Birkhäuser

Monografie Matematyczne

Volume 75

Starting in the 1930s with volumes written by such distinguished mathematicians as Banach, Saks, Kuratowski, and Sierpinski, the original series grew to comprise 62 excellent monographs up to the 1980s. In cooperation with the Institute of Mathematics of the Polish Academy of Sciences (IMPAN), Birkhäuser now resumes this tradition to publish high quality research monographs in all areas of pure and applied mathematics.

Antonio José Guirao • Vicente Montesinos
Václav Zizler

Renormings in Banach Spaces

A Toolbox

Birkhäuser

Antonio José Guirao
Departamento de Matemática Aplicada
Universitat Politècnica de València
Valencia, Spain

Vicente Montesinos
Departamento de Matemática Aplicada
Universitat Politècnica de València
Valencia, Spain

Václav Zizler
University of Alberta
Edmonton, AB, Canada

ISSN 0077-0507 ISSN 2297-0274 (electronic)
Monografie Matematyczne
ISBN 978-3-031-08654-0 ISBN 978-3-031-08655-7 (eBook)
https://doi.org/10.1007/978-3-031-08655-7

This book is published under the imprint Birkhäuser, www.birkhauser-science.com by the registered company Springer Nature Switzerland AG
The registered company address is: Gewerbestrasse 11, 6330 Cham, Switzerland

Dedicated to the memory of Zdeněk Frolík
who managed to unite the classical Čech topological tradition
with the activities of the new Banach space workshop in Prague

A mathematician is a person who can find analogies between theorems. A better mathematician is one who can see analogies between proofs and the best mathematician can notice analogies between theories. One can imagine that the ultimate mathematician is one who can see analogies between analogies.

Stefan Banach

Contents

Preface

Introduction

Banach spaces provide a framework for modern mathematical analysis, as they blend classical analysis, geometry, topology and linearity. This makes Banach space theory a wonderful, elegant, and active research area in Mathematics. Many problems in modern linear and nonlinear analysis are of infinite-dimensional nature. Renormings represent a big and important part. They deal with structural properties, geometry, differential calculus etc. In this area, there are many important open problems.

In this text, "differentiability" and "smoothness" are synonymous terms, as they are "rotundness" and "strict convexity". These and related concepts will be properly defined in Chapters 4 and 5, respectively.

Questions concerning the supply of smooth functions on Banach spaces are of crucial importance in differential calculus. Smooth functions are usually obtained from equivalent smooth norms and smooth norms are in turn often constructed by using dual rotund norms. The existence of equivalent smooth or rotund norms or nontrivial smooth functions on a particular Banach space depends on its structure and has a profound impact on its geometry. For example, a separable Banach space that admits a Fréchet smooth function with bounded support must have a separable dual (Theorems 276, 283, and 284); if this function is, moreover, infinitely differentiable, then the space must contain a copy of c_0 or ℓ_p (Theorem 516). If the space does not contain c_0 and has a C^2-smooth function with bounded support, it is superreflexive (Theorem 453). If for a compact space K, its Cantor derivative $K^{(\omega_1)}$ is empty, then $C(K)$ admits a C^∞-smooth norm (Theorem 813). However, there is K such that $K^{(\omega_1)}$ is a singleton and $C(K)$ does not admit any Gâteaux differentiable norm [Hay99] (cf., e.g., [DeGoZi93, Chapter VII], [Ziz03, Theorem VII.6.4], and Theorem 529).

This text attempts to provide a modern and concise guide to some basic renorming results and techniques of Banach spaces and their applications, and discusses some streams of contemporary research in this area. We tried to keep a reader-friendly relaxed form to enjoy the text. For understanding the relations of concepts, we provide here as many counterexamples as possible. We made an effort to write the text in an informative manner, briefly pointing out main ideas and motivations for the results and the connections between them. The experienced reader may skip some detailed explanations in the introductory parts of the text. We include them to provide for some training in techniques used in this area, specially for young students. In many instances, specially in more advanced topics, we prefer to sketch the construction and motivate the result pointing out the main idea behind, better than to provide all details of a proof. In case that it is missing at all, we refer then either to the original paper or to books where the results appear in full.

This style enables us to discuss much wider spectrum of things and their relations than it could go to the regular book form. This allows us to include observations and comments that may help the reader, suggesting ways to complete the information provided here.

The text can be considered as a commented guide to optional graduate courses in Banach spaces. It starts with a review of basic tools and continues with several rather independent sections for the instructor's choice. The course would discuss many motivating open problems, and thus provides material for a research seminar as well.

The central topic of the text consists in renorming Banach spaces. This means to endow the given space with an equivalent norm that enjoys extra properties:

> If $\|\cdot\|$ is a norm on a Banach space X, another norm $|||\cdot|||$ is **equivalent** to $\|\cdot\|$ if two positive numbers α and β exist such that $\alpha\|x\| \leq |||x||| \leq \beta\|x\|$ for all $x \in X$.

Observe that the topology on X induced by the norm does not change when such new norm is adopted, although the metric properties may be really different. The renorming theory has thus many applications in the geometry of Banach spaces and in problems in analysis on them, and it is an active, large area of research in Banach spaces.

Note that in a given finite-dimensional space, all norms are equivalent (see Proposition 3 below). On the other hand, on any given infinite-dimensional Banach space $(X, \|\cdot\|)$ there is always a norm $|||\cdot|||$ that is not equivalent to $\|\cdot\|$ (see again Proposition 3).

Usually, any good result in renorming has many corollaries and it ensures many applications, since it is often stronger than the mere required particular property. Consider in this direction the Krein–Milman Theorem 7, differentiability of convex functions, and some other properties discussed below in the text.

In analysis in infinite dimensions, some problems require further techniques for their solution. For example, though every finite convex function on a finite-dimensional space is necessarily continuous (see Theorem 52), note that every infinite-dimensional Banach space admits a discontinuous linear functional (see the paragraph after the proof of Theorem 49). Bounded linear functionals do not necessarily attain their maxima on closed balls (as an easy example, observe that if B_{c_0} is the closed unit ball of the space c_0 and f is the continuous linear functional on c_0 defined by $f(x) = \sum \frac{1}{2^i} x_i$ for $x = (x_i)$, then the supremum of f over B_{c_0} is 1, but it is not difficult to show that for every $x \in B_{c_0}$, $f(x) < 1$. Therefore there is no closest point to the origin in this $f^{-1}(1)$. Also, there is no extreme point in B_{c_0}. Another example of a continuous linear functional that does not attain its supremum on the closed unit ball of a Banach space is given after Theorem 12).

An infinite-dimensional Rolle's theorem does not hold true (see, e.g., [FHHMZ11, Exercise 7.25]), and so new variational principles are needed. We will meet among others Ekeland's, compact, and smooth variational principles here (see Chapter 11).

Let us see an example of how renormings can help, typically, in some of these questions. Let $(X, \|\cdot\|_1)$ be a reflexive Banach space and let $\|\cdot\|_2$ be an equivalent strictly convex norm on X (its existence is guaranteed by a theorem of S.L. Troyanski (Theorem 218 below). By a result of K.S. Lau (Theorem 260 below), there exists an interior point c_1 of $B_{\|\cdot\|_1}$ from which a point $y_1 \in S_{\|\cdot\|_1}$ at farthest $\|\cdot\|_2$-distance exists. Then it is not difficult to see that y_1 is an extreme point of $B_{\|\cdot\|_1}$. We shall see in the text that, in a similar way, J. Lindenstrauss proved that all closed bounded convex sets in reflexive spaces do have an extreme point with extra properties (Theorem 259), by using M.I. Kadets and S.L. Troyanski renorming by locally uniformly convex norms (a property better than simple strict convexity). Similarly, if x_0 is a point of Fréchet smoothness of the second norm, then x_0 is also a Fréchet smooth point of the first norm. This, via a dualization of the Kadets–Troyanski result and the Lindenstrauss method, gives Fréchet smooth points on balls in reflexive spaces (see Item 5 in Subsection 5.3.2). These results of M.I. Kadets, Troyanski, and Lindenstrauss have been one of the most important results in this part of Banach space theory. Let us note that these are in a sense the best possible results in this direction; as we shall see later, measure theory can hardly be used here. It is the (miraculous) power of the concept of differentiability that allows for refraining from the requirement of reflexivity in the cases mentioned above. This is one of the goals of the smooth variational principle (Theorem 289) we will meet in this text.

Another thing we would like to mention at the beginning is that M.I. Kadets proved that if X is a Banach space with a Schauder basis $\{e_i, f_i\}$, then X can be renormed so that if $\|x_n\| \to \|x\|$ and $f_i(x_n) \to f(x)$ for each i, then $\|x_n - x\| \to 0$ (Corollary 177). So this norm shares this property with the Hilbertian norm. He used this approach for solving the long-standing problem of M. Fréchet and S. Banach whether that all infinite-dimensional separable Banach spaces are homeomorphic (Theorem 186). We shall discuss it in this text.

Let us briefly explicit here our point of view about the actual situation of renorming:

1. The present understanding of the basics is rather satisfactory in the separable case for local uniform rotundity, C^1-Fréchet and Gâteaux smoothness.
2. The understanding of basics in local uniform rotundity and C^1-Fréchet smoothness is relatively satisfactory for the class of weakly compactly generated spaces and their relatives.

3. The understanding of Gâteaux smoothness is not yet satisfactory in nonseparable spaces.
4. The understanding of higher order smoothness is not yet satisfactory in separable or nonseparable spaces.
5. There is a huge potential for more research in applications of renormings in linear and nonlinear analysis, in particular in Convexity Theory and Approximation.

There are many techniques for providing equivalent norms in specific Banach spaces or in wide classes of them. The state of affairs in Renorming Theory up to 1993 was presented in [DeGoZi93]. Later on, some remarkable contributions appeared, and some of them are presented in [MOTV09]. Now it is almost impossible to collect all sources and references in this area. Some are provided in the bibliography at the end of this manual. Among them we may add to the previously mentioned references, for example, [BeLi00], [BoVa09], [Di75], [Fab97], [FHHMPZ01], [FHHMZ11], [Go01b], [GoLaZi14], [HHZ96], [HMVZ07], [HáJo14], [HáMoZi12], [Meg98], [PelBe79], and [Ziz03]. Only a tiny part of the theory and techniques of renorming was discussed in [FHHMZ11] and [HMVZ07]. This is why we prepared this text, where we try to upgrade the information, including also some examples, counterexamples, and application of renorming. A big portion of the results are mentioned here for the first time in book form. Emphasis will be done on basic techniques, so that the reader may use them —or convenient variations—for constructing norms with various required properties.

When presenting and using the various relevant concepts we shall stress the relationship between them. Quite often results —even important ones— come without a detailed proof (only a sketch of the underlying ideas is provided), and many of them actually without proof at all (in those cases we shall always refer to sources). There are four main reasons for skipping some proofs: First, if incorporated, then the length of the text would have been unpractical. Second, that nowadays almost —if not all— sources are available at a click, and what is missing is, maybe, a map to search and locate what the interested reader is looking for. Third, that we prefer to focuss instead on the underlying ideas that may enlighten the panorama, especially stressing what is missing in some other texts like, e.g., [FHHMZ11], [DeGoZi93], [HMVZ07]. And finally, that we have in mind that the most efficient streamlined proof is, sometimes, not the best for a future use. In other words, that reading the original —sometimes distant— sources is, quite often, not only an exercise in mathematical history, but a must for true understanding (read the masters!). Let us quote here N.H. Abel: "It appears to me that if one wants to make progress in mathematics one should study the masters and not the pupils" (N.H. Abel (1802–1829), quoted from an unpublished source by O. Ore in [Ore74, p. 138]). The Abstract of H.M. Edwards in [Edw81], that we reproduce here, insists on this matter: "It is as good an idea to read the masters now as it was in Abel's time. The best mathematicians know this and do it all the time. Unfortunately, students of mathematics normally spend their early years using textbooks (which may be, but usually aren't, written by masters) and taking lecture courses which are self-contained and make little or no reference to the primary literature of the subject. The students are left to discover on their own the wisdom of Abel's advice. In this they are being cheated".

This way the text may help the reader to get an overview of the whole body of the renorming theory better than to focus on detailed explanations— trying to summarize results on individual classes of spaces—. Thus, we can often afford to discuss more than one proof of the given result. For example, we discuss several proofs of the fact that non-trivial Fréchet differentiable functions cannot live in a space where there exists a convex continuous function that is nowhere Fréchet differentiable: Kurzweil's, Ekeland's, Whitfield's, Borwein–Preiss, and Deville's methods. We show several alternative proofs of the Bishop–Phelps Theorem 14. We discuss several methods in constructing projections in nonseparable spaces: Lindenstrauss', Plichko's, Valdivia's, Orihuela's, and Troyanski's methods, just to name a few authors involved. For relatives of weakly compactly generated spaces we consider both Troyanski et al. and Godefroy et al. transfer methods. Note they both involve a very sophisticated use of the interplay of convexity and compactness. We will see in the text that each one of them is used many times in this area. In the other direction, we use in several circumstances recent results to provide streamlined proofs of some older ones. This approach also enables us to effectively summarize results concerning a given individual property.

In any case, to make this text easily readable, we shall try to provide all the required definitions, make comments on them, and give hints to significant results —especially if the material is not easily accessible— trying to be to some extent self-contained. We hope that the text may be interesting to experts in Banach spaces in general, as a second reading on renormings, and that may help in supervising PhD theses on this subject. To assist the reader, we prepared organized lists of properties of particular spaces and in Chapter 51 a list of coordinates of various counterexamples. In Chapter 52 a list of easier problems to start independent work on this subject is included.

Of course, some basic knowledge of Banach space theory is required, as contained in basic undergraduate courses in functional analysis and in the introductory chapters of [FHHMZ11]. For a more extensive background, the reader may consult, for instance, the introductory article by Johnson and Lindenstrauss [JoLi03], [FHHMZ11], [LinTz77], [Di84], and [AlKa06]. In particular, it is an asset if the reader is more or less familiar with some standard known implications between the very basic notions of rotundity, smoothness and their duality, as contained, e. g., in [DeGoZi93], [FHHMZ11, Chapters 7, 8]. A certain knowledge of convex functions and, in particular, norms, is certainly needed. Practicing on this is a must. We may suggest the reader to look, for example, to [BoVa09], [FHHMZ11], [HMVZ07], [Phe93], [Gil82], and [MoZiZi15].

A basic knowledge of set theory and a certain background in topology are a must. We may suggest to consult [BessPe75], [Du66], [Eng85], [Kam50], and [Kel55].

There are many crucial open problems in the renorming area. Some of them are discussed, e. g., in [DeGoZi93], [FHHMZ11], [HMVZ07], [GuiMoZi16], and in [HáJo14]. A **bump function** (or just a **bump**) on a Banach space X is a function with nonempty bounded support. If ϕ is a C^k-smooth function on the real numbers such that $\phi(t) = 1$ if $t \in [-1, 1]$ and $\phi(t) = 0$ if $|t| \geq 2$, and $\|\cdot\|$ is a C^k-smooth norm —outside the origin— on X, then the composition $\phi(\|x\|)$ is a C^k-smooth bump. We will see that there are nonseparable Banach spaces that admit C^∞-smooth bumps but do not admit Gâteaux differentiable norms (Theorem 817). We will show that the sup norm of $C[0, 1]$ is nowhere Fréchet differentiable (Proposition 280) and that this space admits no Fréchet differentiable bump (Theorem 288). We will see that this is unlike finite-dimensional spaces, where Rademacher's Theorem 438 holds saying that any real-valued Lipschitz function on a finite-dimensional Banach space is Fréchet differentiable on a dense set of points. We will also meet the remarkable Preiss' Theorem 443 that on Asplund spaces —in particular, on Hilbert spaces— every real-valued Lipschitz function is Fréchet differentiable at points of a dense set.
Let us note in passing, for example, that the following problems are open.

1. Assume that a separable space X admits a C^2-smooth bump. Does it admit a C^2-smooth norm?
2. If X is an Asplund space, does X admit a C^1-smooth bump?
3. If X admits a C^1-smooth bump, does it allow for smooth approximations of continuous functions?
4. It is unknown if every norm on ℓ_2 can be approximated by real-analytic norms.
5. From the general theory, it should be noted that, amazingly, it is unknown whether every nonseparable Banach space has a infinite-dimensional separable quotient, see [GuiMoZi16, Problem 32].

This and many other open problems, famous and less famous, are discussed in this text. A list of them appears in Chapter 52. For a larger collection of open problems in the general theory of Banach spaces, we suggest to look at [GuiMoZi16].

A survey of the present situation in the area of the weakly compactly generated spaces and their relatives can be found in [FGMZ04] and [FGHZ03].

The text is divided into three parts.

(I) An introductory course in renorming that covers the basic necessary material in convexity and norms needed in this area, including the most basic renorming techniques. To make the text accessible to not-experienced readers, we explain here even basic concepts —like norming subspaces, Fenchel duality and convolution— and refer to basic results.

(II) An intermediate course in renorming covering more advanced techniques, together with applications to differentiability of convex functions and extremal structure.

(III) Advances and developments in renormings, covering contemporary research material concerning the connections to various structural properties of the spaces.

The text also shows a little bit on history and personal memories regarding when and where some of the results have been obtained, since we think that this, too, helps understanding the subject, and can provide for a moral support and encouragement: For example, the Czech and Spanish schools started from the beginning in the seventies of the last century, and now they are renowned centers of Banach spaces.

At the end of this text we provide for the reader's convenience a concise list giving the coordinates for examples and counterexamples discussed here. This information is doubled, in a telegraphic way, in several of the carefully prepared indices (General, Symbols, Authors, Renormings and Impossible Renormings), allowing for locating the information needed. We also provide a seizable list of references.

Summing up, this text is directed mainly to young research personnel in the area and can be considered as a supplement to [FHHMPZ01], [FHHMZ11], [HMVZ07], [Fab97], and [GuiMoZi16].

We thank our Institutions: The Mathematical Institute of the Czech Academy of Sciences and the Faculty of Mathematics and Physics at Charles University in Prague (Prague, Czech Republic), the University of Alberta (Edmonton, Alberta, Canada), the Real Academia de Ciencias Exactas, Físicas y Naturales of Spain (Madrid, Spain), the Universitat Politècnica de València and its Instituto de Matemática Pura y Aplicada (Valencia, Spain), for support and providing excellent working conditions. To Canadian, German, Czech and Spaniard grants, in particular Grants MTM2017-83262-C2-1-P (Spain), MTM2017-83262-C2-2-P (Spain) and PID2021-122126NB-C33 (Spain), for supporting this research over the years, and to our colleagues and friends for contributions and discussions on the subject of this work. We thank the staff of Springer Verlag, and especially Ms. Dorothy Mazlum (Birkhauser) and Mr. Marc Strauss (Springer), for their interest in this text.

The first named author wants to express his gratitude to the Murcia School of Functional Analysis (Spain), in particular to Professors Gabriel Vera, José Orihuela, Stanimir Troyanski, and the late Bernardo Cascales. His thanks also extend to Professors Petr Hájek from Prague (Czech Republic), and Robert Deville, from Bordeaux (France).
The second named author wants to acknowledge the debt with the late Professor Manuel Valdivia, founder of the Spanish School of Functional Analysis. His

overwhelming contribution to Mathematics, to his many students and to the intellectual mathematical life in Valencia (Spain) deserves a very special place in this list of acknowledgement. He also wants to thank Professors Václav Zizler (Prague, Czech Republic, and Edmonton, Canada), Marián Fabian (Prague, Czech Republic), Petr Hájek (Prague, Czech Republic), and Gilles Godefroy (Paris, France) for support, cooperation, and friendship.
The third named author uses this opportunity to thank all organizers of the annual International Czech Winter and Paseky Schools on Abstract Analysis (last year was its forty eighth meeting), where many important results from renormings were presented even for the first time —for example, the solution to the scalar-plus-compact famous open problem (Theorem 474 below)—. Special appreciation goes to the founder of these Schools, the late Professor Zdeněk Frolík, of the Czech Academy of Sciences, to whom this book is devoted. The appreciation goes for thousands of hours of selfless work of the whole team of mathematicians in Prague contributing to the success of these Schools (this was needed as they wanted to keep a warm and friendly working atmosphere in these workshops; Schools are placed in Czech mountain resorts to achieve a good condition for team work). The focus has been always directed to young participants. The Schools were open even for graduate students. Special appreciation goes to Professor Kamil John, from Prague Mathematical Institute of the Czech Academy of Sciences, as it was he who brought to the Prague group the project of systematically study the linear topological structure of nonseparable Banach spaces and started to have good results in this area. The work of this group in Prague would not have been possible without a strong moral support from the Professors Czesław Bessaga and Aleksander Pełczyński from the Warsaw School. He would also like to express his appreciation to the late Professor Jiří Jelínek, who introduced him to Functional Analysis, and to the late Professor Josef Kolomý of the Charles University in Prague, who put him and Marián Fabian on the track of Renorming Theory in the 60's of the last century. Special thanks go to the previously mentioned Professors Marián Fabian (Prague, Czech Republic), Gilles Godefroy (Paris, France), Petr Hájek (Prague, Czech Republic), and Kamil John (Prague, Czech Republic), for their long-life mathematical cooperation and friendship. He also thanks Professors Franz Hering (Dortmund, Germany), John Whitfield (Thunder Bay, Canada), and again Gilles Godefroy (Paris, France) for their crucial help with emigrating Zizler family to Canada in 1983. Special thanks of the third named author go to University of Alberta in Edmonton for keeping his MathSciNet account at present. We thank many our colleagues for their criticism on the text.

Above all, we thank our wives, Marta, Danuta, and Jarmila, respectively, for their moral support and encouragement. This book is also dedicated to them.

We would be happy if this text helped in motivating some young researchers to choose Renorming Theory as their main work field, and/or help some team of specialists prepare a new textbook on renormings. When preparing this text, we have had in mind also students that may like renormings and are staying

at universities that do not specialize in Banach spaces. For them we included a discussion of some folklore results needed in this area. At the end of the text we include many easy (and some not easy) problems for them to start with. Due to the ample bibliography included, Banach space specialists in general can use it as a reference list in this area. We hope that the books [DeGoZi93], [Fab97], [HMVZ07], [MOTV09], [FHHMZ11], and the present one, could offer a reasonable overview of the present state of renormings and their applications.
We wish readers a pleasant time using this book.

Notation

We follow the standard notation as in [LinTz77], [DuSch], or [FHHMZ11], for example. In any case, symbols will be properly defined when they will appear, and a list of them is included as a special index. Here we shall only mention some adopted conventions that maybe are not of general use: For example, if f is a function from a set A into a set B, and if $b \in B$, we shall denote the set $\{x \in A : \; f(x) = b\}$ simply by $f(x) = b$. This will be used, in particular, in many of our pictures. Thus, $f(x) = 0$ will denote the kernel $\{x \in A : \; f(x) = 0\}$ of a function f from A into a space. If nothing is said on the contrary, by the word "space" we shall have in mind an infinite-dimensional real normed space (usually denoted by X). If X^* denotes its dual space, then the action of an element $x^* \in X^*$ on an element $x \in X$ will be indistinctly denoted by $x^*(x)$ or $\langle x, x^* \rangle$. We find it convenient, when dealing with higher dual spaces, to consider in the previous evaluation notation that "even" duals are "on the left" and "odd" duals "on the right". In this way, $\langle x, x^* \rangle$ will not change order if x is considered as an element in X^{**} via the canonical linear isometric injection j from X into X^{**}; accordingly, it will be written $\langle j(x), x^* \rangle$ or, in a more simple way, $\langle x, x^* \rangle$ if there is no risk of misunderstanding. By the way, and having in mind the existence of the mapping j, X will always be considered as a linear subspace of X^{**}.

Often we say that a Banach space is, say, strictly convex if its norm is strictly convex, etc. Sometimes we say "separable Banach space" and mean "separable infinite-dimensional Banach space". Often we say "a function", meaning a "real valued function". The word "operator" refers to a linear mapping between two spaces. When we say that a space "admits a norm" we mean an *equivalent* norm.

Vectors x in a (generalized) sequence space will be denoted by $(x_\gamma)_{\gamma \in \Gamma}$ or $(x(\gamma))_{\gamma \in \Gamma}$, indistinctly (sometimes the notation will be shortened to (x_γ) if the index set is understood). The first or the second notation will be used according to notational convenience. Sequences (or generalized sequences) will be denoted by $\{x_n\}_{n=1}^\infty$, or by $\{x_n\}_{n \in \mathbb{N}}$ (respectively, $\{x_\gamma\}_{\gamma \in \Gamma}$ for general ordered sets Γ of indices).

Typography

To be spotted easily, definitions are included in special boxes on a gray background. Since the subject of this text is renormings, results explicitly on this matter are also included in boxes again on gray background.
To help focusing on the essential, some explanations can be avoided at a first reading; for them, a particular environment that we call "annotation", is used,

➡ as in this example. They are preceded by a special symbol for the reader's convenience. This resource will be used for some proofs, complementary material, and detours. Skipping them should not damage much a first approach.

Historical notes will be included in a special environment headed by H

Remarks end by the symbol ®

Proofs end by the symbol □

Exercises end by the symbol ◇

Examples end by the symbol ✠

The list of open questions in Chapter 52 follows the same typographical convention than the results along the text: Those directly related to renorming are included in special boxes on a gray background. Not all the items in the list come from the text. For those that do we provide the right location.

Part I

An Introductory Course in Renorming

Chapter 1

Norms, normed spaces, Banach spaces

In this chapter we will set the basic definitions —and some results, with their references— that will be used along the text. We include them here in order to avoid repetitions later on. For further information, the reader may access any of the texts mentioned in the references.

1.1 Norms, normed and Banach spaces

Definition 1. *A* **norm** *on a real vector space* X *is a function* $\|\cdot\|$ *from* X *into* $\mathbb{R}$ *that satisfies, for all* $x, y \in X$*, and* $\lambda \in \mathbb{R}$*,*
(i) $\|x\| \geq 0$,
(ii) $\|x\| = 0$ *if, and only if,* $x = 0$,
(iii) $\|\lambda x\| = |\lambda|.\|x\|$, *and*
(iv) $\|x + y\| \leq \|x\| + \|y\|$ *(the* **triangle inequality***).*
A real vector space X *with a norm* $\|\cdot\|$ *is said to be a* **(real) normed space***, and we speak of the normed space* $(X, \|\cdot\|)$*. A sequence* $\{x_n\}_{n=1}^{\infty}$ *in* X $\|\cdot\|$-**converges to an element** $x \in X$ *if* $\|x_n - x\| \to 0$*. A sequence* $\{x_n\}_{n=1}^{\infty}$ *in* X *is* $\|\cdot\|$**-Cauchy** *if for every* $\varepsilon > 0$ *there exists* $n_0 \in \mathbb{N}$ *such that* $\|x_n - x_m\| < \varepsilon$ *for every* $n, m \geq n_0$*. A normed space* $(X, \|\cdot\|)$ *is a* **(real) Banach space** *if every Cauchy sequence in* X *is* $\|\cdot\|$*-convergent to some point in* X *(i.e., if the space is* **complete***).*

In this text, we shall deal with *real* normed spaces. Thus, the term "normed space" ("Banach space") will mean "real normed space" (respectively, "real Banach space"). Although many concepts defined below make sense in the context of normed spaces —without the hypothesis of completeness— and again several results do not depend on this requirement, Banach spaces will be the natural framework in which we shall work. Only in a few instances we shall define the concepts or formulate the results in the more general context of general (real) normed spaces (in a few instances we shall refer to the theory of locally convex spaces; a brief introduction will be presented in Section 2.1).
Thus, X, or more precisely, $(X, \|\cdot\|)$, will be a real Banach space if nothing is said to the contrary. The term "space" without other qualifications will refer to a real Banach space. If nothing is said on the contrary, the term "subspace" refers to a linear subspace.

A. J. Guirao et al., *Renormings in Banach Spaces*, Monografie Matematyczne 75,
https://doi.org/10.1007/978-3-031-08655-7_1

A norm $\|\cdot\|$ on X defines a **distance** d: Precisely, $d(x_1, x_2) := \|x_1 - x_2\|$ for $x_1, x_2 \in X$. This function d satisfies all the requirements of an abstract distance: (i) $d(x_1, x_2) \geq 0$, (ii) $d(x_1, x_2) = 0$ if, and only if, $x_1 = x_2$, (iii) $d(x_1, x_2) = d(x_2, x_1)$, and (iv) $d(x_1, x_3) \leq d(x_1, x_2) + d(x_2, x_3)$, for all $x_1, x_2, x_3 \in X$. The distance defined by a norm has some extra features: For example, $d(x_1, x_2) = d(x_1 + x_3, x_2 + x_3)$, or $d(\lambda x_1, \lambda x_2) = \lambda d(x_1, x_2)$, for all $x_1, x_2, x_3 \in X$ and $\lambda \geq 0$. The space $(X, \|\cdot\|)$, with the distance defined above, is a particular instance of a metric space, and we may speak of "proximity": Two points x_1 and x_2 in X are "close" when $d(x_1, x_2)$ (i.e., $\|x_1 - x_2\|$) is small. The set of points in X whose distance to a given point $x_0 \in X$ is equal to some constant $\varepsilon > 0$ is called the **sphere centered at** x_0 **and having radius** ε. The set of points in X whose distance to a given point $x_0 \in X$ is less than or equal to some constant $\varepsilon > 0$ is, by definition, the **closed ball** $B(x_0, \varepsilon) := \{x \in X : \ \|x - x_0\| \leq \varepsilon\}$ **centered at** x_0 **and having radius** ε. The set $\operatorname{Int} B(x_0, \varepsilon) := \{x \in X : \ \|x - x_0\| < \varepsilon\}$ is called the **open ball centered at** x_0 **and having radius** ε. We usually denote by B_X (or by $B_{(X, \|\cdot\|)}$ if we want to be more precise) the set $B(0, 1)$ (i.e., $\{x \in X : \ \|x\| \leq 1\}$, the closed unit ball), by $\operatorname{Int} B_X$ or $\operatorname{Int} B_{(X, \|\cdot\|)}$ the open unit ball, and by $S_X := \{x \in X : \ \|x\| = 1\}$ the **unit sphere**. A set $S \subset X$ is said to be **bounded** if there exists some $n \in \mathbb{N}$ such that $S \subset B(0, n)$. The class $\mathcal{O}$ of all open sets defined by the metric d is a **topology** on X: A set $O \subset X$ is **open** whenever, for every $x_0 \in O$, there exists $\varepsilon > 0$ such that $B(x_0, \varepsilon)$ is contained in O (note that the empty set is considered also to be open). This topology, that comes from a norm $\|\cdot\|$ in X, shall be denoted by $\mathcal{T}_{\|\cdot\|}$, and we shall speak about the **norm topology**, or the topology **defined by the norm**. The terminology is consistent: A closed ball is a closed set, an open ball is an open set; the **topological interior** (i.e., the biggest open set contained in a given set) of the closed ball $B(x_0, \varepsilon)$ is the open ball $\operatorname{Int} B(x_0, \varepsilon)$, and the **topological closure** (i.e., the smallest closed set containing a given set) of the open ball $\operatorname{Int} B(x_0, \varepsilon)$ is the closed ball $B(x_0, \varepsilon)$.

Clearly, a sequence $\{x_n\}_{n \in \mathbb{N}}$ in X is $\mathcal{T}_{\|\cdot\|}$-convergent to $x \in X$ if, and only if, $\|x - x_n\| \to 0$. The topological concepts in the topology $\mathcal{T}_{\|\cdot\|}$ can be described by using sequences: For example, the complement of an open subset of X is said to be **closed**; it turns out that A is closed in $(X, \mathcal{T}_{\|\cdot\|})$ if, and only if, $x \in A$ whenever $\{x_n\}_{n \in \mathbb{N}}$ is a sequence in A that $\mathcal{T}_{\|\cdot\|}$-converges to x.

Note that $\|\cdot\|$, as a function from X into $\mathbb{R}$, is $\mathcal{T}_{\|\cdot\|}$-continuous. Indeed, a consequence of the triangle inequality (iv) in Definition 1 is that

$$\big|\|x\| - \|y\|\big| \leq \|x - y\|, \text{ for all } x, y \in X.$$

Therefore, if $\{x_n\}_{n \in \mathbb{N}}$ is a sequence in X that $\mathcal{T}_{\|\cdot\|}$-converges to $x \in X$, then $\big|\|x\| - \|x_n\|\big| \leq \|x - x_n\| \to 0$.

To simplify some expressions, topological concepts in the $\mathcal{T}_{\|\cdot\|}$-topology will sometimes be written by using the prefix $\|\cdot\|$- instead of the more complicated $\mathcal{T}_{\|\cdot\|}$-, as referring to "$\|\cdot\|$-convergence" instead of "$\mathcal{T}_{\|\cdot\|}$-convergence". Both ways will be used indistinctly.

1.2 Equivalent norms

1.2.1 Definition

The concept of equivalent norms has been already presented in the Introduction. Let us repeat it here since it is the central object of our study. A wider treatment of the basic concept of equivalence of norms will be done in Chapter 3. At this stage we need just its definition:

Definition 2. *If* $(X, \|\cdot\|)$ *is a normed space, another norm* $|||\cdot|||$ *on* X *is said to be* **equivalent** *(to* $\|\cdot\|$*) whenever there exist two positive constants* $c \le C$ *such that*

$$c\|x\| \le |||x||| \le C\|x\|, \text{ for all } x \in X. \tag{1.1}$$

To be able to construct equivalent norms on a given Banach space that enjoy some desired properties —essentially rotundity and/or differentiability to some degrees, see Chapters 4 and 5 for definitions— is the main purpose of this text.

1.3 Linear operators and linear functionals. Duality, weak topologies

If $(X, \|\cdot\|_X)$ and $(Y, \|\cdot\|_Y)$ are normed spaces, a linear mapping (also called an **operator**) $T : X \to Y$ is continuous if, and only if, it is **bounded**, i.e., if TB_X is a bounded set. It is equivalent to say that there exists $C \ge 0$ such that $\|Tx\|_Y \le C\|x\|_X$ for all $x \in X$. The minimum constant C with this property is said to be the **norm of** T, denoted by $\|T\|$. In other words, $\|T\| = \sup\{\|Tx\|_Y : x \in B_X\}$. The set $\mathcal{B}(X, Y)$ of all continuous linear operators from X into Y, with the algebraic operations sum and product by a scalar, is a vector space. When endowed with the norm $\|\cdot\|$ defined above, $\mathcal{B}(X, Y)$ becomes a normed space. If $(Y, \|\cdot\|_Y)$ is complete, then it is easy to see that so it is $(\mathcal{B}(X, Y), \|\cdot\|)$:

➡ First, notice that a Cauchy sequence is always bounded. Second, if $\{f_n\}_{n\in\mathbb{N}}$ is a Cauchy sequence in $(\mathcal{B}(X, Y), \|\cdot\|)$, and $x \in X$, then $\{f_n(x)\}_{n\in\mathbb{N}}$ is a Cauchy sequence in Y, hence convergent (say to $f(x)$). The so-defined mapping $f : X \to Y$ is clearly linear, and the boundedness of $\{f_n\}_{n\in\mathbb{N}}$ readily gives that f is bounded (i.e., $f(B_X)$ is a bounded set in Y), hence continuous.

The converse also holds: If $(\mathcal{B}(X, Y), \|\cdot\|)$ is complete, and $X \neq \{0\}$, then $(Y, \|\cdot\|_Y)$ is complete, too:

➡ Let $x_0 \in S_X$ and $f_0 \in S_{X^*}$ be such that $f_0(x_0) = 1$ (the existence of f_0 is guaranteed by the Hahn–Banach extension theorem, see below in this section). Let $\{y_n\}_{n\in\mathbb{N}}$ be a Cauchy sequence in Y. For $n \in \mathbb{N}$, let us define a mapping $\phi_n \in \mathcal{B}(X,Y)$ as $\phi_n(x) := f_0(x)y_n$, for $x \in X$. Note that $\phi_n(x_0) = y_n$ for all n. Since $\|\phi_n - \phi_m\| \leq \|y_n - y_m\|$ for $n, m \in \mathbb{N}$, the sequence $\{\phi_n\}_{n\in\mathbb{N}}$ is $\|\cdot\|$-Cauchy. Thus, it converges to some function $\phi \in \mathcal{B}(X,Y)$. In particular, $y_n = \phi_n(x_0) \to \phi(x_0)$.

A particular instance of the previous situation is when $(Y, \|\cdot\|_Y)$ is just $(\mathbb{R}, |\cdot|)$. The space $X^* := \mathcal{B}(X,\mathbb{R})$, called the **dual space** of X. Elements in X^* are called **continuous linear forms**. We shall use the symbol $\|\cdot\|^*$ for the defined norm on X^*, referred to as the **dual norm** (dual to $\|\cdot\|$). If there is no risk of misunderstanding, sometimes we shall use $\|\cdot\|$ instead. By the previous observation, $(X^*, \|\cdot\|^*)$ is always a Banach space, even in the case that X is not complete. This shows, in particular, that every finite-dimensional normed space is complete.

➪ Indeed, notice first that every linear mapping between two finite-dimensional normed spaces is continuous —after all, its action can be described by matrix multiplication—. If $(X, \|\cdot\|)$ is finite-dimensional, and $\{e_i\}_{i=1}^n$ is an algebraic basis, the set $\{e_i^*\}_{i=1}^n$, where $e_i^*(e_j) = \delta_{i,j}$, $i,j = 1,2,\ldots,n$, is an algebraic basis of X^*. The mapping $T : (X, \|\cdot\|) \to (\mathcal{B}(X^*,\mathbb{R}), \|\cdot\|)$ given by $Tx(f) = f(x)$ for all $f \in X^*$ is clearly a linear isometry. It is onto: Given $\phi \in \mathcal{B}(X^*,\mathbb{R})$, put $\phi(e_i^*) = a_i$ for $i = 1,2,\ldots,n$. Then clearly $Tx = \phi$, where $x := \sum_{i=1}^n a_i e_i$. The result follows from the completeness of $(\mathcal{B}(X^*,\mathbb{R}), \|\cdot\|)$.

Notice that $\|x^*\|^* = \sup\{\langle x, x^*\rangle : x \in B_{(X,\|\cdot\|)}\}$. We shall sometimes shorten this expression by writing, instead, $\sup\langle B_{(X,\|\cdot\|)}, x^*\rangle$. The same applies to similar formulas as long as it will not induce any misunderstanding.

If X and Y are normed spaces, and $T : X \to Y$ is a linear operator, we may naturally define an operator $T^* : Y^* \to X^*$ (called its **adjoint**) by the formula $\langle x, T^*x^*\rangle = \langle Tx, x^*\rangle$ for $x \in X$ and $x^* \in X^*$. It is easy to see that T^* is continuous in case that T is continuous, and that $\|T^*\| = \|T\|$.

We say that X and X^* form a **dual pair** (denoted by $\langle X, X^*\rangle$), or that they are **in duality**. Technically speaking, this means that there exists a **bilinear** (i.e., linear in each variable) **form** b from $X \times X^*$ into $\mathbb{R}$ such that, if $b(x,x^*) = 0$ for all $x^* \in X^*$, then $x = 0$ and, conversely, if $b(x,x^*) = 0$ for all $x \in X$, then $x^* = 0$. In our case, this bilinear form b is just $b(x,x^*) := x^*(x)$. Due to the intrinsic symmetry involved, we shall often use $\langle x, x^*\rangle$ instead of $x^*(x)$. Note that the definition of the norm $\|\cdot\|^*$ on X^* shows that

$$|\langle x, x^*\rangle| \leq \|x\|.\|x^*\|^*, \quad \text{for all } x \in X \text{ and } x^* \in X^*. \tag{1.2}$$

The reason for the existence of nonzero continuous linear forms on any normed space $X \neq \{0\}$ is the fundamental Hahn–Banach extension theorem (see, e.g., [Kothe69, §17.3] and [FHHMZ11, Chapter II, Theorem 2.1]), that says that *if X is a vector space endowed with a* **sublinear function** p (i.e., a mapping $p : X \to \mathbb{R}$ such that $p(x_1 + x_2) \leq p(x_1) + p(x_2)$ and $p(\lambda x) = \lambda p(x)$ for $x_1, x_2, x \in X$ and

$\lambda \geq 0$), *then* any *linear form* f *defined on a subspace* Y *of* X *such that* $f(y) \leq p(y)$ *for all* $y \in Y$ *has an extension* $\widehat{f} : X \to \mathbb{R}$ *such that* $\widehat{f}(x) \leq p(x)$ *for all* $x \in X$ (the main point consists in defining a suitable extension to a subspace of X that contains Y as a 1-codimensional subspace; then Zorn's Lemma is applied to get the sought extension to X). Thus, if $(X, \|\cdot\|)$ is a normed space, by taking $p := \|\cdot\|$ we get that any continuous linear form f on a subspace Y has a **norm-preserving extension** (i.e., the norm of f and the norm of its extension coincide) $\widehat{f}$ to X. A consequence is that X^* does not reduce to $\{0\}$ if $X \neq \{0\}$. Even more precise: Given $x \in S_X$, there exists $x^* \in S_{X^*}$ such that $\langle x, x^* \rangle = 1$.

On a normed space X we may define other topologies besides $\mathcal{T}_{\|\cdot\|}$, defined above. One of the most useful is the **weak topology** $w(X, X^*)$ (often denoted just by w): A basis of neighborhoods of 0 in this topology is given by the finite intersections of open halfspaces $\{x \in X : \langle x, x^* \rangle < \varepsilon\}$ (where $x^* \in X^*$ and $\varepsilon > 0$). A basis of neighborhoods of any other point $x \in X$ is obtained just by translating to x the previous given basis.

Some features of the weak topology derive from this description:
(i) The weak topology $w = w(X, X^*)$ is indeed a topology on X, and it is clearly Hausdorff.
(ii) All members of the given basis of w-neighborhoods of any point are convex sets (a subset C of a vector space X is said to be **convex** if $\lambda x + (1-\lambda)y \in C$ for all $x, y \in C$ and $0 \leq \lambda \leq 1$. It is equivalent to say that $\sum_{i=1}^n \lambda_i x_i \in C$ whenever $n \in \mathbb{N}$, $x_i \in C$, $\lambda_i \geq 0$ for all $i = 1, 2, \ldots, n$, and $\sum_{i=1}^n \lambda_i = 1$). By the way, given a subset A of a vector space E, the smallest convex set containing A is called the **convex hull** of A, it is denoted by $\operatorname{conv}(A)$. It is simple to show that $\operatorname{conv}(A) = \{\sum_{i=1}^n \lambda_i a_i : n \in \mathbb{N},\ \lambda_i \geq 0,\ \sum_{i=1}^n \lambda_i = 1,\ a_i \in A,\ i = 1, 2, \ldots, n\}$. We refer to [FHHMZ11, Chapters I and II] for more on the basic concepts on convexity.
(iii) Every member of the given basis of w-neighborhoods of any point is $\mathcal{T}_{\|\cdot\|}$-open. Thus, the topology w on X is **coarser** than the topology $\mathcal{T}_{\|\cdot\|}$ (i.e., every w-open subset of X is $\mathcal{T}_{\|\cdot\|}$-open). This fact will be denoted by $w \prec \mathcal{T}_{\|\cdot\|}$.

➪ *A short note on nets.* As we mentioned above, the norm topology $\mathcal{T}_{\|\cdot\|}$ on X can be described by using convergence of sequences. Behind this is the fact that this topology is induced by a metric. When this is not the case, a topology can still be described by using convergence, only that in general sequences are not enough. We need the concept of a **net**: It is a mapping r from a partially ordered set of indices $(I, \leq)$ (i.e., $\leq$ is reflexive and transitive) that is upward directed (two indices have a common successor) into X. Elements $r(i)$ will just be denoted by x_i, and we shall speak about the net $\{x_i : i \in I, \leq\}$ (or in a shorter form $\{x_i\}_{i \in I}$) if there is no risk of misunderstanding. The net $\{x_i\}_{i \in I}$ **converges to some** $x \in X$ whenever, given an arbitrary neighborhood U of x, there is $i_0 \in I$ such that $x_i \in U$ for all $i \geq i_0$.

The weak topology w on X can be described by saying that a net $\{x_i\}_{i\in I}$ converges to $x \in X$ if, and only if, $\{\langle x_i, x^*\rangle\}_{i\in I}$ converges to $\langle x, x^*\rangle$ for each $x^* \in X^*$. Since every $x^* \in X^*$ is a $\mathcal{T}_{\|\cdot\|}$-continuous function, this shows again that $\mathcal{T}_{\|\cdot\|}$-convergence of a net to a point implies w-convergence (to the same point), what amounts to say that the topology w is coarser than the topology $\mathcal{T}_{\|\cdot\|}$, as it was already noticed.
Sequences, used above, are particular instances of nets: The partially ordered index set I is just $\mathbb{N}$, endowed with its natural order.
We refer, e.g., to [Kel55], [Du66], and [Eng85] for more on nets, needed in the rest of this text.

➩ *On the norm and weak topologies.* To see how different these two topologies can be on infinite-dimensional spaces —in finite-dimensional ones they coincide, see Section 1.5 below—, note that *every* neighborhood of 0 in the basis for the w-topology above (and so *every* w-neighbourhood of 0) is unbounded (after all, it always contains a straight line, see Figure 1.1). As a consequence (always in infinite-dimensional spaces), $0 \in \overline{S_X}^w$. However, the same convex sets are closed for the $\|\cdot\|$- and the w- topologies, a consequence of the Hahn–Banach Separation Theorem.

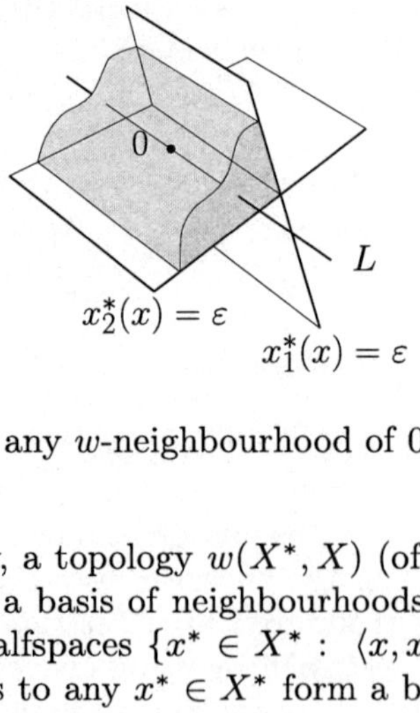

Figure 1.1: If $\dim X = \infty$, any w-neighbourhood of 0 contains a straight line L

In a kind of symmetric way, a topology $w(X^*, X)$ (often denoted just by w^*) on X^* is defined by providing a basis of neighbourhoods of 0: It consists on all the finite intersections of the halfspaces $\{x^* \in X^* : \langle x, x^*\rangle < \varepsilon\}$ (where $x \in X$ and $\varepsilon > 0$). Then, the translates to any $x^* \in X^*$ form a basis of neighborhoods of x^* in (X^*, w^*).

➩ *On the w^*-convergence of nets.* A net $\{x_i^*\}_{i\in I}$ in X^* is w^*-convergent to some $x^* \in X^*$ precisely when the net $\{\langle x, x_i^*\rangle\}_{i\in I}$ converges to $\langle x, x^*\rangle$ for any $x \in X$.

The importance of the topology w^* on X^* derives from the fact that $B_{(X^*,\|\cdot\|^*)}$ is w^*-compact (this is the Alaoglu–Bourbaki theorem, see, e.g., [FHHMZ11, Theorem

3.37], a consequence of Tychonoff's theorem on the compactness of an arbitrary product of compact topological spaces).

Due to the "symmetry" evidenced by the notation $\langle x, x^* \rangle$ for the action of $x^* \in X^*$ on $x \in X$, we may also consider that x acts on X^* (precisely, $x(x^*) = \langle x, x^* \rangle$ for $x^* \in X^*$). Equation (1.2) above shows that $x : X^* \to \mathbb{R}$ is continuous, so X can be considered as a subspace of the space dual to $(X^*, \|\cdot\|^*)$ (called the **second dual space**, or just the **bidual**, and denoted by X^{**}). The (linear) mapping $j : X \to X^{**}$ that to $x \in X$ associates $j(x) : X^* \to \mathbb{R}$ given by $\langle j(x), x^* \rangle = \langle x, x^* \rangle$ for all $x^* \in X^*$, is an algebraic isomorphism into, and the Hahn–Banach theorem quoted above shows that j is, indeed, an isometry when X is endowed with its norm $\|\cdot\|$ and X^{**} with the **bidual norm** $\|\cdot\|^{**}$, defined as the dual norm of $\|\cdot\|^*$. In this text, we shall always identify $(X, \|\cdot\|)$ with a linear subspace of $(X^{**}, \|\cdot\|^{**})$, precisely the image of X under the mapping j above.
Higher dual spaces are defined recursively ($(X^{***}, \|\cdot\|^{***})$ is the dual space of $(X^{**}, \|\cdot\|^{**})$, and so on). In order to avoid cumbersome notations, we shall often omit the superscript when dealing with norms in higher dual spaces —if there is no possibility of misunderstanding—. As we mentioned in the Introduction, when working with higher dual spaces, we shall locate the even duals (X, X^{**}, ...) "to the left", and odd duals (X^*, X^{***}, ...) "to the right". So, for example, the action of $x \in X$ or $x^{**} \in X^{**}$ on $x^* \in X^*$ is represented as $\langle x, x^* \rangle$ and $\langle x^{**}, x^* \rangle$, respectively, what reminds that x can be considered as an element in X^{**}.

If it happens that $j(X) = X^{**}$, then the space X is said to be **reflexive**. Notice that every reflexive normed space is always complete (i.e., a Banach space), due to the previous observation that the dual space of a normed space is always complete. The Alaoglu-Bourbaki theorem mentioned above shows that X is reflexive if, and only if, (B_X, w) is a compact topological space. It is worth to mention that a space X can be linearly isomorphic (even linearly isometric) to X^{**} without being reflexive (see, e.g., the James' space J in [FHHMZ11, Definition 4.43 and Exercise 4.70]; see also Section 20.3 below).

Given a subset U of X, the set

$$U^\circ := \{x^* \in X^* : \langle x, x^* \rangle \leq 1 \text{ for all } x \in U\} \tag{1.3}$$

is called the **polar set of** U. If $V \subset X^*$, then its **prepolar** is defined as

$$V_\circ := \{x \in X : \langle x, x^* \rangle \leq 1 \text{ for all } x^* \in V\}. \tag{1.4}$$

The **absolute polar** of U is the set $\{x^* \in X^* : |\langle x, x^* \rangle| \leq 1 \text{ for all } x \in U\}$. The **absolute prepolar** of a set $V \subset X^*$ is defined accordingly. Note that the polar and the absolute polar of a **balanced** set U coincide (a set $U \subset X$ is called **balanced** if $\lambda U \subset U$ for all $\lambda \in [-1, 1]$). A subset of X is said to be **absolutely convex** if it is simultaneously convex and balanced. It is equivalent to say that $\sum_{i=1}^n \lambda_i x_i \in U$ whenever $n \in \mathbb{N}$, $x_i \in U$ for $i = 1, 2, \ldots, n$, and $\sum_{i=1}^n |\lambda_i| \leq 1$. The smallest

absolutely convex set that contains a given set S is the **absolutely convex hull of** S, and is denoted by $\Gamma(S)$.
Note that the polar U° of an arbitrary subset U of X is always w^*-closed and convex, and contains 0. Its absolute polar is w^*-closed and absolutely convex. The set $(U^\circ)_\circ$ is the closed convex hull $\overline{\text{conv}}\,(U\cup\{0\})$ of $U\cup\{0\}$, i.e., the smallest closed convex set that contains $U\cup\{0\}$. If we consider absolute polars instead, we get the closed absolutely convex cover $\overline{\Gamma(U)}$ of U, i.e., the smallest closed absolutely convex set that contains U. The set $U^{\circ\circ}$ $(\subset X^{**})$ is the set $\overline{\text{conv}}^{\,w^*}(U\cup\{0\})$, and a similar result works for the absolute polars. Those results are usually grouped under the name of **the bipolar theorem** (see, e.g., [FHHMZ11, Theorem 3.38]) and all of them follow from the Hahn–Banach Separation Theorem. Note that, in particular,

$$\boxed{(B_X)^\circ = B_{X^*}, \text{ and } (B_{X^*})_\circ = B_X.} \tag{1.5}$$

Moreover,

$$\boxed{B_{X^{**}} = (B_X)^{\circ\circ} = \overline{B_X}^{w^*},} \tag{1.6}$$

hence B_X is w^*-dense in $B_{X^{**}}$.

1.4 A few basic examples

We list here some examples of Banach spaces that appear naturally as building blocks for the theory. For details, see, e.g., [FHHMZ11, Sections 1.1 and 2.2].

The space c_0 consists of all **null** real sequences $x = (x_n)$, i.e., those sequences (x_n) that satisfy $\lim_n x_n = 0$. It is endowed with the norm $\|\cdot\|_\infty$ defined by $\|x\|_\infty = \sup\{|x_n| : n \in \mathbb{N}\}$, where $x = (x_n) \in c_0$. The space $(c_0, \|\cdot\|_\infty)$ is a Banach space.

If $1 \le p$, then the space ℓ_p consists of all **p-absolutely summable** real sequences $x = (x_n)$, i.e., $\sum_{n=1}^\infty |x_n|^p < \infty$. It is endowed with the norm $\|x\|_p := (\sum_{n=1}^\infty |x_n|)^{1/p}$, where $x = (x_n) \in \ell_p$. The space $(\ell_p, \|\cdot\|_p)$ is a Banach space. A separable real **Hilbert space** is a Banach space linearly isometric to $(\ell_2, \|\cdot\|_2)$. Sometimes a Hilbert space will be denoted by the letter H.

Nonseparable versions of c_0 and ℓ_p for $1 \le p < \infty$ (denoted $c_0(\Gamma)$ and $\ell_p(\Gamma)$, respectively, for an arbitrary uncountable set Γ) are defined similarly. For example, $c_0(\Gamma)$ is the vector space of all function $x : \Gamma \to \mathbb{R}$ such that for every $\varepsilon > 0$ the set $\{\gamma \in \Gamma : |x(\gamma)| > \varepsilon\}$ is finite. It is equipped with the sup norm. In the same way, $\ell_2(\Gamma)$ is the vector space of all functions $x : \Gamma \to \mathbb{R}$ such that $\sup\{(\sum_{\gamma\in F} x(\gamma)^2)^{1/2} : F \subset \Gamma \text{ finite}\} < \infty$, equipped with the norm given by this supremum. Note that functions (i.e., vectors) in those two last vector spaces have countable support (see, e.g., [FHHMZ11, Section 1.2]).

The space ℓ_∞ consists of all **bounded** real sequences $x = (x_n)$, i.e. $\sup\{|x_n| : n \in \mathbb{N}\} < \infty$. It is endowed with the norm $\|x\|_\infty := \sup\{|x_n| : n \in \mathbb{N}\}$. The space $(\ell_\infty, \|\cdot\|_\infty)$ is a Banach space, and c (i.e., the space of all convergent real sequences) and c_0 are (closed) subspaces. The space $\ell_\infty(\Gamma)$ for a general index set Γ is defined similarly.

If K is a compact Hausdorff topological space, then the space $C(K)$ consists of all continuous real-valued functions defined on K. It is endowed with the norm $\|f\|_\infty := \sup\{|f(x)| : x \in K\}$. The space $(C(K), \|\cdot\|_\infty)$ is a Banach space. See, e.g., the references to the space $C(K)$ in [FHHMZ11, Chapters 1, 2, and 3].
If $(X, \|\cdot\|_X)$ is a normed space, it often happens then that its dual space $(X^*, \|\cdot\|_X^*)$ can be naturally mapped by a linear isometry ϕ onto another easily identified Banach space $(Y, \|\cdot\|_Y)$. Thus, X and Y form a dual pair $\langle X, Y\rangle$ where the bilinear form $b : X \times Y \to \mathbb{R}$ is defined as $b(x, y) := x^*(x)$ for $x \in X$ and $x^* \in X^*$ such that $\phi(x^*) = y$. In this case it is natural to identify $(X^*, \|\cdot\|_X^*)$ with $(Y, \|\cdot\|_Y)$, and write $\langle x, y\rangle$ instead of $b(x, y)$, for $x \in X$ and $y \in Y$. Several instances of this procedure are exhibited below.
(i) If $(X, \|\cdot\|_X) := (c_0, \|\cdot\|_\infty)$, then $(Y, \|\cdot\|_Y) = (\ell_1, \|\cdot\|_1)$.
(ii) If $(X, \|\cdot\|_X) := (\ell_1, \|\cdot\|_1)$, then $(Y, \|\cdot\|_Y) = (\ell_\infty, \|\cdot\|_\infty)$.
(iii) If, for some $1 < p < \infty$, we have $(X, \|\cdot\|_X) := (\ell_p, \|\cdot\|_p)$, then $(Y, \|\cdot\|_Y) = (\ell_q, \|\cdot\|_q)$, where $p^{-1} + q^{-1} = 1$. Those spaces are reflexive.
In all cases (i), (ii), and (iii), the bilinear form on $X \times Y$ is $\langle x, y\rangle = \sum_{k=1}^\infty x_k y_k$, where $x = (x_k) \in X$ and $y = (y_k) \in Y$.

Since $(\ell_2, \|\cdot\|_2)$ is linearly isometric to its dual space, the bilinear form $\langle x, y\rangle$ for $x \in \ell_2$ and $y \in \ell_2$ $(= \ell_2^*)$ becomes an **inner product** on ℓ_2 (thus, on the separable real Hilbert space H). Nonseparable Hilbert spaces are linearly isometric to spaces $(\ell_2(\Gamma), \|\cdot\|_2)$ for uncountable Γ, and the previous considerations about the separable case apply to this one, too.

For a Hausdorff compact space K, the dual space of $(X, \|\cdot\|_X) := (C(K), \|\cdot\|_\infty)$ is linearly isometric to the space $(Y, \|\cdot\|_Y) := (M(K), v)$ of all regular Borel measures μ on K endowed with the **total variation norm** $v(\mu) = \overline{v}(K) - \underline{v}(K)$, where $\overline{v}(K) := \sup\{\mu(S) : S \subset K,\ S \text{ Borel}\}$, and $\underline{v}(K) := \inf\{\mu(S) : S \subset K,\ S \text{ Borel}\}$ (for a reference see, e.g., [Hal50, VI.28]). Each element $k \in K$ defines an element $\delta_k \in C(K)^*$ whose action on functions $f \in C(K)$ is the evaluation at k, i.e., $\langle f, \delta_k\rangle = f(k)$ for $f \in C(K)$. Note that $\|\delta_k\|_\infty^* = 1$, and $\|\delta_{k_1} - \delta_{k_2}\|_\infty^* = 2$ for $k_1, k_2 \in K$, $k_1 \neq k_2$. In the aforementioned linear isometry, δ_k corresponds to the measure (denoted again by δ_k) that to a Borel set $B \subset K$ that contains k associates the weight 1, and 0 in case that $k \notin B$. The space K is homeomorphic to the subset $\{\delta_k : k \in K\}$ of $C(K)^*$ endowed with the restriction of the w^* topology on $C(K)^*$. We refer to any of the two interpretations of the element δ_k as the **Dirac delta** (of the element k). For details see, e.g., [FHHMZ11, Lemma 3.116].

In the rest of the book, and when nothing will be said on the contrary, the classical sequence spaces c_0, ℓ_p for $1 \leq p \leq \infty$, their uncountable counterparts, and the function space $C(K)$ will be endowed with the norms $\|\cdot\|_\infty$, $\|\cdot\|_p$, and $\|\cdot\|_\infty$ defined above, respectively.

1.5 Finite-dimensional spaces

A **finite-dimensional (real) normed space** $(F, \|\cdot\|)$ (of algebraic dimension a natural number n) is a vector space having a finite algebraic basis $\{e_1, e_2, \dots, e_n\}$ $(\subset F)$ and endowed with a norm $\|\cdot\|$. Algebraically, F is linearly isomorphic to $\mathbb{R}^n$ (to $x := \sum_{k=1}^n a_k e_k \in F$, where $a_k \in \mathbb{R}$ for $k = 1, 2, \dots, n$, we associate the element $(a_1, a_2, \dots, a_n) \in \mathbb{R}^n$). Thus, a finite-dimensional normed space can be just represented as the space $(\mathbb{R}^n, \|\cdot\|)$ for some norm $\|\cdot\|$ there.

We shall often use the n-finite-dimensional (for some $n \in \mathbb{N}$) counterparts, denoted ℓ_p^n, of the spaces ℓ_p defined in Section 1.4 above. Those spaces can be seen to be linearly isometric to finite-dimensional subspaces of the corresponding space $(\ell_p, \|\cdot\|_p)$. Note that $(\ell_2^n, \|\cdot\|_2)$ is precisely the usual ***n*-dimensional Euclidean space**.

Finite-dimensional normed spaces $(F, \|\cdot\|)$ have a special feature that infinite-dimensional ones do not posses: B_X is compact in the topology $\mathcal{T}_{\|\cdot\|}$ (see Proposition 3 below). This characterization has a profound impact on the behaviour of finite-dimensional normed spaces; some straightforward consequences are listed after its proof; some other will appear once and again along this text. Proposition 3 is a particular case of a result formulated for topological vector spaces (see Theorem 6 below). The reason for giving here this "restricted" version lies in its purely normed-space proof, in contrast with the topological vector space one there.

Proposition 3. *Let $(X, \|\cdot\|)$ be a normed space. Then, the following statements are equivalent:*
(i) *X is finite-dimensional.*
(ii) *All norms on X are equivalent. In particular, all norms on X are equivalent to the norm $\|\cdot\|_1$ defined as $\|x\|_1 := \sum_{i=1}^n |x_i|$, where $x = (x_i)_{i=1}^n \in X$ in some basis $\{z_1, \dots, z_n\}$ of X* (see Section 1.4 above).
(iii) *B_X is $\mathcal{T}_{\|\cdot\|}$-compact.*

Proof. (i)⇒(ii) Assume that $\dim X = n$. Fix an algebraic basis $\{z_i\}_{i=1}^n$ of X, and naturally identify X with $\mathbb{R}^n$. Thus, the problem has been translated to prove that an arbitrary norm $\|\cdot\|$ on $\mathbb{R}^n$ is equivalent to $\|\cdot\|_1$. For simplicity, put $B := B_{(\mathbb{R}^n, \|\cdot\|)}$ and $B_1 := B_{(\mathbb{R}^n, \|\cdot\|_1)}$. Let $\{e_i\}_{i=1}^n$ be the canonical basis of $\mathbb{R}^n$. Let $C := \max\{\|e_i\| : i = 1, 2, \dots, n\}$ (> 0) (see Figure 1.2).
If $x = \sum_{i=1}^n x_i e_i \in \mathbb{R}^n$, then

$$\|x\| = \left\|\sum_{i=1}^n x_i e_i\right\| \le \sum_{i=1}^n |x_i|.\|e_i\| \le C \sum_{i=1}^n |x_i| = C\|x\|_1.$$

Thus $\|\cdot\| \leq C\|\cdot\|_1$, so $\|\cdot\|$ is continuous on the space $(\mathbb{R}^n, \|\cdot\|_1)$ (i.e., ℓ_1^n). Notice that B_1 is $\mathcal{T}_{\|\cdot\|_1}$-compact.

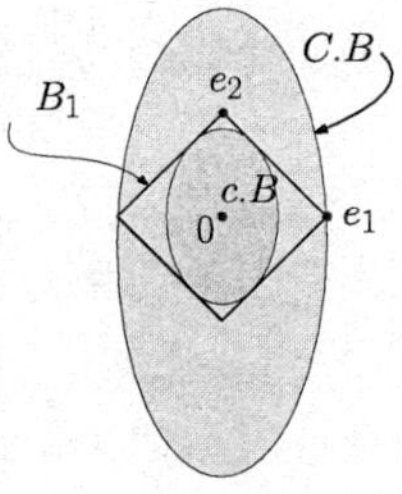

Figure 1.2: The construction in the proof of (i)⇒(ii) (Proposition 3)

➡ Indeed, if $\{x_k\}_{k\in\mathbb{N}}$ is a sequence in B_1, then for each $i \in \{1, 2, \ldots, n\}$ the sequence $\{x_k(i)\}_{k=1}^{\infty}$ is bounded, so it has a convergent subsequence. A simple diagonal procedure gives a coordinatewise convergent subsequence of $\{x_k\}_{k=1}^{\infty}$ (to a limit $x = (x(i))_{i=1}^n$). Clearly, $x \in B_1$ and the convergence is in the norm $\|\cdot\|_1$. Thus, B_1 is $\mathcal{T}_{\|\cdot\|_1}$-compact.

In particular, the unit sphere S_1 of ℓ_1^n is $\mathcal{T}_{\|\cdot\|_1}$-compact. By Weierstrass' theorem, the function $\|\cdot\|$ attains its (positive) minimum c on S_1. Therefore $c \leq \|\frac{x}{\|x\|_1}\|$ for all $x \neq 0$ in X. All together, $c\|x\|_1 \leq \|x\| \leq C\|x\|_1$ for all $x \in X$, and this gives the equivalence of $\|\cdot\|$ and $\|\cdot\|_1$.
(ii)⇒(iii) In the proof of the previous implication we showed (independently of (i)) that B_1 is $\mathcal{T}_{\|\cdot\|_1}$-compact. Notice that $B \subset (1/c)B_1$. Given a sequence $\{x_k\}_{k=1}^{\infty}$ in B, it is in $(1/c)B_1$, hence it has a $\|\cdot\|_1$-convergent subsequence. The equivalence of $\|\cdot\|$ and $\|\cdot\|_1$ shows that it clearly $\|\cdot\|$-converges to an element in B. Thus, B is $\mathcal{T}_{\|\cdot\|}$-compact.
(iii)⇒(i) Note first that if $(X, \|\cdot\|)$ is a normed space and Y is a proper closed subspace of X, the Hahn–Banach Separation Theorem gives a nonzero continuous linear form f that vanishes on Y. Without loss of generality, we may assume that $\|f\| = 1$, so given and fixed some $0 < \varepsilon < 1$ there exists $x_0 \in B_{(X,\|\cdot\|)}$ such that $f(x_0) > 1 - \varepsilon$. Thus, $\|x_0 - y\| \geq |f(x_0) - f(y)| = f(x_0) > 1 - \varepsilon$ for all $y \in Y$ (this is, essentially, Riesz's Lemma, check, e.g., [FHHMZ11, Lemma 1.37], see also Figure 1.3; its statement is reproduced below).

Lemma 4 (Riesz). *Let $(X, \|\cdot\|)$ be a normed space, and let Y be a proper closed linear subspace of X. If $0 < \varepsilon < 1$, then there exists $x_\varepsilon \in S_X$ such that $\|x_\varepsilon - y\| > 1 - \varepsilon$ for every $y \in Y$.*

➡ If Y in Riesz's statement is finite-dimensional, then we can find $x \in S_X$ such that $\|x - y\| \geq 1$ for all $y \in Y$. Indeed, apply Lemma 4 to Y and a finite-dimensional subspace Y^+ of X that contains Y. For every $n \in \mathbb{N}$ we can find

$x_n \in S_{Y^+}$ such that $\|x_n - y\| > 1 - 1/n$ for all $y \in Y$. Since S_{Y^+} is compact, there exists a $\|\cdot\|$-cluster point $x \in S_{Y^+}$ of the sequence $\{x_n\}_{n=1}^{\infty}$. This point x satisfies the requirement.

If X is infinite-dimensional, the fact that every finite-dimensional subspace is closed —it is complete, see Section 1.3 above— together with an inductive argument and the note above give a sequence $\{x_n\}_{n=1}^{\infty}$ in $B_{(X,\|\cdot\|)}$ that is 1-separated (i.e., $\|x_n - x_m\| \geq 1$ for all $n \neq m$ in $\mathbb{N}$), showing that $B_{(X,\|\cdot\|)}$ is not compact. This proves (iii)⇒(i). □

Riesz's Lemma in infinite-dimensional normed spaces has been significantly extended. The reader is advised to check Chapter 40 below for this, as well as for other related results.

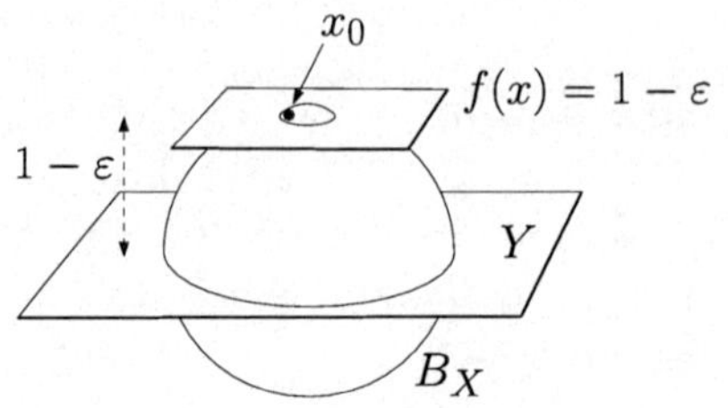

Figure 1.3: Riesz's Lemma

Let us collect here some of the important facts regarding finite-dimensional normed spaces spread across previous lines: (i) Every linear mapping between two finite-dimensional normed spaces is continuous (use matrix representation). (ii) Every finite-dimensional normed space is complete (essentially due to the completeness of $\mathbb{R}$, see Section 1.3). (iii) Every finite-dimensional normed space is reflexive. (iv) Finite-dimensionality is characterized by the compactness of the closed unit ball (Proposition 3). (iv) Finite-dimensionality is characterized by the equivalence of all norms defined on the space (Proposition 3). Other characterizations of finite-dimensionality by using tools from a more abstract setting will be provided in Theorem 6 and Corollary 16 below.

Remark 5. A concrete example of a discontinuous linear functional f on a (infinite-dimensional) normed space $(X, \|\cdot\|)$ is given by $f(p) := p'(0)$, where X is the space of all polynomials in $[-1, 1]$ endowed with the maximum norm, and $p \in X$. The proof is standard, see, e.g., [Phe93, Example 1.4 (a)]. ®

Chapter 2

Some basic definitions and tools

In this short chapter we shall provide some basic information on the theory of locally convex spaces and related topics that will be needed in the sequel. This may unify the approach to some results in Banach spaces, avoiding unnecessary repetitions. For a more complete treatment we advice the reader to look at [Kothe69], [Jarc81], and, for a concise exposition, [RoRo64]. In the second part we shall state the fundamental Krein–Milman theorem on the existence of extreme points of nonempty compact convex subsets of a vector space. This theorem links two quite different objects: extreme points of convex sets and the compactness of the set in some locally convex topology on the space —this is one of the reasons for working in the locally convex context, even if we are interested in Banach spaces—. Finally, we shall collect, for further reference, some important theorems that will appear ubiquitously in the rest of the text. We refer, too, for these and other topics, to [FHHMZ11, Chapter 3].

2.1 A short utility-grade approach to locally convex spaces

Several topologies on a Banach space already appeared in this text ($\mathcal{T}_{\|\cdot\|}$, w, w^* —the last one in a dual space—). In order to unify previous and forthcoming treatments it is convenient to present —briefly— a useful structure related to a linear space E. The purpose is to define a topology $\mathcal{T}$ on E that interacts with the algebraic structure. We shall always assume that the topologies to be considered are Hausdorff. Precisely, $(E, \mathcal{T})$ is called a **topological vector space** if the mappings $(x, y) \to x + y$ from $E \times E$ into E and $(\lambda, x) \to \lambda.x$ from $\mathbb{R} \times E$ into E, are both continuous. If $(E, \mathcal{T})$ is a topological vector space, its **dual space** E^* (or, to be more precise, $(E, \mathcal{T})^*$) is the linear space of all continuous linear forms $f : E \to \mathbb{R}$ defined on E. We want to use duality tools that depend on the existence of non-trivial continuous linear forms $f : E \to \mathbb{R}$. Thus, relying on the Hahn–Banach theorem (mentioned in Section 1.3 above), our interest is the case of 0 having a basis of convex neighborhoods (and also, by translation, any other point $x \in E$). If this is the case, the space $(E, \mathcal{T})$ is said to be a **locally convex space**.
An equivalent way to define a locally convex topology on a linear space E is to provide a system of seminorms $\mathcal{P} = \{p_i : i \in I\}$. A **seminorm** p on E is a mapping from E into $\mathbb{R}$ that satisfies (i) $p(x) \geq 0$, (iii) $p(\lambda x) = |\lambda| p(x)$, and (iv) $p(x + y) \leq p(x) + p(y)$ for all $x, y \in E$ and $\lambda \in \mathbb{R}$ (i.e., all the requirements but (ii) of a norm in Definition 1). The way the family $\mathcal{P}$ defines a topology $\mathcal{T}_{\mathcal{P}}$ on E is the following: A basis of neighborhoods of 0 is given by the sets

A. J. Guirao et al., *Renormings in Banach Spaces*, Monografie Matematyczne 75,
https://doi.org/10.1007/978-3-031-08655-7_2

$\{U_{F,\varepsilon} : F \subset I, F \text{ finite}, \varepsilon > 0\}$, where $U_{F,\varepsilon} := \{x \in E : p_i(x) < \varepsilon, i \in F\}$. A basis of neighborhoods of any other point $x \in E$ is obtained by translation. As mentioned, we work only with Hausdorff topologies, so we shall always request that the family $\mathcal{P}$ should **separate points**, i.e., if for some $x \in E$ we have $p(x) = 0$ for all $p \in \mathcal{P}$, then $x = 0$. The family $\mathcal{P}$ is said to **define** the topology $\mathcal{T}_{\mathcal{P}}$. A net $\{x_i\}_{i\in I}$ in E is $\mathcal{T}_{\mathcal{P}}$-convergent to 0 if, and only if, $p(x_i) \to 0$ for each $p \in \mathcal{P}$.

Dual pairs were introduced in Section 1.3 above in the context of normed spaces. More generally, a **dual pair** $\langle E, F\rangle$ is a couple (E, F) of vector spaces together with a bilinear form $\langle\cdot,\cdot\rangle : E \times F \to \mathbb{R}$ (i.e., a mapping that is linear in each variable separately) that **separates points** (i.e., $\langle e, f\rangle = 0$ for all $e \in E$ implies $f = 0$, and $\langle e, f\rangle = 0$ for all $f \in F$ implies $e = 0$). Note that each $f \in F$ defines a linear form on E by the formula $f(e) = \langle e, f\rangle$ for $e \in E$. Symmetrically, each $e \in E$ defines a linear form on F by the formula $e(f) = \langle e, f\rangle$ for $f \in F$. If $(E, \mathcal{T})$ is a locally convex space, we say that the topology $\mathcal{T}$ on E is **compatible with the dual pair** $\langle E, F\rangle$ if $(E, \mathcal{T})^* = F$. It is a consequence of the Hahn–Banach theorem that all topologies on E compatible with the dual pair $\langle E, F\rangle$ define the same family of convex closed subsets of E (Mazur's theorem, see, e.g., [FHHMZ11, Theorem 3.45]).

Given a dual pair $\langle E, F\rangle$, several topologies related to this structure can be defined on E (symmetrically, on F). Those are natural extensions of the previously defined w and w^* topologies on a normed space and its dual space, respectively. On E, the topology $w(E, F)$ is defined by the system $\{p_f : f \in F\}$ of seminorms $p_f(e) := |\langle e, f\rangle|$, $e \in E$, $f \in F$. Due to the separation property of the bilinear form, $w(E, F)$ is a Hausdorff topology. The topological vector space $(E, w(E, F))$ is locally convex. The topology $w(E, F)$ is the coarsest locally convex topology on E compatible with the dual pair $\langle E, F\rangle$. According to the previous observations, a net $\{e_i\}_{i\in I}$ in E is $w(E, F)$-convergent to 0 if, and only if, $\langle e_i, f\rangle \to 0$ for each $f \in F$. The finest locally convex topology on E that is compatible with the dual pair $\langle E, F\rangle$ is the **Mackey topology** $\mu(E, F)$ (the Mackey–Arens theorem, see, e.g., [Kothe69, §21.4(2)]). It is defined by the system $\{p_K : K \in \mathcal{K}\}$ of seminorms $p_K(e) := \sup_{k\in K}\langle e, k\rangle$, $e \in E$, where $\mathcal{K}$ is the family of all absolutely convex and $w(F, E)$-compact subsets of F.

If $\langle E, F\rangle$ is a dual pair, a subspace E_0 of E is $w(E, F)$-dense in E if, and only if, it **separates points of** F (i.e., $\langle e_0, f\rangle = 0$ for all $e_0 \in E_0$ implies $f = 0$). If this is the case, the restriction of the bilinear form $\langle\cdot,\cdot\rangle$ on $E \times F$ to $E_0 \times F$ shows that $\langle E_0, F\rangle$ is still a dual pair.

Every topological vector space $(E, \mathcal{T})$ can be naturally seen as a dense linear subspace of a complete topological vector space $(\widetilde{E}, \widetilde{\mathcal{T}})$ —its **completion**—. Indeed, a topological vector space is a particular instance of a uniform space. As such, it has a completion, that is readily seen to be —under the operations defined by (uniform) extension of the operations on E— a (complete) topological vector space. For details see, e.g., [Kothe69, §15.3].

Regarding finite-dimensional topological vector spaces, let us mention the following classical result (its "normed space version" is Proposition 3 above). A subset P of a topological vector space $(E, \mathcal{T})$ is said to be **precompact** if for every neighborhood V of 0 there exists a finite set $F \subset E$ such that $P \subset F+V$. The set P is precompact if, and only if, its closure in the completion of $(E, \mathcal{T})$ is compact.

Theorem 6. *Let $(E, \mathcal{T})$ be a (real) Hausdorff topological vector space. Then, the following are equivalent:*
(i) *E is finite-dimensional.*
(ii) **[Tychonoff]** *E is linearly homeomorphic to $\mathbb{R}^n$, where $n := \dim(E)$ and $\mathbb{R}^n$ is endowed with its usual topology.*
(iii) *E has a precompact neighborhood U of 0.*

➡ (The proof below follows closely the one in [Jarc81, Theorem 3.5.6])
(i)⇒(ii) Let $\{e_k\}_{k=1}^n$ be an algebraic basis of E, and let $\Phi : \mathbb{R}^n \to E$ be the linear isomorphism defined as $\Phi(\mathbf{a}) = \sum_{k=1}^n a_k e_k$, where $\mathbf{a} = (a_k)_{k=1}^n \in \mathbb{R}^n$. The fact that $\mathcal{T}$ is a topological vector space topology shows that Φ is continuous. In order to prove that Φ^{-1} is also continuous, we shall proceed by induction on n. For $n = 1$ this is trivial. Assume that it has been proved for $n-1$, and that $\dim(E) = n$. For $k \in \{1, 2, \ldots, n\}$, put $E_k := \operatorname{span}\{e_i : i = 1, 2, \ldots, n,\ i \neq k\}$. The mapping $\Phi_k : \mathbb{R}^{n-1} \to E_k$ defined as $\Phi_k(\mathbf{a}) = \sum_{i=1,\ i\neq k}^n a_i e_i$ for $\mathbf{a} = (a_i)_{i=1,\ i\neq k}^n \in \mathbb{R}^{n-1}$ is, by the induction hypothesis, a linear homeomorphism. This shows, in particular, that E_k is complete, hence closed. If π_k denotes the canonical projection of $\mathbb{R}^n$ onto its k-th coordinate, then the mapping $u_k := \pi_k \circ \Phi^{-1} : E \to \mathbb{R}$ is a linear form whose kernel is E_k. This proves that u_k is continuous. Since this is true for every $k = 1, 2, \ldots, n$, we get that Φ^{-1} is continuous.
(ii)⇒(iii) is trivial, having in mind the Bolzano–Weierstrass Theorem.
To prove that (iii)⇒(i), notice that $\{2^{-n}U\}_{n=1}^\infty$ is a basis of neighborhoods of 0. Indeed, given a neighborhood V of 0, find another neighborhood W of 0 such that $W+W \subset V$ and $\lambda W \subset V$ for all λ with $|\lambda| \le 1$ (for details, see, e.g., [Kothe69, §15.1(2)]). There exists a finite set $F \subset E$ such that $U \subset F + W$. Find $n \in \mathbb{N}$ such that $F \subset 2^n W$. Thus, $U \subset 2^n W + W \subset 2^n W + 2^n W \subset 2^n V$. This shows that $2^{-n}U \subset V$. Since U is precompact, there exists a finite subset M of E such that $U \subset M + (1/2)U$. Let $F := \operatorname{span}(M)$. Thus, $U \subset F + (1/2)U$, and so $(1/2)U \subset F + (1/4)U$. It follows that $U \subset F + (1/4)U$. Iterate this argument to show $U \subset F + 2^{-n}U$ for all $n \in \mathbb{N}$. By (ii), F is closed, and the fact that $U \subset \bigcap_{n=1}^\infty (F + 2^{-n}U)$ shows that $U \subset F$. This implies $F = E$. □

Examples of locally convex spaces are
(i) Normed spaces $(X, \|\cdot\|)$. A basis of neighborhoods of 0 is given by the family of closed balls $\{(1/n)B_X : n \in \mathbb{N}\}$ —alternatively, the family consisting of the single seminorm (in this case, a norm) $\|\cdot\|$ defines the norm topology $\mathcal{T}_{\|\cdot\|}$.

(ii) The spaces $(X, w(X, X^*))$ and $(X^*, w(X^*, X))$ as defined above, where X is a normed space.
(iii) More generally, if $(X, \|\cdot\|)$ is a normed space and $Y \subset X^*$ is a linear subspace of X^* that separates points (i.e., Y is $w(X^*, X)$-dense in X^*), then the restriction of the bilinear form $\langle\cdot,\cdot\rangle$ to $X \times Y$ makes $\langle X, Y\rangle$ a dual pair, so we may define $w(X, Y)$ on X according to the previous observations. The dual space of $(X, w(X, Y))$ is then Y. If a subspace Y of X separates points of X^* (i.e., Y is $w(X, X^*)$-dense in X, what amounts to say that Y is $\mathcal{T}_{\|\cdot\|}$-dense in X), then the restriction of the bilinear form $\langle\cdot,\cdot\rangle$ to $Y \times X^*$ shows that $\langle Y, X^*\rangle$ is a dual pair, so we may consider the topology $w(X^*, Y)$ on X^*. The dual space of $(X^*, w(X^*, Y))$ is Y.
(iv) If $(X, \|\cdot\|)$ is a normed space, then $\mathcal{T}_{\|\cdot\|}$ agrees with the Mackey topology $\mu(X, X^*)$ (see, e.g., [Kothe69, §21.5(3)]). If X is not reflexive, then $\mu(X^*, X)$ and $\mathcal{T}_{\|\cdot\|^*}$ are different: The dual space of $(X^*, \mu(X^*, X))$ is X, while the dual space of $(X^*, \|\cdot\|^*)$ is X^{**}.

2.2 Extreme points

If C is a convex subset of a Banach space, a point $e \in C$ is said to be an **extreme point of** C if there are no $x, y \in C$, $x \neq y$, such that $c = (x+y)/2$. It follows from the parallelogram identity in a real Hilbert space H (see Subsection 5.2.1 below) that for $x, y \in B_H$, $x \neq y$, we have $\|(x+y)/2\|^2 \leq 1 - \|(x-y)/2\|^2$, hence every $x \in S_H$ is an extreme point of B_H. In a linear space, the convex hull C of a finite set P of points (i.e., a **polytope**) has extreme points (sometimes called **vertices** of the polytope), all of them in P —in fact, C is the convex hull of the set of its extreme points—. In $(\mathbb{R}^n, \|\cdot\|_1)$, the closed unit ball has $2n$ extreme points (precisely, points $\pm e_k$, $k = 1, 2, \dots, n$), and in $(\mathbb{R}^n, \|\cdot\|_\infty)$, its closed unit ball has 2^n extreme points (precisely, points $x = (x_k)_{k=1}^n$, where $|x_k| = 1$ for all $k = 1, 2, \dots, n$). Similar descriptions hold for the spaces $(\ell_1, \|\cdot\|_1)$ and $(\ell_\infty, \|\cdot\|_\infty)$, see below. After the previous examples, it is somehow against the intuition that there are closed convex and bounded subsets of a Banach space without extreme points (the closed unit ball of $(c_0, \|\cdot\|_\infty)$ is such, see the box below). Behind this "strange" behaviour is the absence of a locally convex topology on the space that makes the set compact (see Theorem 7).

➪ Indeed, every point in $B_{(c_0, \|\cdot\|_\infty)}$ can be written as the middle point of two other points in $B_{(c_0, \|\cdot\|_\infty)}$, so there are no extreme points of $B_{(c_0, \|\cdot\|_\infty)}$. Figure 2.1 gives the reason for it.

In the presence of compactness the situation is different:

Theorem 7 (Krein–Milman, see, e.g., [FHHMZ11], Theorem 3.65). *Let $(E, \mathcal{T})$ be a locally convex space, and let K be a nonempty compact and convex subset of E. Then K is the closed convex hull of the set* Ext (K) *of all its extreme points (in particular,* Ext (K) *is nonempty).*

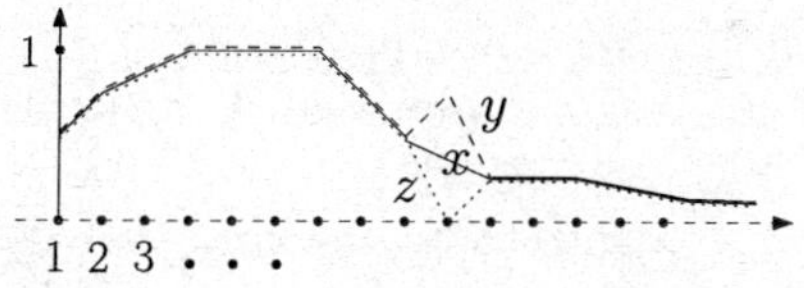

Figure 2.1: $x \in B_{(c_0, \|\cdot\|_\infty)}$ is the middle point of y and z, both in $B_{(c_0, \|\cdot\|_\infty)}$

An example that the closure in Theorem 7 is needed is in [FHHMZ11, Exercise 3.149]. In contrast, there is no need to compute the closure in the finite-dimensional case (see Item (ii) in the next paragraph).

Theorem 7 has some striking consequences. We mention here just a few:
(i) If K is a convex subset of a locally convex space $(E, \mathcal{T})$, and it turns out that K is compact in some compatible topology —the least demanding requirement is to ask for weak compactness—, then K can be "reproduced" as the closure of the convex hull of the set of its extreme points —and the closure can be computed in any compatible topology, say in $\mathcal{T}$, as we noted above (Mazur's theorem).
(ii) Every closed convex and bounded subset of a finite-dimensional space is the convex hull of the set of all its extreme points. For a more precise statement check Carathéodory's theorem (see, e.g., [FHHMZ11, Proposition 3.70]). A consequence of this last result is that the convex hull of a compact subset in a finite-dimensional Banach space is compact.
(iii) If K is a w^*-closed, convex and bounded subset of a dual Banach space, then K is the w^*-closed convex hull of the set of all extreme points. This shows, in view of the absence of extreme points of the closed unit ball of $(c_0, \|\cdot\|_\infty)$ (see the introduction to this section), that $(c_0, \|\cdot\|_\infty)$ cannot be linearly isometric to a dual space. In view of the description of the set of extreme points of $S_{C[0,1]}$ (see Example 4 in this section) the same happens to $(C[0,1], \|\cdot\|_\infty)$.
(iv) The closed unit ball of a reflexive Banach space is the closed convex hull of the set of all its extreme points.

The following is a useful kind of converse to Theorem 7:

Theorem 8 (Milman, see, e.g., [FHHMZ11], Theorem 3.66). *Let $(E, \mathcal{T})$ be a locally convex space, and let K be a nonempty compact and convex subset of E. If K is the closed convex hull of a subset A, then* Ext $(K) \subset \overline{A}$.

In this direction, let us mention here the following beautiful important result of Grothendieck:

Theorem 9 (Grothendieck). (i) *Let X be a Banach space. Let $A \subset X$ be totally bounded. Then, there exists a sequence $\{x_n\}$ in X such that $x_n \to 0$ and $A \subset \{\sum_{n=1}^\infty \lambda_n x_n : \lambda_n \geq 0$ for all $n \in \mathbb{N}$, and $\sum_{n=1}^\infty \lambda_n = 1\}$.*

(ii) *Let* $\{x_n\}_{n=1}^\infty$ *be a* w*-null sequence in a sequentially complete locally convex space. Then, the closed absolutely convex hull* Γ *of* $\{x_n : n \in \mathbb{N}\}$ *is* w*-compact, and* $\Gamma = \{\sum_{n=1}^\infty \lambda_n x_n : \sum_{n=1}^\infty |\lambda_n| \leq 1\}$.

Proof. For (i) see, e.g., [FHHMZ11, Exercise 1.69]. The first part of the conclusion in (ii) follows from Theorem 7. For a proof of the second part see, e.g., [Kothe69, §20.10(6)]. □

Let us list some examples (and some consequences):

1. $\operatorname{Ext}(B_{(c_0,\|\cdot\|_\infty)}) = \emptyset$, as it was proved above. Then —and this was also mentioned there—, then $(c_0, \|\cdot\|_\infty)$ is not linearly isometric to any dual Banach space. See, e.g., [FHHMZ11, Exercise 3.131].

2. $\operatorname{Ext}(B_{(\ell_1,\|\cdot\|_1)}) = \{\pm e_n : n \in \mathbb{N}\}$, where e_n is the n-th vector of the canonical basis of ℓ_1. See also, e.g., [FHHMZ11, Exercise 3.135].

3. $\operatorname{Ext}(B_{(\ell_\infty,\|\cdot\|_\infty)}) = \{x = x(n) \in S_{(\ell_\infty,\|\cdot\|_\infty)} : x(n) = \pm 1, \text{ for all } n \in \mathbb{N}\}$. See also, e.g., [FHHMZ11, Exercise 3.129].

4. $\operatorname{Ext}(B_{(C[0,1],\|\cdot\|_\infty)}) = \{1, -1\}$, where 1 denotes the constant function with value 1, and -1 the constant function with value -1. As in the example of $(c_0, \|\cdot\|_\infty)$ mentioned above, $(C[0,1], \|\cdot\|_\infty)$ is not linearly isometric to any dual Banach space. This has been already mentioned after Theorem 7. See also, e.g., [FHHMZ11, Exercise 3.133].

5. $\operatorname{Ext}((B_{C(K)^*}, \|\cdot\|_\infty^*)) = \{\pm\delta_k : k \in K\}$, where K is a compact space, and $\delta_k \in C(K)^*$ is the Dirac delta associated to $k \in K$ (see Section 1.4 above). Note, too, that $(c_0, \|\cdot\|_\infty)$ cannot be linearly isometric to any $(C(K), \|\cdot\|_\infty)$ space.

6. Every completely regular Hausdorff topological space T can be embedded into a compact topological space of the form $[0,1]^\Gamma$:

   ➪ Indeed, the mapping $x \to \delta_x$ from T into $B_{CB(T)^*}$, where $CB(T)$ denotes the Banach space of all continuous bounded functions from T into $\mathbb{R}$ endowed with the supremum norm, is a topological homeomorphism when $B_{CB(T)^*}$ carries the w^*-topology. This shows that T can be seen as a subspace of the compact topological space $[-1,1]^{B_{CB(T)^*}}$ in its product topology (and so of some $[0,1]^\Gamma$, as claimed).

   Thus, its closure there is a **compactification** of T (i.e., a compact topological space where T is dense). If L is locally compact and not compact, a simpler way to build a compactification is to adjoin a single point $\infty \notin L$ and to define sets in $\widetilde{L} := L \cup \{\infty\}$ to be open if either they lie in L and they are open there, or they contain ∞ and their complement in $\widetilde{L}$ (a subset of L) is compact in L. In this way $\widetilde{L}$ becomes a compact space where L is dense.

As usual, $C(\widetilde{L})$ denotes the space of all real-valued continuous functions on $\widetilde{L}$ endowed with the supremum norm (see the previous item). $C_0(L)$ denotes the subspace consisting of all functions $f \in C(\widetilde{L})$ **vanishing at** ∞, i.e., such that $f(\infty) = 0$. If $f \in C_0(L)$, then $f|_L$ is a continuous function on L such that for every $\varepsilon > 0$ there exists a compact subset K of L with $|f(x)| < \varepsilon$ for all $x \in L \setminus K$. Conversely, such a function can be extended to an element in $C_0(L)$ by letting $f(\infty) = 0$. Note that $C_0(L) = \ker \delta_\infty$, where $\delta_\infty \in C(\widetilde{L})^*$ is the Dirac delta of the element ∞. Thus, $C_0(L)^* = C(\widetilde{L})^*/\text{span}\,\{\delta_\infty\}$. Let $q : C(\widetilde{L})^* \to C(\widetilde{L})^*/\text{span}\,\{\delta_\infty\}$ be the canonical quotient mapping. Every extreme point of $B_{C_0(L)^*}$ is the image under q of an extreme point of $B_{C(\widetilde{L})^*}$, i.e., of an element $\pm\delta_{\tilde{l}}$ for some $\tilde{l} \in \widetilde{L}$ (see again the previous item).

➡ This holds for any normed space $(X, \|\cdot\|)$ and a closed subspace $Y \subset X$. The dual space Y^* is linearly isometric to $X^*/Y^\perp$ endowed with the quotient norm of the norm $\|\cdot\|^*$. If $q : X^* \to Y^*$ is the canonical quotient mapping, and y^* is an extreme point of B_{Y^*}, then $\{x^* \in S_{X^*} : q(x^*) = y^*\}$ is a w^*-compact convex and nonempty set. By Theorem 7, it has an extreme point x^*, that is easily seen to be an extreme point of B_{X^*}. Moreover, $q(x^*) = y^*$.

Since $q(\delta_\infty) = 0$, necessarily $\tilde{l} \neq \infty$. Let us show that, conversely, every point $q(\pm\delta_{l_0})$, for $l_0 \in L$, is an extreme point of $B_{C_0(L)^*}$. It is enough to argue for $q(\delta_{l_0})$ (for $q(-\delta_{l_0})$ the argument is similar). The proof follows the same idea used to show that for a compact space K and $k \in K$, δ_k is an extreme point of $B_{C(K)^*}$ (see Item 5 above; see also, e.g., [FHHMZ11, Lemma 3.116]). Fix $l_0 \in L$. There exists a basis $\mathcal{U}$ of compact neighborhoods of l_0. Given $U \in \mathcal{U}$, find a continuous function $f_U : L \to [0,1]$ such that $f_U(l_0) = 1$ and $f_U|_{L\setminus U} \equiv 0$. Then $f_U \in C_0(L)$. We get $q(\delta_{l_0}) \in F := \bigcap_{U\in\mathcal{U}}\{\mu \in B_{C_0(L)^*} : \langle f_U, \mu\rangle = 1\}$. The set F is nonempty, convex, and w^*-compact, so it contains an extreme point (that is also an extreme point of $B_{C_0(L)^*}$, i.e., an element of the form $q(\pm\delta_l)$ for some $l \in L$). Since $f_U(L) \subset [0,1]$ for all $U \in \mathcal{U}$, it must be of the form $q(\delta_l)$. From the fact that $\mathcal{U}$ is a basis of neighborhoods of l_0 and that $f_U(l) = 1$ for every $U \in \mathcal{U}$, we must have $l = l_0$. This proves that $q(\delta_{l_0})$ is an extreme point of $B_{C_0(L)^*}$. This happens for every $l_0 \in L$. Thus, finally, $\text{Ext}\, B_{C_0(L)^*} = \{q(\pm\delta_l) : l \in L\}$. Those elements are just the evaluation and minus the evaluation, respectively, of functions $f \in C_0(L)$ at a given point $l \in L$.

Natural applications of the concept of extreme point are to the theory of optimization. We need the basic definition of a convex function:

Definition 10. *Let X be a vector space, and let $C \subset X$ be a nonempty convex subset. A function $f : C \to \mathbb{R}$ is said to be* **convex** *whenever $f(\lambda x + (1-\lambda)y) \leq \lambda f(x) + (1-\lambda)f(y)$ for every $x, y \in C$ and $\lambda \in [0,1]$.*

Observe that a function $f : C \to \mathbb{R}$ is convex if, and only if, its **epigraph** $\operatorname{epi}(f) := \{(x, r) \in C \times \mathbb{R} : f(x) \leq r\}$ is a convex set.

We just mention here two results:

(i) *If X is a normed space, C is a convex and w-compact subset of X, and $x^* \in X^*$, then x^* attains its supremum on C at an extreme point of C.* Indeed, if M is the supremum of x^* on C, then $C \cap \{x \in X : f(x) = M\}$ is a convex closed subset of C. It has extreme points by Theorem 7, that are readily seen to be extreme points of C.

(ii) *If for some $n \in \mathbb{N}$, C is a convex compact subset of $\mathbb{R}^n$ and $f : C \to \mathbb{R}$ is a convex continuous function, then f attains its supremum on C at an extreme point* (the Bauer Maximum Principle) (see, e.g., [MoZiZi15, Exercise 13.523], see also Figure 2.2). For the proof, note first that if f is restricted to an open segment I of C, and attains it supremum on I at some point, then it must be constant on I. Thus, if f attains its supremum on C at a point x_0 that is interior (relatively to span (C)) of C, it must be constant on C and we are done. On the contrary, the point has a supporting hyperplane H. If $H \cap C = \{x_0\}$, then x_0 is an extreme point. Otherwise, repeat the argument on $C \cap H$, so reducing the dimension of the space. The conclusion is achieved by induction.

A strengthening of the notion of extreme point is in the next definition:

Definition 11. *A point $x_0 \in S_X$ is said to be a (w-)* **strongly extreme point of** B_X *if given two sequences $\{y_n\}_{n=1}^{\infty}$ and $\{z_n\}_{n=1}^{\infty}$ in B_X such that $(y_n + z_n) \to 2x_0$, then $y_n \to x_0$ (respectively, $y_n \xrightarrow{w} x_0$).*

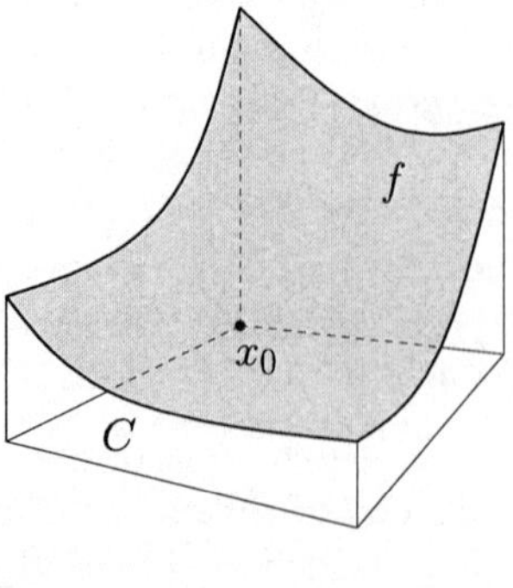

Figure 2.2: A convex function on a compact convex set in $\mathbb{R}^n$ attains its supremum at an extreme point

2.3 Some basic results in Banach space theory

Theorem 12 (Banach Open Mapping and Closed Graph theorems). (See, e.g., [FHHMZ11, Theorems 2.25 and 2.27]) *Let X and Y be Banach spaces. A linear mapping $T : X \to Y$ with closed graph $G(T) := \{(x, Tx) : x \in X\} \subset X \times Y$ is continuous. If, moreover, T is onto, then T is open (i.e., the image of an open subset of X is open in Y).*

In particular, if X and Y are Banach spaces, an $T : X \to Y$ is linear, continuous, one-to-one, and onto, then T is a linear isomorphism.

Let X be a Banach space, and let M be a nonempty subset of X. We say that $x_0^* \in X^*$ **attains its supremum on** M if there exists $x_0 \in M$ such that $\langle x_0, x_0^* \rangle = \sup\{\langle x, x_0^* \rangle : x \in M\}$. In the particular case that $M := B_X$, we say then that x^* **attains its norm**. The set of all elements in X^* that attain their norm is denoted by $NA(X, \|\cdot\|)$ or, if the norm $\|\cdot\|$ is understood, by $NA(X)$. In general, $NA(X)$ is a proper subset of X^*. For example, the element $x^* := (1 - 1/n)_{n=1}^\infty \in \ell_\infty$ does not belong to $NA(\ell_1, \|\cdot\|_1)$.

Assume that a Banach space $(X, \|\cdot\|)$ is finite-dimensional. Then, by Proposition 3, B_X is $\|\cdot\|$-compact, and so every continuous linear functional $x^* \in X^*$ attains its norm. More generally, if X is reflexive (see Section 1.3), (B_X, w) is w-compact, and again every $x^* \in X^*$ (a w-continuous mapping) attains its norm. In both cases we have $NA(X) = X^*$. It is a striking fundamental result that the converse is true:

Theorem 13 (James). (See, e.g., [FHHMZ11, Theorem 3.130]) *A Banach space X is reflexive if, and only if, $NA(X) = X^*$. More generally, a weakly closed subset M of X is w-compact if, and only if, every $x^* \in X^*$ attains its supremum on M.*

A proof in the separable case, based on the concept of James boundary and on Simons inequality (Lemma 537 and Theorem 538), will be provided in Chapter 28 below. There we shall mention some landmarks to the final version of this important theorem.

Spaces $(X, \|\cdot\|)$ where $NA(X)$ was a $\|\cdot\|^*$-dense subset of X^* where called **subreflexive**. An important result is that every Banach space is such:

Theorem 14 (Bishop–Phelps, [BiPh61]). (See, e.g., [FHHMZ11, Theorem 7.41]) *If $(X, \|\cdot\|)$ is a Banach space, then $NA(X)$ is norm-dense in X^*. More precisely, if C is a nonempty closed convex and bounded subset of X, then the set of all continuous linear functionals on X that attain their maximum on C is norm-dense in X^*.*

A proof via the Smooth Variational Principle for separable —or, more generally, weakly compactly generated— Banach spaces is in [GuiMoZi18]. This shows that also the Bishop–Phelps theorem can be proved by the "good witch" Baire category theorem. Here we include the very interesting original proof (which preceded, and in fact, motivated, Ekeland's formulation of his Variational Principle (see Theorems 286, 287, and Subsection 11.7.2):

Proof. Assume $f \in S_{X^*}$ and $\varepsilon > 0$ be given. By the Hyperplane Parallel Lemma (see, e.g., [FHHMZ11, Exercise 2.13]), it suffices to find $g \in S_X^*$ such that $|g(x)| \leq 1$ for all $x \in T := \{x \in X : \; f(x) = 0, \; \|x\| \leq 2\varepsilon^{-1}\}$ and for which there is $x \in S_X$ such that $g(x) = 1$. Let C be the convex hull of the union of the sets T and B_X (see Figure 2.3). Suppose that there is $x_0 \in B_X$ which is also in the boundary of C. Since C has nonempty interior, by the Hahn–Banach Separation Theorem, there is $g \in S_{X^*}$ such that $\sup\{g(x) : \; x \in C\} = g(x_0)$. It follows that $\|x_0\| = 1$ and thus $g(x_0) \leq 1$. On the other hand, $g(x_0) \geq \sup_{B_X} g = 1$. Thus $g(x_0) = 1 = \|x_0\|$.

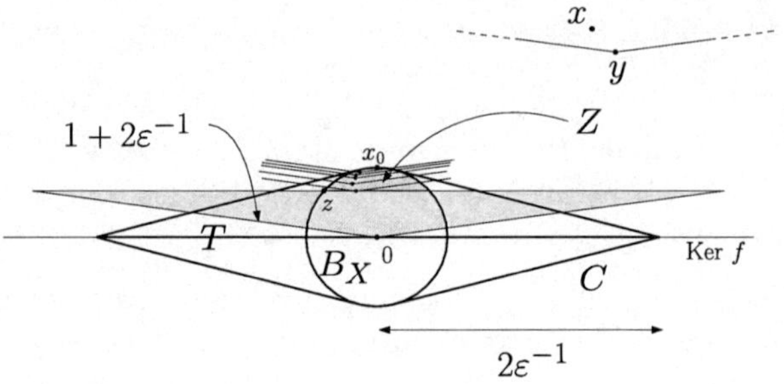

Figure 2.3: The construction in the proof of Theorem 14. In light gray, the cone giving the partial order $\prec$. Note that $y \prec x$

Thus it remains to show that $B_X \cap \text{bdry}C = \emptyset$. To this end, choose $z \in B_X$ such that $f(z) > 0$. If $f(z) = 1$, there is nothing to prove. Otherwise put $K := (f(z))^{-1}(1 + 2\varepsilon^{-1})$. Define a partial ordering on the set $Z := \{x \in B_X : \; f(x) \geq f(z)\}$ as follows: We say that $y \prec x$ whenever

$$\text{(i) } f(y) < f(x), \text{ and}$$
$$\text{(ii) } \|x - y\| \leq K\big(f(x) - f(y)\big)$$

(geometrically, this means that x is in a cone pointed at y as in the picture). You may wish to note that there are points $x \in Z$ such that $z \prec x$. Suppose now that W is a totally ordered subset of Z (some points are represented in Figure 2.3). By (i), the net of real numbers $\{f(x) : \; x \in W, \; \prec\}$ is bounded above and monotonically increasing, so it converges to its supremum. From (ii) it follows that W is a Cauchy net. By the completeness of X , W converges to a point $y \in B_X$. It follows that y is an upper bound for W. By Zorn's Lemma, there exists a maximal element $x_0 \in Z$. Since $x_0 \in B_X \subset C$, we need only to show that $x_0 \in \text{bdry } C$. If not, then x_0 is in the interior of C and there is $\alpha > 0$ such that $x_0 + \alpha z \in C$ (see Figure 2.4). From the definition of C, we then have that there exist $y \in B_X$, $x \in T$, and $\lambda \in [0, 1]$, such that $x_0 + \alpha z = \lambda y + (1 - \lambda)x$. Then $f(z) \leq f(x_0) < f(x_0 + \alpha z) = \lambda f(y) \leq f(y)$, so that $y \in Z$. Also, $y - x_0 = (1 - \lambda)(y - x) + \alpha z$.

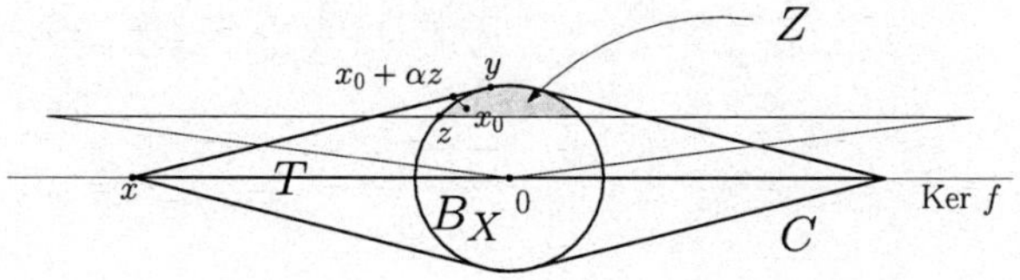

Figure 2.4: If $x_0 \notin$ bdry C in the proof of Theorem 14, then searching for a contradiction

Thus

$$\begin{aligned}\|y - x_0\| &\leq (1-\lambda)\|y - x\| + \alpha \leq (1-\lambda)(\|y\| + \|x\|) + \alpha \\ &\leq (1-\lambda)(1+2\varepsilon^{-1}) + \alpha \leq (1-\lambda+\alpha)(1+2\varepsilon^{-1}).\end{aligned}$$

On the other hand, $f(y-x_0) = (1-\lambda)f(y)+\alpha f(z) \geq (1-\lambda+\alpha)f(z)$, so $\|y-x_0\| \leq K\big(f(y) - f(x_0)\big)$. This shows that $x_0 \prec y$, a contradiction, so $x_0 \in$ bdry C, and the proof is complete. □

A variant of Theorem 14 is due to Bollobás (see Figure 2.5):

Theorem 15 (Bishop–Phelps–Bollobás, [Bol99], Theorem 8.11)**.** (See also, e.g., [FHHMZ11, Exercise 7.53]) *Let $(X, \|\cdot\|)$ be a Banach space, and let $x_0 \in S_X$ and $x_0^* \in S_{X^*}$ such that*

$$|\langle x_0, x_0^*\rangle - 1| < \delta := \frac{1}{4}\varepsilon^2,$$

where $0 < \varepsilon < 1$. Then there exists $x_1 \in S_X$ and $x_1^* \in S_{X^*}$ such that $\langle x_1, x_1^*\rangle = 1$, $\|x_1^* - x_0^*\| \leq \varepsilon$, and $\|x_1 - x_0\| \leq \varepsilon$.

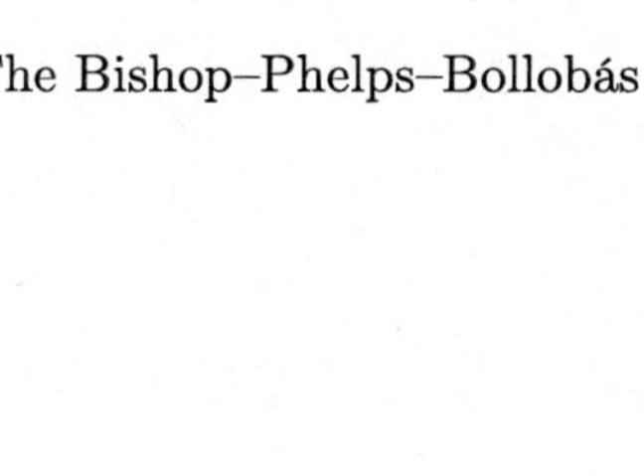

Figure 2.5: The Bishop–Phelps–Bollobás Theorem 15

Check for updates

Chapter 3

Equivalent norms

In this chapter we discuss the basic notion of an equivalent norm (Definition 2 above) in Section 3.1, and for finite-dimensional spaces, in Section 3.2. We consider then several basic constructions that provide equivalent norms. The first one works in the dual space, and generates the new equivalent norm on the original space (Section 3.3). The second one considers norming subspaces of the dual (Sections 3.4 and 3.5). The third one combines, in a way or another, some convex bodies in the space to produce an equivalent norm there (Section 3.6).

3.1 On the definition of equivalent norms

The basic concept of equivalent norms has been presented in Definition 2. The ubiquitous appearance of equivalent norms in this text advices to record some alternatives to the definition. **In the items below $\|\cdot\|$ and $|||\cdot|||$ are norms on a vector space X. They are equivalent if, and only if, each of the following items —and then all of them— holds**:

1. The geometric counterpart to (1.1) is the following:

$$c.B_{(X,|||\cdot|||)} \subset B_{(X,\|\cdot\|)} \subset C.B_{(X,|||\cdot|||)}. \tag{3.1}$$

2. Both norms provide the same family of convergent sequences.

   ➪ Indeed, if (1.1) holds, then obviously a sequence $\{x_n\}_{n\in\mathbb{N}}$ is $\|\cdot\|$-convergent (say to x) if, and only if, it is $|||\cdot|||$-convergent (to x again). Assume, on the other side, that (1.1) does not hold. In particular, for $n\in\mathbb{N}$ we have $(1/n)B_{(X,|||\cdot|||)} \not\subset B_{(X,\|\cdot\|)}$ or $B_{(X,\|\cdot\|)} \not\subset nB_{(X,|||\cdot|||)}$. One of the alternative holds for a subsequence $\{n_k\}$. Assume the first one. Then there exists a sequence $\{x_k\}_{k\in\mathbb{N}}$ in $(1/n_k)B_{(X,|||\cdot|||)}$ such that $\|x_k\|>1$ for all $k\in\mathbb{N}$. Thus, $\{\sqrt{n_k}x_k\}$ is $|||\cdot|||$-convergent to 0 (we say that the sequence is $|||\cdot|||$-**null**), while $\|\sqrt{n_k}x_k\| \geq \sqrt{n_k}$ for all $k\in\mathbb{N}$, so $\{\sqrt{n}_k x_{n_k}\}_{k\in\mathbb{N}}$ does not $\|\cdot\|$-converge. If, instead, the other part of the alternative holds for a subsequence $\{n_k\}_{k\in\mathbb{N}}$, then there exists a sequence $\{x_k\}_{k\in\mathbb{N}}$ in $B_{(X,\|\cdot\|)}$ such that $|||x_k|||>n_k$ for all $k\in\mathbb{N}$. Thus, $\{(1/\sqrt{n}_k)x_k\}_{k\in\mathbb{N}}$ is $\|\cdot\|$-null while $|||(1/\sqrt{n}_k)x_k|||>\sqrt{n_k}$ for all $k\in\mathbb{N}$.

3. The topologies $\mathcal{T}_{\|\cdot\|}$ and $\mathcal{T}_{|||\cdot|||}$ associated to $\|\cdot\|$ and $|||\cdot|||$, respectively, coincide. This can be seen by looking at Item 2 above, since two metrizable topologies coincide as soon as they provide the same convergent sequences.

A. J. Guirao et al., *Renormings in Banach Spaces*, Monografie Matematyczne 75,
https://doi.org/10.1007/978-3-031-08655-7_3

4. The identity mapping $I : (X, \|\cdot\|) \to (X, |||\cdot|||)$ is continuous and its inverse is also continuous.

   ➪ (i) If both $(X, \|\cdot\|)$ and $(X, |||\cdot|||)$ are Banach spaces, and $\|\cdot\| \leq |||\cdot|||$, then the Open Mapping Theorem 12 above shows that the two norms are equivalent.

   (ii) We may ask what happens when there exists an arbitrary linear, one-to-one, onto, and continuous mapping $T : (X, \|\cdot\|) \to (X, |||\cdot|||)$ with a continuous inverse.

   (a) If both $(X, \|\cdot\|)$ and $(X, |||\cdot|||)$ are Banach spaces, then the Baire category theorem shows that the set $T(B_{(X,\|\cdot\|)})$ is the closed unit ball of a norm on X equivalent to $|||\cdot|||$.

   (b) If the completeness condition is removed, it may happen that two norms $\|\cdot\|$ and $|||\cdot|||$ on a vector space X make $(X, \|\cdot\|)$ and $(X, |||\cdot|||)$ linearly topologically isomorphic and yet the two norms are not equivalent. As an example, let c_{00} be the subspace of c_0 consisting of all vectors with finite support. For $x = (x_k) \in c_{00}$, put $\|x\| := \sum_{k=1}^{\infty} k|x_{2k-1}| + \sum_{k=1}^{\infty} |x_{2k}|$, and $|||x||| := \sum_{k=1}^{\infty} |x_{2k-1}| + \sum_{k=1}^{\infty} k|x_{2k}|$. By looking at the canonical basis of c_{00} it turns out that the two norms are not equivalent. However, the mapping $T : (c_{00}, \|\cdot\|) \to (c_{00}, |||\cdot|||)$ that interchanges odd and even coordinates is a linear topological isomorphism. This example was provided by J. Bonet (personal communication).

5. Inequalities (1.1) and inclusions (3.1) are equivalent to the following inequalities and inclusions concerning the dual norms and closed dual unit balls, respectively, a consequence of the bipolar theorem.

$$c|||x^*|||^* \leq \|x^*\|^* \leq C|||x^*|||^*, \text{ for all } x^* \in X^*. \tag{3.2}$$

   In other words,

$$c.B_{(X^*,\|\cdot\|^*)} \subset B_{(X^*,|||\cdot|||^*)} \subset C.B_{(X^*,\|\cdot\|^*)}. \tag{3.3}$$

6. The dual spaces of $(X, \|\cdot\|)$ and $(X, |||\cdot|||)$ coincide (as vector spaces; obviously, the respective dual norms are different as soon as the two norms $\|\cdot\|$ and $|||\cdot|||$ are).

   ➪ Indeed, if $\|\cdot\|$ and $|||\cdot|||$ are equivalent, both topologies $\mathcal{T}_{\|\cdot\|}$ and $\mathcal{T}_{|||\cdot|||}$ coincide, and so they do the continuous linear forms on $(X, \|\cdot\|)$ and on $(X, |||\cdot|||)$. On the other hand, the topology of a metrizable locally convex space coincide with the Mackey topology (see, e.g., [Kothe69, §21.5(3)]). Thus, $\mathcal{T}_{\|\cdot\|} = \mathcal{T}_{|||\cdot|||}$, and we can apply Item 3 above.

The following elementary construction provides, in every normed space $(X, \|\cdot\|)$, an equivalent norm $|||\cdot|||$ which is not a multiple of the original norm (see Figure 3.1): Let $x_0 \notin B_{(X,\|\cdot\|)}$. The set $\operatorname{conv}(B_{(X,\|\cdot\|)} \cup \{x_0\} \cup \{-x_0\})$ is absolutely convex, bounded, and closed, so it is the closed unit ball of a norm $|||\cdot|||$ on X. Since $B_{(X,\|\cdot\|)} \subset B_{(X,|||\cdot|||)} \subset \|x_0\|.B_{(X,\|\cdot\|)}$, the norm $|||\cdot|||$ is an equivalent norm.

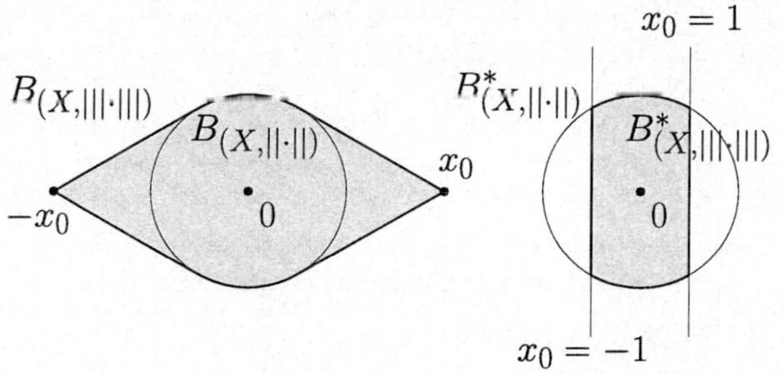

Figure 3.1: The closed unit ball of the norm $|||\cdot|||$ and its dual (both in gray)

3.2 Finite-dimensionality and equivalent norms

Regarding our central subject —renorming— a basic fact is that all norms on a finite-dimensional vector space F are equivalent (see, e.g., [FHHMZ11, Proposition 1.36] and Proposition 3 above, where in fact it is stated that this property characterizes the finite-dimensional Banach spaces). As we mentioned in Items 3 and 2 in Section 3.1 above, another way to formulate this is to say that all norms on X define the same topology —or, alternatively, they provide the same convergent sequences.

Thus, from the topological point of view, there is nothing more to say: There is one and only one topology on F that is defined by a norm (and so all finite-dimensional Banach spaces are linearly topologically isomorphic to the Euclidean space $(\mathbb{R}^n, \|\cdot\|_2)$, where n is the algebraic dimension of F). However, from the geometric point of view, the differences when using a norm or another can be dramatic: Uniqueness and location of points of best approximation, recovering the closed unit ball from a minimal amount of information (points), the possibility to use differential calculus —or not— in optimization or other problems or, if the use of a norm is imposed by the particular problem treated, the possibility to approximate the given norm by "better" (infinitely many times differentiable, for example) norms with a prescribed accuracy, and to provide bounds for the committed error in any of the subsequent computations.

Proposition 3 above provided two statements equivalent to the finite-dimensionality of the space in terms of the compactness of its closed unit ball and the equivalence of all norms on it (Figure 3.2 illustrates the reason why the $\|\cdot\|_2$, $\|\cdot\|_1$, and $\|\cdot\|_\infty$-norms on $\mathbb{R}^2$ are equivalent, in view of Item 1 in Section 3.1 above).

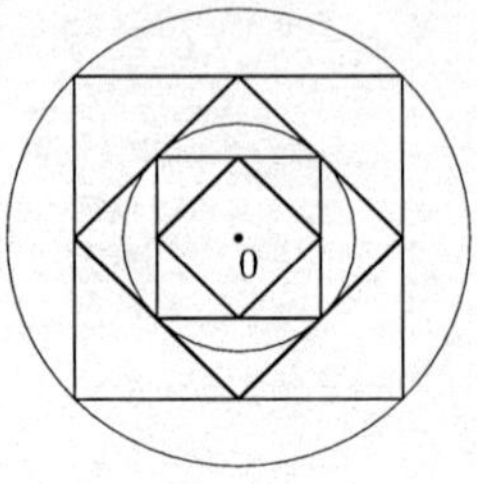

Figure 3.2: All norms in $\mathbb{R}^2$ are equivalent (thin, Euclidean norm; heavier, the ℓ_∞-norm; fat, the ℓ_1-norm

The next result lists some other equivalences to the finite-dimensionality of a normed space X, now in terms of the coincidence of topologies on X.

Corollary 16. *Let $(X, \|\cdot\|)$ be a normed space. Then, the following are equivalent:*
(i) *X is finite-dimensional.*
(ii) *The topologies w and $\mathcal{T}_{\|\cdot\|}$ on X coincide.*
(iii) *There is only one Hausdorff topological vector space topology on X.*

Proof. (i)⇒(ii) It was mentioned in Section 1.3 above that $w \prec \mathcal{T}_{\|\cdot\|}$. Proposition 3 shows that B_X is $\mathcal{T}_{\|\cdot\|}$-compact. Since w is Hausdorff, it follows that $w = \mathcal{T}_{\|\cdot\|}$.
(ii)⇒(i) follows from the fact that $0 \in \overline{S_X}^w$ if X is infinite-dimensional (see one of the annotations in Section 1.3).
(i)⇔(iii) was proved in Theorem 6 above. □

3.3 Dual equivalent norms

3.3.1 Norms on the dual versus dual norms

A natural device to build an equivalent norm on a Banach space $(X, \|\cdot\|)$ is to define first an equivalent norm on X^*. This may sound a little awkward at first. However, the reader is aware that some constructions are easier in the dual of a Banach space than on the space (just remember that some compactness of the closed dual unit ball is available for free —the Alaoglu–Bourbaki theorem—). The procedure has some extra bonus: Often, good properties of the dual norm "induce" good properties of the original norm. Essentially, this is the content of the Šmulyan Lemma (see Lemma 54 and Propositions 82 and 91 below). And, last but not least, it is much easier to analytically construct rotund norms than to geometrically construct smooth norm.

It is necessary, however, to hint at an initial problem: If $(X, \|\cdot\|)$ is a Banach space, not every equivalent norm $|||\cdot|||$ on X^* is the dual norm of an equivalent norm on X (if this is the case, we say that $|||\cdot|||$ is a **dual norm**). Examples of this situation are in the annotation below.

The rest of this section touches on some of these aspects.

To start with, note that a necessary condition for $|||\cdot|||$ on X^* being a dual norm is that its closed unit ball should be w^*-closed (in fact, w^*-compact; it shall be proved in Proposition 17 that in fact this condition is also sufficient). The following example builds on the necessity: Let the equivalent norm on ℓ_∞ be $|||x||| := \|x\|_\infty + \limsup_n |x(n)|$, for $x = (x(n)) \in \ell_\infty$. The sequence $\{x_m\}_{m\in N}$, where $x_m := \chi_{\{1,2,\ldots,m\}}$ for $m \in \mathbb{N}$, belongs to $B_{(\ell_\infty,|||\cdot|||)}$, it w^*-converges to the constant function $x = 1$ in ℓ_∞, and $|||x||| = 2$, so $B_{(\ell_\infty,|||\cdot|||)}$ is not w^*-closed and thus $|||\cdot|||$ is not a dual norm.

➭ *The framework for the previous example.* An equivalent non-dual norm on X^{**}. As soon as a Banach space $(X, \|\cdot\|)$ is not reflexive, it is possible to define an equivalent norm on X^{**} that is not a dual norm. This is the general setting for the previous example (there, $(X, \|\cdot\|) := (c_0, \|\cdot\|_\infty)$, and the non-dual equivalent norm was defined on c_0^{**} $(= \ell_\infty)$). To simplify the notation, we shall write at this moment $\|\cdot\|$ for the norm on a space X and their dual, as well as for higher dual norms. Notice first that X^* and $X^\perp := \{x^\perp \in X^{***} : \langle x, x^\perp\rangle = 0 \text{ for all } x \in X\}$ are subspaces of X^{***}. A result of Dixmier (see, e.g., [FHHMZ11, Exercise 4.7]) shows that $(X^{***}, \|\cdot\|)$ is linearly topologically isomorphic to $X^* \oplus X^\perp$. Take $0 \neq x^\perp \in X^\perp$ (in our preceding example the element $x^\perp \in c_0^\perp$ $(\subset \ell_\infty^*)$ was defined by $\langle x, x^\perp\rangle := \limsup_n x(n)$ for $x := (x(n)) \in \ell_\infty$). We mentioned above (a consequence of the bipolar theorem) that $B_{X^{**}} = \overline{B_X}^{w(X^{**},X^*)}$. Let $x_0^{**} \in S_{(X^{**},\|\cdot\|)}$ be such that $\langle x_0^{**}, x^\perp\rangle \neq 0$, and let $\{x_i\}_{i\in I}$ be a net in $B_{(X,\|\cdot\|)}$ that $w(X^{**}, X^*)$-converges to x_0^{**}. Define a norm $|||\cdot|||$ on X^{**} by $|||x^{**}||| := \|x^{**}\| + |\langle x^{**}, x^\perp\rangle|$ for all $x^{**} \in X^{**}$. Clearly, it is an equivalent norm on $(X^{**}, \|\cdot\|)$. Note that $|||x_i||| = \|x_i\| \leq 1$ for all $i \in I$, although $|||x_0^{**}||| = \|x_0^{**}\| + |\langle x_0^{**}, x^\perp\rangle| > 1$. This shows that $B_{(X^{**},|||\cdot|||)}$ is not $w(X^{**}, X^*)$-closed, so $|||\cdot|||$ is not a dual norm.

An equivalent non-dual norm on X^*.
(i) In Item 1 in Subsection 3.4.4 below, we shall modify the previous construction to build an equivalent non-dual norm in X^* for every Banach space X that contains an isomorphic copy of ℓ_1.
(ii) It is enough to rely on a result of W.J. Davis and W.B. Johnson [DaJo73] that characterizes reflexivity in terms of equivalent norms (see (vii) in Proposition 31 below) and observe, in view of Proposition 17 below, that their condition happens if, and only if, every equivalent norm on X^* is a dual norm. Thus, it is enough to consider any non-reflexive Banach space.

The property of $B_{(X^*,|||\cdot|||)}$ exhibited in the previous example characterizes, in fact, dual norms. This characterization will be needed so often that we may present it here in the form of a simple result (for one of the equivalent definitions of **lower semicontinuity** of an extended-valued function defined on a topological space see the proof of (ii)⇔(iv) in Proposition 17 below; a function $f: T \to [-\infty, +\infty]$ is **upper semicontinuous** whenever $-f$ is lower semicontinuous):

Proposition 17. *Let X be a Banach space, and let $|||\cdot|||$ be an equivalent norm on X^*. Put $B := B_{(X^*,|||\cdot|||)}$ and $S := S_{(X^*,|||\cdot|||)}$. Then the following statements are equivalent:*
(i) *$|||\cdot|||$ is a dual norm.*
(ii) *B is w^*-closed.*
(iii) *$\overline{S}^{w^*} \subset B$.*
(iv) *The mapping $|||\cdot||| : X^* \to \mathbb{R}$ is w^*-lower semicontinuous.*

➡ *Proof of Proposition* 17. Indeed, (i)⇒(ii) is a consequence of the Alaoglu–Bourbaki theorem, and (ii)⇒(i) follows from the bipolar theorem (indeed, $B_\circ$ is the closed unit ball of an equivalent norm on X; its dual unit ball is B, and so $|||\cdot|||$ is a dual norm).
That (ii)⇔(iv) follows from the fact that a function f from a topological space T into $\mathbb{R} \cup \{-\infty, +\infty\}$ is lower semicontinuous if, and only if, all sets $\{t \in T : f(t) \le r\}$, $r \in \mathbb{R}$, are closed.
Obviously, (ii)⇒(iii).
In order to prove (iii)⇒(ii), assume first that X is infinite-dimensional. Thus, $B \subset \overline{S}^{w^*}$ thanks to the fact that, in this case, every w^*-neighbourhood of any point $b^* \in B$ is unbounded, so it cuts S (see the note at the annotation in page 8); this shows that $(\overline{S}^{w^*} \subset)\ B \subset \overline{S}^{w^*}$, hence $B = \overline{S}^{w^*}$ and (ii) holds. Assume no that X is finite-dimensional. Then X^* is also finite-dimensional. By Proposition 3 and Corollary 16, the topologies w^* and $\mathcal{T}_{|||\cdot|||}$ coincide, so B is w^*-closed. □

Another useful sufficient condition for an equivalent norm $|||\cdot|||$ on the dual of a Banach space to be a dual norm, in terms of the coincidence of the w and the w^* topologies on $S_{|||\cdot|||}$ will be given in Proposition 93 and its Corollary 94. It enlightens Proposition 17 above.

Remark 18. 1. Assume that $(X, \|\cdot\|)$ is a Banach space and that $|||\cdot|||$ is a norm on X^* equivalent to $\|\cdot\|^*$. According to (3.2), the dual norm $|||\cdot|||^*$ and the bidual norm $\|\cdot\|^{**}$ on X^{**} are equivalent, and so there are their restriction to X. Certainly, $\|\cdot\|^{**}$ induces on X the original norm $\|\cdot\|$. Summarizing, the norm on X induced by $|||\cdot|||^*$ (denoted again $|||\cdot|||^*$) is equivalent to the original norm $\|\cdot\|$. However, the dual norm on X^* of the norm $|||\cdot|||^*$ on X is not, necessarily, $|||\cdot|||$. This happens (see Proposition 17 above) if, and only if, $|||\cdot|||$ on X^* is w^*-lower semicontinuous. In other words, this happens if, and only if, $B_{(X^*,|||\cdot|||)}$ is w^*-closed, or $|||\cdot|||$ on X^* is, from the beginning, a dual norm.

2. Closely related to the previous observation, note that if $(X, \|\cdot\|)$ is a Banach space and $|||\cdot|||$ is a norm on X^{**} that is equivalent to $\|\cdot\|^{**}$, then $|||\cdot|||$ is a bidual norm (i.e., there exists an equivalent norm $|\cdot|$ on X such that $|\cdot|^{**} = |||\cdot|||$) if, and only if, $\overline{B_{(X^{**},|||\cdot|||)} \cap X}^{w^*} = B_{(X,|||\cdot|||)}$. Indeed, $|||\cdot|||$ induces an equivalent norm $|\cdot|$ on X, $B_{(X,|\cdot|)} = B_{(X^{**},|||\cdot|||)} \cap X$, and the bipolar theorem shows that its bidual unit ball is just $\overline{B_{(X,|\cdot|)}}^{w^*}$. ®

➪ In this section we considered equivalent norms on X^* that may or may not be the dual norm of an equivalent norm on X. The reader should be warned that an alternative definition of a "dual" norm may raise some delicate questions. We say that a Banach space $(X_0, |||\cdot|||)$ is an **isometric predual** of a Banach space $(X, \|\cdot\|)$ whenever $(X_0^*, |||\cdot|||^*)$ is linearly isometric to $(X, \|\cdot\|)$. If this is the case, then $(X_0, |||\cdot|||)$ can be viewed as a (closed) linear subspace of $(X^*, \|\cdot\|^*)$ endowed with the restriction of the norm $\|\cdot\|^*$. This is so because $(X_0^{**}, |||\cdot|||^{**})$ is linearly isometric to $(X^*, \|\cdot\|^*)$. In this sense, $\|\cdot\|$ on X should be named a "dual norm". To elucidate the existence —and, if this is the case, the uniqueness, to be defined below— of isometric preduals is not easy in general. The following proposition, a characterization of Banach spaces having an isometric predual, has a simple formulation; however, its use for deciding on the matter in case of particular examples may be tricky.

Proposition 19. *A Banach space $(X, \|\cdot\|)$ has an isometric predual if, and only if, there exists $S \subset X^*$ that separates points of X and such that $(B_X, w(X, S))$ is compact. If this happens, then $(\overline{\operatorname{span}}^{\|\cdot\|^*}(S), \|\cdot\|^*)$ is an isometric predual of $(X, \|\cdot\|)$.*

Proof. One direction is clear: If $X_0 \subset X^*$ is a predual, then $S := X_0$ satisfies the requirements. Assume now that such a set S exists. Put $X_0 := \overline{\operatorname{span}}^{\|\cdot\|^*}(S)$ $(\subset X^*)$ and endow X_0 with the restriction of the norm $\|\cdot\|^*$. Its dual space X_0^* is linearly isometric to $X^{**}/X_0^{\perp}$ (endowed with the quotient norm). Let $q : X^{**} \to X^{**}/X_0^{\perp}$ the canonical quotient map. Its restriction to X is one-to-one, due to the fact that S separates points of X. Moreover, $q(B_{X^{**}}) = B_{X_0^*}$. Since q is w^*-w^*-continuous, $q(B_X)$ is w^*-dense in $B_{X_0^*}$ and $w(X_0^*, S)$-compact. A simple argument shows that it is also $w(X_0^*, X_0)$-compact, hence $q(B_X) = B_{X_0^*}$. □

Looking at Proposition 19, we may notice that there are Banach spaces without an isometric predual. One example is $(c_0, \|\cdot\|_\infty)$. Indeed, $B_{(c_0,\|\cdot\|_\infty)}$ cannot be compact in any locally convex topology on c_0, due to the absence of extreme points of $B_{(c_0,\|\cdot\|_\infty)}$ and the Krein–Milman Theorem 7.
The property of having an isometric predual is a metric property (i.e., a

Banach space $(X, \|\cdot\|)$ may have an isometric predual and an equivalent norm $|||\cdot|||$ such that $(X, |||\cdot|||)$ lacks an isometric predual. Indeed, let $(X, \|\cdot\|)$ be a nonreflexive Banach space such that X^{**} is separable. Certainly, $(X^*, \|\cdot\|^*)$ has an isometric predual. It is known that $(X^*, \|\cdot\|^*)$ has an equivalent Fréchet differentiable norm $|||\cdot|||$ (Theorem 276) . This norm $|||\cdot|||$ cannot be a dual norm (since X is not reflexive, Theorem 105). Thus, $|||\cdot|||$ on X^* is nicely smooth (see Definition 33 and the statement immediately after it). Assume that $(X^*, |||\cdot|||)$ has an isometric predual. According to Proposition 19, there exists a $|||\cdot|||^*$-closed and w^*-dense subspace N of X^{**} whose dual space is linearly isometric to $(X^*, |||\cdot|||)$ and such that $(B_{(X^*,|||\cdot|||)}, w(X^*, N))$ is compact. In particular, N is 1-norming for $(X^*, |||\cdot|||)$ (see Definition 22 below). Necessarily $N = X^{**}$, what forces $B_{(X^*,|||\cdot|||)}$ to be w-compact. This is a contradiction with the fact that X is not reflexive.

The following result gives a characterization of the existence of an isometric predual for Asplund spaces (see Definition 58 below). The mapping $j_X : X \to X^{**}$ denotes the canonical embedding. Given a Banach space $(X, \|\cdot\|)$, put $R_X := \{x^{**} \in X^{**} : \ker x^{**} \text{ is a 1-norming subspace of } X^* \text{ for } X\}$ (for the concept of an 1-norming subspace see again Definition 22 below).

Proposition 20 (Godefroy, [Go78]). *Let X be an Asplund space. The following are equivalent:*
(i) *X has an isometric predual.*
(ii) $X^{**} = j(X) \oplus R_X$.

It should be noticed that R_X is a w^*-closed linear subspace of X^{**} in case that X is Asplund [Go78, Lemme 3].

Certainly $(X, \|\cdot\|)$ is an isometric predual of the Banach space $(X^*, \|\cdot\|^*)$. However, it may happen that other isometric preduals do exist. In [Go78] the following concept is introduced: If $(X, \|\cdot\|)$ is a Banach space, it is said that $(X, \|\cdot\|)$ is the **unique isometric predual** of $(X^*, \|\cdot\|^*)$ if given any Banach space $(X_0, |||\cdot|||)$ such that $(X_0^*, |||\cdot|||^*)$ is linearly isometric to $(X^*, \|\cdot\|^*)$ and any linear isometry $I : (X^*, \|\cdot\|^*) \to (X_0^*, |||\cdot|||^*)$, we have $I^*(j_{X_0}(X_0)) = j_X(X)$, where I^* is the adjoint mapping to I (see Figure 3.3). It is clear that if X is the unique isometric predual of X^*, then every Banach space X_0 such that X_0^* is linearly isometric to X^* is linearly isometric to X.

Theorem 21 (Godefroy, [Go78], Theorem 10). *In each of the following cases, X is the unique isometric predual of X^*:*
(i) *X^* does not have isomorphic copies of ℓ_1.*
(ii) *X has the Radon–Nikodým property* (see Definition 61 and Chapter 29 below).
(iii) *X is locally uniformly rotund* (see Definition 86 below).

(iv) *X is separable and (B_X, w) is a Baire space* (see page 424 below).

Examples of spaces that have not unique isometric preduals include ℓ_1 (the spaces c and c_0 are isometric preduals, although they are not linearly isometric) and $C(K)$, where K is an infinite compact metric space. If K' is another compact metric space of the same cardinality, the spaces $C(K)^*$ and $C(K')^*$ are linearly isometric. However, $C(K)$ and $C(K')$ are not linearly isometric if K is not homeomorphic to K'. Thus, if K is a metric compact space, then $C(K)$ is never the unique isometric predual of $C(K)^*$ (see [Go78]).
Examples of spaces X being the unique isometric predual of X^* are subspaces of weakly compactly generated dual spaces (see Definitions 131 and 132 below) —in particular, separable dual spaces—, spaces X whose second dual is weakly compactly generated (related to this, see Problem 3 in Section 52.4 below), separable spaces that are G_δ-subsets of (X^{**}, w^*), the space $\ell_1(\Gamma)$ (see [Go78]) and the space $L_1(\mu)$ (Grothendieck), von Neumann algebras (Sakai), as well as the space H^∞ (Ando and Wojtaszczyk) (see the definition in page 437) and the James' quasi-reflexive space J (Brown and Ito) (for references, see [Go78] and [BroIt80]; see also some notes after Theorem 412 in Chapter 20 below, and [FHHMZ11, Definition 4.43]).

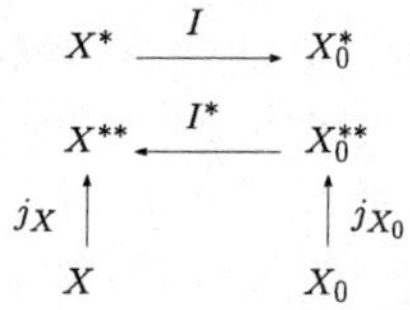

Figure 3.3: Diagram for the uniqueness of the predual

3.4 Equivalent norms and norming subspaces

3.4.1 Norming and 1-norming subspaces

If $(X, \|\cdot\|)$ is a Banach space, then, for $x^* \in X^*$ we have $\|x^*\|^* = \sup\{\langle x, x^*\rangle : x \in B_X\}$ (by definition). Since $B_{X^{**}} = \overline{B_X}^{w^*}$ and each $x^* \in X^*$ is certainly w^*-continuous on X^{**}, we have $\|x^*\|^* = \sup\{\langle x^{**}, x^*\rangle : x^{**} \in B_{X^{**}}\}$. We say then that X is a **norming** (more precisely, **1-norming**) subspace of X^{**} (for $(X^*, \|\cdot\|^*)$). We may translate this situation to a Banach space X and suitable subspaces of its dual space X^* in order to get equivalent norms on X.

Definition 22. *Let $(X, \|\cdot\|)$ be a Banach space. Let N be a subspace of X^* (not necessarily closed), and let $B_N := B_{X^*} \cap N$. The subspace N is said to be* **norming** *(for $(X, \|\cdot\|)$) if*

$$\|x\|_N := \sup\langle x, B_N\rangle, \ x \in X, \tag{3.4}$$

*is an equivalent norm on X. More precisely, N is said to be α-***norming**, *for some $\alpha \geq 1$, if $(1/\alpha)\|x\| \leq \|x\|_N$ $(\leq \|x\|)$ for every $x \in X$.*

The **Dixmier characteristic** $r(N)$ of N is defined in the following way:

$$r(N) := \max\{\alpha^{-1} : \ \alpha \geq 1, \ (1/\alpha)\|x\| \leq \|x\|_N, \text{ for all } x \in X\}$$
$$\left(= \inf\left\{\frac{\|x\|_N}{\|x\|} : \ x \in X, \ x \neq 0\right\}\right). \tag{3.5}$$

Obviously, a norming subspace $N \subset X$ becomes 1-norming for the equivalent norm $\|\cdot\|_N$.

Let $(X, \|\cdot\|)$ be a Banach space and N a subspace of X^*. If N is a norming subspace, notice that in order to define the equivalent norm $\|\cdot\|_N$ on X, we may always assume that N is $\|\cdot\|$-closed.

➩ *To define $\|\cdot\|_N$, the subspace N can be taken to be $\|\cdot\|$-closed.* In order to avoid a cumbersome notation here, we shall denote again by $\|\cdot\|$ the norm on X^* dual to $\|\cdot\|$. Observe that if N is a subspace of X^*, then $\overline{B_N}^{\|\cdot\|} = B_{\overline{N}^{\|\cdot\|}}$. Indeed, $\overline{B_N}^{\|\cdot\|} \subset B_{\overline{N}^{\|\cdot\|}}$. On the other hand, if $x^* \in B_{\overline{N}^{\|\cdot\|}}$ and $x^* \neq 0$, then there exists a sequence $\{x_n^*\}_{n\in\mathbb{N}}$ in N that $\|\cdot\|$-converges to x^*. The sequence $\{x_n^*/\|x_n^*\|\}_{n\in\mathbb{N}}$ belongs to B_N and $\|\cdot\|$-converges to $x^*/\|x^*\|$, hence $x^* \in \|x^*\|.\overline{B_N}^{\|\cdot\|}$. Since $\|x^*\| \leq 1$, we get $x^* \in \overline{B_N}^{\|\cdot\|}$. If alternatively $x^* = 0$, then obviously $x^* \in \overline{B_N}^{\|\cdot\|}$. This proves the statement, Since $\|x\|_N = \sup\langle x, \overline{B_N}^{\|\cdot\|}\rangle$ for $x \in X$, the assertion follows. □

3.4.2 Separating subspaces

We mentioned in Section 1.3 that if $(X, \|\cdot\|)$ is a Banach space, then $\langle X, X^*\rangle$ is a dual pair, and that a (not necessarily closed) subspace Y of X^* separates points of X if, and only if, it is w^*-dense. This statement is obviously equivalent to the assertion that $\|\cdot\|_Y$, defined as in (3.4) above (for $N = Y$), is in fact a (not necessarily equivalent) norm on X. Notice that $\|\cdot\|_Y$, as the supremum of a family of $w(X,Y)$-continuous linear mappings, is $w(X,Y)$-lower semicontinuous. Naturally, $\|\cdot\|_Y$ is called the **Y-lower semicontinuous envelope of** $\|\cdot\|$. Again in case that Y separates points of X, a simple consequence of the bipolar theorem is that

$$B_{(X,\|\cdot\|_Y)} = \overline{B_{(X,\|\cdot\|)}}^{w(X,Y)} \tag{3.6}$$

(in particular, $B_{(X,\|\cdot\|_Y)}$ is $w(X,Y)$-closed, also a consequence of the $w(X,Y)$-lower semicontinuity of the norm $\|\cdot\|_Y$).

If N ($\subset X^*$) is a norming subspace, then $\|\cdot\|_N$ is a norm (in fact, an equivalent norm), hence N is w^*-dense in X^*. All together, among the subspaces Y of X^* that separate points of X, the norming subspaces N are those for which the norm $\|\cdot\|_N$ is in fact an equivalent norm (and $\|\cdot\|_N = \|\cdot\|$ if, and only if, N is 1-norming). There are examples of Banach spaces $(X, \|\cdot\|)$ and separating non-norming subspaces of X^*.

➪ *An example of a separating non-norming subspace of X^*.* The subspace

$$\{x = (x_n) \in \ell_1 : \; x_{n_k} = (1/n_k) \sum_{m \in I_{n_k}} x_m, \; k \in \mathbb{N}\}$$

is $w(\ell_1, c_0)$-dense and not norming; here, $\{I, I_{n_1}, I_{n_2}, \ldots\}$ is a splitting of $\mathbb{N}$ into infinite pairwise disjoint sets, and $I = \{n_1, n_2, \ldots\}$). For details, see, e.g., [FHHMZ11, Exercise 3.92].

Theorems 24 and 25 below settle the question of the existence of separating non-norming subspaces of X^*. First, we need a definition:

Definition 23. *A Banach space X is said to be* **quasi-reflexive** *whenever* $\dim(X^{**}/X) < \infty$.

An example of a quasi-reflexive Banach space is the space J of James mentioned in Section 1.3 (see Section 20.3 below, and, e.g., [FHHMZ11, Definition 4.43]). Petunin [Pet64] and Singer [Sin63] showed the following result:

Theorem 24 (Petunin [Pet64], Singer [Sin63]). *If X is a quasi-reflexive Banach space, then every w^*-dense subspace of X^* is norming.*

The proof is presented in [FHHMZ11, Exercise 5.5]. In the other direction, Davis and Lindenstrauss proved the following result:

Theorem 25 (Davis, Lindenstrauss [DaLi72]). *For every non-quasi-reflexive Banach space X, the space X^* contains a w^*-dense non-norming subspace.*

Theorems 24 and 25 together show the following characterization of quasi-reflexivity:

Corollary 26. *A space X is quasi-reflexive if, and only if, every w^*-dense subspace of X^* is norming*

For a characterization of reflexive spaces in terms of norming subspaces of the dual see Proposition 31; see also Remark 27 below.
The proof of the aforementioned theorem of Davis and Lindenstrauss shows the following result, that is of independent interest: *If X is not quasireflexive, then there is an infinite-dimensional subspace Y of X such that X/Y is not quasireflexive.*

Remark 27. Any device that would produce norming subspaces N of X^* will become a tool for getting equivalent norms on a Banach space X. However, not every equivalent norm on a Banach space $(X, \|\cdot\|)$ is the $\|\cdot\|_N$-norm associated to a certain norming subspace N of X^*. For example, let $(X, \|\cdot\|)$ be a reflexive space. If N is a closed norming subspace of X^*, then $N = X^*$, hence $\|\cdot\|_N = \|\cdot\|$. Indeed, N, being a norming subspace, separates points of X, so it is w^*-dense in X^*. Since $w^* = w$ on X^*, N is w-dense, and this is equivalent to say that N is $\|\cdot\|$-dense. Due to the fact that N is $\|\cdot\|$-closed, we get $N = X^*$, as claimed. This shows that no equivalent norm on X can be defined via (3.4) for some (closed) proper norming subspace N of X^*, simply because there is no such object in the dual. Ⓡ

3.4.3 Characterizing norming subspaces

In view of Remark 27, it is useful to provide tools for getting norming subspaces of the dual of a given Banach space X; this is the purpose of Subsection 3.4.4. Here we characterize norming and 1-norming subspaces among the class of separating subspaces Y of X^*. This will be done either by looking at the ball $B_{(X,\|\cdot\|_Y)} \subset X$, or at the ball $B_Y \subset X^*$, or even at the orthogonal subspace $Y^\perp$ of Y in X^{**}.

In all the statements below, $(X, \|\cdot\|)$ is a Banach space, B_X its closed unit ball, Y a separating subspace of X^* (in other terms, a w^*-dense subspace of X^*), $B_Y := B_{X^*} \cap Y$, and $\|x\|_Y := \sup\{\langle x, y^*\rangle : y^* \in B_Y\}$.

1. **In terms of balls in** X (recall, see (3.6), that $B_{(X,\|\cdot\|_Y)} = \overline{B_{(X,\|\cdot\|)}}^{w(X,Y)}$):

   Lemma 28. *Let X be a Banach space, and $Y \subset X^*$ a separating subspace.*

   (i) *Y is norming if, and only if, $B_{(X,\|\cdot\|_Y)}$ is bounded. More precisely, if $\alpha \geq 1$ then Y is α-norming if, and only if ($B_X \subset$) $B_{(X,\|\cdot\|_Y)} \subset \alpha B_X$* (see Figure 3.4).

   (ii) *In particular, Y is 1-norming if, and only if, $B_X = B_{(X,\|\cdot\|_Y)}$, if, and only if, B_X is $w(X,Y)$-closed, and if, and only if, $\|\cdot\|$ is $w(X,Y)$-lower semicontinuous.*

   Proof. See, e.g., [FHHMZ11, Exercise 3.90]. □

2. **In terms of balls in** X^*, we have, dually, that (see Figure 3.5)

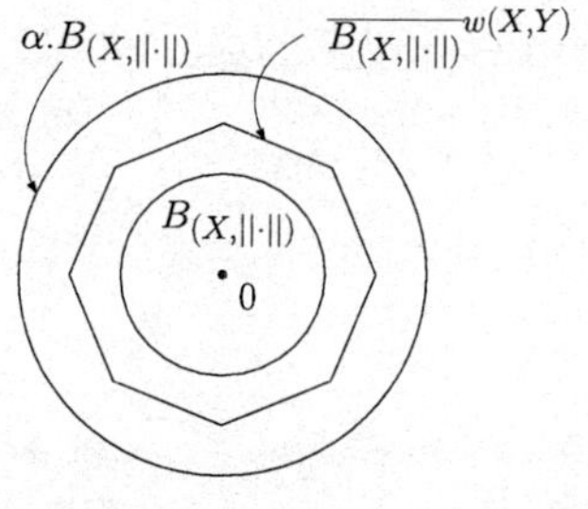

Figure 3.4: Y is norming if, and only if, $\overline{B_{(X,\|\cdot\|)}}^{w(X,Y)}$ is bounded

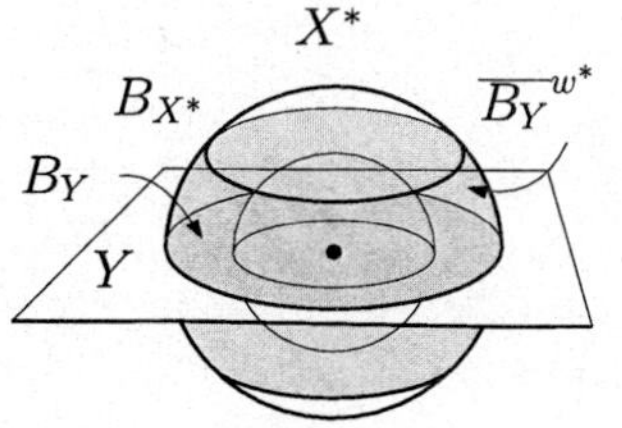

Figure 3.5: A norming subspace $Y \subset X^*$ (in dark gray, B_Y, in light gray its w^*-closure; it contains a multiple of B_{X^*})

Lemma 29. *Let X be a Banach space, and $Y \subset X^*$ a separating subspace.*

(i) *Y is norming if, and only if, $\overline{B_Y}^{w^*}$ contains a multiple of B_{X^*}. More precisely, if $0 < \alpha \leq 1$ then Y is $(1/\alpha)$-norming if, and only if, $\alpha B_{X^*} \subset \overline{B_Y}^{w^*}$.*

(ii) *In particular, Y is 1-norming if, and only if, $\overline{B_Y}^{w^*} = B_{X^*}$.*

3. **In terms of orthogonal subspaces**, let Y, as above, be a separating subspace of X^*, and $Y^\perp$ be the subspace of X^{**} orthogonal to Y. Observe that $X \cap Y^\perp = \{0\}$, hence $X + Y^\perp$ is an algebraic direct sum in X^{**}. We have (see, e.g., [FHHMZ11, Exercise 5.3])

Lemma 30. *Let X be a Banach space, and $Y \subset X^*$ a separating subspace.*

(i) *Y is norming if, and only if, $X + Y^\perp$ is closed in X^{**} (so $X + Y^\perp$ is a topological direct sum in X^{**}, i.e., the projection P (respectively, $I - P$) from $X + Y^\perp$ onto X (respectively, $Y^\perp$) is continuous). To be more precise, for $\alpha \geq 1$, the space Y is α-norming if, and only if, $\|P\| \leq \alpha$.*

(ii) *Y is 1-norming if, and only if, $X+Y^\perp$ is closed in X^{**} and the projection $P : X + Y^\perp \to X$ has norm 1.*

Proof. We shall only prove the statement about the norm of the projection in (i): Assume first that Y is α-norming. Given $x \in X$ and $y^{\perp} \in Y^{\perp}$ ($\subset X^{**}$),

$$\|x\| \le \alpha\|x\|_Y = \alpha\|x + y^{\perp}\|_Y \le \alpha\|x + y^{\perp}\|,$$

hence $\|P\| \le \alpha$.

Assume now that $\|P\| \le \alpha$. Let $|||\cdot|||$ be the norm on Y induced by $\|\cdot\|^*$. Notice that $(Y^*, \overline{|||\cdot|||^*})$ is linearly isometrically isomorphic to $(X^{**}/Y^{\perp}, \overline{\|\cdot\|^{**}})$, where $\overline{\|\cdot\|^{**}}$ is the norm induced by $\|\cdot\|^{**}$ on the quotient $X^{**}/Y^{\perp}$. Let $q : X^{**} \to X^{**}/Y^{\perp}$ be the canonical quotient mapping. Thus, for $x^{**} \in X^{**}$ we have

$$\overline{\|q(x^{**})\|^{**}} = |||q(x^{**})|||^* = \sup\{|\langle x^{**}, y\rangle| : \ y \in B_Y\} = \|x^{**}\|_Y.$$

Notice, too, that by the definition of the quotient norm,

$$\overline{\|q(x^{**})\|^{**}} = \inf\{\|x^{**} + y^{\perp}\|^{**} : \ y^{\perp} \in Y^{\perp}\}.$$

For $x^{**} = x \in X$ we have $\|x\| = \|P(x + y^{\perp})\| \le \alpha\|x + y^{\perp}\|$, i.e., $\|x + y^{\perp}\| \ge (1/\alpha)\|x\|$, for every $y^{\perp} \in Y^{\perp}$. Thus, $\inf\{\|x + y^{\perp}\| : \ y^{\perp} \in Y^{\perp}\} \ge (1/\alpha)\|x\|$. All together, we get $\|x\|_Y \ge (1/\alpha)\|x\|$, and so Y is α-norming. □

3.4.4 Getting norming subspaces

We collect here some results that provide norming and 1-norming subspaces.

1. Let X be a non-reflexive Banach space. If $x^{**} \in X^{**} \setminus X$, then $\ker x^{**}$ is a proper, closed and norming subspace of X^* (see, e.g., [FHHMZ11, Exercise 3.88]). Moreover, for such an x^{**} we have that $\ker x^{**}$ is 1-norming if, and only if, $\|x^{**} + x\| \ge \|x\|$ for all $x \in X$ (see, e.g., [FHHMZ11, Exercise 5.4]).

As a byproduct of the previous result, we may compare the quasireflexive case described in Subsection 3.4.2 above with the reflexive one. Despite that the notion of a Fréchet differentiable norm will be given later (see Definition 42 below) we include here Proposition 31 for getting a more complete picture of reflexivity in terms of norming subspaces.

Proposition 31. *Let $(X, \|\cdot\|)$ be a Banach space. Then, the following statements are equivalent:*
(i) *X is reflexive.*
(ii) *Every norming subspace of X^* is 1-norming.*
(iii) *$\|\cdot\|$ is $w(X,N)$-lower semicontinuous for every norming subspace N of X^*.*
(iv) *$B_{(X,\|\cdot\|)}$ is $w(X,N)$-closed for every norming subspace N of X^*.*
(v) *There are no proper closed norming subspaces of X^*.*
(vi) *For every norming subspace N of X^* there exists an equivalent Fréchet differentiable and $w(X,N)$-lower semicontinuous norm on X.*
(vii) (Davis, Johnson, [DaJo73]) *For every equivalent norm $|||\cdot|||$ on X^*, $B_{(X^*,|||\cdot|||)}$ is w^*-closed.*

Proof. That (ii) to (iv) are equivalent is almost trivial. (i)⇒(iv) follows from the w-compactness of B_X, and for (iv)⇒(i) see [FHHMZ11, Exercise 3.91]. The argument for proving (i)⇒(v) was given in Remark 27. The result above shows that $\ker x^{**}$ ($\subset X^*$) is a proper closed and norming subspace of X^*; thus, (v)⇒(i). If X is reflexive, there is no proper norming subspace of X^*. Troyanski's renorming Theorem 218 gives an equivalent LUR norm on X^*, and thus its predual norm on X is Fréchet differentiable (Proposition 91 below). This shows that (i)⇒(vi). If N is a $\|\cdot\|$-norming subspace of X^* and $|||\cdot|||$ is a $w(X,N)$-lower semicontinuous Fréchet differentiable equivalent norm on X, note that N is also $|||\cdot|||$-norming. The argument in Section 3.5 gives $N = X^*$, and the reflexivity of X follows from (v). Notice that (i)⇒(vii) is obvious. Assume finally that (vii) holds. Let $x^{**} \in S_{X^{**}}$. The set $B_n := B_{(X^*,\|\cdot\|^*)} \cap \{x^* \in X^* : |\langle x^{**}, x^*\rangle| \le 1/n\}$ is the closed unit ball of an equivalent norm on X^* (see Figure 3.6). According to the hypothesis, it is w^*-closed. Since $B := \bigcap_{n=1}^{\infty} B_n = B_{(X^*,\|\cdot\|^*)} \cap \ker x^{**}$, we get that B is w^*-closed. Thus, x^{**} is w^*-continuous (see, e.g., [FHHMZ11, Corollary 3.94]), i.e., $x^{**} \in X$. See also [FHHMZ11, Exercise 3.120]. □

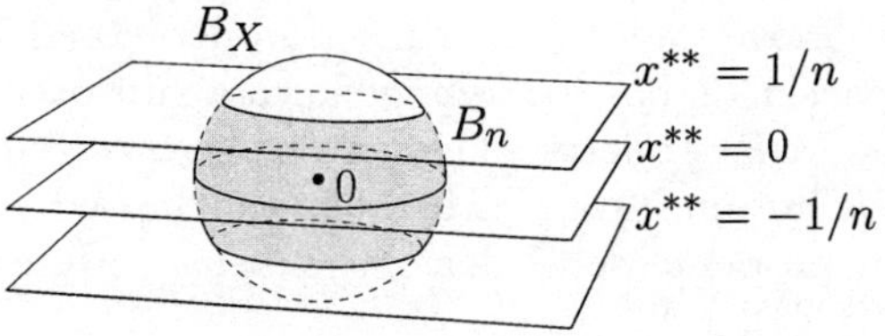

Figure 3.6: The sets in the proof of (vii)⇒(i) in Proposition 31

Remark 32. (i) In view of Proposition 17 above, (vii) in Proposition 31 is equivalent to say that all equivalent norms on X^* are in fact dual norms.

(ii) If the word "norming" is replaced by "1-norming" in Proposition 31 (v), then we get, instead of reflexivity, the concept of a nicely smooth norm (see Definition 33 below). ®

In view of (vii) in Proposition 31, given a non-reflexive Banach space there always exists on X^* an equivalent norm that is not a dual norm (see Section 3.3). Based on the norming character of the kernel of any $x^{**} \in X^{**} \setminus X$, we may give explicit examples of norms on X^* that are not dual norms.

➪ *An equivalent non-dual norm in X^* for some spaces X.* Let X be a separable Banach space that contains an isomorphic copy of ℓ_1. By [DeGoZi93, Theorem III.2.5], then X has an equivalent octahedral norm $\|\cdot\|$ (for this notion, see [DeGoZi93, Definition III.2.1]). This shows the existence of $x^{**} \in X^{**} \setminus \{0\}$ such that $\|x^{**}+x\| = \|x^{**}\| + \|x\|$ for every $x \in X$ ([GoKa89] and, e.g., [DeGoZi93, Lemma III.2.1]). In particular, $\|x^{**} + x\| \geq \|x\|$ for every $x \in X$. In view of the statement at the beginning of this number, the subspace $N := \ker x^{**}$ of X^* is a proper (closed) 1-norming subspace for X (and so $\overline{B_{(N,\|\cdot\|)}}^{w^*} = B_{(X^*,\|\cdot\|)}$). Define on X^* the equivalent norm $|||x^*||| := \|x^*\| + |\langle x^{**}, x^*\rangle|$, for $x^* \in X^*$. Find $x_0^* \in S_{(X^*,\|\cdot\|)}$ such that $\langle x^{**}, x_0^*\rangle \neq 0$ and let $\{x_i^*\}_{i\in I}$ be a net in $B_{(N,\|\cdot\|)}$ that w^*-converges to x_0^*. Note that $|||x_i^*||| = \|x_i^*\| \leq 1$ for all i. However, $|||x_0^*||| = \|x_0^*\| + |\langle x^{**}, x_0^*\rangle| > 1$. Thus, $B_{(X^*,|||\cdot|||)}$ is not w^*-closed. By Proposition 17, $|||\cdot|||$ on X^* is not a dual norm.

2. Let Y be a closed subspace of a Banach space X. Let $q : X^* \to X^*/Y^\perp$ be the canonical quotient mapping. If N is a norming subspace of Y^* $(= X^*/Y^\perp)$, then $q^{-1}(N)$ is a norming subspace of X^* (see, e.g., [FHHMZ11, Exercise 5.6]).

➪ If N is a 1-norming subspace of Y^* we may not always conclude that $q^{-1}(N)$ $(\subset X^*)$ would be also 1-norming. Indeed, let Y be a separable Banach space that contains an isomorphic copy of ℓ_1. We proved in the annotation above that Y^* contains a proper closed 1-norming subspace N. The space Y is linearly isometric to a subspace of a Banach space X with the MIP property (Theorem 569 below), and X is then nicely smooth (Definition 33 and Theorem 582 below). Thus, X^* does not contain any proper closed 1-norming subspace. Note that $q^{-1}(N) \neq X^*$, since $q(q^{-1}(N)) \subset N$.

3. Another useful fact (see, e.g., [FHHMZ11, Exercise 4.34]) is that, *in case $(X, \|\cdot\|)$ is a separable Banach space with a Schauder basis $\{e_n, e_n^*\}_{n=1}^\infty$, the subspace $F :=$* span *$\{e_n^* : n \in \mathbb{N}\}$ of X^* is norming. For $n \in \mathbb{N}$, let $P_n : X \to X$ be the canonical projection on* span *$\{e_1, \ldots, e_n\}$. If the basis is monotone*

(i.e., its basis constant is 1 or, in other words, if $\|P_n(x)\| \le \|P_{n+1}(x)\|$ for all $n \in \mathbb{N}$ and all $x \in X$), *then* F *is* 1*-norming.*

➪ We provide here the proof of this result: Let $|||x||| := \sup_n \|P_n(x)\|$ for $x \in X$. Then $|||\cdot|||$ is an equivalent norm on X and $|||P_n||| = 1$, so $|||P_n^*|||^* = 1$, for every $n \in \mathbb{N}$ (see, e.g., [FHHMZ11, Lemma 4.9 and Exercise 4.27]). Given $x^* \in X^*$ such that $|||x^*|||^* \le 1$, we have $|||P_n^*(x^*)|||^* \le 1$, $P_n^*(x^*) \in \operatorname{span}\{e_1^*, \ldots, e_n^*\}$, and $P_n^*(x^*) \xrightarrow{w^*} x^*$. Hence

$$\overline{F \cap B_{(X^*,|||\cdot|||^*)}}^{w^*} = B_{(X^*,|||\cdot|||^*)}.$$

Thus F is 1-norming for $(X, |||\cdot|||)$. Since $|||\cdot|||$ is an equivalent norm on X, F is a norming subspace for $(X, \|\cdot\|)$, as we wanted to show.

For the last part, observe that $P_n(x) \to x$ and, if the basis is monotone, $\|P_n(x)\| \le \|P_{n+1}(x)\|$ for all $n \in \mathbb{N}$. Thus, the norm $|||\cdot|||$ defined above coincides in this case with $\|\cdot\|$, and the conclusion follows.

In the former proof, note that $|||P_n||| = 1$ for all $n \in \mathbb{N}$ ensures that for $x^* \in B_{(X^*,|||\cdot|||^*)}$, the w^*-approximating sequence $\{P_n^*(x^*)\}$ is in $B_{(X^*,|||\cdot|||^*)}$. For getting the result for $\|\cdot\|$ is then enough that $\|P_n\| \to 1$ (and so $\|P_n^*\|^* \to 1$). Thus, a slight modification of the former proof allows for the following extension: *Let* $(X, \|\cdot\|)$ *be a Banach space with a Schauder basis* $\{e_n, e_n^*\}_{n=1}^{\infty}$. *Let* $\{P_n\}$ *be the sequence of associated projections. Assume that* $\|P_n\| \to 1$. *Then the subspace* $F := \operatorname{span}\{e_n^* : n \in \mathbb{N}\}$ *is* 1*-norming.*

➪ Indeed, we have $\|P_n^*\| \to 1$. Given $x^* \in B_{X^*}$, the sequence $\{P_n^*(x^*)\}$ converges to x^* in the w^*-topology, and so it does the sequence $\{P_n^*(x^*)/\|P_n^*\|\}$. This last sequence is in $F \cap B_{X^*}$, so F is 1-norming.

The former observations regarding Schauder bases should be compared with Theorem 828 below. Observe, too, that the converse of the statement in Item 3 fails: The canonical basis of the space ℓ_2 can be perturbed, according to the Krein–Milman–Rutman–Valdivia result (see [FHHMZ11, Theorem 4.23]), to get a non-monotone Schauder basis with functional coefficients spanning $\overline{\operatorname{span}}\{e_n^* : n \in \mathbb{N}\} = \ell_2^*$, obviously 1-norming for ℓ_2.

4. Let K be a compact topological space. For $f \in C(K)$, $\|f\|_\infty = \langle f, \delta_k \rangle$ for some $k \in K$, where δ_k is the Dirac delta associated to k (see Section 1.4 above). Thus, the space $\operatorname{span}\{\delta_k : k \in K\}$ is a 1-norming subspace of $C(K)^*$ for $(C(K), \|\cdot\|_\infty)$ (in particular, w^*-dense in $C(K)^*$). More generally, if $D \subset K$ is a dense subset of K, then $\operatorname{span}(D)$ is a 1-norming subspace of $C(K)^*$. Regarding the first assertion, observe, too, the following more precise statement: $\overline{\operatorname{conv}}^{w^*}\{\pm\delta_k : k \in K\} = B_{C(K)^*}$. This is a consequence of the fact that $\{\pm\delta_k : k \in K\}$ is the set of all extreme points of $B_{C(K)^*}$ (see Section

2.2, Example 5) and the Krein–Milman Theorem 7 (see also [FHHMZ11, Theorem 3.65]). It is not always true that $\overline{\text{conv}}^{\|\cdot\|_\infty^*}\{\pm\delta_k : k \in K\} = B_{C(K)^*}$. Indeed, note that every element $\mu \in \overline{\text{span}}^{\|\cdot\|_\infty^*}\{\delta_k : k \in K\}$ is separably supported, although there are compact spaces K such that not every Borel regular measure on it is such. It is a result of G. Godefroy (Theorem 539 below) that if $JB \subset (S_{X^*}, \|\cdot\|^*)$ is a separable James boundary for a Banach space X (a **James boundary** is a subset JB of S_{X^*} such that every $x \in X$ attains its supremum on B_{X^*} at a point in JB, see Chapter 28), then JB is a **strong James boundary** (i.e., $\overline{\text{conv}}^{\|\cdot\|^*}(JB) = B_{X^*}$). This provides a proof of James' compactness theorem for the separable case. For this proof and for more information on James boundaries and James' compactness theorem, see again Chapter 28.

5. We can provide some extra information related to Item 4 above. If K is an uncountable compact space, then $(C(K)^*, \|\cdot\|_\infty^*)$ is not separable. Indeed, in the opposite case, $(\delta(K), \|\cdot\|_\infty^*)$ would also be separable. However, $\|\delta_{k_1} - \delta_{k_2}\|_\infty^* = 2$ for every $k_1, k_2 \in K$, $k_1 \neq k_2$, and a countable subset of $\delta(K)$ cannot be, then, $\|\cdot\|_\infty^*$-dense in $\delta(K)$. On the other side, if K is countable, then, since $JB := \{\pm\delta_k : k \in K\}$ is a ($\|\cdot\|_\infty^*$-separable) James boundary, it is a strong James boundary (see again Theorem 539 below), and this implies that $\overline{\text{span}}^{\|\cdot\|_\infty^*} JB = C(K)^*$, hence $(C(K)^*, \|\cdot\|_\infty^*)$ is separable. Thus, $C(K)$ is Asplund (Definition 58). By Theorem 274 below, K is **scattered**, i.e., every closed subspace $L \subset K$ has an isolated point in L. Since $C(K)$ is separable, it follows from [FHHMZ11, Lemma 3.102] or, more directly, from the fact that $(\delta(K), w^*)$ is metrizable, that K is metrizable. For a more direct approach to these and other related results see [FHHMZ11, Section 14.3].

3.5 Some other results related to norming subspaces

For motivating Definition 33 below, we mention here a simple result pertaining to differentiability theory, to be considered in Chapter 4: If $(X, \|\cdot\|)$ is a Banach space, and $\|\cdot\|$ is Fréchet differentiable (see Definition 42 below), then there is no proper closed 1-norming subspace of X^*.

➪ Indeed, assume that $N \subset X^*$ is such. If $x_0^* \in S_{X^*} \cap NA(X)$ (say $\langle x_0, x_0^*\rangle = 1$ for some $x_0 \in S_X$), the Fréchet differentiability of $\|\cdot\|$ at x_0 shows that the diameter of the slices $\{x^* \in B_{X^*} : \langle x_0, x^*\rangle > 1 - 1/n\}$ goes to 0 as $n \to \infty$. Moreover, $x_0^* \in \overline{B_N}^{w^*}$. All together, $x_0^* \in N$. If x_0^* does not attains its norm, the Bishop–Phelps Theorem 14 and the argument above conclude, again, that $x_0^* \in N$. □

This behaviour of a Fréchet differentiable norm suggests the following concept:

Definition 33. *The norm* $\|\cdot\|$ *on a Banach space* $(X, \|\cdot\|)$ *is said to be* **nicely smooth** *(***NS***, for short) if there is no proper closed* 1*-norming subspace of* X^*.

Thus, the statement above shows that *a Fréchet differentiable norm is nicely smooth* (a stronger statement is in Theorem 582 below). We observed above (see (v) in Proposition 31), that the norm of every reflexive Banach space is nicely smooth. Nicely smooth norms will be discussed in Chapter 34.

3.6 The interplay between norms and their unit balls

This and the next sections provide tools for defining norms associated to selected subsets of a vector space. Of course, the objective will be to ensure that the so-defined norm will enjoy one or more of the desired properties —enhanced convexity or differentiability, for example— and that the new norm is equivalent and, maybe, close enough to the old one. Another purpose of the present section will be to provide analytic tools for computing dual norms. Finally, we shall explore how to combine several norms into new ones by manipulating their closed unit balls.

The detailed explanation of these two sections in not necessary for following the rest of the text; however, it will be beneficial.

3.6.1 Minkowski functionals

If $(X, \|\cdot\|)$ is a Banach space, it is clear that $\|x\| = \inf\{\lambda > 0 : \ x \in \lambda B_X\}$ for all $x \in X$. This trivial observation relating $\|\cdot\|$ and its closed unit ball suggests the definition of the so-called Minkowski functional of a subset of a vector space.

Given an **absorbent** subset B of X (i.e., a set such that for every $x \in X$ there exists $\lambda > 0$ such that $x \in \lambda B$) the following function is well defined (and is finite everywhere):

$$\boxed{p_B(x) := \inf\{\lambda > 0 : \ x \in \lambda B\}.} \tag{3.7}$$

The function p_B is called the **Minkowski functional of** B. It is clearly non-negative and positively homogeneous (i.e., $p_B(rx) = rp_B(x)$ for $x \in X$ and $r > 0$). If B is, moreover, convex, then p_B is subadditive. It is easily seen to be a seminorm in case that B is, additionally, balanced. The set B does not contain rays from 0 if, and only if, p_B is a norm. If $(X, \|\cdot\|)$ is a Banach space and B is an absolutely convex subset of X, then p_B is an equivalent norm if, and only if, B is bounded and has a nonempty topological interior. In other words, this happens if, and only if, B is contained in a multiple of B_X and contains another multiple of B_X.

A subset of X that is closed, convex and has a nonempty interior is called a **convex body**.

3.6.2 Support functions

Another way to associate a function —this time on X^*— to a certain subset of a normed space $(X, \|\cdot\|)$ is the following: Given a nonempty set $C \subset X$, its **support function** σ_C is defined, for every $x^* \in X^*$, as

$$\boxed{\sigma_C(x^*) := \sup\{x^*(x) : x \in C\},} \tag{3.8}$$

a function that depends on the particular properties of the set C. It is defined on X^*, and takes values in $(-\infty, +\infty]$. Note that $\|\cdot\|^* = \sigma_{B_X}$.
The classical tools of convex analysis —to be explored below— provide alternative ways to compute efficiently norms and their dual norms —and their corresponding balls—. As a byproduct, we shall deduce formulas for new norms obtained from blending several convex bodies.

Properties of the support function

Let us list in the following proposition some properties of the support function σ_C defined in (3.8). This little catalogue will spare repeating once and again some computations that will often appear in the rest of the text. We refer to [Rock96] for a complete treatment of support functions.

Proposition 34. *Let X be a normed space, A, B subsets of X, $\{C_i\}_{i\in I}$ a nonempty family of nonempty subsets of X, and $t \geq 0$. Then:*
(i) σ_A *is a convex function.*
(ii) $\sigma_A = \sigma_{\overline{A}}$.
(iii) *If $A \subset B$ then $\sigma_A \leq \sigma_B$. The converse is true whenever A and B are both convex and closed.*
(iv) $\sigma_A = \sigma_{\overline{\text{conv}}\,(A)}$.
(v) $\sigma_{A+B} = \sigma_A + \sigma_B$.
(vi) $\sigma_{\bigcup_i C_i} = \sup\{\sigma_{C_i} : i \in I\}$.
(vii) $\sigma_{A\cup B} = \max\{\sigma_A, \sigma_B\}$.
(viii) $\sigma_{tA} = t\sigma_A$.

Proof. (i) follows easily from the fact that σ_A is defined as the pointwise supremum of a family of affine functions.
(ii) is a consequence of the continuity of every $x^* \in X^*$.
(iii) follows straightforwardly from the definition. The second part of the statement is a consequence of the Hahn–Banach Separation Theorem (see Section 1.3).
(iv) That $\sigma_A \leq \sigma_{\overline{\text{conv}}\,(A)}$ follows from (iii) above. That $\sigma_{\text{conv}\,(A)}(x^*) \leq \sigma_A(x^*)$ is a consequence of the convexity of x^*. Use then (ii) above.

(v) can be checked as follows: take $x^* \in X^*$

$$\begin{aligned}\sigma_{A+B}(x^*) &= \sup\{x^*(x) : x \in A+B\} \\ &= \sup\{x^*(x_a) + x^*(x_b) : x_a \in A, x_b \in B\} \\ &= \sup\{x^*(x_a) : x_a \in A\} + \sup\{x^*(x_b) : x_b \in B\} \\ &= \sigma_A(x^*) + \sigma_B(x^*).\end{aligned}$$

(vi) We have

$$\begin{aligned}\sigma_{\bigcup_{i\in I} C_i}(x^*) &= \sup\{x^*(x) :\ x \in \cup_i C_i\} \\ &= \sup\{x^*(x) :\ \exists i \in I \text{ with } x \in C_i\} \\ &= \sup_{i\in I}\{\sup\{x^*(x) :\ x \in C_i\}\} \\ &= \sup_{i\in I}\{\sigma_{C_i}(x^*)\}.\end{aligned}$$

(vii) follows, obviously, from (vi), and (viii) should be clear. □

Remark 35. Notice that Proposition 34 does not include the item $\sigma_{A\cap B}$. The reason is that the result needs a new operation, the so-called **infimal convolution** of two functions, denoted □, whose definition and basic properties will be the subject of Subsection 3.6.5 below. For A and B the unit balls of two norms on a vector space X, the result and the computation are explicit there. ®

3.6.3 Polarity

The concepts of polar and absolute polar (prepolar, absolute prepolar) of a set in a normed space (in the dual of a normed space, respectively), were introduced in Section 1.3 (see equations (1.3) and (1.4) above). Easy computations lead to the following results pertaining to subsets A and B of a normed space X and a nonzero real number α: (i) $(\alpha A)^\circ = (1/\alpha)A^\circ$, (ii) if $A \subset B$, then $A^\circ \supset B^\circ$, (iii) $A \subset A^{\circ\circ}$, and (iv) $[(A^\circ)_\circ]^\circ = A^\circ$. If $\{C_i :\ i \in I\}$ is a collection of convex weakly closed subsets of a normed space X containing 0, we have

$$\left(\bigcap_i C_i\right)^\circ = \overline{\mathrm{conv}}\bigcup_i C_i^\circ. \tag{3.9}$$

The dual formula holds for arbitrary subsets S_i of X, i.e.,

$$\left(\bigcup_i S_i\right)^\circ = \bigcap_i S_i^\circ. \tag{3.10}$$

Equalities (1.5) and (1.6) above computing the polar of a ball are useful to understand the geometry of the unit ball —and its dual and bidual ones— without explicitly computing the norms. It is worth mentioning that the support function of a set $S \subset X$ is the Minkowski functional of its polar S°.

➡ Note first that for a nonempty subset S of a Banach space X and $x^* \in X^*$, the two following statements are equivalent:
(i) x^* is bounded above on S.
(ii) S° absorbs x^* (i.e., there exists $\lambda > 0$ such that $x^* \in \lambda S^\circ$).
Indeed, if (i) holds, then $\langle s, x^*\rangle \leq \lambda$ for all $s \in S$ and some λ, that we can choose to be greater than 0. Thus, $\langle s, x^*/\lambda\rangle \leq 1$, so $x^*/\lambda \in S^\circ$, and $x^* \in \lambda S^\circ$. This shows (ii). On the other hand, if (ii) is true, there exists $\lambda > 0$ such that $x^* \in \lambda S^\circ$. Thus, $\langle s, x^*/\lambda\rangle \leq 1$ for all $s \in S$, i.e., $\langle s, x^*\rangle \leq \lambda$ for all $s \in S$, and this is (i).
The argument above shows that in case that x^* is bounded above on S, then $\sigma_S(x^*) \leq p_{S^\circ}(x^*)$. If, moreover, $\sigma_S(x^*) > 0$, we get $\sigma_S(x^*) = p_{S^\circ}(x^*)$.

3.6.4 Fenchel conjugation

A device used in Convexity Theory that, in a sense, plays the rôle of the duality theory in Linear Algebra, is the so-called Fenchel conjugation. When applied to the norm of a Banach space provides the indicator function of the closed dual unit ball (see Item 1 below in this paragraph for the definition, and see also Item 2 there), and conversely, when applied to the indicator function of the closed unit ball gives the dual norm (see Item 3 in this paragraph). Another example is equation 3.17 below, that provides a way to compute $\|\cdot\|^*$ by using two ingredients: The elementary function $\varphi_2(t) := |t|^2/2$ (see equation (3.15)), and the Fenchel conjugate of a convenient function. We provide here a straightforward approach to this tool, presenting only what is needed for our purposes.

Let X be a Banach space and $f\colon X \to (-\infty, +\infty]$ be a function. Let $\mathrm{Dom}(f)$ be the set of all $x \in X$ such that $f(x) < +\infty$ —sometimes called the **effective domain of** f—. A function f is said to be **proper** if $\mathrm{Dom}(f) \neq \emptyset$. The **Fenchel conjugate** of f is defined for each $x^* \in X^*$ by

$$\boxed{f^*(x^*) = \sup\{\langle x, x^*\rangle - f(x) :\ x \in X\}} \tag{3.11}$$

(see Figure 3.7, where the value $f^*(x^*)$ is computed for a particular $x^* \in X^*$).

Remark 36. The notation f^* for the Fenchel conjugate of a function f is widely accepted. In case that $f : X \to \mathbb{R}$ is the norm, this convention will certainly may lead to some misunderstandings, since *the Fenchel conjugate of the function* $\|\cdot\|$ *is not the dual norm* (see Item 2 below). We suggest to avoid this kind of problems by denoting

$$\boxed{FC\|\cdot\| \text{ the Fenchel conjugate of a norm } \|\cdot\|,}$$

and sticking to $\|\cdot\|^*$ for denoting the dual norm. ®

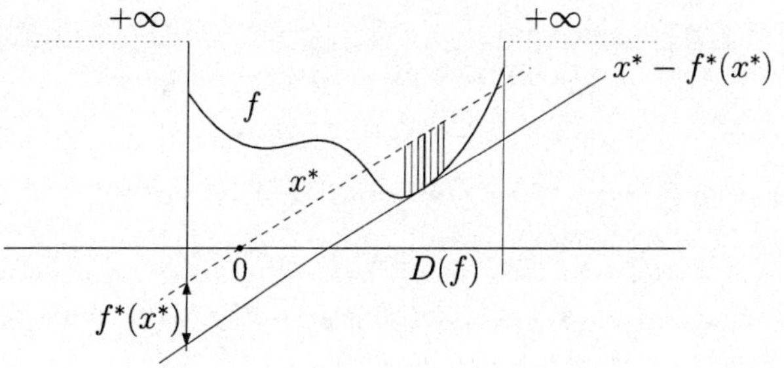

Figure 3.7: Computing the Fenchel conjugate $f^*(x^*)$

Clearly, the three following properties are equivalent:
(i) The function f is proper.
(ii) For every $x^* \in X^*$, $f^*(x^*) > -\infty$.
(iii) There exists $x^* \in X^*$ such that $f^*(x^*) > -\infty$.
In the case that f is proper, the two following properties are equivalent:
(i) The function f^* is proper.
(ii) There exists $x^* \in X^*$ and $t \in \mathbb{R}$ such that $f(x) \geq \langle x, x^* \rangle - t$ for all $x \in X$.

If X is the dual of a Banach space Y, we can define $f_* = f^* \circ j : Y \to [-\infty, +\infty]$, where $j\colon Y \to X^*$ is the canonical isometric embedding.

➪ Fenchel conjugation is treated at length, for example in [Rock96] and [BoVa09]. Here we shall only mention the properties that will be used in the rest of this text. Most of these have a simple proof. Below, f, f_i, and g are arbitrary functions from a normed space X into $(-\infty, +\infty]$, for $i \in I$.
(i) f^* is convex and w^*-lower semicontinuous.
(ii) $(f^*)_*$ is the greatest convex w-lower semicontinuous function dominated by f (i.e., $(f^*)_*(x) \leq f(x)$ for all $x \in X$). If X is a dual space, then $(f_*)^*$ is the greatest convex w^*-lower semicontinuous function dominated by f.
(iii) $f^*(0) = -\inf\{f(x) : \in X\}$.
(iv) If $f \leq g$, then $f^* \geq g^*$, and, in case that X is a dual space, $f_* \geq g_*$.
(v) $(\inf_{i\in I} f_i)^* = \sup_{i\in I} f_i^*$, and $(\sup_{i\in I} f_i)^* \leq \inf_{i\in I} f_i^*$.
(vi) If $t > 0$, then $(tf)^*(x^*) = t(f^*((1/t)x^*))$ for all $x^* \in X^*$.
(vii) For all $t \in \mathbb{R}$, $(f + t)^*(x^*) = f^*(x^*) - t$, for $x^* \in X^*$.
(viii) If $x_0 \in X$ and $g(x) := f(x - x_0)$ for $x \in X$, then $g^*(x^*) = f^*(x^*) + \langle x_0, x^* \rangle$ for $x^* \in X^*$.
(ix) If f is quadratic-homogeneous, that is, if $f(tx) = t^2 f(x)$ for all $x \in X$ and $t \in \mathbb{R}$, then f^* is also quadratic-homogeneous and $(tf)^* = (1/t)f^*$.
(x) (The Fenchel–Young inequality) $f(x) + f^*(x^*) \geq \langle x, x^* \rangle$, for $x \in X$ and $x^* \in X^*$.

(To be convinced that the second inequality in (v) can be strict, observe that the infimum of two convex functions may fail to be convex.)

Some examples

1. **The Fenchel conjugate of an affine function**: Given $x_0^* \in X^*$, define $f(x) = \langle x, x_0^* \rangle$ for $x \in X$. Then

$$f^*(x^*) = \begin{cases} 0 & \text{if } x^* = x_0^*, \\ +\infty & \text{otherwise.} \end{cases}$$

It is convenient, then, to introduce the so-called indicator function: If S is a subset of a normed space X, its **indicator function** $i_S : X \to \{0, +\infty\}$ is defined as

$$i_S(x) := \begin{cases} 0 & \text{if } x \in S, \\ +\infty & \text{otherwise.} \end{cases} \tag{3.12}$$

Thus, for the function $f(x) = \langle x, x_0^* \rangle$ for $x \in X$ defined above, $f^* = i_{\{x_0^*\}}$. More generally, if for some $x_0^* \in X^*$ and $a \in \mathbb{R}$ we define $f(x) = \langle x, x_0^* \rangle + a$ for $x \in X$, then it follows from (vii) above that $f^* = i_{\{x_0^*\}} - a$.

➪ If $\{S_\gamma : \gamma \in \Gamma\}$ is a collection of subsets of X, note that $i_{\bigcup_{\gamma\in\Gamma} S_\gamma} = \inf_{\gamma\in\Gamma}\{i_{S_\gamma}\}$, and $i_{\bigcap_{\gamma\in\Gamma} S_\gamma} = \sup_{\gamma\in\Gamma}\{i_{S_\gamma}\}$. Note, too, that $i_{A\cap B} = i_A + i_B$.

Observe that the indicator function of a set $S \subset X$ is lower semicontinuous if, and only if, S is closed.

2. **The Fenchel conjugate of the norm**: If $f = \|\cdot\|$ ($= p_{B_X}$, where p_B denotes the Minkowski functional of a set B), then $f^* = i_{B_{X^*}}$.

➪ Note: Although this was already mentioned in Remark 36, we stress here that the dual norm $\|\cdot\|^*$ *is not* the Fenchel conjugate of the norm (just look above). We agreed to write $FC\|\cdot\|$ for the Fenchel conjugate of a norm $\|\cdot\|$. Thus,

$$\boxed{FC\|\cdot\| = i_{B_{X^*}}} \tag{3.13}$$

3. **The Fenchel conjugate of the indicator function of the closed unit ball**:

$$\boxed{(i_{B_X})^* = \|\cdot\|^*} \tag{3.14}$$

4. **Another way to get $\|\cdot\|^*$**: An alternate way to Item 3 above for getting the dual norm without using Minkowski functionals or indicator functions is the following: Let us consider, for $p \in (1, +\infty)$, the function $\varphi_p \colon \mathbb{R} \to \mathbb{R}$ defined by

$$\boxed{\varphi_p(t) := |t|^p/p, \text{ for every } t \in \mathbb{R},} \tag{3.15}$$

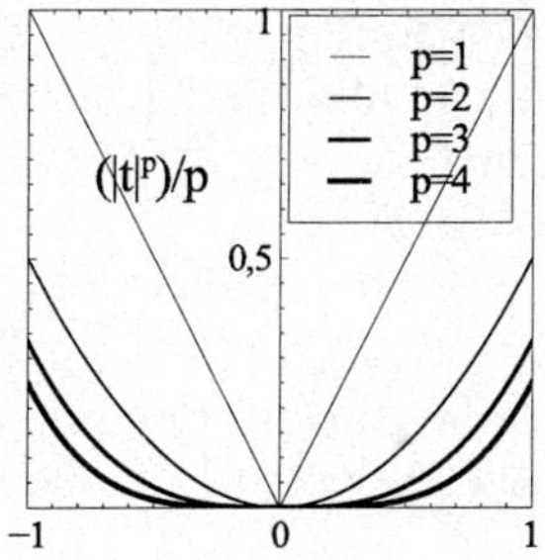

Figure 3.8: The function φ_p for several p's

(see Figure 3.8). It is easy to see that $\varphi_p^* = \varphi_q$, where $p^{-1} + q^{-1} = 1$. In particular, the function φ_2 is self-conjugate, in the sense that $\varphi_2^* = \varphi_2$.

Let $f\colon \mathbb{R} \to \mathbb{R}$ be a non-negative function with $f(0) = 0$ and set $g = f(\|\cdot\|)$. Thus, for $x^* \in X^*$,

$$\begin{aligned} g^*(x^*) &= \sup_{x\in X}\{\langle x, x^*\rangle - f(\|x\|)\} = \sup_{r\geq 0}\sup_{\|x\|=r}\{\langle x, x^*\rangle - f(r)\} \\ &= \sup_{r\geq 0}\Big\{\sup_{\|x\|=r}\{\langle x, x^*\rangle\} - f(r)\Big\} = \sup_{r\geq 0}\{r\|x^*\|^* - f(r)\} \\ &= f^*(\|x^*\|^*). \end{aligned} \tag{3.16}$$

This computation implies, in particular, that if $g = \varphi_2(\|\cdot\|)$, then $g^* = \varphi_2(\|\cdot\|^*)$ or, in other words,

$$\boxed{\|x\| = \sqrt{2g(x)} \text{ for } x \in X \;\Rightarrow \|x^*\|^* = \sqrt{2g^*(x^*)} \text{ for } x^* \in X^*.} \tag{3.17}$$

3.6.5 The infimal convolution of two convex functions

The following device, the infimal convolution of two functions, is a well-known operation usual in Convex Theory. It plays an important rôle in manipulating norms, as it will become evident in Section 3.7, in particular in questions regarding duality. Let us present this concept and some basic properties.

Definition 37. *Let f and g be two convex and w-lower semicontinuous functionals on X. Define their* **infimal convolution** *(sometimes shortened to* **inf-convolution***) as*

$$\boxed{(f\,\square\, g)(x) := \inf\left\{\frac{f(x+y)+g(x-y)}{2} : \; y \in X\right\},\; x \in X.} \tag{3.18}$$

H The version of the inf-convolution above is, according to Day (see [Day73, Definition VII.§2.5]), due to Asplund [Asp67]. Another version is due to Moreau: $(f \,\Box\, g)(x) := \inf\{f(x-y) + g(y) : \ y \in X\}$, for $x \in X$. Note that $(fMg)(x) = 2(fAg)(x/2)$ for all $x \in X$, where M and A stand for the Moreau and Asplund versions (respectively) of the infimal convolution.

➡ For a definition of the infimal convolution, some restrictions on the framework are needed. Two examples are worth to be considered:

1. If X is a Banach space and x^* and y^* are two distinct elements in X^*, note that

$$\begin{aligned}(x^*\Box y^*)(x) &= \inf_{y\in X} (1/2)(\langle x+y, x^*\rangle + \langle x-y, y^*\rangle) \\ &= \inf_{y\in X} (1/2)(\langle x, x^*+y^*\rangle + \langle y, x^*-y^*\rangle) = -\infty,\end{aligned}$$

   for every $x \in X$.

2. If X is a non-reflexive Banach space, let $x_0^* \in S_{X^*}$ be an element that does not attain its norm. Let $H_1 := \{x \in X : \ \langle x, x_0^*\rangle = 1\}$, $f := i_{H_1}$ and $g := i_{B_X}$, where i denotes the indicator function. Due to the fact that both H_1 and B_X are w-closed, the functions f and g are w-lower semicontinuous (w-lsc, for short). However, $f\Box g$ is not w-lower semicontinuous, since $f\Box g = i_D$, where $D := \{x \in X : \ 0 < \langle x, x_0^*\rangle < 1\}$ (see Figure 3.9).

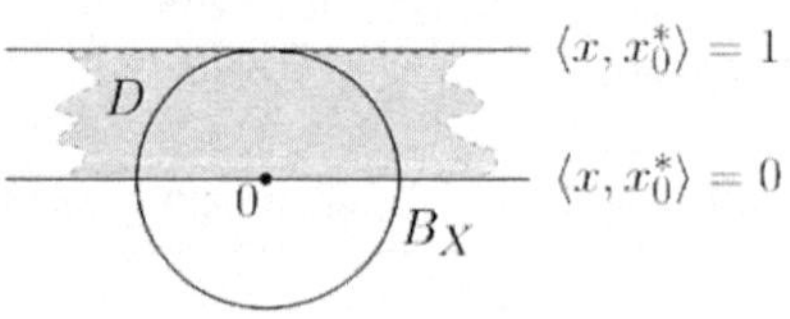

Figure 3.9: The infimal convolution of two lsc functions may fail to be lsc

To avoid the inconveniences arising from the two previous examples, we shall introduce the following restriction when dealing with the infimal convolution of two real-valued w-lower semicontinuous functions f and g defined on a Banach space X:

> When dealing with the infimal convolution $f\Box g$ of two convex functions f and g as in Definition 37, we shall always request the existence of an affine function $h := a + x^*$ (with $a \in \mathbb{R}$ and $x^* \in X^*$) such that $h \le f$ and $h \le g$.

Some important properties of the inf-convolution are listed in the following proposition:

Proposition 38. *Let f and g be two convex and w-lower semicontinuous functions on a Banach space X. Then,*
(i) *$f \,\Box\, g$ is convex and w-lower semicontinuous.*
(ii) *$f \,\Box\, g \leq (f+g)/2$.*
(iii) *$(f \,\Box\, g)^* = (f^* + g^*)/2$ and, dually, $((f+g)/2)^* = f^* \,\Box\, g^*$.*

Proof. Let us show just the first in (iii) above: Fix $x^* \in X^*$, then

$$\begin{aligned}
(f \,\Box\, g)^*(x^*) &= \sup_{x\in X}\{x^*(x) - (f \,\Box\, g)(x)\} \\
&= \sup_{x\in X}\sup_{y\in X}\{x^*(x) - (f(x+y) + g(x-y))/2\} \\
&= \sup_{x,y\in X}\{(x^*(x+y) - f(x+y))/2 + (x^*(x-y) - g(x-y))/2\} \\
&= \sup_{u,v\in X}\{(x^*(u) - f(u))/2 + (x^*(v) - g(v))/2\} \\
&= \sup_{u\in X}\{(x^*(u) - f(u))/2\} + \sup_{v\in X}\{(x^*(v) - g(v))/2\} \\
&= (f^*(x^*) + g^*(x^*))/2.
\end{aligned}$$ □

At Figure 3.10 we plot four functions, $f(x) := x^2$, $g(x) := x$, $f \,\Box\, g$, and $(f+g)/2$ for illustrating some of those properties.

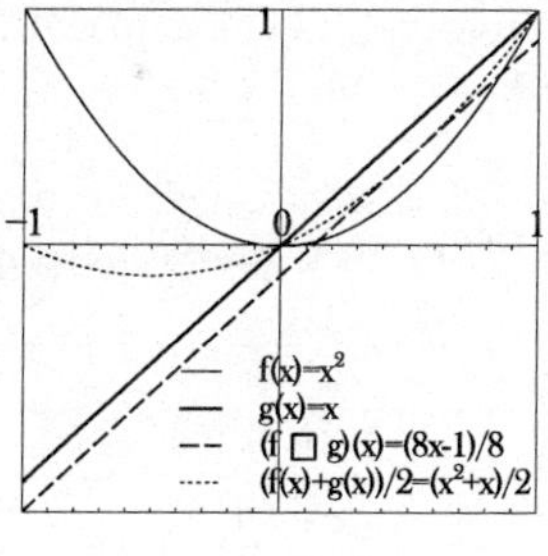

Figure 3.10: Functions f, g, $f \,\Box\, g$, and $(f+g)/2$

An example: The support function of the intersection of two balls

Let $\|\cdot\|_1$ and $\|\cdot\|_2$ be two norms on a vector space X. Observe that the set $B := B_{\|\cdot\|_1} \cap B_{\|\cdot\|_2}$ is absorbing and absolutely convex, and does not contain any ray from 0, so $\|\cdot\| := p_B$ is a norm on X (see Subsection 3.6.1). Then,

$$\sigma_{(B_{\|\cdot\|_1} \cap B_{\|\cdot\|_2})} = \sigma_{B_{\|\cdot\|_1}} \Box \sigma_{B_{\|\cdot\|_2}} \; (= \|\cdot\|_1^* \Box \|\cdot\|_2^*). \tag{3.19}$$

Indeed, and letting $B_i := B_{\|\cdot\|_i}$ for $i = 1, 2$,

$$\begin{aligned}\|\cdot\|^* &= (\iota_{B_1\cap B_2})^* \text{ (equation (3.14))}\\ &= ((\iota_{B_1} + \iota_{B_2})/2)^* \text{ (this is obvious)}\\ &= (\iota_{B_1})^* \,\square (\iota_{B_2})^* \text{ ((iii) in Proposition 38)}\\ &= \|\cdot\|_1^* \,\square\, \|\cdot\|_2^* \text{ (equation (3.14)).}\end{aligned}$$

□

3.7 Norm and ball arithmetics; duality

This section should be considered a (we hope useful) bench work to perform some basic operations on the raw material of norms, their dual ones, and their corresponding unit balls. We follow in part [Yu12].
Table 3.1 (that reproduces Table 1 in [Yu12]) serves as a guide to our program and summarizes what will be mentioned below. It is formulated for a finite family of norms $\{\|\cdot\|_i\}_{i=1}^n$ on a vector space X. In the following subsections we shall justify the validity of the entries there for the case of two norms (the general case will follow easily). By the formula $+_i\|\cdot\|_i$ we mean $\sum_{i=1}^n \|\cdot\|_i$, by $+_iB_i$ we mean the algebraic sum of the sets $\{B_i\}_{i=1}^n$, and by $\overline{+_iB_i}$ the closure of $+_iB_i$. The same applies to the other symbols. Positions (2.ii) and (4.ii) mentions a closure, and so a topology. Only in these cases we shall assume that the norms $\|\cdot\|_i$ are mutually equivalent, and the topology involved is the one defined by those equivalent norms.

Table 3.1: A table of norm operations, balls, and dual norms

		(i)	(ii)		(iii)
	operation	formula	ball	dual operation	dual formula
(1)	$+$	$+_i\|\cdot\|_i$	$\sharp_i B_{\|\cdot\|_i}$	$\sharp$	$\sharp_i\|\cdot\|_i^*$
(2)	$\sharp$	$\sharp_i\|\cdot\|_i$	$\overline{+_i B_{\|\cdot\|_i}}$	$+$	$+_i\|\cdot\|_i^*$
(3)	$\vee$	$\bigvee_i \|\cdot\|_i$	$\bigcap_i B_{\|\cdot\|_i}$	$\square$	$\square_i\|\cdot\|_i^*$
(4)	$\square$	$\square_i\|\cdot\|_i$	$\overline{\text{conv}} \bigcup_i B_{\|\cdot\|_i}$	$\vee$	$\bigvee_i \|\cdot\|_i^*$

We shall follow the suggested order for arithmetics of a finite number of norms:
(1) Sum ($+$).
(2) Inverse summation ($\sharp$) (Definition 39 below).
(3) Supremum ($\vee$).
(4) Infimal convolution ($\square$) (Definition 37 above).
At each level, we shall explore (i) the norm arithmetics, (ii) the resulting unit balls, and finally (iii) their dual operation. We shall address each entry in Table 3.1 by its coordinates. For example, (2.iii) means dealing with the sum of the dual norms.

Thus, in the following subsections, X will denote a vector space over the real numbers, and $\|\cdot\|_1$ and $\|\cdot\|_2$ two norms on X.

3.7.1 Sum of two norms

(1.i) Formula

Put

$$\|\cdot\| := \|\cdot\|_1 + \|\cdot\|_2. \tag{3.20}$$

It is obvious that $\|\cdot\|$ is also a norm on X. Observe that in case that $\|\cdot\|_1$ and $\|\cdot\|_2$ are equivalent norms, the norm $\|\cdot\|$ is equivalent to both of them.

(1.ii) Ball

We have (and the following equation contains the definition of the operation $\sharp$ —the inverse summation— for two sets)

$$\boxed{B_{(\|\cdot\|_1+\|\cdot\|_2)} = \bigcup_{\lambda\in[0,1]} \left(\lambda.B_{\|\cdot\|_1} \cap (1-\lambda).B_{\|\cdot\|_2}\right) =: B_{\|\cdot\|_1} \sharp B_{\|\cdot\|_2}.} \tag{3.21}$$

➪ Let $\|\cdot\| := \|\cdot\|_1 + \|\cdot\|_2$. If $x \in B_{\|\cdot\|}$, then $0 \le \|x\|_1 + \|x\|_2 \le 1$ (thus $0 \le \lambda := \|x\|_1 \le 1$), $x \in \lambda.B_{\|\cdot\|_1}$, $x \in \|x\|_2 B_{\|\cdot\|_2} \subset (1-\lambda)B_{\|\cdot\|_2}$, and this shows $B_{\|\cdot\|} \subset \bigcup_{\lambda\in[0,1]}(\lambda.B_{\|\cdot\|_1} \cap (1-\lambda).B_{\|\cdot\|_2})$. The reverse inclusion is obvious. (Notice that $\lambda.B_{\|\cdot\|_1} \cap (1-\lambda).B_{\|\cdot\|_2} = \{0\}$ whenever $\lambda = 0$ or $\lambda = 1$; observe, too, that $\bigcup_{\lambda\in[0,1]} \left(\lambda.B_{\|\cdot\|_1} \cap (1-\lambda).B_{\|\cdot\|_2}\right)$ is a closed set.)

(1.iii) Duality

We have

$$\boxed{(\|\cdot\|_1 + \|\cdot\|_2)^* = \|\cdot\|_1^* \sharp \|\cdot\|_2^*,} \tag{3.22}$$

where $\sharp$ is the inverse summation (see Definition 39 below).

➪ Indeed, having in mind (3.21),

$$\begin{aligned}
(\|\cdot\|_1 + \|\cdot\|_2)^* &= \sigma_{\bigcup_{t\in[0,1]}(tB_{\|\cdot\|_1}\cap(1-t)B_{\|\cdot\|_2})} \\
&= \sup_{t\in[0,1]} \left\{\sigma_{\left(tB_{\|\cdot\|_1}\cap(1-t)B_{\|\cdot\|_2}\right)}\right\} \\
&= \sup_{t\in[0,1]} \left\{\sigma_{tB_{\|\cdot\|_1}} \square\, \sigma_{\left((1-t)B_{\|\cdot\|_2}\right)}\right\} \qquad \text{(equation (3.19))} \\
&= \sup_{t\in[0,1]} \left\{(t\sigma_{B_{\|\cdot\|_1}}) \square ((1-t)\sigma_{B_{\|\cdot\|_2}})\right\} \\
&= \sup_{t\in[0,1]} \left\{(t\|\cdot\|_1^*) \square ((1-t)\|\cdot\|_2^*)\right\}.
\end{aligned} \tag{3.23}$$

Notice now that

$$\begin{aligned}\|x^*\|_1^* \,\square\, \|x^*\|_2^* &= \inf\{\tfrac{\|x^*+y^*\|_1^*+\|x^*-y^*\|_2^*}{2} : \ y^* \in X^*\} \\ &= \inf\{\|x^*/2 + y^*/2\|_1^* + \|x^*/2 - y^*/2\|_2^* : \ y^* \in X^*\} \\ &= \inf\{\|x_1^*\|_1^* + \|x_2^*\|_2^* : \ x_1^* + x_2^* = x^*\}.\end{aligned}$$

Thus, $\sup_{t\in[0,1]}\{(t\|\cdot\|_1^*)\,\square\,((1-t)\|\cdot\|_2^*)\} = \|x^*\|_1 \sharp \|x^*\|_2$. Carrying this into (3.23), we get (3.22). □

3.7.2 Inverse summation of two norms

(2.i) Formula

In view of the definition of the operation $\sharp$ **for sets** (see (3.21)), it is natural to define a similar operation **for norms**, also denoted by $\sharp$, and called the **inverse summation** of two norms:

Definition 39. *Let* $\|\cdot\|_1$ *and* $\|\cdot\|_2$ *be two norms on a vector space* X. *Their* **inverse summation** *is given by*

$$\|x\|_1 \sharp \|x\|_2 := \sup_{\lambda\in[0,1]} \inf_{x=x_1+x_2} \lambda\|x_1\|_1 + (1-\lambda)\|x_2\|_2 = \inf_{x=x_1+x_2} \|x_1\|_1 \vee \|x_2\|_2. \tag{3.24}$$

➪ Equality in equation (3.24) is a straightforward consequence of Sion's Minimax Theorem [Sio58] (an extension of the classical von Neumann's Minimax Theorem): *Let K be a compact convex subset of a topological vector space, and let C be a convex subset of another topological vector space. Let $f : K\times C \to \mathbb{R}$ be such that*
(i) *$f(k,\cdot)$ is upper semicontinuous and quasiconcave on C, for all $k \in K$, and*
(ii) *$f(\cdot, c)$ is lower semicontinuous and quasiconvex on K, for all $c \in C$. Then*

$$\min_{k\in K}\sup_{c\in C} f(k,c) = \sup_{c\in C}\min_{k\in K} f(k,c). \tag{3.25}$$

(A function $g : S \to \mathbb{R}$ defined on a convex set S of a vector space is said to be **quasiconvex** (**quasiconcave**) if each sublevel set $\{x \in S : \ g(x) \le \alpha\}$ is a convex set (respectively, if $-g$ is a quasiconvex function).)
Indeed, take $K := [0,1]$ and $C := \{(x_1, x_2) \in X \times X : \ x_1 + x_2 = x\}$, and put $f(\lambda,(x_1,x_2)) := -(\lambda\|x_1\|_1 + (1-\lambda)\|x_2\|_2)$, $\lambda \in \mathbb{R}$, $(x_1,x_2) \in C$. Observe that $f(\lambda,\cdot)$ is clearly concave for $\lambda \in [0,1]$ on $(x_1,x_2) \in C$ (and continuous if X carries the product topology $\mathcal{T}_{\|\cdot\|_1} \times \mathcal{T}_{\|\cdot\|_2}$), and linear in $\lambda \in \mathbb{R}$. Now, equality in (3.24) follows from (3.25), since $f(\cdot,(x_1,x_2))$ attains its minimum

on $[0,1]$ at $\{0,1\}$, for each $(x_1, x_2) \in C$.

The reader may get a fair idea of the reason for equality in (3.24) by looking at Figure 3.11: Each level corresponds to a choice of $x_1, x_2 \in X$ such that $x = x_1 + x_2$ (so moving up or down means to choose different such couples (x_1, x_2)), while going from left to right means letting λ in $\lambda\|x_1\|_1 + (1 - \lambda)\|x_2\|_2$ move from 0 to 1.

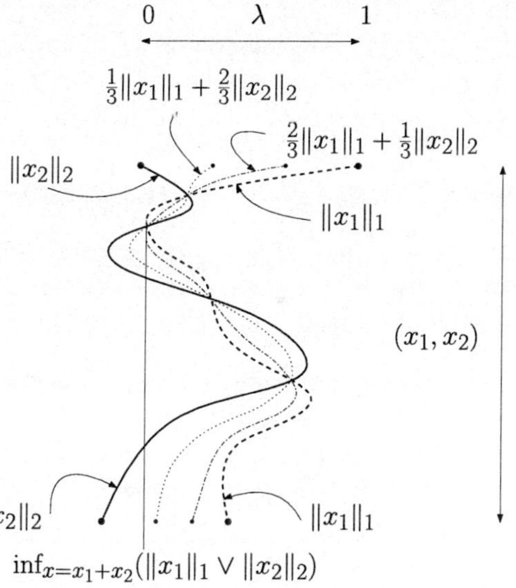

Figure 3.11: Proving the equality in equation (3.24). "Lines" for four λ's are drawn ($\lambda = 0, 1/3, 2/3, 1$, respectively)

Remark 40. Assume now that $\|\cdot\|_1$ and $\|\cdot\|_2$ on X are equivalent norms, say $a\|x\|_2 \le \|x\|_1 \le b\|x\|_2$ for all $x \in X$ and some $0 < a < b$. We claim that the set $B := \{x \in X : \|x\|_1 \sharp \|x\|_2 \le 1\}$ is closed. To prove this, observe first that, for a sequence $\{x_n\}_{n=1}^{\infty}$ in X, we have $\|x_n\|_1 \sharp \|x_n\|_2 \to 0$ if, and only if, $\|x_n\|_1 \to 0$. Indeed, if for some $\varepsilon > 0$ and some $n \in \mathbb{N}$ we have $\|x_n\|_1 \sharp \|x_n\|_2 < \varepsilon$, then we can find y_n and z_n in X such that $x_n = y_n + z_n$, $\|y_n\|_1 < \varepsilon$ and $\|z_n\|_2 < \varepsilon$. Thus, $\|x_n\|_1 \le \|y_n\|_1 + \|z_n\|_1 \le \varepsilon(1+b)$. This shows that $\|x_n\|_1 \to 0$. The proof of the sufficient condition is also easy. Assume now that $\{s_n\}_{n=1}^{\infty}$ is a sequence in B that converges to $x \in X$. For $n \in \mathbb{N}$, find y_n and z_n in X such that $s_n = y_n + z_n$ and $\|y_n\|_1 + \|z_n\|_2 < 1 + 1/n$. Since $x = s_n + (x - s_n) = y_n + z_n + (x - s_n)$, we have $\|x\|_1 \sharp \|x\|_2 \le \|y_n + (x - s_n)\|_1 \vee \|z_n\|_2 \le (\|y_n\|_1 + \|x - s_n\|_1) \vee \|z_n\|_2$. Since this is true for every $n \in \mathbb{N}$ and $\|x - s_n\|_1 \to 0$, we get $\|x\|_1 \sharp \|x\|_2 \le 1$, as claimed. Ⓡ

(2.ii) Ball

Here we assume that $\|\cdot\|_1$ and $\|\cdot\|_2$ are equivalent norms. The closure in (3.26) refers to the common norm topology on X.

$$\boxed{B_{(\|\cdot\|_1 \sharp \|\cdot\|_2)} = \overline{B_{\|\cdot\|_1} + B_{\|\cdot\|_2}}.} \tag{3.26}$$

➪ In the argument that follows, we shall use the description of $\sharp$ given by the third member in (3.24). If for some $x \in X$ we have $\|x\|_1 \sharp \|x\|_2 \leq 1$ and we provide an arbitrary $\varepsilon > 0$, then there exists x_1, x_2 in X such that $x = x_1 + x_2$, and $\|x_1\|_1 \vee \|x_2\|_2 < 1 + \varepsilon$. Thus, $\|x_i\|_i < 1 + \varepsilon$, i.e., $x_i \in (1+\varepsilon)B_{\|\cdot\|_i}$, for $i = 1, 2$, and so $x \in (1+\varepsilon)(B_{\|\cdot\|_1} + B_{\|\cdot\|_2})$. Since this is true for every $\varepsilon > 0$, we get $x \in \overline{B_{\|\cdot\|_1} + B_{\|\cdot\|_2}}$. This shows that $B_{\|\cdot\|_1 \sharp \|\cdot\|_2} \subset \overline{B_{\|\cdot\|_1} + B_{\|\cdot\|_2}}$. On the other hand, if $x \in B_{\|\cdot\|_1} + B_{\|\cdot\|_2}$ we get $x = x_1 + x_2$, where $x_i \in B_{\|\cdot\|_i}$, $i = 1, 2$. Thus, $\|x_1\|_1 \vee \|x_2\|_2 \leq 1$, and so $\|x\|_1 \sharp \|x\|_2 \leq 1$. We get then $B_{\|\cdot\|_1} + B_{\|\cdot\|_2} \subset B_{\|\cdot\|_1 \sharp \|\cdot\|_2}$. Since $B_{\|\cdot\|_1 \sharp \|\cdot\|_2}$ is closed (see Remark 40 above), we get the conclusion. Observe, too, that formula (3.26) shows that $\|\cdot\|_1 \sharp \|\cdot\|_2$ is, in fact, a norm (equivalent to both $\|\cdot\|_1$ and $\|\cdot\|_2$).

(2.iii) Duality

Since the closed unit ball B of the norm $\|\cdot\| := \|\cdot\|_1 \sharp \|\cdot\|_2$ is given by (3.26), it should be clear that $\|\cdot\|^*$, as the supremum on B, is

$$\boxed{(\|\cdot\|_1 \sharp \|\cdot\|_2)^* = \|\cdot\|_1^* + \|\cdot\|_2^*}.$$

3.7.3 Supremum of two norms

(3.i) Formula

The supremum $\|\cdot\|_1 \vee \|\cdot\|_2$ of two norms $\|\cdot\|_1$ and $\|\cdot\|_2$ on X is again a norm on X.

(3.ii) Ball

The following equality is obvious:

$$\boxed{B_{(\|\cdot\|_1 \vee \|\cdot\|_2)} = B_{\|\cdot\|_1} \cap B_{\|\cdot\|_2}.} \tag{3.27}$$

(3.iii) Duality

We have

$$\boxed{(\|\cdot\|_1 \vee \|\cdot\|_2)^* = \|\cdot\|_1^* \square \|\cdot\|_2^*.} \tag{3.28}$$

➪ Indeed,

$$\begin{aligned}(\|\cdot\|_1 \vee \|\cdot\|_2)^* &= (i_{B_1\cap B_2})^* \text{ (equations (3.27) abd (3.14))}\\ &= ((i_{B_1}+i_{B_2})/2)^* \text{ (this is obvious)}\\ &= (i_{B_1})^* \square (i_{B_2})^* \text{ ((iii) in Proposition 38)}\\ &= \|\cdot\|_1^* \square \|\cdot\|_2^* \text{ (equation (3.14))} \qquad (3.29)\end{aligned}$$

3.7.4 The infimal convolution of two norms

(4.i) Formula

The infimal convolution of two convex functions was introduced in Definition 37. For two norms $\|\cdot\|_1$ and $\|\cdot\|_2$ on X it reads (see the proof of (3.22)):

$$(\|\cdot\|_1 \square \|\cdot\|_2)(x) := \inf_{x=x_1+x_2} (\|x_1\|_1 + \|x_2\|_2),\ x \in X.$$

From the obvious fact that $(\|\cdot\|_1\square\|\cdot\|_2)(\lambda.x) = |\lambda|.(\|\cdot\|_1\square\|\cdot\|_2)(x)$ for every $\lambda \in \mathbb{R}$ and $x \in X$, and from Proposition 38, it follows that $\|\cdot\|_1\square\|\cdot\|_2$ is a seminorm on X. If $(\|\cdot\|_1\square\|\cdot\|_2)(x) = 0$ and $\varepsilon > 0$ we can find a sequence $(x_{1,n}, x_{2,n})_{n=1}^{\infty}$ in $X \times X$ with $x_{1,n} + x_{2,n} = x$ for all $n \in \mathbb{N}$, and such that $\|x_{i,n}\|_i \to 0$ as $n \to \infty$, for $i = 1, 2$. It follows that $x = 0$, so $\|\cdot\|_1\square\|\cdot\|_2$ is in fact a norm.
It is worth to notice that the epigraph of the norm $\|\cdot\|_1\square\|\cdot\|_2$ is the closed convex hull of the epigraph of $\|\cdot\|_1 \wedge \|\cdot\|_2$. This is a simple consequence of the description of its closed unit ball in (3.30) below.

(4.ii) Ball

As we mentioned above, here we assume that the two norms $\|\cdot\|_1$ and $\|\cdot\|_2$ are equivalent. The closure in (3.30) refers to the common topology defined.
We have

$$\boxed{B_{(\|\cdot\|_1\square\|\cdot\|_2)} = \overline{\text{conv}}\,(B_{\|\cdot\|_1} \cup B_{\|\cdot\|_2}).} \qquad (3.30)$$

➪ Let $\|\cdot\| := \|\cdot\|_1\square\|\cdot\|_2$, and let $x \in X$ such that $\|x\| \le 1$. Given $n \in \mathbb{N}$, we can find $x_{1,n}, x_{2,n} \in X$ such that $x = x_{1,n} + x_{2,n}$ and $\|x_{1,n}\|_1 + \|x_{2,n}\|_2 \le 1 + 1/n$. If $a_n := \|x_{1,n}\|_1$, we have $x_{1,n} \in a_n B_{\|\cdot\|_1}$ and $x_{2,n} \in (1 - a_n + 1/n) B_{\|\cdot\|_2}$, for $n \in \mathbb{N}$. This shows $(1+1/n)^{-1} x \in \text{conv}\,(B_{\|\cdot\|_1} \cup B_{\|\cdot\|_2})$. Since this is true for every $n \in \mathbb{N}$, we get $x \in \overline{\text{conv}}\,(B_{\|\cdot\|_1} \cup B_{\|\cdot\|_2})$.
On the other hand, if $x \in \text{conv}\,(B_{\|\cdot\|_1} \cup B_{\|\cdot\|_2})$ then $x = \lambda b_1 + (1-\lambda) b_2$, for $b_i \in B_{\|\cdot\|_i}$, $i = 1, 2$. Thus, $\|x\| \le 1$. Since $B_{\|\cdot\|}$ is closed, we get $\overline{\text{conv}}\,(B_{\|\cdot\|_1} \cup B_{\|\cdot\|_2}) \subset B_{\|\cdot\|}$. All together, (3.30) holds. □

(4.iii) Duality

We have

$$\boxed{(\|\cdot\|_1 \square \|\cdot\|_2)^* = \max\{\|\cdot\|_1^*, \|\cdot\|_2^*\}}. \tag{3.31}$$

➪ Indeed, by using the properties of the support function and the description of the unit ball in (3.30), we have

$$\begin{aligned}(\|\cdot\|_1 \square \|\cdot\|_2)^* &= \sigma_{\overline{\text{conv}}(B_1 \cup B_2)} \\ &= \sigma_{B_1 \cup B_2} = \max\{\sigma_{B_1}, \sigma_{B_2}\} = \max\{\|\cdot\|_1^*, \|\cdot\|_2^*\}.\end{aligned}$$

3.7.5 A basic result

The following result summarizes the previous observations made for the case of two norms. Its proof follows the pattern of the arguments in the previous Subsections. The conjugate index to $p = 1$ is $q = +\infty$. In this case, the second formula in (3.32) naturally becomes $\inf_{x^*=\sum_{i=1}^n x_i^*} \sup\{\|x_i^*\|_i^* : i = 1, 2, \ldots, n\}$, while the first formula in (3.33) becomes $\inf_{x=\sum_{i=1}^n x_i} \sup\{\|x_i\|_i : i = 1, 2, \ldots, n\}$ (due to the fact that $\lim_{x\to+\infty}(a^x + b^x)^{1/x} = \max\{a, b\}$ for $a > 0$ and $b > 0$).

Theorem 41. *Let $\{\|\cdot\|_i\}_{i=1}^n$ be a finite family of norms on a normed space X. Let $p \geq 1$, and let $q \geq 1$ be its conjugate index, i.e., $1/p + 1/q = 1$. Then*
(i) *The dual norm to the norm*

$$\left(\sum_{i=1}^n \|x\|_i^p\right)^{1/p} \quad \textit{is} \quad \inf_{x^*=\sum_{i=1}^n x_i^*} \left(\sum_{i=1}^n \|x_i^*\|_i^{*q}\right)^{1/q}. \tag{3.32}$$

(ii) *The dual norm to the norm*

$$\inf_{x=\sum_{i=1}^n x_i} \left(\sum_{i=1}^n \|x_i\|^q\right)^{1/q} \quad \textit{is} \quad \left(\sum_{i=1}^n \|x^*\|_i^{*p}\right)^{1/p}. \tag{3.33}$$

3.7.6 Dealing with seminorms

Let $p\colon X \to \mathbb{R}$ be a continuous seminorm on the Banach space X (for the definition of a seminorm see Section 2.1 above). Denote by Z_p the kernel of p (it is a closed subspace of X) and by B_p its unit ball (i.e., $B_p = \{x \in X : p(x) \leq 1\}$). Note that p is a norm if, and only if, $Z_p = \{0\}$. The quotient space $X_p := X/Z_p$ can be endowed with the norm defined by $\|\widehat{x}\|_p = p(x)$, where $x \in \widehat{x}$.

➪ That $\|\cdot\|_p$ is well defined is clear: If x and y belong to $\hat{x}$, then $x - y \in Z_p$. Thus, $p(x) = p(y)$. That $\|\cdot\|_p$ so defined on X_p is indeed a norm follows right away from this and the definition.

Only in this subsection we shall denote by $Q_p\colon X \to X_p$ the canonical quotient map (not the be confused with a later manipulation of a seminorm defined in (5.2)). Observe that $Q_p(B_p)$ is the closed unit ball of $(X_p, \|\cdot\|_p)$. The dual space of $(X_p, \|\cdot\|_p)$ is $Z_p^\perp$. Thus, $x^* \in Z_p^\perp$ if, and only if, x^* is bounded on B_p. In other words, *$Z_p^\perp$ is the effective domain of σ_{B_p}*. Note, too, that $Z_p \subset B_p$. The support function σ_{B_p} is not bounded in case that $Z_p \neq \{0\}$. Moreover, $\|F\|_p^* = \sigma_{B_p}(Q_p^* F)$ for every $F \in Z_p^\perp$.

Given $f\colon \mathbb{R} \to \mathbb{R}$ a non-negative function with $f(0) = 0$, let us define the function $g := f \circ p : X \to \mathbb{R}$. Then, for $z^* \in Z_p^\perp$ we have

$$\begin{aligned} g^*(z^*) &= \sup_{x\in X}\{\langle x, z^*\rangle - f(p(x))\} = \sup_{r\geq 0}\sup_{p(x)=r}\{\langle x, z^*\rangle - f(r)\} \\ &= \sup_{r\geq 0}\Big\{\sup_{p(x)=r}\{\langle x, z^*\rangle\} - f(r)\Big\} = \sup_{r\geq 0}\{r\sigma_{B_p}(z^*) - f(r)\} \\ &= f^*(\sigma_{B_p}(z^*)), \end{aligned}$$

since it is clear that $\sup_{p(x)=r}\langle x, z^*\rangle = r\sup_{x\in B_p}\langle x, z^*\rangle$. This computation generalizes the one we obtained for norms in Subsection 3.6.4, Item 4. However, it has a main drawback: $g^*(x^*) = +\infty$ for every $x^* \notin Z_p^\perp$ (since $\sigma_{B_p}(x^*) = +\infty$ in such a case, see above). On the other hand, g^* can be written in terms of the dual norm $\|\cdot\|_p^*$. Indeed, we may define $\widehat{g} : X_p \to \mathbb{R}$ as $\widehat{g} = f \circ \|\cdot\|_p$. Observe that $\widehat{g}\circ Q = g$, and also that, for $z^* \in Z_p^\perp$,

$$\begin{aligned} (\widehat{g})^*(z^*) &= \sup_{\widehat{x}\in X_p}\{\langle \widehat{x}, z^*\rangle - \widehat{g}(\widehat{x})\} \\ &= \sup_{x\in X}\{\langle x, z^*\rangle - g(x)\} = g^*(z^*). \end{aligned}$$

We may apply Subsection 3.6.4, Item 4 to get $(\widehat{g})^*(z^*) = f^*(\|z^*\|_p^*)$. Thus, $g^*(z^*) = f^*(\|z^*\|_p^*)$ for every $z^* \in Z_p^\perp$.

Fix $r \in (0, \infty)$. Let p_1 and p_2 be two continuous seminorms on X such that $\|x\| = (p_1(x)^r + p_2(x)^r)^{1/r}$ defines a norm on X. Then, as before, $\varphi_r(\|\cdot\|) = \varphi_r \circ p_1 + \varphi_r \circ p_2$ (where φ_r was defined in (3.15), and thus

$$\varphi_s(\|\cdot\|^*) = 2[(\varphi_s \circ \sigma_{B_{p_1}})\Box(\varphi_s \circ \sigma_{B_{p_2}})](\cdot/2)$$

where s is the adjoint of r. Equivalently

$$\|\cdot\|^* = 2^{1/s}\left[(\sigma_{B_{p_1}})^s \Box (\sigma_{B_{p_2}})^s\right]^{1/s}(\cdot/2).$$

Of course this should be explained, since the support functions in the two previous expressions are not bounded. Given $x^* \in X^*$, since $Z_{p_1} \cap Z_{p_2} = \{0\}$ we can find $x_i^* \in Z_{p_i}^\perp$ such that $x^* = x_1^* + x_2^*$. Then, the vector $y^* = (x_1^* - x_2^*)$ satisfies $x^* + y^* \in Z_{p_1}^\perp$ and $x^* - y^* \in Z_{p_2}^\perp$ so,

$$\|x^*\|^* = \inf\left\{\left[(\sigma_{B_{p_1}}(z_1^*))^s + (\sigma_{B_{p_2}}(z_2^*))^s\right]^{1/s} : \; z_i^* \in Z_{p_i}^\perp,\; i = 1, 2,\; z_1^* + z_2^* = x^*\right\}.$$

We can write this expression in terms of the dual norms of the corresponding quotient spaces X_{p_i}:

$$\|x^*\|^* = \inf\Big\{\big[(\|F_1\|_{p_1}^*)^s + (\|F_2\|_{p_2}^*)^s\big]^{1/s} : \\ F_i \in X_{p_i}^*,\ i=1,2,\ Q_{p_1}^* F_1 + Q_{p_2}^* F_2 = x^*\Big\}.$$

A different approach

Let us show a different approach to the computation of dual norms by using simply classical results of Banach Space Theory that will give us the result, even for a countable number of seminorms.

Let $(X_1, \|\cdot\|_1)$ and $(X_2, \|\cdot\|_2)$ be two Banach spaces. Given $p \in (1,\infty)$ the p-sum of X_1 and X_2, denoted by $X_1 \oplus_p X_2$, is defined as the linear space $X_1 \times X_2$ together with the norm $\|(x_1,x_2)\|_p := (\|x_1\|_1^p + \|x_2\|_2^p)^{1/p}$ for every $(x_1,x_2) \in X_1 \times X_2$. It is easy to see that the dual of $X_1 \oplus_p X_2$ is isometric to $X_1^* \oplus_q X_2^*$ where $p^{-1} + q^{-1} = 1$.

➡ Let, for $i = 1,2$, $T_i\colon X_i \to X_1 \oplus_p X_2$ be the canonical inclusion and $P_i := T_i^*$. Let $\Phi\colon (X_1 \oplus_p X_2)^* \to X_1^* \times X_2^*$ be defined for any $f \in (X_1 \oplus_p X_2)^*$ by $\Phi(f) = (P_1 f, P_2 f)$. Let $(x_n^i)_n \in S_{X_i}$ be such that $\langle x_n^i, P_i f\rangle \to \|P_i f\|_i$. Set $z_n := (\|P_1 f\|_1^{q-1} x_n^1, \|P_2 f\|_2^{q-1} x_n^2)$. Then, for $n \in \mathbb{N}$,

$$\langle z_n, f\rangle \le \|f\|_p^* \|z_n\|_p = \|f\|_p^* \left(\|P_1 f\|_1^q + \|P_2 f\|_2^q\right)^{1/p},$$

and since $\|P_1 f\|_1^q + \|P_2 f\|_2^q = \lim_n \langle z_n, f\rangle$, we obtain

$$\left(\|P_1 f\|_1^q + \|P_2 f\|_2^q\right)^{1/q} \le \|f\|_p^*,$$

which implies $\|\Phi(f)\| \le \|f\|_p^*$. On the other hand, given $(x_1^*, x_2^*) \in X_1^* \oplus_q X_2^*$ we can define $f \in (X_1 \oplus_p X_2)^*$ by $f(x_1,x_2) = x_1^*(x_1) + x_2^*(x_2)$. Clearly $\Phi(f) = (x_1^*, x_2^*)$ and $\|f\|_p^* \le \|(x_1^*, x_2^*)\|_q$ by Hölder's inequality. Since $\|f\|_p^* \le \|\Phi(f)\|_q \le \|f\|_p^*$, we get that Φ is an isometry between $(X_1 \oplus_p X_2)^*$ and $X_1^* \oplus_q X_2^*$.

Let X be a Banach space, $p \in (1,+\infty)$, and, for $i = 1,2$, let $\|\cdot\|_i$ be a continuous seminorm on X such that the function

$$\|\cdot\| := (\|\cdot\|_1^p + \|\cdot\|_2^p)^{1/p}$$

is an equivalent norm on X. Set $Z_i := \{x \in X : \|x\|_i = 0\}$ and $X_i := X/Z_i$ for $i = 1,2$. Endow X_i with the norm obtained canonically from $\|\cdot\|_i$ and consider the quotient maps $Q_i\colon X \to X_i$, $i = 1,2$. The map $\varphi\colon (X, \|\cdot\|) \to X_1 \oplus_p X_2$ defined by $\varphi(x) = (Q_1 x, Q_2 x)$ is an isometry onto $\varphi(X)$. Then, the dual $(X, \|\cdot\|)^*$ is isometric to $(\varphi(X), \|\cdot\|_p)^*$, which is in turn isometric to $(X_1 \oplus_p X_2)^*/\varphi(X)^\perp$.

This chain of isometries shows that

$$\|x^*\|^* = \inf\left\{[(\|x_1^*\|_1^*)^q + (\|x_2^*\|_2^*)^q]^{1/q} : \; Q_1^* x_1^* + Q_2^* x_2^* = x^*\right\}$$

In particular, when both seminorms are norms, we have that $Q_1 = Q_2 = \mathrm{Id}_X$, and the dual norm is

$$\|x^*\|^* = \inf\{((\|y^*\|_1^*)^q + (\|x^* - y^*\|_2^*)^q)^{1/q} : \; y^* \in X^*\},$$

which is the same formula we obtained by using the Fenchel conjugation.

➬ The previous argument can be performed for a countable number of continuous seminorms $(\|\cdot\|_n)_n$ keeping in mind that the dual of a p-sum of Banach spaces is a q-sum of its corresponding duals (where $p^{-1} + q^{-1} = 1$). Indeed, if $\|\cdot\| = (\sum_n \|\cdot\|_n^p)^{1/p}$ defines a norm on X, then its dual norm can be computed as follows:

$$\|x^*\|^* = \inf\left\{\left(\sum_n (\|x_n^*\|_n^*)^q\right)^{1/q} : x_n^* \in X_n^*, \; \sum_n Q_n^* x_n^* = x^*\right\},$$

where Q_n is the canonical quotient from X to $X_n := X/Z_n$ (where Z_n is the zero set of the seminorm $\|\cdot\|_n$) endowed with the corresponding norm.

3.7.7 The predual norm of a p-sum of w^*-lower semicontinuous seminorms

Let $(\|\cdot\|_n)_n$ be a sequence of continuous and w^*-lsc seminorms on the dual X^* of a Banach space X such that $\|\cdot\| = (\sum_n \|\cdot\|_n^p)^{1/p}$ is a norm on X^*. Then, of course, $\|\cdot\|$ is a dual norm. From the previous subsection we know how to compute the dual norm $\|\cdot\|^*$ but, clearly, this will be the bidual norm of the predual norm $\|\cdot\|_*$ on X. Since X is isometric to $\pi(X) \subset X^{**}$, where $\pi\colon X \to X^{**}$ is the canonical isometric embedding, the predual norm can be computed as

$$\|x\|_* = \inf\left\{\left(\sum_n (\|x_n^*\|_n^*)^q\right)^{1/q} : \; x_n^* \in X_n^*, \; \sum_n Q_n^* x_n^* = \pi(x)\right\}$$

using again the same notation as in the previous subsection. Observe that this formula is highly inefficient since we need to compute the infimum over sequences of elements in X^{**}. An intrinsic representation of $\|\cdot\|_*$ (that is, an expression depending just on the behaviour of X) would be of interest.

Chapter 4

Basic differentiability in Banach spaces

Differentiability and rotundity are related concepts for convex functions. In a sense, they are "in duality" (whatever it means at this moment). So it is difficult to prioritize one or the other. One is more "analytical", the other more "geometrical". We choose to start with differentiability; despite this, it will be necessary to mention from the beginning some rotundity terms and to use some geometric arguments. Rotundity will be the subject of Chapter 5 below.

Differentiability of a function means, essentially, the possibility to approximate the given function, at least locally, by a linear one. In the setting of real-valued functions f on a nonempty open interval (a, b) in $\mathbb{R}$, the approximating function at a point $x_0 \in (a, b)$ has a graph that is the tangent line to the graph of f at $(x_0, f(x_0))$ (by the way, a necessary condition for the approximation is the continuity of the function at the given point, due to the fact that a linear function on $\mathbb{R}$ is continuous). If this is done for real-valued functions defined on a Banach space X, linear functionals should appear naturally in the description (as approximating functions or, geometrically, defining tangent planes to the graph). We shall consider, for a matter of economy, the case of a general real-valued function on a Banach space —mostly convex or Lipschitz functions— and then the particular case of its norm: Its geometric and analytic properties have a profound impact on the underlying structure of the Banach space and the behaviour of the functions defined on it.

Thus, this chapter will be organized as follows: Section 4.1 will present the basic facts on differentiability of real-valued functions on a Banach space. We shall describe first, for getting some intuition, the finite-dimensional case, and only after that the general case. We shall consider Lipschitz functions, then convex functions and finally, even risking some redundance, the case of a norm. To have a reservoir of examples, the canonical norms in classical spaces will be considered. Section 4.2 will introduce the concept of strong Gâteaux differentiability, and Section 4.3 some uniformities in differentiability. Finally, Section 4.4 will describe a typical tool in renorming: How to use level sets of (differentiable) functions to define differentiable equivalent norms.

A. J. Guirao et al., *Renormings in Banach Spaces*, Monografie Matematyczne 75,
https://doi.org/10.1007/978-3-031-08655-7_4

4.1 Gâteaux and Fréchet differentiability

4.1.1 The basic definitions

The concept of Gâteaux differentiability at a point refers to the existence of a "tangent plane" to the graph of a function. Fréchet differentiability asks for a "uniform" (in directions) approximation between both of them, the tangent plane and the graph. Precisely, this is recorded in the following

Definition 42. *Let $(X, \|\cdot\|)$ and $(Y, |||\cdot|||)$ be two normed spaces, and U be a nonempty open subset of X. We say that a function $f : U \to Y$ is* **Gâteaux differentiable (or Gâteaux smooth) at a point** $x_0 \in U$ *if for every $h \in X$ the following limit (called, if $h \neq 0$, the* **directional derivative of** f **at** x_0 **in the direction** h*, and sometimes denoted $D_h f(x_0)$) exists and defines an element $f'(x_0) \in \mathcal{B}(X,Y)$, i.e., a linear and continuous mapping from X into Y:*

$$(D_h f(x_0) =)\quad f'(x_0)(h) := \lim_{t\to 0} \frac{f(x_0 + th) - f(x_0)}{t}, \tag{4.1}$$

where $t \neq 0$ is small enough to have $x_0 + th \in U$. The continuous linear operator $f'(x_0)$ ($\in \mathcal{B}(X,Y)$) is called the **Gâteaux derivative** *or* **Gâteaux differential** *of f at x_0.*
In case that the limit (4.1) is uniform in $h \in S_X$ we say that f is **Fréchet differentiable** *(or* **Fréchet smooth***)* **at** x_0.

Definition 43. *Let $f : U \to Y$ be a function, and let $x_0 \in U$. Assume that f is Fréchet differentiable at points of a neighborhood $W \subset U$ of x_0. If the mapping $x \to f'(x)$ from W into $\mathcal{B}(X,Y)$ is continuous we say that f is* C^1**-smooth***. If, moreover, it is Fréchet differentiable at x_0 we say that f is* **twice Fréchet differentiable at** x_0*. In this case, $(f')'(x_0)$ (an element in $\mathcal{B}(X, \mathcal{B}(X,Y))$) is denoted by $f''(x_0)$ or, alternatively, by $f^{(2)}(x_0)$. Notice that the space $\mathcal{B}(X, \mathcal{B}(X,Y))$ can be canonically identified with $\mathcal{B}(X^2, Y)$. If $f^{(2)}$ is continuous in W we say that f is C^2-smooth. Inductively we can define the concept of* k**-times Fréchet differentiability of** f **at** x_0 *(the corresponding mapping from W into $\mathcal{B}(X^k, Y)$ is denoted by $f^{(k)}$) and the* C^k**-smoothness of** f *for any $k \in \mathbb{N}$. If f is C^k-smooth on its domain for every $k \in \mathbb{N}$ we say that f is* C^∞**-smooth**.
We denote by $C^k(U,Y)$ the set of all C^k-smooth mappings from U into Y for $k \in \mathbb{N} \cup \{\infty\}$.

There are relatively few differentiability results for arbitrary continuous functions from general Banach spaces to general Banach spaces. The Mean Value Theorem is one example:

➡ *The Mean Value Theorem*

Theorem 44 (Mean Value Theorem). *Let $f : U \to Y$ be a Gâteaux differentiable function defined on a non-empty open convex subset U of a normed space X with values in a normed space Y. If $x_1, x_2 \in U$, then*

$$\|f(x_1) - f(x_2)\| \leq \|x_1 - x_2\|. \sup_{t\in[0,1]} \|f'(x_1 + t(x_2 - x_1))\|. \tag{4.2}$$

Proof. Find $y_0^* \in S_{Y^*}$ such that $\langle f(x_2) - f(x_1), y_0^* \rangle = \|f(x_2) - f(x_1)\|$. Define a mapping $r : (-\varepsilon, 1+\varepsilon) \to Y$ as $r(t) := f\big(x_1 + t(x_2 - x_1)\big)$, where $\varepsilon > 0$ is such that $x_1 + t(x_2 - x_1) \in U$ for $t \in (-\varepsilon, 1+\varepsilon)$. Observe that $r : (-\varepsilon, 1+\varepsilon) \to Y$ is differentiable and $r'(t) = f'\big(x_1 + t(x_2 - x_1)\big)(x_2 - x_1)$ for $t \in (1-\varepsilon, \varepsilon)$. Let $g := y_0^* \circ r : (-\varepsilon, 1+\varepsilon) \to \mathbb{R}$. The Chain Rule shows that $g'(t) = \langle r'(t), y_0^* \rangle$ for $t \in (-\varepsilon, 1-\varepsilon)$. An application of the one-variable Mean Value Theorem of the classical Calculus gives $t_0 \in (0,1)$ such that $g(1) - g(0) = g'(t_0)$, i.e.,

$$(\|f(x_2) - f(x_1)\| =) \ \langle f(x_2) - f(x_1), y_0^* \rangle = \langle f'\big(x_1 + t_0(x_2 - x_1)\big), y_0^* \rangle.$$

This shows (4.2) above. □

Extensions can be found, for example, in [Car67, Theorem 3.1.2], or in [Dieu69, Theorem 8.5.1].

Let us consider the following properties of a function f defined on a nonempty open subset U of a Banach space at a point $x_0 \in U$:
(i) Continuity.
(ii) Existence of all directional derivatives.
(iii) Gâteaux differentiability.
(iv) Fréchet differentiability.
Clearly, (iv)⇒(iii)⇒(ii). Certainly, (iv)⇒(i). No other implication holds without extra assumptions.
Even in the finite-dimensional setting, the concepts of Gâteaux and Fréchet differentiability may differ. The following example is typical in this context: Define $f : \mathbb{R}^2 \to \mathbb{R}$ by $f(x_1, x_2) := (x_1^4 x_2)/(x_1^6 + x_2^3)$ for $x_1^6 + x_2^3 \neq 0$ and $f(x_1, x_2) = 0$ otherwise. This function is Gâteaux differentiable at $(0,0)$; however, it is not continuous there (so, in particular, it is not Fréchet differentiable at $(0,0)$).
The following result shows that under the continuity of the derivative at a given point, Gâteaux differentiability and Fréchet differentiability coincide at this point (compare with Lemma 48 below):

Theorem 45. *Let $f : U \to Y$ be a Gâteaux differentiable function, where U is a nonempty open subset of a Banach space and Y is a Banach space, and let $x_0 \in U$. If the mapping $x \to f'(x)$ from U into $\mathcal{B}(X, Y)$ is continuous at x_0, then f is Fréchet differentiable at x_0.*

➡ *Proof.* Fix $v \in S_X$ and define the mapping $g(t) := f(x_0 + tv) - f(x_0) - tf'(x_0)(v)$ for $t \in [-\delta, \delta]$, where $\delta > 0$ satisfies $[x_0+\delta v, x_0-\delta v] \subset U$. Note that $g(0) = 0$. A simple application of the Mean Value Theorem 44 for $t \in [-\delta, \delta]$ shows that $\|g(t) - g(0)\| \leq |t|. \sup_{s\in(0,1)} \|f'(x_0 + stv)(v) - f'(x_0)(v)\|$. The continuity of $x \to f'(x)$ at x_0 and the fact that $v \in S_X$ is arbitrary give the result. □

The following is an interesting separable-reduction result. It plays an important rôle in some of the arguments below.

Lemma 46 (Preiss, [Prei84]). *Let X and Y be Banach spaces, G a nonempty open subset of X, $f : G \to Y$ be a continuous function and V be a separable subspace of X. Then there is a separable subspace W of X that contains V and such that f is Fréchet differentiable at every point x of $W \cap G$ where the restriction of f to W is Fréchet differentiable.*

In the next subsections, we shall briefly describe the situation regarding differentiability when f is either Lipschitz or convex (and, in the last case when, in particular, f is the norm). In each of them, we shall distinguish between the finite and the infinite-dimensional cases.

4.1.2 Lipschitz functions I

A function f defined on a subset D of a normed space X with values in a Banach space Y is said to be **Lipschitz** if there exists a constant C (called the **Lipschitz constant of** f) such that $\|f(x_1) - f(x_2)\| \leq C\|x_1 - x_2\|$, for every $x_1, x_2 \in D$ (we say then that f is C-**Lipschitz**).
If x_0 is an interior point of D, then f is said to be **locally Lipschitz at** x_0 if there exists a neighborhood $U \subset D$ of x_0 where f is Lipschitz. If D is open, f is said to be **locally Lipschitz** if it is locally Lipschitz at every point of D.
The norm of a Banach space is an example of a Lipschitz function. The Mean Value Theorem 44 ensures the Lipschitz character of any continuous real-valued function defined on an interval $[a, b]$ such that it is differentiable on (a, b) and has a bounded Gâteaux derivative there. Theorem 49 below shows that a convex function defined on a convex open subset of a Banach space is locally Lipschitz at points where it is continuous (a condition that is automatically satisfied when the space is finite-dimensional, see Theorem 52).

The following sufficient condition for a function to be Lipschitz is a consequence of the Mean Value Theorem 44 above.

Proposition 47. *Let f be a Gâteaux differentiable function defined on an open bounded subset U of a Banach space X, with values in a Banach space Y. Assume that the set of Gâteaux derivatives of f at points of U is a bounded subset of $\mathcal{B}(X, Y)$. Then f is Lipschitz on U.*

For a kind of converse to Proposition 47 see Remark 51 below.

Finite-dimensional Banach spaces

To motivate Lemma 48 and Theorem 52 below, let us consider the following function defined on $\mathbb{R}^2$:

$$f(x,y) = \begin{cases} \frac{x^3}{x^2+y^2} & \text{if } (x,y) \neq (0,0), \\ 0 & \text{otherwise.} \end{cases}$$

This function is continuous on $\mathbb{R}^2$. At every point $\mathbf{x} \in \mathbb{R}^2$, f has all directional derivatives $D_{\mathbf{v}}f(\mathbf{x})$, and $D_{\mathbf{v}}f(\mathbf{0}) = f(\mathbf{v})$ for $\mathbf{v} \neq \mathbf{0}$. Since $\mathbf{v} \to D_{\mathbf{v}}f(\mathbf{0})$ is not linear, f is not Gâteaux differentiable at $\mathbf{0}$ (of course, it is Gâteaux differentiable at every nonzero point). It is easy to see that there exists $M > 0$ such that $|\nabla f(\mathbf{x})| \leq M$ for all $\mathbf{x} \in \mathbb{R}^2$, where $\nabla f(\mathbf{x}) = (D_{(1,0)}f(\mathbf{x}), D_{(0,1)}f(\mathbf{x}))$ is the gradient of f at $\mathbf{x}$.

➡ Indeed, $\nabla f(0,0) = (1,0)$, while

$$\nabla f(x,y) = \left(\frac{x^4 + 3x^2y^2}{(x^2+y^2)^2}, \frac{-2x^3y}{(x^2+y^2)^2} \right), \text{ for } (x,y) \neq (0,0).$$

In the last case, and putting $x = r\cos\theta$ and $y = r\sin\theta$ for some $r > 0$ and $\theta \in [0, 2\pi)$, we get $\nabla f(x,y) = (\cos^4\theta + 3\cos^2\theta\sin^2\theta, -2\cos^3\theta\sin\theta)$, and the conclusion follows.

The function f is even Lipschitz. Indeed, notice that $|f(\mathbf{x}) - f(\mathbf{0})| = |f(\mathbf{x})| \leq \|\mathbf{x}\|$ for all $\mathbf{x}$. Thus, if $0 \in [\mathbf{x}, \mathbf{y}]$, then $\|f(\mathbf{x}) - f(\mathbf{y})\| \leq \|f(\mathbf{x}) - f(\mathbf{0})\| + \|f(\mathbf{0}) - f(\mathbf{y})\| \leq \|\mathbf{x}\| + \|\mathbf{y}\| = \|\mathbf{x} - \mathbf{y}\|$. If, on the contrary, $\mathbf{0} \notin [\mathbf{x}, \mathbf{y}]$, this follows from the Mean Value Theorem 44 and the inequality $\|\nabla f(\mathbf{x})\| \leq M$ above.

The following results shows that for Lipschitz functions in finite-dimensional spaces, the concepts of Gâteaux and Fréchet differentiability agree (compare with Theorem 45 and the example preceding it):

Lemma 48. *Let X be a finite-dimensional Banach space, and f be a Lipschitz map from an open set U in X into a possibly infinite-dimensional Banach space Y. If f is Gâteaux differentiable at some $x_0 \in U$, then f is Fréchet differentiable at x_0.*

Proof. Fix $\varepsilon > 0$, and choose a finite ε-net for S_X, i.e., a subset $\{u_i\}_{i=1}^m$ of S_X such that for any $u \in S_X$ there is $i \in \{1, 2, \ldots, m\}$ with $\|u - u_i\| < \varepsilon$. Since f is Gâteaux differentiable at x_0 and the set $\{u_i : i = 1, 2, \ldots, m\}$ is finite, there exists $t_0 > 0$ such that

$$\|f(x_0 + tu_i) - f(x_0) - tf'(x_0)(u_i)\| \leq \varepsilon|t|$$

for all $t \in (-t_0, t_0)$ and for all $i \in \{1, 2, \ldots, m\}$. It follows that for every $u \in S_X$,

$$\|f(x_0 + tu) - f(x_0) - tf'(x_0)u\| \leq (C + \|f'(x_0)\| + 1)\varepsilon|t|$$

for all $t \in (-t_0, t_0)$, where C is the Lipschitz constant of f. □

➡ The proof of Lemma 48 cannot be reduced to an application of Theorem 45. There are Lipschitz differentiable functions whose derivative is discontinuous. An example is the function $f : \mathbb{R} \to \mathbb{R}$ defined as $f(x) := x^2 \sin(1/x)$ for $x \neq 0$ and $f(0) = 0$.

Arbitrary Banach spaces

In infinite-dimensional Banach spaces we cannot expect a version of Lemma 48: The norm is a Lipschitz function, and every separable Banach space has an equivalent Gâteaux differentiable norm (see Subsection 9.2.2 and, in particular, Proposition 156). If the space is not Asplund (see Definition 58 below), no equivalent norm on it can be Fréchet differentiable (see Remark 59 and (ii) in Theorem 275).

4.1.3 Convex functions

Let us repeat here the definition of a convex function (see Definition 10 above): A function $f : C \to \mathbb{R}$, where C is a convex subset of a vector space, is said to be **convex** if $f(\lambda x + (1-\lambda)y) \leq \lambda f(x) + (1-\lambda)f(y)$ for every $x, y \in C$ and $0 \leq \lambda \leq 1$. The norm of a normed space is a convex function. More generally, any seminorm on a vector space is a convex function.

A basic, simple, and useful property of a convex function defined on an interval $[a,b] \subset \mathbb{R}$ is that

$$\frac{f(t_1)-f(t_0)}{t_1-t_0} \leq \frac{f(t_2)-f(t_0)}{t_2-t_0} \leq \frac{f(t_2)-f(t_1)}{t_2-t_1} \tag{4.3}$$

for $a \leq t_0 \leq t_1 \leq t_2 \leq b$ (see Figure 4.1 below). This behaviour implies the existence of (finite) lateral derivatives at any point of (a,b), in particular the continuity of f on (a,b).

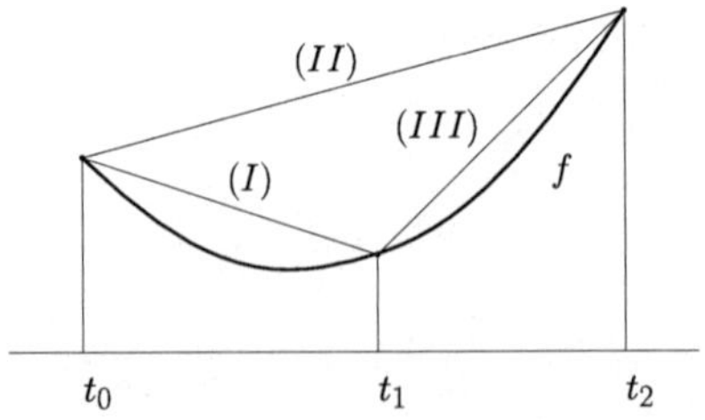

Figure 4.1: Slope (I)≤ slope (II)≤ slope (III) for a convex function

This property of convex functions in one variable readily shows the existence of all one-sided directional derivatives (see equations (4.5) and (4.6) below) for convex functions defined on convex open subsets U of an arbitrary normed space X: Indeed, notice first that if $x_0 \in U$, $h \in X$, and $\varepsilon > 0$ is such that $B(x_0, \varepsilon) \subset U$, then the "slope" function $t \to (1/t)[f(x_0 + th) - f(x_0)]$, defined on $(-\varepsilon, 0) \cup (0, \varepsilon)$, satisfies (see Figure 4.2)

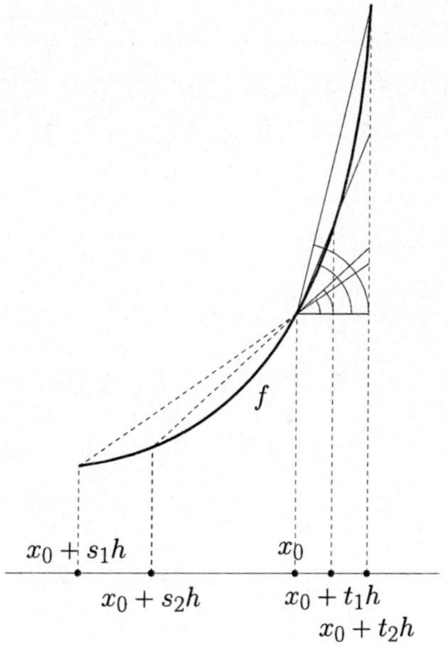

Figure 4.2: For a convex function, the slopes (all positive in the picture) increase

$$\begin{aligned}\frac{f(x_0+s_1h)-f(x_0)}{s_1} &\le \frac{f(x_0+s_2h)-f(x_0)}{s_2} \le \frac{f(x_0+t_1h)-f(x_0)}{t_1} \\ &\le \frac{f(x_0+t_2h)-f(x_0)}{t_2}, \quad \text{for } -\varepsilon < s_1 < s_2 < 0 < t_1 < t_2 < \varepsilon. \qquad (4.4)\end{aligned}$$

This shows, in particular, the existence of the one-sided directional derivatives

$$D_h^- f(x_0) := \lim_{t \to 0-} \frac{f(x_0+th)-f(x_0)}{t} \qquad (4.5)$$

$$\text{and } D_h^+ f(x_0) := \lim_{t \to 0+} \frac{f(x_0+th)-f(x_0)}{t}, \qquad (4.6)$$

i.e., the one-sided directional derivatives (in any direction h) of f at x_0, both exist finite at any point $x_0 \in U$. Moreover, $D_h^- f(x_0) \le D_h^+ f(x_0)$.

Arbitrary Banach spaces

A function $f : U \to \mathbb{R}$, where U is an open subset of a Banach space X, is said to be **locally bounded above (below) at a point** $x_0 \in U$ if there exists a neighborhood $V(x_0)$ of x_0 contained in U and such that f is bounded above (respectively, below) on $V(x_0)$. We say that f is **locally bounded at** x_0 if it is locally bounded above and below at x_0.

Theorem 49. *Let f be a convex function defined on a nonempty open subset U of a Banach space X, and let $x_0 \in U$. Then, the following are equivalent:*
(i) *f is continuous at x_0.*
(ii) *f is locally bounded above at x_0.*
(iii) *f is locally Lipschitz at x_0.*

Proof. (See, e.g., [FHHMZ11, Lemma 7.3])
(i)⇒(ii) is obvious.
(ii)⇒(iii) Fix $x_0 \in U$. By the assumption, there exists $\delta > 0$ and $M \in \mathbb{R}$ such that $f(x) \leq M$ for $x \in B(x_0, 2\delta) \subset U$. Let $x, y \in B(x_0, \delta)$, $x \neq y$. Put $d := \|x - y\|$, and let $z := y + (\delta/d)(y - x)$ (see Figure 4.3).

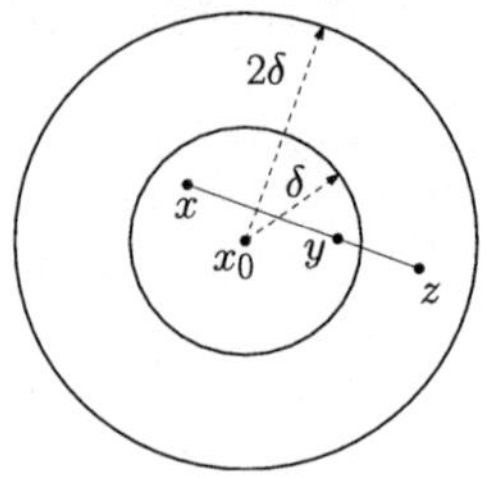

Figure 4.3: The construction in the proof of Theorem 49

Clearly, $z \in B(x_0, 2\delta)$. Note that $y = \big(d/(d+\delta)\big)z + \big(\delta/(d+\delta)\big)x$. Since f is convex, $f(y) \leq \big(d/(d+\delta)\big)f(z) + \big(\delta/(d+\delta)\big)f(x)$. Thus,

$$f(y) - f(x) \leq \big(d/(d+\delta)\big)(f(z) - f(x)) \leq (d/\delta)2M = (2M/\delta)\|x - y\|,$$

as we wanted to show.
(iii)⇒(i) is obvious. □

Certainly, continuity of a convex function defined on an infinite-dimensional Banach space is not guaranteed. In fact, on every such a space there is always a discontinuous linear form (if $\{e_n : n \in \mathbb{N}\}$ is a linearly independent subset of S_X, the linear extension of the mapping f that to each e_n associates n is unbounded, hence discontinuous, see Section 1.3). This is in contrast with the behaviour of convex functions defined on a finite-dimensional space (see Theorem 52 below).

Inequalities (4.3), (4.4), and the existence of the limits (4.5) and (4.6) for a given $h \in X$ show that, for a convex function f, the fact that it has a directional derivative (4.1) at x_0 in the direction h is equivalent to the fact that the limit

$$\lim_{t\to 0} \frac{f(x_0+th)+f(x_0-th)-2f(x_0)}{t} \tag{4.7}$$

exists with value 0. A corresponding statement holds when uniformity in h of both limits (4.1) and (4.7) is considered.

Theorem 50. *Let f be a convex function defined on a nonempty open convex subset U of a Banach space X, and let $x_0 \in U$. If f is continuous at x_0 and the directional derivatives at x_0 exists for all the directions, then f is Gâteaux differentiable at x_0.*

➪ *Proof.* (The first part of the proof here can be made simpler; however, we prefer this version to assist the proof of Theorem 52 below.) Fix $\alpha_1, \alpha_2 \in [0,1]$ such that $\alpha_1+\alpha_2=1$. Let h_1 and h_2 be arbitrary elements in X. In all the expressions below, $s<0<t$ are such that the vectors computed by using α_1, α_2, s, and t, belong to U. The convexity of f shows that (see Figure 4.4)

$$\begin{aligned} &f\big(x_0+t(\alpha_1h_1+\alpha_2h_2)\big)\\ &\quad= f(\alpha_1x_0+\alpha_1th_1+\alpha_2x_0+a_2th_2)\\ &\quad\le \alpha_1 f(x_0+th_1)+\alpha_2 f(x_0+th_2). \end{aligned} \tag{4.8}$$

Thus, for $t>0$, we get from (4.8)

$$\begin{aligned} &\frac{f\big(x_0+t(\alpha_1h_1+\alpha_2h_2)\big)-f(x_0)}{t}\\ &\quad\le \alpha_1\frac{f(x_0+th_1)-f(x_0)}{t}+\alpha_2\frac{f(x_0+th_2)-f(x_0)}{t}. \end{aligned} \tag{4.9}$$

If in (4.8) we divide by $s<0$ instead, we get

$$\begin{aligned} &\frac{f\big(x_0+s(\alpha_1h_1+\alpha_2h_2)\big)-f(x_0)}{s}\\ &\quad\ge \alpha_1\frac{f(x_0+sh_1)-f(x_0)}{s}+\alpha_2\frac{f(x_0+sh_2)-f(x_0)}{s}. \end{aligned} \tag{4.10}$$

Putting equations (4.9) and (4.10) together, and having in mind that the right-hand side (left-hand side) in (4.9) (respectively in (4.10)) converges to $\alpha_1 f'(x_0)(h_1)+\alpha_2 f'(x_0)(h_2)$ when $t\to 0$ (respectively, when $s\to 0$), it follows that there exists $f'(x_0)(\alpha_i h_1+\alpha_2h_2)$ $(=\alpha_1 f'(x_0)(h_1)+\alpha_2 f'(x_0)(h_2))$. Since obviously $f'(x_0)$ is homogeneous, we get that $f'(x_0)$ is linear. The fact that

f is locally Lipschitz (Theorem 49) shows that for t small enough,

$$\left|\frac{f(x_0+th)-f(x_0)}{t}\right| \leq \frac{K\|th\|}{|t|} = K\|h\|,$$

hence $|f'(x_0)(h)| \leq K\|h\|$ for $h \in X$, $h \neq 0$, where K is the Lipschitz constant of f in a neighborhood of x_0. Thus, $f'(x_0) \in X^*$, and f is Gâteaux differentiable at x_0. □

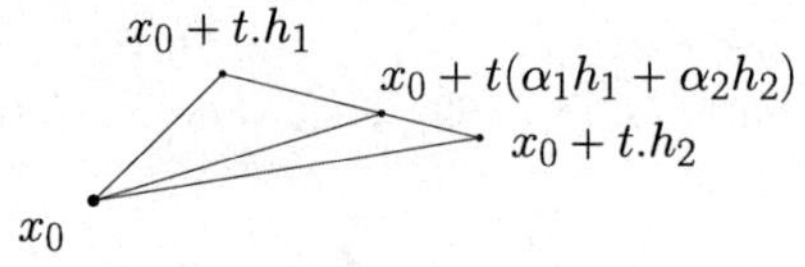

Figure 4.4: On the proof of Theorem 50

If f is a convex function defined on an open convex subset U of a Banach space, $x_0 \in U$, $x_0 + h \in U$, and $0 < t < 1$, the fact that $x_0 + th = (1-t)x_0 + t(x_0+h)$ and the convexity of f show that, in case that f is Gâteaux differentiable at x_0,

$$\big(f'(x_0)(h) =)\big)\ \ f'(x_0)(x_0+h) - f'(x_0)(x_0) \leq f(x_0+h) - f(x_0). \tag{4.11}$$

Geometrically (see Figure 4.5) equation (4.11) says that the graph of f stays above the "tangent plane" $x \to f(x_0) + f'(x_0)(x - x_0)$. For a general convex continuous function $f : U \to \mathbb{R}$ —not necessarily Gâteaux differentiable— the set of all $x^* \in X^*$ that satisfy (4.11), i.e., such that

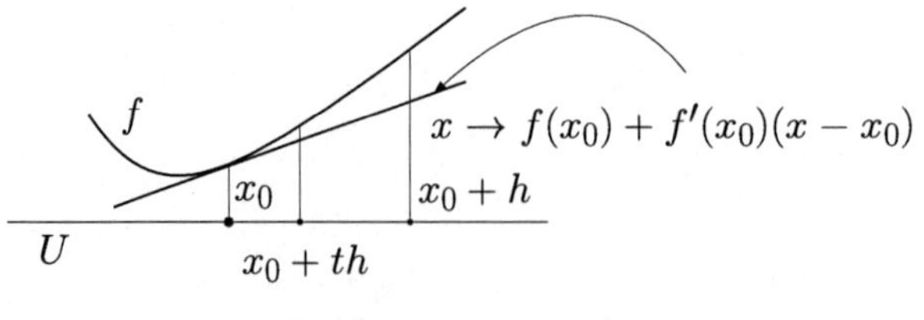

Figure 4.5: The graph of a Gâteaux differentiable convex function is above the tangent line (the tangent plane for higher dimensions)

$$\big(x^*(h) = \big)\ x^*(x_0+h) - x^*(x_0) \leq f(x_0+h) - f(x_0) \tag{4.12}$$

for h small enough to have $x_0 + h \in U$, is called the **subdifferential of f at** x_0, and is denoted by $\partial f(x_0)$ (see Figure 4.6). It is always nonempty thanks to the Hahn–Banach Separation Theorem applied to the epigraph of f and the point $(x_0, f(x_0))$. The multivalued mapping ∂f is always $\|\cdot\|$-w^*-**upper semicontinuous**,

i.e., for every $x_0 \in U$ and any w^*-open subset W of X^* that contains $\partial f(x_0)$, there exists $\varepsilon > 0$ such that $\partial f(B(x_0, \varepsilon)) \subset W$ (see, e.g., [Phe93, Proposition 2.5]; for the case of the norm this is proved in Subsection 4.1.4 below).
If f is convex and Gâteaux differentiable at a point x_0, (4.11) shows that $f'(x_0) \in \partial f(x_0)$ (in fact, $\partial f(x_0) = \{f'(x_0)\}$). If f is, moreover, Fréchet differentiable at $x_0 \in U$, then ∂f is $\|\cdot\|$-$\|\cdot\|$-upper semicontinuous at x_0, in the sense that given $\varepsilon > 0$ there exists $\delta > 0$ such that $\|x - x_0\| < \delta$ implies $\partial f(x) \subset B(x_0^*, \varepsilon)$, where $x_0^* = f'(x_0)$ (see, e.g., [Phe93, Lemma 2.6]).

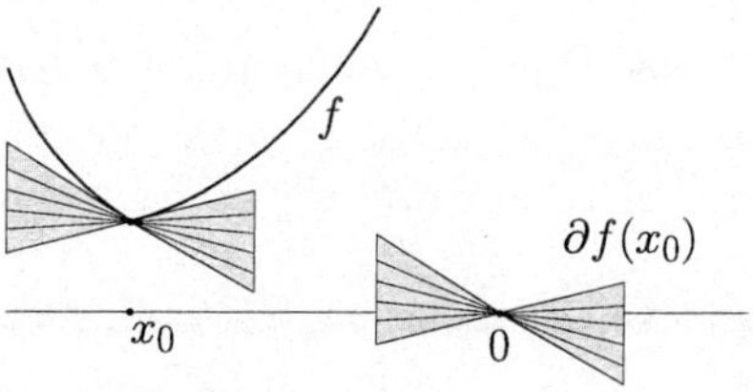

Figure 4.6: The graphs of elements in the set $\partial f(x_0)$

Remark 51. Observe that for a convex and C-Lipschitz real-valued function defined on a nonempty open convex subset U of a Banach space X, where $C > 0$ is a constant, and $x_0 \in U$, we have $\partial f(x_0) \subset C.B_{X^*}$, a consequence of the definition of the subdifferential (see equation (4.12) above). Ⓡ

Finite-dimensional Banach spaces

We consider convex functions f defined on a nonempty convex open subset U of $\mathbb{R}^n$. By a **partial derivative** of f at a point $x_0 \in U$ we mean the directional derivative of f at x_0 in the direction of an element of the canonical basis of $\mathbb{R}^n$.

Theorem 52. *Let f be a convex function defined on a nonempty open convex subset U of $\mathbb{R}^n$. Then*
(i) *f is continuous on U.*
(ii) *f is locally Lipschitz on U.*
(iii) *If all the partial derivatives of f at a point $x_0 \in U$ exist, then f is Fréchet differentiable at x_0.*

Proof. (Sketch) (i) We may assume that $\mathbb{R}^n$ is equipped with the norm $\|\cdot\|_1$. The continuity of f was proved for $n = 1$ right after the definition of convexity. For $n = 2$ fix a point $x_0 \in U$. Assume, without loss of generality, that $x_0 = 0$ and $f(x_0) = 0$. Find $\varepsilon > 0$ such that $B(0, \varepsilon) \subset U$. Put $M(\varepsilon) := \max\{f(\varepsilon e_i) : I = 1, 2\}$ (see Figure 4.7).
Let $x \in B(0, \varepsilon)$. By convexity, $0 = f(0) \le f(x) + f(-x)$, hence

$$\Big(M(\varepsilon) \ge \Big)\ f(-x) \ge -f(x) \ge M(\varepsilon). \tag{4.13}$$

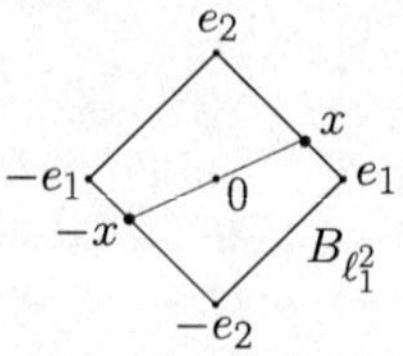

Figure 4.7: Checking continuity at 0 in $\mathbb{R}^2$

Thus, $-M(\varepsilon) \leq f(x) \leq M(\varepsilon)$ for all $x \in B(0, \varepsilon)$. We may rely on Theorem 49 to get that f is continuous at 0. Alternatively, the 1-dimensional case shows that $M(\varepsilon) \to 0$ as $\varepsilon \to 0$, and (4.13) concludes the continuity of f at 0. The argument for $n > 2$ is similar.
(ii) This is, e.g., in [MoZiZi15, Proposition 1032]. In fact, it follows from (i) and Theorem 49.
(iii) From Theorem 50 above and its proof we get that the existence of all partial derivatives at a point $x_0 \in U$ implies the existence of all directional derivatives at x_0 due to the fact that the space is finite-dimensional. We get, too, the linearity and continuity of the derivative. Thus, f is Gâteaux differentiable at x_0. The Fréchet differentiability of f follows then from Lemma 48. □

4.1.4 The case of the norm

We refer to $\partial\|\cdot\|$ as to the **duality mapping**. Its upper semicontinuity can be checked by Lemma 53 below.

Given $x^* \in X^*$ and $\delta > 0$, the **slice** defined by x^* and δ of the ball B_X is the set $S(B_X, x^*, \delta) := \{x \in B_X : \langle x, x^* \rangle > \|x^*\| - \delta\}$ (see Figure 4.8).

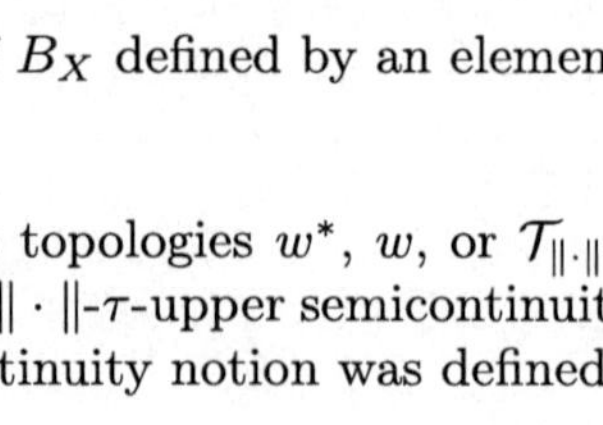

Figure 4.8: A slice of B_X defined by an element $x^* \in X^*$ and $\delta > 0$

➪ If τ denotes one of the topologies w^*, w, or $\mathcal{T}_{\|\cdot\|^*}$ on X^*, the following is a characterization of the $\|\cdot\|$-τ-upper semicontinuity of the duality mapping at some $x_0 \in X$. This continuity notion was defined in Subsection 4.1.3 above.

Lemma 53 (Giles, Gregory, Sims, [GiGrSi78b]). *If X is a Banach space, the duality mapping $\partial\|\cdot\|$ is $\|\cdot\|$-τ-upper semicontinuous at some $x_0 \in X$ if, and only if, for each τ-neighborhood N of 0 in X^*, $\partial\|\cdot\|(x_0)+N$ contains a slice of B_{X^*} determined by x_0.*

Proof. If $\partial\|\cdot\|(x_0)+N$ contains $S(B_{X^*},x_0,\delta)$ and $x \in B(x_0,\delta)\cap S_X$, then $|\langle x_0,x^*\rangle - 1| = |\langle x_0 - x, x^*\rangle| < \delta$ for every $x^* \in \partial\|\cdot\|(x)$, hence $\partial\|\cdot\|(x) \subset S(B_{X^*},x_0,\delta)$ $(\subset \partial\|\cdot\|(x_0)+N)$, and so $\partial\|\cdot\|$ is $\|\cdot\|$-τ-upper semicontinuous at x_0.
Conversely, there exists $\delta' > 0$ such that $\partial\|\cdot\|\big(B(x_0;\delta')\cap S_X\big) \subset \partial\|\cdot\|(x_0)+N/2$. Find $0<\delta<\delta'$ such that $\delta B_{X^*} \subset N/2$. If $x^* \in S(B_{X^*},x_0,\delta^2/4)$ then $|\langle x_0,x^*\rangle - 1| < \delta^2/4$. By the Bishop–Phelps–Bollobás Theorem 15 we get $x_1 \in S_X$ and $x_1^* \in \partial\|\cdot\|(x_1)$ such that $\|x_1 - x_0\| < \delta$ and $\|x_1^* - x^*\| < \delta$. Thus, $x_1 \in B(x_0,\delta')\cap S_X$, so $\partial\|\cdot\|(x_1) \subset \partial\|\cdot\|(x_0)+N/2$. Since $x^* \in \partial\|\cdot\|(x_1)+\delta B_{X^*}$, we get $x^* \in \partial\|\cdot\|(x_0)+N$. This shows, finally, that $S(B_{X^*},x_0,\delta^2/4) \subset \partial\|\cdot\|(x_0)+N$. □

Note that, for $\tau := w^*$, the duality mapping $\partial\|\cdot\|$ satisfies the condition in Lemma 53 at any point (so $\partial\|\cdot\|$ is always $\|\cdot\|$-w^*-upper semicontinuous). Indeed, If $x_0 = 0$ then $\partial\|\cdot\|(x_0) = B_{X^*}$, and the result holds trivially. If $x_0 \neq 0$ and N is a w^*-open neighborhood of 0, assume that no slice of B_{X^*} defined by x_0 is in $\partial\|\cdot\|(x_0)+N$. Thus, there exists a sequence $\{x_n^*\}$ in $B_{X^*} \setminus \big(\partial\|\cdot\|(x_0)+N\big)$ such that $\langle x_0,x_n^*\rangle \to 1$. It has a w^*-cluster point $x_0^* \in B_{X^*} \setminus \big(\partial\|\cdot\|(x_0)+N\big)$. However, $\langle x_0,x_0^*\rangle = 1$, hence $x_0^* \in \partial\|\cdot\|(x_0)$, a contradiction.

The $\|\cdot\|$-w^*-upper semicontinuity of $\partial\|\cdot\|$ at x_0 means, in the case that $\|\cdot\|$ is Gâteaux differentiable at x_0, that $x_n \xrightarrow{\|\cdot\|} x_0$ implies $x_n^* \xrightarrow{w^*} \partial\|\cdot\|(x_0)$, where $x_n^* \in \partial\|\cdot\|(x_n)$ for all n. This follows from the proof above and the fact that $\partial\|\cdot\|(x_0)$ is a singleton.

The norm is never Gâteaux differentiable at 0. In the future, when we shall say that $\|\cdot\|$ **is Gâteaux (Fréchet) differentiable** we shall mean that it is so at every point $0 \neq x_0 \in X$. By homogeneity, it is enough to check differentiability at every point $x_0 \in S_X$. If $0 \neq x_0 \in X$, then the existence for every $h \in X$ of the limit in (4.1) implies (see Theorem 50) the linearity and continuity of the function $\|\cdot\|'(x_0)$ (now an element in S_{X^*}). The convexity of the norm shows that f is Gâteaux (Fréchet) differentiable at x_0 if, and only if, for every $h \in X$ the following limit exists (respectively, if the limit exists uniformly in $h \in S_X$) and is 0 (see (4.7) above):

$$\lim_{t\to 0} \frac{\|x_0 + th\| + \|x_0 - th\| - 2\|x_0\|}{t} = 0. \tag{4.14}$$

It is simple to show that, for $0 \neq x_0 \in X$, we have $\partial\|\cdot\|(x_0) = \{x^* \in B_{X^*} : \langle x_0,x^*\rangle = \|x_0\|\}$, while $\partial\|\cdot\|(0) = B_{X^*}$ (another way to recall that $\|\cdot\|$ is never Gâteaux differentiable at 0). Several instances of the set $\partial\|\cdot\|(x)$ can be seen in

Figure 4.12 below.

We reproduce here the following crucial result (see, e.g., for the statement and an elementary proof, [DeGoZi93, Theorem I.1.4 and Corollary I.1.5], and [FHHMZ11, Corollary 7.22], where a version that includes differentiability of the dual norm is stated and proved, see Lemma 142 below) pertaining the Gâteaux and Fréchet differentiability of the norm $\|\cdot\|$ of a normed space $(X, \|\cdot\|)$. It is illustrated in Figure 4.12. (The version for an arbitrary convex continuous function defined on a nonempty open convex subset of a Banach space can be found in, e.g., [FHHMZ11, Theorems 7.15 and 7.17].)

Lemma 54 (Šmulyan). *The norm $\|\cdot\|$ of a Banach space X is, at a point $x_0 \in X$, Gâteaux differentiable if, and only if, $\partial\|\cdot\|(x_0) = \{x_0^*\}$ for some $x_0^* \in S_{X^*}$ (and $x_0^* = \|\cdot\|'(x_0)$). It is Fréchet differentiable there if, and only if,* $\operatorname{diam} S(B_{X^*}, x_0, \delta) \to 0$ *as* $\delta \to 0$.

➡ Despite that the proof of Lemma 54 can be found in most texts (several references precede its statement) it is worth to provide a sketch for it (it tries to convey its geometric meaning, and it will be presented in such a way that trivial changes will provide the proof for any continuous convex function defined on a nonempty convex open subset of X): Work on $X \times \mathbb{R}$. Notice that the continuity and convexity of $\|\cdot\|$ ensures a closed convex epigraph $\operatorname{epi}(\|\cdot\|) := \{(x,t) : \ t \geq \|x\|, \ x \in X\}$ with a nonempty interior. Thus, the Hahn–Banach Separation Theorem for $(x_0, \|x_0\|)$ and $\operatorname{epi}(\|\cdot\|)$ provides at least an element in $\partial\|\cdot\|(x_0)$.
(i) *The Gâteaux-differentiability version.* Assume that $\|\cdot\|$ is Gâteaux differentiable at x_0 (see Figure 4.10). Clearly, $\|\cdot\|'(x_0) \in \partial\|\cdot\|(x_0)$. Let $x^* \in \partial\|\cdot\|(x_0)$. If $s < 0 < t$ and $h \in X$,

$$\frac{\|x_0 + sh\| - \|x_0\|}{s} \leq x^*(h) \leq \frac{\|x_0 + th\| - \|x_0\|}{t},$$

and the Gâteaux differentiability of $\|\cdot\|$ at x_0 gives $x^*(h) = \|\cdot\|'(x_0)(h)$. Thus, $x^* = \|\cdot\|'(x_0)$ and so $\partial\|\cdot\|(x_0) = \{\|\cdot\|'(x_0)\}$.
Assume now that $\|\cdot\|$ is not Gâteaux differentiable at x_0 (see Figure 4.11). Thus, there exists $h \in S_X$ and $\alpha \in \big(D_h^-\|\cdot\|(x_0), D_h^+\|\cdot\|(x_0)\big)$, where D_h^- and D_h^+ are the one-sided directional derivatives, defined by (4.5) and (4.6), respectively. Define a function g on $x_0 + \operatorname{span}\{h\}$ as $g(x_0 + th) = \|x_0\| + t\alpha$, for $t \in \mathbb{R}$. It is simple to show that $\operatorname{graph}(g) \cap \operatorname{epi}(\|\cdot\|) = \{(x_0, \|x_0\|)\}$. The Hahn–Banach Separation Theorem gives a closed hyperplane H in $X \times \mathbb{R}$ such that $(x_0, \|x_0\|) + H$ separates $\operatorname{graph}(g)$ and $\operatorname{epi}(\|\cdot\|)$. It is easy to show that H is the graph of an element $x_0^* \in \partial\|\cdot\|(x_0)$. Each $\alpha \in \big(D_h^-\|\cdot\|(x_0), D_h^+\|\cdot\|(x_0)\big)$ defines a different x_0^*, so $\partial\|\cdot\|(x_0)$ does not reduce to a singleton.
(ii) *The Fréchet differentiability version.* Assume first that $\|\cdot\|$ is Fréchet

differentiable at $x_0 \in S_X$. Fix $\varepsilon > 0$. Find $\delta > 0$ such that

$$\left| \frac{\|x_0 + th\| - \|x_0\|}{t} - \langle h, x_0^* \rangle \right| < \varepsilon, \text{ for all } h \in S_X, \ 0 < |t| < \delta. \tag{4.15}$$

In order to find $\operatorname{diam} S(B_{X^*}, x_0, \varepsilon\delta)$, we shall compute $\|x^* - x_0^*\|$ for $x^* \in S(B_{X^*}, x_0, \varepsilon\delta)$ by using the so-called Parallel Hyperplane Lemma (see, e.g., [FHHMZ11, Exercise 2.13]; see Figure 4.9 for a sketch). Let $h \in S_X$ such that $\langle h, x_0^* \rangle = 0$. We shall prove that $|\langle h, x^* \rangle| < 2\varepsilon$. This will conclude that $\|x^* - x_0^*\| \leq 4\varepsilon$.
Note that

$$\langle h, x^* \rangle = \frac{\langle x_0 + th - x_0, x^* \rangle}{t}. \tag{4.16}$$

If $t > 0$, then it follows from (4.15) —having in mind that $\langle h, x_0^* \rangle = 0$— and (4.16) that

$$\langle h, x^* \rangle \leq \frac{\|x_0 + th\| - 1 + \varepsilon\delta}{t} = \frac{\|x_0 + th\| - 1}{t} + \frac{\varepsilon\delta}{t} < \varepsilon + \frac{\varepsilon\delta}{t}.$$

Since this is true for every $0 < t < \delta$, we get $\langle h, x^* \rangle < 2\varepsilon$. If, on the contrary, $t < 0$, we get in a similar way that $\langle h, x^* \rangle > -2\varepsilon$. This proves that $\operatorname{diam} S(B_{X^*}, x_0, r) \to 0$ as $r \downarrow 0$, and the conclusion follows.
Assume now that $\operatorname{diam} S(B_{X^*}, x_0, \delta) \to 0$ as $\delta \downarrow 0$. Then $\partial\|\cdot\|(x_0)$ is a singleton, say $\{x_0^*\}$. If $\|\cdot\|$ is not Fréchet differentiable at x_0, then there exists $\varepsilon > 0$ and sequences $\{t_n\}_{n\in\mathbb{N}}$ in $(0, +\infty)$ and $\{h_n\}_{n\in\mathbb{N}}$ in S_X such that $t_n \to 0$ and

$$\|x_0 + t_n h_n\| - \|x_0\| - \langle t_n h_n, x_0^* \rangle > t_n \varepsilon, \text{ for all } n \in \mathbb{N}. \tag{4.17}$$

Pick $x_n^* \in \partial\|\cdot\|(x_0 + t_n h_n)$ for $n \in \mathbb{N}$. Then

$$\langle x_0 - (x_0 + t_n h_n), x_n^* \rangle \leq \|x_0\| - \|x_0 + t_n h_n\|, \text{ for all } n \in \mathbb{N}. \tag{4.18}$$

Thus, it follows from (4.17) and (4.18) that $\langle h_n, x_n^* - x_0^* \rangle \geq \varepsilon$, hence $\|x_n^* - x_0^*\| \geq \varepsilon$, for all $n \in \mathbb{N}$, a contradiction with the fact that $\langle x_0, x_n^* \rangle \to 1$. □

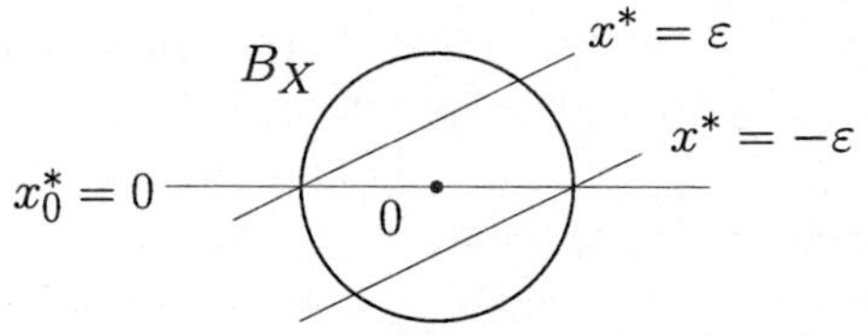

Figure 4.9: If this happens, then $\|x^* - x_0^*\| \leq 2\varepsilon$ or $\|x^* + x_0^*\| \leq 2\varepsilon$

In other words, Gâteaux differentiability of $\|\cdot\|$ at some point $x_0 \in X$, with $\|\cdot\|'(x_0) = x_0^*$, means that x_0, as a mapping on X^*, **exposes** B_{X^*} at x_0^*, while Fréchet differentiability there means that x_0^* is **strongly exposed** by x_0, according to the next definition. Here, $S(A, x^*, \delta)$ denotes a **slice** of the set $A \subset X$ defined by $x^* \in X^*$ and $\delta > 0$. It was defined for the closed unit ball of X at the beginning of Subsection 4.1.4. Precisely, $S(A, x^*, \delta) := \{x \in A : \langle x, x^* \rangle > \sigma_A(x^*) - \delta\}$.

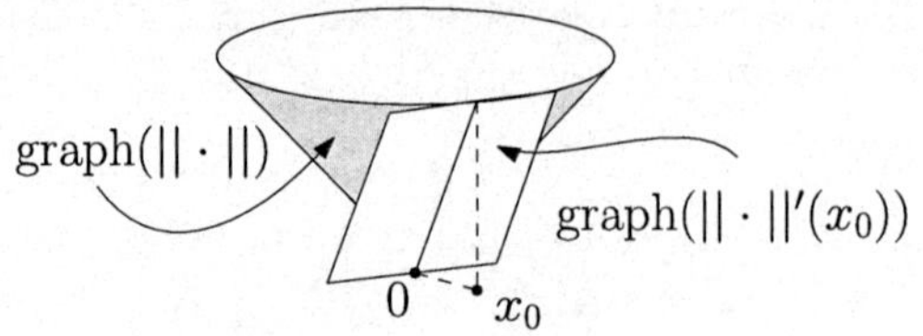

Figure 4.10: The graph of $\|\cdot\|'(x_0)$ is tangent to the graph of $\|\cdot\|$ at $(x_0, \|x_0\|)$ (differentiability case in the proof of Lemma 54)

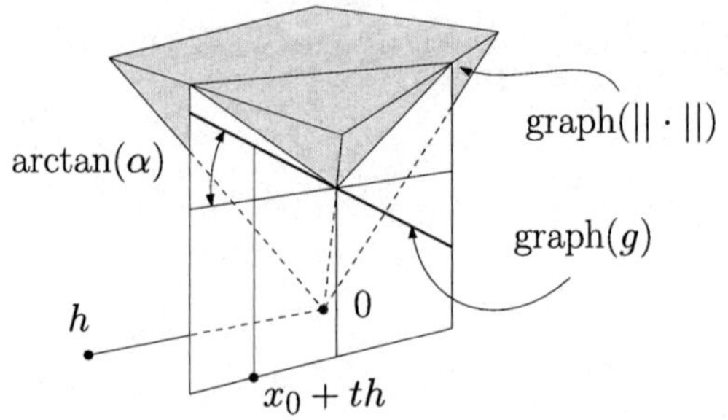

Figure 4.11: Two closed convex sets (epi ($\|\cdot\|$) and graph (g)) to be separated (non-differentiability case in the proof of Lemma 54)

Definition 55. *Let $(X, \|\cdot\|)$ be a Banach space. A point x_0 in a nonempty subset A of X is an* **exposed** *(***strongly exposed***) point of A whenever there exists $x^* \in X^*$ such that $\langle x, x^* \rangle < \langle x_0, x^* \rangle$ for all $x \in A$, $x \neq x_0$ (whenever* $\operatorname{diam} S(A, x^*, \delta) \to 0$ *as $\delta \to 0$, respectively). We say then that x_0* **is exposed by** *x^* (respectively, that x_0* **is strongly exposed by** *x^*).*

Obviously, every strongly exposed point is exposed, and every exposed point of a convex set is an extreme point of it. If a nonempty closed convex set C in a Banach space is $\|\cdot\|$-compact, a simple compactness argument shows that every exposed point of C is in fact strongly exposed.

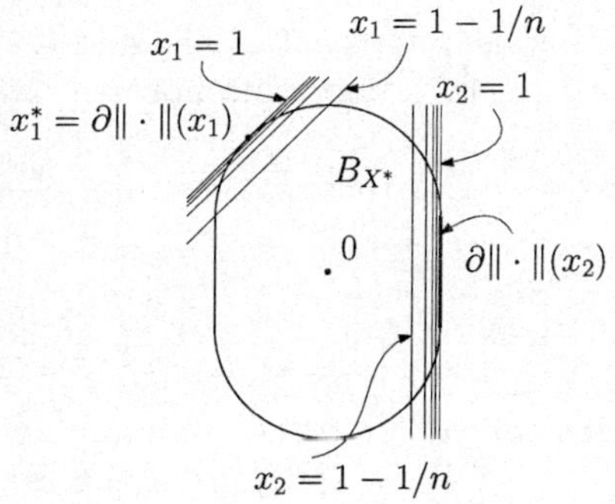

Figure 4.12: The norm is differentiable (Fréchet and Gâteaux) at $x_1 \in S_X$, but not at $x_2 \in S_X$

Theorem 56 (Straszewicz, [Str35]). *Every norm-compact convex set in a Banach space is the closed convex hull of its exposed points.*

Proof. A nice self-contained proof is, e.g., in [FoLiPh01, Proposition 4.3]. □

Theorem 57 (Corson, [Cor65]). *There is a convex compact set in* $\mathbb{R}^3$ *whose exposed points form a set* E_p *of first category in itself, i.e.,* E_p *is the union of a countable number of closed sets each of which has no interior with respect to* E_p.

This should be compared with [FHHMZ11, Lemma 3.73] concerning the G_δ property of extreme points in the metrizable case.

Some examples

➪ Most of the examples below can be checked by using Šmulyan Lemma 54 for computing $\partial\|\cdot\|(x)$ at a certain point x and, if this is the case, the character of Gâteaux (Fréchet) differentiability of the given norm there.

1. *Finite-dimensional spaces.* A consequence of Lemma 54 is that the concepts of Fréchet and Gâteaux differentiability of a norm coincide in finite-dimensional spaces. This follows, too, from Lemma 48 above.

2. *Hilbert spaces.* If (H, b) is a real Hilbert space, where $b : H \times H$ is the associated bilinear form, then the norm induced by b, i.e., $\|x\| = \sqrt{b(x,x)}$ for $x \in H$, is Fréchet differentiable (we shall mention later that it is even better than this: It is **uniformly Fréchet differentiable**, see Definition 70 below). Indeed, the parallelogram identity easily shows that $\operatorname{diam} S(B_{H^*}, x, \delta) \leq 2\sqrt{\delta(2-\delta)}$ for $x \in S_H$ and $\delta > 0$, and the result follows from Lemma 54. Moreover, $\|\cdot\|'(x) = D(x)/\|x\|$ for all $0 \neq x \in H$, where $D(x)(y) := b(x, y)$ for all $y \in H$.

3. *$C(K)$ spaces.* If K is a compact space, and $C(K)$ is endowed with the supremum norm $\|\cdot\|_\infty$, then this norm is Gâteaux differentiable at $(0 \neq)\ f \in C(K)$ if, and only if, $M_f := \{k \in K : \ |f(k)| = \|f\|_\infty\}$ is a singleton $\{k_0\}$ (and then $\delta_{k_0} = \|\cdot\|'_\infty(f)$, where $\delta_{k_0} \in C(K)^*$ is the Dirac delta at k_0). The same is true —with the same proof— for $(C_0(L), \|\cdot\|_\infty)$, where $C_0(L)$ is the space of all real-valued bounded continuous functions on L and vanishing at infinity, being L a locally compact topological space (see Section 2.2 above). The norm $\|\cdot\|_\infty$ is Fréchet differentiable at $f \in C_0(L)$ if, and only if, in addition the point k_0 is isolated. This follows easily from the Šmulyan test (Lemma 54) for one implication and the definition of $\|\cdot\|'_\infty$ for the other.
4. *The space c_0.* As a particular case of Item 3 above, we get that $\|\cdot\|_\infty$ in $c_0 := c_0(\mathbb{N})$ $(= C_0(\mathbb{N})$, where $\mathbb{N}$ carries the discrete topology, see Section 2.2 above) is Gâteaux differentiable at $(0 \neq)\ x = (x_n) \in c_0$ if, and only if, the set $M_x := \{n \in \mathbb{N} : \ |x(n)| = \|x\|_\infty\}$ is a singleton, say $\{n_0\}$. If this is the case, then $\partial\|\cdot\|_\infty(x) = \{e_{n_0}\}$ if $x(n_0) = \|x\|_\infty$, or $\partial\|\cdot\|_\infty(x) = \{-e_{n_0}\}$ if $x(n_0) = -\|x\|_\infty$, where e_n denotes the n-th element of the canonical basis of $\ell_1(\mathbb{N})$. It is easy to see that $\|\cdot\|_\infty$ on c_0 is Fréchet differentiable precisely where it is Gâteaux differentiable.
5. *The space ℓ_∞.* The norm $\|\cdot\|_\infty$ on ℓ_∞ is Gâteaux differentiable at some $(0 \neq)\ x = (x_n) \in \ell_\infty$ if, and only if, the set $M_x := \{n \in \mathbb{N} : \ |x_n| = \|x\|_\infty\}$ is a singleton, say n_0, and $|x_{n_0}| > \sup\{|x_n| : \ n \in \mathbb{N} \setminus \{n_0\}\}$. If this is the case, then $\partial\|\cdot\|_\infty(x) = e_{n_0}$ (i.e., the n_0-th element in the canonical basis of ℓ_1) if $x(n_0) = \|x\|_\infty$, or $-e_{n_0}$ if $x(n_0) = -\|x\|_\infty$. The norm $\|\cdot\|_\infty$ is Fréchet differentiable precisely where it is Gâteaux differentiable.
6. *The space ℓ_1.* The norm $\|\cdot\|_1$ on $\ell_1 := \ell_1(\mathbb{N})$ is Gâteaux differentiable precisely at points $(0 \neq)\ x = (x_n)$ such that $x_n \neq 0$ for all $n \in \mathbb{N}$. Thus, for an uncountable set Γ of indices, the norm $\|\cdot\|_1$ on $\ell_1(\Gamma)$ is nowhere Gâteaux differentiable. The norm $\|\cdot\|_1$ on ℓ_1 is nowhere Fréchet differentiable.

Two examples of a norm with no points of Gâteaux differentiability

Item 6 in the previous annotation shows that the canonical norm $\|\cdot\|_1$ on $\ell_1(\Gamma)$, for uncountable Γ, is nowhere Gâteaux differentiable. Let us provide two other examples of this phenomenon:

1. The canonical norm $\|\cdot\|_\infty$ (i.e., the essential supremum) on $L_\infty[0,1]$ is nowhere Gâteaux differentiable (see, e.g., [FHHMZ11, Exercise 7.41]) (in particular, this space cannot be linearly isometric to the space ℓ_∞ (see Item 5 in the previous annotation), although both spaces are linearly isomorphic (see, e.g., [FHHMZ11, Exercises 4.42 and 5.91])).

2. Let $\|\cdot\|_\infty$ be the supremum norm on ℓ_∞. Put $p(x) := \limsup |x_n|$, for $x = (x_n) \in \ell_\infty$. Let $\|x\| := \|x\|_\infty + p(x)$, $x \in \ell_\infty$. This norm was considered in Section 3.3, where it was proved that it is an equivalent non-dual norm. The norm $\|\cdot\|$ is nowhere Gâteaux differentiable (see, e.g., [FHHMZ11, Exercise 7.42]).

4.1.5 The set of points of differentiability of general functions on a Banach space

The fact that a convex function f defined on an interval (a, b) in $\mathbb{R}$ has an at most countable set of points of non-differentiability —a consequence of the non-decreasing character of the lateral derivatives, see Subsection 4.1.3— makes it natural to ask about how many points of Gâteaux or Fréchet differentiability a locally Lipschitz function or, more particularly, a continuous convex function on a Banach space has. There are many different situations. We mention here some results in this direction. We follow in part the summary presented by D. Preiss of the joint work [LPT12] in [Prei10]. A property defined on a space of measure happens **almost everywhere** when the set of points where it fails has measure zero.

1. There are continuous nowhere differentiable functions $f : \mathbb{R} \to \mathbb{R}$ (B. Bolzano, and, with a detailed proof, K. Weierstrass in 1875).

2. Every monotone function from $\mathbb{R}$ into $\mathbb{R}$ is differentiable almost everywhere. This is the celebrated Lebesgue Differentiation Theorem (see, e.g., [MoZiZi15, Theorem 424]). As a consequence, every Lipschitz function from $\mathbb{R}$ into $\mathbb{R}$ is differentiable almost everywhere (indeed, every Lipschitz function is absolutely continuous, hence of bounded variation, what implies that f is the difference of two increasing functions; see, e.g., [MoZiZi15, Subsection 4.5.7]). This result was extended by Rademacher to Lipschitz functions from $\mathbb{R}^n$ into $\mathbb{R}$ (see Theorem 438 below).

3. We mentioned in Subsection 4.1.4 (see also Theorem 172 below) that the canonical norm $\|\cdot\|_1$ on ℓ_1 is nowhere Fréchet differentiable. Spaces for which this behaviour does not happen are called **Asplund spaces**. We shall discuss this later (see Section 11.6). The present section deals with differentiability of functions defined on a Banach space. It will be then natural to provide the precise definition of Asplundness in terms of differentiability —that, by the way, was the original Asplund's definition of this concept, named by him **strong differentiability**—. This is done in Subsection 11.6.1 below. However, the following equivalent Definition 58 is clearly more handleable and will accordingly be mostly used in this text:

Definition 58. *A Banach space X is called an* **Asplund space** *if every separable subspace of X has a separable dual.*

Remark 59. Observe that if the norm $\|\cdot\|$ of a Banach space $(X, \|\cdot\|)$ is Fréchet differentiable, then X is Asplund.

➪ Indeed, it will be clearly enough to prove the result for a separable Banach space $(X, \|\cdot\|)$. Note that (B_{X^*}, w^*) is metrizable (and so separable); let D be a countable w^*-dense subset of B_{X^*}. If $NA(X)$ denotes the set of all norm-attaining elements in X^* and $x^* \in NA(X) \cap S_{X^*}$ attains its norm at $x \in S_X$, then there exists a sequence $\{d_n^*\}_{n\in\mathbb{N}}$ in D that w^*-converges to x^*. In particular, $\langle x, d_n^* \rangle \to \langle x, x^* \rangle$. The Šmulyan Lemma 54 implies that $\|d_n^* - x^*\| \to 0$. This shows that $\overline{D}^{\|\cdot\|^*}$ contains $NA(X) \cap S_{X^*}$. Since this last set is $\|\cdot\|^*$-dense in S_{X^*} by the Bishop–Phelps Theorem 14, S_{X^*} (and so X^*) is $\|\cdot\|^*$-separable. □

A streamlined argument —still relying on the Bishop–Phelps Theorem 14— for the previous statement uses the $\|\cdot\|$-$\|\cdot\|$-continuity of the (single-valued) subdifferential mapping $\partial\|\cdot\|$ (see the notes after Theorem 50), what ensures the $\|\cdot\|$-density in $NA(X)$ of the set $\partial\|\cdot\|(D)$. Ⓡ

We shall discuss the following result in Chapter 22.

Theorem 60 (Preiss, [Prei90]). *Every real-valued Lipschitz function on an Asplund space has points of Fréchet differentiability.*

4. The set of points of differentiability of a Lipschitz function on the real line need not be residual. Indeed, for every Lebesgue-null set $A \subset \mathbb{R}$ there exists a Lipschitz function on $\mathbb{R}$ not differentiable at points of A ([Zah46], see also [FHHMZ11, Exercise 7.34]).

5. Rademacher's Theorem 438 below does not hold for Lipschitz mappings with values in an infinite-dimensional space: The following function from $\mathbb{R}$ into c_0 is Lipschitz and nowhere differentiable:

$$f(t) := \left(\frac{\sin(t)}{1}, \frac{\sin(2t)}{2}, \frac{\sin(3t)}{3} \dots \right).$$

The Lipschitz property of f can be checked by using the Mean Value Theorem 44 on each coordinate of the range. Note that the only candidate for the derivative $f'(t)$ is $(\cos(t), \cos(2t), \cos(3t), \dots)$, that does not define a mapping $\mathbb{R} \to c_0$.

Spaces for which this pathology does not happen are called spaces with the **Radon–Nikodým property (RNP)**, closely related to the class of Asplund spaces defined above (see Definition 58 above and Theorem 547 below). The following definition gives an alternative description of this property in terms of strongly exposed points (see Definition 55), and should be compared with the statement of Theorem 259. Later, in Chapter 29, still an equivalent definition (there in terms of vector-valued measures, Definition 542) will be provided, and Theorem 544 will list a collection of statements equivalent to the RNP property of a Banach space.

Definition 61. *A Banach space X is said to have the* **Radon–Nikodým property** (**RNP**, *for short) if every closed convex bounded set in X is the closed convex hull of the set of its strongly exposed points.*

The following result should be compared with Theorem 443 and Corollary 444 below.

Theorem 62 (Aronszajn, Christensen, Mankiewicz)**.** (For the complete statement see, e.g., [FHHMZ11, Theorem 11.21]) *Every Lipschitz function from a separable Banach space into a space with the RNP is Gâteaux differentiable at points of a dense subset of X.*

However, in the direction of the Fréchet differentiability, no "almost everywhere" result is known.

6. In every nonseparable reflexive space, there are convex continuous functions whose set of points of Gâteaux differentiability is not G_δ (though it must be residual) [HSZ98].

7.

Definition 63. *A Banach space X is called a* **weak Asplund space** *if every convex continuous function on it is Gâteaux differentiable at each point of a dense G_δ subset of X.*

If $(X, \|\cdot\|)$ is a *separable* Banach space and $f : X \to \mathbb{R}$ is a continuous convex function, then *the set of points of Gâteaux differentiability is itself a dense G_δ subset of X* (Mazur, see Theorem 362 below; see also, e.g., [FHHMZ11, Theorem 8.2] and [Phe93, Theorem 1.20]). Thus, every separable Banach space is a weak Asplund space.

There is a non-weak Asplund space that is a **Gâteaux differentiability space**, i.e., every convex continuous function on it is Gâteaux differentiable at the points of a dense set [MoSo06].

8. If $(X, \|\cdot\|)$ is an arbitrary Banach space and $f : X \to \mathbb{R}$ is a continuous convex function, then the set of points where f is Fréchet differentiable is always a G_δ (it can be empty, as the example of the $\|\cdot\|_1$-norm in ℓ_1 shows, see Subsection 4.1.4, Item 6). For a proof see, e.g., [DeGoZi93, Remark after Theorem I.5.2] or [FHHMZ11, Lemma 7.45].

9.

Proposition 64. *If f is a convex function on a nonempty open convex subset U of $\mathbb{R}^n$, then f is Fréchet differentiable at points of a dense G_δ set in U.*

Proof. (Sketch) We propose several approaches to this result:

(i) We can use Mazur's theorem (see Item 7 above, see also, e.g., [DeGoZi93, Theorem I.4.12] or [MoZiZi15, Theorem 1033]) for the Gâteaux differentiability, and then Theorem 52 above.

(ii) The result is also a consequence of the Smooth Variational Principle (Theorem 289 below). Details can be seen, e.g., in the proof of [FHHMZ11, Corollary 7.44].

(iii) By Theorem 52, f is locally Lipschitz. Rademacher's Theorem 438 shows that every point has a neighborhood where the set of points of Fréchet differentiability of f is dense. This shows that the set of point of Fréchet differentiability of f in U is dense. It is always a G_δ-set (see Item 8 above).

(iv) The result appears as a particular case of Theorem 443, in view of Proposition 3, the fact that the Euclidean norm on $\mathbb{R}^n$ is Fréchet differentiable, Theorem 437, and Item 8 above. □

10. Connected with the Gâteaux and Fréchet differentiability of Lipschitz functions, see again the fundamental Theorem 443 below.

4.2 Strengthening Gâteaux differentiability

If the norm $\|\cdot\|$ of a Banach space $(X, \|\cdot\|)$ is Gâteaux differentiable, the (single-valued) duality mapping $x \to \|\cdot\|'(x)$ from S_X into X^* (see Subsection 4.1.4) is $\|\cdot\|$-w^*-continuous (see Subsection 4.1.3). A stronger differentiability notion is then to ask in addition for its $\|\cdot\|$-w-continuity, leading to a concept introduced in [DiFa74]:

Definition 65. (See [Di75, page 31]) *The norm $\|\cdot\|$ on a Banach space X is called* **very smooth at a point** $x_0 \in X$ **(VS)** *if it is Gâteaux differentiable at x_0 and the duality mapping is $\|\cdot\|$-w-upper semicontinuous at x_0. It is said to be* **very smooth** *if it is very smooth at each point $x \in X$, $x \neq 0$.*

The norm $\|\cdot\|$ of a Banach space can be Gâteaux differentiable at some $x_0 \in S_X$ and yet $\|\cdot\|^{**}$ can fail to be Gâteaux differentiable at x_0 (see the examples below in this same section). The following strengthening of the Gâteaux differentiability seems then natural: The norm $\|\cdot\|$ of a Banach space X is said to be **strongly Gâteaux differentiable at a point** x_0 if $\|\cdot\|^{**}$ in X^{**} is Gâteaux differentiable at x_0. (If $\|\cdot\|$ is strongly Gâteaux differentiable at any point of the sphere of X, it is said that it is **strongly Gâteaux differentiable** (SG).) It turns out that the two notions —very smoothness and strong Gâteaux differentiability— coincide.

➪ *Proof of the previous assertion.* Assume first that $\|\cdot\|^{**}$ is Gâteaux differentiable at some $x_0 \in S_X$ (with derivative $x_0^* \in S_{X^*}$, since x_0 attains it supremum on B_{X^*}). Thus, $(X^{***} \supset)\ \partial\|\cdot\|^{**}(x_0) = \{x_0^*\} \in S_{X^*}$. This shows that $\|\cdot\|$ is Gâteaux differentiable at x_0 and, since the duality mapping $\partial\|\cdot\|^{**} : X^{**} \to 2^{X^{***}}$ is always $\|\cdot\|$-$w(X^{***}, X^{**})$-upper semicontinuous, we get that $\partial\|\cdot\| : X \to 2^{X^*}$ is $\|\cdot\|$-$w(X^*, X^{**})$-upper semicontinuous, as claimed.
In order to prove the reverse implication, we may use Lemma 53. Assume then that $\|\cdot\|$ is Gâteaux differentiable at $x_0 \in S_X$ (with derivative $x_0^* \in S_{X^*}$) and that $\partial\|\cdot\|$ is $\|\cdot\|$-w-upper semicontinuous at x_0. According to the aforementioned lemma, given an arbitrary w-neighbourhood $N := \{x^* \in X^* : |\langle x_i^{**}, x^*\rangle| \leq \varepsilon,\ i = 1, 2, \ldots, n\}$ of 0 in X^* (where $x_i^{**} \in X^{**}$ for $i = 1, 2, \ldots, n$ and $\varepsilon > 0$) we can find a (closed) slice $S(B_{X^*}, x_0, \delta)$ contained in $\{x_0^*\} + N$. Thus, the corresponding slice $S(B_{X^{***}}, x_0, \delta)$ of $B_{X^{***}}$ (just the $w(X^{***}, X^{**})$-closure of $S(B_{X^*}, x_0, \delta)$) is contained in $\{x_0^*\} + M$, where

$$M := \{x^{***} \in X^{***} : |\langle x_i^{**}, x^{***}\rangle| \leq \varepsilon : i = 1, 2, \ldots, n\}.$$

Since the family of all such M's is a $w(X^{***}, X^{**})$-basis of neighbourhoods of 0 in X^{***}, we get that $\partial\|\cdot\|^{**}(x_0) = \{x_0^*\}$, as we wanted to show.

The $\|\cdot\|$-w-continuity of the single-valued mapping $\partial\|\cdot\|$ for a very smooth norm $\|\cdot\|$ on a Banach space X, together with the argument given in Remark 59, show that *every space* $(X, \|\cdot\|)$ *where the norm is very smooth is Asplund* (Theorem 372 below). The description of the set of points of Gâteaux differentiability in the space c_0 and in the space ℓ_∞, respectively, shows that $\|\cdot\|_\infty$ on c_0 is strongly Gâteaux differentiable at every point where it is Gâteaux differentiable. Given a Banach space $(X, \|\cdot\|)$, *if* $(X^*, \|\cdot\|^*)$ *is very smooth then* $(X, \|\cdot\|)$ *is reflexive* (see Theorem 411 below). Any of the two previous observations implies that $\|\cdot\|_1$ on ℓ_1 is not a very smooth norm.

4.3 Uniformities in differentiability

Uniformities (in differentiability and convexity, to be discussed below) are quantified by using the so-called modulus of smoothness and modulus of convexity, respectively. These two functions are defined here for future references:

Definition 66 (Day, [Day43]). *Let* $(X, \|\cdot\|)$ *be a Banach space. For* $\tau > 0$, *we define the* **modulus of smoothness of** $\|\cdot\|$ *by* (see Figure 4.13)

$$\rho_X(\tau) = \sup\left\{\frac{\|x+y\| + \|x-y\|}{2} - 1 :\ x, y \in X,\ \|x\| = 1,\ \|y\| = \tau\right\}. \quad (4.19)$$

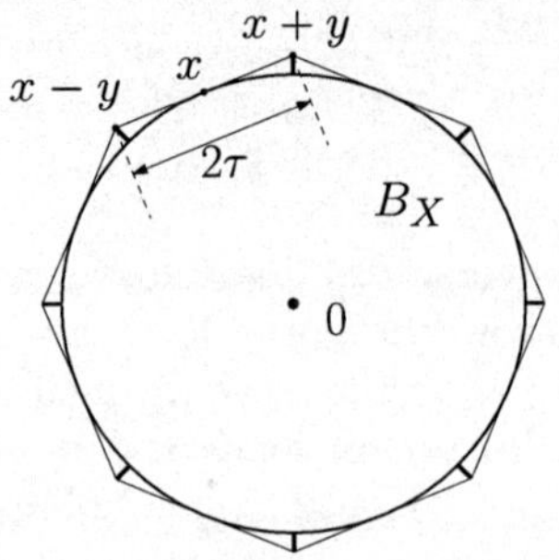

Figure 4.13: For the ℓ_2-norm, the modulus of smoothness is the length of one of the heavier segments

Definition 67 (Clarkson, [Clar36]). *The* **modulus of convexity** *of the norm* $\|\cdot\|$ *on a Banach space* X *is the function on* $(0, 2]$ *given by*

$$\delta_X(\varepsilon) = \inf\left\{1 - \left\|\frac{x+y}{2}\right\| :\ x, y \in B_X,\ \|x - y\| \geq \varepsilon\right\}. \quad (4.20)$$

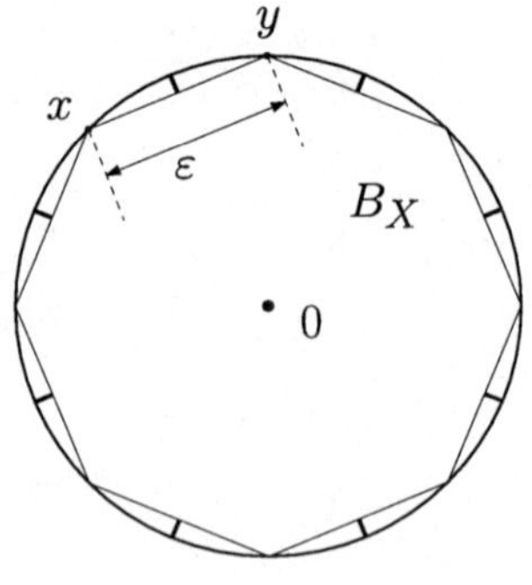

Figure 4.14: For the ℓ_2-norm, the modulus of convexity is the length of one of the heavier segments

Fréchet differentiability is the uniform —with respect to directions— version of the concept of Gâteaux differentiability. There are other important uniform variants of the basic definition:

(a) Keeping the direction and asking in (4.1) for uniformity on the "base points" on the sphere:

Definition 68. *The norm on a Banach space is called* **uniformly Gâteaux differentiable (UG)** *if for every $h \in X$ the limit* (4.1) *exists uniformly on $x_0 \in S_X$.*

This is equivalent to the fact that

$$\lim_{t \to 0} \frac{\|x+th\| + \|x-th\| - 2\|x\|}{t} = 0 \tag{4.21}$$

uniformly for $x \in S_X$ (see (4.14) above).

(b) Asking for uniformity on directions of the one-sided derivative:

Definition 69. *The norm $\|\cdot\|$ on X is called* **strongly subdifferentiable (SSD) at** $x \in X$ *if the one-sided limit*

$$\lim_{t \to 0+} \frac{\|x+th\| - \|x\|}{t}$$

exists uniformly for $h \in S_X$. It this happens for all $x \in X$, we say that the norm is **strongly subdifferentiable (SSD) on** X.

This concept will be considered again in Chapter 30 below, where a geometric description in terms of the size of slices from B_{X^*} in the direction of Šmulyan's Lemma 54 will be given (Theorem 548).

Uniformity in the definition of the derivative can be required not only on directions as in (b) above, but on both directions and base-points on the sphere:

Definition 70. *The norm $\|\cdot\|$ on X is called* **uniformly Fréchet smooth (UF)** *(also called* **uniformly smooth***) if*

$$\lim_{h \to 0} \frac{\|x+h\| + \|x-h\| - 2\|x\|}{\|h\|} = 0, \quad \textit{uniformly for } x \in S_X. \tag{4.22}$$

The geometric counterpart of the previous definition is that for every $\varepsilon > 0$ there exists $\delta > 0$ such that every slice $S(B_{X^*}, x, \delta)$ cut by any $x \in S_X$ off B_{X^*} have a $\|\cdot\|$-diameter less than ε.

We have that $(X, \|\cdot\|)$ is uniformly Fréchet differentiable if, and only if,

$$\lim_{t \to 0} \frac{\rho_X(\tau)}{\tau} = 0,$$

where ρ_X is the modulus of smoothness (see Definition 66 above).

Lemma 71 (Lindenstrauss). *Let $(X, \|\cdot\|)$ be a Banach space, let δ_X be the modulus of convexity of $\|\cdot\|$ and ρ_{X^*} be the modulus of smoothness of the dual norm $\|\cdot\|^*$. Then for every $\tau > 0$,*

$$\rho_{X^*}(\tau) = \sup\left\{\tau\frac{\varepsilon}{2} - \delta_X(\varepsilon) : 0 < \varepsilon \leq 2\right\}.$$

For the proof see, e.g., [FHHMZ11, Lemma 9.8]. The concept of uniform convexity (also called "uniform rotundity") will be introduced in Definition 83.

Corollary 72. *A Banach space is uniformly convex if, and only if, its dual is uniformly smooth.*

Theorem 73 (Clarkson, see, e.g., [FHHMZ11], Theorem 9.3). *If (Ω, μ) is a measure space and $p > 1$, then the canonical norm of $L_p(\mu)$ is uniformly convex.*

Thus we have

Theorem 74. *If (Ω, μ) is a measure space and $1 < p < \infty$, then the canonical norm of $L_p(\mu)$ is both uniformly convex and uniformly smooth.*

Proposition 75. *If H is a Hilbert space, then its modulus of convexity is*

$$\delta_H(\varepsilon) = 1 - \sqrt{1 - \frac{\varepsilon^2}{4}},\ \varepsilon \in (0, 2], \tag{4.23}$$

and its modulus of smoothness is

$$\rho_H(\tau) = \sqrt{1 + \tau^2} - 1,\ \tau > 0. \tag{4.24}$$

For a proof see, e.g., [Di75, Chapter 3, §3]. The evaluation of the moduli of convexity and of smoothness for several classes of spaces is, e.g., in [Di75, Chapter 3].

The following concept can be understood as the description of being "uniformly non-Fréchet differentiable", and it allows for several important structural properties of Banach spaces (Asplundness, see Theorems 274, 276, and 284, and the non-existence of a Fréchet differentiable bump, see Theorem 283, among others):

Definition 76 (Leach, Whitfield, [LeaWh72]). *Given $\varepsilon > 0$, a norm $\|\cdot\|$ on a Banach space X is said to be* **ε-rough** *if for all $x \in X$,*

$$\limsup_{\|h\| \to 0} \frac{\|x + h\| + \|x - h\| - 2\|x\|}{\|h\|} \geq \varepsilon. \tag{4.25}$$

The norm is **rough** *if it is ε-rough for some $\varepsilon > 0$* (see Figure 4.15).

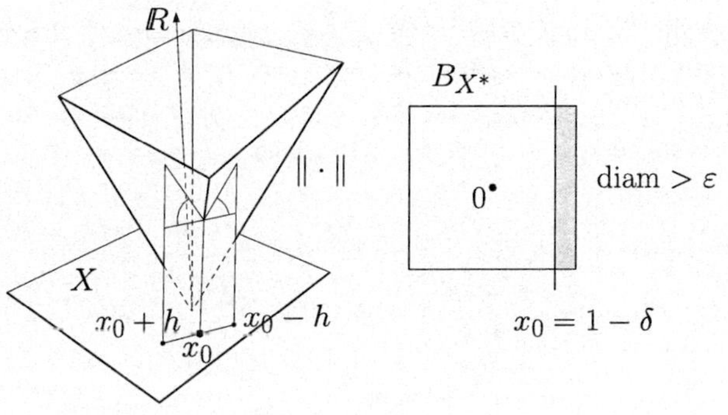

Figure 4.15: Roughness at x_0 (two versions)

A geometric description of roughness is the following (see again Figure 4.15): The norm $\|\cdot\|$ on X is ε-rough if, and only if, $\operatorname{diam} S(B_{X^*}, x, \delta) \geq \varepsilon$ for all $x \in S_X$ and $\delta > 0$ (see Theorem 278 below for this and other geometric descriptions). Examples of rough norms are $\|\cdot\|_\infty$ on $C[0,1]$ (see Proposition 280), and $\|\cdot\|_1$ on $\ell_1(\Gamma)$ for an infinite set Γ (see Proposition 281).

4.4 Norms defined by level sets of differentiable functions

In order to renorm $(\mathbb{R}^2, \|\cdot\|_\infty)$ by a smooth norm with a prescribed approximation to the original $\|\cdot\|_\infty$ norm, we draw a "smooth" absolutely convex body B "close" to $B_{\|\cdot\|_\infty}$, and claim that the Minkowski functional of B is an equivalent smooth norm close to $\|\cdot\|_\infty$ (see Figure 4.16).

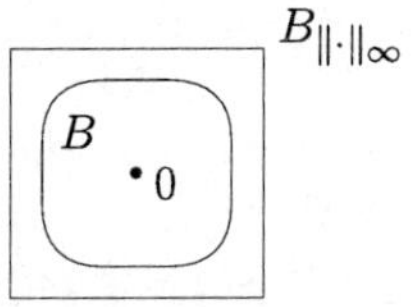

Figure 4.16: Approximating a norm by a smooth norm

Proposition 78 below formalizes this approach. The justification of this procedure is done through the Implicit Function Theorem 77. As it is usual, if f is defined on an open subset of a product space $X \times Y$, and if f is Fréchet differentiable as a function of the variable $x \in X$, its Fréchet derivative at a point $(x_0, y_0) \in G$ is denoted $(\partial f/\partial x)(x_0, y_0)$. Similarly for the Fréchet derivative $(\partial f/\partial y)(x_0, y_0)$ (we hope that the symbol ∂f here does not create any misunderstanding with the use of the same symbol denoting the subdifferential of a function, see page 74).

Theorem 77 (Implicit Function Theorem, see, e.g., [HáJo14], Theorem 81). *Let X be a normed space, Y and Z Banach spaces, $G \subset X \oplus Y$ a nonempty open set, and $f \in C^1(G, Z)$. Let $(x_0, y_0) \in G$ and let $(\partial f/\partial y)(x_0, y_0)$ be an isomorphism of Y onto Z. Then there are open neighbourhoods U of x_0 in X and V of y_0 in Y such that there is a unique mapping $u : U \to V$ satisfying $f\big(x, u(x)\big) = f(x_0, y_0)$ for all $x \in U$. Moreover, this mapping u is C^1-smooth on U, and*

$$u'(x) = -\Big(\frac{\partial f}{\partial y}\big(x, u(x)\big)\Big)^{-1} \circ \frac{\partial f}{\partial x}\big(x, u(x)\big)$$

for all $x \in U$. If f is moreover C^k-smooth for some $k \in \mathbb{N} \cup \{+\infty\}$, then $u \in C^k(U; Y)$.

For a sketch, see Figure 4.17.

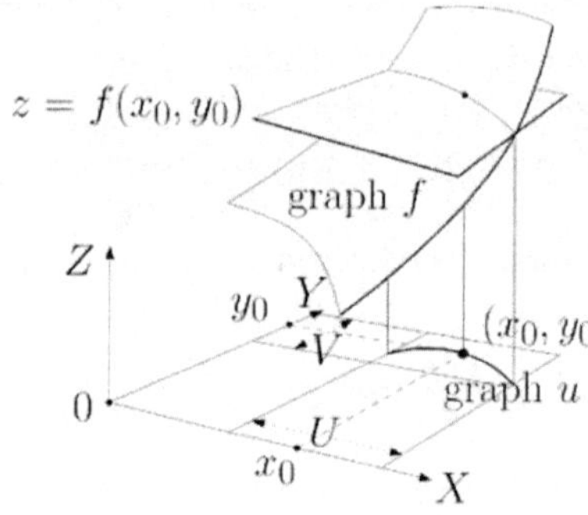

Figure 4.17: The Implicit Mapping Theorem 77

Proposition 78. *Let $(X, \|\cdot\|)$ be a normed space and $D \neq \emptyset$ be an open, absolutely convex subset of X. Let $f : D \to \mathbb{R}$ be an even, convex, and continuous function. Assume the existence of $a > f(0)$ such that the set $B := \{x \in D : \ f(x) \leq a\}$ is a closed and bounded subset of X. If there exists an open set O with $\{x \in D : f(x) = a\} \subset O$ and such that f is C^k-smooth on O for some $k \in \mathbb{N} \cup \{+\infty\}$, then the Minkowski functional p of B is an equivalent C^k-smooth norm on X.*

Proof. (Sketch. For details, see, e.g., [Ru18].) The set B is absolutely convex and closed (see Figure 4.18). Put

$$V := \{(x, \rho) \in X \setminus \{0\} \times (0, +\infty) : \ \rho^{-1}x \in O\}.$$

The set V is open. Define $F : V \to \mathbb{R}$ as $F(x, \rho) := f(\rho^{-1}x)$, a function obviously C^k-smooth on V. We claim that, by using the Implicit Function Theorem 77, the equation $F = a$ gives p as the (unique) C^k-smooth implicit function from $X \setminus \{0\}$ into $\mathbb{R}$ (i.e., $F(x, p(x)) = a$ for all $x \in X \setminus \{0\}$). Indeed, $F(x, p(x)) = f(x/p(x)) \leq a$, since $x/p(x) \in B$. In case that $F(x, p(x)) < a$ for some $x \in X \setminus \{0\}$, there exists,

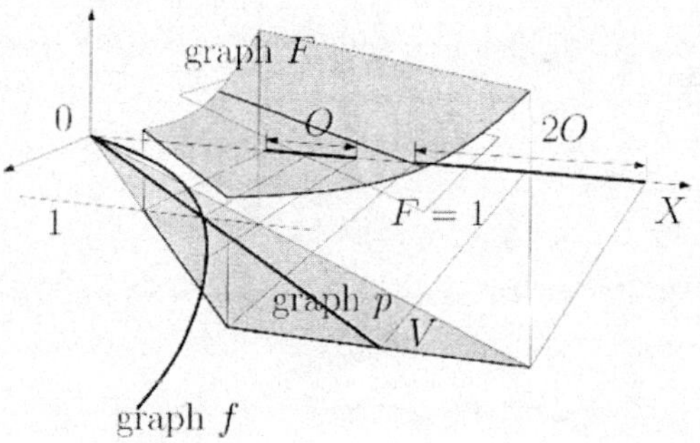

Figure 4.18: The construction in the proof of Proposition 78 (for simplicity, $a = 1$, $f(0) = 0$, and we represent only half of the picture)

by continuity, $\varepsilon > 0$ such that $F(x, p(x) - \varepsilon) = f(x/(p(x) - \varepsilon)) < a$. Thus, $x \in (p(x) - \varepsilon)B$, a contradiction. It follows that $F(x, p(x)) = a$ for $x \in X \setminus \{0\}$. The convexity of f, together with the fact that $f(0) < a$, shows that $p(x)$ is the unique value in $(0, +\infty)$ with $F(x, p(x)) = a$.
The convexity of f shows that

$$\frac{f((1+t)h) - f(h)}{t} \geq f(h) - f(0) \tag{4.26}$$

for every $h \in D$ and $t > 0$ small. Consequently, for $h \in D$ such that $f(h) = a$, we get from (4.26)

$$\langle h, f'(h) \rangle = \lim_{t \to 0+} \frac{f((1+t)h) - f(h)}{t} \geq f(h) - f(0) = a - f(0) > 0. \tag{4.27}$$

It follows from (4.27) that, for $0 \neq h \in D$,

$$\frac{\partial F}{\partial \rho}(h, p(h)) = -\frac{1}{p^2(h)} \left\langle h, f'\left(\frac{h}{p(h)}\right)\right\rangle = -\frac{1}{p(h)} \left\langle \frac{h}{p(h)}, f'\left(\frac{h}{p(h)}\right)\right\rangle \neq 0. \tag{4.28}$$

Applying the Implicit Function Theorem 77 above we deduce that p is C^k-smooth on $X \setminus \{0\}$. □

The following result, close in spirit to Proposition 78, ensures the differentiability of the Minkowski functional defined by a level set of a convex differentiable function.

Proposition 79. *Let $f\colon X \to (-\infty, +\infty]$ be a lsc convex function, $C := \{x : f(x) \leq \alpha\}$ where $f(0) < \alpha$ and $0 \in \mathrm{Int}C$. Assume that f is differentiable Fréchet (resp. Gâteaux) at all x in the boundary of C. Then p_C, the Minkowski functional of C, is differentiable Fréchet (resp. Gâteaux) at all x where $p_C(x) > 0$.*

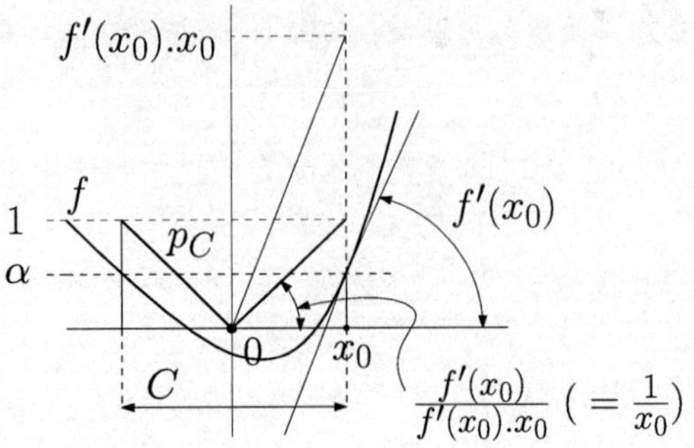

Figure 4.19: How to construct p_C for a level-α set C of f (f and p_C in bold)

➡ (Hint) For the Fréchet case, fix x_0 in the boundary of C. Let

$$x_0^* := f'(x_0)/(\langle x_0, f'(x_0)\rangle).$$

This functional is the derivative of p_C at x_0 (see Figure 4.19, a sketch for a function $f : \mathbb{R} \to \mathbb{R}$). The fact that f is Fréchet differentiable at x_0 gives the Fréchet differentiability of p_C at x_0, as in Proposition 78. For the Gâteaux case, it is possible to use an argument of reduction to finite-dimensional spaces and then consider that Gâteaux differentiability implies in the convex (and then locally Lipschitz) case C^1-differentiability (see Lemma 48 and Theorem 49). An alternative argument uses a proof by Borwein and Vanderwerff [BoVa09] for the Gâteaux case —by using subdifferentials—, that can be extended to any bornology by using theorems of continuity of the subdifferential.

Chapter 5

Basic rotundity

Convexity is one of the fundamental concepts in geometry and analysis on Banach spaces. The function $\|\cdot\|$ is always convex (the triangle inequality). In other terms, the epigraph of $\|\cdot\|$ is a convex set in $X \times \mathbb{R}$ or, equivalently, the closed unit ball is convex. In this chapter we shall explore the most useful variations of convexity — strict, uniform and locally uniform—, their connections with differentiability, and their implications in the topological structure of the space. Only at the end we shall mention some other variants that still play an important rôle in the geometry and structure of Banach spaces.

5.1 Strict convexity

More demanding properties of a norm than merely convexity —that is automatic— entail a more precise behaviour. A useful extra property of a norm is the so-called strict convexity:

Definition 80. *The norm $\|\cdot\|$ on X is said to be* **strictly convex** *(or* **rotund***), (***R***, for short), if its unit sphere contains no nondegenerate straight line segments.*

In Lemma 81 below we shall provide some equivalent formulations to the strict convexity of a norm. This will be completed in Proposition 113.

Lemma 81. *Let $(X, \|\cdot\|)$ be a normed space. The following are equivalent:*
(i) *$\|\cdot\|$ is strictly convex.*
(ii) *If $x, y \in S_X$ satisfy $\|x+y\| = 2$, then $x = y$.*
(iii) *If $x, y \in X$, $x \neq 0$ and $y \neq 0$, satisfy $\|x+y\| = \|x\| + \|y\|$, then $x = \lambda y$ for some $\lambda > 0$.*
(iv) *Every point $x \in S_X$ is an extreme point of B_X.*

Proof. (i)$\Rightarrow$(ii): If $\|z\| < 1$, then $[z, x) \in \mathrm{Int} B_X$, so if $\|(x+y)/2\| = 1$ then it follows that $[x, y] \subset S_X$.
(ii)$\Rightarrow$(iii): Let $\|x+y\| = \|x\| + \|y\|$ for some $x \neq 0$, $y \neq 0$. We may assume that $0 < \|x\| \leq \|y\|$. Then

$$\begin{aligned} 2 &\geq \big\|x/\|x\| + y/\|y\|\big\| \geq \big\|x/\|x\| + y/\|x\|\big\| - \big\|y/\|x\| - y/\|y\|\big\| \\ &= (1/\|x\|)\|x+y\| - \|y\|(1/\|x\| - 1/\|y\|) \\ &= (1/\|x\|)(\|x\| + \|y\|) - \|y\|(1/\|x\| - 1/\|y\|) = 2. \end{aligned}$$

A. J. Guirao et al., *Renormings in Banach Spaces*, Monografie Matematyczne 75,
https://doi.org/10.1007/978-3-031-08655-7_5

Thus $\|x/\|x\| + y/\|y\|\| = 2$. By (ii) we get $x/\|x\| = y/\|y\|$, and the conclusion follows.
(iii)⇒(iv): If $z \in S_X$ is not an extreme point of B_X, then $z = (1/2)(x+y)$ for some $x, y \in B_X$. Thus, $1 = \|z\| \leq (1/2)(\|x\| + \|y\|) \leq 1$, so $\|x+y\| = \|x\| + \|y\|$ and $\|x\| = \|y\| = 1$. From (iii) we get $x = \lambda y$ for some $\lambda > 0$, and so $\lambda = 1$ and $x = y$.
(iv)⇒(i) is obvious. □

There is no apparent connection between differentiability and convexity, since the point x_1 in Figure 5.1 is extreme and the norm is differentiable there, the point x_2 is not extreme and the norm is differentiable there, and the point x_3 is extreme and the norm is not differentiable there.

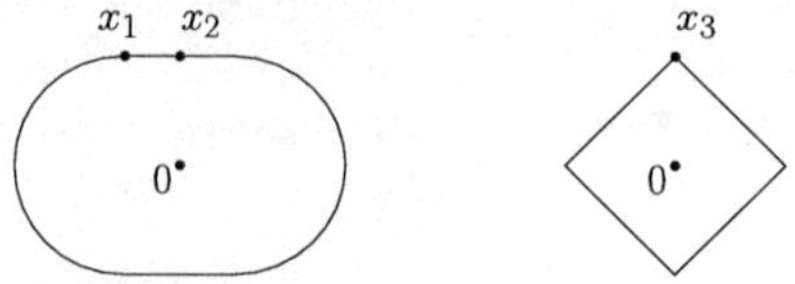

Figure 5.1: Two balls in $\mathbb{R}^2$

However, as we mentioned before, strict convexity and differentiability for a norm are, in a sense, dual concepts. This is reflected in the following basic result, a straightforward consequence of Lemma 54:

Proposition 82. *Let* $(X, \|\cdot\|)$ *be a Banach space.*
(i) *If* $\|\cdot\|^*$ *is strictly convex, then* $\|\cdot\|$ *is Gâteaux differentiable.*
(ii) *If* $\|\cdot\|^*$ *is Gâteaux differentiable, then* $\|\cdot\|$ *is strictly convex.*

Proof. (i) If $\|\cdot\|^*$ is strictly convex, then, for every $0 \neq x \in X$, the face $F_x := \{x^* \in B_{X^*} : \langle x, x^*\rangle = 1\}$ $(= \partial\|\cdot\|(x))$ must be a singleton. Use now Lemma 54.
(ii) If $x \in S_X$ satisfies $x = (y+z)/2$, where $y, z \in B_X$ and $y \neq z$, and $x^* \in S_{X^*}$ satisfies $\langle x, x^*\rangle = 1$, then $\{y, z\} \subset \partial\|\cdot\|^*(x^*)$ $(\subset X^{**})$, and so $\|\cdot\|^*$ is not Gâteaux differentiable at x^*, again by Lemma 54. □

It is worth to mention that the reverse implication in Proposition 82(i) is false. This was proved in [Klee59] and [Tr70]. Even more, there is a Fréchet differentiable equivalent norm on a separable Banach space whose dual norm is not strictly convex [GuiMoZi12] (see Example 719 below).
The impossibility to reverse the implication in (ii) holds even in a strong sense: There are separable Banach spaces $(X, \|\cdot\|)$ with a strictly convex norm whose dual has no equivalent Gâteaux smooth renorming. Indeed, ℓ_∞ admits no equivalent Gâteaux differentiable norm (see also Theorem 805; see also, e.g., [FHHMZ11, Exercise 7.64]), while ℓ_1, being a separable space, admits a strictly convex equivalent norm (see the application in Subsection 9.2.1 and Theorem 154 below). It seems to be open whether the dual space of every space with a Gâteaux differentiable norm can be renormed by a strictly convex (nondual in general) norm

(see. e.g., [Fab97, Notes and Remarks 5.4]. For a full discussion see Section 48.1, Subsection 48.1.2, in particular Theorems 805, 806, and 808, and Corollary 807.

5.2 Uniform convexity

5.2.1 Introduction

In a Hilbert space H the quality of the convexity of its closed unit ball is "the best" we can expect: Thanks to the **parallelogram identity** (i.e., $\|x+y\|^2+\|x-y\|^2 = 2\|x\|^2+2\|y\|^2$ for all $x, y \in H$), it follows that

$$\left\|\frac{x+y}{2}\right\|^2 = 1 - \left\|\frac{x-y}{2}\right\|^2, \quad \text{for } x, y \in S_H.$$

From this, if $\|x\| = \|y\| = 1$ and $\|x-y\| = \varepsilon$, then $\|(x+y)/2\| = \sqrt{1-\varepsilon^2/4}$. If $x^* \in S_{H^*}$ and $x, y \in S(B_H, x^*, \delta)$ $(:= \{h \in B_H : \langle h, x^*\rangle \geq 1-\delta\})$, then we get easily that $\|x-y\| \leq 2\sqrt{\delta(2-\delta)}$, so $\operatorname{diam}(S(B_H, x^*, \delta)) \leq 2\sqrt{\delta(2-\delta)}$. By computing the diameter of the intersection with any 2-dimensional subspace, we get in fact that $\operatorname{diam}(S(B_H, x^*, \delta)) = 2\sqrt{\delta(2-\delta)}$.
The following concept, due to J.A. Clarkson [Clar36], coins under the name "uniform convexity" the special behaviour of the middle point of two separated points on the unit sphere of a Banach space:

Definition 83. *The norm* $\|\cdot\|$ *of a Banach space* X *is called* **uniformly rotund (UR)** *(also called* **uniformly convex (UC)***) if* $\delta_X(\varepsilon) > 0$ *for each* $0 < \varepsilon \leq 2$*, where* δ_X *is the modulus of convexity* (see Definition 67 and (a) in Figure 5.2).

The following proposition lists some straightforward equivalent formulations to the uniform convexity.

Proposition 84. *Let* $(X, \|\cdot\|)$ *be a Banach space. The following statements are equivalent:*
(i) $\|\cdot\|$ *is uniformly convex.*
(ii) *Given two sequences* $\{x_n\}_{n\in\mathbb{N}}$ *and* $\{y_n\}_{n\in\mathbb{N}}$ *in* S_X *such that* $\|x_n+y_n\| \to 2$*, then* $\|x_n-y_n\| \to 0$.
(iii) *Given* $\varepsilon > 0$ *there exists* $\delta > 0$ *such that* $\operatorname{diam} S(B_X, x^*, \delta) < \varepsilon$ *for all* $x^* \in S_X^*$ (see (b) in Figure 5.2).

5.2.2 A consequence of the uniform convexity

Note that, easily, the equivalence (i)⇔(iii) in Proposition 84 above shows that *every UR space is reflexive*: Indeed, observe first that the w^*-closure in X^{**} of a $\|\cdot\|$-small set in X is again $\|\cdot\|$-small in X^{**} (after all, the w^*-closure of a ball in X is a ball with the same center and radius in X^{**}, see Figure 5.3). If $x^{**} \in S_{X^{**}}$,

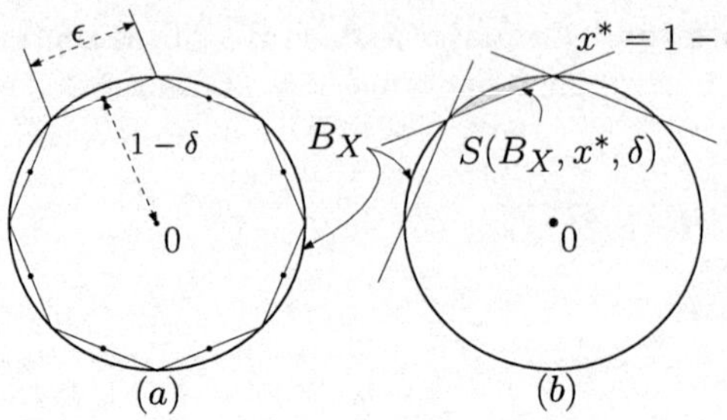

Figure 5.2: A uniformly convex ball: (a) depth of the middle point; (b) diameter of slices

find a $\|\cdot\|$-small slice of $B_{X^{**}}$ defined by an element in S_{X^*} and containing x^{**}. Since every such slice intersects X (indeed, X is w^*-dense in X^{**}), we get that $\operatorname{dist}(x^{**}, X)$ is arbitrarily small.

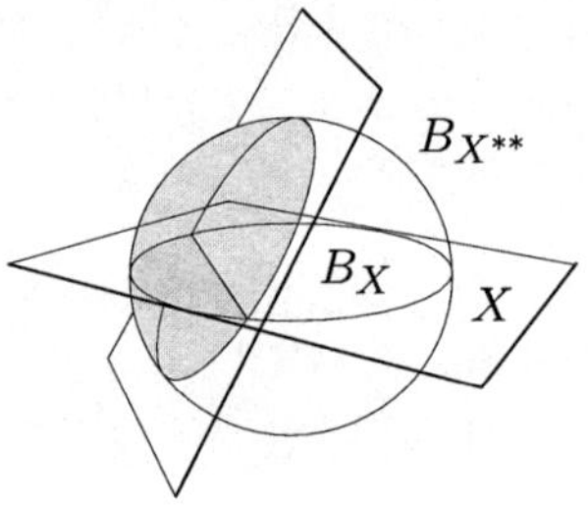

Figure 5.3: If a slice of B_X is small, then the corresponding slice of $B_{X^{**}}$ is small, too

Certainly, James' weak compactness Theorem 13 can be used to provide an alternative proof of the preceding statement: Indeed, under the uniform convexity every $x^* \in X^*$ attains its supremum on B_X. However, the previous proof shows that in this case it is not necessary to rely on such a deep result for proving reflexivity.

5.2.3 A weak version of the uniform convexity

The "weak" variant of the uniform convexity was introduced by V.L. Šmulyan [Smu40]:

Definition 85. *The norm* $\|\cdot\|$ *on* X *is* **weakly uniformly rotund (WUR)** *if* $x_n - y_n \to 0$ *in the weak topology of* X *whenever* $x_n, y_n \in S_X$ *are such that* $\|x_n + y_n\| \to 2$.
The norm of X^* *is* **weak* uniformly rotund (W*UR)** *if* $f_n - g_n \to 0$ *in the* w^**-topology of* X^* *, whenever* $f_n, g_n \in S_{X^*}$ *are such that* $\|f_n + g_n\| \to 2$.

Note that a WUR norm is strictly convex. The same applies to a W*UR dual norm.

5.3 Local uniform convexity

5.3.1 Introduction

The following is one of the most useful concepts in the geometry and analysis of Banach spaces. It was introduced in R. Lovaglia's PHD dissertation [Lov55] under the supervision of the late R.C. James. This notion is used in almost all areas in Banach space theory: Geometry and analysis on Banach spaces, extremal structure of convex sets, differentiability of convex functions, approximation, topology, etc. Used via duality (see Proposition 91), it guarantees Fréchet differentiability on the space, a most important tool for analysis.

Definition 86. *Let* $(X, \|\cdot\|)$ *be a Banach space. Let* $x_0 \in S_X$. *The norm* $\|\cdot\|$ *on* X *is* **locally uniformly rotund (LUR)** *(also called* **locally uniformly convex***)* **at** x_0 *if given* $\varepsilon \in (0, 2]$ *there exists* $\delta = \delta(x_0, \varepsilon) > 0$ *such that* $\|(x_0 + x)/2\| \le 1 - \delta$, *whenever* $x \in S_X$ *is such that* $\|x_0 - x\| \ge \varepsilon$. *If* $\|\cdot\|$ *is LUR at every point of* S_X, *we say that* $\|\cdot\|$ **is LUR**.

An equivalent definition in terms of the so-called **modulus of local uniform convexity** will be given in the proof of Theorem 186.

Obviously, if the norm is LUR then it is strictly convex. As in the case of the uniform convexity, to list some equivalent formulations of the LUR property of the norm $\|\cdot\|$ of a Banach space X at a point $x_0 \in S_X$ can be useful (for the full proof of the equivalences see, e.g., [DeGoZi93, Proposition II.1.2]). This is done in Proposition 87 below:

Proposition 87. *Let* $(X, \|\cdot\|)$ *be a Banach space, and let* $x_0 \in S_X$. *The following statements are equivalent:*
(i) $\|\cdot\|$ *is LUR at* x_0.
(ii) *If* $\{x_n\}_{n\in\mathbb{N}}$ *is a sequence in* S_X *such that* $\|x_0 + x_n\| \to 2$, *then* $\|x_0 - x_n\| \to 0$ (it can be shown easily that we can take sequences $\{x_n\}_{n\in\mathbb{N}}$ in B_X instead).

We will see that many spaces —for example, the separable or the reflexive ones, see Theorems 173 and 218, respectively— can be given an equivalent LUR norm.

On the other hand ℓ_∞ does not admit any equivalent LUR norm (see Theorem 802).

5.3.2 Some consequences of the local uniform convexity

1. **LUR and small slices.** If $\|\cdot\|$ is LUR at $x_0 \in S_X$, and $x_0^* \in \partial\|\cdot\|(x_0)$, then $\operatorname{diam} S(B_X, x_0^*, \delta) \to 0$ as $\delta \to 0$ (see Figure 5.4).

   ➪ Indeed, if this is not the case we may find $\varepsilon > 0$ and a sequence $\{x_n\}$ in B_X such that $\langle x_n, x_0^*\rangle \to 1$ and $\|x_0 - x_n\| \geq \varepsilon$ for all $n \in \mathbb{N}$. Then

$$1 \geq \left\| \frac{x_0 + x_n}{2} \right\| \geq \left\langle \frac{x_0 + x_n}{2}, x_0^* \right\rangle \to 1.$$

   This shows that $\|(1/2)(x_0+x_n)\| \to 1$, while $\|x_0 - x_n\| \geq \varepsilon$ for all $n \in \mathbb{N}$, a contradiction with the LUR property of $\|\cdot\|$ at x_0 (see Proposition 87).

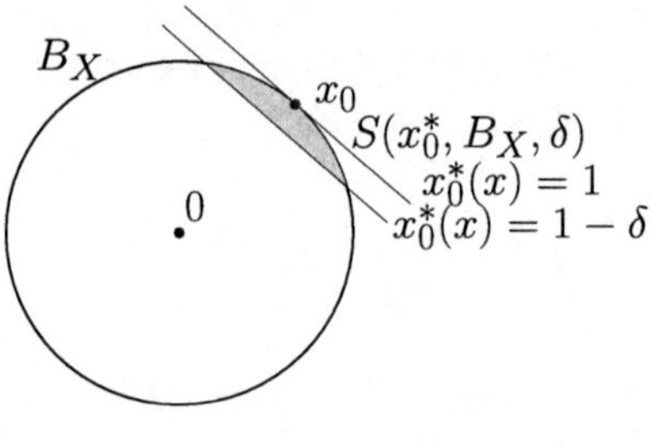

Figure 5.4: If LUR at x_0, then the diameter of the slices goes to 0

2. **LUR and topologies on the sphere.** A quantification of the argument in Item 1 above gives the following proposition (see also Section 5.4).

   Proposition 88. *Assume that the norm $\|\cdot\|$ of a Banach space X is LUR at a point $x_0 \in S_X$, and that D is a w^*-dense subset of B_{X^*}. If a net $\{x_i\}_{i\in I}$ in B_X is $w(X, D)$-convergent to x_0, then it is $\|\cdot\|$-convergent (to x_0).*

   Proof. Let $\varepsilon > 0$. Find $\delta := \delta(x_0, \varepsilon) > 0$ according to Definition 86. By the w^*-density of D in B_{X^*}, there exists $d_0^* \in D$ such that $\langle x_0, d_0^*\rangle > 1 - \delta$. Find $i_0 \in I$ such that $\langle x_i, d_0^*\rangle > 1 - \delta$ for $i \geq i_0$. Then

$$\left\| \frac{x_i + x_0}{2} \right\| \geq \left\langle \frac{x_i + x_0}{2}, d_0^* \right\rangle > 1 - \delta,$$

   hence $\|x_i - x_0\| < \varepsilon$ for $i \geq i_0$, and the conclusion follows. □

The property exhibited in Proposition 90 below plays an important rôle in the geometry of Banach spaces. For later use, the following definition coins a name for it:

Definition 89. *Let* $(X, \|\cdot\|)$ *be a Banach space, and let* τ_1 *and* τ_2 *be two topologies on* X. *We say that* $\|\cdot\|$ *has the* τ_1-τ_2-**Kadets–Klee property** *(for short, that the norm is* τ_1-τ_2-**KK**) *if* τ_1 *and* τ_2 *coincide on* S_X.
We say that $\|\cdot\|$ *has the* τ_1-τ_2-**sequential Kadets–Klee property** *(for short, that the norm is* τ_1-τ_2-**SKK**) *if a sequence in* S_X *is* τ_1*-convergent to a point* $x \in S_X$ *if, and only if, it is* τ_2*-convergent to* x.

In order to simplify the notation, we shall use in some circumstances just the acronym KK (SKK) for denoting the w-$\mathcal{T}_{\|\cdot\|}$-KK property (the w-$\mathcal{T}_{\|\cdot\|}$-SKK property, respectively), if there is no risk of misunderstanding.

The following consequence of Proposition 88 holds:

Proposition 90. *Every LUR norm* $\|\cdot\|$ *on a Banach space has the* $w(X, D)$-$\mathcal{T}_{\|\cdot\|}$*-KK property, for any* w^**-dense subset* D *of* B_{X^*}.

The two properties KK and SKK are not the same. We shall return to this in Section 5.4 (proving that under some circumstances, the two properties coincide) and provide examples in Section 20.3 (see Remarks 404 and 409 below).

3. **LUR in duality**. A very close argument to the one in Item 1 above, together with the geometric characterization of Fréchet differentiability given in Lemma 54, shows the following important result:

   Proposition 91. *Let* X *be a Banach space. Assume that* $x_0 \in S_X$ *and* $x_0^* \in \partial\|\cdot\|(x_0)$. *If the dual norm* $\|\cdot\|^*$ *is LUR at* x_0^**, then the original norm is Fréchet differentiable at* x_0. *In particular, if* $\|\cdot\|^*$ *on* X^* *is LUR, then* $\|\cdot\|$ *on* X *is Fréchet differentiable.*

   H Dual LUR norms were used in 1968 by Asplund ([Asp68]) in his differentiability result on convex functions on spaces with separable dual (see Theorem 272).

   The converse of Proposition 91 fails in general. For an example see Item 2 in Paragraph 9.4.2 below.

4. **Strongly exposed and farthest points**. The classical Krein–Milman Theorem 7 describes any nonempty convex w-compact subset C of a Banach space as the closed convex hull of its extreme points. Lindenstrauss in 1963 [Lind63] (by using norm-attaining operators and the Kadets' LUR renorming —see

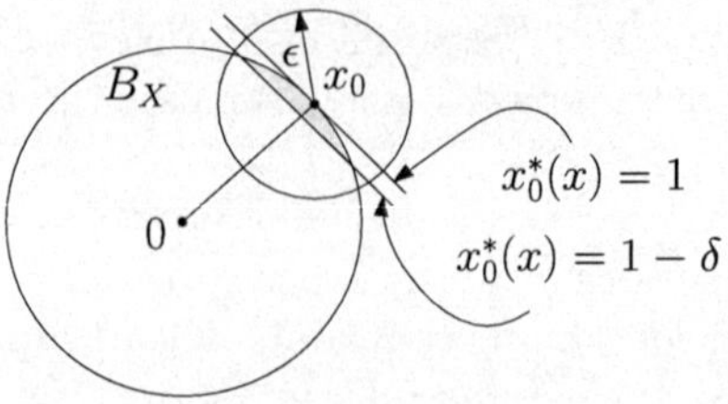

Figure 5.5: The w and $\|\cdot\|$ topologies coincide on S_X if LUR

Theorem 176 below—) showed that *it is possible to use strongly exposed points instead if the space is separable* (for the notion of exposed and strongly exposed points see Definition 55). Later on the separability was removed by Troyanski [Tr71] (see Theorem 259 below). Strongly exposed points are related, in LUR spaces, to another geometrical concept: Given a bounded nonempty set S in X and a point $c \in X$, a point $x_0 \in S$ is said to be a **farthest point in S from c** if $\|x_0 - c\| = \sup\{\|x - c\| : x \in S\}$.

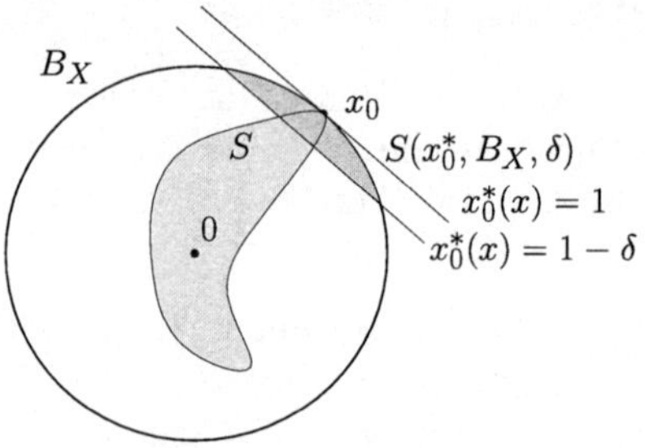

Figure 5.6: x_0 farthest is a strongly exposed point of S if X is LUR

If $(X, \|\cdot\|)$ is a LUR Banach space and S is a nonempty bounded subset of X, then *every farthest point x_0 in S from a certain point $c \in X$ is a strongly exposed point of S*. Indeed, we may always assume that $c = 0$ and that $\|x_0\| = 1$ (see Figure 5.6). If $x_0^* \in S_{X^*}$ is a supporting functional of B_X at x_0, we showed in Item 1 above that the diameter of the slices $S(B_X, x_0^*, \delta)$ goes to 0 as $\delta \to 0$. Thus x_0 is a strongly exposed point of B_X, and so of S. In particular, every point $x_0 \in S_X$ is a strongly exposed point of B_X.

A trivial observation is that although the concept of farthest point depends on the particular norm on X, *this is not the case for the concept of strongly exposed point.* Since Lau's Theorem 260 below guarantees the existence of a substantial number of points c in a Banach space from which a given nonempty convex and w-compact subset of a Banach space X has farthest points, we get then that *in a LUR-renormable Banach space, every nonempty w-compact convex subset of X has a strongly exposed point.* For a more precise statement in this direction see Theorem 259 and, obtained by using the same tools, Corollary 264 below.

5. **Fréchet differentiability in reflexive spaces**. We may combine these observations to prove that *the closed unit ball of every equivalent norm $|||\cdot|||$ in a reflexive Banach space $(X, \|\cdot\|)$ has at least a point of Fréchet differentiability and at least a strongly exposed point* (see Figure 5.7, and check Theorem 259 for the general result). For this, we may assume, by using Theorem 218 below, that $\|\cdot\|$ is LUR. Since $B_{(X,|||\cdot|||)}$ is w-compact, we are then in case of Item 4 above, so $B_{(X,|||\cdot|||)}$ has a strongly exposed point (strongly exposed by, say, $x_0^* \in S_{(X^*,\|\cdot\|^*)}$). Lemma 54 above shows that $|||\cdot|||^*$ is Fréchet differentiable at x_0^*. For the second assertion, interchange the rôles of X and X^*. ®

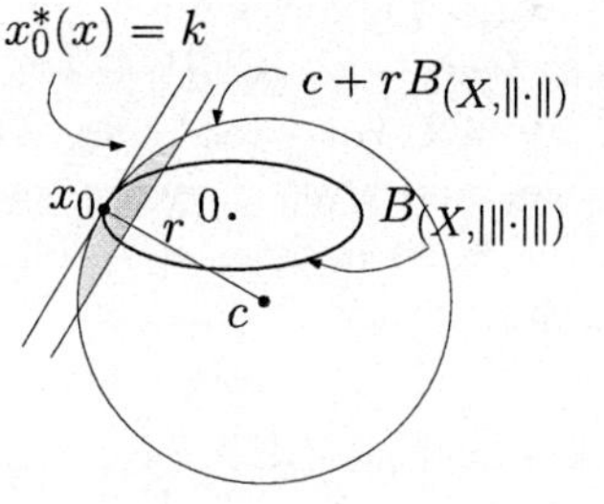

Figure 5.7: x_0 strongly exposed of $B_{(X,|||\cdot|||)}$, $|||\cdot|||^*$ Fréchet differentiable at x_0^*

5.3.3 A weak version of the local uniform rotundity property

The following concept is due to Lovaglia [Lov55], too. The $\|\cdot\|$-convergence of a sequence that characterizes the LUR property (see Proposition 87) is replaced here by w-convergence.

Definition 92. *Let $(X, \|\cdot\|)$ be a Banach space. Let $x_0 \in S_X$. The norm $\|\cdot\|$ on X is* **weakly locally uniformly rotund (WLUR)** *(also called* **weakly locally uniformly convex***)* **at** x_0 *if given a sequence $\{x_n\}_{n\in\mathbb{N}}$ in S_X such that $\|x_0 + x_n\| \to 2$, then $x_n \to x_0$ in the weak topology. If $\|\cdot\|$ is WLUR at every point of S_X, we say that $\|\cdot\|$* **is WLUR**. *Similarly, we say that the norm $\|\cdot\|^*$ on X^* is* **w^*-locally uniformly rotund** *(or* **w^*-locally uniformly convex (W*LUR)***) if for a sequence $\{x_n^*\}_{n\in\mathbb{N}}$ in S_{X^*} and a point $x_0^* \in S_{X^*}$, the condition $\|x_0^* + x_n^*\|^* \to 2$ implies that $x_n^* \to x_0^*$ in the w^*-topology.*

For a characterization of the WLUR (W*LUR) property of the norm in terms of the function $Q_{\|\cdot\|}$ defined in (5.2) below, see Proposition 116 (Proposition 117, respectively). One important consequence of the WLUR property of the norm is that it allows for a LUR renorming of the space (see Theorem 227 below).

5.4 Topological properties of the unit sphere

In Definition 89 above we introduced the KK and SKK properties of the norm of a Banach space, and we proved in Proposition 90 that if the norm of a normed space is LUR, then it is w-$\mathcal{T}_{\|\cdot\|}$-KK. It is an observation due to Troyanski, and based on Rosenthal's ℓ_1-theorem (see , e.g., [FHHMZ11, Section 5.3]), that *in a separable Banach space* $(X, \|\cdot\|)$ *without isomorphic copies of* ℓ_1, *if* $\|\cdot\|$ *is* w-$\mathcal{T}_{\|\cdot\|}$-*SKK then it is* w-$\mathcal{T}_{\|\cdot\|}$-*KK*. In fact, *the result holds also in the nonseparable case.*

➩ Indeed, and arguing by contradiction, assume that there exists $A \subset S_X$ that is closed in $(S_X, \|\cdot\|)$ although for some $x_0 \in S_X \setminus A$ we have $x_0 \in \overline{A}^w$. By Kaplanski's theorem (see, e.g., [FHHMZ11, Theorem 3.54]) there exits a countable set $A_0 \subset A$ such that $x_0 \in \overline{A_0}^w$. Put $Y := \overline{\text{span}}(A_0)$. Then $\ell_1 \not\hookrightarrow Y$, Y is separable, and it is easy to see that the closure of A_0 in $(S_Y, \|\cdot\|)$ is contained in A, so it does not contain x_0, a contradiction with the separable version of the result.

The norm $\|\cdot\|_1$ on ℓ_1 has the w-$\mathcal{T}_{\|\cdot\|_1}$-SKK property due to the Schur property (i.e., every w-convergent sequence in ℓ_1 is $\|\cdot\|_1$-convergent, see, e.g., [FHHMZ11, Theorem 5.36]). Even more, *it has the* w-$\mathcal{T}_{\|\cdot\|_1}$-KK *property* (so it gives an example of a w-$\mathcal{T}_{\|\cdot\|}$-KK norm without the LUR property). In fact, a stronger property holds: $\|\cdot\|_1$ is τ_p-$\mathcal{T}_{\|\cdot\|_1}$-KK, where τ_p denotes the topology of the pointwise convergence on ℓ_1. This holds even for $\ell_1(\Gamma)$, where Γ is an arbitrary set. An argument is provided in the annotation below.

➩ Indeed, let $\{x_i : i \in I, \leq\}$ be a net in $S_{\ell_1(\Gamma)}$ that pointwise converges to some $x \in S_{\ell_1(\Gamma)}$. Figure 5.8 below tries to convey the main idea of the proof. Given $\varepsilon > 0$, find a finite subset F of Γ such that $\sum_F |x(\gamma)| > 1 - \varepsilon$ (so $\sum_{\Gamma\setminus F} |x(\gamma)| < \varepsilon$). Find $i_0 \in I$ such that, for $i \geq i_0$, we have $|x(\gamma) - x_i(\gamma)| < \varepsilon/\text{card}\,(F)$ for $\gamma \in F$. Then, for $i \geq i_0$ and $\gamma \in F$, we have

$$\begin{aligned}\sum_{\gamma\in F} |x_i(\gamma)| &= \sum_{\gamma\in F} |x(\gamma)| + \sum_{\gamma\in F} (|x_i(\gamma)| - |x(\gamma)|) \\ &> 1 - \varepsilon - \varepsilon = 1 - 2\varepsilon,\end{aligned}$$

thus $\sum_{\Gamma\setminus F} |x_i(\gamma)| < 2\varepsilon$. Finally, for $i \geq i_0$,

$$\begin{aligned}\|x - x_i\|_1 &= \sum_F |(x - x_i)(\gamma)| + \sum_{\Gamma\setminus F} |(x - x_i)(\gamma)| \\ &\leq \sum_F |(x - x_i)(\gamma)| + \sum_{\Gamma\setminus F} |x(\gamma)| + \sum_{\Gamma\setminus F} |x_i(\gamma)| < \varepsilon + \varepsilon + 2\varepsilon,\end{aligned}$$

and this shows that $\|x - x_i\|_1 \to 0$. □

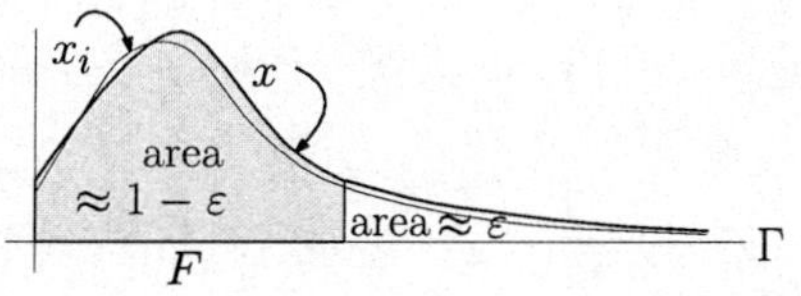

Figure 5.8: The vectors x and x_i in $\ell_1(\Gamma)$ in the proof of the coincidence of topologies

In order to give an example of a Banach space with the w-$\mathcal{T}_{\|\cdot\|}$-SKK property and failing the w-$\mathcal{T}_{\|\cdot\|}$-KK property, (necessarily a Banach space containing an isomorphic copy of ℓ_1, see the previous paragraph) we may rely on Smith's Theorem 399 and Corollary 408 below (see Remark 409). Another option is explained in Remark 404.

If $(X, \|\cdot\|)$ is a Banach space, then the norm $\|\cdot\|^*$ on X^* is w^*-lower semicontinuous. A norm $|||\cdot|||$ on X^* that is equivalent to $\|\cdot\|^*$ is, certainly, w-lower semicontinuous (its closed unit ball is w-closed); it is not, in general, w^*-lower semicontinuous (this happens if, and only if, it is a dual norm, see Proposition 17). It is useful to know that, under some circumstances, the lower semicontinuity of the norm remains when passing to a coarser topology. This is the content of the next result. To formulate it in a way that will apply to a space or to a dual space, let us consider a Banach space $(X, \|\cdot\|)$, two Hausdorff topologies τ_1 and τ_2 on X, and a nonempty subset A of X. We say that the norm $\|\cdot\|$ has the **τ_1-τ_2-KK property with respect to A** if τ_1 and τ_2 coincide on $S_X \cap A$.

Proposition 93. *Let $(X, \|\cdot\|)$ be a Banach space, and let A be a cone in X such that $\overline{A \cap B_X}^{\|\cdot\|} = B_X$. Assume that $\|\cdot\|$ has the τ_1-τ_2-KK property with respect to A for two Hausdorff locally convex topologies τ_1 and τ_2 on X such that both are coarser than $\mathcal{T}_{\|\cdot\|}$. If $\|\cdot\|$ is τ_2-lower semicontinuous, then it is also τ_1-lower semicontinuous.*

Proof. (See Figure 5.9.) Assume that $\|\cdot\|$ is not τ_1-lower semicontinuous. Thus, there exists a net $\{y_i : i \in I, \leq\}$ in B_X that τ_1-converges to some $y \in X$ with $\|y\| > 1$. Find $x \in A$ with $\|x\| = \|y\|$ and such that $\|x - y\| < (\|y\| - 1)/3$. Fix $i \in I$.

Put $x_i := y_i + (x - y)$; observe that $\|x_i\| < \|y\|$. Define a function $f_i : [0, +\infty] \to \mathbb{R}$ as $f_i(t) := \|x + t(x_i - x)\|$, for $t > 0$, and note that $f_i(1) = \|x_i\| < \|y\|$, while $f_i(t) \to +\infty$ as $t \to +\infty$. Since f_i is continuous, there exists $t_i > 1$ such that $f_i(t_i) = \|y\|$. Thus, if $z_i := x + t(x_i - x)$, we have $\|z_i\| = \|y\|$. Observe that $\|x_i - x\| \geq (2/3)(\|y\| - 1)$ and that $(1 <)\ |t_i| \leq 3\|y\|/(\|y\| - 1)$. This happens for all $i \in I$. There exists, then, a convergent subnet of $\{t_i\}_{i\in I}$ (that we may denote again by $\{t_i\}_{i\in I}$). Since $\{x_i\}_{i\in I}$ is τ_1-convergent to x, we get that $\{z_i\}_{i\in I}$ also τ_1-converges to x. Given $i \in I$ and $n \in \mathbb{N}$, there exists $\omega_{i,n} \in A$ such that $\|\omega_{i,n}\| = \|z_i\|$ $(= \|y\|)$ and $\|\omega_{i,n} - z_i\| \leq 1/n$. We endow the set $I \times \mathbb{N}$ with the upward directed preorder $\preceq$ defined by $(i, n) \preceq (j, m)$ if, and only if, $i \leq j$ and $n \leq m$. Let us show that $\{\omega_{i,n} : \ (i, n) \in I \times \mathbb{N} \ \preceq\}$ is a net τ_1-convergent to x. Indeed, given a τ_1-neighborhood U of 0, find another τ_1-neighborhood V of 0 such that $V + V \subset U$. Since $\tau_1 \prec \mathcal{T}_{\|\cdot\|}$, there exists $\varepsilon > 0$ such that $\varepsilon B_X \subset V$. Find $i_0 \in I$ such that $z_i \in x + V$ for all $i \geq i_0$. Let $n_0 \in \mathbb{N}$ such that $1/n_0 < \varepsilon$. Choose $(i, n) \in I \times \mathbb{N}$ such that $(i_0, n_0) \preceq (i, n)$. Then $\omega_{i,n} = z_i + (\omega_{i,n} - z_i) \in x + V + V \subset x + U$, so $\{\omega_{i,n} : \ (i, n) \in I \times \mathbb{N}, \ \preceq\}$ τ_1-converges to x. Since $\omega_{i,n} \in A \cap \|y\| B_X$ for all i, n and $x \in A \cap \|y\| B_X$, the τ_1-τ_2-KK with respect to A property of $\|\cdot\|$ shows that $\{\omega_{i,n} : \ (i, n) \in I \times \mathbb{N}, \ \preceq\}$ τ_2-converges to x. A similar argument (by using now τ_2 neighborhoods instead, and the fact that $\tau_2 \leq \mathcal{T}_{\|\cdot\|}$), shows that $\{z_i\}_{i\in I}$ is τ_2-convergent to x. Since $\{t_i\}_{i\in I}$ is bounded below by 1, we conclude that $\{x_i\}_{i\in I}$ τ_2-converges to x, hence $\{y_i\}_{i\in I}$ τ_2-converges to y. This is false, due to the τ_2-lower semicontinuity of the norm. □

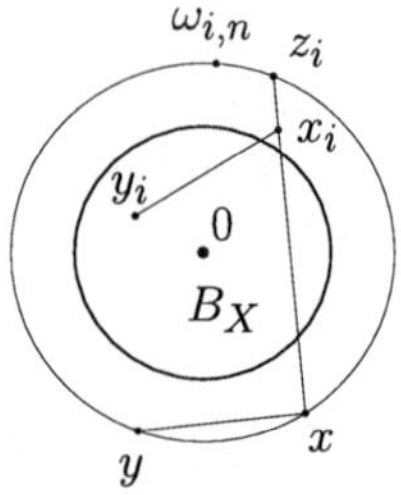

Figure 5.9: The construction in the proof of Proposition 93

The following consequence of Proposition 93 should be applied together with Raja's result [Raj02] showing that if the dual norm of a Banach space has the property w^*-w-KK, then the space X has an equivalent norm whose dual norm is LUR.

Corollary 94. *Let $(X, \|\cdot\|)$ be a Banach space. Let $|||\cdot|||$ be a norm on X^* equivalent to $\|\cdot\|^*$. Assume that the topologies w and w^* on X^* coincide on $S_{(X^*,|||\cdot|||)}$. Then $|||\cdot|||$ is a dual norm.*

➪ *A dual norm without the property in Corollary* 94. The norm $\|\cdot\|_\infty$ on ℓ_∞ is a dual norm (dual to the norm $\|\cdot\|_1$ on ℓ_1), and it has not the w^*-w-KK property. Indeed, denote by $\{e_n^*\}_{n\in\mathbb{N}}$ the canonical basis of c_0 and by $\{e_n\}_{n\in\mathbb{N}}$ the canonical basis of ℓ_1. Let $F \in \ell_\infty^*$ be a $w(\ell_\infty^*, \ell_\infty)$-cluster point of the sequence $\{e_n\}_{n\in\mathbb{N}}$; thus, given $l := (l_n) \in \ell_\infty$ we have that $\langle F, l\rangle$ is a cluster point of the sequence $\{\langle e_n, l\rangle\}_{n\in\mathbb{N}}$, i.e., of the sequence $\{l_n\}_{n\in\mathbb{N}}$. In particular, $F \in c_0^\perp$. Let $s_n^* := \sum_{k=1}^n e_k^* \in c_0$ $(\subset \ell_\infty)$ for $n \in \mathbb{N}$. The sequence $\{s_n^*\}_{n\in\mathbb{N}}$ belongs to S_{ℓ_∞}, clearly $w(\ell_\infty, \ell_1)$-converges to $s^* := (1, 1, \ldots) \in S_{\ell_\infty}$, $\langle F, s_n^*\rangle = 0$ for all $n \in \mathbb{N}$, although $\langle F, s^*\rangle = 1$. Thus, $\{s_n^*\}_{n\in\mathbb{N}}$ is not $w(\ell_\infty, \ell_\infty^*)$-convergent to s^*.

5.5 Variants of uniform convexity and local uniform convexity

Natural weakening of the uniform convexity and many other similar concepts consist in replacing the norm-convergence requirement by other more relaxed ones. We list some below.

5.5.1 2-rotundness and weakly 2-rotundness

Definition 95. *The norm of X is called* **2-rotund (2R)** *(resp.* **weakly 2-rotund (W2R)***) if whenever $\{x_n : n \in \mathbb{N}\} \subset B_X$ is such that* $\lim_{n,m\to\infty} \|x_n + x_m\| = 2$, *then for some $x \in X$, $x_n \to x$ in the norm (resp. in the weak topology).*

5.5.2 Rotundness in directions

The following concept is due to A.L. Garkavi [Gark64] (it is important, for example, in Fixed Point Theory; we shall return to this below):

Definition 96. *The norm $\|\cdot\|$ of a Banach space X is called* **uniformly rotund in every direction (URED)** *if* $\lim \|x_n - y_n\| = 0$ *whenever $x_n, y_n \in S_X$, $n = 1, 2, \ldots$ are such that* $\lim \|x_n + y_n\| = 2$ *and there is $z \in S_X$ such that $x_n - y_n \in$* span z *for each n.*

The following concept is due to M.A. Smith [Sm77]:

Definition 97. *The norm $\|\cdot\|$ of X is said to be* **uniformly rotund in weakly compact sets of directions (URWC)** *if whenever $z \in X$ and $\{x_n\}_{n=1}^\infty$ and $\{y_n\}_{n=1}^\infty$ are sequences in X such that $\|x_n\| \to 1$, $\|y_n\| \to 1$, $\|x_n + y_n\| \to 2$, and $x_n - y_n \to z$ weakly, then $z = 0$.*

Observe that if the norm is URWC , then it is URED. Indeed, assume that two sequences $\{x_n\}_{n=1}^{\infty}$ and $\{y_n\}_{n=1}^{\infty}$ in S_X satisfy $\|x_n + y_n\| \to 2$, and there is $z \in S_X$ such that $(x_n - y_n) \in \operatorname{span} z$. Take an arbitrary subsequence $\{n_k\}_{k=1}^{\infty}$ of $\mathbb{N}$. Since the sequence $\{x_{n_k} - y_{n_k}\}_{k=1}^{\infty}$ is bounded and is contained in a 1-dimensional subspace, it has a convergent subsequence that, according to the definition of URWC, converges to 0. This shows that the sequence $\{x_n - y_n\}_{n=1}^{\infty}$ converges to 0. An alternative argument uses Smith's characterization of URED and URWC norms in Theorem 99. Still another one uses Propositions 118 and 121.

The following results of Smith provide for an elegant useful insight into the concepts around uniform rotundity. We need first a definition.

Definition 98. *If X is a Banach space and $z \in X \setminus \{0\}$ and $0 < \varepsilon \leq 2$, then the* **directional modulus of rotundity** *$\delta(\varepsilon, z)$ is defined by*

$$\delta(\varepsilon, z) = \inf\left\{1 - \left\|\frac{x+y}{2}\right\| : \; x, y \in S_X, \; x - y \in \operatorname{span} z, \; \|x - y\| \geq \varepsilon\right\}.$$

If $A \subset X \setminus \{0\}$, we define (see Figure 5.10)

$$\delta(\varepsilon, A) = \inf\{\delta(\varepsilon, z) : \; z \in A\}.$$

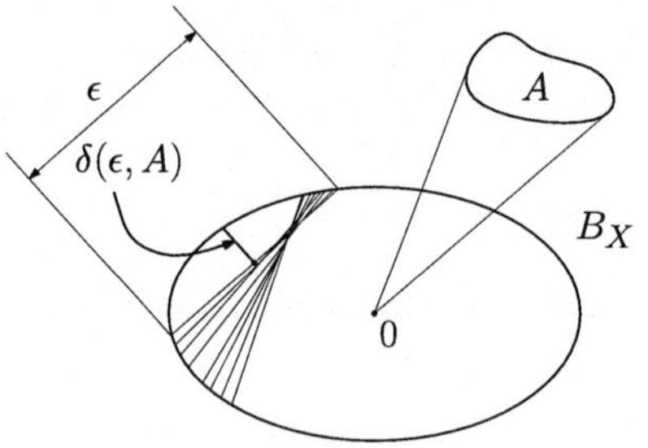

Figure 5.10: The A-directional modulus of rotundity $\delta(\varepsilon, A)$ (Definition 98)

Then we have

Theorem 99 (M.A. Smith, [Sm77]). *Let X be a Banach space. Then:*
(i) *X is UR if, and only if, $\delta(\varepsilon, S_X) > 0$ for all $0 < \varepsilon \leq 2$.*
(ii) *X is WUR if, and only if, $\delta(\varepsilon, A) > 0$ for every weakly closed and bounded subset A of $X \setminus \{0\}$ and for every $0 < \varepsilon \leq 2$.*
(iii) *X is URWC if, and only if, $\delta(\varepsilon, A) > 0$ for every weakly compact set A in $X \setminus \{0\}$ and every $0 < \varepsilon \leq 2$.*
(iv) *X is URED if, and only if, $\delta(\varepsilon, A) > 0$ for every norm compact set A in $X \setminus \{0\}$ and every $0 < \varepsilon \leq 2$.*

Proof. We sketch the proof of (ii). Assume that for some A as in the statement, and for some $\varepsilon > 0$, we have $\delta(\varepsilon, A) = 0$. Choose $z_n \in A$ and $x_n, y_n \in S_X$ such that $x_n - y_n = \alpha_n z_n$ for some $\alpha_n \in \mathbb{R}$, $\|x_n - y_n\| \geq \varepsilon$, and $\|x_n + y_n\| \geq 2 - \frac{1}{n}$. Since $\{z_n : n \in \mathbb{N}\} \subset A$, the sequence $\{z_n\}_{n=1}^{\infty}$ is bounded and bounded away from 0. Since $\varepsilon \leq |\alpha_n| \|z_n\| \leq 2$, the same is true for the sequence $\{|\alpha_n|\}_{n=1}^{\infty}$. Since $z_n \not\to 0$ weakly, it follows that $x_n - y_n \not\to 0$ weakly and thus X is not WUR.
Conversely, if X is not WUR, then there exists $\varepsilon > 0$, $f \in S_{X^*}$ and sequences $\{x_n\}_{n=1}^{\infty}$ and $\{y_n\}_{n=1}^{\infty}$ in S_X such that $\|x_n + y_n\| \to 2$ and $f(x_n - y_n) \geq \varepsilon$ for all n. Let

$$A = \{x \in X : f(x) \geq \varepsilon\} \cap 2B_X.$$

Then A is a bounded weakly closed subset of $X \setminus \{0\}$, $\|x_n - y_n\| \geq \varepsilon$, and $x_n - y_n \in A$. Hence

$$0 = \inf_n \delta(\varepsilon, x_n - y_n) \geq \delta(\varepsilon, A) \geq 0.$$

Thus the condition does not hold.
For (iv) we mention in passing a recurrent pattern: Regarding many geometrical properties, a norm-compact subset of a normed space can be "replaced" by a single point (please, do not take it literally). □

5.5.3 Near uniform convexity

Definition 100. *For a sequence $\{x_n\}_{n=1}^{\infty}$ in a Banach space X, its* **separation** *is defined by* $\operatorname{sep}\{x_n\}_{n=1}^{\infty} = \inf\{\|x_n - x_m\| : m, n \in \mathbb{N},\ m \neq n\}$.

Definition 101. *The norm $\|\cdot\|$ is called* **nearly uniformly convex (NUC)** *if for every $\varepsilon > 0$ there is $\delta < 1$ such that if $\{x_n\}_{n=1}^{\infty}$ is a sequence in B_X such that* $\operatorname{sep}\{x_n\}_{n=1}^{\infty} \geq \varepsilon$, *then there is a point $x \in \operatorname{conv}(\{x_n : n \in \mathbb{N}\})$ with $\|x\| \leq \delta$.*

5.5.4 Midpoint and weak midpoint local uniform convexity

The following concept was introduced by K.W. Anderson [And60]:

Definition 102. *The norm $\|\cdot\|$ is called* **midpoint locally uniformly rotund (MLUR)** *if whenever $x \in S_X$ and $x_n \in X$, $n \in \mathbb{N}$, are such that $\|x + x_n\| \to 1$ and $\|x - x_n\| \to 1$, then $x_n \to 0$.*

For "negative" results on Fréchet differentiability the notion of MLUR can be used (see Theorem 105 below): To see this, first of all let us show that MLUR has an equivalent form as presented in the introduction, namely:

Lemma 103. *The norm $\|\cdot\|$ in X is MLUR if, and only if, whenever $\{x_n\}_{n=1}^{\infty}$ and $\{y_n\}_{n=1}^{\infty}$ are sequences in X and $x \in X$ is such that $\|x_n\| \to \|x\|$, $\|y_n\| \to \|x\|$, and $\|(x_n + y_n)/2 - x\| \to 0$, then $\|x_n - y_n\| \to 0$.*

Proof. For a nonzero vector $z \in X$, put $u(z) := z/\|z\|$. Assume first that $\|\cdot\|$ is MLUR, and let $\{x_n\}_{n=1}^{\infty}$, $\{y_n\}_{n=1}^{\infty}$ and x as in the statement. If $x = 0$ we easily get $x_n \to 0$ and $y_n \to 0$, and so $\|x_n - y_n\| \to 0$. Assume then $x \neq 0$ (hence we may assume $x_n \neq 0$ and $y_n \neq 0$ for all n). Put $z_n := (u(x_n) + u(y_n))/2$ for $n \in \mathbb{N}$. Since $\|u(x) - z_n\| \to 0$, we get $\|u(x) + [u(x_n) - z_n]\| = \|[u(x) - z_n] + u(x_n)\| \to 1$, and $\|u(x) - [u(x_n) - z_n]\| = \|[u(x) + [u(y_n) - z_n]\| = \|[u(x) - z_n] + u(y_n)\| \to 1$. From this it follows that $\|u(x_n) - u(y_n)\| \to 0$. Since $\|x_n\| \to \|x\|$ and $\|y_n\| \to \|x\|$, we get $\|x_n - y_n\| \to 0$.
Assume now that the conclusion of the lemma holds. Let $x \in X$ and $\{x_n\}_{n=1}^{\infty}$ in X be such that $\|x \pm x_n\| \to 1$. Then $[(x + x_n) + (x - x_n)]/2 = x$, so we get $\|2x_n\| \to 0$, i.e., $x_n \to 0$. □

Remark 104. Obviously, in Lemma 103 we may restrict our attention to sequences $\{x_n\}_{n=1}^{\infty}$, $\{y_n\}_{n=1}^{\infty}$ in S_X, and to $x \in S_X$. ®

Theorem 105. *If $(X, \|\cdot\|)$ is a space and $\|\cdot\|^*$ on X^* is Fréchet differentiable, then $\|\cdot\|$ on X is MLUR (and the space X is reflexive).*

Proof. Let $x_n, y_n, z \in S_X$, $n \in \mathbb{N}$, are such that $\|\frac{1}{2}(x_n + y_n) - z\| \to 0$. Let $f \in S_{X^*}$ be such that $f(z) = 1$. Since $f(x_n) \leq 1$ and $f(y_n) \leq 1$ for all $n \in \mathbb{N}$, we have that necessarily $f(x_n) \to 1$ and $f(y_n) \to 1$. Since the dual norm is Fréchet differentiable, from Šmulyan's Lemma 54 we get $\|x_n - y_n\| \to 0$.
For the reflexivity of X, see Theorem 144 below. □

We will also need the following

Theorem 106. *If the norm is LUR, it is MLUR.*

Proof. As in the proof of the implication that the dual norm LUR implies the norm is Fréchet differentiable (Proposition 91), it follows that if the norm is LUR and $z \in S_X$ is such that for some $f \in S_{X^*}$ we have $f(z) = 1$, then f is a point of Fréchet differentiability of the dual norm. Let x_n, $y_n \in S_X$, $n \in \mathbb{N}$, be such that $\|\frac{1}{2}(x_n + y_n) - z\| \to 0$. Then $\frac{1}{2}f(x_n + y_n) \to f(z) = 1$. Since $f(x_n) \leq 1$ and $f(y_n) \leq 1$ for all $n \in \mathbb{N}$, we have that necessarily $f(x_n) \to 1$ and $f(y_n) \to 1$. Since the dual norm is Fréchet differentiable at the point f, from Šmulyan's Lemma 54 we get $\|x_n - y_n\| \to 0$. □

Note that a MLUR norm means that every $x \in S_X$ is a strongly extreme point ([DoHuSm95]) (see Definition 11).

Definition 107. *If in the definition of MLUR we may only conclude that $x_n \to 0$ in the weak topology, we speak then of* **weakly midpoint locally uniformly rotund** *norms (***WMLUR***).*

5.6 Using the square of the norm

The use of the square of the norm in the description of some of its geometric properties is a wonderful device, as it was proved in the 70's of the last century. Maybe the clue for this is the simple inequality (5.3) in Lemma 108 below. It originates in the particular behaviour of the Euclidean norm $\|\cdot\|_2$. Observe that the Parallelogram Identity in a Hilbert space gives $\|x-y\|_2^2 = 2\|x\|_2^2 + 2\|y\|_2^2 - \|x+y\|_2^2$. Moreover, $\big|\|x\|_2 - \|y\|_2\big| \le \|x-y\|_2$, so

$$0 \le (\|x\|_2 - \|y\|_2)^2 \le 2\|x\|_2^2 + 2\|y\|_2^2 - \|x+y\|_2^2. \tag{5.1}$$

Although for a general norm —even less for a seminorm— in a Banach space we cannot expect something like the Parallelogram Identity, it turns out that the inequality (5.1) above holds (see Lemma 108) just by asking that the involved function should be *subadditive*: If X is a Banach space, we say that a function $p : X \to \mathbb{R}$ is **subadditive** whenever $p(x+y) \le p(x) + p(y)$ for all $x, y \in X$.

Due to the fact that the formula $2p^2(x)+2p^2(y)-p^2(x+y)$ for an arbitrary function $p : X \to \mathbb{R}$ —present in inequality (5.1) above for the case of the Euclidean norm— will often appear in subsequent arguments, we codify it by using the following expression:

$$\boxed{\text{For } x, y \in X, \text{ put } Q_p(x,y) := 2p^2(x) + 2p^2(y) - p^2(x+y).} \tag{5.2}$$

The two following lemmata, despite their elementary nature, hint at the crucial rôle played, in many of the computations that follow, by the procedure of squaring a nonnegative function p. The first one considers the case when p is subadditive, and the second one when p is convex (in particular, both apply to a seminorm in a Banach space). They are very similar. Equation (5.4) will appear, in particular, in the relevant Lemma 426 and subsequent computations.

Lemma 108. *Let X be a normed space. Let $p : X \to [0, +\infty)$ be a subadditive function. Then*

$$0 \le [p(x) - p(y)]^2 \le Q_p(x,y) \quad \textit{for all } x, y \in X. \tag{5.3}$$

Proof. This is a consequence of the following computations:

$$\begin{aligned} 0 &\le (p(x) - p(y))^2 = p^2(x) + p^2(y) - 2p(x)p(y) \\ &= p^2(x) + p^2(y) + \left[p^2(x) + p^2(y) - (p(x) + p(y))^2\right] \\ &\le 2p^2(x) + 2p^2(y) - p(x+y)^2 = Q_p(x,y), \end{aligned}$$

where the last inequality follows from the subadditivity of p. □

Lemma 109. *Let X be a normed space. Let $p : X \to [0, +\infty)$ be a convex function. Then*

$$0 \leq \frac{1}{4}[p(x) - p(y)]^2 \leq \frac{1}{2}p^2(x) + \frac{1}{2}p^2(y) - p^2\left(\frac{x+y}{2}\right) \quad \textit{for all } x, y \in X. \tag{5.4}$$

Proof. This is a consequence of the following computations:

$$\begin{aligned} 0 &\leq \frac{1}{4}[p(x) - p(y)]^2 = \frac{1}{4}[p^2(x) + p^2(y) - 2p(x)p(y)] \\ &= \frac{1}{4}\left[p^2(x) + p^2(y) + \left[p^2(x) + p^2(y) - (p(x) + p(y))^2\right]\right] \\ &\leq \frac{1}{4}\left[2p^2(x) + 2p^2(y) - 4p^2\left(\frac{x+y}{2}\right)\right] \\ &= \frac{1}{2}p^2(x) + \frac{1}{2}p^2(y) - p^2\left(\frac{x+y}{2}\right), \end{aligned}$$

where the last inequality follows from the convexity of p. □

5.6.1 Some useful computations using $\|\cdot\|^2$, and some tools

Lemma 110. (See, e.g., [DeGoZi93, Fact II.2.3 (i)]) *If p is a subadditive function from a normed space X into $[0, +\infty)$, and x, x_j, $j = 1, 2, \dots$ are elements in X, then the following are equivalent:*
(i) $\lim_j p(x_j) = p(x)$ *and* $\lim_j p(x + x_j) = 2p(x)$.
(ii) $\lim_j Q_p(x, x_j) = 0$.

Proof. (i)⇒(ii) is trivial. The reverse implication follows directly from (5.3) above. □

Another useful tool in renorming is the following (see, e.g., [DeGoZi93, Fact II.2.3 (ii)]; both in Lemma 111 and in Remark 112 we assume that the sequence $\{\alpha_n\}_{n\in\mathbb{N}}$ is such that the function p is well defined):

Lemma 111. *Let X be a space. Let $\{\alpha_n\}_{n\in\mathbb{N}}$ be a sequence of positive real numbers and $\{p_n\}_{n\in\mathbb{N}}$ a sequence of subadditive functions from X into $[0, +\infty)$. Put*

$$\boxed{p^2 := \sum_n \alpha_n p_n^2.} \tag{5.5}$$

Then, for each $n \in \mathbb{N}$, and $x, y \in X$, we have $Q_{p_n}(x, y) \leq \alpha_n^{-1} Q_p(x, y)$. Thus, if $Q_p(x, y) = 0$ for some $x, y \in X$, then $Q_{p_n}(x, y) = 0$ for $n \in \mathbb{N}$. If $x \in X$ and $\{x_j\}_{j\in\mathbb{N}}$ is a sequence in X such that $\lim_j Q_p(x, x_j) = 0$, then $\lim_j Q_{p_n}(x, x_j) = 0$ for each $n \in \mathbb{N}$.

Proof. Note that $Q_p(x,y) = \sum_{k=1}^{\infty} \alpha_k Q_{p_k}(x,y)$ for all $x, y \in X$ (assuming convergence; if not, both terms are $+\infty$). The non-negativity of each Q_n shows that $Q_{p_n}(x,y) \le \alpha_n^{-1} \sum_{k=1}^{\infty} \alpha_k Q_{p_k}(x,y) = \alpha_n^{-1} Q_p(x,y)$. Note, too, that if for positive numbers a_j^n, $(j, n \in \mathbb{N})$, it happens that $\sum_n \alpha_n a_j^n \to 0$ as $j \to +\infty$, then, for each $n \in \mathbb{N}$, we have $\lim_j a_j^n = 0$. □

Remark 112. Regarding Lemma 111, observe that, due to the fact that the functions $p_n : X \to [0, \infty)$, $n \in \mathbb{N}$, are subadditive, then p *defined by* (5.5) *above is subadditive, too.* In order to show this, put $P(x) := (\sqrt{\alpha_n} p_n(x))_{n=1}^{\infty}$ for $x \in X$. Then $p(x) = \|P(x)\|_2$, where $\|\cdot\|_2$ denotes the canonical norm in ℓ_2. Thus

$$p(x+y) = \|P(x+y)\|_2 \le \|P(x) + P(y)\|_2 \le \|P(x)\|_2 + \|P(y)\|_2 = p(x) + p(y),$$

where the first inequality above comes from the subadditivity of each p_n, and the second one is the triangle inequality of the $\|\cdot\|_2$ norm in ℓ_2. In particular, if each p_n is a **seminorm**, then p is also a seminorm on X. Ⓡ

5.6.2 Describing rotundity

The usefulness of the function Q_p defined in (5.2) above (specially when p is the norm of a Banach space) appears when describing several convexity properties by means of this wonderful device. For later reference, let us list here some of them. We shall include only proofs for the first simple case, and for those requiring a not-completely straightforward argument:

Strict convexity

Proposition 113. *Let* $(X, \|\cdot\|)$ *be a normed space. The following are equivalent:*
(i) $\|\cdot\|$ *is strictly convex.*
(ii) *If* $x, y \in X$ *satisfy* $Q_{\|\cdot\|}(x,y) = 0$, *then* $x = y$.

Proof. (i) ⇒(ii): Equation (5.3) above shows that if $Q_{\|\cdot\|}(x,y) = 0$, then $\|x\| = \|y\|$. Hence we may assume that $x, y \in S_X$; we get $\|x+y\| = 2$ and (ii) implies $x = y$. (ii)⇒(i): If there exists $z \in S_X$ such that $z = (1/2)(x+y)$, $x, y \in S_X$ then $Q_{\|\cdot\|}(x,y) = 0$ and so $x = y$. □

Uniform convexity

Proposition 114. *Let* $(X, \|\cdot\|)$ *be a Banach space. The following are equivalent:*
(i) *The norm* $\|\cdot\|$ *is uniformly convex.*
(ii) *If* $\{x_n\}_{n\in\mathbb{N}}$ *and* $\{y_n\}_{n\in\mathbb{N}}$ *are sequences in* X *such that* $\{x_n\}_{n\in\mathbb{N}}$ *is bounded and* $Q_{\|\cdot\|}(x_n, y_n) \to 0$, *then* $\|x_n - y_n\| \to 0$.

Note that in (ii) above, one of the sequences (and then the other) has to be bounded, see [GuiHá07a].

Local uniform convexity

Proposition 115. *Let $(X, \|\cdot\|)$ be a Banach space. The following are equivalent:*
(i) *The norm $\|\cdot\|$ is locally uniformly convex.*
(ii) *If $x \in X$ and $\{x_n\}_{n\in\mathbb{N}}$ is a sequence in X such that $Q_{\|\cdot\|}(x, x_n) \to 0$, then $\|x - x_n\| \to 0$.*

WUR and WLUR

Proposition 116. *Let $(X, \|\cdot\|)$ be a Banach space.*
(i) *The norm $\|\cdot\|$ is WUR if, and only if, given two sequences $\{x_n\}$ and $\{y_n\}$ in X, $\{x_n\}$ bounded, such that $Q_{\|\cdot\|}(x_n, y_n) \to 0$, then $(x_n - y_n) \xrightarrow{w} 0$.*
(ii) *The norm $\|\cdot\|$ is WLUR if, and only if, given $x \in X$ and a sequence $\{x_n\}$ in X such that $Q_{\|\cdot\|}(x, x_n) \to 0$, then $x_n \xrightarrow{w} x$.*

Proof. This is a consequence of (5.3) (see, e.g., [DeGoZi93, Proposition II.6.2]). □

W*UR and W*LUR

Proposition 117. *Let $(X, \|\cdot\|)$ be a Banach space.*
(i) *The norm $\|\cdot\|^*$ on X^* is W*UR if, and only if, given sequences $\{x_n^*\}_{n\in\mathbb{N}}$ and $\{y_n^*\}_{n\in\mathbb{N}}$ in X^*, $\{x_n^*\}_{n\in\mathbb{N}}$ bounded, and such that $Q_{\|\cdot\|^*}(x_n^*, y_n^*) \to 0$, then $(x_n^* - y_n^*) \xrightarrow{w^*} 0$.*
(ii) *The norm $\|\cdot\|^*$ on X^* is W*LUR if, and only if, given $x^* \in X^*$ and a sequence $\{x_n^*\}_{n\in\mathbb{N}}$ in X^* such that $Q_{\|\cdot\|^*}(x^*, x_n^*) \to 0$, then $x_n^* \xrightarrow{w^*} x^*$.*

URED

Proposition 118. *The norm $\|\cdot\|$ on a Banach space X is URED if, and only if, whenever $x_n, y_n \in X$, $n = 1, 2, \ldots$ are such that $\lim_n Q_{\|\cdot\|}(x_n, y_n) = 0$, $\{x_n\}_{n\in\mathbb{N}}$ is bounded, there is $z \in X \setminus \{0\}$ and real numbers λ_n, $n = 1, 2, \ldots$ which satisfy $x_n - y_n = \lambda_n z$ for each n, then $\lim_n \lambda_n = 0$.*

Proof. We will make use of the following simple assertion in which, for $x, y \in X$, we denote by $[x, y]$ the closed linear segment between x and y, i.e., $[x, y] = \{tx + (1-t)y : 0 \le t \le 1\}$.

Lemma 119. *Let $(X, \|\cdot\|)$ be a Banach space: Let $x_n, y_n \in S_X$, and $z_n \in [x_n, y_n]$, $n \in \mathbb{N}$, be such that $\lim \|z_n\| = 1$ and*

$$\inf_n \big\{ \min\{\|x_n - z_n\|, \|y_n - z_n\|\} \big\} > 0. \tag{5.6}$$

Then $\lim_n \min\{\|tx_n + (1-t)y_n\| : 0 \le t \le 1\} = 1$.

Proof of Lemma 119. Let $z_n = t_n x_n + (1-t_n)y_n$, $t_n \in [0,1]$. In view of (5.6) we can choose $\delta > 0$ such that $\delta < t_n < 1-\delta$ for each n. For every n, choose $f_n \in S_{(X^*,\|\cdot\|^*)}$ such that $f_n(z_n) = \|z_n\|$. Then, for each n,

$$\|z_n\| = f_n(z_n) = t_n f_n(x_n) + (1-t_n) f_n(y_n).$$

Since $\lim \|z_n\| = 1$, $0 < \delta < t_n < 1-\delta$, $f_n(x_n) \le 1$, and $f_n(y_n) \le 1$ for each n, it follows that $\lim f_n(x_n) = \lim f_n(y_n) = 1$. Hence for every $\varepsilon > 0$ there is $n_0 \in \mathbb{N}$ such that $f_n(x_n) > 1-\varepsilon$ and $f_n(y_n) > 1-\varepsilon$ for $n \ge n_0$. Therefore if $n \ge n_0$ and $t \in [0,1]$, then $\|tx_n + (1-t)y_n\| \ge f_n\big(tx_n + (1-t)y_n\big) \ge 1-\varepsilon$. The conclusion follows. □

Another ingredient in the proof of Proposition 118 is the following lemma.

Lemma 120. *Let the norm $\|\cdot\|$ on a Banach space $(X, \|\cdot\|)$ be URED. If $z \in S_X$, $\{x_n\}_{n\in\mathbb{N}}$ and $\{y_n\}_{n\in\mathbb{N}}$ are sequences in B_X such that $\lim_n \|x_n + y_n\| = 2$, and for all $n \in \mathbb{N}$ we have $x_n - y_n = \lambda_n z$, where $\{\lambda_n\}_{n\in\mathbb{N}}$ is a sequence of real numbers, then $\lim_n \lambda_n = 0$.*

Proof of Lemma 120. Assume that there are $x_n, y_n \in B_X$ and $\varepsilon > 0$ such that $\lim \|x_n + y_n\| = 2$, $x_n - y_n = \lambda_n z$, $z \in S_X$ and $|\lambda_n| \ge \varepsilon > 0$ for each n.

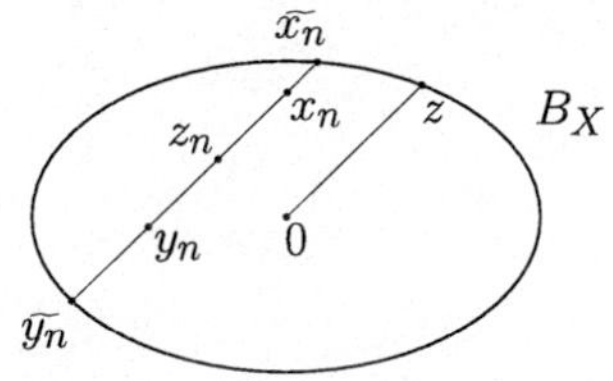

Figure 5.11: The construction in Lemma 120

Let $\widetilde{x}_n, \widetilde{y}_n \in S_X$ be such that $[x_n, y_n] \subset [\widetilde{x}_n, \widetilde{y}_n]$ and put $z_n = (1/2)(x_n + y_n)$ (see Figure 5.11). Then $\lim \|z_n\| = 1$ and $\inf_n \{\min(\|z_n - \widetilde{x}_n\|, \|z_n - \widetilde{y}_n\|)\} \ge (1/2)\inf\{\|x_n - y_n\|\} \ge \varepsilon/2$. It follows from Lemma 119 that $\lim \|(1/2)(\widetilde{x}_n + \widetilde{y}_n)\| = 1$. Since for each n, $\widetilde{x}_n - \widetilde{y}_n = \beta_n z$ for some $|\beta_n| \ge |\lambda_n| \ge \varepsilon$, it follows that $\|\cdot\|$ does not have the URED property, so the proof of Lemma 120 is complete. □

To finish the proof of Proposition 118, let us assume that the norm $\|\cdot\|$ is URED but does not satisfy the condition. Then there are $x_n, y_n \in X$, $n \in \mathbb{N}$ and $\varepsilon > 0$ such that $\lim(2\|x_n\|^2 + 2\|y_n\|^2 - \|x_n + y_n\|^2) = 0$, $\{x_n\}$ is bounded, $x_n - y_n = \lambda_n z$ for some $z \in X \setminus \{0\}$ and $|\lambda_n| \ge \varepsilon$ for each n. Since $\{x_n\}_{n\in\mathbb{N}}$ is bounded and since $\lim_n(\|x_n\| - \|y_n\|) = 0$ (see Lemma 108 below), we may assume without loss of generality that $\lim \|x_n\| = \lim \|y_n\| = 1$. Put, for $n \in \mathbb{N}$, $\delta_n := \big(\max(\|x_n\|, \|y_n\|)\big)^{-1}$ and

$$x'_n := \delta_n x_n \quad \text{and} \quad y'_n := \delta_n y_n, \text{ for } n \in \mathbb{N}.$$

Then for each n, x'_n, $y'_n \in B_X(\|\cdot\|)$, $\lim \|x'_n + y'_n\| = 2$, $x'_n - y'_n = \delta_n \lambda_n z$, and $|\delta_n \lambda_n| > \varepsilon/2$ for large n. Since we assumed that $\|\cdot\|$ is URED we obtained a contradiction. Therefore the norm $\|\cdot\|$ satisfies the condition whenever $\|\cdot\|$ is URED. The converse implication is straightforward. □

Uniform convexity in a weakly compact set of directions

Proposition 121. *The norm $\|\cdot\|$ of a Banach space has property URWC if, and only if, for a bounded sequence $\{x_n\}_{n\in\mathbb{N}}$ in X, another sequence $\{y_n\}_{n\in\mathbb{N}}$ in X such that $x_n - y_n \xrightarrow{w} z$ for some $z \in X$ we have $Q_{\|\cdot\|}(x_n, y_n) \to 0$, then $z = 0$.*

5.7 Extension of rotund norms

Let $(X, \|\cdot\|)$ be a Banach space and $Y \subset X$ a closed subspace. Denote by $\|\cdot\|_Y$ the restriction of $\|\cdot\|$ to Y. If $|||\cdot|||_Y$ is a norm on Y equivalent to $\|\cdot\|_Y$, then there exists an equivalent norm $|||\cdot|||$ on X that extends $|||\cdot|||_Y$, i.e., $|||y||| = |||y|||_Y$ for every $y \in Y$.

➪ We may always assume that $|||\cdot|||_Y \le \|\cdot\|_Y$. Put $B := \overline{\text{conv}}\,(B_{(Y,|||\cdot|||_Y)} \cup B_{(X,\|\cdot\|)})$. B is the closed unit ball of a norm $|||\cdot|||$ on X that satisfies the requirements (see Figure 5.12).

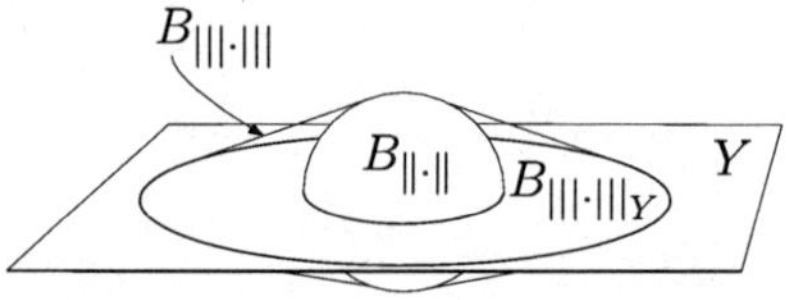

Figure 5.12: Extending a norm

Extra work is needed in case we want that $|||\cdot|||$ would keep some of the convexity or smoothness properties that $\|\cdot\|_Y$ may have.

In 1996, Wee-Kee Tang [WKT96b] improved results by K. John and V. Zizler, R. Deville, G. Godefroy and V. Zizler, and M. Fabian, on extension of rotund norm. We present here his theorem and the proof that shows that the 2-sum technique of Clarkson and Day (see Section 9.2 below) can be used with general convex functions —not only norms— whose sublevel sets are equivalent balls.

Theorem 122 (Wee-Kee Tang, Fabian, [WKT96b]). *Let Y be a closed subspace of a Banach space X. Suppose that both X and Y admit R (UR, LUR, WUR, URED, w-$\mathcal{T}_{\|\cdot\|}$-SKK) norms, say $\|\cdot\|$ and $|\cdot|_Y$, respectively. Then $|\cdot|_Y$ can be extended to a R (UR, LUR, WUR, URED, w-$\mathcal{T}_{\|\cdot\|}$-SKK, respectively) norm on X.*

➪ *Proof.* (Sketch) Let $|\cdot|$ be any extension of $|\cdot|_Y$ (see the previous paragraph). Without loss of generality we may assume that $\|\cdot\| \leq (1/\sqrt{2})|\cdot|$. Define the function $p(\cdot)$ on $B_{(X,|\cdot|)}$ by

$$p^2(x) = |x|^2 + \text{dist}(x, Y)^2 + q(x),$$

where $q(x) := \text{dist}(x, Y)^2 e^{\|x\|^2}$. The function q is convex on $B_{(X,\|\cdot\|)}$, so p is also convex (and symmetric). Its sublevel set of height 1 defines a set B whose Minkowski functional is an equivalent norm $|||\cdot|||$ on X.
We sketch below the proof of the UR case. The other cases are proven similarly:
Let $\{x_n\}_{n\in\mathbb{N}}$ and $\{y_n\}_{n\in\mathbb{N}}$ be two sequence in $S_{(X,|||\cdot|||)}$ whose midpoint goes to the unit sphere. Then $Q_p(x_n, y_n) \to 0$, where Q_p was defined in (5.2). From Lemma 111 we deduce that $Q_{|\cdot|}(x_n, y_n) \to 0$, $Q_{\text{dist}(\cdot,Y)}(x_n, y_n) \to 0$ and $Q_q(x_n, y_n) \to 0$. Lemma 108 shows that

$$\lim_n \big(\text{dist}\,(x_n, Y) - \text{dist}\,(y_n, Y)\big) \to 0, \text{ and}$$

$$\lim_n \left(\text{dist}\left(\frac{x_n + y_n}{2}, Y\right) - \text{dist}\,(x_n, Y)\right) = 0.$$

Assuming that there exists $\varepsilon > 0$ and a subsequence of $\{x_n - y_n\}_{n\in\mathbb{N}}$, denoted again as $\{x_n - y_n\}_{n\in\mathbb{N}}$, such that $|||x_n - y_n||| > \varepsilon$ for all $n \in \mathbb{N}$.
We consider two cases:
(I) $\text{dist}\,(x_n, Y) \to 0$. This gives readily a contradiction with the fact that $|\cdot|_Y$ is UR.
(II) $\limsup_n \text{dist}\,(x_n, Y) = d > 0$. There exists then a subsequence (denoted again $\{x_n\}_{n\in\mathbb{N}}$) such that $\text{dist}\,(x_n, Y) = d$. Thus $Q_q(x_n, y_n) \to 0$, hence $(e^{\|x_n\|^2} + e^{\|y_n\|^2} - 2e^{\|(x_n+y_n)/2\|^2}) \to 0$. The expression of the exponential as a series shows that each summand converges to 0; in particular, $Q_{\|\cdot\|}(x_n, y_n) \to 0$. The uniform rotundity of $\|\cdot\|$ and the last limit gives a contradiction.

In this connection, see also [Fab95].

The result in Theorem 122 fails for Gâteaux differentiability [Ziz89], see Theorem 703. For Fréchet differentiability is open, even in the separable case .

The following theorem is a special case of Theorem 122. Its proof is different; the argument is used in some questions regarding smoothness.

Theorem 123 (K. John, V. Zizler, [JoZi76] see, e.g., [DeGoZi93], Proposition II.8.2). *Let Y be a subspace of a separable Banach space X and let $|\cdot|$ be an equivalent LUR norm on Y. Then $|\cdot|$ can be extended to a LUR norm on X.*

Chapter 6

Some structural properties of Banach spaces

In this chapter we shall introduce, for later references, some structures in a Banach space that allow for vector representation and computations —a coordinate system—, and consequent classification of spaces in such terms. They will also allow for renormings of the Banach space or its dual.

6.1 Biorthogonal systems, Markushevich bases, Schauder bases, and projectional resolutions of the identity

6.1.1 Biorthogonal systems

Definition 124. *If X is a Banach space, a* **biorthogonal system** $\{x_\gamma, f_\gamma\}_{\gamma\in\Gamma}$ *for X is a subset of $X \times X^*$ such that $\langle x_\alpha, f_\beta\rangle = \delta_{\alpha,\beta}$ for all $\alpha, \beta \in \Gamma$, where $\delta_{\alpha,\alpha} = 1$ and $\delta_{\alpha,\beta} = 0$ for $\alpha \neq \beta$. It is called an* **Auerbach system** *if, moreover, $\|x_\gamma\| = \|f_\gamma\| = 1$ for all $\gamma \in \Gamma$.*

For the use of this concept in renormings, see, e.g., Theorem 570.

Definition 125. *A biorthogonal system $\{x_\gamma, f_\gamma\}_{\gamma\in\Gamma}$ for a Banach space X is called* **fundamental** *if $\overline{\text{span}}\,\{x_\gamma : \gamma \in \Gamma\} = X$. It is called* **total** *if $\text{span}\,\{f_\gamma : \gamma \in \Gamma\}$ is w^*-dense in X^*.*

The concept of a fundamental biorthogonal system appears scattered along this text. We shall complete the references to it in Section 45.10 below. It can be successfully used in renorming (see, e.g., Theorems 223 and 648).

6.1.2 Markushevich bases

Definition 126. *A fundamental and total biorthogonal system for a Banach space X is said to be a* **Markushevich basis** *(for short, an* **M-basis***). If the M-basis is, moreover, an Auerbach system, it is called an* **Auerbach basis**.

A. J. Guirao et al., *Renormings in Banach Spaces*, Monografie Matematyczne 75,
https://doi.org/10.1007/978-3-031-08655-7_6

Every separable Banach space has an M-basis (Markushevich, see, e.g., [FHHMZ11, Theorem 4.59]). It is an open problem whether it always has an Auerbach system. The existence of M-bases is guaranteed in a number of wide classes of nonseparable Banach spaces (see, e.g., Theorems 135 and 315, (iii) and (v) in Theorem 774, and (vii), (viii) and (ix) in Theorem 827).
M-bases will be constructed in some nonseparable spaces by transfinite inductive methods based on the existence of projectional resolutions of the identity (see, e.g., Subsections 6.1.4 and 6.1.5, the proof of Theorem 306, and (iii) and (v) in Theorem 774).

The use of M-bases in renorming will appear below (see, e.g., Theorems 218, 222, 314, Example 432, and (v) in Theorem 827).

Remark 127. If a Banach space X has an M-basis $\{x_\gamma, f_\gamma\}_{\gamma\in\Gamma}$, then there exists a continuous linear and one-to-one mapping into $c_0(\Gamma)$. Indeed, if $\{x_\gamma, f_\gamma\}_{\gamma\in\Gamma}$ is an M-basis of X, we may always assume that $\|f_\gamma\| = 1$ for all $\gamma \in \Gamma$. The mapping $T : X \to \ell_\infty(\Gamma)$ given by $Tx = (\langle x, f_\gamma\rangle)$ is clearly linear, bounded, and one-to-one. Moreover, $T(x_\gamma) = e_\gamma$ for all $\gamma \in \Gamma$, so $T(\text{span}\,\{x_\gamma : \gamma \in \Gamma\}) \subset c_0(\Gamma)$, hence $T(X) \subset c_0(\Gamma)$. In particular, the set

$$\Gamma(x) := \{\gamma \in \Gamma : \langle x, f_\gamma\rangle \neq 0\} \tag{6.1}$$

is countable for every $x \in X$. The existence of such a mapping T, together with the possibility to endow $c_0(\Gamma)$ with a "good" norm (see, e.g., Subsections 9.4.1 and 9.4.2, and Sections 10.7 and 21.1, Theorems 505, 506, 810, and 830), and the use of "transfer" techniques (to be described later) allow for renormings of spaces having M-bases or similar structures. ®

The space $\ell_\infty(\Gamma)$, for an infinite set Γ, has a fundamental biorthogonal system (Rosenthal, see, e.g., [HMVZ07, Corollary 4.23], and Theorem 770 below), yet it has no M-basis (Johnson, see, e.g., [HMVZ07, Theorem 5.12]). We advice the reader to check, e.g., [HMVZ07, Chapter 1].

The following definition introduces some particular types of M-bases. They allow for a precise characterization of some important classes of spaces, in the sense that each of them is described by the existence of an M-basis of a certain type. See Theorem 135 below.

Definition 128. *A Markushevich basis* $\{x_\gamma, f_\gamma\}_{\gamma\in\Gamma}$ *for a Banach space* X *is called*
(i) **strong** *if every* $x \in X \setminus \{0\}$ *belongs to* $\overline{\operatorname{span}}\{x_\gamma :\ \gamma \in \Gamma(x)\}$, *where* $\Gamma(x)$ *is given in* (6.1) *above,*
(ii) **shrinking** *if* $\overline{\operatorname{span}}\{f_\gamma :\ \gamma \in \Gamma\} = X^*$,
(iii) **σ-shrinkable** *if* $G := \{x_\gamma :\ \gamma \in \Gamma\} = \bigcup_{n=1}^\infty G_n$ *so that for every neighborhood* U *of* 0 *in* $(X^{**}, \|\cdot\|^{**})$ *and for every* $x \in G$, *there is* $n \in \mathbb{N}$ *such that* $x \in G_n$ *and* $G'_n \subset U$, *where* G'_n *denotes the set of all accumulation points of* G_n *in* (X^{**}, w^*),
(iii) **weakly σ-shrinkable** *if* U *in* (iii) *above is an arbitrary* w^**-neighbourhood of* 0 *in* X^{**}.
(iv) **weakly compact** *(***weakly Lindelöf***) if* $\{x_\gamma :\ \gamma \in \Gamma\} \cup \{0\}$ *is weakly compact (respectively, weakly Lindelöf),*
(v) **norming** *if* $\overline{\operatorname{span}}\{f_\gamma :\ \gamma \in \Gamma\}$ *is a norming subspace of* X^* (see Subsection 3.4.1 above).

Observe that if $\{x_\gamma, f_\gamma\}_{\gamma\in\Gamma}$ is a shrinking M-basis of X such that the set $V := \{x_\gamma :\ \gamma \in \Gamma\}$ is bounded, then V is w-relatively compact.

➪ Indeed, let $x^{**} \in X^{**}$ be a w^*-cluster point of V. Then, $\langle x^{**}, f_\gamma\rangle = 0$ for each $\gamma \in \Gamma$. Since $\{f_\gamma :\ \gamma \in \Gamma\}$ is $\|\cdot\|$-dense in X^*, we get that $x^{**} = 0$.

6.1.3 Schauder bases

The following particular class of M-bases in separable Banach spaces is a central object in the theory: If $(X, \|\cdot\|)$ is a separable Banach space, a **(Schauder) basis** is a sequence $\{x_n\}_{n\in\mathbb{N}}$ of vectors in X such that every $x \in X$ can be uniquely written as $\sum_{n=1}^\infty a_n x_n$, where $\{a_n\}_{n\in\mathbb{N}}$ is a sequence in $\mathbb{R}$ and the convergence is in $\|\cdot\|$. The functions $f_n : X \to \mathbb{R}$ defined by $f_n(x) = a_n$ for $x = \sum_{n=1}^\infty a_n x_n \in X$, $n \in \mathbb{N}$, are called the **functional coefficients** or **coordinate functionals**. They form a bounded set in X^*. The mappings $P_n := \sum_{k=1}^n f_k . x_k : X \to X$ are continuous linear projections (called the **natural projections associated to a basis**), and $\sup\{\|P_n\| : n \in \mathbb{N}\}$ is called the **basis constant**; it is finite. In contrast with the situation with M-bases, not every separable space has a basis (P. Enflo). For information on bases see, e.g., [LinTz77, Section 1.a], or [FHHMZ11, Chapter 4].

Characterizations of structural properties of Banach spaces with bases in terms of their qualities —to be shrinking, boundedly complete, or the combination of both properties, for example— are due to R.C. James. See, e.g., [LinTz77, Section 1.b] or [FHHMZ11, Section 4.3].

Quoting Lindenstrauss and Tzafriri's book [LinTz77, Section 1.c], "the existence of a Schauder basis in a Banach space does not give very much information on the structure of the space. (...) Undoubtedly, the most useful and widely studied special class of bases is that of unconditional bases". This quote refers to the separable situation. We add here that unconditionality is not only a desirable extra property, but a must in the nonseparable case:

Definition 129. *A collection* $\{e_\gamma\}_{\gamma\in\Gamma}$ *of vectors in a separable or nonseparable Banach space* X *is an* **unconditional basis** *for* X *if for every* $x \in X$ *there is a unique family of real numbers* $\{a_\gamma\}_{\gamma\in\Gamma}$ *such that* $x = \sum a_\gamma e_\gamma$ *in the sense that for every* $\varepsilon > 0$ *there is a finite set* $F \subset \Gamma$ *such that* $\|x - \sum_{\gamma\in F'} a_\gamma e_\gamma\| \leq \varepsilon$ *for every finite* $F' \supset F$*. For* $\gamma \in \Gamma$*, the corresponding coordinate functionals* f_γ *is defined as* $f_\gamma(x) = a_\gamma$ *whenever* $x = \sum a_\gamma e_\gamma$*.*

Examples of unconditional bases are the canonical ones in $c_0(\Gamma)$ and $\ell_p(\Gamma)$ for all Γ and for $1 \leq p < \infty$, as well as the Haar basis in $L_p[0,1]$ for $p > 1$ (Paley and Marcinkiewicz, see, e.g., [AlKa06, page 130]). However, $C[0,1]$ does not admit any unconditional basis (see, e.g., [HMVZ07, Corollay 6.51], and the comments after Theorem 790).

➜ Unconditional Schauder bases should not be confused with long Schauder bases —for the definition see, e.g., [HMVZ07, Definition 4.1]. The first ones are —even after reordering— particular cases of the second, although $C[0,\omega_1]$ has a long Schauder basis but no unconditional Schauder basis (see, e.g., [HMVZ07, Exercise 3.25]). Every long Schauder basis is a (strong and bounded) M-basis (see. e.g., [HMVZ07, Fact 4.7]).

In particular, every unconditional basis is a (norming) M-basis (see, e.g., [HMVZ07, Exercise 7.8]). We shall provide more information on unconditional bases in Section 45.9 below.

6.1.4 Projectional resolutions of the identity I

Let $X := c_0(\Gamma)$ for $\Gamma := [0,\mu]$, where μ is some uncountable ordinal number. Let us consider the collection $\{P_\gamma : \omega_0 \leq \gamma \leq \mu\}$ of continuous linear operators $P_\gamma : c_0(\Gamma) \to c_0(\Gamma)$ given by $P_\gamma x(\alpha) = x(\alpha)$ if $1 \leq \alpha \leq \gamma$, $P_\gamma x(\alpha) = 0$ if $\gamma < \alpha \leq \mu$, and $P_\mu := \mathrm{Id}_X$. The family $\{P_\gamma : \omega_0 \leq \gamma \leq \mu\}$ consists of continuous linear projections, and it enjoys the following properties:
(i) $\|P_\alpha\| = 1$,
(ii) $P_\alpha P_\beta = P_\beta P_\alpha = P_\alpha$, if $\omega_0 \leq \alpha \leq \beta \leq \mu$,
(iii) dens $(P_\alpha(X)) \leq$ card (α),
(iv) $\bigcup\{P_{\beta+1}(X) : \omega_0 \leq \beta < \alpha\}$ is dense in $P_\alpha(X)$, for every $\alpha \in (\omega_0, \mu]$, and
(v) $P_\mu = \mathrm{Id}_X$.

The previous collection of projections is the model of what, in general, is called a *projectional resolution of the identity*. On many spaces it is possible to define such a structure. This will be properly defined in Definition 130, and later on will appear in Sections 10.1, 10.2, 10.3, 10.4, 12.5, 45.5, Chapters 12, 46, 47 and Section 48.6. Projectional resolutions of the identity are a powerful device for building special structures —as M-bases, see, e.g., Subsection 6.1.5 and comments there— or proving certain properties by transfinite induction —like the existence of a LUR renorming (see, e.g., Theorem 220).

Let S be an infinite set. Note that the smallest ordinal whose cardinality is $\text{dens}\,(X)$ must be a limit ordinal.

Definition 130. *Let X be a nonseparable Banach space and μ be the smallest ordinal such that its cardinality* $\text{card}\,(\mu) =$ *$\text{dens}\,(X)$. A* **projectional resolution of the identity** *(***PRI***, for short) on X is a collection $\{P_\alpha : \omega_0 \leq \alpha \leq \mu\}$ of projections of X into X that satisfy, for every α such that $\omega_0 \leq \alpha \leq \mu$, the following conditions:*
(i) $\|P_\alpha\| = 1$,
(ii) $P_\alpha P_\beta = P_\beta P_\alpha = P_\alpha$, *if* $\omega_0 \leq \alpha \leq \beta \leq \mu$,
(iii) $\text{dens}\,(P_\alpha(X)) \leq \text{card}\,(\alpha)$,
(iv) $\bigcup\{P_{\beta+1}(X) : \omega_0 \leq \beta < \alpha\}$ *is dense in* $P_\alpha(X)$, *for every* $\alpha \in (\omega_0, \mu]$, *and*
(v) $P_\mu = \text{Id}_X$.

If $\{x_\alpha, f_\alpha\}_{0\leq\alpha\leq\mu}$ is an unconditional basis for a Banach space $(X, \|\cdot\|)$, for an uncountable ordinal μ, then there is an equivalent norm $|||\cdot|||$ on the space such that the natural projections $P_\beta(x) = \sum_{\alpha\leq\beta} f_\alpha(x)x_\alpha$, $\omega_0 \leq \beta \leq \mu$, defined as in the example of $c_0(\Gamma)$ at the beginning of this subsection, form a PRI on $(X, |||\cdot|||)$. Details can be found in Subsection 10.4.1.

We will see later that, for example, every weakly Lindelöf determined space (see Definition 134), in particular every weakly compactly generated space (see Definition 131), has a PRI (see the notes after Definition 207 and Theorem 208), while ℓ_∞ does not (see Section 12.4).

Projectional resolutions of the identity will often appear along this text. Chapter 12 will be devoted to this subject.

6.1.5 Markushevich bases versus projectional resolutions of the identiy

In general, the existence of a PRI does not guarantee the existence of an M-basis. For an example see, e.g., [HMVZ07, Section 5.1].
Sometimes it does: Assume a class $\mathcal{C}$ of Banach spaces with the following property: If $X \in \mathcal{C}$, then there exists a PRI $\{P_\alpha : \omega_0 \leq \alpha \leq \mu\}$ on X such that $(P_{\alpha+1} - P_\alpha)(X) \in \mathcal{C}$. We say then that $\mathcal{C}$ is a **$\mathcal{P}$-class**. Every member X of a $\mathcal{P}$-class has a strong M-basis (obtained by a natural transfinite induction process on the density of X, and starting with the separable case, where the existence of a strong M-basis is guaranteed by a result of Terenzi —see Theorem 224 below). For details see, e.g., [HMVZ07, Theorem 5.1].

In the other direction, the existence of an M-basis —even a strong M-basis in a space of density ω_1— does not guarantee the existence of a PRI. See Theorem 225 below, that shows the result even in a quite dramatic way.

6.1.6 Several classes of Banach spaces defined in terms of Markushevich bases or projectional resolutions of the identity

Various kinds of the shrinking properties of M-bases are closely connected to various types of weak compact generating, as shown in [HMVZ07, Chapter 6].
We list here some relevant definitions in increasing order of generality, and provide alternative descriptions in the language of M-bases, giving the proper references. Finite-dimensional, separable Banach spaces, and Hilbert spaces have already been mentioned. Examples of Banach spaces in those more general classes will be provided in due term, and some relevant aspects of their properties will be considered later on.

Definition 131. *Let $(X, \|\cdot\|)$ be a Banach space. X is said to be* **weakly compactly generated** *(***WCG***, for short), if there exists a weakly compact subset K of X such that* $\overline{\operatorname{span}}(K) = X$.

Definition 132. *Let $(X, \|\cdot\|)$ be a Banach space. X is said to be* **subspace of a weakly compactly generated** *(***SWCG***, for short), if it is linearly isomorphic to a subspace of a WCG Banach space.*

WCG and SWCG spaces will be treated at length in Sections 10.1 and 10.6, and in Chapter 46.

Definition 133. *A Banach space X is called* **Vašák** *(also called* **weakly countably determined, WCD** *for short) if there is a countable collection $\{K_n : n \in \mathbb{N}\}$ of w^*-compact subsets of X^{**} such that for every $x \in X$ and $u \in X^{**} \setminus X$ there is $n_0 \in \mathbb{N}$ for which $x \in K_{n_0}$ and $u \notin K_{n_0}$.*

We devote Section 45.5 to some complements on WCD spaces and their connection to WCG and other classes of Banach spaces.

If Γ is a set, then $\ell_\infty^c(\Gamma)$ denotes the Banach space formed by all elements of $\ell_\infty(\Gamma)$ that are countably supported (endowed with the supremum norm).

Definition 134. *A Banach space X is called* **weakly Lindelöf determined (WLD)** *if, for some set Γ, there is a one-to-one bounded linear operator from X^* into $\ell_\infty^c(\Gamma)$ that is weak*-to-pointwise continuous.*

The subject of WLD spaces will be considered along this book at several instances. An equivalent description of the property of being a WLD space will be given in Theorem 135 below in terms of the existence of an M-basis with additional properties; another one, by looking at its dual closed unit ball, will be mentioned in Section 6.2 below. Still another one, this time in terms of the existence of a particular projectional generator, will be provided after Definition 207.

We have the following chain of implications:

$$\boxed{\text{WCG} \Rightarrow \text{SWCG} \Rightarrow \text{WCD} \Rightarrow \text{WLD}}$$

The first implication is obvious. For the second see, e.g., [HMVZ07, Theorem 6.13], and for the last see Theorem 201 below (see also, e.g., [HMVZ07, Theorems 5.37 and 6.25]). Some of those implications can be seen at a glance once the following characterizations in terms of M-bases are proved:

➡ We collect here characterizations of the previous classes in terms of existing M-bases. Some of them will be discussed later.

Theorem 135. *Let* $(X, \|\cdot\|)$ *be a Banach space. Then,* X *is*
(i) *WCG if, and only if, it admits a weakly compact M-basis.*
(ii) *SWCG if, and only if, it admits a* σ*-shrinkable M-basis.*
(iii) *WCD if, and only if, it admits a weakly* σ*-shrinkable M-basis.*
(iv) *WLD if, and only if, it admits an M-basis* $\{x_\gamma, f_\gamma\}_{\gamma\in\Gamma}$ *that countably supports* X^**, i.e.,* $\operatorname{card}\{\gamma:\ \gamma\in\Gamma,\ \langle x_\gamma, f\rangle \neq 0\} \leq \aleph_0$ *for all* $f \in X^*$.

Proof. (i) See Theorem 235. (ii) [HMVZ07, Theorem 6.13]. (iii) [HMVZ07, Theorem 6.25]. (iv) [HMVZ07, Theorem 5.37] (in fact, *every* M-basis on a WLD space has the property exhibited in (iv); see Example 432 below; see also Theorem 224). □

One influential result was in [JoZi75a], where it was proved that *X admits a shrinking Markushevich basis if both X and* X^* *are subspaces of WCG spaces.* It provided the first nontrivial examples of nonreflexive spaces with shrinking Markushevich bases. See also Theorems 315 and 839 below.

6.1.7 Superreflexivity, finite representability

Definition 136. *A Banach space X is said to be* **superreflexive** *if it admits an equivalent UR norm.*

This was not the original definition of superreflexivity. It was formulated in terms of finite representability (Definition 137 and Item (iv) in Theorem 447 below). The equivalence with the above formulation was established by P. Enflo in [Enf72]. Theorem 447 lists this and some other equivalent formulations.

Every superreflexive space is reflexive. There are examples of reflexive spaces which are not superreflexive (see Theorem 454 below). For this subject we refer to [FHHMZ11, Chapters 4 and 9], and we devote Chapter 24 below to some aspects of the theory of superreflexive spaces.

Definition 137. *A Banach space Y is said to be* **crudely finitely representable in a Banach space** X *if there exists $K > 0$ such that, for every finite-dimensional subspace F of Y there is a linear isomorphism T of F onto $T(F) \subset X$ so that $\|T\|.\|T^{-1}\| < K$. We say that Y is* **finitely representable in a Banach space** X *if for every $\varepsilon > 0$, Y is crudely finitely representable in X with constant $K = 1+\varepsilon$.*

We will see that *a Banach space X is superreflexive if, and only if, each Banach space that is finitely representable in X is necessarily reflexive* (see Chapter 24 and, in particular, see again Theorem 447).

H The concept of finite representability is one of the most important concepts in the modern Banach space theory. It was introduced by R.C. James in [Jame72a] and [Jame72].

6.1.8 Generation and strong generation

A Banach space X is said to be **generated by a Banach space** Y if there exists a continuous linear operator $T : Y \to X$ with dense image.
It is obvious that this is equivalent to say that every point $x \in X$ is "almost absorbed" by TB_Y, in the sense that for every $\varepsilon > 0$ there exists $n \in \mathbb{N}$ such that $x \in nTB_Y + \varepsilon B_X$ (see Section 10.1). It is a simple exercise to prove that this is in turn equivalent to say that every $\|\cdot\|$-compact is almost absorbed by TB_Y.
An important result is given by Theorem 191 below, that ensures that *a Banach space is WCG if, and only if, it is generated by a reflexive Banach space.* It is natural then to explore some other versions of the concept of generation and strong generation: A Banach space is said to be **strongly generated by a Banach space** Y if every w-compact subset of X is almost absorbed by TB_Y.

A Banach space X is said to be **strongly weakly compactly generated** (β**WCG**, for short) if it is strongly generated by a reflexive space.

We shall present some results on this class of Banach spaces in Section 48.11 below.

A Banach space X is said to be **Hilbert generated** (**strongly Hilbert generated**) if it is generated (respectively, strongly generated) by a Hilbert space.

We say that a Banach space X **injects into a Banach space** Y if there is a bounded linear one-to-one operator from X into Y.

In this text we shall be often interested in Banach spaces that inject into classical Banach spaces. See, e.g., Remark 127.

We say that a Banach space is a **DENS space** (**DENS**, for short) if the weak* density of X^* equals to the norm density of X.

We say that a Banach space X has **property C** (of Corson) if any family $\mathcal{A}$ of closed convex sets in X with empty intersection has a countable subfamily $\mathcal{B}$ of $\mathcal{A}$ with empty intersection.

For information on Corson's property C the reader may check, e.g., [FHHMZ11, Chapter 14] and [KKLP13, Chapter 13].

6.2 Some classes of compact topological spaces

We mention here some classes of compact topological spaces that play an important role in Banach space theory. For a wider account of these, see, e.g., [FHHMZ11, Chapter 14].

A compact space K is said to be **Eberlein compact** if, for some set Γ, K is homeomorphic to a compact set in $(c_0(\Gamma), w)$. It turns out that K is an Eberlein compact if, and only if, it is homeomorphic to a weakly compact subset of a Banach space endowed with its weak topology, if, and only if, $(C(K), \|\cdot\|_\infty)$ is WCG (see Definition 131 below). The continuous image of an Eberlein compact is again an Eberlein compact [BRW77]. So, X is a subspace of a WCG space if, and only if, (B_{X^*}, w^*) is an Eberlein compact. For this and related results see Section 10.1 below.

A compact space K is called **uniform Eberlein compact** if it is homeomorphic to a weakly compact set in a Hilbert space endowed with its weak topology.

A compact space K is said to be **Corson compact** if K is homeomorphic to a subset S of $[-1,1]^\Gamma$ that is compact in the pointwise topology and such that for every $x \in S$ we have $\operatorname{card}\{\gamma \in \Gamma : x(\gamma) \neq 0\} \leq \aleph_0$. Notice that a Banach space is WLD if, and only if, (B_{X^*}, w^*) is a Corson compact (see Theorem 833).

Let Λ be a set and Z be a separable metric space. The Banach space $c_1(Z \times \Lambda)$ (or **Mercourakis' space**) is the closed subspace of $\ell_\infty(Z \times \Lambda)$ formed by all functions f that for every compact subset K of Z the restriction of f to $K \times \Lambda$ belongs to $c_0(K \times \Lambda)$. For renorming purposes note that it readily follows that if Z is not totally bounded, then Mercourakis space does contain a copy of ℓ_∞.

The **α-Cantor derivative** $A^{(\alpha)}$ of a topological space A is defined as $A^{(0)} = A$, $A^{(\alpha+1)}$ is the set of all accumulation points of $A^{(\alpha)}$ and, if α is a limit ordinal, then $A^{(\alpha)} = \bigcap_{\beta<\alpha} A^{(\beta)}$.

Definition 138. *Let T be a topological space. A subset S of T is said to be* **(relatively) countably compact** *if every countable subset of S has a cluster point in S (respectively, in T).*
S is said to be **(relatively) sequentially compact** *if every sequence in S has a subsequence that converges to a point in S (respectively, in T).*

Definition 139. *A topological space T is said to be* **angelic** *if every relatively countably compact subset A of T is relatively compact, and, for such a set A, every element $x \in \overline{A}$ is the limit of a sequence in A.*

Observe that every metric space is such, and that the Eberlein–Šmulyan theorem, together with Kaplansky's theorem (see, e.g., [FHHMZ11, Theorem 3.54]), ensures that every Banach space is angelic in its weak topology. For more information on angelic spaces see, e.g., [Fl80, Chapter 3] and [KKLP13, Chapter 4].

6.3 Extra definitions needed

Recall that a function is a **Baire 1 function** if it is a pointwise limit of continuous functions.

Let M be a subset of B_X and let $C \subset X^*$. Then

$$\operatorname{diam}_M(C) := \sup_{f,g\in C} \sup_{x\in M} \{|f(x) - g(x)|\}.$$

Let C be a weak* compact convex subset of X^*. Let $\varepsilon > 0$. We denote

$$D_{(M,\varepsilon)}(C) := \\ \{f \in C : \operatorname{diam}_M(S) > \varepsilon \text{ for every } w^*\text{-open slice } S \text{ of } C \text{ with } f \in C\}.$$

Note that $D_{(M,\varepsilon)}(C)$ is w^*-compact and convex. We can proceed then to form the second "Cantor derivation" of it, to get $D^{(2)}_{(M,\varepsilon)}(C) := D_{(M,\varepsilon)}(D_{(M,\varepsilon)}(C))$ and continue by induction. We say that M has **finite dual ε-index** if $D^{(n)}_{(M,\varepsilon)}B_{X^*} = \emptyset$ for some $n \in \mathbb{N}$. We say that X has **dual dentability index** ω_0 if for each $\varepsilon > 0$, B_X has finite ε-index for every $\varepsilon > 0$. We say that X has **σ-finite dual index** if for every $\varepsilon > 0$ there is a decomposition $B_X = \bigcup_{k=1}^{\infty} M_k^{\varepsilon}$ where each set M_k^{ε} has finite dual ε-index.

The following important concept was introduced and substantially used by Toruńczyk in [Tor73] (see Section 13.2 below).

Definition 140. *We say that a real-valued function* φ *on a Banach space* X **locally depends on finitely many functionals (or coordinates)** *(***LFC***, for short) if for every* $x \in X$ *there is a neighborhood* U *of* x*, functionals* $f_1, f_2, \ldots, f_n \in X^*$*, and a real-valued function* ψ *on* $\mathbb{R}^n$ *such that* $\varphi(z) = \psi(f_1(z), f_2(z), \ldots, f_n(z))$ *for all* $z \in U$*. When* X *is a subspace of* $\ell_\infty(\Gamma)$ *for some set* Γ*, we shall understand that the functionals* f_i*,* $i = 1, 2, \ldots, n$ *are coordinates, i.e.,* $f_i(z) = z(\gamma_i)$ *for* $\gamma_i \in \Gamma$*,* $i = 1, 2, \ldots, n$*.*

We shall devote Chapter 38 to the subject of polyhedral spaces, a concept defined below:

Definition 141. *A Banach space* X *is called* **polyhedral** *if for every finite-dimensional subspace* $F \subset X$ *the ball of* F *is the convex hull of a finite set of points.*

Chapter 7

The use of Šmulyan's tests

We already mentioned some duality results, that we collect here generically under the name **Šmulyan's tests**. These are important, since often is easier to renorm the dual space than the original space to achieve a certain property. For example, to get a renorming of X with a given property, say Gâteaux differentiability, we may produce a strict convex renorming of X^* instead, provided that the new norm on X^* would be an (equivalent) *dual norm.* In Proposition 17 and Corollary 94 we provided several statements for an equivalent norm on the dual of a Banach space to be a dual norm. In general, if $(X, \|\cdot\|)$ is a Banach space and $D \subset X^*$ is a w^*-dense subspace, we may consider on X the Hausdorff locally convex topology $w(X, D)$ of the pointwise convergence on the elements of D. Then, the norm $\|\cdot\|$ is $w(X, D)$-lower semicontinuous if, and only if, B_X is $w(X, D)$-closed.
A simple remark: If a norm $\|\cdot\|$ on a Banach space is $w(X, D)$-lower semicontinuous for some w^*-dense subspace D of X^*, then so it is the (convex) function $\|\cdot\|^p$ for any $p > 0$. This is so since $\{x \in X : \|x\|^p \leq r\} = \{x \in X : \|x\| \leq r^{1/p}\}$.

Before we start on examples, let us mention that a review on the differentiability properties of standard norms of basic classical spaces like $C[0,1]$, c_0, ℓ_1 etc. is in [DeGoZi93, Examples I.1.6] or [FHHMZ11, Exercises 7.36–7.41].

Generally speaking, to deal with smoothness is more challenging than dealing with rotundity, as even definitions concerning smoothness are more subtle. Thus the following remark may be useful: For proving "positive" results on Fréchet differentiability is convenient to use that *it is implied by LUR dual norms* (see Proposition 91 above). On the other hand, for "negative" results on the Fréchet differentiability the notion of MLUR can be used (see Theorem 105).

If we directly try to get a smooth ball by addition of a "smooth convex set" to the closed unit ball, the problem is to get in this way a closed body. This may be done by using compactness arguments. However, a much easier procedure is to work dually by adding a strictly convex function to convexify a convenient function.

7.1 Some comments on the Šmulyan tests

A basic result on differentiability of the norm in Banach spaces is the so-called Šmulyan Lemma 54. Note that it can be rephrased —and slightly enlarged— in the following way (see, e.g., [FHHMZ11, Corollary 7.22]:

A. J. Guirao et al., *Renormings in Banach Spaces*, Monografie Matematyczne 75,
https://doi.org/10.1007/978-3-031-08655-7_7

Lemma 142 (Šmulyan). *Let $(X, \|\cdot\|)$ be a Banach space.*
(i) *The norm $\|\cdot\|$ is Fréchet differentiable at $x \in S_X$ if, and only if, $\|f_n - g_n\| \to 0$ whenever $\{f_n\}_{n\in\mathbb{N}}$ and $\{g_n\}_{n\in\mathbb{N}}$ are sequences in B_{X^*} such that $f_n(x) \to 1$ and $g_n(x) \to 1$, if, and only if, a sequence $\{f_n\}_{n\in\mathbb{N}}$ in B_{X^*} converges in norm whenever $f_n(x) \to 1$.*
(ii) *The norm $\|\cdot\|$ is Gâteaux differentiable at $x \in S_X$ if, and only if, $f_n - g_n \stackrel{w^*}{\to} 0$ whenever $\{f_n\}_{n\in\mathbb{N}}$ and $\{g_n\}_{n\in\mathbb{N}}$ are sequences in B_{X^*} such that $f_n(x) \to 1$ and $g_n(x) \to 1$, if, and only if, there is a unique $f \in B_{X^*}$ such that $f(x) = 1$.*
(iii) *The dual norm $\|\cdot\|^*$ on X^* is Fréchet differentiable at $f \in S_{X^*}$ if, and only if, $\|x_n - y_n\| \to 0$ whenever $\{x_n\}_{n\in\mathbb{N}}$ and $\{y_n\}_{n\in\mathbb{N}}$ are sequences in B_X such that $f(x_n) \to 1$ and $f(y_n) \to 1$, if, and only if, a sequence $\{x_n\}_{n\in\mathbb{N}}$ in B_X converges in norm whenever $f(x_n) \to 1$.*
(iv) *The dual norm $\|\cdot\|^*$ on X^* is Gâteaux differentiable at $f \in S_{X^*}$ if, and only if, $x_n - y_n \stackrel{w}{\to} 0$ whenever $\{x_n\}_{n\in\mathbb{N}}$ and $\{y_n\}_{n\in\mathbb{N}}$ are sequences in B_X such that $f(x_n) \to 1$ and $f(y_n) \to 1$, if, and only if, $\{x_n\}_{n\in\mathbb{N}}$ is weakly Cauchy whenever $\{x_n\}_{n\in\mathbb{N}}$ is a sequence in B_X such that $f(x_n) \to 1$.*

We collect in the next result some of the information spread in this and previous sections.

Theorem 143. *Let $(X, \|\cdot\|)$ be a Banach space.*
(i) *If $\|\cdot\|^*$ is strictly convex, then $\|\cdot\|$ is Gâteaux differentiable.*
(ii) *If $\|\cdot\|^*$ is Gâteaux differentiable then $\|\cdot\|$ is strictly convex (and X does not contain an isomorphic copy of ℓ_1).*
(iii) *If $\|\cdot\|^*$ is LUR, then $\|\cdot\|$ is Fréchet differentiable.*
(iv) *$\|\cdot\|$ is UG if, and only if, $\|\cdot\|^*$ is W* UR.*
(v) *$\|\cdot\|$ is WUR if, and only if, $\|\cdot\|^*$ is UG.*
(vi) *$\|\cdot\|$ is UR if, and only if, $\|\cdot\|^*$ is UF.*
(vii) *$\|\cdot\|$ is UF if, and only if, $\|\cdot\|^*$ is UR.*

Proof. (Sketch)
(i) and the first part of (ii) are in Proposition 82. See also [FHHMZ11, Corollary 7.23]. For the second part of (ii) see, e.g., [FHHMZ11, Corollary 7.26].

(iii) Proposition 91 above.

(iv) Assume $\|\cdot\|$ is UG. Then, given $h \in S_X$ and $\varepsilon > 0$, there exists $\delta > 0$ such that $\|x + th\| + \|x - th\| \le 2 + \varepsilon|t|$ for all $x \in S_X$ and $|t| \le \delta$.
Let $\{f_n\}$ and $\{g_n\}$ be sequences in S_{X^*} such that $\|f_n + g_n\| \to 2$. For $n \in \mathbb{N}$, find $x_n \in S_X$ such that $(f_n + g_n)(x_n) \to 2$. Then $f_n(x_n) \to 1$ and $g_n(x_n) \to 1$. Obviously,

$$f_n(x_n + th) + g(x_n - th) \le \|x_n - th\| + \|x_n - th\| \le 2 + \varepsilon|t| \le 2 + \varepsilon\delta, \ |t| \le \delta,$$

so there exists $n_0 \in \mathbb{N}$ such that, for $n \ge n_0$, $(f_n - g_n)(th) \le 2 + \varepsilon\delta - f_n(x_n) - g_n(x_n) < 2\varepsilon\delta$. In particular, $\delta(f_n - g_n)(h) < 2\varepsilon\delta$ for $n \ge n_0$. This shows that $(f_n - g_n) \to 0$ in the w^*-topology. Thus, $\|\cdot\|^*$ is W*UR.

For the reverse implication, assume that $\|\cdot\|$ is not UG. By convexity, there exist $h \in S_X$, $\varepsilon > 0$, and sequences $\{x_n\}$ in S_X and $\{t_n\}$ in $\mathbb{R}$ with $0 < t_n \to 0$ such that

$$\|x_n + t_n h\| + \|x_n - t_n h\| \geq 2 + \varepsilon t_n, \ n \in \mathbb{N}.$$

Find f_n and g_n in S_{X^*} such that $f_n(x_n + t_n h) = \|x_n + t_n h\|$ and $g_n(x_n - t_n h) = \|x_n - t_n h\|$ for $n \in \mathbb{N}$. Thus,

$$f_n(x_n + t_n h) + g_n(x_n - t_n h) \geq 2 + \varepsilon t_n, \ n \in \mathbb{N}. \tag{7.1}$$

Note that $f_n(x_n) = f_n(x_n + t_n h) - f_n(t_n h) \to 1$: analogously, $g_n(x_n) \to 1$, hence $\|f_n + g_n\| \to 2$. From (7.1) we get

$$(f_n - g_n)(t_n h) \geq 2 + \varepsilon t_n - f_n(x_n) - g_n(x_n) \geq \varepsilon t_n, \ n \in \mathbb{N},$$

hence $(f_n - g_n)(h) \geq \varepsilon$ for all $n \in \mathbb{N}$, and so $\|\cdot\|^*$ is not W*UR.

To show (v) we may follows the proof of (iv) with some minor changes: If $\|\cdot\|^*$ is UG, then $\|\cdot\|$ is WUR by just dualizing the argument above. If, to prove the reverse implication, we assume that $\|\cdot\|^*$ is not UG, we may find, as above, $h^* \in S_{X^*}$, $\varepsilon > 0$, and sequences $\{f_n\}$ in S_{X^*} and $\{t_n\}$ in $\mathbb{R}$ with $0 < t_n \to 0$ such that

$$\|f_n + t_n h^*\|^* + \|f_n - t_n h^*\|^* \geq 2 + \varepsilon t_n, \ n \in \mathbb{N}.$$

The only change in the rest of the argument is that sequences $\{x_n\}$ and $\{y_n\}$ are chosen in S_X with $(f_n + t_n h^*)(x_n) > \|f_n + t_n h^*\|^* - t_n/n$ and $(f_n - t_n h^*)(y_n) > \|f_n - t_n h^*\|^* - t_n/n$ for all $n \in \mathbb{N}$, and we proceed accordingly. □

The following theorem collects some results of a structural type: The existence of a norm with a certain property implies an isomorphic property of the space.

Theorem 144. *Let $(X, \|\cdot\|)$ be a Banach space.*
(i) *If $\|\cdot\|$ is UR, then X is reflexive.*
(ii) *If $\|\cdot\|$ is UF, then X is reflexive.*
(iii) *If $\|\cdot\|^*$ is Fréchet differentiable, then X is reflexive.*
(iv) (Hájek [Há96a]) *If $\|\cdot\|$ is WUR, then X is Asplund.*

Observe that (iii) in Theorem 144 can be strengthened in the following way (already mentioned in Section 4.2): If $\|\cdot\|^*$ is very smooth (see Definition 65 above), then X is reflexive (Theorem 411 below).
For (iv) in Theorem 144 see [FHHMZ11, Theorem 8.12].

7.2 An application of the Šmulyan lemma to w^*-convergence

The following is just a reformulation of (ii) in Lemma 142 (Figure 7.1 hints at the underlying argument). A straightforward consequence is Corollary 147 below.

Proposition 145. *Let $(X, \|\cdot\|)$ be a Banach space. Assume that $\|\cdot\|$ is Gâteaux differentiable at some point $x_0 \in S_X$. Then, if $\{x_0^*\} = \partial\|\cdot\|(x_0)$ and $\{x_n^*\}_{n\in\mathbb{N}}$ is a sequence in B_{X^*} such that $\langle x_0, x_n^*\rangle \to \langle x_0, x_0^*\rangle$, then $x_n^* \to x_0^*$ in the w^*-topology.*

Notice that the previous proposition could be formulated for nets, too.

Remark 146. Notice that a consequence of the Bishop–Phelps Theorem 14 is that, for a Banach space $(X, \|\cdot\|)$ such that the norm is Gâteaux differentiable, given any $x_0^* \in S_{X^*}$ there exists a sequence $\{x_n^*\}_{n=1}^{\infty}$ of derivatives of the norm that $\|\cdot\|$-converges to x_0. Indeed, it is enough to observe that every norm-attaining functional in S_{X^*} is the derivative of the norm at the point where the functional attains its norm. Ⓡ

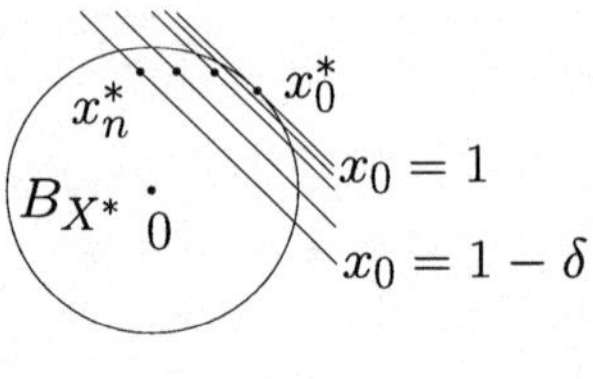

Figure 7.1: Getting w^*-convergence under Gâteaux differentiability

Proposition 145 and the Fréchet version of Lemma 54 give the following result, that allows for some variants, replacing $\mathcal{T}_{\|\cdot\|}$ by some other topologies on X^*:

Corollary 147. *Let $(X, \|\cdot\|)$ be a Banach space. Assume that $\|\cdot\|$ is Gâteaux differentiable at some point $x_0 \in S_X$ and that $\|\cdot\|^*$ is w^*-$\mathcal{T}_{\|\cdot\|^*}$-SKK. Then $\|\cdot\|$ is Fréchet differentiable at x_0.*

Proof. Let $x_0^* \in S_{X^*}$ be the derivative of $\|\cdot\|$ at x_0, and let $\{x_n\}_{n\in\mathbb{N}}$ be a sequence in B_{X^*} such that $\langle x_0, x_n^*\rangle \to \langle x_0, x_0^*\rangle$ $(= 1)$. Proposition 145 shows that $\{x_n^*\}_{n\in\mathbb{N}}$ is w^*-convergent to x_0^*. Since $\|x_n^*\|^* \to 1$, we may assume that $\|x_n^*\| = 1$ for all $n \in \mathbb{N}$. By the assumption, $x_n^* \to x_0^*$ in the norm topology. This shows, by (i) in Lemma 142, that $\|\cdot\|$ is Fréchet differentiable at x_0. □

Let us finish this section by mentioning that the norm on a Banach space is C^1-smooth in the Fréchet sense if it is Fréchet differentiable. This directly follows from Lemma 53 and the Šmulyan's characterization of Fréchet differentiability of $\|\cdot\|$ at a point (Lemma 54).

Chapter 8

Asplund averaging I

Let X be a Banach space and $\|\cdot\|_1$, $\|\cdot\|_2$ be two equivalent norms on X. Assume that $\|\cdot\|_i$ satisfies certain property $\mathcal{P}_i$, $i = 1, 2$. It is important to elucidate whether a new equivalent norm $\|\cdot\|$, satisfying both properties $\mathcal{P}_1$ and $\mathcal{P}_2$ simultaneously, can be defined on X.
This will be done here by following the original technique developed by E. Asplund. Later on, in Section 10.8, we shall present another technique based on Baire's category theorem. Both procedures are important. The second one for its strength and elegance, the first one for providing useful geometric arguments.

8.1 Asplund original approach

Asplund [Asp67] successfully developed an averaging procedure to achieve the goal mentioned at the introduction to this chapter. His technique is based on the Fenchel duality and the operation of inf-convolution defined in Subsections 3.6.4 and 3.6.5.
We recall that $\|\cdot\|_1$ and $\|\cdot\|_2$ are two equivalent norms on a Banach space X. Assume without loss of generality that $\|\cdot\|_1 \leq \|\cdot\|_2 \leq K\|\cdot\|_1$, for some $K > 1$. Let $\varphi_2\colon \mathbb{R} \to \mathbb{R}$ be the mapping defined in (3.15), i.e., $\varphi_2(t) := |t|^2/2$ for every $t \in \mathbb{R}$. Set

$$\boxed{g_0 := \varphi_2(\|\cdot\|_1), \text{ and } f_0 := \varphi_2(\|\cdot\|_2).} \tag{8.1}$$

Then g_0 and f_0 are convex, quadratic-homogeneous, and $g_0 \leq f_0$. We have $g_0^* = \varphi_2(\|\cdot\|_1^*)$ and $f_0^* = \varphi_2(\|\cdot\|_2^*)$ (see Subsection 3.6.4, Item 4). Setting

$$\boxed{C := K^2 - 1} \tag{8.2}$$

we rewrite the previous inequalities as $g_0 \leq f_0 \leq (1 + C)g_0$. Fix $n \in \mathbb{N} \cup \{0\}$ and assume that the functions f_n and g_n have been defined in such a way that they are convex, quadratic-homogeneous, and

$$g_n \leq f_n \leq (1 + 4^{-n}C)g_n.$$

Then set

$$\boxed{g_{n+1} := f_n \,\Box\, g_n \text{ and } f_{n+1} := (f_n + g_n)/2}$$

(where $\Box$ denotes the infimal convolution, defined in (3.18)), which are convex, quadratic-homogeneous, and satisfy

$$\boxed{g_n \leq g_{n+1} \leq f_{n+1} \leq f_n} \tag{8.3}$$

A. J. Guirao et al., *Renormings in Banach Spaces*, Monografie Matematyczne 75,
https://doi.org/10.1007/978-3-031-08655-7_8

(the first inequality by the convexity of g_n, for the second see (ii) in Proposition 38, and the third one is obvious). These new functions satisfy the following inequality:

$$g_{n+1} \leq f_{n+1} \leq \left(1 + 4^{-(n+1)}C\right) g_{n+1}.$$

➡ The first inequality is in (8.3). Let us prove the second one. From the induction hypothesis we know that $f_{n+1} \leq f_n$ and that $f_{n+1} \leq a_n g_n$ where $a_n = (1 + 2^{-2n-1}C)$. Fix x and y in X. Then, due to the quadratic homogeneity of f_{n+1} and its convexity, it holds that

$$\begin{aligned} f_n(x+y) + g_n(x-y) &\geq f_{n+1}(x+y) + f_{n+1}(x-y)/a_n = \\ &= f_{n+1}(x+y) + a_n f_{n+1}((x-y)/a_n) \geq \\ &\geq (1+a_n) f_{n+1}(2x/(1+a_n)) = 4f_{n+1}(x)/(1+a_n). \end{aligned}$$

Therefore, by taking the infimum over $y \in X$ we deduce that

$$g_{n+1}(x) \geq 2f_{n+1}(x)/(1+a_n).$$

which is the desired inequality $f_{n+1} \leq (1 + 4^{-(n+1)}C) g_{n+1}$.

This induction process gives us two sequences of convex quadratic-homogeneous functions $\{f_n\}$ and $\{g_n\}$ such that, for every $n \in \mathbb{N} \cup \{0\}$, it holds

$$\boxed{g_n \leq f_n \leq (1 + 4^{-n}C) g_n.} \tag{8.4}$$

Then, in view of (8.3) and (8.4), both sequences converge uniformly on any bounded set to a certain convex quadratic-homogeneous continuous function h such that

$$\boxed{g_0 \leq g_1 \leq \cdots \leq g_n \leq \cdots \leq h \leq \cdots \leq f_m \leq \cdots \leq f_1 \leq f_0.}$$

The Fenchel duality gives that

$$f_0^* \leq f_1^* \leq \cdots \leq f_m^* \leq \cdots \leq h^* \leq \cdots \leq g_n^* \leq \cdots \leq g_1^* \leq g_0^*.$$

The convex quadratic-homogeneous continuous function h defines an equivalent norm $\|\cdot\|$ on X as

$$\|x\| = \sqrt{2h(x)} \text{ for } x \in X. \tag{8.5}$$

➡ This can be done by using Proposition 78. Indeed, notice first that f_0 and g_0 are obviously (convex) even and continuous. Moreover, for any $a > 0$, the sets $\{x \in X : \ f_0(x) \leq a\}$ and $\{x \in X : \ g_0(x) \leq a\}$ are bounded. If $x \in X$ satisfies $(f_0 \Box g_0)(x) \leq a$, then there exists $y \in X$ such that $(1/2)(f_0(x+y) + g_0(x-y)) \leq 2a$, and so $g_0(x+y) + g_0(x-y) \leq 4a$. By convexity, $g_0(x) \leq 2a$; thus, x belongs to the bounded set $\{x \in X : \ g_0(x) \leq 2a\}$.

Thus, for $k = 0$ the function f_1 fulfills all the requirements in Proposition 78. A simpler argument shows that this happens also to g_1. By induction, this holds for any f_n and g_n, and so also for h. Put $B = \{x \in X :\ h(x) \leq 1/2\}$; the (equivalent) norm $\|\cdot\|$ was defined in the aforementioned proposition as the Minkowski functional of B. Since $\|x\| = \inf\{\lambda \geq 0 :\ x \in \lambda B\}$, and for $\lambda > 0$ we have $x/\lambda \in B$ if, and only if, $h(x/\lambda) \leq 1/2$, it follows from the quadratic-homogeneity of h that this happens if, and only if, $h(x) \leq \lambda^2/2$, i.e., $\sqrt{2h(x)} \leq \lambda$. This shows that $\|x\| = \sqrt{2h(x)}$ for $x \in X$.

We shall say that $\|\cdot\|$ **has been obtained from $\|\cdot\|_1$ and $\|\cdot\|_2$ by Asplund averaging**. We obtain that

$$\boxed{\|\cdot\|_1 \leq \|\cdot\| \leq \|\cdot\|_2.}$$

Note that, by (8.5) and (3.17),

$$\boxed{h = \varphi_2(\|\cdot\|),\ \text{(and so } h^* = \varphi_2(\|\cdot\|^*)).} \tag{8.6}$$

Observe that $f_0^* \leq h^* \leq g_0^*$.

The following theorem of Asplund gives the crucial property of the norm $\|\cdot\|$ in (8.5) defined from $\|\cdot\|_1$ and $\|\cdot\|_2$ by Asplund averaging. The use of the expression $Q_{\|\cdot\|}$, defined in (5.2) above, will shorten some of the computations that follow. We reproduce here its formula for the reader's convenience:

$$\boxed{\text{For } x, y \in (X, \|\cdot\|),\ Q_{\|\cdot\|}(x, y) := 2\|x\|^2 + 2\|y\|^2 - \|x + y\|^2.} \tag{8.7}$$

Theorem 148 (Asplund, [Asp67]). *The norm $\|\cdot\|$ obtained by Asplund averaging of two equivalent norms $\|\cdot\|_1$ and $\|\cdot\|_2$ such that $\|\cdot\|_1 \leq \|\cdot\|_2 \leq K\|\cdot\|_1$ satisfies, for every $n \in \mathbb{N}$, $x, y \in X$ and $x^*, y^* \in X^*$,*

$$Q_{\|\cdot\|}(x, y) \geq 2^{-n} Q_{\|\cdot\|_2}(x, y) - 2^{-2n+1} C\left(\|x\|_2^2 + \|y\|_2^2\right) \tag{8.8}$$

and

$$Q_{\|\cdot\|^*}(x^*, y^*) \geq 2^{-n} Q_{\|\cdot\|_1^*}(x^*, y^*) - 2^{-2n+1} C\left((\|x^*\|_1^*)^2 + (\|y^*\|_1^*)^2\right). \tag{8.9}$$

where $C = K^2 - 1$, and the function $Q_{\|\cdot\|}$ is defined in (8.7).

Proof. First of all, by the inequalities (8.4) we know that

$$0 \leq f_n - h \leq f_n - g_n \leq 4^{-n} C f_0, \text{ for } n \in \mathbb{N} \cup \{0\}.$$

Define the function $h_n := f_n - 2^{-n} f_0$, for $n \in \mathbb{N}$. Then

$$h_n + (2^{-n} - 4^{-n} C) f_0 \leq h \leq h_n + 2^{-n} f_0, \text{ for } n \in \mathbb{N}. \tag{8.10}$$

Observe that

$$2h_n = (2f_n - f_{n-1}) + (f_{n-1} - 2^{-(n-1)}f_0) = g_{n-1} + h_{n-1}, \text{ for } n \in \mathbb{N},$$

and thus

$$h_n = \sum_{j=1}^{n} \frac{1}{2^{n+1-j}} g_j,$$

so h_n is convex as a positive linear combination of convex functions, for every $n \in \mathbb{N}$. Let x, y belong to X. Then, by using the previous inequalities, the convexity of h_n, and the fact that each h_n is quadratic homogeneous, we have

$$\begin{aligned}
Q_{\|\cdot\|}(x,y) &= 4h(x) + 4h(y) - 2h(x+y) \text{ (from (8.6))} \\
&\geq 4\Big(h_n(x) + (2^{-n} - 4^{-n}C)f_0(x) + h_n(y) + (2^{-n} - 4^{-n}C)f_0(y)\Big) \\
&\quad - 2\big(h_n(x+y) + 2^{-n}f_0(x+y)\big) \text{ (from (8.10))} \\
&= 4\left(h_n(x) + h_n(y) - \frac{h_n(x+y)}{2}\right) \\
&\quad + 4\Big((2^{-n} - 4^{-n}C)\big(f_0(x) + f_0(y)\big)\Big) - 2^{-n+1}f_0(x+y) \\
&= 8\left(\frac{h_n(x) + h_n(y)}{2} - h_n\left(\frac{x+y}{2}\right)\right) \\
&\quad + 4.2^{-n}\left(f_0(x) + f_0(y) - 2f_0\left(\frac{x+y}{2}\right)\right) - 4^{-n+1}C\big(f_0(x) + f_0(y)\big) \\
&\geq 2^{-n}Q_{\|\cdot\|_2}(x,y) - 2^{-2n+2}C\big(f_0(x) + f_0(y)\big) \\
&= 2^{-n}Q_{\|\cdot\|_2}(x,y) - 2^{-2n+1}C\big(\|x\|_2^2 + \|y\|_2^2\big) \\
&\quad \text{(from the convexity of } h_n \text{ and (8.1)).}
\end{aligned}$$

The dual inequality follows easily by making the same computations with the Fenchel conjugates of the g's and f's, and where the rôle of f_0 is now played by g_0^*. □

8.2 Applications

Theorem 148, combined with the description of several convexity properties of the norm and its dual —a source of smoothness on the space— by using the function Q defined in (8.7) (see Propositions 113, 115, 116, and 118), allows for a number of results in the direction mentioned at the introduction of Section 8.1.
This will be done through some examples. We shall follow the notation in Section 8.1. Recall that $\|\cdot\|_1$ and $\|\cdot\|_2$ are two equivalent norms on a Banach space X:

1. Assume that $\|\cdot\|_2$ and $\|\cdot\|_1^*$ are both strictly convex. Then $\|\cdot\|$ and $\|\cdot\|^*$ are both strictly convex, too, so in particular $\|\cdot\|$ is simultaneously strictly convex and Gâteaux smooth.

➡ Indeed, let $x, y \in X$ be such that $Q_{\|\cdot\|}(x,y) = 0$. From (8.8) we get ($0 \leq$) $Q_{\|\cdot\|_2}(x,y) \leq 2^{-n+1}C(\|x\|_2^2 + \|y\|_2^2)$, for all $n \in \mathbb{N}$. Hence $Q_{\|\cdot\|_2}(x,y) = 0$ and, by the fact that $\|\cdot\|_2$ is strictly convex, we get $x = y$ (see Proposition 113). This shows that $\|\cdot\|$ is strictly convex. A similar argument, now using (8.9), shows that $\|\cdot\|^*$ is strictly convex. Lemma 54 ends the proof. □

2. Assume that both $\|\cdot\|_2$ and $\|\cdot\|_1^*$ are locally uniformly rotund. Then $\|\cdot\|$ and $\|\cdot\|^*$ are locally uniformly rotund, too, so $\|\cdot\|$ is simultaneously locally uniformly rotund and Fréchet differentiable.

   ➡ Indeed, let $x, x_m \in X$, for $m \in \mathbb{N}$, such that $Q_{\|\cdot\|}(x, x_m) \to 0$. From (8.8) we get, for all $m, n \in \mathbb{N}$,

$$Q_{\|\cdot\|_2}(x, x_m) \leq 2^n Q_{\|\cdot\|}(x, x_m) + 2^{-n+1}C(\|x\|_2^2 + \|x_m\|_2^2). \tag{8.11}$$

   It follows from (5.3) that the sequence $\{\|x_m\|\}_{m=1}^{\infty}$ (and hence also $\{\|x_m\|2\}_{m=1}^{\infty}$) is bounded. Fix $\varepsilon > 0$ and find $n \in \mathbb{N}$ such that $2^{-n+1}C(\|x\|_2^2 + \|x_m\|_2^2) < \varepsilon/2$ for all $m \in \mathbb{N}$. There exists $m_0 \in \mathbb{N}$ such that, for $m \geq m_0$, $2^n Q_{\|\cdot\|}(x, x_m) < \varepsilon/2$. Hence, from (8.11) we get $Q_{\|\cdot\|_2}(x, x_m) < \varepsilon$ for $m \geq m_0$. Thus, $Q_{\|\cdot\|_2}(x, x_m) \to 0$. From the local uniform rotundity of $\|\cdot\|_2$ we get $\|x_m - x\| \to 0$. This shows the local uniform rotundity of $\|\cdot\|$ (see Proposition 115). A similar argument, now by using (8.9) instead, shows that $\|\cdot\|^*$ is locally uniformly rotund. Lemma 54 concludes that $\|\cdot\|$ is Fréchet differentiable. □

3. Assume that both $\|\cdot\|_2$ and $\|\cdot\|_1^*$ are uniformly rotund (thus, the space is reflexive, see Subsection 5.2.2). Then $\|\cdot\|$ and $\|\cdot\|^*$ are both uniformly rotund, too, hence $\|\cdot\|$ is simultaneously uniformly rotund and uniformly Fréchet differentiable (see (vii) in Theorem 143). The argument follows the one in Item 2 above combined with the characterization of uniform rotundity in Proposition 114. Notice that the boundedness of one of the sequences $\{x_n\}_{n=1}^{\infty}$ or $\{y_n\}_{n=1}^{\infty}$ to be used is needed.

4. A similar argument to the one used in the proof of Item 2 above shows the following results (we need Propositions 116 and 117 both in (i) and (ii), and Proposition 118 for (iii)):

   (i) If $\|\cdot\|_2$ is WUR and $\|\cdot\|_1^*$ is W*UR, then $\|\cdot\|$ is WUR and $\|\cdot\|^*$ is W*UR (so $\|\cdot\|$ is WUR and uniformly Gâteaux differentiable, see, e.g., [FHHMZ11, Exercise 7.30]).

   (ii) If $\|\cdot\|_2$ is WLUR and $\|\cdot\|_1^*$ is W*LUR, then $\|\cdot\|$ is WLUR and $\|\cdot\|^*$ is W*LUR.

(iii) If $\|\cdot\|_2$ is URED and $\|\cdot\|_1^*$ is URED, too, then $\|\cdot\|$ is URED and $\|\cdot\|^*$ is URED.

5. Some other combinations are possible. Let us mention, for example, that if $\|\cdot\|_2$ is strictly convex and $\|\cdot\|_1^*$ is W*UR, then $\|\cdot\|$ is strictly convex and $\|\cdot\|^*$ is W*UR (so $\|\cdot\|$ is also UG, see Theorem 143 above). The proof follows from Theorem 148, the characterization of strict convexity in Proposition 113, the characterization of W*UR in Proposition 117, and again a simple convergence argument.

It is an open problem whether a Banach space X admits an equivalent LUR and Fréchet differentiable norm as soon as it admits an equivalent LUR norm and an equivalent Fréchet differentiable norm.

Chapter 9

Tools for renorming

This chapter focuses on separable Banach spaces. Despite this, some non-separable results are included, mostly because they provide basic tools for renorming. For example, Day's norm on $\ell_\infty(\Gamma)$ is presented in Subsection 9.4.1, although at this stage we shall only proof its rotundity character (Theorem 170) —its LUR property will be given in Corollary 243—. Theorem 171 provides another equivalent norm on $c_0(\Gamma)$ whose dual norm is LUR (and so, the original norm on $c_0(\Gamma)$ is Fréchet differentiable).
Section 9.5 presents the crucial Kadets' deep renorming theorem (Theorems 173, 175, and 176) that provides every separable Banach space with an equivalent LUR norm. This was really a cornerstone in the whole renorming theory —aimed originally at proving the homeomorphic character of any two infinite-dimensional separable Banach spaces, a result due to M.I. Kadets—, and motivating, among other things, the fundamental renorming theorem of WCG spaces by Kadets' former student Troyanski (Theorem 218 below).

In this chapter we shall discuss some of the basic renorming techniques:
• A strictly convex renorming (built in two different ways) for Banach spaces in a certain class that includes the separable spaces and also ℓ_∞ (Propositions 152 and 161).
• A strictly convex dual renorming of X^* —i.e., the equivalent norm defined on X^* is a dual norm, and it is strictly convex— (and so a Gâteaux smooth renorming of the space) for Banach spaces in a class that include the separable spaces (Proposition 156).
• Combining the two previous techniques, we get an equivalent norm on a separable space that is simultaneously strictly convex and Gâteaux differentiable (Proposition 158).
• A W*UR (W*LUR) dual renorming for spaces in a class that includes the separable spaces (Proposition 162).
• A WUR renorming for the separable Asplund spaces (Proposition 164).
• Uniformly rotund in every direction (uniformly rotund in a compact set of directions) renormings for some classes of Banach spaces (Proposition 167 and 168).

9.1 Two simple facts

We shall often need the following simple result, whose straightforward proof is omitted:

A. J. Guirao et al., *Renormings in Banach Spaces*, Monografie Matematyczne 75,
https://doi.org/10.1007/978-3-031-08655-7_9

Proposition 149. *Let X_i be Banach spaces, $i = 1, 2$. Let $T : X_1 \to X_2$ be a continuous linear mapping. Then $\ker T = (T^* X_2^*)_\perp$, and $\ker T^* = (TX_1)^\perp$. Thus,*
(i) T is one-to-one if, and only if, $T^* X_2^*$ is w^*-dense in X_1^*, and
(ii) T^* is one-to-one if, and only if, TX_1 is $\|\cdot\|$-dense in X_2.

Some of the procedures in this chapter to renorm a space X rely on the existence of continuous linear operators from or to a "good" space (as ℓ_2). Thus, we shall include here the following elementary result, that shall be frequently used in this text from now on:

Proposition 150. *Let X be a Banach space.*
(i) If X is separable, then there exists a continuous linear operator $T : \ell_2 \to X$ with dense range.
(ii) If (X^*, w^*) is separable, then there exists a one-to-one continuous linear operator $T : X \to \ell_2$.

Proof. (i) Let $\{x_n : n \in \mathbb{N}\}$ be a dense subset of S_X. Let $T : \ell_2 \to X$ be defined as $Tb := \sum_{n=1}^{\infty} (1/n) b_n x_n$, whenever $b = (b_n) \in \ell_2$. Fix $n \leq m$ in $\mathbb{N}$. Note that, by the Cauchy–Schwarz inequality,

$$\left\| \sum_{k=n}^{m} \frac{1}{k} b_k x_k \right\| \leq \sum_{k=n}^{m} \frac{1}{k} |b_k| \leq \left(\sum_{k=n}^{m} \frac{1}{k^2} \right)^{1/2} \left(\sum_{k=n}^{m} b_k^2 \right)^{1/2}. \tag{9.1}$$

This shows that the series $\sum (1/n) b_n x_n$ is Cauchy, hence convergent, and so T is well defined. From (9.1) we get that $\|Tb\| \leq (\sum_{n=1}^{\infty} 1/n^2)^{1/2} \|b\|_2$, hence T is (linear and) continuous. Note that $\{x_n : n \in \mathbb{N}\} \subset T\ell_2$, hence $T\ell_2$ is dense in X.
(ii) Let $\{x_n^* : n \in \mathbb{N}\}$ be a w^*-dense subset of X^*. We may assume that $x_n^* \neq 0$ for all $n \in \mathbb{N}$. Define $T : X \to \ell_2$ as

$$Tx = \left(\frac{\langle x, x_n^* \rangle}{n \|x_n^*\|^*} \right), \text{ for } x \in X.$$

Then, for $n \leq m$ in $\mathbb{N}$,

$$\left(\sum_{k=n}^{m} \left(\frac{\langle x, x_k^* \rangle}{k \|x_k^*\|^*} \right)^2 \right)^{1/2} \leq \left(\sum_{k=n}^{m} \frac{1}{k^2} \right)^{1/2} \|x\|. \tag{9.2}$$

Thus, T is well defined. From (9.2) we get that T is (linear and) continuous. If $Tx = 0$ then $\langle x, x_n^* \rangle = 0$ for all $n \in \mathbb{N}$. The w^*-density of $\{x_n^* : n \in \mathbb{N}\}$ in X^* concludes that $x = 0$; thus, T is one-to-one. □

Remark 151. Observe that if a Banach space X is separable, then (X^*, w^*) is separable, too. There are nonseparable Banach spaces —like ℓ_∞, use Goldstine's theorem— whose dual is w^*-separable. Ⓡ

9.2 An elementary strictly convex renorming

9.2.1 A strictly convex renorming of the space

The following simple proposition is one of the earliest results in the area of renorming. Clarkson's contribution to renorming has been substantial. We shall encounter his methods again and again in this text (see, e.g., Subsection 9.3.1 and Chapter 18 below). In [Clar36, Theorem 9] the reader may find the seminal idea both behind Proposition 152 and Proposition 161 below.

Proposition 152 (Clarkson, [Clar36]). *Assume that $(X_i, \|\cdot\|_i)$, $i = 1, 2$, are Banach spaces such that $(X_2, \|\cdot\|_2)$ is strictly convex, and that there exists a continuous linear and one-to-one operator $T : X_1 \to X_2$. Then*

$$\boxed{|||x_1||| := \|x_1\|_1 + \|Tx_1\|_2,\ x_1 \in X_1,} \tag{9.3}$$

defines a strictly convex equivalent norm on X_1.

Proof. To check that $|||\cdot|||$ is an equivalent norm is easy.
Let $x_1 \neq 0$ and $y_1 \neq 0$ in X_1 such that $|||x_1+y_1||| = |||x_1|||+|||y_1|||$, i.e., $\|x_1+y_1\|_1 + \|Tx_1+Ty_1\|_2 = \|x_1\|_1+\|Tx_1\|_2+\|y_1\|_1+\|Ty_1\|_2$. Since $\|x_1+y_1\|_1 \leq \|x_1\|_1+\|y_1\|_1$ and $\|Tx_1 + Ty_1\|_2 \leq \|Tx_1\|_2 + \|Ty_1\|_2$, we get in particular $\|Tx_1 + Ty_1\|_2 = \|Tx_1\|_2 + \|Ty_1\|_2$. The injectivity of T gives $Tx_1 \neq 0$ and $Ty_1 \neq 0$. By (iii) in Lemma 81 we get $Tx_1 = \lambda Ty_1$ for some $\lambda > 0$, i.e. $T(x_1) = T(\lambda y_1)$. Again by the injectivity of T, $x_1 = \lambda y_1$. A new appeal to (iii) in Lemma 81 gives that $|||\cdot|||$ is strictly convex. □

Remark 153. A different proof of Proposition 152 (by using the square of the norm instead) will be given below (see Proposition 161). Yet, the present technique may be applied to several situations, for example when dealing with norm-attaining operators. Ⓡ

An application

Theorem 154. *If $(X, \|\cdot\|)$ is a Banach space with a w^*-separable dual (in particular, if $(X, \|\cdot\|)$ is separable), then X has an equivalent strictly convex norm.*

Proof. This follows from (ii) in Proposition 150 and from Proposition 152. □

Remark 155. 1. Theorem 154 applies also to some nonseparable spaces, like, e.g., ℓ^∞, as it was noticed in Remark 151.

2. Of course, Theorem 154 applies to Banach spaces X whose dual unit ball is w^*-separable (since then (X^*, w^*) is also separable). However, it is worth to mention that there are spaces X such that (X^*, w^*) is separable, although for no equivalent norm its dual unit ball is w^*-separable [JoLi74]. Ⓡ

9.2.2 A strictly convex renorming of the dual space

By using the previous technique, we may provide also a dual version of Proposition 152, getting a strictly convex renorming of the dual space, with the extra property that the norm so defined is a dual norm. This will give a Gâteaux smooth renorming of the original space, in view of Proposition 82 above.

Proposition 156 (Day, [Day55]). *Let* $(X_i, \|\cdot\|_i)$ *be a Banach space,* $i = 1, 2$. *Let* $T : X_2 \to X_1$ *be a linear and continuous mapping such that* TX_2 *is dense in* X_1. *If* $\|\cdot\|_2^*$ *on* X_2^* *is strictly convex, then* $|||\cdot|||^*$ *defined on* X_1^* *as*

$$\boxed{|||x_1^*|||^* := \|x_1^*\|_1^* + \|T^*x_1^*\|_2^*, \; x_1^* \in X_1^*} \tag{9.4}$$

is a strictly convex dual equivalent norm on X_1^* *(so the predual norm* $|||\cdot|||$ *on* X_1 *is a Gâteaux differentiable equivalent norm).*

Proof. By Proposition 149, $T^* : X_1^* \to X_2^*$ is one-to-one. It follows from Proposition 152 that the (equivalent) norm $|||\cdot|||^*$ is strictly convex. Note that T^* is w^*-w^*-continuous, and that $\|\cdot\|_1^*$ and $\|\cdot\|_2^*$ are, certainly, w^*-lower semicontinuous. Thus, $|||\cdot|||^*$ is w^*-lower semicontinuous, too, hence a dual norm (see Proposition 17). By Proposition 82, its predual norm $|||\cdot|||$ on X_1 is Gâteaux differentiable. □

An example of the application of this technique will be given below and in the proof of Proposition 232.

The following geometric construction gives the predual norm $|||\cdot|||$ on X_1 to the norm $|||\cdot|||^*$ on X_1^* defined in Proposition 156, and it is inspired in the techniques and results in Subsection 3.7.4 (there, only sum of balls were considered; however, the arguments can be adapted to the situation described in (9.5) below). Denote by B_i the closed unit ball of $(X_i, \|\cdot\|_i)$, $i = 1, 2$. Put

$$B := \overline{B_1 + TB_2} \; (\subset X_1). \tag{9.5}$$

Then B is closed convex, balanced, and $B_1 \subset B \subset (1 + \|T\|)B_1$, so B is the closed unit ball of an equivalent norm on X_1. We claim that its dual norm is just $|||\cdot|||^*$ defined by (9.4). Indeed, for $x_1^* \in X_1^*$,

$$\begin{aligned} |||x_1^*|||^* &= \sup\{\langle b, x_1^*\rangle : \; b \in B\} = \sup\{\langle b, x_1^*\rangle : \; b \in B_1 + TB_2\} \\ &= \sup\{\langle b_1, x_1^*\rangle : \; b_1 \in B_1\} + \sup\{\langle Tb_2, x_1^*\rangle : \; b_2 \in B_2\} \\ &= \|x_1^*\|_1^* + \sup\{\langle b_2, T^*x_1^*\rangle : \; b_2 \in B_2\} = \|x_1^*\|_1^* + \|T^*x_1^*\|_2^*. \end{aligned} \tag{9.6}$$

An application

Theorem 157 (Day, [Day55]). *If* X *is separable, then* X^* *admits an equivalent strictly convex dual norm. Thus,* X *admits an equivalent Gâteaux differentiable norm.*

➪ It follows from (i) in Proposition 150 and Proposition 156. □

This result will be improved in Subsection 9.2.3 below.

9.2.3 A strictly convex and Gâteaux smooth renorming of the space

We may combine the two previous procedures in one by using Asplund averaging (see Chapter 8). Indeed, the conclusion of Proposition 158 below follows by using Theorem 154, Theorem 157, and Item 1 in Section 8.2. Here we present an easy version of it.

Proposition 158 (Klee). *Let $(X_1, \|\cdot\|_1)$ be a Banach space. Assume that $\|\cdot\|_1$ is Gâteaux smooth, and that there exists a continuous linear and one-to-one mapping from $(X_1, \|\cdot\|_1)$ into a Banach space $(X_2, \|\cdot\|_2)$ such that $\|\cdot\|_2$ is simultaneously Gâteaux smooth and strictly convex. Then X_1 has an equivalent norm $|||\cdot|||$ that is simultaneously Gâteaux smooth and strictly convex.*

Proof. Define $|||\cdot|||$ on X_1 by equation (9.3). Proposition 152 shows that $|||\cdot|||$ is an equivalent strictly convex norm on X_1. Since $\|\cdot\|_2$ is Gâteaux smooth and T is a linear mapping, the Chain Rule shows that $x \to \|Tx\|_2$ is also Gâteaux smooth, and so it is $|||\cdot|||$ on X_1. □

An application

Theorem 159 (Klee, [Klee59]). *Every separable Banach space X has an equivalent norm $|||\cdot|||$ that is simultaneously Gâteaux smooth and strictly convex.*

Proof. We shall provide two proofs of this result:
(i) Theorem 157 shows that there exists an equivalent Gâteaux smooth norm $\|\cdot\|_1$ on X. Moreover, (ii) in Proposition 150 gives a linear continuous and one-to-one mapping $T: X \to \ell_2$. Obviously, the Euclidean norm $\|\cdot\|_2$ on ℓ_2 is simultaneously strictly convex and Gâteaux smooth. Proposition 158 applies to provide an equivalent norm $|||\cdot|||$ on X that is simultaneously strictly convex and Gâteaux smooth.
(ii) Theorem 154 provides an equivalent strictly convex norm $\|\cdot\|_1$ on X. Theorem 157 gives an equivalent strictly convex norm $\|\cdot\|_2$ on X^*. It is enough to use now the Asplund averaging technique (see Item 1 in Section 8.2 above). □

9.3 Transfer methods by using the square of the norm

Let us see how the lemmata in Subsection 5.6.1 can be used in renorming:
Let P be a property of a seminorm p that can be described by using the formula (5.2), i.e., $Q_p(x, y) := 2p^2(x) + 2p^2(y) - p^2(x + y)$, to get a connection between x and y (for example, see (ii) in Proposition 113 for a norm $p := \|\cdot\|$ and the

property of being strictly convex). Assume that the formula $p^2 = \sum_{n=1}^{\infty} \alpha_n p_n^2$ for seminorms p_n's and positive numbers α_n's defines the function p. Then, according to Lemma 111, the "Q_p behaviour" of p passes to *all* p_n's. If the sought connection is obtained by checking one of those p_n's, we are done, and p has property P. This powerful technique will appear in many of the following sections.

9.3.1 Clarkson's method

One of the earliest methods for renorming is due to Clarkson [Clar36] —another one, also due to him, was recorded in Proposition 152 above—. It fits precisely into the scheme presented in the introduction to the present section, although it will involve only two seminorms instead of a full sequence of them: Let $(X_1, \|\cdot\|_1)$ and $(X_2, \|\cdot\|_2)$ be two Banach spaces, and assume that $(X_2, \|\cdot\|_2)$ enjoys some property tested via the expression (5.2). Let $T : (X_1, \|\cdot\|_1) \to (X_2, \|\cdot\|_2)$ be a continuous linear mapping. On X_1 we consider, besides the original norm $p_1 := \|\cdot\|_1$, the seminorm $p_2(x) := \|Tx\|_2$, $x \in X_1$. Define then $|||\cdot||| := p$ as in (5.5) in Lemma 111 (letting $p_n := 0$ for $n = 3, 4, \ldots$), i.e.:

$$\boxed{p^2(x) = p_1^2(x) + p_2^2(x), \quad \text{i.e.,} \quad |||x|||^2 = \|x\|_1^2 + \|Tx\|_2^2, \quad x \in X_1.} \tag{9.7}$$

Note that $|||\cdot|||$ is an equivalent norm on X_1 (for the subadditivity, see Remark 112 above).
Following this scheme, the quality of the new norm $|||\cdot|||$ is tested on $\|\cdot\|_2$. Unfortunately, this term is written as $\|Tx\|_2^2$ (and it is not in general a norm). In order to "return" to $(X_1, |||\cdot|||)$ we shall need injectivity of T on one hand, and some kind of compactness on the other, according to the following remark:

Remark 160. The reason for the last statement is the following basic fact: If K and S are topological spaces, K is compact, and $f : K \to S$ is a one-to-one continuous mapping, then $f(K)$ is compact and f is a homeomorphism from K onto $f(K)$. In particular, if $\{f(k_i)\}$ is a net in $f(K)$ that converges to $s \in S$, then $\{k_i\}$ converges to some $k \in K$ and $f(k) = s$. Ⓡ

If the sought geometric condition of the new norm depends, as in the case of strict convexity, only on two points, we can dispense the compactness requirement. Proposition 113 above is an example. This result is formally the same as Proposition 152 (only the new norm is different). The reader may wonder why to appeal to a new method when an apparently simpler one was available. The answer is that for strict convexity, indeed, there is no need for another approach. However, for renorming with some extra important properties, the former technique in Section 9.2 is not powerful enough. This is the case when, instead, the sought property depends on the convergence of a sequence (see, e.g., Proposition 117)—and then compactness is a must—. For example, the important locally uniformly rotundness property (LUR, for short) is formulated by using sequences (see

Proposition 115). For the norm defined via (9.8), and for a sequence $\{x_n\}_{n\in\mathbb{N}}$ in X_1 such that $Q_{|||\cdot|||}(x, x_n) \to 0$ (see equation (8.7)), we get, in case that $(X_2, \|\cdot\|_2)$ is LUR, that $\|Tx_n - Tx\|_2 \to 0$. However, there is no guarantee that $\{x_n\}_{n\in\mathbb{N}}$ will converge, and so no LUR conclusion for $(X_1, |||\cdot|||)$ can be reached this way. This is why a more sophisticated way to renorm $(X_1, \|\cdot\|_1)$ is needed. This will be accomplished, for separable spaces, by the fundamental Theorem 173. A sufficient condition for making the transfer scheme for LUR renorming work in a general setting will be given in Theorem 221. Similar situation happens with URED norms, etc.
The sought compactness —in the w^*-topology— for sequences is guaranteed if the to-be-renormed-space is a dual space X_1^* and the sequence is bounded; in the case of the weak topology of X_1 —and still for a bounded sequence— the procedure consists in going to the bidual X_1^{**} (the requirement will be now the injectivity of T^{**}). This will be made explicit in the next few items.

Clarkson's method, as we mentioned, is modelled according to the following scheme:

$(X_1, \|\cdot\|_1) \xrightarrow{T} (X_2, \|\cdot\|_2)$, where T is linear, one-to-one, and bounded.
Consider $(X_1, |||\cdot|||)$, where $|||\cdot|||$ is defined as in (9.7), i.e.,

$$|||x|||^2 = \|x\|_1^2 + \|Tx\|_2^2. \tag{9.8}$$

For the right application of the suggested renorming method to several geometric properties of a norm $\|\cdot\|$, we provided in Subsection 5.6.2 alternative characterizations of those in terms of the expression $Q_{\|\cdot\|}$ in (8.7). The following ones were included there: For the space X, strict convexity (R) (Proposition 113), weak uniform convexity (WUR) and weak local uniform convexity (WLUR) (Proposition 116), uniform convexity in every direction (URED) (Proposition 118), and uniform convexity in a compact set of directions (URWC) (Proposition 121). For the dual space, w^*-uniform convexity (W*UR) and w^*-local uniform convexity (W*LUR) (Proposition 117).

Strictly convex renorming

Strictly convex renorming was achieved thanks to the simple technique in Section 9.2 above. The same goal may be attained by using Clarkson's method described here.

As we already mentioned, the following result is formally identical to Proposition 152. The only change is that the equivalent norm defined in its proof is different from the one given by equation (9.3).

Proposition 161. *Let $(X_i, \|\cdot\|_1)$, $i = 1, 2$, be Banach spaces. Let $T : X_1 \to X_2$ be a linear, continuous and one-to-one mapping. If $(X_2, \|\cdot\|_2)$ is strictly convex, then the norm $|||\cdot|||$ on X_1 given by (9.7) is strictly convex, too.*

Proof. If for $x, y \in X_1$ we have $Q_{|||\cdot|||}(x, y) = 0$, then (directly or by taking $x_n := y$ for all $n \in \mathbb{N}$ and applying Lemma 111) we get $Q_{\|\cdot\|_2}(Tx, Ty) = 0$. Since $\|\cdot\|_2$ is (R), we get $Tx = Ty$, and due to the fact that T is one-to-one, we finally obtain $x = y$. By Proposition 113, $|||\cdot|||$ is strictly convex. □

Applications. The previous technique provides alternative strictly convex renormings for the class of spaces with w^*-separable dual spaces (see Subsection 9.2.1). We recover Theorem 154 above:
If X is a Banach space with a w^*-separable dual space, then X has an equivalent strictly convex norm.

➩ A linear continuous and one-to-one operator from X into ℓ_2 was constructed in Proposition 150. Instead of using Proposition 152 we apply now Proposition 161. The result is the same. The built equivalent norm is not.

In particular, *every separable Banach space admits a strictly convex equivalent norm*, and *this happens too, for the space ℓ_∞*.

Dual w^*-renormings

For bounded sequences in the dual of a Banach space, w^*-relative compactness is guaranteed. We need w^*-continuity and injectivity of the transfer mapping T^*. We are interested here in the W*UR and W*LUR properties introduced in Definition 85.

Obviously, for a dual norm we have W*UR⇒W*LUR⇒R.

Proposition 162. *Let $(X_i, \|\cdot\|_1)$, $i = 1, 2$, be Banach spaces. Let $T : X_2 \to X_1$ be a linear and continuous mapping such that $T^* : X_1^* \to X_2^*$ is one-to-one (equivalently, TX_2 is dense in X_1). If $(X_2^*, \|\cdot\|_2^*)$ is W*UR (W*LUR), then the norm $|||\cdot|||^*$ defined on X_1^* as $(|||x_1^*|||)^2 = (\|x_1^*\|_1^*)^2 + (\|T^*x_1^*\|_2^*)^2$ is a dual equivalent W*UR norm (respectively, W*LUR). Thus, in the first case, X_1 has an equivalent UG norm.*

Proof. We shall prove the W*UR part. The proof of the other one is similar. Since T^* is w^*-w^*-continuous, $|||\cdot|||^*$ is w^*-lower semicontinuous, and so it is a dual norm (see Proposition 17). Let $\{x_n^*\}_{n\in\mathbb{N}}$ be a bounded sequence in X_1^*, and $\{y_n^*\}_{n\in\mathbb{N}}$ another sequence in X_1^* such that $Q_{|||\cdot|||^*}(x_n^*, y_n^*) \to 0$. Then, by Lemma 108, the sequence $\{y_n^*\}_{n\in\mathbb{N}}$ is also bounded, and $Q_{\|\cdot\|_2^*}(T^*x_n^*, T^*y_n^*) \to 0$. This shows that $(T^*x_n^* - T^*y_n^*) \overset{w^*}{\to} 0$. Remark 160 gives $(x_n^* - y_n^*) \overset{w^*}{\to} 0$. Finally, Proposition 117 shows that $|||\cdot|||$ is W*UR. The UG property of the predual norm $|||\cdot|||$ on X follows from Theorem 143. □

Applications. The following consequence of Proposition 162 extends Theorem 157 (since every W*UR norm in a dual space X^* is strictly convex):

Theorem 163. *If X is separable, then X^* has an equivalent W*UR dual norm (hence X has an equivalent UG norm).*

➡ *Proof.* By Proposition 150, there exists a mapping $T : \ell_2 \to X$ such that $T\ell_2$ is $\|\cdot\|$-dense in X. By Proposition 149, $T^* : X^* \to \ell_2^*$ $(= \ell_2)$ is one-to-one. The conclusion follows from Proposition 162, and the last statement from Theorem 143.

Weak UR and weak LUR renormings

For bounded sequences in X, only w^*-relative compactness in X^{**} is guaranteed. Accordingly, we should request injectivity of T^{**}, where T is the "transfer" mapping.

Proposition 164. *Let $(X_i, \|\cdot\|_i)$, $i = 1, 2$, be Banach spaces. Let $T : X_1 \to X_2$ be a linear and continuous mapping such that $T^{**} : X_1^{**} \to X_2^{**}$ is one-to-one. If $(X_2, \|\cdot\|_2)$ is WUR (WLUR), then the norm $|||\cdot|||$ on X_1 defined by (9.7) is an equivalent WUR (respectively, WLUR) norm.*

Proof. We only prove the WUR part. The proof of the WLUR part is similar. Let $\{x_n\}_{n=1}^\infty$ and $\{y_n\}_{n=1}^\infty$ be sequences in X_1, and assume that $\{x_n\}_{n=1}^\infty$ is bounded. If $Q_{|||\cdot|||}(x_n, y_n) \to 0$, it follows from (5.3) that $\|x_n\|_1 - \|y_n\|_1 \to 0$ (in particular, $\{y_n\}_{n=1}^\infty$ is also bounded), and from Lemma 111, $Q_{\|\cdot\|_2}(Tx_n, Ty_n) \to 0$. By Proposition 116, $(Tx_n - Ty_n) \overset{w}{\to} 0$. The set $\{x_n, y_n : n \in \mathbb{N}\}$ is w^*-relatively compact in X_1^{**}, and T^{**} is one-to-one. Remark 160 shows that $(x_n - y_n) \overset{w}{\to} 0$. This proves, again by Proposition 116, that $|||\cdot|||$ is also WUR. □

Application.

Theorem 165. *If X^* is separable, then X has an equivalent WUR norm.*

➡ Let $\{x_n^* : n \in \mathbb{N}\}$ be a dense subset of $S_{(X^*,\|\cdot\|^*)}$. Define the mapping $T : (X, \|\cdot\|) \to (\ell_2, \|\cdot\|_2)$ as $Tx := ((1/n)\langle x, x_n^*\rangle)_{n=1}^\infty$ for $x \in X$. Clearly, T is well defined, bounded, and linear. Note that $T^* e_n^* = (1/n)x_n^*$ for $n \in \mathbb{N}$, where e_n^* is the n-th vector of the canonical basis of ℓ_2^* $(= \ell_2)$. Thus, $T^*\ell_2^*$ is $\|\cdot\|$-dense in X^*, hence T^{**} is one-to-one (see Proposition 149). The conclusion follows from Proposition 164, as $(\ell_2, \|\cdot\|_2)$ is obviously WUR.

Remark 166. Every space with a WUR norm is Asplund [Há96a] (see Theorem 144 above). Thus, the class of separable Banach spaces with an equivalent WUR norm is, precisely, the class of spaces with a separable dual. ®

Directional renormings

V. Zizler [Ziz71] and M.A. Smith [Sm78] proved statements for the properties URED and URWC, respectively, similar to those in Propositions 161 and 164 above.

1. **Uniform convexity in every direction**

Proposition 167 (V. Zizler, [Ziz71], Proposition 14). *Let $(X_i, \|\cdot\|_1)$, $i = 1, 2$, be Banach spaces. Let $T : X_1 \to X_2$ be a linear, one-to-one and continuous mapping. If $\|\cdot\|_2$ is URED, then the norm $|||\cdot|||$ on X_1 given by (9.7) is URED, too.*

Proof. Assume that $\|\cdot\|_2$ is URED. Let $\{x_n\}_{n=1}^{\infty}$ be a bounded sequence in X_1 and $z \in X_1 \backslash \{0\}$. Put $y_n := x_n - \lambda_n z$ for some $\lambda_n \in \mathbb{R}$, $n \in \mathbb{N}$. Assume that $Q_{|||\cdot|||}(x_n, y_n) \to 0$. Then, $Q_{\|\cdot\|_2}(Tx_n, Ty_n) \to 0$. Since $\{Tx_n\}_{n=1}^{\infty}$ is bounded, $Ty_n = Tx_n - \lambda_n Tz$, and $Tz \neq 0$, we get $\lambda_n Tz = Tx_n - Ty_n \to 0$, hence $\lambda_n \to 0$, and so $x_n - y_n \to 0$. This proves that then $|||\cdot|||$ is also URED. □

2. **Uniform convexity in a compact set of directions**

Proposition 168 (M.A. Smith, [Sm78]). *Let $(X_i, \|\cdot\|_1)$, $i = 1, 2$, be Banach spaces. Let $T : X_1 \to X_2$ be a linear, one-to-one and continuous mapping. If $\|\cdot\|_2$ is URWC, then the norm $|||\cdot|||$ on X_1 given by (9.7) is URWC, too.*

Proof. Let $\{x_n\}_{n=1}^{\infty}$ be a bounded sequence in X_1, $\{y_n\}_{n=1}^{\infty}$ a sequence in X_1, and $z \in X_1$ such that $x_n - y_n \xrightarrow{w} z$. Assume that $Q_{|||\cdot|||}(x_n, y_n) \to 0$. Then, $Q_{\|\cdot\|_2}(Tx_n, Ty_n) \to 0$. Since $\{Tx_n\}_{n=1}^{\infty}$ is bounded, it follows that $Tx_n - Ty_n \xrightarrow{w} Tz$. Thus, we get $Tz = 0$, hence $z = 0$. By using Proposition 168, this proves that $|||\cdot|||$ is also URWC. □

3. **An application.** The following result is a consequence of Propositions 168 and 150, and the observation after Definition 97:

Corollary 169. *If X is a Banach space such that (X^*, w^*) is separable, then X has an equivalent norm which is URWC (in particular, URED).*

9.4 Some "ad-hoc" norms on classical spaces

9.4.1 Day's norm I: Day's norm on $\ell_\infty(\Gamma)$

Crucial for renorming theory has been Day's discovery of the norm $\|\cdot\|_D$ on $\ell_\infty(\Gamma)$ now known as **Day's norm.** On $c_0(\Gamma)$ it is a lattice (i.e., $|x(\gamma)| \le |y(\gamma)|$ for all $\gamma \in \Gamma$ implies $\|x\|_D \le \|y\|_D$ for $x, y \in c_0(\Gamma)$) LUR norm equivalent to the supremum norm $\|\cdot\|_\infty$, a result due to J. Rainwater (see Corollary 243 below, where the full proof, based on a technical lemma in [MOTV09], will be presented; check also Section 21.1). Here we shall just prove Day's result (Theorem 170 below), showing that $\|\cdot\|_D$ on $c_0(\Gamma)$ is strictly convex.

Day's norm $\|\cdot\|_D$ on $\ell_\infty(\Gamma)$ is defined for $x = \{x(\gamma)\}$ in the following way: First, for $n \in \mathbb{N}$ let us introduce the seminorm $\|\cdot\|_n$ on $\ell_\infty(\Gamma)$ given by

$$\|x\|_n^2 = \sup\left\{\sum_{\gamma\in N} x(\gamma)^2 : N \subset \Gamma,\ \operatorname{card}(N) = n\right\}. \tag{9.9}$$

Then put, for $x \in \ell_\infty(\Gamma)$,

$$\|x\|_D := \left(\sum_{n=1}^{\infty} \frac{1}{2^n}\|x\|_n^2\right)^{\frac{1}{2}}. \tag{9.10}$$

Note that $\|\cdot\|_D$ is a norm on $\ell_\infty(\Gamma)$, obviously equivalent to $\|\cdot\|_\infty$. Figure 9.1 depicts one-fourth of the closed unit ball for $\|\cdot\|_D$ in $\mathbb{R}^2$.

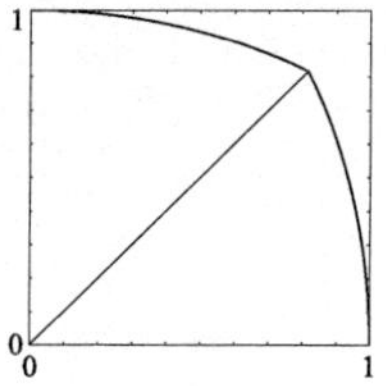

Figure 9.1: The closed unit ball of Day's norm in the positive quadrant of $\mathbb{R}^2$

For a one-step formula that gives Day's norm on $c_0(\Gamma)$ write (see equation (9.12) below):

$$|x|_D := \left(\sum_{n=1}^{\infty} \frac{1}{2^{n-1}} x(\gamma_n)^2\right)^{\frac{1}{2}}, \tag{9.11}$$

defined on $\ell_\infty^c(\Gamma)$ —the space of all real and bounded sequences on Γ with countable support— where $\{\gamma_n\}_{n\in\mathbb{N}}$ is an enumeration of the support of x in such a way that $|x(\gamma_j)| \ge |x(\gamma_{j+1})| > 0$ for every $j \in \mathbb{N}$.

➪ Note that the particular association of coefficients to coordinates in (9.11) maximizes the result. Indeed, given two sequences $\{a_n\}$ and $\{b_n\}$ of positive numbers with $a_n \geq a_{n+1}$ and $b_n \geq b_{n+1}$ for all $n \in \mathbb{N}$ and such that $\sum_n a_n b_n$ converges, we have $\sum_n a_n b_n \geq \sum_n a_{p(n)} b_{q(n)}$ for any couple of permutations p and q of $\mathbb{N}$. To show this, put $I_m := \{1, 2, \ldots, m\}$, and $R_m = \mathbb{N} \setminus I_m$ for $m \in \mathbb{N}$. Note first that, by reordering, we may always assume that p is the identity. If the result fails, then there exists $m \in \mathbb{N}$ such that $\sum_{n=1}^m a_n b_n < \sum_{n=1}^m a_n b_{q(n)}$. Certainly, q can fail to be a permutation of I_m. However, by redefining q on $I_m \cap q^{-1}(R_m)$ to become a permutation q' of I_m, we get $\sum_{n=1}^m a_n b_n < \sum_{n=1}^m a_n b_{q(n)} \leq \sum_{n=1}^m a_n b_{q'(n)}$, so we may always assume that $q|_{I_m}$ is a permutation of I_m. Now, every permutation of a finite set is a composition of transpositions. Since obviously $(a_1 b_1 + a_2 b_2) \geq (a_1 b_2 + a_2 b_1)$, we get a contradiction.
As a complement to this observation, note too that given real numbers a, b, α, β such that $a > b > 0$ and $\alpha > \beta > 0$, we have $\alpha a + \beta b > \beta a + \alpha b$.

The expression (9.11) gives a norm on $\ell_\infty^c(\Gamma)$. Although it is written as an "ℓ_2-sum", the selection of coordinates and its order depend completely on x and it is in no way fixed. By looking at the restriction of both norms to $c_{00}(\Gamma)$ (the space of functions on Γ having a finite support), we can prove that

$$\boxed{\|\cdot\|_D = |\cdot|_D \text{ on } c_0(\Gamma).} \tag{9.12}$$

➪ Along this box, $\|\cdot\|_n$ refers to the seminorms defined in (9.9) for $n \in \mathbb{N}$. Observe first that $\|\cdot\|_1 = \|\cdot\|_\infty$, where $\|\cdot\|_\infty$ denotes the supremum norm on $\ell_\infty(\Gamma)$. Thus, $|x|_D^2 \leq \|x\|_1^2(1 + 1/2 + 1/2^2 + \ldots) = 2\|x\|_1^2$. Note, too, that $(1/2)\|x\|_1^2 \leq \|x\|_D^2$. It follows that $|\cdot|_D \leq 2\|\cdot\|_D$. It is easy to prove that both norms are equivalent to $\|\cdot\|_\infty$ on $c_0(\Gamma)$. Since $c_{00}(\Gamma)$ is norm dense in $c_0(\Gamma)$, to verify the coincidence of $\|\cdot\|_D$ and $|\cdot|_D$ on $c_0(\Gamma)$ we just need to check this on $c_{00}(\Gamma)$. So, let us fix $x \in c_{00}(\Gamma)$ and write it as $x = \sum_{n=1}^k a_n e_{\gamma_n}$, where the γ_n's are all different and $|a_j| \geq |a_{j+1}| > 0$ for all $1 \leq j < k$. Then,

$$|x|_D = \left(\sum_{n=1}^{k} \frac{1}{2^{n-1}} a_n^2 \right)^{1/2}.$$

On the other hand, $\|x\|_r^2 = \sum_{j=1}^s a_j^2$, where $s = \min(r, k)$. Fix $m \geq k$. Then

$$\begin{aligned} \sum_{n=1}^{m} \frac{1}{2^n} \|x\|_n^2 &= \sum_{n=1}^{k-1} \frac{1}{2^n} \left(\sum_{j=1}^{n} a_j^2 \right) + \sum_{n=k}^{m} \frac{1}{2^n} \left(\sum_{j=1}^{k} a_j^2 \right) \\ &= \sum_{j=1}^{m} \frac{1}{2^j} a_1^2 + \sum_{j=2}^{m} \frac{1}{2^j} a_2^2 + \cdots + \sum_{j=k}^{m} \frac{1}{2^j} a_k^2. \end{aligned}$$

Taking limits when m goes to ∞, we obtain that

$$\|x\|_D^2 = a_1^2 + \frac{1}{2}a_2^2 + \cdots + \frac{1}{2^{k-1}}a_k^2 = |x|_D^2,$$

as we wanted to show.

As we mentioned above, the following theorem is the original result of Day.

Theorem 170 (Day, [Day55]). *For every set Γ, Day's norm is strictly convex on $c_0(\Gamma)$.*

Proof. If $x = (x_\gamma)$, then a sequence $\{\gamma_j\}$ in Γ such that $|x_{\gamma_1}| \geq |x_{\gamma_2}| \geq \ldots$ is called an **appropriate sequence for** x. We have that if $\|x+y\|_D = \|x\|_D + \|y\|_D$, $\|x\|_D = \|y\|_D = 1$ and $\{\gamma_j\}$ is an appropriate sequence for $x+y$, then, the subadditive property of the $\|\cdot\|_2$ norm in ℓ_2 gives

$$2 = \|x+y\|_D = \left(\sum_{j=1}^{\infty} \frac{(x+y)_{\gamma_j}^2}{2^{j-1}}\right)^{\frac{1}{2}} \leq \left(\sum_{j=1}^{\infty} \frac{x_{\gamma_j}^2}{2^{j-1}}\right)^{\frac{1}{2}} + \left(\sum_{j=1}^{\infty} \frac{y_{\gamma_j}^2}{2^{j-1}}\right)^{\frac{1}{2}}.$$

Assume that $\{\gamma_j\}$ is not an appropriate sequence for x. Thus, for some $j \in \mathbb{N}$ we have $|x|_{\gamma_j} < |x|_{\gamma_{j+1}}$. By transposing γ_j and γ_{j+1} we get another sequence (denoted $\{\gamma_j'\}$) such that, by the observation in page 152, $\left(\sum_{j=1}^{\infty} \frac{x_{\gamma_j}^2}{2^{j-1}}\right)^{\frac{1}{2}} < \left(\sum_{j=1}^{\infty} \frac{x_{\gamma_j'}^2}{2^{j-1}}\right)^{\frac{1}{2}}$. So

$$\begin{aligned}
2 &\leq \left(\sum_{j=1}^{\infty} \frac{x_{\gamma_j}^2}{2^{j-1}}\right)^{\frac{1}{2}} + \left(\sum_{j=1}^{\infty} \frac{y_{\gamma_j}^2}{2^{j-1}}\right)^{\frac{1}{2}} \\
&< \left(\sum_{j=1}^{\infty} \frac{x_{\gamma_j'}^2}{2^{j-1}}\right)^{\frac{1}{2}} + \left(\sum_{j=1}^{\infty} \frac{y_{\gamma_j}^2}{2^{j-1}}\right)^{\frac{1}{2}} \leq \left(\sum_{j=1}^{\infty} \frac{x_{\rho_j}^2}{2^{j-1}}\right)^{\frac{1}{2}} + \left(\sum_{j=1}^{\infty} \frac{y_{\gamma_j}^2}{2^{j-1}}\right)^{\frac{1}{2}} \leq 2,
\end{aligned}$$

where $\{\rho_j\}$ is a appropriate sequence for x (again by the observation in page 152), a contradiction. This shows that $\{\gamma_j\}$ is an appropriate sequence for x. Similarly we argue for y. Thus $\{\gamma_j\}$ is an appropriate sequence for both x and y. By the fact that $\|\cdot\|_2$ in ℓ_2 is rotund, we then get $x = y$. □

The LUR property of Day's norm is carefully discussed in Section 10.7 below; alternatively, we refer to [Di75, Section 4.1], [DeGoZi93, Theorem II.7.3], and [MOTV09, Proposition 3.38].

Several general renorming results included in this text produce, in the particular case of $c_0(\Gamma)$, LUR equivalent norms on this last space.

This is the case of Theorem 222. Since the canonical basis of $c_0(\Gamma)$ is obviously strong (see Definition 128), we can accordingly define a LUR norm on $c_0(\Gamma)$ (observe that the second proof we offer of Theorem 222 does not use Day's norm $\|\cdot\|_D$).
Another possibility is to use Deville's Master Lemma 426 adapted to the space $c_0(\Gamma)$: a sketch of the construction of the LUR norm there is presented in Section 21.1 below.
A third approach is to use Theorem 229 together with the result that $c_0(\Gamma)$ has an equivalent norm whose dual norm is LUR (Theorem 171). Observe that the canonical injection $I : \ell_1(\Gamma) \to c_0(\Gamma)$ is $\|\cdot\|_1$-$\|\cdot\|_\infty$ continuous, and it has obviously dense image. If $\{l_j\}_J$ is a w^*-null net in $\ell_1(\Gamma)$, then $\langle x, l_j\rangle \to 0$ for each $x \in c_0(\Gamma)$. Thus, in particular, $\langle Il_j, l\rangle \to 0$ for each $l \in \ell_1(\Gamma)$, and this shows that $\{Il_j\}_J$ is a w-null net in $c_0(\Gamma)$. Thus, I is w^*-w-continuous. The existence of an equivalent LUR norm on $c_0(\Gamma)$ then follows from Theorem 229.
Still another approach is to use Theorem 337. Observe once more that $\ell_1(\Gamma)$ has a dual LUR equivalent norm (Theorem 171); finally, use Theorem 337.

For more on variants on Day's norm we refer, e.g., to [HMVZ07, Section 3.5], [DeGoZi93, Section II.7], [HáJo09], [HáPro14], [Fab97], and [MOTV09].

9.4.2 Some other particular norms on $c_0(\Gamma)$

In Proposition 115 we provided a characterization of the LUR property of a norm $\|\cdot\|$ in terms of the function $Q_{\|\cdot\|}$. Let us define two other —besides the supremum and Day's— equivalent norms on $c_0(\Gamma)$.

1. The first one has a LUR dual norm:

Theorem 171 (Troyanski, [Tr72]). *The space $(c_0(\Gamma), \|\cdot\|_\infty)$ has an equivalent norm $|||\cdot|||$ whose dual norm is LUR. Thus, $|||\cdot|||$ on $c_0(\Gamma)$ is Fréchet differentiable.*

Proof. (Sketch, see , e.g., [FHHMZ11, Lemma 13.26] and [DeGoZi93, Theorem II.7.4]) We follow the procedure in Subsection 9.3.1 (Proposition 162): Let $I : \ell_2(\Gamma) \to c_0(\Gamma)$ be the identity mapping. The adjoint $I^* : \ell_1(\Gamma) \to \ell_2^*(\Gamma)$ ($= \ell_2(\Gamma)$) is again the identity mapping. The sought norm $|||\cdot|||$ on $c_0(\Gamma)$ is the predual of the norm $|||\cdot|||^*$ on $\ell_1(\Gamma)$ defined as in (9.7), i.e.,

$$\boxed{(|||x|||^*)^2 := \|x\|_1^2 + \|I^*x\|_2^2 \ (= \|x\|_1^2 + \|x\|_2^2) \ \text{ for } x \in \ell_1(\Gamma),} \tag{9.13}$$

where $\|\cdot\|_i$ is the canonical norm in $\ell_i(\Gamma)$, $i = 1, 2$. Proposition 162 gives that $|||\cdot|||^*$ is a dual W*LUR norm. A special behaviour of the $\|\cdot\|_1$ norm guarantees that $|||\cdot|||^*$ is in fact LUR (see below). For the Fréchet differentiability of the norm $|||\cdot|||$ see Proposition 91. □

➪ Let us show that $|||\cdot|||^*$ defined in (9.13) is in fact LUR. In $\ell_1(\Gamma)$ take x and a sequence $\{x_n\}_{n\in\mathbb{N}}$ such that $Q_{|||\cdot|||^*}(x,x_n)\to 0$ (see the notation in equation (5.2)). According to Lemma 111, we have $Q_{\|\cdot\|_i}(x,x_n)\to 0$ for $i=1,2$. This shows that $\|x_n\|_i \to_n \|x\|_i$, $i=1,2$. Since $\ell_2(\Gamma)$ is LUR, we get from Proposition 115 that $\|x-x_n\|_2\to 0$; in particular, $x_n(\gamma)\to x(\gamma)$ for all $\gamma\in\Gamma$.

Use now the fact that on $S_{(\ell_1(\Gamma),\|\cdot\|_1)}$ the topology of the pointwise convergence and the topology induced by $\|\cdot\|_1$ agree (see the argument in page 104 and modify it slightly to suit the present situation, in which $\|x_n\|_1\to\|x\|_1$ ($\neq 0$)) to conclude that $\|x-x_n\|_1\to 0$. All together, $|||x-x_n|||^*\to 0$, and so $|||\cdot|||^*$ is LUR by Proposition 115. □

2. The mapping $I^*:\ell_1(\Gamma)\to\ell_2(\Gamma)$ defined in the proof of Theorem 171 allows, via Proposition 152, to define another strictly convex dual norm $\|\cdot\|$ on $\ell_1(\Gamma)$. Precisely, put

$$\boxed{\|x\|=\|x\|_1+\|x\|_2 \text{ for } x\in\ell_1(\Gamma).} \tag{9.14}$$

Therefore, on $c_0(\Gamma)$, the predual norm $\|\cdot\|_*$ of $\|\cdot\|$ is Gâteaux differentiable (see Proposition 82).

However, the norm on $\ell_1(\Gamma)$ defined by (9.14) is not LUR.

➪ (The argument presented here is for $\ell_1(\mathbb{N})$. It can be extended with a similar proof to $\ell_1(\Gamma)$ for an infinite index set Γ.) Indeed, if $x:=(1/2,0,\ldots)$ and $x_n=(0,1/n,\ldots,1/n,0,\ldots)$, where $1/n$ is repeated n times, then $\|x\|=1$, $\|x_n\|\to 1$, and $\|x+x_n\|\to 2$, while $\|x-x_n\|\geq 1$ for $n\in\mathbb{N}$.

Despite this, $(\ell_1,\|\cdot\|)$ has the w^*-$\mathcal{T}_{\|\cdot\|}$-KK property, i.e., on $S_{(\ell_1(\Gamma),\|\cdot\|)}$ the w^* and the norm topologies coincide.

➪ Indeed, if $\{x_i\}_{i\in I}$ is a net in $S_{(\ell_1(\mathbb{N}),\|\cdot\|)}$ that w^*-converges to some $x\in S_{(\ell_1(\mathbb{N}),\|\cdot\|)}$, by passing to a subnet if necessary we may assume that $\|x_i\|_1\to L$ for some $L\in(0,1)$ (so $\|x_i\|_2\to 1-L$). By the Alaoglu-Bourbaki theorem we have $\|x\|_1\leq L$ and $\|x\|_2\leq(1-L)$. This shows that indeed $\|x\|_1=L$ and $\|x\|_2=1-L$. The argument in the proof of Theorem 171 shows then that $\|x-x_i\|_1\to 0$. Note that then $\|x-x_i\|_2\to 0$, so $\|x-x_i\|\to 0$.

It follows from Corollary 147 that its predual norm $\|\cdot\|_*$ on $c_0(\mathbb{N})$ is, in fact, Fréchet differentiable. This gives and example of a Fréchet differentiable norm whose dual norm is not LUR.

9.4.3 Phelps' norm on ℓ_1

Theorem 172 below answered a question of Mazur in [Maz33].

Theorem 172 (Phelps, [Ph60a]). *There exists an equivalent dual norm on ℓ_1 that is Gâteaux differentiable and nowhere Fréchet differentiable.*

Proof. (Sketch) We shall follow the procedure in Subsection 9.2.2. Define a continuous linear mapping $T : \ell_2 \to \ell_1$ such that $Te_i = (1/2^{i/2})e_i$ for $i \in \mathbb{N}$. Thus, the norm $|||\cdot|||^*$ on ℓ_∞ defined in Proposition 156, i.e.,

$$|||f|||^* = \|f\|_\infty + \|T^* f\|_2^* = \|f\|_\infty + \left(\sum_{i=1}^{\infty} \frac{1}{2^i}|f_i|^2\right)^{\frac{1}{2}} \tag{9.15}$$

for $f \in \ell_\infty$ is a dual strictly convex equivalent norm (notice that $\|f\|_\infty \le |||f|||^* \le 2\|f\|_\infty$ for $f \in \ell_\infty$). We will show that the (Gâteaux differentiable) predual norm $|||\cdot|||$ in ℓ_1 is nowhere Fréchet differentiable by using Šmulyan Lemma 54. For this, let $x = (x_i) \in S_{(\ell_1,|||\cdot|||)}$ and $f := \partial|||\cdot|||(x)$. Define a sequence $\{f_n\}_{n=1}^\infty$ in ℓ_∞ such that each f_n differs from f only at one coordinate, $f_n(x) \to f(x) = 1$ and $|||f_n||| \to |||f||| = 1$. Figure 9.2 tries to convey the idea of the construction (each dot is the only perturbation of f to get f_n, and the procedure depends on (a) $f \in c_0$ or (b) $f \notin c_0$, with $|f(i)| > 2\varepsilon$ for some $\varepsilon > 0$ and infinitely many coordinates in the second case). Then $|||f - f_n||| > 1/8$ in the first case and $|||f - f_n||| > \varepsilon$ in the second, for $n \in \mathbb{N}$. Šmulyan's Lemma 54 shows that $|||\cdot|||$ is not Fréchet differentiable at x. For a complete proof, see, e.g., [Phe93, Example 1.14(c)].
Observe that the norm $|||\cdot|||$ on ℓ_1 is a dual norm. Indeed, according to Item 2 in Remark 18 above, it suffices to show that $B_{(\ell_\infty,|||\cdot|||^*)} = \overline{B_{(c_0,|||\cdot|||^*)}}^{w^*}$, where $|||\cdot|||^*$ on c_0 is the restriction to c_0 of the norm $|||\cdot|||^*$ on ℓ_∞. Let $f = (f_n) \in B_{(\ell_\infty,|||\cdot|||^*)}$. For $n \in \mathbb{N}$, let $P_n : \ell_\infty \to c_0$ be the canonical projection onto $\overline{\operatorname{span}}\{e_1, \ldots, e_n\}$. Clearly, $|||P_n f||| \le 1$ for all $n \in \mathbb{N}$ (see formula (9.15)), and $P_n f \to f$ in the $w(\ell_\infty, \ell_1)$-topology. The conclusion follows. □

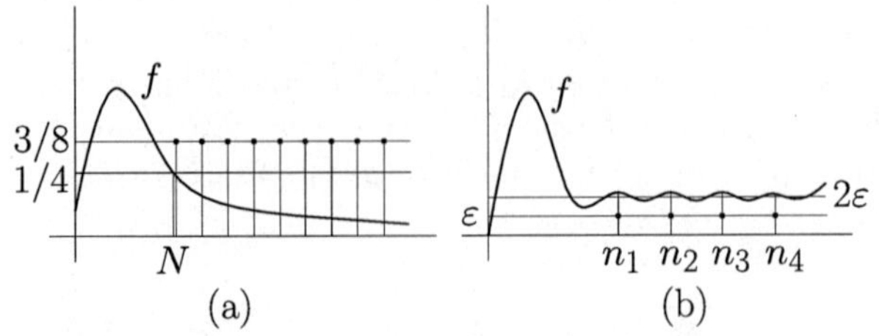

Figure 9.2: Perturbing f to get f_n in the proof of Theorem 172

In Theorem 370 below, a characterization of those separable Banach spaces having a Gâteaux differentiable norm that is nowhere Fréchet differentiable will be discussed.

9.5 Locally uniformly convex norms in separable spaces: Kadets renorming and homeomorphism theorems

9.5.1 Kadets renorming theorem

One of the most important result in the area of LUR renorming has been Kadets' Theorem 173 below. The original Kadets' result has a more precise formulation, stated as Theorem 176. We shall reproduce Kadets' ingenious proof, based on a wonderful interplay between compactness and convexity. His argument —worth to be studied carefully— turned out to be crucial in this area, and gave birth to many variants and some improvements. We present also a streamlined proof of Theorem 173 due to Davis and Johnson [DaJo73]. Kadets' main purpose for his renorming theorem was Theorem 186 below on the homeomorphism of all separable infinite-dimensional Banach spaces.

Theorem 173 (M.I. Kadets, see, e.g., [FHHMZ11], Theorem 8.1). *Every separable Banach space admits an equivalent LUR norm.*

Proof [DaJo73] (sketch of the basic idea; the complete proof is given below). Denote the norm of X by $\|\cdot\|$. Let $\{f_i\}_{i=1}^{\infty}$ be a w^*-dense sequence in the dual sphere and $\{F_i\}_{i=1}^{\infty}$ be an increasing sequence of finite-dimensional subspaces of X such that $\bigcup_{i=1}^{\infty} F_i$ is dense in X.
Let us define an equivalent norm $|||\cdot|||$ on X by

$$|||x|||^2 = \|x\|^2 + \sum_{i=1}^{\infty} \frac{1}{2^i} f_i^2(x) + \sum_{i=1}^{\infty} \frac{1}{2^i} \operatorname{dist}^2(x, F_i). \tag{9.16}$$

This is the desired norm. Indeed, it follows that if $Q_{|||\cdot|||}(x, x_n) \to 0$ (where $Q_{|||\cdot|||}$ was defined by (8.7)), then the same holds for all summands in (9.16) (see Lemma 111), and thus, by Lemma 110, $f_i(x_n) \to_n f_i(x)$ for each i and $\operatorname{dist}(x_n, F_i) \to_n \operatorname{dist}(x, F_i)$ for each i.
Then the compactness of the balls in finite dimensions gives the relative norm compactness of $\{x_n : n \in \mathbb{N}\}$, and since $f_i(x_n) \to_n f_i(x)$ for each i, from the coincidence on norm-compact sets of the norm topology and the topology of the pointwise convergence on the set $\{f_i : i \in \mathbb{N}\}$, we get that $|||x_n - x||| \to 0$. The conclusion follows from Proposition 115. $\square$

To appreciate the depth of Kadets' Theorem 173, consult, e.g., Theorem 452 below (see [DeGoZi93a]): Renormed by an equivalent LUR norm, the behaviour of the associated pointwise modulus of convexity at each point of the unit sphere provides a test for superreflexivity of any separable space.

We record now the original result of Kadets (Theorem 176 below; we follow [Milm71, Theorem 1.6]). This proof goes deeper inside the present problem. First we have an intermediate result (Theorem 175). The new norm is defined by using a sequence of "moduli of convexity", and the proof of the uniform behaviour of a pointwise convergent sequence relies on the Arzelà–Ascoli theorem. Let us recall here some other definitions needed: If (X,ρ) and (Y,π) are metric spaces, a family F of functions from X into Y is said to be **equicontinuous** whenever given $x \in X$ and $\varepsilon > 0$ there exists $\delta(x,\varepsilon) > 0$ such that $\pi(f(x), f(z)) < \varepsilon$ for all $z \in X$ with $\rho(x,z) < \delta$ and all $f \in F$. A family F of functions from X into Y is said to be **uniformly equicontinuous** if given $\varepsilon > 0$ there exists $\delta(\varepsilon) > 0$ such that $\pi(f(x), f(z)) < \varepsilon$ for all $f \in F$ whenever $x, z \in X$ satisfy $\rho(x,z) < \delta$. If X is compact, then every equicontinuous family in $C(X,Y)$ is uniformly equicontinuous. The Arzelà–Ascoli theorem says that *if K is a compact metric space and $(X, \|\cdot\|)$ is a Banach space, a subset F of $C(K,X)$ is $\|\cdot\|_\infty$-relatively compact if, and only if, F is equicontinuous and for every $k \in K$, the set $F(k) := \{f(k) :\ f \in F\} \subset X$ is $\|\cdot\|$-relatively compact in X* (see, e.g., [Dieu69, Theorem 7.5.7]).

Remark 174. Let us observe the following fact: *Assume that (K,ρ) is a compact metric space, D a dense subset of K, and F a subset of $C(K)$ with the following property: For every $\varepsilon > 0$ there exists $\delta > 0$ such that for all $d_1, d_2 \in D$ with $\rho(d_1,d_2) < \delta$ we have $|f(d_1) - f(d_2)| < \varepsilon$ for every $f \in F$. Then F is uniformly equicontinuous.*

➪ Indeed, fix $\varepsilon > 0$ and find $\delta > 0$ accordingly. Let $k_i \in K$ for $i = 1,2$ such that $\rho(k_1,k_2) < \delta$. Find r such that $\rho(k_1,k_2) < r < \delta$. Fix $f \in F$. Since f is continuous, we can find $\delta_1 \in (0,(1/2)(r - \rho(k_1,k_2)))$ such that $\rho(z_i,k_i) < \delta_1$ implies $|f(z_i) - f(k_i)| < \varepsilon$ for $i = 1,2$. Find $d_i \in D$ such that $\rho(d_i,k_i) < \delta_1$ for $i = 1,2$. Then

$$\begin{aligned}\rho(d_1,d_2) &\le \rho(d_1,k_1) + \rho(k_1,k_2) + \rho(k_2,d_2) < \delta_1 + \rho(k_1,k_2) + \delta_1 \\ &= \rho(k_1,k_2) + 2\delta_1 < \rho(k_1,k_2) + r - \rho(k_1,k_2) = r < \delta.\end{aligned}$$

So, $|f(d_1) - f(d_2)| < \varepsilon$. It follows that

$$|f(k_1) - f(k_2)| \le |f(k_1) - f(d_1)| + |f(d_1) - f(d_2)| + |f(d_2) - f(k_2)| < 3\varepsilon.$$

Since this holds for every $f \in F$, it follows that F is uniformly equicontinuous.

®

As we mentioned above, the following is an intermediate result in the way to the important Theorem 176 —that in turn was devised for proving Kadets' fundamental homeomorphism Theorem 186—. Recall that the LUR property of a norm

$\|\cdot\|$ implies the $w(X,D)$-$\mathcal{T}_{\|\cdot\|}$-KK property, for any w^*-dense subset D of B_{X^*} (see Proposition 90 above).

Theorem 175 (M.I. Kadets, see [Milm71], Theorem 1.6)**.** *Let $(X, \|\cdot\|)$ be a separable Banach space and N be a norming subspace of X^*. Then there is an equivalent norm $\|\cdot\|_1$ on X that is $w(X,N)$-lower semicontinuous (so N is $\|\cdot\|_1$-1-norming) and has property $w(X,N)$-$\mathcal{T}_{\|\cdot\|_1}$-SKK.*

Proof. The w^* topology on B_{X^*} is metrizable. Let ρ be a metric that induces on B_{X^*} the w^*-topology. Put $B_N := B_{X^*} \cap N$.
For $x \in X$ as a function on B_N, we introduce its modulus of uniform continuity: For $\delta > 0$,

$$\boxed{\omega(x,\delta) = \sup_{f,g\in B_N,\ \rho(f,g)\le\delta} |f(x) - g(x)|.} \tag{9.17}$$

Obviously, $x \to \omega(x,\delta)$ is a seminorm on X. It is $w(X,N)$-lower semicontinuous, since it is the supremum of a family of $w(X,N)$-continuous functions. By choosing δ conveniently in (9.17) we may define a sequence of seminorms on the space X:

$$\omega_0(x) = \sup_{f\in B_N} |f(x)|;\quad \omega_m(x) = \omega\left(x, \frac{1}{m}\right),\quad m \in \mathbb{N},\ x \in X.$$

Observe that ω_0 is an equivalent norm on X, say $\eta\|x\| \le \omega_0(x)$ $(\le \|x\|)$ for $x \in X$, where η is the Dixmier characteristic of N (see equation (3.5)), and that ω_0 is also $w(X,N)$-lower semicontinuous. Note, too, that

$$(2\|x\| \ge)\ \omega_m(x) \ge \omega_{m+1}(x)\ (\ge 0),\ \text{ for all } m \in \mathbb{N}\cup\{0\} \text{ and all } x \in X. \tag{9.18}$$

Moreover, for $x \in X$, the function x restricted to (B_{X^*}, ρ) is uniformly continuous, hence

$$\lim_{m\to\infty} \omega_m(x) = 0,\ \text{ for all } x \in X. \tag{9.19}$$

Define a new norm on X by

$$\boxed{\|x\|_1 = \sum_{m=0}^{\infty} 2^{-m}\omega_m(x).} \tag{9.20}$$

Observe that, for all $x \in X$,

$$\eta\|x\| \le \|x\|_1 \le 3\|x\|,$$

so $\|\cdot\|_1$ is an equivalent norm. Note that $\|\cdot\|_1$ is the supremum of the sequence of partial sums in (9.20), hence it is $w(X,N)$-lower semicontinuous.
We will show that the norm $\|\cdot\|_1$ satisfies $w(X,N)$-$\mathcal{T}_{\|\cdot\|_1}$-SKK. For this let $x \in X$ and a sequence $\{x_n\}_{n\in\mathbb{N}}$ in X be such that $\|x_n\|_1 = \|x\|_1 = 1$ and $x_n \to x$ in

the $w(X,N)$-topology. The $w(X,N)$-lower semicontinuity of ω_m, $m = 0,1,2,\ldots$, shows that

$$\omega_m(x) \leq l_m := \liminf_n \omega_m(x_n), \; m = 0,1,2,\ldots \tag{9.21}$$

From this it follows that

$$\lim_{n\to\infty} \omega_m(x_n) = \omega_m(x), \quad m = 0,1,2,\ldots \tag{9.22}$$

➪ Indeed, define a mapping $\psi : X \to \ell_1$ by $\psi(x) = (2^{-m}\omega_m(x))_{m=0}^\infty$. Given an arbitrary subsequence $\{x_{n_k}\}$ of $\{x_n\}$, a diagonal procedure gives a further subsequence $\{x_{n_{k_j}}\}$ such that $2^{-m}\omega_m(x_{n_{k_j}})$ converges to $2^{-m}l_m$ for $m = 0,1,2,\ldots$. To avoid a cumbersome notation, the subsequence $\{n_{k_j}\}$ will be denoted simply by $\{n\}$. Thus, the sequence $\{\psi(x_n)\}$ pointwise converges to $(2^{-m}l_m)_{m=0}^\infty$, Since the sequence $\{\psi(x_n)\}$ is in S_{ℓ_1}, it has a $w(\ell_1, c_0)$-cluster point in B_{ℓ_1}. This cluster point must be $(2^{-m}l_m)_{m=0}^\infty$. Thus, $(2^{-m}l_m)_{m=0}^\infty \in B_{\ell_1}$, and the sequence $\{\psi(x_n)\}$ is $w(\ell_1, c_0)$-convergent to $(2^{-m}l_m)_{m=0}^\infty$. We must have $\omega_m(x) = l_m$ for all $m = 0,1,2,\ldots$, since (9.21) holds, $\|\psi(x)\|_1 = 1$, and $(2^{-m}l_m)_{m=0}^\infty \in B_{\ell_1}$. Due to the fact that the subsequence $\{n_k\}$ was arbitrary, (9.22) holds.

From (9.22) it follows that the set $\{x_n : n \in \mathbb{N}\}$, as a subset of $C((\overline{B_N}^{w^*}, w^*))$ (i.e., the space of all real-valued functions defined on the compact metric space $(\overline{B_N}^{w^*}, w^*)$), is bounded and uniformly equicontinuous.

➪ Indeed, the boundedness is obvious (after all, $\{x_n\}_{n\in\mathbb{N}}$ is a bounded sequence). To show the uniform equicontinuity, let $\varepsilon > 0$ be given. Find $m_0 > 0$ so that $\omega_m(x) < \varepsilon/2$ for all $m \geq m_0$ (see (9.19); here x is the $w(X,N)$-limit of the sequence $\{x_n\}$). Then we can choose, by (9.22), $n_0 = n_0(m_0, \varepsilon)$ so that

$$|\omega_{m_0}(x_n) - \omega_{m_0}(x)| < \varepsilon/2, \quad \text{for } n \geq n_0.$$

Thus we have

$$(0 \leq)\ \omega_{m_0}(x_n) < \varepsilon \text{ for } n \geq n_0. \tag{9.23}$$

Keep n_0 fixed and find $m \geq m_0$ such that, thanks to (9.19), $\omega_m(x_n) < \varepsilon$ for $n = 1,2,\ldots,n_0 - 1$. Due to (9.18) we still have (9.23) for m instead of m_0. All together, we have

$$\omega_m(x_n) = \omega\left(x_n, \frac{1}{m}\right) < \varepsilon \quad \text{for all } n \in \mathbb{N}.$$

This gives, in view of Remark 174, the uniform equicontinuity of the set $\{x_n : n \in \mathbb{N}\}$ as a subset of $C((\overline{B_N}^{w^*}, w^*))$.

By the Arzelà–Ascoli theorem, the sequence $\{x_n\}_{n\in\mathbb{N}}$ is $\|\cdot\|_\infty$-relatively compact in $C((\overline{B_N}^{w^*}, w^*))$. The topology on $C((\overline{B_N}^{w^*}, w^*))$ of the pointwise convergence on the set B_N is Hausdorff and coarser than the topology induced by $\|\cdot\|_\infty$, so $\|x_n - x\|_\infty \to 0$ in $C((\overline{B_N}^{w^*}, w^*))$. Note that $\overline{B_N}^{w^*}$ contains a positive multiple of $B_{(X^*,\|\cdot\|)}$, due to the fact that N is norming. This shows that $\|x_n - x\| \to 0$, hence $\|x_n - x\|_1 \to 0$, so $\|\cdot\|_1$ has property $w(X,N)$-$\mathcal{T}_{\|\cdot\|_1}$-SKK, as we wanted to show. □

Theorem 175 provides the basic ingredient for the following important result:

Theorem 176 (M.I. Kadets, see, e.g., [Milm71]). *Let $(X, \|\cdot\|_1)$ be a separable Banach space and $N \subset X^*$ be a norming subspace. Then there is an equivalent $w(X,N)$-lower semicontinuous norm $\|\cdot\|$ on X (so N is $\|\cdot\|$-1-norming) that has simultaneously properties LUR and $w(X,N)$-$\mathcal{T}_{\|\cdot\|}$-KK.*

Proof. As we mentioned at the beginning of the proof of Theorem 175, (B_{X^*}, w^*) is (compact and) metrizable, hence (hereditarily) separable. In particular, B_N is w^*-separable. Find a w^*-dense subset $\{\varphi_k : k \in \mathbb{N}\}$ of B_N. Put $N_1 := \text{span}\,\{\varphi_k : k \in \mathbb{N}\}$ $(\subset N)$. Thus, N_1 is a norming subspace of X^*. By Theorem 175 we may assume that the original norm $\|\cdot\|_1$ is $w(X,N_1)$-lower semicontinuous and has property $w(X,N_1)$-$\mathcal{T}_{\|\cdot\|_1}$-SKK.
To define the sought equivalent norm $\|\cdot\|$ on X we follow the technique in (9.8): Let $T(x) := (2^{-k/2}\varphi_k(x))_{k=1}^\infty$ for $x \in X$. Then $T : X \to \ell_2$ is linear, one to one, and bounded. Put $\|x\|^2 := \|x\|_1^2 + \|Tx\|_2^2$ for $x \in X$, i.e.,

$$\boxed{\|x\|^2 = \|x\|_1^2 + \sum_{k=1}^{\infty} 2^{-k}\varphi_k^2(x).} \tag{9.24}$$

Note that $\|\cdot\|$ is $w(X,N_1)$-lower semicontinuous. This is due to two facts: First, $\|\cdot\|_1$ has this property. Second, each function φ_k is obviously $w(X,N_1)$-continuous on $B_{(X,\|\cdot\|)}$; the Weierstrass M-test shows that $J := \sum_{k=1}^\infty 2^{-k}\varphi_k^2$ is a $w(X,N_1)$-continuous function on $B_{(X,\|\cdot\|)}$ as well. The 1-norming character of N_1 (and then of N) follows from what was said in Section 3.4.
In order to prove the theorem, we will first show that the norm $\|\cdot\|$ has property $w(X,N)$-$\mathcal{T}_{\|\cdot\|}$-SKK. So, let

$$\|x_n\| = \|x\| = 1,\ x_n \overset{w(X,N_1)}{\longrightarrow} x,$$

i.e., $\lim_{n\to\infty}\varphi_k(x_n) = \varphi_k(x)$, $k = 1, 2, \ldots$ By the previous observation, $J(x_n) \to J(x)$. Since $\|y\|_1^2 = \|y\|^2 - J(y)$ for all $y \in X$, we get $\|x_n\|_1 \to \|x\|_1$. The $w(X,N_1)$-$\mathcal{T}_{\|\cdot\|}$-SKK property of $\|\cdot\|_1$ concludes that $\|x_n - x\|_1 \to 0$, and so $\|x_n - x\| \to 0$.
We now will show that the norm $\|\cdot\|$ is LUR. Let $Q_{\|\cdot\|}(x, x_n) \to 0$ (see the definition of $Q_{\|\cdot\|}$ in (8.7)). From Lemma 111 we get $Q_{\|\cdot\|_1}(x, x_n) \to 0$ and $Q_{\varphi_k}(x, x_n) \to_n 0$ for $k \in \mathbb{N}$. From Lemmas 108 and 110 we get $\|x_n\|_1 \to \|x\|_1$, and $\varphi_k(x_n) \to \varphi_k(x)$

for all $k \in \mathbb{N}$, i.e., $x_n \overset{w(X,N_1)}{\longrightarrow} x$. Since $\|\cdot\|_1$ has property $w(X,N_1)$-$\mathcal{T}_{\|\cdot\|_1}$-SKK, we get $\|x_n - x\|_1 \to 0$, hence $\|x - x_n\| \to 0$.
The conclusion that $\|\cdot\|$ has property $w(X,N)$-$\mathcal{T}_{\|\cdot\|}$-KK follows from the LUR character of $\|\cdot\|$ and Proposition 90. □

We list now some consequences of the previous results.

Since the space generated by the coefficients of a Schauder basis in separable spaces is norming (see Subsection 3.4.4, Item 3, and [FHHMZ11, Exercise 4.34]) the following corollary holds. It was used by Kadets in his proof that all infinite-dimensional separable Banach spaces are homeomorphic.

Corollary 177 (M.I. Kadets, see [Milm71] or [BessPe75], Theorem VI.3.3). *Let X be a separable Banach space with a Schauder basis $\{e_n, e_n^*\}_{n=1}^\infty$. Let $F :=$ span $\{e_n^* : n \in \mathbb{N}\}$. Then X admits an equivalent norm $|||\cdot|||$ with the following properties:.*
(i) *The basis is monotone in the norm $|||\cdot|||$ (in particular, $|||\cdot|||$ is $w(X,F)$-lower semicontinuous or, in other terms, F is 1-norming for $|||\cdot|||$). Moreover, for $n \in \mathbb{N}$ and $a_1, \ldots, a_{n+1} \in \mathbb{R}$,*

$$\left|\left|\left|\sum_1^n a_k e_k\right|\right|\right| < \left|\left|\left|\sum_1^{n+1} a_k e_k\right|\right|\right|, \text{ if } a_{n+1} \neq 0. \tag{9.25}$$

(ii) *The norm $|||\cdot|||$ is $w(X,F)$-$\mathcal{T}_{|||\cdot|||}$-KK.*
(iii) *The norm $|||\cdot|||$ is LUR*

Proof.

1. We follow the hint in [FHHMZ11, Exercise 8.9]: First produce, by a standard method (see Subsection 10.4.1, see also, e.g., [FHHMZ11, Lemma 4.9]), an equivalent norm $\|\cdot\|$ on X in which the Schauder basis is monotone. Then, modify the formula for the norm in (9.16) in the following way (where P_i is the i-th canonical projection associated to the basis):

$$|||x|||^2 = \|x\|^2 + \sum_i \frac{1}{2^i}(x_i^*)^2(x) + \sum_i \frac{1}{2^i}\|x - P_i(x)\|^2. \tag{9.26}$$

   The argument in the first proof of Theorem 173, now for $|||\cdot|||$ defined by (9.26), shows that $|||\cdot|||$ is LUR. This is (iii). Thus, from Proposition 90 we get (ii).

2. The basis $\{e_n, e_n^*\}_{n=1}^\infty$ is clearly monotone for the norm $|||\cdot|||$ (it is obvious that $|||\sum_{i=1}^n a_i e_i||| \leq |||\sum_{i=1}^m a_i e_i|||$ for scalars $a_1, \ldots, a_m$, $n \leq m$). This implies that F is $|||\cdot|||$-1-norming (see Subsection 3.4.4, Item 3).

➪ Notice that, for $n \in \mathbb{N}$, $x \in X$ and $y \in P_nX$, we have $\|x - P_n(x)\| \le \|x - y\| + \|y - P_n(x)\| = \|x - y\| + \|P_n(y) - P_n(x)\| \le 2\|x - y\|$, hence $\operatorname{dist}(x, P_nX) \le \|x - P_n(x)\| \le 2 \operatorname{dist}(x, P_nX)$ (due to the fact that the basis is monotone for $\|\cdot\|$). Since $\{P_nX\}_{n=1}^{\infty}$ is an increasing sequence of finite-dimensional subspaces of X and $\bigcup_{n=1}^{\infty} P_nX$ is dense in X, formula (9.16) for the norm would have been enough to produce a LUR norm. However, formula (9.26) has the extra feature that now the basis is monotone for $|||\cdot|||$.

3. Observe, too, that (9.25) is a consequence of the monotonicity of the basis for the norm $|||\cdot|||$ and its strict convexity. Indeed, fix $n \in \mathbb{N}$ and a finite sequence $\{a_i\}_{i=1}^{n+1}$ in $\mathbb{R}$ such that $a_{n+1} \neq 0$. Put $x := \sum_{i=1}^{n} a_i e_i$, and $y := \sum_{i=1}^{n+1} a_i e_i$. Note that $(1/2)(x + y) = \sum_{i=1}^{n} a_i e_i + (a_{n+1}/2)e_{n+1}$. By the monotonicity of the basis we have $|||x||| \le |||y|||$. Assume that $|||x||| = |||y|||$. Then, again by monotonicity,
$$(|||y||| =)\ \ |||x||| \le |||(1/2)(x + y)|||,$$
and, obviously,
$$|||(1/2)(x + y)||| \le |||x|||\ \ (= |||y|||),$$
so we get, all together, $|||x||| = |||(1/2)(x + y)||| = |||y|||$. We reach a contradiction with the strictly convexity of $|||\cdot|||$. □

9.5.2 Some consequences of Kadets' renorming theorem

Theorem 176, when applied to a (separable) dual space, gives the following result:

Corollary 178 (M.I. Kadets). *Let $(X, |||\cdot|||)$ be a Banach space. Assume that X^* is separable. Then X has an equivalent norm $\|\cdot\|$ whose dual norm $\|\cdot\|^*$ is LUR (and so its predual norm in X is Fréchet differentiable).*

Proof. The space X is a (1-) norming subspace for $(X^*, |||\cdot|||)$. Theorem 176, applied to the separable space X^* and to $N := X$, gives a LUR equivalent norm on X^* that is w^*-lower semicontinuous —hence a dual norm, denoted by $\|\cdot\|^*$— (see Proposition 17). Its predual norm $\|\cdot\|$ has the sought property (the Fréchet differentiability coming from Proposition 91). □

If, instead, Theorem 176 is applied to a (separable) $C(K)$ space, we get

Corollary 179 (M.I. Kadets). *Assume that K is a compact metric space . Then $C(K)$ admits a pointwise lower semicontinous LUR norm $\|\cdot\|$ with the property that on the unit sphere of $\|\cdot\|$, the coordinatewise and norm topologies coincide.*

Proof. We just need to notice that $(C(K), \|\cdot\|_\infty)$ is separable and the linear hull N of $\{\delta_k : k \in K\}$ in $C(K)^*$ is (1-) norming for the norm $\|\cdot\|_\infty$. The result follows from Theorem 176. □

Having in mind that the space generated by the functional coefficients of any Schauder basis is norming (see Section 3.4.4, Item 3), we have the following corollary.

Corollary 180 (M.I. Kadets, Vanderwerff, [Milm71], [Vand92a]). *Let X be a separable Banach space with a Schauder basis $\{e_i, f_i\}_{i=1}^\infty$. Then there is an equivalent norm $\|\cdot\|$ on X that is LUR and Fréchet differentiable at any point of* span $\{e_i : i \in \mathbb{N}\}$ *and moreover, if $x_n, x \in S_{(X,\|\cdot\|)}$, $n \in \mathbb{N}$, are such that for each $i \in \mathbb{N}$, $f_i(x_n) \to_n f_i(x)$, then $\|x_n - x\| \to 0$.*

Actually, J. Vanderwerff proved the following surprising theorem.

Theorem 181 (Vanderwerff, [Vand92a]). *Any normed linear space of countable dimension admits an equivalent Fréchet differentiable norm.*

Theorem 181 should be compared with the fact that the dual of a separable Banach space with Fréchet differentiable norm is separable (see Remark 59 above and Theorem 276 below).

Through the work of J. Lindenstrauss ([Lind63]) and E. Asplund ([Asp68]), Theorem 173 led to new results on the extremal structure of convex sets and differentiability of convex functions (see, e.g., [FHHMZ11, Chapters 7 and 11].

Both proofs of the existence of a LUR renorming of any separable Banach space given here use compactness arguments. As we saw, the original proof of Kadets relies on the Arzelà–Ascoli theorem (see Theorems 175 and 176), while Davis and Johnson proof ([DaJo73]) used (see Theorem 173) the compactness of the balls in finite-dimensional spaces (see also, e.g., [FHHMZ11, Theorem 8.1]). Some other subsequent proofs, useful in other contexts, can be found in the literature, see for example [DeGoZi93, Chapter 2, Section 8, and Chapter 8]). This helps, for example, in questions involving smoothness.

Davis and Johnson technique of constructing LUR norms (see again the proof of Theorem 173) is quite flexible. For example it allowed for some variants, where the distance function in equation (9.16) was replaced by projections on finite-dimensional subspaces (later on used in cases where differentiability is needed). See, for example, Corollary 177 and, in particular, equation (9.26).

9.5.3 Kadets homeomorphism theorem for separable spaces

In the proof of Kadets' Theorem 186 on the homeomorphism of all infinite-dimensional separable Banach spaces, Theorem 173 together with its Corollary

177—devised precisely for such a purpose— and the two following results were crucial.

The first one is the Bartle–Graves Selection Principle. For a proof see, e.g., [FHHMZ11, Corollary 7.56], [DeGoZi93, Lemma VII.3.2], or [BessPe75, Corollary II.7.1].

Theorem 182 (Bartle–Graves Selection Principle). *If X is a Banach space, and Y is a closed linear subspace of X, then there exists a continuous positively homogeneous mapping $\phi : X/Y \to X$ such that $q \circ \phi = Id_{X/Y}$, where $q : X \to X/Y$ is the canonical quotient mapping. Moreover, given $\varepsilon > 0$ the mapping ϕ can be taken so that $\phi(B_{X/Y}) \subset (1+\varepsilon)B_X$.*

The following is a consequence of Theorem 182 that shall be used in the proof of Kadets' Homeomorphism Theorem 186 below. The symbol $\sim$ means "homeomorphic to".

Corollary 183. *Let X be a Banach space, and let $Y \subset X$ be a closed linear subspace. Then $X \sim Y \times X/Y$.*

Proof of Corollary 183. (Sketch) Define a mapping $P : X \to X$ by $Px = x - \phi(qx)$ for $x \in X$ (we follow the notation in Theorem 182). Prove first that $P(X) \subset Y$. Then define $\psi : X \to Y \times X/Y$ by $\psi(x) = (Px, q(x))$ for $x \in X$. Show first that ψ is one-to-one, and then that it is onto (given $(y, q(x)) \in Y \times (X/Y)$, put $x_0 := y + \phi(q(x))$; show that $\psi(x_0) = (y, q(x))$). Thus, $\psi^{-1}(y, q(x)) = y + \phi(q(x))$ for $y \in Y$ and $x \in X$. Obviously, ψ and ψ^{-1} are both continuous. □

Remark 184. Theorem 182 can be obtained from Michael's selection theorem (see, e.g., [FHHMZ11, Theorem 7.53]). This is due to the fact that the multivalued mapping $\Phi : X/Y \to 2^X$ that to a class $\overline{x} \in X/Y$ associates the closed set $x + Y \subset X$, where $x \in \overline{x}$, is lower semicontinuous, and the space X/Y, being a Banach space, is paracompact. ®

The second one is due to Bessaga and Pełczyński:

Theorem 185 (Bessaga, Pełczyński, [BessPe60], Proposition VI.9.2). *If all infinite-dimensional separable Banach spaces with Schauder basis are homeomorphic to ℓ_2, then all infinite-dimensional separable Banach spaces are homeomorphic to ℓ_2.*

Proof. (We follow M.I. Kadets in [Kad67]) We will consider products of Banach spaces X_n in the sense of c_0, i.e., of the form $(\sum X_n)_{c_0}$. We denote by $=$ a linear isomorphism and, as we mentioned above, by $\sim$ a homeomorphism.

➪ To be more precise, given two Banach spaces X and Y, we consider on the product $X \times Y$ the norm $\|(x,y)\| := \max\{\|x\|, \|y\|\}$. Then $(X \times Y, \|\cdot\|)$ becomes a Banach space. More generally, we may define the c_0-sum of a sequence $\{X_n\}_{n=1}^{\infty}$ of Banach spaces: Put $(\sum X_n)_{c_0} := \{x = (x_n) : x_n \in X_n \text{ for } n \in \mathbb{N}, \lim_n \|x_n\| = 0\}$, endowed with the norm $\|x\| = \sup_n \|x_n\|$.

It is simple to prove that $((\sum X_n)_{c_0}, \|\cdot\|)$ is a Banach space. Observe that $(\sum_n (X \times Y))_{c_0} = X \times (\sum (Y \times X))_{c_0}$ via the linear isomorphism

$$((x_1, y_1), (x_2, y_2), \ldots) \to (x_1, (y_1, x_2), (y_2, x_3), \ldots).$$

It is also simple to prove that $c_0 = c_0 \times c_0$, and that, more generally, $(\sum c_0)_{c_0} = c_0$.

To continue with the proof of Theorem 185, let us consider an arbitrary infinite-dimensional separable Banach space X. Observe that

$$C := C[0,1] \sim X \times C/X, \tag{9.27}$$

and that

$$X \sim Y \times X/Y \tag{9.28}$$

for any closed subspace Y of X with a Schauder basis.

➡ Indeed, every separable Banach space is linearly isometric to a subspace of C, and so $C \sim X \times C/X$ by Corollary 183 above. Every Banach space contains a basic sequence. Its closed linear span is then a subspace with a Schauder basis, and we can use again Corollary 183 above.

By the hypothesis,

$$Y \sim C \sim c_0. \tag{9.29}$$

By using (9.28) and (9.29), together with the observations at the beginning of the proof, we have

$$\begin{aligned} X &\sim Y \times X/Y \sim c_0 \times X/Y \\ &\sim (c_0 \times c_0) \times X/Y \sim c_0 \times (c_0 \times X/Y) \sim c_0 \times X \sim C \times X. \end{aligned} \tag{9.30}$$

Then we get, from (9.29) and (9.27), and again by the notes at the beginning of the proof,

$$\begin{aligned} C &\sim c_0 \sim \Big(\sum c_0\Big)_{c_0} \sim \Big(\sum C\Big)_{c_0} \\ &\sim \Big(\sum (X \times C/X)\Big)_{c_0} \sim X \times \Big(\sum (C/X \times X)\Big)_{c_0} \sim X \times C. \end{aligned} \tag{9.31}$$

Taking (9.30) and (9.31) into account, we get $X \sim C$ and $C \sim \ell_2$ as it has a basis. So, $X \sim \ell_2$. □

Kadets' homeomorphism theorem is the following fundamental result:

Theorem 186 (M.I. Kadets, [Kad67]). *All infinite-dimensional separable Banach spaces are mutually homeomorphic.*

The proof can also be found in [BessPe75]. An easier argument for the reflexive case is in [FHHMZ11, Theorem 12.46]. Here we proceed only with a sketch of the proof:

Proof. (Sketch) We shall construct a homeomorphism of a separable infinite-dimensional Banach space X to ℓ_2. In view of Theorem 185, it is enough to consider that X has a Schauder basis $\{e_i, f_i\}_{i=1}^{\infty}$. Let $\|\cdot\|$ be the norm on X from Corollary 177. Put, for $x \in S_X$ and $0 \leq \delta \leq 1$ (see Figure 9.3),

$$G(x,\delta) = \{z \in B_X : \ \|\lambda x + (1-\lambda)z\| \geq 1-\delta \text{ for } 0 \leq \lambda \leq 1\},$$
$$\varepsilon(x,\delta) = 2^{-1}\sup\{\|x-z\| : \ z \in G(x,\delta)\}.$$

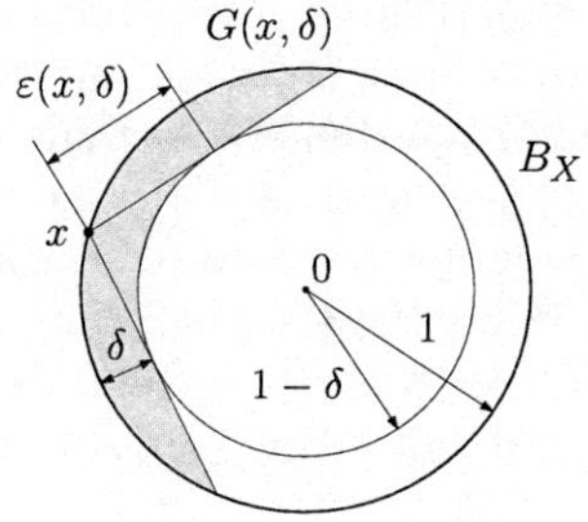

Figure 9.3: In gray, the set $G(x,\delta)$

For convenience we put

$$\varepsilon(x,\delta) = 1 \text{ for } x \in S_X \text{ and for } \delta > 1.$$

➪ Bessaga and Pełczyński call the function $\varepsilon(x,\delta)$ the **modulus of local uniform convexity**. Lovaglia [Lov55] used, instead, the following version: Given $x \in B_X$ and $\varepsilon > 0$, the **modulus of local convexity** is the function $\delta(x,\varepsilon) := \inf\{1 - \|(x+y)/2\| : \ y \in B_X, \ \|x-y\| \geq \varepsilon\}$. Clearly, $(X, \|\cdot\|)$ is LUR if, and only if, $\delta(x,\varepsilon) > 0$ for every $\varepsilon > 0$ and every $x \in S_X$. From Figure 9.3 it should be clear that $(X, \|\cdot\|)$ is LUR if, and only if, $\varepsilon(x,\delta) > 0$ for every $\delta > 0$ and $x \in S_X$.

For $x \in X$ and $n \in \mathbb{N}$, let $P_n(x) := \sum_{k=1}^{n} f_k(x)e_k$, $P_0(x) := 0$; put

$$L(x) = \bigcup_{n=1}^{\infty}\{y \in X : \ y = (1-\lambda)P_{n-1}(x) + \lambda P_n(x); \ 0 \leq \lambda \leq 1\},$$
$$\varphi(x) = \varepsilon(\|x\|^{-1}\cdot x, 1-\|x\|) \ \text{ for } 0 < \|x\| \leq 1, \ \varphi(0) = 1,$$
$$\Phi(x) = \inf\{\varphi(y) : \ y \in L(x)\} \ \text{ for } \|x\| \leq 1,$$
$$F(x) = \|x\|^2\Big(1 - \Phi(x)\Big)^2 \text{ for } \|x\| \leq 1.$$

Define a map h from the unit sphere of X endowed with the norm $\|\cdot\|$ onto the unit sphere of ℓ_2 by

$$h(x) = \Big(|F(P_n(x)) - F(P_{n-1}(x))|^{\frac{1}{2}} \operatorname{sgn} f_n(x)\Big)_{n=1}^{\infty}.$$

Finally, we define the homeomorphism H from X onto ℓ_2 by

$$H(x) = \begin{cases} 0 \text{ for } x = 0, \\ \|x\| \cdot h\left(\frac{x}{\|x\|}\right) \text{ for } x \neq 0. \end{cases}$$ □

H Brouwer's homeomorphism theorem says that, *if n and m are two positive integers such that $n \neq m$, then $\mathbb{R}^n$ and $\mathbb{R}^m$ are not homeomorphic* (see, e.g., [Du66, Chapter 16], and [FHHMZ11, Theorem 12.43]). It is a consequence of Brouwer's theorem on invariance of domains in $\mathbb{R}^m$. From Brouwer's fixed point theorem it follows easily that, for $n \in \mathbb{N}$, $B_{\mathbb{R}^n}$ and $\mathbb{R}^n$ are not mutually homeomorphic. This is in contrast with the infinite-dimensional case: If X is an infinite-dimensional Banach space, then X, B_X, and S_X are mutually homeomorphic [BessPe75, Proposition VI.6.1 and Theorem VI.6.2].

The historical notes that follow concerning the separable case come from the detailed account given by Bessaga and Pełczyński in [BessPe75, VI §10]:

The question of the topological equivalence of all separable infinite-dimensional Banach spaces was first posed in [Fre28, pages 94–96] and in [Ban32, page 233].

In [Maz29] it was shown that all L_p-spaces and λ_p-spaces, $1 \leq p < \infty$, are mutually topologically equivalent.

It was proved in [Kad53] that c_0 is topologically equivalent to ℓ_2.

It was proved in [Kad55] that all separable infinite-dimensional uniformly convex spaces are topologically equivalent.

It was proved in [Kad58] that all reflexive separable infinite-dimensional spaces are topologically equivalent.

For the final solution see [Kad66] and [Kad67].

The non-separable situation was considered by many authors. Troyanski [Tr67] proved that $c_0(\Gamma)$ and $\ell_1(\Gamma)$ are homeomorphic for all sets Γ.

Then Bessaga [BessPe60] proved that all reflexive Banach spaces have this property (see also [BessPe75, Theorem VII.4.1]).

Nowadays, Kadets' homeomorphism Theorem 186 is a special case of the result of Toruńczyk [Tor81] that all infinite-dimensional Banach spaces of the same density are mutually homeomorphic. (Toruńczyk's proof is beyond the scope of most textbooks in Banach spaces).

It would be interesting to see whether Bessaga's approach can be extended to, for example, WCG spaces (see Definition 131).

For a more complete information we refer, e.g., to [FHHMZ11, Chapter 12].

Kadets' homeomorphism Theorem 186 should be compared with the following theorem of O. H. Keller:

Theorem 187 (Keller). *All compact convex infinite-dimensional sets in ℓ_2 are mutually homeomorphic.*

For a proof see, e.g., [FHHMZ11, Subsection 12.3.5].

➩ Let us briefly comment here on the original Keller's result: A **Keller space** is a compact convex subset of a topological vector space affinely homeomorphic to an infinite-dimensional compact convex subset of ℓ_2. If C is an infinite-dimensional compact convex subset of a topological vector space such that there exists a sequence of continuous linear functionals that separates points of C, then C is a Keller space (so it is, in particular, every infinite-dimensional compact convex and metrizable subset of a locally convex space). Keller's Theorem 187 shows then that *all Keller spaces are mutually homeomorphic* (in particular, all of them are homeomorphic to the Hilbert cube $[-1,1]^{\mathbb{N}}$ endowed with the product topology). The proof of Theorem 187 consists in showing, first, that every closed convex subset C of ℓ_2 is homeomorphic to a closed **elliptically convex** subset EC of ℓ_2 (i.e., a convex set EC such that every continuous linear functional on ℓ_2 that is bounded above and non-constant on EC attains its supremum on EC at a single point). This step is in a sense reminiscent of M.I. Kadets' instrumental result showing the existence of an equivalent LUR norm on every separable Banach space and, more precisely, for a separable Banach space with a Schauder basis, Corollary 177 —another clear similitude between the two results, Keller's and Kadets', is the use of ℓ_2 as the reference space—. The second and last step consists in proving that all infinite-dimensional elliptically convex subsets of ℓ_2 are mutually homeomorphic via a geometric argument. For a more complete information on the topic of Keller Homeomorphism Theorem 187 we refer, e.g., to [FHHMZ11, Section 12.3].

Note that, unlike Kadets' homeomorphism Theorem 186, a nonseparable Keller-type homeomorphism theorem does not hold true (see [Av09]).

9.5.4 Some statements concerning homeomorphic normed spaces

We include here some results on homeomorphic normed spaces that may help to understand some of the consequences of Kadets' homeomorphism Theorem 186.

Theorem 188 (Mazur, Sternbach, [MazSt33]). *If X is a Banach space and Y is a linear subspace of X that is G_δ, then Y is closed in X.*

Proof. Let $Y_1 = \overline{Y}$. Then Y is G_δ in the Banach space Y_1. We will show that $Y = Y_1$. Indeed, if $y_0 \in Y_1 \setminus Y$, then the translate $Y + y_0$ would be a dense G_δ set in Y_1 and $(Y + y_0) \cap Y = \emptyset$, a contradiction with Baire's theorem. □

Theorem 189 (Lavrentieff, Sierpiński). *If X is a Banach space and φ is a homeomorphism of X into a Banach space Y, then $\varphi(X)$ is a G_δ set in Y.*

Proof. See [BessPe75, Theorem I.3.2 and Corollary I.3.2]. □

Corollary 190. *A non-complete normed space Y cannot be homeomorphic to a Banach space X.*

Proof. Let φ be a homeomorphism of a Banach space X onto a noncomplete normed space Y. Then by Theorem 189, Y is a G_δ set in the completion of Y, which is a Banach space. Thus, by Theorem 188, Y must be closed in its completion, a contradiction. □

Note that there is an uncountable collection of pairwise nonhomeomorphic separable normed spaces (see [PelBe79, page 277, Notes after Theorem 13.9]) and it is an open problem whether every normed space is homeomorphic to an inner product (non complete in general) space [PelBe79, page 278, Notes after Theorem 13.9]. For additional information on homeomorphisms, the reader may consult, e.g., [GoLaZi14] and [BeLi00].

Part II

An Intermediate Course in Renorming

Chapter 10

Locally uniformly convex renorming of nonseparable spaces

This chapter deals with renormings —for the main part, LUR renormings— of the more "accessible" nonseparable Banach spaces (weakly compactly generated —in particular, reflexive— and related classes of Banach spaces). A basic tool was introduced by J. Lindenstrauss —later elaborated by D. Amir and J. Lindenstrauss himself—, the so-called "projectional resolution of the identity", building in the space a "staircase" of complemented subspaces that reaches, through a well-ordered number of steps, the whole of it, and allowing for convenient inductive arguments (Theorems 200, 201, 203, 208, 210, 211, 216 and 220). We present, because of the intrinsic value of the construction, the original Lindenstrauss' approach and then, in a subsequent stage, two new approaches, the first by using a kind of "volleyball" technique, the contribution of several authors, and the second by using Deville's Master Lemma 426. A fundamental step based on the Amir–Lindenstrauss method was made by Troyanski, renorming a weakly compactly generated Banach space by a LUR norm (Theorem 218). His idea was pushed forward —the so-called (linear and nonlinear) transfer technique— by several authors, and was included, in an almost final form, in two monographs, [DeGoZi93] and [MOTV09]. We present Troyanski's ideas in detail, as they are a source of more recent developments.

We then shall show some important applications of renorming to structural questions of Banach spaces. For example, that every weakly compact and convex subset of a Banach space X can be reproduced just by using the family of all its strongly exposed points —a result of J. Lindenstrauss and S. Troyanski, see Theorem 259 below—, or that there are plenty of points from which the farthest distance to a given weakly compact subset of X is attained (Lau, see Theorem 260 below).

A fundamental result that justifies the introduction of the important class of the strong differentiability spaces —nowadays called Asplund spaces— is that spaces such that their duals have a dual LUR norm belong to this class. Since X is an Asplund space if, and only if, its dual has the Radon–Nikodým property (see Theorem 547), it is justified, starting from renorming theory, to study those classes, the Asplund spaces on one hand and spaces with the RNP on the other. Rough norms —and bumps— become an important tool for this study. Another one is the use of the so-called variational principles.

Until now, we were satisfied with ensuring the existence of an equivalent norm with better convexity or better differentiability properties than the original norm. Now, we ask for an approximation, as good as we wish, of those improved norms

A. J. Guirao et al., *Renormings in Banach Spaces*, Monografie Matematyczne 75,
https://doi.org/10.1007/978-3-031-08655-7_10

to the original one. A tool for succeeding is the so-called "partitions of the unity".

10.1 Weakly compactly generated spaces I: Some introductory results

As we mentioned in the introduction to this chapter, maybe the most "useful" class of non-separable Banach spaces is the one formed by the so-called "weakly compactly generated" Banach spaces (Definition 131 above). Those are spaces containing a weakly compact and linearly dense subset K (we say that K "**generates**" the space, or that K is a "**generator**" of the space). All the reflexive spaces belong to this class —since their closed unit ball is a generator of the space—, as well as the spaces $L_1(\mu)$ for a finite measure μ (look at the image of $B_{L_2(\mu)}$ under the natural injection $L_2(\mu) \to L_1(\mu)$), more generally the spaces $L_1(\mu)$ for a σ-finite measure μ, too, or the space $c_0(\Gamma)$ for any nonempty set Γ (the image of $B_{\ell_2(\Gamma)}$ under the natural injection $\ell_2(\Gamma) \to c_0(\Gamma)$ is a w-compact generator).

Krein's theorem (see, e.g., [FHHMZ11, Theorem 3.133]) allows for a generator to be absolutely convex and closed. Under this assumption, the following are easily seen to be equivalent: (i) K is a generator of X. (ii) K **almost absorbs** any point $x \in X$ (i.e., given $x \in X$ and $\varepsilon > 0$, there exists $n \in \mathbb{N}$ such that $\{x\} \subset nK + \varepsilon B_X$). (iii) K almost absorbs any $\|\cdot\|$-compact subset of X. (iv) K almost absorbs any $\|\cdot\|$-null sequence in X.

➩ (Sketch of the argument) For (i)⇒(ii), any $x \in \operatorname{span} K$ is written

$$x = s\sum_{i=1}^{m}(\lambda_i/s)x_i \in sK \subset nK, \quad \text{where } s := \sum_{i=1}^{m}|\lambda_i| \leq n,$$

due to the fact that K is absolutely convex. For (ii)⇒(iii), just observe that for $\varepsilon > 0$ and a $\|\cdot\|$-compact subset M of X, we have $M \subset \bigcup_{i=1}^{n} B(x_i, \varepsilon)$ for some $\{x_1, \ldots, x_n\} \subset X$. (iii)⇒(iv) and (iv)⇒(i) are obvious.

H The class of WCG spaces and their relatives was created by J. Lindenstrauss and his collaborators in Israel and the U. S. A. (mainly by H. Corson and D. Amir) in the sixties. It is a wide class of nonseparable spaces (some subclasses were mentioned in the introduction to this section).

The more general class of Vašák spaces originated in Prague (Vašák's thesis), Paris (Talagrand), and Tomsk (Gul'ko) in the seventies. The step from WCG to Vašák spaces was made in search of a concept that would: (i) In the old golden language "be elegant", and yet have most properties of the more complicated class of the weakly K-analytic spaces (for the definition see, e.g., [DeGoZi93, Definition after Lemma VI.6.6]), (ii) be more stable than WCG, (iii) be easy to work with, and (iv) in the now contemporary language, promise to create new jobs, i.e. new theorems. Vašák spaces were introduced by Vašák in [Vas81].

Yet the even more general class of WLD spaces originated in Greece (Argyros, Mercourakis), Spain (Valdivia, in [Vald88], Orihuela, and Cascales), and Ukraine (Plichko) in the eighties.

All these classes share in a sense properties of spaces with Schauder bases (for example, their elements have PRI's, see again the notes after Definition 207 and Theorem 208, and they have strong Markushevich bases, see Definition 128 and Theorem 224).

Along this chapter we shall prove or just mention some results on WCG spaces that turn out to be crucial for the theory of such spaces. We may start by an important "factorization" theorem whose proof does not depend on additional material but Krein's theorem on the weakly compact character of the absolutely convex and closed hull of a weakly compact set in a Banach space (see, e.g., [FHHMZ11, Thm. 3.133]).

Theorem 191 (Davis, Figiel, Johnson, Pełczyński, [DFJP74]). (see, e.g., Thm. 13.22 in [FHHMZ11]) *Let X be a WCG Banach space generated by a weakly compact set K. Then there exists a reflexive Banach space R and a bounded linear one-to-one operator $T : R \to X$ such that $K \subset T(B_R)$.*

An obvious reformulation of Theorem 191 is that a Banach space X is WCG if, and only if, there exists a reflexive Banach space R and a bounded linear mapping $T : R \to X$ such that TB_R is a generator of X (note that, by passing to $R/\ker(T)$, we may always assume that T is one-to-one).
Still an equivalent formulation —that justifies naming the result a "factorization theorem"— is that every weakly compact operator S from a Banach space Y into a Banach space X "factors" through a reflexive Banach space, in the sense that there exists a reflexive Banach space R and continuous linear operators $S_0 : Y \to R$ and $T : R \to X$ such that $S = T \circ S_0$. Indeed, the space $X_0 := \overline{\text{span}}\{S(B_Y)\}$ is a WCG space generated by $\overline{S(B_Y)}$. Apply Theorem 191 to get a reflexive Banach space R and a bounded linear operator $T : R \to X_0$, and recall that it is possible to choose T one-to-one. It is enough to put $S_0(y) = T^{-1}(S(y))$ for $y \in Y$.

The proof of the following result relies, to some extent, on the separable 1-complementability property of WCG Banach spaces (see the proof of Theorem 208 below).

Theorem 192 (Valdivia, [Vald77]). *Let X be a Banach space such that X^{**}/X is separable. Then X is isomorphic to $R \oplus S$, where R is a reflexive Banach space and S is a separable Banach space. Moreover, if Y is a separable subspace of X, then Y^{**} is separable. In particular, S^{**} is separable.*

Remark 193. The separability of S^{**} comes from the fact that $X^{**} = R \oplus S^{**}$ and the existence of a natural linear isomorphism between the separable space X^{**}/X and S^{**}/S. Since S is separable, it follows that S^{**} is separable, too. Now, if Y is a separable subspace of X, and if P_1 (P_2) is the canonical projection from X onto R (respectively, onto S), we have $Y \subset P_1 Y \oplus P_2 Y$, hence $Y^{**} \subset P_1 Y \oplus (P_2 Y)^{**}$. Since $P_1 Y$ and $(P_2 Y)^{**}$ are both separable (the last one thanks to the separability of S^{**}), the conclusion follows. ®

Corollary 194 (Valdivia, [Vald77]). *If X^{**}/X is separable, then X, X^*, and X^{**} are all of them WCG.*

Another straightforward consequence is the following theorem:

Theorem 195 (Kuo, [Kuo75]). *If X is a Banach space, and X^{**}/X is separable, then X^{**} has the RNP.*

Proof. The argument of the proof is similar to the one in Remark 193. We follow the notation in Theorem 192. Let $Y \subset X^* = R^* \times S^*$ be a separable subspace. Let P_1 (P_2) be the canonical projection from X^* onto R^* (respectively, onto S^*). Then $Y \subset P_1 Y \times P_2 Y$, a separable subspace of X^* whose dual is $(P_1 Y)^* \oplus (P_2 Y)^*$. This last space is separable, since $P_1 Y$ is separable and reflexive, and $(P_2 Y)^* \subset S^{**}$ (remember that S^{**} is separable). Thus, X^* is Asplund, hence X^{**} has the RNP (see Theorem 274 below). □

10.2 Lindenstrauss–Troyanski tools for renorming, and developments

It was the approach of Davis and Johnson (see Theorem 173), that allowed Troyanski to nontrivially extend in [Tr71] Kadets' Theorem 173 above to the large class of the WCG Banach spaces (see Section 10.4 below and, in particular, Theorem 218). This was a result that helped open the whole area of renormings of nonseparable spaces overall. In Troyanski's result, Day's norm was used in a very subtle way combining both M.I. Kadets' and Day's techniques. This was allowed by using the Lindenstrauss' method of projectional resolutions of the identity. It used the fact that the Day norm $\|\cdot\|_D$ (see Subsection 9.4.1 and Corollary 243) "survives" its LUR properties when carefully perturbed with (weighted) distances to proper finite-dimensional subspaces.

Thus, the crucial ingredient was a tool —called a **projectional resolution of the identity (PRI)**, see Definition 130—, introduced by J. Lindenstrauss in his pioneering work [Lind64b], [Lind66b], which provides a beautiful crystal clear geometric argument.

10.2.1 A basic Lindenstrauss lemma

The construction of a PRI on some classes of nonseparable Banach spaces is based on the following fundamental and extremely important result:

Lemma 196 (Lindenstrauss, [Lind66b]). *Let $(X, \|\cdot\|)$ be a WCG Banach space generated by a (absolutely convex and) weakly compact set K. Put $N := \operatorname{span}(K)$. Let A be a finite-dimensional subspace of N, and E a finite-dimensional subspace of N^* $(= X^*)$. Let $n, q \in \mathbb{N}$ and let $\varepsilon > 0$. Then there is a finite-dimensional subspace $Z_{\varepsilon,n,q} \subset X$ containing A such that for every subspace L of X containing A with $\dim L/A = n$ and $p_K \leq q\|\cdot\|$ on L —where p_K denotes the Minkowski functional of K—, there is an operator $T : L \to Z_{\varepsilon,n,q}$ such that $Ta = a$ for every $a \in A$, $|\,\|Tx\| - \|x\|\,| < \varepsilon\|x\|$, $|p_K(Tx) - p_K(x)| < \varepsilon\|x\|$ and $|\langle Tx - x, x^*\rangle| < \varepsilon\|x^*\|.\|x\|$ for all $x \in L$ and $x^* \in E$.*

Proof. (Sketch) The construction in the proof of this result has a nice geometric description (we follow [JoZi72]; see also [Day73, III §5, proof of Theorem 1]): Figure 10.1 depicts the case $n = 1$ (the general case follows similarly, or it can be obtained from this by finite induction). The idea is to "move" $\mathcal{L}$, the set of all subspaces $(L, \|\cdot\|)$ of X that contain A as a hyperplane and satisfy the other requirements in the statement, to the single structure $(A \times \mathbb{R}, \|\cdot\| \oplus_1 |\cdot|)$ (where $\|\cdot\|$ denotes here the norm $\|\cdot\|$ restricted to A) by using the linear mapping $T_L(a + tb_L) := (a, t)$, where $a \in A$, $t \in \mathbb{R}$, and b_L and $-b_L$ are two chosen contact points between B_L and the two support hyperplanes $f = 1$ and $f = -1$ to B_L, being $f : L \to \mathbb{R}$ a continuous linear mapping in S_{L^*} with $f|_A = 0$. Note that if $x = a + tb_L \in B_L$, then $1 \geq \|x\| \geq |f(a + tb_L)| = |t|$.
All the resulting norms $\|(a,t)\|_L := \|a + tb_L\|$ on $A \times \mathbb{R}$ form an equicontinuous (hence relatively compact) set $\{\|\cdot\|_L|_{B_{A\times\mathbb{R}}} : L \in \mathcal{L}\}$ of functions defined on the closed unit ball of $(A \times \mathbb{R}, \|\cdot\| \oplus_1 |\cdot|)$ (in other words, the associated balls $T_L(B_L)$ are all in the "diabolo" set D, drawn in gray in Figure 10.1), hence relatively compact. The set of all seminorms $(p_K)_L$ on $A \times \mathbb{R}$ defined as $(p_K)_L(a,t) = p_K(a + tb_L)$, is also relatively compact (note that $p_K \leq q\|\cdot\|$ on L), and so it is the set $\{x_L^* : x^* \in B_E\}$ of functionals on $A \times \mathbb{R}$, where $x_L^*(a,t) = x^*(a + tb_L)$. All together, we get a relatively compact set of norms, seminorms, and functionals indexed by $\mathcal{L}$, and so having a finite ε-net indexed by a finite subset $\{L_i\}_{i=1}^m$ of $\mathcal{L}$. Put $Z_{\varepsilon,n,q} := \operatorname{span} \bigcup_{i=1}^m L_i$, The sought mapping $T : L \to Z_{\varepsilon,n,q}$ is the composition $T_{L_i}^{-1} \circ T_L$, where $T_{L_i}(B_{L_i})$ is the closest ball of this type to $T_L(B_L)$. Given $x = a + tb_L \in B_L$, we have $|\,\|Tx\| - \|x\|\,| = |\,\|(a,t)\|_{L_i} - \|(a,t)\|_L\,| < \varepsilon$, $|\,p_K(Tx) - p_K(x)\,| = |(p_K)_{L_i}(a,t) - (p_K)_L(a,t)\,| < \varepsilon$, and

$$|\langle Tx - x, x^*\rangle| = |t\langle b_{L_i} - b_L, x^*\rangle| < t\varepsilon \leq \varepsilon$$

for $x^* \in B_E$. This shows that $T : L \to Z_{\varepsilon,n,q}$ has the sought properties. □

Remark 197. If $(X, \|\cdot\|)$ is a Banach space and K is a w-compact linearly dense subset of X (that we may always assume absolutely convex and closed thanks to

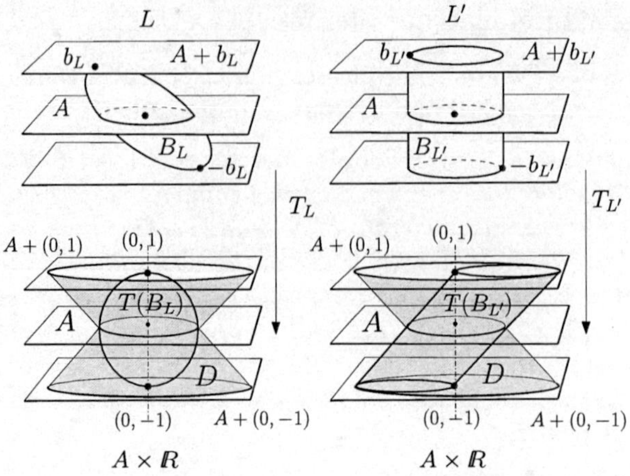

Figure 10.1: The proof of Lemma 196 for $n = 1$ (and two spaces L and L') in $\mathcal{L}$

Krein's theorem, see, e.g., [FHHMZ11, Theorem 3.133]), we may define a norm $\|\cdot\|_K$ on X^* as $\|x^*\|_K := \sup\{|\langle k, x^*\rangle| : k \in K\}$. A continuous linear operator $T : (X^*, \|\cdot\|^*) \to (X^*, \|\cdot\|^*)$ is w^*-w^*-continuous (and so T is the adjoint S^* of an operator $S : X \to X$) if there exists a constant $C \geq 0$ such that $\|T\|_K \leq C$. Ⓡ

Indeed, let us consider the adjoint operator $T^* : X^{**} \to X^{**}$. We claim that $T^*(K) \subset X$. To prove the claim, let $x^* \in K^\circ$ ($\subset X^*$). Thus, $\|x^*\|_K \leq 1$. It follows that $|\langle k, Tx^*\rangle| \leq \|Tx^*\|_K \leq C\|x^*\|_K \leq C$, and so $|\langle T^*k, x^*\rangle| \leq C$ for all $k \in K$. This shows that $T^*k \in CK^{\circ\circ} = C\overline{K}^{w(X^{**},X^*)} = CK$ for all $k \in K$, and the claim is proved.
Since $T^* : (X^{**}, \|\cdot\|^{**}) \to (X^{**}, \|\cdot\|^{**})$ is also continuous, we obtain that $T^*(X) \subset X$. The restriction to X of T^* is the sought operator $S : X \to X$.

Observe that if X is not reflexive, then $\|\cdot\|_K$ cannot be an equivalent norm on X^*. Indeed, if for some $a > 0$ we have ($aK^\circ =$) $aB_{\|\cdot\|_K} \subset B_{(X^*,\|\cdot\|^*)}$, then ($(1/a)K =$) $(1/a)K^{\circ\circ} \supset B_{(X^{**},\|\cdot\|^{**})}$, and we get that X is reflexive. Ⓡ

H Remark 197 enabled to work in the dual of a Banach space and to use projections in the original space. This opened the way to the concept of a Vašák space (see Definition 133 and Section 10.1 above).

10.2.2 An application to reflexive spaces

If X is reflexive, we can take $K := B_X$ (so $N = X$) and $q := 1$ in Lemma 196. Let A be a finite-dimensional subspace of X, and $\{x_j^* : 1 \leq j \leq m\} \subset S_{X^*}$. Find, for

$n \in \mathbb{N}$ and $\varepsilon = 1/n$, a finite-dimensional subspace $Z_n := Z_{1/n,n,1} \supset A$ according to Lemma 196. Put $Z := \overline{\operatorname{span}} \bigcup_{n=1}^{\infty} Z_n$. Given $n \in \mathbb{N}$ and a finite-dimensional subspace $L \supset A$ of X such that $\dim L/A = n$, Lemma 196 provides a bounded linear operator $T_L : L \to Z$ with $\|T_L\| \leq 1 + 1/n$ such that $T_L a = a$ for all $a \in A$ and $\|T^* x_j^* - x_j^*\| < 1/n$ for $1 \leq j \leq m$. Extend —nonlinearly— T_L to X by letting $\widetilde{T}_L(x) = 0$ for all $x \in X \setminus L$. Let us consider the family $\mathcal{L}$ of all finite-dimensional subspaces of X that contain A, partially ordered by inclusion. Note that $\widetilde{T}_L(B_X) \subset 2B_Z$ for all $L \in \mathcal{L}$. The net $\{\widetilde{T}_L|_{B_X}\}$ is in $(2B_Z)^{B_X}$ (a compact space when B_Z is endowed with the weak topology and the product with the pointwise topology), and so it has a cluster point $T \in (2B_Z)^{B_X}$. Given $x \in B_X$ and $n \in \mathbb{N}$, we may always consider a finite-dimensional subspace $L_0 \in \mathcal{L}$ with $\dim(L_0/A) = n$ and such that $x \in L_0$. This shows that T so defined gives a linear norm-1 operator $T : X \to Z$ with $Ta = a$ for all $a \in A$ and $T^* x_j^* = x_j^*$ for $1 \leq j \leq m$.

This proves, then, the following version of Lemma 196:

Lemma 198. *Let X be a reflexive Banach space. Let $A \subset X$ be a finite-dimensional subspace, $m \in \mathbb{N}$, and $\{x_j^* : j = 1, 2, \dots, m\}$ a subset of X^*. Then, there exists a separable subspace Z that contains A, and a norm-1 linear operator $T : X \to Z$ such that $Ta = a$ for all $a \in A$, and $T^* x_j^* = x_j^*$ for all $j = 1, 2, \dots, m$.*

A finite-induction process —that we describe below— gives, now, the following result:

Theorem 199. *Let X be a reflexive space. Let Y be a separable subspace of X and E a separable subspace of X^*. Then there exists a separable subspace W of X such that $Y \subset W$, and a norm-1 projection P from X onto W such that $P^* x^* = x^*$ for all $x^* \in E$.*

Proof. (Sketch) Put $Y_0 := Y = \overline{\operatorname{span}}\{y_n^0 : n \in \mathbb{N}\}$, and $E = \overline{\operatorname{span}}\{e_n^* : n \in \mathbb{N}\} \subset X^*$. We get, by the previous construction, a norm-1 linear operator $T^1 : X \to Y^1 = \overline{\operatorname{span}}\{y_i^1 : i \in \mathbb{N}\}$ fixing $\{y_1^0\}$ and such that $(T^1)^*$ fixes $\{e_1^*\}$. By induction we get a norm-1 linear operator $T^n : X \to Y^n = \overline{\operatorname{span}}\{y_i^n : i \in \mathbb{N}\}$ fixing $\{y_i^k : 0 \leq k \leq n-1, 1 \leq i \leq n\}$ and such that $(T^n)^*$ fixes $\{e_j^* : 1 \leq j \leq n\}$. If $W := \overline{\operatorname{span}} \bigcup_{n=0}^{\infty} Y^n$ $(\supset Y)$, the limit of a convergent subnet of $\{T^n\}$ in the previous sense gives a norm-1 linear *projection* T from X onto W such that $T^* x^* = x^*$ for all $x^* \in E$. □

In particular, we get the following result:

Theorem 200 (Lindenstrauss). *Any reflexive Banach space has an infinite-dimensional separable 1-complemented subspace.*

We shall prove later on (see the proof of Theorem 208 below) that the so-called **separable 1-complementation property** exhibited in Theorem 199 for the class of reflexive spaces —every separable subspace of X is contained in a separable and

1-complemented subspace of X— still holds for the larger class of WCG Banach spaces, a result due to Amir and Lindenstrauss.

In this direction, see also [FHHMZ11, Theorem 13.52] and [FHHMZ11, Exercise 13.2], due to Valdivia.

10.2.3 A glimpse of the weakly compactly generated case and Hahn–Banach extension operators

H For WCG spaces the Lindenstrauss method was adapted by Amir and Lindenstrauss in [AmLi68] (see the fundamental Theorem 201 below). It allows for a decomposition of the space X into pieces $(P_{\alpha+1} - P_\alpha)(X)$, where each P_α is a norm-1 projection on X, and the subsequent use of transfinite induction in checking a certain property of the space (see also [DeGoZi93], [FHHMZ11], [HMVZ07], and [Fab97]). This seminal result was later extended to other more general classes of spaces. First of all, K. John in Prague modified the proof of Lindenstrauss result by removing the use of eigenvector and matrices, and thus adapted it to situations where one can effectively use the smoothness of the space, and to subspaces of WCG, even if they are not WCG (see [Ros74] for counterexamples). See a series of papers by K. John and V. Zizler from that time, e.g., [JoZi72] and [JoZi74a]. It turned out that for WCG relatives one can work on the dual space, if one additional norm is involved, namely the supremum over the weakly compact generating set (see Remark 197). This approach allowed Vašák in [Vas81] to adapt the Lindenstrauss' construction (see, e.g., [FHHMZ11, Definition 13.5]) to the nowadays named Vašák spaces (see Definition 133 and Section 10.1 above).

Thus we get

Theorem 201 (Vašák, [Vas81], Valdivia, [Vald88]). *Every nonseparable WCD Banach space is WLD and admits a PRI. Thus every WCD space admits an injection into some $c_0(\Gamma)$ and a LUR equivalent norm.*

Proof. (Sketch) First, by using finite intersections of the sets in the definition of a WCD space (see Definition 133) it was proven in Vašák's work (see [FHHMZ11, Exercise 13.29]) that these sets can be taken to be absolutely convex and having a nonempty interior. Then Vašák was able to use K. John's refinement of the Amir–Lindenstrauss construction to ensure that these sets are all of them preserved when applying Tychonoff's theorem, as it was done in Lindenstrauss' construction in X^{**}. This way, by the properties of these sets, it is possible to get that X is preserved. Thus, Lindenstrauss' method gives a PRI on X. Furthermore, by using the Lindelöf property of subsets of the metrizable Baire space $\mathbb{N}^{\mathbb{N}}$ we get that X is Lindelöf in its weak topology (as Vašák showed). The reason why an injection into $c_0(\Gamma)$ exists is in Remark 127 and (iii) and (iv) in Theorem 135. For the existence of an equivalent LUR norm see, e.g., Theorem 220 below. For other details see [FHHMZ11, Chapter 13] and [HMVZ07, Theorem 5.44]. □

Let us mention now the following interesting property connected with Lemma 196 (see, e.g., [SiYo89]). First note the following definition:

If M is a subspace of a Banach space X, then M^* is naturally a quotient of X^*. The quotient mapping $X^* \mapsto M^*$ is the restriction mapping R_M that to $f \in X^*$ associates $f|_M$ ($\in M^*$). If $f \in M^*$, then put

$$H_M(f) := \{\tilde{f} \in X^*, \ \|\tilde{f}\| = \|f\|, \ \text{ and } R_M(\tilde{f}) = f\}; \tag{10.1}$$

It is a non-empty —due to the Hahn–Banach theorem— w^*-compact and convex subset of X^*. A selector $T : M^* \to X^*$ of R_M such that $Tf \in H_M(f)$ for all $f \in M^*$ is called a **Hahn–Banach extension function for** M. It is a norm-preserving function. It is a natural question whether T can be chosen to be linear (if this is the case, T becomes a linear isometry).

Some observations are in order: If $T : M^* \to X^*$ is a linear Hahn–Banach extension function for some subspace M of X, then (i) $P := TR_M : X^* \to X^*$ is a norm-1 projection with range $T(M^*)$ and kernel $M^{\perp}$, and (ii) T^* is a norm-1 projection of X^{**} onto M^{**} (observe that M^{**} is isometrically isomorphic to the closure of M in (X^{**}, w^*), that $T^* : X^{**} \to M^{**}$ is w^*-w^*-continuous, and that obviously $T^*m = m$ for all $m \in M$). On the other hand, (iii) if M is the range of a norm-1 projection $P : X \to M$, then $P^* : M^* \to X^*$ is a linear Hahn-Banach extension operator for M.

Kakutani's theorem [Kaku39] says that *if X is a finite-dimensional Banach space with* $\dim X \geq 3$ *such that every subspace of X is the range of a norm-1 projection, then X is isometrically isomorphic to a Hilbert space* . Later on, Lindenstrauss and Tzafriri proved that if every closed subspace of a Banach space X is complemented, then X is isomorphic to a Hilbert space (see Theorem 446 below, and, e.g., [FHHMZ11, Theorem 6.16]).
From Kakutani's theorem it is possible to obtain the following result:

Theorem 202 (Kakutani, [Kaku39]). *Every subspace of a Banach space X admits a linear Hahn–Banach extension operator if and only if X is isometrically isomorphic to a Hilbert space.*

Proof. Indeed, let N be an arbitrary 3-dimensional subspace of a Banach space X with the property in the statement, and let M be a 2-dimensional subspace of N. Since for M there is a Hahn–Banach extension operator $T : M^* \to X^*$, it follows from the observation (ii) above that $T^* : X^* \to M^{**}$ $(= M)$ is a norm-1 projection. Its restriction to N is a norm-1 projection from N to M, and Kakutani's theorem on finite-dimensional spaces, mentioned before the statement of Theorem 202, concludes that N is isometrically isomorphic to a Hilbert space. This finishes the proof. □

Based on Lemma 196 we have

Theorem 203 (Sims, Yost, [SiYo89]). *Let N be a subspace of a Banach space X. Then there is a subspace M of X such that $M \supset N$,* $\operatorname{dens} M = \operatorname{dens} N$, *and M admits a linear Hahn–Banach extension operator .*

Corollary 204 (Heinrich, Mankiewicz, Sims, Yost, [HeiMa82], [SiYo89]). *If X is a Banach space then X^* contains a 1-complemented subspace of density less than or equal to the continuum c.*

This follows from taking N separable in Theorem 203, finding M accordingly, and noting that then $\operatorname{dens} M^* \leq 2^{\aleph_0} = c$. The sought 1-complemented subspace of X^* is then TM^*, where T is the Hahn–Banach extension operator for M.

This should be compared with Lindenstrauss' result that *every infinite-dimensional complemented subspace of ℓ_∞ is isomorphic to ℓ_∞* ([LinTz77, Theorem 2.a.7]).

Corollary 205. *If X is an Asplund space then X^* contains a 1-complemented separable subspace.*

We will see that this result can be much strengthened (Theorem 320). A particular case of Corollary 205 is Theorem 200 above.

Related to the problem of the Hahn–Banach extension, it is also natural to ask for its uniqueness. In other words, to ask for conditions on a closed subspace M of a Banach space X to ensure the uniqueness of the norm-preserving extension —i.e., whenever H_M defined in (10.1) is a singleton— or even conditions on the space for ensuring that *all* closed subspaces have the uniqueness extension property. This will be the subject of Section 45.1 below.

10.3 Projectional resolutions of the identity II: A tool

After the fundamental contributions of Amir and Lindenstrauss and others to the theory of weakly compactly generated Banach spaces, a new method for constructing PRI's was discovered that, instead of compactness, used a tool called a **projectional generator**. This is a way to assign to elements in X^* countable subsets of X that, in a sense, *support* them. With this in hand, it proceeded similarly to the Lindenstrauss' method described above. While this method is not as fine as Lindenstrauss' one it still suffices for many purposes.
For better understanding, let us first present a simple but crucial lemma that sets the basic ingredients for having a single norm-1 projection:

Lemma 206 (See, e.g., [Fab97], Lemma 6.1.1). *Let X be a Banach space. Let V be a subspace of X and Y be a subspace of X^*. Assume that Y 1-norms V and V in turn separates the points of $\overline{Y}^{w^*}$, i.e. $V^\perp \cap \overline{Y}^{w^*} = \{0\}$. Then there is a projection $P : X \to X$ such that $\|P\| = 1$, $PX = V$, $P^{-1}(0) = Y_\perp$ and $P^*X^* = \overline{Y}^{w^*}$.*

Proof. (i) Observe that $V + Y_\perp$ is dense in X. Indeed, if $\langle v + y, x^*\rangle = 0$ for all $v \in V$ and all $y \in Y_\perp$, then $x^* \in \overline{Y}^{w^*} \cap V^\perp$ (due to the fact that $(Y_\perp)^\perp = \overline{Y}^{w^*}$), so $x^* = 0$ since V separates points of $\overline{Y}^{w^*}$.
(ii) Note that $V \cap Y_\perp = \{0\}$, since Y norms V. Thus, $V + Y_\perp$ is an algebraic direct sum, say $D := V \oplus Y_\perp$.
(iii) Let us see that $V \oplus Y_\perp$ is a topological direct sum. Let $P : D \to V$ be one of the associated projections. Given $x = v + y \in D$, with $v \in V$ (and $y \in Y_\perp$), we have

$$\|Px\| = \sup |\langle Px, B_Y\rangle| = \sup\langle x, B_Y\rangle \le \|x\|,$$

(the first equality since Y 1-norms V, and the second equality since $x - Px \in Y_\perp$) hence P is continuous and has norm 1.
(iv) As a consequence, D is closed (so $D = X$). Indeed, D is complete in X. For a direct proof, note that if $v_n + y_n \to x$, where $v_n \in V$ and $y_n \in Y_\perp$ for all $n \in \mathbb{N}$, then $v_n = Pv_n \to Px$, so $y_n \to x - Px \in Y_\perp$ (since $Y_\perp$ is closed), and so $x = Px + (x - Px) \in D$.
(v) Finally, if $y^* \in \overline{Y}^{w^*}$, for $x \in X$ we have

$$\langle x, P^*y^*\rangle = \langle Px, y^*\rangle = \langle x, y^*\rangle$$

(the second equality since $x - Px \in Y_\perp$), hence $y^* = P^*y^*$. □

Here comes the definition of a projectional generator.

Definition 207 (Orihuela, Valdivia, [OrVa90]). *Let X be a Banach space, W be a 1-norming subset of X^* such that $\overline{W}$ is linear, and $\Phi : W \to 2^X$ be a countably-valued mapping such that for every nonempty subset B of W such that $\overline{B}$ is linear, we have $\Phi(B)^\perp \cap \overline{B}^{w^*} = \{0\}$. The couple (W, Φ) is called a* **projectional generator (PG)** *on X. The projectional generator is called* **full** *if $W = X^*$.*

The projectional generator works similarly to the Mazur exhaustion method (see, e.g., [LinTz77, Theorem 1.a.5]), but it gives projections on the whole space under, for example, the assumptions that distinguished points of X^* have a countable support on distinguished subsets of X. The device of a projectional generator is at hand in many of the most natural examples of Banach spaces:

(i) If X is separable and $\{x_n : n \in \mathbb{N}\}$ is a dense subset of X, take $W := X^*$ and $\Phi(x^*) := \{x_n : n \in \mathbb{N}\}$ for $x^* \in X^*$. Obviously, (X^*, Φ) is a projectional generator on X.
(ii) More generally, if X is WCG and $K \subset X$ is absolutely convex, w-compact, and linearly dense, then (X^*, Φ) is a PG, where $\Phi(x^*) = \{x\}$ and x is an element in K where x^* attains its supremum. Indeed. Let B be a set as in the

statement. The fact that $\overline{B}^{\|\cdot\|}$ is a linear subspace (hence a convex set) shows that $\overline{B}^{w^*} = \overline{B}^{\mu^*}$, where $\mu^* := \mu(X^*, X)$ is the Mackey topology on X^* for the dual pair $\langle X, X^*\rangle$, i.e., the topology on X^* of the uniform convergence on the family of all absolutely convex and w-compact subsets of X (the Mackey–Arens Theorem, see Section 2.1 above). Let $x^* \in \Phi(B)^\perp \cap \overline{B}^{w^*}$ and $\Phi(x^*) = \{x\}$. Assume that $0 < 3\varepsilon := \langle x, x^*\rangle$. Find $b^* \in B \cap (x^* + \varepsilon K^\circ)$, and let $\Phi(b^*) = \{b\}$. Then $|\langle x, x^* - b^*\rangle| \leq \varepsilon$, so $\langle x, b^*\rangle \geq 2\varepsilon$. Since $\langle x, b^*\rangle \leq \langle b, b^*\rangle$ we get $\langle b, b^*\rangle \geq 2\varepsilon$. Note that $|\langle b, b^* - x^*\rangle| \leq \varepsilon$. This shows $0 = \langle b, x^*\rangle > \varepsilon$. a contradiction. Thus, $\langle x, x^*\rangle = 0$. Due to the fact that K is linearly dense, we finally get $x = 0$. See also [HMVZ07, Proposition 3.43].
(iii) X is WLD if, and only if, X has a full PG (see, e.g, [HMVZ07, Theorem 5.44]).
(iv) For PG's on Vašák spaces we refer to [DeGoZi93, Chapter 6] and [FHHMZ11, Exercise 13.31], this last one due to Valdivia, where he used Jayne–Rogers selectors instead of the mappings in the first reference.
(v) If the norm $\|\cdot\|$ on a Banach space X is Fréchet differentiable, then the couple (X, ϕ), where ϕ is the mapping $\phi : X \to X^*$ that sends 0 to 0 and $x \in X \setminus \{0\}$ to $\partial\|\cdot\|(x)$, is a projectional generator on X^*. See, e.g., [Fab97, Section 8.2]; the result and its proof as presented there are extremely instructive.
(vi) More generally, for the existence of projectional generators in duals of Asplund spaces we refer to [Fab97, Section 8.2].

The importance of the concept of a projectional generator in structure and renorming of Banach spaces lies in the following result, just a consequence of the ingenuity of the definition of a PG:

Theorem 208. *If X admits a PG, then it admits a PRI.*

For a proof see, e.g., [HMVZ07, Theorem 3.42].

➡ Maybe a sketch of the way to get the "first" —of a transfinite-induction process— norm-1-projection on a (infinite-dimensional) WCG Banach space can be instructive: Fix an (infinite-dimensional) separable subspace V_0 of a WCG Banach space X (generated by an absolutely convex and w-compact subset K), and let $S_0 := \{x_{0,n} : n \in \mathbb{N}\}$ be a dense subset of S_{V_0}. Let N_0 be a countable subset of S_{X^*} that 2^{-1}-norms S_0 (i.e., $\sup|\langle x_{0,n}, N_0\rangle| > 1 - 2^{-1}$). Put $Y_0 := \text{span}\,(N_0)$ and let $D_0 := \{x^*_{0,n} : n \in \mathbb{N}\}$ be a countable dense subset of Y_0. Each $x^*_{0,n}$ attains its supremum on K (say at $x_{1,n}$). Put $S_1 := S_0 \cup \{x_{1,n} : n \in \mathbb{N}\}$, and find a countable subset N_1 that 2^{-2}-norms S_1. Continue this way. The subspaces $V := \overline{\text{span}}\,(\bigcup_{n=0}^\infty S_n)$ of X and $Y := \overline{\text{span}}\,(\bigcup_{n=0}^\infty Y_n)$ of X^* satisfy the requirements of Lemma 206.
Indeed, that Y 1-norms V is clear. Without loss of generality, we may assume that $K \subset B_X$. Let $x^* \in V^\perp \cap \overline{Y}^{w^*}$. We shall prove that $x^* = 0$ (the argument

is similar to the one used in (ii) at the annotation after Definition 207). Assume $x^* \neq 0$. Then $0 < \sup\langle K, x^*\rangle$ $(= 5\varepsilon)$. As above, $\overline{Y}^{w^*} = \overline{Y}^{\mu(X^*,X)}$. In particular, there exists $y^* \in Y$ such that $\sup|\langle K, x^* - y^*\rangle| \leq \varepsilon$ (and so $\sup\langle K, y^*\rangle \geq 4\varepsilon$). Find some $n \in \mathbb{N}\cup\{0\}$ and $y_0^* \in D_n$ such that $\|y^* - y_0^*\|^* \leq \varepsilon$, hence $3\varepsilon \leq \sup\langle K, y_0^*\rangle$ $(= \langle x_0, y_0^*\rangle$ for some $x_0 \in S_{n+1} \cap K)$. However, $\langle x_0, x^*\rangle = 0$, hence $|\langle x_0, y^*\rangle| \leq \varepsilon$ and so $|\langle x_0, y_0^*\rangle| \leq 2\varepsilon$, a contradiction. Thus, V $(\supset V_0)$ is the (separable) range of a norm-1 projection on X whose kernel is $Y_\perp$. Observe that, in particular, we proved that *every separable subspace of a WCG Banach space X is contained in a separable* 1*-complemented subspace of X* (see Lindenstrauss' Theorems 199 and 200 for the reflexive case and Corollary 205 above for the Asplund case). This property is called the **separable 1-complementation property**.

A partially ordered set $(\Gamma, \leq)$ is called **up-directed** if given $s \in \Gamma$ and $t \in \Gamma$, there is $u \in \Gamma$ such that $t \leq u$ and $t \leq u$. It is called **σ-complete** if every increasing sequence in it has a supremum.

Definition 209 (Kubiś, [Kub09]). *A* **projectional skeleton** *in a Banach space X is a family $(P_s : s \in \Gamma)$ of norm-*1 *projections in X for some up-directed σ-complete partially ordered set Γ such that*
(i) *P_sX is separable for every $s \in \Gamma$,*
(ii) $X = \bigcup_{s\in\Gamma} P_sX$
(iii) *$P_t \circ P_s = P_s \circ P_t = P_s$ whenever $s, t \in \Gamma$ are such that $s \leq t$.*
(iv) *Given a sequence $s_1 < s_2 < \ldots$ in Γ, and $t := \sup_{n\in\mathbb{N}} s_n$ $(\in \Gamma)$, then $P_tX = \overline{\bigcup_{n\in\mathbb{N}} P_{s_n}X}$.*

Theorem 210 (Kubiś, [Kub09]). *Every Banach space that admits a projectional generator admits also a projectional skeleton.*

For an elementary proof, see [CuFa17].

Theorem 211 (Kubiś, [Kub09]). *Every Banach space that admits a projectional skeleton admits also a PRI.*

The proof is elementary (see the reference above).

Observe then that ℓ_∞ does not admit a projectional skeleton, since ℓ_∞ does not admit a PRI (see Section 12.4).

Theorem 212 (Kalenda, [Kal]). *Every Banach space that admits a projectional skeleton, admits also a strong Markushevich basis.*

Proposition 213 (Kubiś, [Kub09]). *Let X be a Banach space with density $\aleph_1$. Then X admits a PRI if, and only if, it admits a projectional skeleton.*

Note that the space $\ell_2(2^{\mathfrak{c}}) \oplus \ell_\infty$ has a PRI (see, e.g., [HMVZ07, Section 5.1]) but it does not admit a projectional skeleton, as otherwise it would admit a strong

Markushevich basis (see Theorem 212 above) and thus a LUR equivalent norm (see Theorem 222 below), which is not the case as ℓ_∞ does not admit a LUR norm (Corollary 804 below). It seems not clear if the existence of a projectional skeleton implies the existence of a projectional generator (see [Kal]).

➡ Projectional skeletons can be used to characterize several classes of Banach spaces. We provide here some examples:

Theorem 214 (Kubiś, [Kub09]). *A Banach space $(X, \|\cdot\|)$ is weakly Lindelöf determined if, and only if, there exists a projectional skeleton $\{P_{\gamma\colon\ \gamma\in\Gamma}\}$ on $(X, \|\cdot\|)$ such that $\bigcup\{P_\gamma^* X^* :\ \gamma \in \Gamma\} = X^*$.*

A somehow streamlined proof can be found in [FaMo18]. A characterization of the class of the weakly compactly generated spaces in similar terms requires a couple of definitions: First, if $A \subset X$ ia a nonempty bounded set, then we denote $\rho_A(x^*, y^*) := \sup\{|\langle x, x^* - y^*\rangle| :\ x \in A\}$ for $x^*, y^* \in X^*$. Given $\varepsilon \geq 0$, a projectional skeleton $\{P_\gamma :\ \gamma \in \Gamma\}$ on X is called **A-ε-shrinking** if for every sequence $\gamma_1 \leq \gamma_2 \leq \ldots$ in Γ and for every $x^* \in X^*$ we have

$$\limsup_{j\to\infty} \rho_A(P^*_{\gamma_j} x^*, P^*_{\sup \gamma_i} x^*) \leq \varepsilon \|x^*\|;$$

if $\varepsilon = 0$, we say just **A-shrinking**.

Theorem 215 (Fabian, Montesinos, [FaMo18]). *A Banach space $(X, \|\cdot\|)$ is weakly compactly generated if, and only if, there exists a bounded closed absolutely convex and linearly dense set $A \subset X$, and a projectional skeleton $\{P_\gamma :\ \gamma \in \Gamma\}$ on $(X, \|\cdot\|)$ which is A-shrinking and satisfies $P_\gamma(A) \subset A$ for every $\gamma \in \Gamma$.*

For additional information on projectional skeletons, the reader may consult, e.g., [KKLP13, Chapter 17].

Projectional skeletons should be compared with the following crucial result proved by Troyanski in his fundamental result on LUR renorming of WCG spaces [Tr71] (see Theorem 218 below). It holds for WLD spaces as well.

Theorem 216 (Troyanski). *Let X be a WCG space. Then there is a family of projections $\{P_\gamma : \omega \leq \gamma \leq \mu\}$ (where μ is the first ordinal of cardinal* dens X*) such that, letting $R_\gamma := (P_{\gamma+1} - P_\gamma)/(\|P_{\gamma+1}\| + \|P_\gamma\|)$,*
(i) *$P_\gamma P_\delta = P_\delta P_\gamma = P_\gamma$, if $\omega \leq \gamma \leq \delta \leq \mu$.*
(ii) *$P_\omega(X)$ and $R_\gamma(X)$ are separable for $\omega \leq \gamma < \mu$.*
(iii) *$P_\mu = Id_X$.*
(iv) *For every $x \in X$, $\{\|R_\gamma x\| :\ \gamma \in [\omega, \mu)\} \in c_0([\omega, \mu))$.*
(v) *For every $x \in X$ and $\gamma \in [\omega, \mu)$, $P_\gamma x \in \overline{\operatorname{span}}(\{R_\delta(x) :\ \omega \leq \delta < \mu\} \cup \{P_\omega(x)\})$.*

For the proof we refer, e.g., to [HMVZ07, Theorem 3.46].

10.4 Troyanski's renorming theorem, and extensions

A starting landmark in this area was the pioneering Troyanski's results in [Tr68] and [Tr71]. We strongly recommend the reader to study both seminal papers.

Theorem 217 (Troyanski, [Tr68]). *Any Banach space of arbitrary density having an unconditional basis admits an equivalent LUR norm.*

Theorem 218 (Troyanski, [Tr71]). *If X is a WCG space, then it admits a LUR norm.*

Corollary 219. *If X is reflexive, then X admits a Fréchet differentiable norm.*

Proof of Corollary 219. It follows by using Theorem 218 and (iii) in Theorem 143. □

We shall mention several approaches to the proof of the important Theorem 218: In Subsection 10.4.1 we sketch the first one (and we provide a complete proof for the case that X has a transfinite Schauder basis). In Subsection 10.4.2 we offer two proofs: The first one by combining Theorems 221 and 224 in Theorem 222. The second one still uses Theorem 224, although now we construct, from a strong M-basis and without appealing to Theorem 221, the sought LUR renorming.

10.4.1 Troyanski's original proof

First, we sketch the main idea of Troyanski's original proof [Tr71] of Theorem 218. As it was mentioned, it uses Amir–Lindenstrauss projectional resolutions of the identity and a very subtle attachment of distance functions to finite dimensions which, when combined with the use of Day's norm, gives the result. It significantly uses also Theorem 216. To be a little more precise, we sketch here the construction of the norm in the case that $(X, \|\cdot\|)$ has an unconditional basis $\{e_\gamma, f_\gamma\}_{0\leq\gamma\leq\mu}$ (see Definition 129), where μ is an uncountable ordinal. Define $|||\cdot|||$ on X by $|||x||| := \sup_{0\leq\gamma\leq\mu} \|\sum_{i=0}^{\gamma} f_i(x)e_i\|$ for $x \in X$. Then $|||\cdot|||$ is an equivalent norm on X, and each of the projections $P_\gamma : X \to X$ given by $P_\gamma(x) = \sum_{i=0}^{\gamma} f_i(x)e_i$ satisfies $|||P_\gamma||| = 1$. It is now obvious that $\{P_\gamma\}_{0\leq\gamma\leq\mu}$ is a PRI on $(X, |||\cdot|||)$, and $\dim(P_{\alpha+1} - P_\alpha)(X) = 1$ for all $\alpha < \mu$. From now on we assume X endowed with the norm $|||\cdot|||$.

Put $\Gamma := [0, \mu)$. Let $\mathcal{A}_n$ be the family of all finite subsets of Γ with no more that n elements. For $x \in X$ and $A \in \mathcal{A}_n$, let $E_n^A(x) = \operatorname{dist}(x, \operatorname{span}\{e_\alpha\}_{\alpha\in A})$ and $F_n^A(x) = \sum_{\alpha\in A} |f_\alpha(x)|$. Put $G_n(x) = \sup\{E_n^A(x) + nF_n^A(x) : \ A \in \mathcal{A}_n\}$, $G_0(x) = |||x|||$. Let $\Delta = \{0, -1, -2, \dots\} \cup \Gamma$ and define $\Phi : X \to c_0(\Delta)$ by $\Phi(x)(n) = 2^{-n}G_n(x)$ for $n \in \{0, -1, -2, \dots\}$, $\Phi(x)(\alpha) = |f_\alpha(x)|$ for $\alpha \in \Gamma$. Finally, define an equivalent norm $\|x\|$ on X by $\|x\| = \|\Phi(x)\|_D$ where $\|\cdot\|_D$ is Day's norm on $c_0(\Delta)$ (see Subsection 9.4.1). Then $\|\cdot\|$ is the sought norm on X.

Let us briefly outline the main idea of the proof that $\|\cdot\|$ is a LUR norm on X. Let $x_n, x \in X$ be such that $2\|x_n\|^2 + 2\|x\|^2 - \|x + x_n\|^2 \to 0$. Let $\varepsilon > 0$. Find $n \in \mathbb{N}$ and $A' \in \mathcal{A}_n$ such that $E_n^{A'}(x) < \varepsilon$. Assume without loss of generality that

$$\sup\{|f_\alpha(x)| : \alpha \notin A'\} < \min\{|f_\alpha(x)| : \alpha \in A'\}. \tag{10.2}$$

Due to the term n in the definition of G_n's, if m is big enough and $A \in \mathcal{A}_m$ is so chosen that $G_m(x) - (E_m^A(x) + mF_m^A(x)) < \varepsilon$ with $A \in \mathcal{A}_m$, then, *necessarily*, $A \supset A'$ and thus $E_m^A(x) < \varepsilon$. This is because $\{f_\alpha(x)\} \in c_0(\Gamma)$. From the LUR property of Day's norm on $c_0(\Omega)$ it follows that $F_n^A(x_k) \to F_n^A(x)$ for all $A \in \bigcup \mathcal{A}_n$ and that $G_n(x_k) \to G_n(x)$ for all n. As the topology of the coordinatewise convergence in X is Hausdorff, in order to prove that $\|x_k - x\| \to 0$ it suffices to show that $\{x_k\}$ is relatively norm compact in X. The latter is seen from the fact that for large k, $E_m^A(x_k) \leq G_m(x_k) - mF_m^A(x_k) \leq G_m(x) + \varepsilon - mF_m^A(x) + \varepsilon \leq E_m^A(x) + 3\varepsilon \leq 4\varepsilon$.□

By introducing additional parameters into the preceding proof, it can be achieved that in the notation above, $A \supset A'$ can be replaced by $A = A'$. This fact is appropriately called by Haydon "rigidity condition". It was first used in the following version of Troyanski's Theorem:

Theorem 220 (Zizler, [Ziz84]). *Let X have a PRI (P_α) such that each $(P_{\alpha+1} - P_\alpha)X$ admits a LUR norm. Then X admits a LUR norm.*

This theorem is proved in [DeGoZi93, Proposition VII.1.6].

Let us mention that a proof of Troyanski result in which the use of Day's norm is replaced by Deville's Master Lemma 426 is explained in [DeGoZi93, Section VII.1], where Deville's lemma is Lemma 1.1.

10.4.2 Moltó–Orihuela–Troyanski transfer method

H The ingenious concept of the proof of the main result in [Tr71] (see Subsection 10.4.1 above), that consisted of a clever interplay of the properties of Day's norm with the distance function to finite-dimensional properly chosen subspaces, created Theorems 220 and 221, as well as Theorem 231 below.

Later on, Troyanski's proof generated the following elegant important result (see also [MOTV09, Theorems 1.6 and 1.7]).

Theorem 221 (Moltó, Orihuela, Troyanski, [MOT97]). *Let Y be a LUR Banach space. Let T be a bounded linear operator from a Banach space X into Y such that for every $x \in X$ there exists a separable subspace Z_x of X with*

$$x \in \overline{\text{span}} \bigcup_{n=1}^{\infty} Z_{x_n}.$$

whenever $\lim \|Tx_n - Tx\| = 0$. *Then X admits an equivalent LUR norm.*

The condition on the mapping T stated in Theorem 221 turns out to be equivalent to the so-called "co-σ-continuity" for mappings between metric spaces (see Definition 430 below) when applied to normed spaces.

We shall see how Theorem 221 provides a proof of Theorem 222 below, a result that extends Theorem 218. We shall also include another proof of it that does not rely on Theorem 221. For a straightforward application of Theorem 222 see Theorem 224 below. Observe, too, that it applies to all Banach spaces that form a $\mathcal{P}$-class (see Subsection 6.1.3), to dual spaces of Asplund spaces ((v) in Theorem 774), and to spaces with projectional skeletons (Theorem 212), among others.

Theorem 222 (Troyanski, see, e.g., [HMVZ07], Theorem 3.48). *Every space with a strong Markushevich basis admits an equivalent LUR norm.*

Proof 1. (Sketch, by using Theorem 221) Let $\{b_i, b_i^*\}_{i\in I}$ be a strong M-basis of X. We may always assume that $\|b_i^*\| = 1$ for all $i \in I$. We saw in Remark 127 that, in this case, the evaluation mapping $Tx(i) := \langle x, b_i^*\rangle$, $x \in X$, $i \in I$, maps X into $c_0(I)$ (in particular, given $x \in X$, the set $I(x) := \{i \in I : \langle x, b_i^*\rangle \neq 0\}$ is countable). Note that $c_0(I)$ has an equivalent LUR norm $\|\cdot\|_D$ (Corollary 243). For $x \in X$ define the separable space $Z_x := \overline{\text{span}}\{b_i : i \in I(x)\} \subset X$. Let $\{y_n\}_{n\in\mathbb{N}}$ be a sequence in X such that $Ty_n \overset{\|\cdot\|_D}{\longrightarrow} Tx$. If $i \notin \bigcup_{n=1}^{\infty} I(y_n)$, then $\langle y_n, b_i^*\rangle = 0$ for all $n \in \mathbb{N}$, so $\langle x, b_i^*\rangle = 0$, i.e., $i \notin I(x)$. This shows $I(x) \subset \bigcup_{n=1}^{\infty} I(y_n)$. Thus, due to the fact that the M-basis is strong, $x \in \overline{\text{span}}\{b_i : i \in I(x)\} \subset \overline{\text{span}}\{b_i : i \in \bigcup_{n=1}^{\infty} I(y_n)\} = \overline{\text{span}} \bigcup_{n=1}^{\infty} Z_{y_n}$, and we may apply Theorem 221. □

Proof 2. (See [HMVZ07, proof of Theorem 3.48]; the reader will find the similarities between this proof and the one provided for Theorem 218 in case of the existence of an unconditional basis, given in Subsection 10.4.1). Let $\{b_i; b_i^*\}_{i\in I}$ be a strong M-basis of a Banach space X. Assume, without loss of generality, that $\|b_i^*\| \leq 1$ for all $i \in I$. Define, for a finite set $A \subset I$ and some $x \in X$,

$$F_A(x)^2 := \sum_{i\in A} |\langle x, b_i^*\rangle|^2,$$

$$D_A(x) := \|\cdot\|\text{-dist}\,(x, \{b_i : i \in A\}).$$

We have $F_A(x) \leq \sqrt{\operatorname{card} A}\,\|x\|$ and $D_A(x) \leq \|x\|$ for all $x \in X$, so both F_A and D_A are continuous seminorms on X. Given positive integers l and n, let

$$G^2_{l,n}(x) := \sup\{lF^2_A(x) + D^2_A(x) : \ \operatorname{card} A \leq n, \ A \subset I\}, \quad x \in X.$$

Then $G_{l,n}$ is a continuous seminorm with $G_{l,n}(x) \leq \sqrt{nl+1}\|x\|$ for all $x \in X$. Finally, define an equivalent norm $|||\cdot|||$ on X by

$$|||x|||^2 := \|x\|^2 + \sum_{l,n=1}^{\infty} \frac{1}{nl+1}\frac{1}{2^{l+n}} G^2_{l,n}(x), \ x \in X.$$

We shall prove that $|||\cdot|||$ is LUR.
We introduced the expression Q_p in (5.2) for a subadditive function p. To ease the notation, we shall write $Q(p,x,y)$ instead (with the same meaning) along the proof here. Let $x \in X$ and $\{x_k\}_{k=1}^{\infty}$ be a sequence in X such that

$$Q(|||\cdot|||, x, x_k) \to 0. \tag{10.3}$$

We shall prove that $|||x - x_k||| \to 0$.
The strategy will be the following. First, we will check that $\{x_k : \ k \in \mathbb{N}\}$ is a $\|\cdot\|$-relatively compact subset of X by proving that it is as close as we wish to a certain norm-compact subset of X. Second, we will see that $\langle x_k, b^*_i\rangle \to_k \langle x, b^*_i\rangle$ for every $i \in I$. Finally, the coincidence on the norm-compact set $\overline{\{x_k : \ k \in \mathbb{N}\} \cup \{x\}}^{|||\cdot|||}$ of the topologies $|||\cdot|||$ and $w(X, \{b^*_i : \ i \in I\})$ (i.e., the topology of the pointwise convergence on the set $\{b^*_i : \ i \in I\}$) will give the result.
To accomplish it, let $I(x)$ be the set defined in the first proof, and put $I(x) = \{i_1, i_2, \ldots\}$, where $|\langle x, b^*_{i_1}\rangle| \geq |\langle x, b^*_{i_2}\rangle| \geq \ldots$ (observe that $|\langle x, b^*_{i_n}\rangle| \neq 0$ for each $n \in \mathbb{N}$). Fix $\varepsilon > 0$. The M-basis $\{b_i; b^*_i\}_{i \in I}$ is strong, so we can find $m \in \mathbb{N}$ such that for $B := \{i_1, i_2, \ldots, i_m\}$,

$$D_B(x) < \varepsilon. \tag{10.4}$$

By increasing m if necessary, we may always assume that $|\langle x, b^*_{i_m}\rangle| > |\langle x, b^*_{i_{m+1}}\rangle|$, and this gives $\delta > 0$ such that

$$\sup\{F^2_A(x) : \quad A \neq B, \ \operatorname{card} A \leq m\} + \delta < F^2_B(x). \tag{10.5}$$

We can now find $l \in \mathbb{N}$ such that

$$l\delta - \|x\|^2 > 0. \tag{10.6}$$

By the definition of $G_{l,m}(x+x_k)$, we can find finite sets $A_k \subset I$ such that $\operatorname{card} A_k \leq m$, for all $k \in \mathbb{N}$, and

$$(0 \leq)\ c_k := G^2_{l,m}(x+x_k) - lF^2_{A_k}(x+x_k) - D^2_{A_k}(x+x_k) \xrightarrow{k} 0. \tag{10.7}$$

We shall verify the following threefold *Claim:*

(*) $A_k = B$ for big enough k.
(**) $F_B(x_k) \to_k F_B(x)$ and $F_B(x + x_k) \to_k 2F_B(x)$.
(***) $D_B(x_k) \to_k D_B(x)$.

Before proving this Claim, let us note the following consequences:

(a) From (***), and recalling inequality (10.4), there exists $k_0 = k_0(\varepsilon) \in \mathbb{N}$ such that $D_B(x_k) < 2\varepsilon$ for $k \geq k_0$; noticing that $\{x_k\}_{k=1}^{\infty}$ is a bounded sequence (see Lemma 108), this provides a closed bounded subset S_ε of the finite-dimensional space $\{b_i : i \in B\}$ (hence S_ε is a compact set) such that dist $(x_k, S_\varepsilon) < 2\varepsilon$ for $k \geq k_0$. This gives another compact set in the finite-dimensional space $\{x_1, \ldots, x_{k_0}, b_i : i \in B\}$ with the same property. Since this holds for every $\varepsilon > 0$, *the set* $\{x_k : k \in \mathbb{N}\}$ *is thus a* $|||\cdot|||$*-relatively compact set in* X.

(b) Given $j \notin B$ and $k \geq k_0$, we have $\langle x_k, b_j^* \rangle = \langle x_k - y, b_j^* \rangle + \langle y, b_j^* \rangle = \langle x_k - y, b_j^* \rangle$ for every $y \in \{ b_i : i \in B\}$. From $D_B(x_k) < 2\varepsilon$, we get, in particular, $|\langle x_k, b_j^* \rangle| < 2\varepsilon$ for $k \geq k_0$.

(c) To deal with the case $j \in B$, define an isomorphism T from $\{b_i : i \in B\}$ onto $\ell_2(B)$ by $T(\sum_{i\in B} \alpha_i b_i) = (\alpha_i)_{i\in B}$. From (**) and the LUR property of ℓ_2, we get $\|T(x_k) - T(x)\|_2 \to 0$, which implies $\langle x_k, b_j^* \rangle \to_k \langle x, b_j^* \rangle$ for $j \in B$.
This proves that $\{x_k\}_{k=1}^{\infty}$ is $w(X, \{ b_i^* : i \in I\})$-convergent to x, and so $|||\cdot|||$-convergent, as we wanted to show.

Thus, it remains to prove the threefold Claim.
Let us start by establishing (*).
Fix $k \in \mathbb{N}$ and assume that $A_k \neq B$. Then, due to (10.5), we have $F^2_{A_k}(x) + \delta < F^2_B(x)$. From this, and having in mind that $D^2_{A_k}(x) \leq \|x\|^2$ (so $D^2_{A_k}(x) - D^2_B(x) \leq \|x\|^2$), we have

$$\begin{aligned} G^2_{l,m}(x) &\geq lF^2_B(x) + D^2_B(x) \\ &> lF^2_{A_k}(x) + l\delta + D^2_B(x) \geq lF^2_{A_k}(x) + l\delta - \|x\|^2 + D^2_{A_k}(x) \\ &= (l\delta - \|x\|^2) + \left(lF^2_{A_k}(x) + D^2_{A_k}(x)\right). \end{aligned} \tag{10.8}$$

Thus, for every $k \in \mathbb{N}$ such that $A_k \neq B$,

$$\begin{aligned} &Q(G_{l,m}, x, x_k) \\ &\quad > 2\Big\{(l\delta - \|x\|^2) + \Big(lF^2_{A_k}(x) + D^2_{A_k}(x)\Big)\Big\} + 2\Big\{lF^2_{A_k}(x_k) + D^2_{A_k}(x_k)\Big\} \\ &\quad -\Big\{lF^2_{A_k}(x + x_k) + D^2_{A_k}(x + x_k) + c_k\Big\} \\ &\quad = 2(l\delta - \|x\|^2) + lQ(F_{A_k}, x, x_k) + Q(D_{A_k}, x, x_k) - c_k \end{aligned} \tag{10.9}$$

(here we used (10.8), the definition of $G^2_{l,m}(x_k)$ and (10.7), in this order). If k is big enough, we reach a contradiction, thanks to the following reasons:

- $Q(G_{l,m}, x, x_k) \to_k 0$ (recall (10.3) and Lemma 111 above),

- $2(l\delta - \|x\|^2) > 0$ (inequality (10.6)),
- the two expressions in the last line of (10.9) involving Q are nonnegative (see equation (5.3)), and
- $c_k \to 0$ when $k \to \infty$ (see (10.7)).

This proves claim (∗).
From now on, and without loss of generality, we may assume that $A_k = B$ for all $k \in \mathbb{N}$. We have

$$\begin{aligned} 0 \overset{k}{\leftarrow} & Q(G_{l,m}, x, x_k) \\ & \geq 2\Big(lF_B^2(x) + D_B^2(x)\Big) + 2\Big(lF_B^2(x_k) + D_B^2(x_k)\Big) \\ & -\Big(c_k + lF_B^2(x + x_k) + D_B^2(x + x_k)\Big) \\ & = lQ(F_B, x, x_k) + Q(D_B, x, x_k) - c_k. \end{aligned}$$

So, from Lemma 110 above we get (∗∗) and (∗∗∗) and the Claim is proved. This concludes the proof. □

The analysis of the second proof of Theorem 222 gives the following result:

Theorem 223 (Granero, Jiménez-Sevilla, Moreno, [GJM04]). *If $\{x_\alpha, f_\alpha\}$ is a fundamental biorthogonal system in X, then the normed space* span $\{x_\alpha\}$ *admits a LUR norm.*

In order to show how Theorem 221 extends the fundamental Theorem 218, we mention the following result:

Theorem 224 (Terenzi, see, e.g., [HMVZ07], Corollary 5.2). *Every WLD space admits an strong Markushevich basis.*

This was preceded by the result that every separable Banach space has a bounded biorthogonal sequence which is also a Steinitz basis [Ter94] (see an alternative proof in [Ver00]); see also [HMVZ07, Theorem 1.36]. Thus, every WLD space has a LUR renorming.

Theorem 225 (Plichko, Yost, [PlYo01], see also [HMVZ07], Theorem 5.7). *There is a Banach space X of density ω_1 that admits a strong Markushevich basis but admits no PRI in any equivalent norm.*

Regarding the next result, see also Theorem 775 below.

Theorem 226 (See [MOTV09], Corollary 3.24). *If X is Asplund and X^* admits a dual W*LUR norm, then X^* admits a dual LUR norm.*

Theorem 227 (Moltó, Orihuela, Troyanski, Valdivia). ([MOTV99], [MOTV09, Corollary 3.23]) *If X has a WLUR norm, then it admits a LUR norm.*

10.4.3 Godefroy–Fabian transfer results

Fabian's proof of Theorem 218 is given in [FHHMZ11, Theorem 13.27]. The source for Fabian's proof was the original "transfer" method of G. Godefroy ([Go82], [GTWZ83]), which is explained geometrically in [GTWZ83] and is known among the collaborators as the "Embassy Lemma" (Theorem 229 below). In fact, Fabian's proof is the analytical version of Godefroy's clever technique.

H The transfer method was first presented by Godefroy in a Czech Winter School in the early 80's, and created the second revolution in the theory of nonseparable renormings, the first being Lindenstrauss' and Troyanski's.

Here is the main idea of Godefroy's argument for proving Theorem 218, see [GTWZ83]. Along the proof it will be needed that $c_0(\Gamma)$ has an equivalent norm $\|\cdot\|_0$ whose dual norm (denoted again by $\|\cdot\|_0$) is LUR (Theorem 171). Let $(X, \|\cdot\|)$ be a WCG Banach space. Let $T : X^* \to c_0(\Gamma)$ be the one-to-one bounded linear weak*-weak continuous operator from X^* into some $c_0(\Gamma)$ constructed by Amir and Lindenstrauss, see Theorem 236 below and, e.g., [FHHMZ11, Theorem 13.20]. The adjoint operator $L := T^* : \ell_1(\Gamma) \to X$ is w^*-w-continuous and has a dense image. Let B be the closed unit ball for the norm $\|\cdot\|_0$ on $\ell_1(\Gamma)$, and let $K = T^*(B) \subset X$. Thus K is weakly compact and $\overline{\text{span}}(K) = X$.
Let the function ψ on X be defined by

$$\psi(x)^2 = \|x\|^2 + \sum_{p,q\in\mathbb{N}} \frac{1}{2^{p+q}} \text{dist}^2\left(x, \frac{p}{q}K\right). \tag{10.10}$$

Then the Minkowski functional of the set $\{x : \psi(x) \leq 1\}$ is an equivalent norm $|||\cdot|||$ on X. Moreover, from the weak compactness of K and from the LUR property of $\|\cdot\|_0$ on $\ell_1(\Gamma)$, it can be deduced (in a subtle way) that $|||\cdot|||$ is LUR on X.

Note that the weak compactness of K in the preceding argument is absolutely crucial.

Remark 228. Let N be a norming subspace of X^*. Let $\|\cdot\|_N$ be the equivalent norm defined by N (see formula (3.4) above). We mentioned in Subsection 3.4.2 that $\|\cdot\|_N$ is $w(X, N)$-lower semicontinuous. Thus, we may assume from the beginning that $\|\cdot\|$ on X is $w(X, N)$-lower semicontinuous. Note, too, that $x \to \text{dist}\,(x, K)$ is a $w(X, N)$-lower semicontinuous seminorm on X: Indeed, $\{x \in X : \text{dist}\,(x, K) \leq 1\} = K + B_X$ is $w(X, N)$-closed, since B_X is $w(X, N)$-closed and K is $w(X, N)$-compact, due to the fact that the topologies w and $w(X, N)$ coincide on K.

This shows that each summand in (10.10) is $w(X,N)$-lower semicontinuous. Formula (10.10) gives ψ on any bounded set as the supremum of a sequence of $w(X,N)$-lower semicontinuous functions on X, so we get that ψ on any bounded set is $w(X,N)$-lower semicontinuous. This shows that $\{x:\ \psi(x)\le 1\}$ is $w(X,N)$-closed, and so $|||\cdot|||$ is $w(X,N)$-lower semicontinuous, too. ®

Godefroy's technique described above can be formulated as follows:

Theorem 229 (Godefroy, Troyanski, Whitfield, Zizler, [GTWZ83]). (see also [DeGoZi93], Theorem II.2.1 and [Fab91]) *Let X and Y be two Banach spaces such that Y is a dual space that admits a dual LUR norm and let $L : Y \to X$ be a bounded linear operator with dense image.*
(i) *If L is w^*-w-continuous, then X admits a LUR norm.*
(ii) *If X is a dual space and L is w^*-w^*-continuous, then X admits a dual LUR norm.*

The proof of Theorem 229 presented in [DeGoZi93] is from [Fab91].

Some consequences of Theorem 229 and some remarks

1. Theorem 229 implies Troyanski's Theorem 218. Indeed (and following the notation in Theorem 229), we may use for Y the space $\ell_1(\Gamma)$ with the norm given in Theorem 171 and L being the dual operator to the injection of X^* into $c_0(\Gamma)$ given in Theorem 236 below.

2. Theorem 229 gives also the proof of the following result, a pioneering one at that time.

Corollary 230 ([GTWZ84]). *Let X be a Banach space. If X^* is a subspace of a WCG space, then X admits an equivalent LUR norm whose dual norm is also LUR.*

Proof. Assume that X^* is a subspace of the WCG Banach space Y. Let $j : X^* \to Y$ be the canonical injection and $q : Y^* \to X^{**}$ its adjoint, i.e., the canonical quotient mapping. Since $X \subset X^{**}$ is a (1-) norming subspace, the space $N := q^{-1}(X)$ is a subspace of Y^* that is norming for Y, see Item 2 in Subsection 3.4.4 and [FHHMZ11, Exercise 5.6] (see Figure 10.2). The proof of Theorem 218 given at the beginning of this paragraph and Remark 228 here ensure the existence of a LUR equivalent norm $|||\cdot|||$ on Y that is $w(Y,N)$-lower semicontinuous. Its restriction $|||\cdot|||$ to X^* gives an equivalent $w(X^*,X)$-lower semicontinuous norm on X^* (i.e., a dual norm) that is LUR (see [GTWZ84]). Nowadays we can use Theorem 337 to conclude that X then admits an equivalent LUR norm. The Asplund averaging method (see Item 2 in Section 8.2) concludes the proof. For the original approach that does not use Theorem 337 see [GTWZ84] or [DeGoZi93, Theorem VII.2.7] and the results quoted there. □

$$
\begin{array}{ccc}
q^{-1}(X) =: N \subset & Y^* & \qquad Y \\
& q \downarrow & \qquad \uparrow j \\
& X^{**} & \qquad X^* \\
& \cup & \\
& X &
\end{array}
$$

Figure 10.2: The proof of Corollary 230

3. Note that nowadays Corollary 230 follows from Theorem 746 below.
4. Note that Theorem 171 above cannot be obtained from Corollary 230 in case that Γ is uncountable.

➡ Indeed, the space $\ell_1(\Gamma)$ for uncountable Γ cannot be a subspace of a WCG Banach space. This is due to the fact that if a Banach space X is a subspace of a WCG Banach space, then every convex continuous function on X has at least a point of Gâteaux differentiability (something that can be obtained from the Smooth Variational Principle, Theorem 289). However, the canonical norm $\|\cdot\|_1$ of $\ell_1(\Gamma)$ is nowhere Gâteaux differentiable (see, e.g., [Phe93, Example 1.4(b)]).

An alternative argument is the following: A Banach space X is a subspace of a WCG Banach space if, and only if, (B_{X^*}, w^*) is an Eberlein compact (Theorem 239 below). If $\ell_\infty^c(\Gamma)$ denotes the subspace of $\ell_\infty(\Gamma)$ of all vectors with a countable support, the set $B_{\ell_\infty^c(\Gamma)}$ is w^*-dense in $B_{\ell_\infty(\Gamma)}$. If Γ is uncountable, not all points in $B_{\ell_\infty(\Gamma)}$ are w^*-limits of sequences in $B_{\ell_\infty^c(\Gamma)}$ (those limit points have a countable support), a contradiction with the angelic character of any Eberlein compact space (see Section 6.2).

The following is an important useful result. The method of its proof was motivated by Troyanski's original proof of Theorem 218 in the direction of Theorem 221. It is worth to mention that the property of having a strictly convex renorming is not a three-space property (see (xii) in Theorem 655).

Theorem 231 (Godefroy, Troyanski, Whitfield, and Zizler, [GTWZ]). *Let X be a Banach space and Y be a subspace of X.*
(i) (See, e.g., [DeGoZi93, Proposition VII.3.3]) *If Y admits a R norm, and X/Y admits a LUR norm, then X admits a R norm.*
(ii) (See, e.g., [DeGoZi93, Theorem VII.3.1]) *If both Y and X/Y admit LUR norms, then X admits a LUR norm.*

Item (i) in Theorem 231 can be nowadays obtained from Theorem 870 below.

The development of Troyanski's and Godefroy–Fabian techniques generated most of the progress in this part of the renorming theory since. There is a good book [MOTV09] on the development of the transfer technique, in its full strength and use.

It would be good if the method of proof of Theorem 231 can be used to get a new proof of the three-space property for UR renorming (see (viii) in Theorem 655).

10.5 Two more renormings for weakly compactly generated spaces

The following technique, described in Proposition 232 below, and working for WCG spaces, turns out to be very flexible. It is the core of several renorming results for this and related classes of Banach spaces. It is based on two important results ((i) The factorization Theorem 191 of Davis, Figiel, Johnson, and Pełczyński, and (ii) Troyanski's renorming Theorem 218 applied to the reflexive case), and the simple transfer technique —in some of its variants— that has been presented in Section 9.2 above. It produces a Gâteaux norm on every WCG space, and it has the extra property that it does not change the set $NA(X, \|\cdot\|)$ of norm-attaining functionals (see the definition in Section 2.3).

If $(X, \|\cdot\|)$ is a WCG Banach space and $K \subset X$ is an (absolutely convex, by Krein's theorem, see, e.g., [FHHMZ11, Theorem 3.133]) weakly compact subset that generates X, (i) above provides a reflexive Banach space $(R, |\cdot|)$ and a bounded linear and one-to-one operator $T : R \to X$ such that $K \subset T(B_R)$ (so T has a dense range), while (ii) equips R^* with an equivalent norm with additional convexity properties. This norm on R will be the one assumed there from the beginning.

So, the scheme to be used below is the following

$$\begin{array}{ccc} (R, |\cdot|) & \xrightarrow{T} & (X, \|\cdot\|) \\ (R^*, |\cdot|^*) & \xleftarrow{T^*} & (X^*, \|\cdot\|^*) \end{array}$$

Proposition 232 (Fabian, Montesinos, Zizler, [FaMoZi02-b]). *Let $(X, \|\cdot\|)$ be a subspace of a WCG Banach space. Then there exists an equivalent norm $|||\cdot|||$ on X such that, for every norming subspace N of X^*, the N-lower semicontinuous envelope $|||\cdot|||_N$ of $|||\cdot|||$ has a strictly convex dual norm (and so $|||\cdot|||_N$ is Gâteaux differentiable). Moreover, $NA(X, \|\cdot\|) = NA(X, |||\cdot|||)$.*

Proof. First, note that it is enough to prove the result for a WCG Banach space. Indeed, if X is a subspace of a WCG Banach space Z, and Y is a norming subspace

of X^*, then $q^{-1}(Y)$ is a norming subspace of Z, where $q : Z^* \to Z^*/X^\perp$ denotes the canonical quotient mapping (see Item 2 in Subsection 3.4.4 above). The restriction of a $q^{-1}(Y)$-lower semicontinuous equivalent Gâteaux smooth norm on Z to X gives the sought Y-lower semicontinuous norm on X.
Define an equivalent norm $|||\cdot|||$ on X by letting $B_{(X,|||\cdot|||)} := B_{(X,\|\cdot\|)} + T(B_{(R,|\cdot|)})$. If N is a norming subspace of X^*, the closed unit ball for the N-lower semicontinuous envelope $|||\cdot|||_N$ of $|||\cdot|||$ (i.e., $\overline{B_{(X,|||\cdot|||)}}^{w(X,N)}$, see Subsection 3.4.2) satisfies

$$B_{(X,|||\cdot|||_N)} = \overline{B_{(X,\|\cdot\|)} + T(B_{(R,|\cdot|)})}^{w(X,N)}$$
$$= \overline{B_{(X,\|\cdot\|)}}^{w(X,N)} + T(B_{(R,|\cdot|)}) = B_{(X,\|\cdot\|_N)} + T(B_{(R,|\cdot|)}),$$

due to the fact that $T(B_{(R,|\cdot|)})$ is $w(X,X^*)$-compact, hence $w(X,N)$-compact.
It follows that $|||x^*|||_N^* = \|x^*\|_N^* + |T^*x^*|^*$ for every $x^* \in X^*$ (see Subsection 3.7.4), so $|||\cdot|||_N^*$ is strictly convex (indeed, by Proposition 149, T^* is one-to-one; use now Proposition 152). Thus, by Proposition 82, $|||\cdot|||_N$ is Gâteaux differentiable.
The proof of the last statement is also simple: The supremum of a linear functional on the algebraic sum of two sets is the sum of the suprema on each of them (see (v) in Proposition 34). It is enough now to observe that every element in X^* attains its supremum on the w-compact set $T(B_{(R,|\cdot|)})$. Let us give the details. To simplify the notation, put $\sup x^*(A) := \sup\{\langle a, x^*\rangle : a \in A\}$, for $x^* \in X^*$ and $A \subset X$. If $x_0^* \in NA(X, \|\cdot\|)$, then there exists $x_0 \in B_{(X,\|\cdot\|)}$ such that $\sup x_0^*(B_{(X,\|\cdot\|)}) = \langle x_0, x_0^*\rangle$. There exists $r_0 \in B_{(R,|\cdot|)}$ such that $\sup x_0^*(TB_{(R,|\cdot|)}) = \langle Tr_0, x_0^*\rangle$. It is obvious that $z_0 := x_0 + Tr_0 \in B_{(X,|||\cdot|||)}$ satisfies $\sup x_0^*(B_{(X,|||\cdot|||)}) = \langle z_0, x_0^*\rangle$, so $x_0^* \in NA(X, |||\cdot|||)$. Conversely, if $x_0^* \in NA(X, |||\cdot|||)$, there exists $z_0 \in B_{(X,|||\cdot|||)}$ such that

$$\langle z_0, x_0^*\rangle = \sup x_0^*(B_{(X,|||\cdot|||)})$$
$$(= \sup x_0^*(B_{(X,\|\cdot\|)}) + \sup x_0^*(TB_{(R,|\cdot|)}) = \sup x_0^*(B_{(X,\|\cdot\|)}) + \langle Tr_0, x_0^*\rangle).$$

If $z_0 = x_0 + Tr_1$, where $x_0 \in B_{(X,\|\cdot\|)}$ and $r_1 \in B_{(R,|\cdot|)}$, then clearly $x_0^*(B_{(X,\|\cdot\|)}) = \langle x_0, x_0^*\rangle$, hence $x_0^* \in NA(X, \|\cdot\|)$. □

Corollary 233. *Let X be a Banach space such that X^* is WCG. Then X has an equivalent norm $|||\cdot|||$ such that $|||\cdot|||^{**}$ is strictly convex.*

The factorization Theorem 191 can be used together with the basic transfer method to get some renorming results. For example, if X is a Banach space and S is a bounded subset of X, we say that the dual norm $\|\cdot\|$ on X^* is **S-LUR** if for every $x^* \in S_{X^*}$, the fact that a sequence $\{x_n^*\}$ in S_{X^*} satisfies $\|x^* + x_n^*\| \to 2$ implies that $x_n^* \xrightarrow{S} x^*$ (i.e., $\sup_{s\in S}\{|\langle s, x^* - x_n^*\rangle| \to 0\}$). Then we have

Proposition 234. *Let $(X, \|\cdot\|)$ be a WCG Banach space, and let K be a weakly compact subset of X that generates X. Then there exists on X^* a dual equivalent norm that is K-LUR.*

Proof. By Theorem 191 there exists a reflexive Banach space $(R, |\cdot|)$ (whose dual norm can be taken to be LUR, by Theorem 218, see also [FHHMZ11, Thm. 11.22]) and a one-to-one continuous linear operator $T : R \to X$ such that $TB_R \supset K$. Put $|||x^*|||^2 = \|x^*\|^2 + |T^*x^*|^2$ for every $x^* \in X^*$, where $T^* : X^* \to R^*$ is the adjoint operator of T and $|\cdot|$ denotes also the dual norm of $|\cdot|$ in R^*. Then $|||\cdot|||$ is an equivalent dual norm in X^* (that it is an equivalent norm is clear; that it is a dual norm follows from the observation at the beginning of Chapter 7: it is clearly w^*-lower semicontinuous). To see that $|||\cdot|||$ is, indeed, K-LUR, use Proposition 115.

➡ Let $\{x_n^*\}$ be a sequence in $S_{(X^*,|||\cdot|||)}$, and let $x^* \in S_{(X^*,|||\cdot|||)}$ be such that

$$2|||x|||^2 + 2|||x_n^*||| - |||x^* + x_n^*|||^2 \to 0.$$

Then Lemma 110 implies that $2|T^*x|^2 + 2|T^*x_n^*| - |T^*x^* + T^*x_n^*|^2 \to 0$. The LUR property of $|\cdot|$ in R^* concludes that $|T^*x^* - T^*x_n^*| \to 0$, and this implies that $x_n^* \xrightarrow{K} x^*$ (i.e., $x_n^* \to x^*$ uniformly on K). □

10.6 Weakly compactly generated spaces II

Weakly compactly generated spaces (WCG) were introduced in Subsection 6.1.6 (see Definition 131).
Let us start with some review of well-known things we will need for presenting examples on WCG and Vašák, their renormings and injections. We refer to, e.g., [DeGoZi93], [Fab97], [FHHMPZ01], and [FHHMZ11, Chapters 13 and 14] for explaining the material.

First, concerning injections we will need the following few results. Note that they are of interest only in nonseparable spaces (see, e.g., [FHHMZ11, Exercise 5.75].

Theorem 235 (Amir, Lindenstrauss, see, e.g., [FHHMZ11], Theorem 13.16)**.** *Every WCG space admits a weakly compact Markushevich basis.*

Theorem 236 (Amir, Lindenstrauss [AmLi68], see, e.g., [FHHMZ11], Theorem 13.20)**.** *Let X be a Banach space. Then X is WCG if and only if there is a bounded linear one-to-one weak*-weak continuous operator from X^* into some $c_0(\Gamma)$.*

As a consequence, if X is a WCG space, then (B_{X^*}, w^*) is an Eberlein compact space. See also Theorem 239 below.

Corollary 237. *Every WCG space admits a norm whose dual norm is R. In particular, every WCG space admits a Gâteaux smooth norm.*

Theorem 238 (Amir, Lindenstrauss, [AmLi68])**.** (see, e.g., [FHHMZ11], Theorem 14.9) *Let K be a compact space. Then the following are equivalent.*

(i) *K is an Eberlein compact.*
(ii) *$C(K)$ is WCG.*
(iii) *$(B_{C(K)^*}, w^*)$ is an Eberlein compact.*

Note that a consequence of Theorem 236 above and the result of Benyamini, Rudin, and Wage [BRW77], that the continuous image of an Eberlein compact space is again Eberlein compact, is the following result:

Theorem 239. *A Banach space X is a subspace of a WCG space if, and only if, (B_{X^*}, w^*) is an Eberlein compact.*

So, from Theorem 238 we get:

Corollary 240 (see [FHHMZ11], Chapters 13 and 14). *If X is a subspace of a WCG space, then $C\big((B_{X^*}, w^*)\big)$ is WCG.*

The following result solved a long-standing open problem in the area, first formulated by K. John (see [GuiMoZi16, Problem 111]). A related problem was raised by Godefroy (see [GuiMoZi16, Problem 112]): Is every Asplund space with a norming M-basis WCG? Hilbert generated spaces were defined in Subsection 6.1.8 above. Such spaces are precisely those whose closed dual unit ball, endowed with the w^*-topology, is a uniform Eberlein compact.

Theorem 241 (Hájek, [Há20]). *There exists a WCG Banach space —in fact, a Hilbert generated Banach space— with no norming M-basis.*

Godefroy's problem formulated above has been also solved in the negative by P. Hájek, T. Russo, J. Somaglia, and S. Todorčević [HRST20]: *There exists an Asplund space with a norming Markushevich basis that is not weakly compactly generated.*

10.7 Day's norm II: Its locally uniformly convex behaviour on $c_0(\Gamma)$

In Subsection 9.4.1 we introduced Day's norm $\|\cdot\|_D$ on $\ell_\infty(\Gamma)$, a fundamental tool in renorming theory of nonseparable spaces. We proved there that it is rotund on $c_0(\Gamma)$. A group of mathematicians mainly connected to the University of Washington, Seattle, that published several papers under the pseudonym "John Rainwater", proved that, in fact, Day's norm is LUR on $c_0(\Gamma)$ for any nonempty set Γ (see Corollary 243 below). For a "biography" of the aforesaid fictitious mathematician see [Phe02]. The short self-contained complete argument for the LUR character uses Abel's partial summation from classical analysis, and can be found in, e.g., [DeGoZi93, Theorem II.7.3]. Following [MOTV09], we present here another approach to the LUR property of Day's norm. It uses convexity arguments and gives more precise results. The crucial inequalities needed are in Lemma 242.

As we mentioned, Corollary 243 is the main result in this direction. We shall articulate the proof by using the $Q(\cdot,\cdot)$-scheme introduced in equation (5.2). Let us show the details below.
For $x \in \ell^\infty(\Gamma)$ and $\varepsilon > 0$ set

$$\boxed{L_\varepsilon(x) := \{\gamma \in \Gamma : \ |x(\gamma)| \geq \varepsilon\}} \tag{10.11}$$

and

$$\boxed{\eta_\varepsilon(x) := \varepsilon^2 - \sup\{x(\gamma)^2 : \ \gamma \notin L_\varepsilon(x)\}.} \tag{10.12}$$

Given a finite set $A \subset \Gamma$ we consider the seminorm $\|x\|_A = \left(\sum_{\gamma\in A} x(\gamma)^2\right)^{1/2}$. Observe that, for $x, y \in X$, and having in mind that $\|\cdot\|_A$ is the ℓ_2-norm on $\ell_2(A)$,

$$Q_{\|\cdot\|_A}(x,y) = \|x-y\|_A^2 \ (\geq 0), \tag{10.13}$$

see the beginning of Section 5.6.

Lemma 242 ([MOTV09]). *Let $\varepsilon > 0$, and let $x \in \ell^\infty(\Gamma)$ be such that $L = L_\varepsilon(x)$ is finite. Let $\eta = \eta_\varepsilon(x) > 0$, $\ell = \operatorname{card}(L)$, and $\delta = \min\{(\varepsilon - \sqrt{\varepsilon^2 - \eta})^{1/2}, \eta/2\}$. Let $y \in \ell^\infty(\Gamma)$ such that $Q_{\|\cdot\|_D}(x,y) < 2^{-\ell-1}\delta$. Then $\|x-y\|_\infty < 3\varepsilon$.*

➪ By Lemmas 108 and 111, and the definition (9.10) of Day's norm we have

$$(\|x\|_n - \|y\|_n)^2 \leq Q_{\|\cdot\|_n}(x,y) < 2^{n-\ell-1}\delta, \text{ for } n \in \mathbb{N}. \tag{10.14}$$

The proof of Lemma 242 is divided into two cases.

(i) First assume that $\boxed{L = \emptyset}$ (so $\eta = \varepsilon^2 - \|x\|_1^2$ and $\ell = 0$). Then $L_\varepsilon(y) = \emptyset$. Indeed, otherwise we have, from (10.14), $\varepsilon^2 - \|x\|_1^2 \leq (\|y\|_1 - \|x\|_1)^2 < \delta \leq \eta/2 = (\varepsilon^2 - \|x\|_1^2)/2$, a contradiction. Therefore $\|x-y\|_\infty \leq \|x\|_\infty + \|y\|_\infty \leq 2\varepsilon$, and this proves the statement in this case.

(ii) Assume now that $\boxed{L \neq \emptyset}$ (so $\ell \geq 1$). Let $M' \subset \Gamma$ of cardinality $\ell+1$ such that

$$\|x+y\|_{\ell+1}^2 - \delta < \|x+y\|_{M'}^2. \tag{10.15}$$

We claim first that there exists $M \subset M'$ such that card $(M) = \ell$ and

$$\|x+y\|_\ell^2 - \delta < \|x+y\|_M^2. \tag{10.16}$$

If not, we should have $\|x+y\|_{M'\setminus\{\gamma'\}}^2 \leq \|x+y\|_\ell^2 - \delta$ for every $\gamma' \in M'$. This shows that $\left(x(\gamma') + y(\gamma')\right)^2 > \|x+y\|_{\ell+1}^2 - \|x+y\|_\ell^2$ for every $\gamma' \in M'$. Put

$$r := \inf\left\{\left(x(\gamma') + y(\gamma')\right)^2 : \ \gamma' \in M'\right\} - \left(\|x+y\|_{\ell+1}^2 - \|x+y\|_\ell^2\right) \ (> 0).$$

Find now $F \subset \Gamma$ such that card $(F) = \ell$ and $\|x+y\|_\ell^2 - r/2 < \|x+y\|_F^2$. Since there are $\ell+1$ elements in M', at least one of them, say μ, is not in F. Then, putting $F' := F \cup \{\mu\}$, we have

$$\begin{aligned}\|x+y\|_{F'}^2 &= \|x+y\|_F^2 + \big(x(\mu)+y(\mu)\big)^2 \\ &> \|x+y\|_\ell^2 - r/2 + r + \left(\|x+y\|_{\ell+1}^2 - \|x+y\|_\ell^2\right) \\ &= \|x+y\|_{\ell+1}^2 + r/2,\end{aligned}$$

a contradiction. Let $M' := M \cup \{\mu\}$.
Note that, from (10.16),

$$\begin{aligned}Q_{\|\cdot\|_M}(x,y) &= 2\|x\|_M^2 + 2\|y\|_M^2 - \|x+y\|_M^2 \\ &< 2\|x\|_\ell^2 + 2\|y\|_\ell^2 - \|x+y\|_\ell^2 + \delta = Q_{\|\cdot\|_\ell}(x,y) + \delta. \qquad (10.17)\end{aligned}$$

Note, too, that necessarily $\boxed{M = L}$. Indeed, and since card $(L) =$ card (M), otherwise there exit $\alpha \in L \setminus M$ and $\beta \in M \setminus L$. Then $\eta \le \varepsilon^2 - x(\beta)^2 \le x(\alpha)^2 - x(\beta)^2$. Put $N := (\{\alpha\} \cup M) \setminus \{\beta\}$. We get, by using (10.16) and (10.13) in this order,

$$\begin{aligned}\|x+y\|_\ell^2 \le \|x+y\|_M^2 + \delta &\le 2(\|x\|_M^2 + \|y\|_M^2) + \delta \\ &= 2\left(\|x\|_N^2 - (x(\alpha)^2 - x(\beta)^2) + \|y\|_M^2\right) + \delta \\ &\le 2\left(\|x\|_\ell^2 + \|y\|_\ell^2\right) + \delta - 2\eta,\end{aligned}$$

or, in other words, $Q_{\|\cdot\|_\ell}(x,y) \ge 2\eta - \delta$. From (10.14) we get $Q_{\|\cdot\|_\ell}(x,y) < \delta/2$, so altogether $(3/4)\delta > \eta$, a contradiction with the fact that $\delta \le \eta/2$. Now, having in mind that $M = L$, and from (10.13), (10.17), and (10.14), in this order, we get

$$\|x-y\|_L^2 = Q_{\|\cdot\|_L}(x,y) \le Q_{\|\cdot\|_\ell}(x,y) + \delta < 3\delta/2 \le \eta \le \varepsilon^2,$$

or, in other words,

$$\sup\{|x(\gamma) - y(\gamma)| : \ \gamma \in L\} < \varepsilon. \qquad (10.18)$$

Let us now show that

$$\sup\{|x(\gamma) - y(\gamma)| : \ \gamma \notin L\} < 2\varepsilon. \qquad (10.19)$$

We have two possibilities:
(i) In case that $|y(\gamma)| < 2\varepsilon$ for every $\gamma \in \Gamma \setminus L$ we have $|x(\gamma) - y(\gamma)| \le |x(\gamma)| + |y(\gamma)| < 3\varepsilon$ for every $\gamma \in \Gamma \setminus L$, and we are done.

(ii) Assume now that there is some $\gamma \in \Gamma \setminus L$ such that $|y(\gamma)| \geq 2\varepsilon$. Recall (see equation (10.15)) that there exists $\mu \notin L$ such that

$$\|x+y\|^2_{\ell+1} - \delta < \|x+y\|^2_{L\cup\{\mu\}}.$$

Find $\lambda \in \Gamma \setminus L$ such that

$$|y(\lambda)| \geq \max\{|y(\mu)|, 2\varepsilon\} \tag{10.20}$$

(if $|y(\mu)| < 2\varepsilon$, put $\lambda = \gamma$; otherwise, put $\lambda = \mu$). If the statement (10.19) is false, then $|y(\lambda)| - |x(\mu)| > 2\varepsilon - \varepsilon = \varepsilon$. Then, by (10.13),

$$\begin{aligned} Q_{\|\cdot\|_{\ell+1}}(x,y) &> 2\left(\|x\|^2_{L\cup\{\mu\}} + \|y\|^2_{L\cup\{\lambda\}}\right) - \|x+y\|^2_{L\cup\{\mu\}} - \delta \\ &= \|x-y\|^2_L + 2(x(\mu)^2 + y(\lambda)^2) - (x(\mu)+y(\mu))^2 - \delta. \end{aligned}$$

Now, since $\|x-y\|_L \geq 0$ we obtain, in view of (10.20), (10.12), and the definition of δ, that

$$\begin{aligned} Q_{\|\cdot\|_{\ell+1}}(x,y) &\geq x(\mu)^2 + y(\lambda)^2 - 2|x(\mu)||y(\lambda)| - \delta \\ &= (|y(\lambda)| - |x(\mu)|)^2 - \delta > \varepsilon^2 - \delta \geq \eta - \delta \geq \delta, \end{aligned}$$

which yields again a contradiction. Thus, (10.19) holds. By this and (10.18), we get that $\|x-y\|_\infty < 3\varepsilon$, as we wished to prove. □

Now, it follows that

Corollary 243 (J. Rainwater). *Day's norm $\|\cdot\|_D$ is LUR on $c_0(\Gamma)$.*

Proof. (We follow the notation in Lemma 242.) If $x \in S_{(c_0(\Gamma),\|\cdot\|_D)}$, note that for $\varepsilon > 0$ the set $L := L_\varepsilon(x)$ is finite and $\eta := \eta_\varepsilon(x) > 0$. If $\{y_n\}$ is a sequence in $B_{(c_0(\Gamma),\|\cdot\|_D)}$ such that $Q_{\|\cdot\|_D}(x, y_n) \to 0$, then there exists $n_0 \in \mathbb{N}$ such that, for $n \geq n_0$, $Q_{\|\cdot\|_D}(x, y_n) < 2^{-\ell-1}\delta$, hence $\|x - x_n\|_\infty < 3\varepsilon$ for $n \geq n_0$. Due to the fact that $\|\cdot\|_D$ and $\|\cdot\|_\infty$ are equivalent norms on ℓ_∞, the conclusion follows from the characterization of LUR norms in Proposition 115. □

Remark 244. A careful examination of the proof above shows that for any Γ, the bidual norm of Day's norm on $\ell_\infty(\Gamma)$ is LUR at any point of $c_0(\Gamma)$ (see, for details, [DeGoZi93, Theorem II.7.3]). ®

10.8 Asplund averaging II

In Section 8.1 we presented the original Asplund averaging technique for combining two equivalent norms in a single norm sharing the properties of both of them.

Nowadays we may use the following technique: If X is a Banach space, denote by $N(X)$ the set of all equivalent norms on X equipped with the topology of uniform convergence on the closed unit ball of X. It is a Baire space (i.e., the intersection of a countable family of G_δ dense subsets is dense), and so the Baire Category methods can be used there.

Theorem 245 ([FaZaZi81], see also [DeGoZi93, Theorem II.4.1]). *If a Banach space X admits a LUR (respectively UR norm, respectively URED norm) then the set of LUR norms (respectively UR, respectively URED norms) contains a dense G_δ set in $N(X)$.*

Proof. (Sketch) Here is a brief hint how the proof goes in the LUR case. It was motivated in a Honors undergraduate similar project at Charles University in Prague, and used some results of V. Klee: Let r_0 be a LUR norm on X. For $p \in N(X)$ and $j \in \mathbb{N}$ put

$$G(p,j) := \{q \in N(X) : \ \sup\{|p^2(x) + j^{-1}r_0^2(x) - q^2(x)| : \ x \in B_X\} < j^{-2}\}.$$

For $k \in \mathbb{N}$ put

$$G_k := \bigcup \{G(p,j) : \ p \in N(X), \ j \geq k\}.$$

For all $k \in \mathbb{N}$, the set G_k is open and dense in $N(X)$. Put finally

$$G := \bigcap_k G_k.$$

By the Baire category theorem, the set G is (G_δ and) dense. It is enough to show that each element of G is LUR. This is done by using the numbers j^{-1} and j^{-2} in the definition of $G(p,j)$. □

The original proof of Asplund in [Asp67], presented in Section 8.1 above, was constructive and elaborated. The Baire category proof we sketched follows a general rule: "If you have difficulty to find one point having property P try the Baire category arguments to show that most of points have property P".

In this connection we must mention the following variant of a result of Bossard, Godefroy, and Kaufman:

Theorem 246 (Bossard, Godefroy, and Kaufman, [BoGoKa96]). *If X is a separable infinite-dimensional Banach space then the set of all equivalent LUR norms on X is not Borel in $N(X)$.*

By using the Asplund averaging principle we have:

Theorem 247 (Troyanski). *Every reflexive Banach space admits an equivalent norm that is at the same time LUR and Fréchet differentiable* (see, e.g., [FHHMZ11, Chapter 13]).

Here is an elementary proof of the following version:

Theorem 248 ([JoZi79]). *Assume that a Banach space X admits an equivalent norm $\|\cdot\|_1$ that is LUR and an equivalent norm $\|\cdot\|$ whose dual norm is LUR. Then X admits an equivalent norm $|||\cdot|||$ which is LUR and Fréchet differentiable on X.*

The (easy) proof of this result illustrates a version of Asplund averaging in an important special case. It was presented in [JoZi79]:

Proof. For $n \geq 3$, let $\|f\|_n^* = (\|f\|_1^{*2} + n^{-1}\|f\|^{*2})^{\frac{1}{2}}$ for $f \in X^*$. Each $\|\cdot\|_n^*$ is clearly a LUR equivalent norm on X^*, dual to some norm $\|\cdot\|_n$ on X. Furthermore, $\lim_n \|x\|_n = \|x\|_1$ uniformly on bounded sets of X. Since each $\|\cdot\|_n^*$ is LUR, the norm $\|\cdot\|_n$ is Fréchet differentiable. Consider the norm $|||x||| := (\sum_{n=3}^{\infty} 2^{-n}\|x\|_n^2)^{\frac{1}{2}}$. This is an equivalent norm on X. Since the differentials $(\|\cdot\|_n^2)'$ of $\|\cdot\|_n^2$ are uniformly bounded on bounded sets of X, the norm $|||\cdot|||$ is Gâteaux differentiable and the differential $(|||\cdot|||^2)'$ is norm-norm continuous as all $(\|\cdot\|_n^2)'$ are. Thus $|||\cdot|||$ is Fréchet differentiable. To see that $|||\cdot|||$ is LUR, suppose that $x_j,\ x \in X$ and $\lim_j Q_{|||\cdot|||}(x, x_j) = 0$. Then the same is true for any $\|\cdot\|_n$, and since $\{x_j\}$ is then necessarily bounded and $\lim_n \|x\|_n = \|x\|_1$ uniformly on bounded sets, we have $\lim_j Q_{\|\cdot\|_1}(x, x_j) = 0$. So, by the LUR property of the norm $\|\cdot\|_1$, we have $\lim_j \|x_j - x\|_1 = 0$. □

So, using the above, we can formulate the following result:

Theorem 249. *Let X be a separable Banach space. Then X admits a norm that is LUR, URED, and UG.*
If X^* is separable, then X admits a norm that is LUR and URED, and its dual norm is LUR.

As we already mentioned in Section 8.2, it is unknown if in full generality the following holds true: If a Banach space X admits a LUR norm and also a Fréchet differentiable norm, then X admits a norm that shares both these properties.
It is open if in $C[0, \omega_1]$, the set of all Fréchet differentiable norms is dense in the set of all equivalent norms. However, there is in $C[0, \omega_1]$ a Fréchet differentiable LUR norm [HáPro14].
The use of Asplund averaging for getting common moduli of convexity of power type is in [Go01b, page 804]. We will see in Chapter 39 that the Asplund averaging does not work for higher-order smoothness.

Chapter 11

Norm-attaining operators, variational principles, and Asplund spaces

In this chapter we shall review some results on extreme points, points of differentiability, farthest points and norm-attaining operators. All those concepts are related, and play an important rôle in the geometry of a Banach space. We chose to approach them chronologically as they came in history. Then we shall present some basic facts on Asplund spaces. The topic of Variational Principles will cover Ekeland's and some variants, the Smooth Variational Principle and the Compact one.

11.1 Lindenstrauss' extremal structure of ℓ_1

The classical Krein–Milman Theorem 7 states that every nonempty convex weakly compact subset K of a Banach space X has an extreme point. It follows from the Hahn–Banach Separation Theorem and the fact that an extreme point of a "face" of K is, in turn, an extreme point of K, that K is then the closed convex hull of the set of all its extreme points. If the existence of at least one extreme point for every closed convex and bounded subset of X is guaranteed, then the apparently stronger property of reproducing such sets by their extreme points holds even without weak compactness assumptions, as Lindenstrauss showed in the following interesting result:

Lemma 250 (Lindenstrauss). *Let X be a Banach space. Then the following are equivalent:*
(i) *Every bounded, closed and convex subset of X has an extreme point.*
(ii) *Every bounded, closed and convex subset of X is the closed convex hull of its extreme points.*

Proof. Clearly (ii) implies (i). Assume that (i) holds and let K be a bounded, closed and convex subset of X. Let K_0 be the closed convex hull of its extreme points. If $K_0 \neq K$, then by the Bishop–Phelps Theorem 14 above, there is $f \in X^*$ and $y \in K$ such that

$$f(y) = \sup_{x \in K} f(x) > \sup_{x \in K_0} f(x).$$

The set $K_1 = \{x \in K, f(x) = f(y)\}$ is a closed face of K (i.e., an intersection of K with a supporting hyperplane) that is disjoint from K_0. The set K_1 has an extreme point that is thus an extreme point of K, a contradiction. □

A. J. Guirao et al., *Renormings in Banach Spaces*, Monografie Matematyczne 75,
https://doi.org/10.1007/978-3-031-08655-7_11

A space with the (equivalent) properties stated in Lemma 250 is ℓ_1. This is again the contribution of J. Lindenstrauss. The result answered a question by M.A. Rieffel, and nowadays follows from the fact that ℓ_1 has the RNP. We still present it here for its beauty:

Theorem 251 (Lindenstrauss, [Lind66a]). *In the space ℓ_1, every bounded closed and convex subset is the closed convex hull of the set of its extreme points.*

Proof. Let K be a closed convex and bounded subset of ℓ_1. By Lemma 250 we have only to show that it has an extreme point.

➡ We need the following crucial lemma:

Lemma 252 (Lindenstrauss, [Lind66a]). *Let K be a nonempty closed convex and bounded subset of ℓ_1 and let $\varepsilon > 0$. Then there is a closed face F of K and $n \in \mathbb{N}$ such that for all $x = (x_1, x_2, \dots) \in F$ we have $\sum_{i>n} |x_i| < \varepsilon$.*

Proof. Let $M = \sup\{\|x\| : x \in K\}$ and choose $y \in K$ such that $\|y\| > M - \varepsilon/4$. Let $n \in \mathbb{N}$ be such that $\sum_{i=1}^{n} |y_i| > M - \varepsilon/2$ and let $f \in \ell_1^*$ be defined by

$$f(x_1, x_2, \dots) = \sum_{i=1}^{n} \operatorname{sgn}(y_i) x_i,$$

where $\operatorname{sgn} t = 1$, if $t \geq 0$ and $\operatorname{sgn} t = -1$ if $t < 0$. By the Bishop–Phelps theorem, there is $g \in \ell_1^*$ such that $\|f - g\| < \varepsilon/4M$ and $F := \{x \in K : g(x) = \sup\{g(u) : u \in K\}\}$ is not empty. We claim that F has the required properties. Indeed, let $x \in F$. Then

$$\begin{aligned}\sum_{i=1}^{n} |x_i| &\geq f(x) \geq g(x) - \|f - g\|.\|x\| \geq g(y) - \|f - g\|.\|x\| \\ &\geq f(y) - \|f - g\|(\|x\| + \|y\|) \geq M - \varepsilon/2 - \varepsilon/2.\end{aligned}$$

Hence, since $\|x\| \leq M$, we get $\sum_{i>n} |x_i| \leq \varepsilon$. □

We complete now the proof of Theorem 251. By Lemma 252 there is a sequence $\{F_i\}_{i=1}^{\infty}$ of closed faces of K and an increasing sequence of integers $\{n_i\}_{i=1}^{\infty}$ such that F_{i+1} is a face of F_i and $x = (x_1, x_2, \dots) \in F_i$ implies that $\sum_{j>n_i} |x_j| \leq 1/i$. Let $\{y^i\}_{i=1}^{\infty}$ be a sequence of points in ℓ_1 such that $y^i \in F_i$ for each i. Then it follows immediately that $\{y^i : i \in \mathbb{N}\}$ is totally bounded. Since K is closed, $F := \bigcap_{i=1}^{\infty} F_i$ is a nonempty compact face of K. By the Krein–Milman Theorem 7 there is an extreme point of F, which is then an extreme point of K. □

By a careful use of Kadets' result on the homeomorphism of all separable conjugate spaces (see, e.g., [FHHMZ11, Theorem 12.46]), Bessaga and Pełczyński extended Lindenstrauss' Theorem 251 to separable dual spaces. This was a major breakthrough at that time.

Theorem 253 (Bessaga, Pełczyński, [BessPe66]). *Every bounded closed convex set in a separable dual Banach space is the closed convex hull of the set of all its extreme points.*

11.2 Lindenstrauss' results on norm-attaining operators

In Section 2.3 we introduced the set $NA(X, \|\cdot\|)$ of all the norm-attaining functionals on a Banach space $(X, \|\cdot\|)$. This extends to operators between two Banach spaces.

Definition 254. *Let $T \in \mathcal{B}(X, Y)$, and let C be a nonempty subset of X. We say that T* **attains its norm over** *C if there is $x_0 \in C$ such that $\|Tx_0\| = \sup\{\|Tx\| : x \in C\}$. In case $C = B_X$, we just say that T* **attains its norm**, *or that T is a* **norm-attaining operator**.

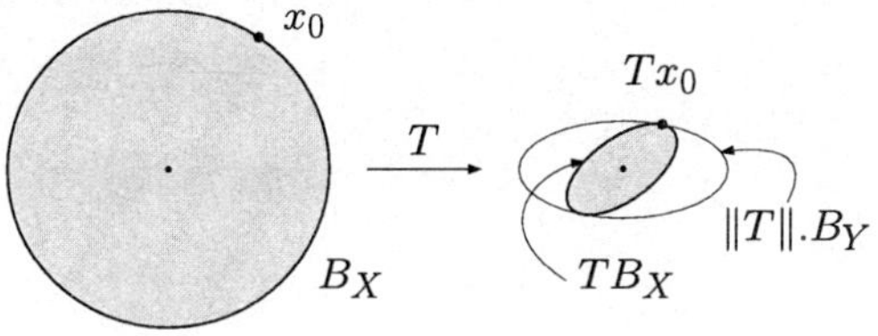

Figure 11.1: T attains its norm at x_0

Note that, in contrast with the case of norm-attaining functionals, a continuous linear operator from a Banach space X into a Banach space Y does not, in general, attain its norm over a non-empty w-compact subset K of X. The reason is that $T(K)$, although certainly w-compact, should not always have a point $y \in T(K)$ at farthest distance from 0 (see Theorem 260 below for a related positive result). A particular striking example —T is an operator on the Hilbert space $(\ell_2, \|\cdot\|_2)$ and K is the closed unit ball B_{ℓ_2} of $(\ell_2, \|\cdot\|_2)$, a w-compact set— is the following (see also [FHHMZ11, Exercise 8.58]): Take a strictly decreasing null sequence $\{a_n\}$ in $(0,1)$, and define $S : \ell_2 \to \ell_2$ by $S(x_1, x_2, \ldots) = (a_1x_1, a_2x_2, \ldots)$. Then put $T := I - S$, where $I : \ell_2 \to \ell_2$ is the identity operator. It is easy to show that $\|T(x)\|_2 \leq 1$ for every $x \in B_{\ell_2}$. Moreover, $T(e_n) = (1 - a_n)e_n$ for $n \in \mathbb{N}$, hence $\|T\| = 1$. Since $\|T(x_1, x_2, \ldots)\|^2 = \sum_{n=1}^{\infty} x_n^2 + \sum_{n=1}^{\infty} a_n^2 x_n^2 - 2\sum_{n=1}^{\infty} a_n x_n$, we get $\|T(x)\|_2 < 1$ for $\|(x_1, x_2 \ldots)\|_2 \leq 1$, so T does not attain its norm.
This is why Theorem 256 below stands as an important result in the area.

Lemma 255 (Lindenstrauss, [Lind63]). *Let X be a reflexive Banach space, Y be a Banach space and T be a bounded operator from X into Y. Then T attains its norm if and only if there is a sequence $\{x_k\}_{k=1}^{\infty}$ in S_X and a sequence $\{y_j^*\}_{j=1}^{\infty}$ in*

S_{Y^*} such that

$$|\langle Tx_k, y_j^* \rangle)| \geq \|T\| - \frac{1}{j} \quad \textit{for all } j \leq k.$$

Proof. Let x be a weak limit point of the sequence $\{x_k\}_{k=1}^{\infty}$. Then, for every j, $(\|T\| \geq \|Tx\| \geq)$ $|\langle Tx, y_j^* \rangle| \geq \|T\| - \frac{1}{j}$ and hence $\|Tx\| = \|T\|$.
Assume now that T attains its norm. There exists $x \in B_X$ such that $\|Tx\| = \|T\|$. Obviously, we may assume that $x \in S_X$. Define $x_k = x$ for $k \in \mathbb{N}$. Find $y_j^* \in S_{Y^*}$ such that $|\langle Tx, y_j^* \rangle| > \|T\| - 1/j$, for $j \in \mathbb{N}$. This shows the other implication. □

Theorem 256 (Lindenstrauss, [Lind63], see, e.g., [FHHMZ11, Theorem 7.47]). *Let X and Y be Banach spaces and C be a weakly compact convex set in X. Then the set of all operators in $\mathcal{B}(X, Y)$ that attain the supremum of their norms over C is norm-dense in $\mathcal{B}(X, Y)$.*

Proof of the case $C := B_X$ and X reflexive. Let $\|T\| = 1$ and $0 < \varepsilon < \frac{1}{3}$ be given. Choose a decreasing sequence $\{\varepsilon_k\}$ of positive numbers such that

$$2\sum_{i=1}^{\infty} \varepsilon_i < \varepsilon, \qquad 2\sum_{i=k+1}^{\infty} \varepsilon_i < \varepsilon_k^2, \qquad \varepsilon_k < \frac{1}{10k}, \qquad k = 1, 2, \ldots$$

Next choose inductively sequences $\{T_k\}$, $\{x_k\}$ and $\{f_k\}$ satisfying

$$\begin{aligned} T_1 &= T, \\ \|T_k x_k\| &\geq \|T_k\| - \varepsilon_k^2, \|x_k\| = 1, \\ f_k(T_k x_k) &= \|T_k x_k\|, \|f_k\| = 1, \\ T_{k+1}x &= T_k x + \varepsilon_k f_k(T_k x) \cdot T_k x_k, \quad x \in X. \end{aligned}$$

Then T_k converge in norm to an operator $\hat{T}$ satisfying $\|\hat{T} - T\| \leq \varepsilon$ and $\|\hat{T} - T_j\| \leq \varepsilon_{j-1}^2$ for $j = 2, 3, \ldots$ Moreover,

$$|f_j(\hat{T}x_k)| \geq \|\hat{T}\| - \frac{1}{j}.$$

Therefore $\hat{T}$ attains its norm in view of Lemma 255 above. Note that $\hat{T} - T$ is compact. □

The following example is in striking contrast with Theorem 256:

Example 257 (Lindenstrauss, [Lind63], see also [FHHMZ11, Exercise 8.59]). Let Y be the space c_0 endowed with its standard $\|\cdot\|_\infty$ norm, and let Z be the space c_0 endowed with Day's LUR $\|\cdot\|_D$ norm (see Section 10.7). Let $X = Y \oplus Z$ under the norm $\|(x, y)\| = \max(\|x\|_\infty, \|y\|_D)$. Let T_0 be an isomorphism from Y onto Z and finally let T from X into X be defined by $T(x, y) = (0, T_0(y))$. Then T is not in the closure of the set of all norm-attaining operators from X into X. ✠

➡ This is due to the "beloved" properties of c_0 in its standard norm $\|\cdot\|_\infty$, in particular that the ball does not have any extreme point (see Section 2.2) and so that the norm-attaining functionals are only the finitely supported ones.

11.3 Norm-attaining operators and strongly exposed points

We shall present the proof of Theorem 259 below only in a particular case of a weakly compact and convex subset of a Banach space, namely the closed unit ball of a reflexive space. It contains the main idea of the general result (to construct a linear isomorphism into a LUR space, see Lemma 258). The method of proof allows for a clear geometrical picture.

Lemma 258. *Let X and Y be Banach spaces. Assume that Y is LUR, and that there exists a linear isomorphism T from X into Y that attains its norm at some point $x \in B_X$. Then x is a strongly exposed point of B_X.*

Proof. Let $g \in Y^*$ satisfy $\|g\| = 1$ and $g(Tx) = \|T\|$ (see Figure 11.2). We shall prove that x is strongly exposed by the functional $f := g \circ T \in X^*$. Suppose that $g(Tx_n) \to g(Tx)$ $(= \|T\|)$, with $\|x_n\| \leq 1$. Then clearly $\|Tx + Tx_n\| \to 2\|T\|$ and hence, by our assumption on Y, $\|Tx_n - Tx\| \to 0$. Since T is an isomorphism, $\|x_n - x\| \to 0$, and this concludes the proof. □

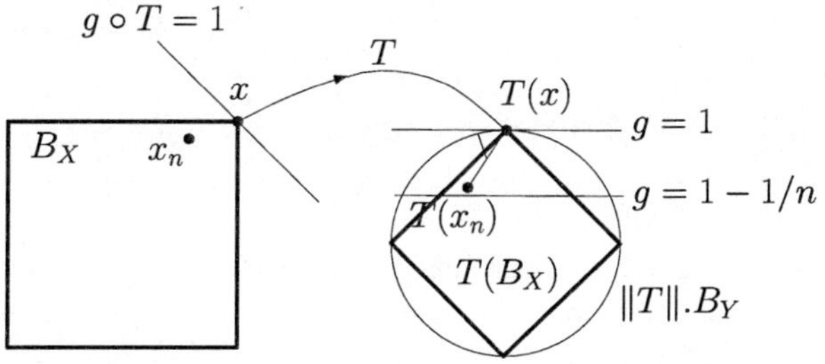

Figure 11.2: The proof of Lemma 258

Theorem 259 (Lindenstrauss, Troyanski, see, e.g., [FHHMZ11, Theorem 8.13]). *Let W be a weakly compact convex set in a Banach space X. Then W is the closed convex hull of its strongly exposed points.*

Proof. (For $W := B_X$, the closed unit ball of a reflexive space X [Lind63], see Figure 11.3) Let us assume that C is the closed convex hull of the set of all strongly exposed points of B_X, where X is a reflexive space. Assume that $C \neq B_X$. Then there is an $f \in S_{X^*}$ and $\delta > 0$ such that $|f(x)| < 1 - \delta$ for $x \in C$.

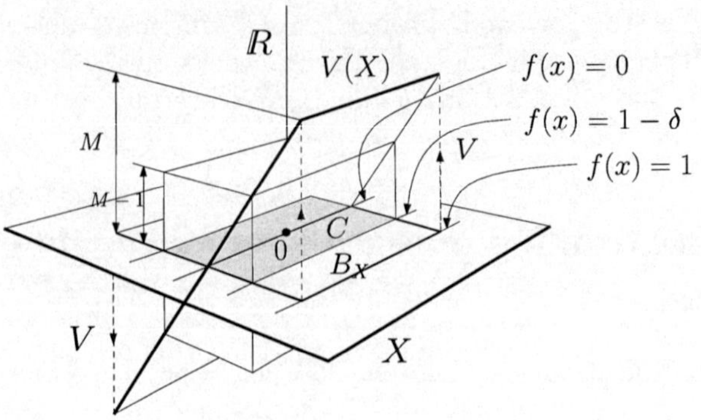

Figure 11.3: The construction in the proof of Theorem 259

Let $|||\cdot|||$ be a LUR norm on X which is equivalent to the given norm $\|\cdot\|$ and such that $|||x||| \leq \|x\|$ for every $x \in X$. Let Y be the space $X \oplus \mathbb{R}$ with the norm $\|(x,r)\| = (|||x|||^2 + r^2)^{\frac{1}{2}}$, $x \in X$, $r \in \mathbb{R}$. Then $(Y, \|\cdot\|)$ is LUR. Let V be the operator from $(X, \|\cdot\|)$ into $(Y, \|\cdot\|)$ defined by $Vx = (x, Mf(x))$, where $\frac{2}{\delta} < M < \frac{2}{\delta^2}$. Then V is an isomorphism into Y, and the same is true for every operator sufficiently close to V. Find $x_0 \in S_X$ such that $f(x_0) = 1$. Then

$$\|V(x_0)\|^2 = \|(x_0, M)\|^2 = |||x_0|||^2 + M^2 \geq M^2, \text{ hence } \|V\| \geq M.$$

On the other hand,

$$\|Vx\|^2 = |||x|||^2 + M^2|f(x)|^2 \leq 1 + M^2(1-\delta)^2 \leq (M-1)^2 \quad \text{for } x \in C.$$

It follows that operators sufficiently close to V that attain their norm (see Theorem 256 above) cannot do it at a point belonging to C. To conclude the proof it is enough to apply Lemma 258 to such an operator to find a strongly exposed point of B_X outside C, a contradiction. □

An alternative proof for the closed unit ball B_X of a reflexive Banach space X is the following: Since X is LUR renormable (see Theorem 218 above), B_X has at least a strongly exposed point (see Subsection 5.3.2, Item 4). Let C be the closed convex hull of the set of all strongly exposed points of B_X. Assume that $C \neq B_X$. Let $x^* \in S_{X^*}$ such that $\sup |\langle C, x^* \rangle| < \alpha$ for some $\alpha < 1$. Since X^* is an SDS space (see Theorem 272 below), the set F of all functionals in X^* where $\|\cdot\|^*$ is Fréchet differentiable is dense. Find $x_0^* \in F$ with $\|x^* - x_0^*\|^*$ small enough. The point $x_0 \in B_X$ where x_0^* attains its norm (a strongly exposed point by the Šmulyan Lemma 54) is not in C, a contradiction. □

11.4 Farthest points

A theorem that in some cases can replace Theorem 259 is the following result of K.S. Lau:

Theorem 260 (Lau [Lau75], see, e.g., [FHHMZ11, Theorem 8.19]). *Let C be a weakly compact set in a Banach space X. Then the set of points that have farthest points in C contains a G_δ dense subset of X.*

Proof of Theorem 260. (Sketch) The **farthest distance function** for the set C is defined for $x \in X$ by

$$r(x) = \sup\{\|x - y\| : \ y \in C\}.$$

This function, as a supremum of a family of convex and 1-Lipschitz continuous functions, is convex and 1-Lipschitz. In particular, $\partial r(x) \subset B_{X^*}$ for every $x \in X$, where ∂r denotes the subdifferential of r (see Remark 51 above). A point $c \in C$ is a **farthest point from** $x \in X$ whenever $r(x) = \|x - c\|$.
The core of the proof is the following statement: *The set*

$$G = \{x \in X : \ \sup\{x^*(x - z) : z \in C\} = r(x) \quad \textit{for every } x^* \in \partial r(x)\}$$

is G_δ dense in X. Furthermore, if $x \in G$, then x has a farthest point in C.
This combines the use of the Lindenstrauss'idea from [Lind63] with a Baire category argument. □

Here is how Lau's result may help in getting at least one point of Fréchet differentiability of any norm on a reflexive space. The basic argument is as follows (see Figure 11.4): Assume that C is the unit ball of a reflexive Banach space $(X, |||\cdot|||)$ such that $|||\cdot|||$ is Fréchet smooth (the existence of such a norm is a consequence of the existence of a —obviously dual— LUR norm in the reflexive space X^* —see Theorem 218, its Corollary 219, and Theorem 247—). Let $\|\cdot\|$ be an equivalent norm on X. Let $x_0 \in X$ have a $\|\cdot\|$-farthest point $y_0 \in C$. Then y_0 is a point where $|||\cdot|||$ is Fréchet differentiable, and so it is $\|\cdot\|$ at y_0, as we may check directly, or use instead again the Šmulyan's Lemma 54. This is done in detail in the next annotation.

➡ There is a useful argument —that can be applied to the precedent situation— that ensures Gâteaux (Fréchet) differentiability of a convex function at a point when it is dominated by a function that is Gâteaux (respectively Fréchet) differentiable at that point. The geometrical clue is Figure 11.5, and the analytic counterpart is based in [FHHMZ11, Lemma 7.4]. This lemma reads: *If a convex function f is defined on an open convex subset U of a Banach space X, and f is continuous at some $x_0 \in U$, then f is Gâteaux (Fréchet) differentiable at x_0 if, and only if,*

$$\lim_{t\to 0} \frac{f(x_0 + th) + f(x_0 - th) - 2f(x_0)}{t} = 0 \textit{ (resp. uniformly on } h \in S_X\textit{).} \tag{11.1}$$

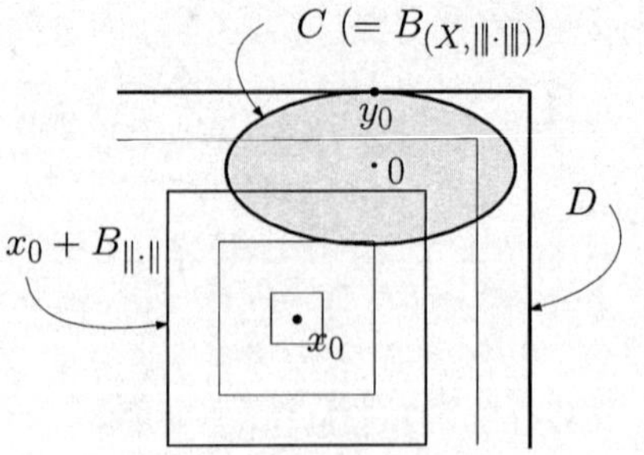

Figure 11.4: The norm $|||\cdot|||$ has a point of Fréchet differentiability

Thus, we readily get the following consequence (see again Figure 11.5): *Let f and g be two convex functions on an open convex subset U of a Banach space X, and let $x_0 \in U$. Assume that (i) both functions are continuous at x_0, (ii) $f(x_0) = g(x_0)$, (iii) $g(x) \leq f(x)$ for all $x \in U$, and (iv) f is Gâteaux (Fréchet) differentiable at x_0. Then g is Gâteaux (respectively, Fréchet) differentiable at x_0.* The argument of the proof uses that, for all $h \in X$ such that $x_0 + h \in U$, and for all $t \in (0,1)$,

$$\frac{g(x_0+th)+g(x_0-th)-2g(x_0)}{t} \leq \frac{f(x_0+th)+f(x_0-th)-2f(x_0)}{t}$$

and that

$$\frac{F(x_0+th)+F(x_0-th)-2F(x_0)}{t} \geq 0$$

for any convex function F on U.

Here is another equivalent approach to the same result (again, Figure 11.5 shows the statement): Let $f : X \to (-\infty, +\infty]$ be a proper (i.e., not identically equal to $+\infty$) function defined on a Banach space X. Fix x_0 in the domain of f (i.e., $f(x_0) < +\infty$), and $\varepsilon \geq 0$. The **ε-subdifferential** (see, e.g., [FHHMZ11, Definition 7.11] of f at x_0 is the set $\partial f_\varepsilon(x_0)$ of all $x^* \in X^*$ such that $f(x) \geq f(x_0) + \langle x - x_0, x^* \rangle$ for all $x \in X$. The 0-subdifferential of f at x_0 is simply denoted by $\partial f(x_0)$. Then, if $f(x_0) = g(x_0)$ and $f(x) \geq g(x)$ for all $x \in X$, we have, clearly, $\partial_\varepsilon g(x_0) \subset \partial_\varepsilon f(x_0)$ for every $\varepsilon \geq 0$. It is simple to prove (see, e.g., [FHHMZ11, Theorems 7.17 and 7.15]) that *if f is a convex continuous function defined on a nonempty open subset U of a Banach space X, then f is Gâteaux (Fréchet) differentiable at x_0 if, and only if, $\partial f(x_0)$ is a singleton (respectively, if, and only if, $\|x_n^* - x_0^*\| \to 0$ whenever $x_n^* \in \partial_{\varepsilon_n} f(x_0)$ for $n \in \mathbb{N}$, where $\varepsilon_n \searrow 0$).* The observation about the inclusion of the ε-subdifferentials of f and g at x_0 shows, as a consequence, the previous result of Gâteaux (respectively, Fréchet) differentiability of g at x_0.

Let us see now how this result can be applied to the differentiability of $\|\cdot\|$ at y_0 (see Figure 11.4). Let p_D be the Minkowski functional of D, where $D := x_0 + \|x_0 - y_0\| B_{\|\cdot\|}$. This is a convex function. It is continuous, due to the fact that D has a nonempty interior. Since $C \subset D$, we have $p_D \leq |||\cdot|||$ on X. Moreover, $P_D(y_0) = |||y_0||| = 1$. Since $|||\cdot|||$ is Fréchet differentiable, we get, by the statement above, that p_D is Fréchet differentiable at y_0. Define a function $h : X \times (\mathbb{R}\setminus 0) \to \mathbb{R}$ by $h(x,r) := p_D(x_0 + \frac{1}{r}(x - x_0)) - 1$, where $x \in X$ and $r \in \mathbb{R} \setminus \{0\}$. Note that $h(y_0, 1) = 0$, and that h is Fréchet differentiable at $(y_0, 1)$. We have, for $x \in X$ and $r \neq 0$,

$$dh_{(x,r)} = \left(\frac{1}{r} d(p_D)_{(x_0 + (x - x_0)/r)}, -\frac{1}{r^2} \left\langle x - x_0, d(p_D)_{(x_0 + (x - x_0)/r)} \right\rangle \right).$$

In particular,

$$dh_{(y_0,1)} = (d(p_D)_{y_0}, \langle y_0 - x_0, d(p_D)_{y_0} \rangle),$$

so

$$\frac{\partial h}{\partial r}(y_0, 1) = \langle y_0 - x_0, d(p_D)_{y_0} \rangle \neq 0.$$

The Implicit Function Theorem 77 gives a neighborhood U of y_0 in X, a $\delta > 0$ (that can be taken less than 1), and a function $r : U \to \mathbb{R}$ such that r is Fréchet differentiable at y_0 and

$$\{(x,r) \in U \times (1 - \delta, 1 + \delta) : \ h(x,r) = 0\} = \{(x, r(x)) : \ x \in U\}.$$

If $x \in U$ we get $h(x, r(x)) = 0$, i.e., $p_D(x_0 + (x - x_0)/r(x)) = 1$, what amounts to $\|(x - x_0)/r(x)\| = 1$, i.e., $r(x) = \|x - x_0\|$. Since r is Fréchet differentiable at y_0, this shows that $\|\cdot\|$ is Fréchet differentiable at y_0, as claimed.

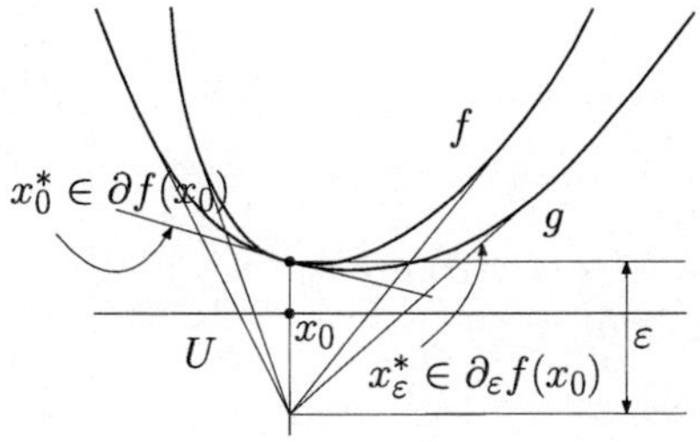

Figure 11.5: If f is differentiable at x_0, so it is g

Theorem 260 has been preceded by [Ed66] and [Asp66].

Note that Theorem 260 characterizes weakly compact sets in the following sense:

Theorem 261 (Deville, Zizler, [DeZiz88]). *Let X be a Banach space and C be a bounded closed convex subset in X that is not weakly compact. Then there is an equivalent norm $|\cdot|$ on X with respect to which the set of points in X having $|\cdot|$-farthest points in C is not dense in X.*

The proof uses James' weak compactness Theorem 13.

Proposition 263 below is a consequence of Theorem 260. It requires a property of the space (the **Mazur intersection property**) that has been widely investigated:

Definition 262. *The norm $\|\cdot\|$ of a Banach space X is said to have the* **Mazur intersection property (MIP)** *if each bounded closed convex set in X is the intersection of closed balls.*

We shall devote Chapter 33 below to this property.

Proposition 263 (Edelstein, Lau, see [FHHMZ11, Proposition 8.20]). *Let X be a Banach space with the MIP. Let C be a nonempty weakly compact convex set in X. Then C is the closed convex hull of the subset of C consisting of points that are farthest points to some points in X.*

Working with Fréchet differentiable LUR norms, we get the following direct corollary (yet, check Theorem 259 above):

Corollary 264. *Let $(X, \|\cdot\|)$ be a Banach space with a separable dual. Then every nonempty weakly compact convex set in X is the closed convex hull of the set of all its strongly exposed points.*

Proof. (Sketch) The space X has an equivalent Fréchet differentiable LUR norm $|||\cdot|||$. Endowed with $|||\cdot|||$, X has the MIP (see Theorem 568). We apply Proposition 263 to get that a given nonempty closed convex and weakly compact subset C of X is the closed convex hull of the set of all $|||\cdot|||$-farthest points. We noticed in Item 4 in Subsection 5.3.2 that every such farthest point in a LUR space $(X, |||\cdot|||)$ is in fact a strongly exposed point of C in $(X, |||\cdot|||)$, hence in $(X, \|\cdot\|)$. □

In this direction we should note the following result.

Proposition 265 (Deville, Zizler, [DeZiz88], see also, e.g., [FHHMZ11, Exercise 8.44]). *Let a set in ℓ_1 be defined by*

$$C := \left\{ (a_n) : \sum_{n=1}^{\infty} (|a_n| + |a_n|^2) \leq 1 \right\}.$$

Then C is a w^*-compact convex subset of ℓ_1 that has no farthest points in the standard norm of ℓ_1.

Proposition 266 (Deville, Zizler, [DeZiz88]). *Let X have the RNP and C be a nonempty w^*-compact set in X^*. Then the set of all points that have farthest points in C in the canonical dual norm contains a dense G_δ set in X^*.*

While there exist nonreflexive Banach spaces and some norms on them in which all bounded closed convex sets possess farthest points [Ziz73], the following seems to be an open problem:

Problem. Is it true that in every nonreflexive Banach space there is a norm in which some closed convex bounded set has no farthest point? (see, e.g., [DeGoZi93, Open problems. Problem III.4]).

In this direction, let us mention a famous problem in the area: It is not known whether the KMP property (Definition 545) implies the RNP property (Definition 61) . We shall return to this question after Definition 299 (see page 241). It is also unknown whether the RNP property implies the existence of an equivalent LUR norm.

11.5 Properties α and β. Uniformly strongly exposed points

J. Lindenstrauss and W. Schachermayer introduced the following notion ([Lind63], [Sch76]) in the study of norm-attaining operators:

Definition 267. (a) *A Banach space X is said to have* **property** α *if there are $0 \leq \lambda < 1$ and a family $\{x_\alpha, x_\alpha^*\}_{\alpha\in I}$ in $S_X \times S_{X^*}$ such that $\langle x_\alpha, x_\alpha^*\rangle = 1$, $|\langle x_\alpha, x_\beta^*\rangle| \leq \lambda$ for all $\alpha, \beta \in I$, $\alpha \neq \beta$, and $B_X = \overline{\Gamma(\{x_\alpha : \alpha \in I\})}$, where $\Gamma(S)$ denotes the absolutely convex hull of a set S.*
(b) *A Banach space X is said to have* **property** β *if there are $0 \leq \lambda < 1$ and a family $\{x_\alpha, x_\alpha^*\}_{\alpha\in I}$ in $S_X \times S_{X^*}$ such that $\langle x_\alpha, x_\alpha^*\rangle = 1$, $|\langle x_\alpha, x_\beta^*\rangle| \leq \lambda$ for all $\alpha, \beta \in I$, $\alpha \neq \beta$, and $\{x_\alpha^* : \alpha \in I\}$ is 1-norming for X.*

J. Lindenstrauss showed in [Lind63] (see, e.g., [HMVZ07, Theorem 8.25]) that if X has property α and Y is an arbitrary Banach space, then the norm-attaining operators from X into Y are norm-dense in the space of all operators.

A typical example of a space with property α is ℓ_1. We have the following interesting

Theorem 268 (Schachermayer, [Sch67]). *Every WCG Banach space can be renormed to have property α.*

Proof. (Sketch, and only the case $X := c_0$) Choose $\lambda < 1$. Let $\{z_n\}_{n\in\mathbb{N}}$ be a dense sequence in the unit ball of c_0. Define x_n by

$$x_n(i) = z_n(i) \text{ if } i \neq n, \qquad x_n(i) = \frac{1}{\lambda} \text{ if } i = n.$$

Let B be the absolutely convex hull of $\{x_n : n \in \mathbb{N}\}$ and let $|||\cdot|||$ be the Minkowski functional of B. Clearly

$$|||\cdot||| \geq \lambda\|\cdot\|.$$

On the other hand, if $\|x\| \leq 1$, fix a subsequence $\{x_{n_j}\}_{j\in\mathbb{N}}$ with $n_1 < n_2 < n_3 < \dots$ such that $x_{n_j} \to x$. For $m \in \mathbb{N}$, let

$$u_m = (x_{n_1} + \ldots + x_{n_m})/m.$$

Then $u_m \in B$ and $u_m \to x$. Therefore $|||\cdot||| \leq \|\cdot\|$. Now let $y_n = \lambda e_n$ for $n \in \mathbb{N}$, where e_n is the n-th coordinate vector in ℓ_1. Then $|||y_n||| = 1$, $y_n(x_n) = 1$, while $|y_n(x_m)| \leq \lambda$ if $n \neq m$. □

The spaces c_0, ℓ_∞, and finite-dimensional spaces with polyhedral unit balls have property β (see Definition 141).

Properties α and β are in duality: $(X, \|\cdot\|)$ has property α if, and only if, $(X^*, \|\cdot\|^*)$ has property β, where, in addition to the defining properties there, the family $\{x^{**}_\alpha\}_{\alpha\in I}$ is in X [Sch67]. The behaviour of the dual pair $\langle x_0, \ell_1\rangle$ prompted C. Finet and W. Schachermayer to introduce the following strengthening of the property β:

Definition 269. *The Banach space X has* **property β in the strong sense** *if X has property β and $B_{X^*} = \overline{\Gamma(\{x^*_\alpha : \alpha \in I\})}^{\|\cdot\|}$.*

Theorem 270 (Finet, Schachermayer, [FiSch89]). *Every separable Asplund space can be renormed to have property β in the strong sense.*

Observe that in Definition 269, the family $\{x^*_\alpha : \alpha \in I\}$ is w^*-**uniformly strongly exposed** in S_{X^*}: A family $\{x_\alpha : \alpha \in I\}$ in S_X is said to be **uniformly strongly exposed** if for every $\varepsilon > 0$ there exists $\delta > 0$ and a family $\{f_\alpha : \alpha \in I\}$ in S_{X^*} such that $\langle x_\alpha, f_\alpha\rangle = 1$ and $\|x - x_\alpha\| < \varepsilon$ whenever $x \in B_X$ and $\langle x, f_\alpha\rangle > 1 - \delta$, for all $\alpha \in I$. If we are in X^* and the family of functionals $\{f_\alpha : \alpha \in I\}$ is in X, we speak of a w^*-**uniformly strongly exposed** family.

In this direction also the following result can be mentioned.

Theorem 271 (Zizler, [Ziz75]). *Let $(X, \|\cdot\|)$ be an Asplund space and assume that the duality mapping is directionally $\|\cdot\|$-lower semicontinuous. Then there is a w^*-strongly exposed family in S_{X^*} that is $\|\cdot\|$-dense in S_{X^*}.*

11.6 Asplund spaces I

11.6.1 Introduction

Every real-valued convex function f defined on a nonempty open convex interval in $\mathbb{R}$ is continuous on it. This also holds for a real-valued convex function defined on a nonempty open convex subset of $\mathbb{R}^n$ (see Theorem 52).

This automatic continuity is no longer true for infinite-dimensional spaces (indeed, there is always a linear everywhere discontinuous function defined on any infinite-dimensional normed space). This was already mentioned after Theorem 49.
Any real-valued convex function defined on an open interval in $\mathbb{R}$ is differentiable at every point in the interval but at most a countable number of them. This is no longer true for convex functions on $\mathbb{R}^n$ (think about $f(x) := \max\{\|x\| - 1, 0\}$ for $x \in \mathbb{R}^2$, see Figure 11.6).

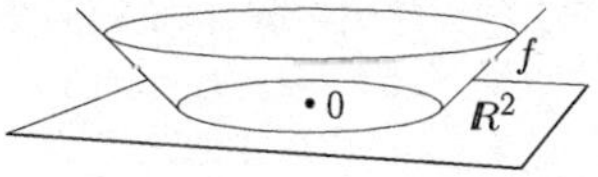

Figure 11.6: f is not differentiable at uncountably many points

Mazur's Theorem 362 was the first result concerning generic differentiability of convex functions in infinite-dimensional spaces (i.e., differentiability at points of a dense G_δ subset of its domain). In the language introduced by E. Asplund, Mazur's theorem says that every separable Banach space is a **weak differentiability space** (for short, **WDS**), i.e., a space X such that every continuous convex function is Gâteaux differentiable at points of a dense G_δ subset of X.
The class of WDS (nowadays called **weak Asplund spaces**, see Definition 63) is strictly larger than the class of what Asplund called **strong differentiability spaces** (**SDS**, for short) (nowadays called **Asplund spaces**). Asplund defined SDS by replacing "Gâteaux differentiability" by "Fréchet differentiability" in the definition of WDS. The norm $\|\cdot\|_1$ on the space ℓ_1 is nowhere Fréchet differentiable (see Subsection 4.1.4 above), hence ℓ_1 is not an Asplund space (for a more precise information see Proposition 281 below). However, ℓ_1 is, by Mazur's theorem, a WDS.

➡ There is no need in the definition of an SDS to add the requirement that the set of points where a continuous convex functions $f : X \to \mathbb{R}$ is Fréchet differentiable contains a G_δ dense set, since this set (it can be empty) is automatically G_δ. See Item 4.1.5.8.
This is not so in the case of Gâteaux differentiability. See, e.g., [CobKen85] and [Tal79].
Note that the concepts of being WDS or SDS are isomorphic in nature.

H Lindenstrauss [Lind63] proved that a reflexive Banach space admitting an equivalent Fréchet differentiable norm is an SDS.
Then Asplund, just by using the Baire category theorem and convex techniques, and before any smooth variational principle, produced a self-contained proof to the following result (the R version of it should be also noted).

Theorem 272 (Asplund, [Asp68]). (i) *If a Banach space X has an equivalent norm such that its dual norm is LUR, then X is an SDS.*
(ii) *If a Banach space X has an equivalent norm such that its dual norm is R, then X is a weak Asplund space.*

➪ Asplund reduces the proof to the case when the convex continuous function to be proved Fréchet (Gâteaux) differentiable at points of a G_δ-dense subset of X is, moreover, Lipschitz. This is done by considering convex functions $g_n(x) := n\|x\| + n$ for $x \in X$, and then functions $f_n := f \wedge g_n$, $n \in \mathbb{N}$. Given x in the effective domain of f, these functions agree with f at a neighborhood of x for n big enough (see Figure 11.7).
Thus, f is supposed to be convex, continuous and Lipschitz. Put

$$h(x) = \frac{1}{2}\|x\|^2, \ x \in X.$$

Since f is Lipschitz, the open sets F_n defined by

$$F_n := \Big\{ x : \ f(x) - \frac{1}{p}\Big(h\big(p(x+y)\big) - \frac{1}{n}\Big) \\ > \sup \Big(f(z) - \frac{1}{p} h\big(p(x+y)\big) \ : \ z \in X \Big) \text{ for some } y \in X \text{ and } p > 0 \Big\}$$

are dense in X. Baire's theorem ensures that

$$F := \bigcap_{n=1}^{\infty} F_n$$

is a dense G_δ subset of X. Under the assumptions on the dual norm Asplund shows that f is Fréchet (respectively Gâteaux) differentiable at each point of F.

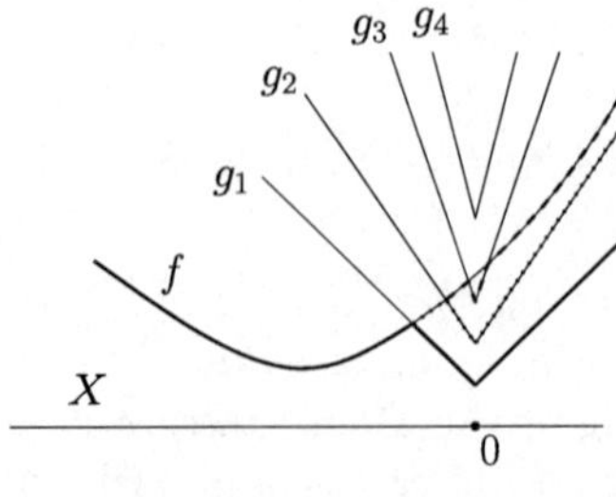

Figure 11.7: Functions g_n and f_n in Asplund's proof

H The third named author is proud to say that, as a graduate student, he personally fed the coal to the stove that heated the room where Professor Asplund had his lecture on Theorem 272 at Prague Mathematical Institute in 1969.

Later on, Preiss, Phelps, and Namioka proved the following important extension of (ii) in Theorem 272:

Theorem 273 (Preiss, Phelps, Namioka, [PrPhNa90]). *Every Banach space that admits a Gâteaux differentiable norm is a weak Asplund space.*

Actually, it suffices to assume that the Banach space admits a Lipschitz and Gâteaux differentiable bump ([DeGoZi91] and [Fos92]). See also [Fab97, Theorems 5.2.2 and 5.3.1].
The proof of Theorem 273 uses the important technique called "Banach–Mazur game". We strongly recommend the student to read the full argument in order to have a glimpse of it. More in this direction can be found in [Phe93].
The result in Theorem 273 was also proven, without relying on Banach–Mazur games, in [Rib92].
Related to Theorem 273, see also the result mentioned in Item 7, Subsection 4.1.5 that we repeat here: *There is a non-weak Asplund space that is a Gâteaux differentiability space* [MoSo06].

➬ In this connection note that $C[0,\omega_1]$ has an equivalent C^∞-norm [Hay99] (see Theorem 336 below), but does not admit any norm with rotund dual norm [Tal86] (see, e.g., [DeGoZi93, Theorem VII.5.2(ii)]).
It is an open question whether every subspace of a weak Asplund space is again weak Asplund. It is also unknown whether every weak Asplund space X^* can be renormed by a rotund not necessarily dual norm, see, e.g., [Fab97, 5.4. Notes and Remarks]. As we mentioned in xxii of the Introduction, it is not known whether every Asplund space admits a C^1-differentiable bump.

11.6.2 Some equivalences

H An intensive research followed the work reflected above, eventually reaching the equivalence (i)⇔(ii) in Theorem 274 below: *A Banach is a strong differentiability space if, an only if, every separable subspace has a separable dual.* Nowadays it is customary to use this separability assumption —we followed this trend— as the definition of an Asplund space (see Definition 58 above).

Theorem 274 below lists some of the known equivalences for a general Banach space to be Asplund, and is the contribution of many mathematicians. We shall properly provide the attributions in the sketch of the proof. An alternative approach to Asplundness, by using the so-called Szlenk indices, can be found in [Go01b] (see also Chapter 43). In the separable case it is possible to add other equivalences (Theorem 276 below), whose validity for nonseparable Banach spaces remains open. This is why we think convenient to present the general and the separable case successively. We provide a sketch of proofs, and some details will be discussed later.

The general case (equivalences)

The concept of a rough norm was introduced in Definition 76.

Theorem 274. *Let X be a Banach space. Then the following are equivalent.*
(i) *X is an Asplund space.*
(ii) *X is a SDS.*
(iii) *Every nonempty and norm-closed convex bounded subset of X^* is the norm-closed convex hull of its strongly exposed points, i.e., X^* has RNP.*
(iv) *X does not admit a rough norm.*
(v) *Every Lipschitz function on X is Fréchet differentiable at a dense set of points.*
If $X = C(K)$ for a compact space K, then the former assertions are equivalent to
(vi) *K is scattered.*

Proof. That (ii)⇒(iv) is trivial. The equivalence of (i), (ii), and (iii) is proved in [FHHMZ11, Theorem 11.8]. The equivalence of (ii) and (iv) is proved in [DeGoZi93, Theorem I.5.3]. For (i)⇒(v) see Theorem 443 below. For proving (v)⇒(ii) observe that every convex continuous function is locally Lipschitz (see Theorem 49 and, e.g., [Phe93, Proposition 1.6] or [FHHMZ11, Lemma 7.3]), and that every Lipschitz function on a subset of a Banach space can be extended to a Lipschitz function defined on the whole space (see Theorem 437 below, see also, e.g., [FHHMZ11, Exercise 2.18]). For the proof of (vi)⇔(i) see, e.g., [FHHMZ11, Theorem 14.25].□

The general case: some sufficient conditions for Asplundness

We list some sufficient conditions for a space X to be Asplund. Whether some of them are also necessary is still open, as we shall mention in due moment. The reader will notice that some are included in others. Still, we shall record them separately, as the genealogy or the arguments used in their proof are different.

Theorem 275. *Each of the following conditions is sufficient for a Banach space X to be Asplund:*
(i) [Asplund [Asp68]] *There exists on X an equivalent norm whose dual norm is LUR.*
(ii) *There exists on X a Fréchet differentiable equivalent norm.*
(iii) [Ekeland, Lebourg [EkLe76]] *There exists on X a Lipschitz C^1-smooth bump.*
(iv) [Ekeland, Lebourg [EkLe76]] *There exists on X a Fréchet differentiable bump.*

Proof. (modulus Theorem 274.) (i) is in Theorem 272. For (ii) note Remark 59. For (iii) see Corollary 290 and Remark 291. For (iv), observe Theorem 283. □

We mentioned after Theorem 272 above that it is open whether every Asplund space admits a C^1-smooth bump. Note also that it is open whether (iv) implies (iii) in Theorem 275. It is open whether the sufficient condition (iv) there is also necessary, i.e., whether every Asplund space must admit a Fréchet differentiable bump.

The separable case

Theorem 276. (see, e.g., [FHHMZ11, Theorem 8.6]) *Let X be a separable Banach space. Then the following are equivalent.*
(i) *X^* is separable*
(ii) *X admits a norm whose dual norm on X^* is LUR.*
(iii) *X admits a Fréchet differentiable norm.*
(iv) *X admits a Fréchet differentiable bump.*
(v) *X admits no rough norm.*
(vi) *X is an SDS.*
(vii) *Every continuous convex function on X is Fréchet differentiable at least at one point.*
(viii) *X admits a WUR norm.*

Proof of Theorem 276.
(i)⇒(ii) is Corollary 178.
(ii)⇒(iii) is Proposition 91.
(iii)⇒(iv) is trivial by using the compositions of functions (in $\mathbb{R}$ there exists a Fréchet differentiable bump).
(iv)⇒(v): Theorem 283.
(v)⇒(i): Theorem 284.
So we get that (i) to (v) are equivalent (notice that we proved in Remark 59 that (iv)⇒(i)).
(vi)⇒(v) is trivial.
(iv)⇒(vi) follows via the Smooth Variational Principle (see Theorem 289 below and [FHHMZ11, Lemma 7.45]), or by the direct beautiful self-contained proof in [PrZa84] (see, e.g., [FHHMZ11, Exercise 8.17]).
Thus all the statements from (i) to (vi) are equivalent.
Now, trivially, (vii)⇒(v), and (vi)⇒(vii).
Concerning (viii), look at (iv) in Theorem 144 and the fact (above) that if X^* is separable, then X admits a WUR norm (see Proposition 164 and its application). If X is separable and it admits a WUR norm, then its dual norm is UG (see (v) in Theorem 276). Thus, X^* is a Vašák space (Lemma 752), hence X^* is DENS (see (v) in Theorem 745). In other words, the w^*-density of X^{**} equals the norm-density of X^*. Since X is w^*-dense in X^{**} and X is separable, we get that X^* is norm-separable. □

Note that there exist separable nonAsplund spaces in which all continuous convex Gâteaux differentiable functions are Fréchet differentiable at a dense sets of points ([DGHZ87] and [DeGoZi93, Chapter III]).

Separable reduction

The proof of the following result shows a method to check the SDS condition on a Banach space X by looking only at its separable subspaces.

Lemma 277. *Let X be a Banach space. Assume that a continuous convex function $f : X \to \mathbb{R}$ has the property that the set D of all points of Fréchet differentiability is not dense in X. Then there exists a separable subspace Y of X where the restriction of f has the same property.*

Proof. There is a nonempty open ball $B \subset X$ in which there is no point of Fréchet differentiability of f. For $n \in \mathbb{N}$ let

$$G_n := \left\{ x \in B : \inf_{\delta>0} \sup_{y \in S_X} \frac{f(x+\delta y) + f(x-\delta y) - 2f(x)}{\delta} < \frac{1}{n} \right\}.$$

It is easy to show that the sets G_n are open for all n (see, e.g., [FHHMPZ01, Lemma 8.23]). From the convexity of f we have that $\emptyset = D \cap B = \bigcap G_n$. As B is a Baire space, we have that G_n is not dense in B for some n. Thus there is n and an open ball U in B such that $G_n \cap U = \emptyset$. In order to find Y we may proceed as follows: Choose $x_1 \in U$. As $x_1 \notin G_n$, from the monotonicity of the quotients we have that there is a sequence $\{y_j^1\}_j$ in S_X such that for all $\delta > 0$,

$$\sup_j \frac{f(x_1+\delta y_j^1) + f(x_1-\delta y_j^1) - 2f(x_1)}{\delta} \geq \frac{1}{2n}.$$

Let Y_1 be the closed linear span of x_1 and $\{y_j^1\}_j$. If an increasing sequence of separable closed subspaces $Y_1, Y_2, \ldots, Y_k$ has been constructed, let $\{x_{k,i}\}_i$ be a countable dense subset of $Y_k \cap U$ and for each i choose a sequence $\{y_j^i\}_j$ in S_X such that for all $\delta > 0$,

$$\sup_j \frac{f(x_{k,i}+\delta y_j^i) + f(x_{k,i}-\delta y_j^i) - 2f(x_{k,i})}{\delta} \geq \frac{1}{2n}.$$

Let Y_{k+1} the closed linear span of Y_k and $\{y_j^i\}_{i,j}$. Let $Y := \overline{\bigcup Y_k}$. Then $\{x_{k,i}\}_{k,i}$ is dense in $U \cap Y$. Moreover, each $x_{k,i}$ is not in $\widetilde{G}_{2n}$, where the latter set is the set G_{2n} defined for the restriction of f to Y. As $\widetilde{G}_{2n}$ is open in Y, we have that $U \cap Y$ is disjoint from $\widetilde{G}_{2n}$ and thus the restriction of f to Y is Fréchet differentiable at no point of $U \cap Y$ (see, for details, [FHHMPZ01, Exercise 8.16]). □

H The paper [Res64] contains apparently the first published proof of the fact that for a separable Banach space X, (i) and (iii) in Theorem 276 are equivalent.

11.6.3 Some tools and proofs on rough norms

Being an SDS exclude the possibility to have an equivalent norm which is nowhere Fréchet differentiable. A particular instance of such a norm is what was called a "rough norm" (see Definition 76). It was described there as "uniformly non-Fréchet differentiable". It is remarkable that the absence of such an equivalent norm characterizes Asplundness; this is the equivalence (i)⇔(iv) in Theorem 274. We now discuss the notion of roughness in more detail, not only because of this characterization, but also being important in understanding smoothness and optimization in Banach spaces.
We start by listing in the following result some equivalences of roughness. Some examples of rough norms will be provided after the proof of this theorem. We present its proof in detail as the methods used there are of general interest in various different situations.

Theorem 278. *Let $(X, \|\cdot\|)$ be a Banach space. Then the following are equivalent.*
(i) *The norm is rough.*
(ii) *There is $\varepsilon > 0$ such that for every $x \in S_X$ there are sequences $\{f_n\}$ and $\{g_n\}$ in S_{X^*} with $\lim f_n(x) = \lim g_n(x) = 1$ and $\limsup \|f_n - g_n\| > \varepsilon$.*
(iii) *There is $\varepsilon > 0$ such that for every $x \in S_X$ and every $\delta > 0$ there are $x_1, x_2 \in X$ and $u \in S_X$ such that $\|x_i - x\| < \delta$ for $i = 1, 2$, and $\|x_2\|'(u) - \|x_1\|'(u) > \varepsilon$, where $\|z\|'(u)$ denotes the* **one-sided Gâteaux differential of $\|\cdot\|$ at z in the direction** u, *i.e.,*

$$\|z\|'(u) = \lim_{t\to 0+} \frac{\|z + tu\| - \|z\|}{t}.$$

(iv) *There is $\varepsilon > 0$ such that for every $x \in S_X$ and for every $\delta > 0$ there is a $v \in X$ with $\|v\| \le 2$ such that*

$$\|x + tv\| \ge \|x\| + \varepsilon|t| - \delta \quad \textit{for any } |t| \le 1. \tag{11.2}$$

Remark 279. From the proof below we shall get the following:
(a) If the norm $\|\cdot\|$ is ε-rough, then (ii) holds for this ε, and conversely.
(b) Let ε be as in (ii), and choose any $0 < \varepsilon' < \varepsilon$. Then (iii) concludes the existence, for every $x \in S_X$ and $\delta > 0$, of $x_i \in X$ with $\|x - x_i\| < \delta$, $i = 1, 2$, and $u \in S_X$ such that $\|x_2\|'(u) - \|x_1\|'(u) > \varepsilon'$.
(c) If $\varepsilon' > 0$ satisfies that for every $x \in S_X$ and every $\delta > 0$ there are $x_i \in X$ with $\|x - x_i\| < \delta$ for $i = 1, 2$, and $u \in S_X$ such that $\|x_2\|'(u) - \|x_1\|'(u) \ge \varepsilon'$, then we get $v \in 2B_X$ such that $\|x + tv\| \ge \|x\| + (\varepsilon'/2)|t| - \delta$ for any t such that $|t| \le 1$.
(d) If (iv) holds for some $\varepsilon > 0$, then the norm $\|\cdot\|$ is ε-rough. ®

Proof of Theorem 278. The equivalence of (i) and (ii) is proved in [FHHMZ11, Proposition 8.4].

(ii)⇒(iii): Let ε be given by (ii). Put $l := \limsup \|f_n - g_n\|$ $(> \varepsilon)$, and let $\eta_0 = (1/2)(l - \varepsilon)$. Fix $x \in S_X$ and $\delta > 0$. There are sequences $\{f_n\}$ and $\{g_n\}$ in S_{X^*} such that $f_n(x) \to 1$, $g_n(x) \to 1$, and $u_n \in S_X$ with $(f_n - g_n)(u_n) > l - \eta_0$ for infinitely many $n \in \mathbb{N}$. So, if $0 < \eta < \min\{\delta, \eta_0\}$ is an arbitrary number, there exists $n = n(x, \eta) \in \mathbb{N}$ such that we have $f_n(x) > 1 - \frac{\eta^2}{4}$, $g_n(x) > 1 - \frac{\eta^2}{4}$, and $(f_n - g_n)(u_n) > l - \mu_0$. Thus, for $t > 0$,

$$\|x + tu_n\| \geq f_n(x + tu_n) > 1 - \frac{\eta^2}{4} + tf_n(u_n) = \|x\| - \frac{\eta^2}{4} + tf_n(u_n),$$

hence

$$\frac{\|x + tu_n\| - \|x\|}{t} > f_n(u_n) - \frac{\eta^2}{4t}.$$

Looking at Figure 11.8 we easily get

$$\|x + tu_n\|'(u_n) \geq \frac{\|x + tu_n\| - \|x\|}{t} \quad \left(> f_n(u_n) - \frac{\eta^2}{4t}\right) \tag{11.3}$$

for all $t > 0$. Similarly, for $t > 0$,

$$\|x - tu_n\| \geq g_n(x - tu_n) > 1 - \frac{\eta^2}{4} - tg_n(u_n) = \|x\| - \frac{\eta^2}{4} - tg_n(u_n),$$

hence

$$\|x\| - \|x - tu_n\| < \frac{\eta^2}{4} + tg_n(u_n).$$

Observe again Figure 11.8 to notice that

$$\|x - tu_n\|'(u_n) \leq \frac{\|x\| - \|x - tu_n\|}{t} \quad \left(< g_n(u_n) + \frac{\eta^2}{4t}\right),$$

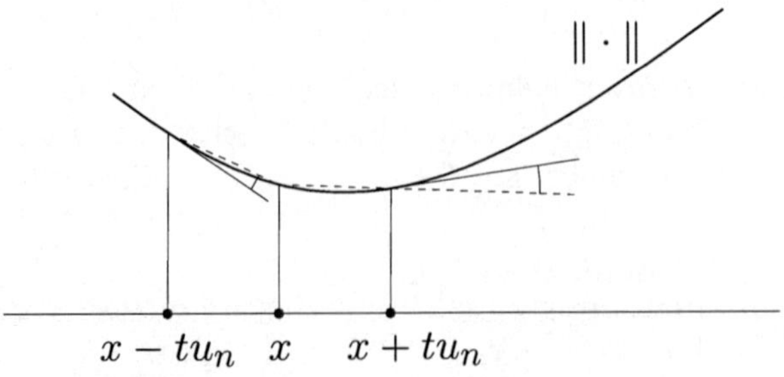

Figure 11.8: One-side tangents (in full) and secants (dashed) to the graph of $\|\cdot\|$

so

$$-\|x - tu_n\|'(u_n) > -g_n(u_n) - \frac{\eta^2}{4t}. \tag{11.4}$$

Then, choosing $t \in (0, \eta)$ such that $\frac{\eta^2}{2t} < \eta$ $(< \eta_0)$, we finally have, by adding (11.3) and (11.4),

$$\|x + tu_n\|'(u_n) - \|x - tu_n\|'(u_n) > (f_n - g_n)(u_n) - \frac{\eta^2}{2t} > l - \eta_0 - \eta_0 = \varepsilon, \tag{11.5}$$

so (iii) holds for $x_1 := x - tu_n$, $x_2 := x + tu_n$, and $u := u_n$.
(iii)⇒(iv):

➪ The idea is the following: Given $x \in S_X$ where $\|\cdot\|$ is not Fréchet differentiable, we should find a direction v such that the graph of the mapping $t \mapsto \|x + tv\|$ will look almost like a symmetric upward cone with vertex at 0 and a generatrix of slope ε (see Figure 11.9). We know that direction u produces at x_1 and x_2 slopes d_1 and d_2, respectively, when moving in the positive sense ($t > 0$). Then we perturb u by using x and the average $m := (1/2)(d_1 + d_2)$ to get $v := u - mx$. In the picture, $d_1 < 0$, $d_2 > 0$, and $m < 0$. In order to simplify the figure, we only depicted $x + tv$ for $t > 0$ (so $s := 1 - tm \in (1, 2)$). In bold, we represented the slopes d_1, d_2, and the resulting cone in the modified direction v. Let us write down the details:

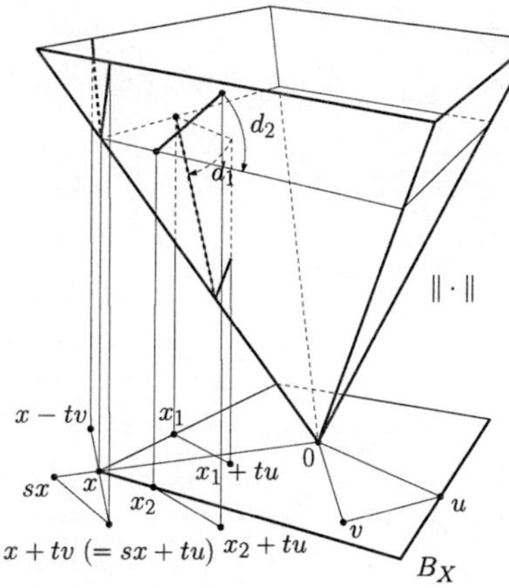

Figure 11.9: Looking for a direction v for the graph of $\|\cdot\|$ at x being (almost) an upward symmetric cone

Find $\varepsilon > 0$ according to (iii). Thus, given $x \in S_X$ and $\delta > 0$, we can find $x_1, x_2 \in X$ such that $\|x - x_i\| < \delta/4$ for $i = 1, 2$, and $u \in S_X$ such that, if $d_i = \|x_i\|'(u)$ for $i = 1, 2$, then $d_2 - d_1 \geq \varepsilon$. Note that $|d_i| \leq 1$ for $i = 1, 2$. Put $m = (d_1 + d_2)/2$.

We still have $|m| \le 1$. Let $v = u - mx$. Observe that $\|v\| \le 2$. Fix $t \in (0,1)$, and let $s = 1 - tm$; note that $0 < s < 2$. Let us compute (11.2):

$$\begin{aligned}
\|x+tv\| &= \|sx+tu\| = s\left\|x+\frac{t}{s}u\right\| \\
&= s\left\|\left(x_2+\frac{t}{s}u\right)+(x-x_2)\right\| \ge s\left(\left\|x_2+\frac{t}{s}u\right\| - \|x-x_2\|\right) \\
&= s\left(\left\|x_2+\frac{t}{s}u\right\| - \|x_2\|\right) + s\|x_2\| - s\|x-x_2\| \\
&= t\frac{\|x_2+(t/s)u\|-\|x_2\|}{t/s} + s\|x\| + s(\|x_2\|-\|x\|) - s\|x-x_2\| \\
&\ge td_2 + s\|x\| + s(\|x_2\|-\|x\|) - s\|x-x_2\| \\
&= td_2 + \|x\| - tm\|x\| + s(\|x_2\|-\|x\|) - s\|x-x_2\| \\
&= \|x\| + t(d_2-m) + s(\|x_2\|-\|x\|) - s\|x-x_2\| \\
&> \|x\| + t\frac{d_2-d_1}{2} - s\frac{\delta}{4} - s\frac{\delta}{4} > \|x\| + t\frac{\varepsilon}{2} - \delta.
\end{aligned} \tag{11.6}$$

If now $t \in (-1,0)$, we still have $0 < s < 2$. Use x_1 instead of x_2 in (11.6) to get, similarly,

$$\begin{aligned}
\|x+tv\| &= \|sx+tu\| = s\left\|x+\frac{t}{s}u\right\| \\
&= s\left\|\left(x_1+\frac{t}{s}u\right)+(x-x_1)\right\| \ge s\left(\left\|x_1+\frac{t}{s}u\right\| - \|x-x_1\|\right) \\
&= s\left(\left\|x_1+\frac{t}{s}u\right\| - \|x_1\|\right) + s\|x_1\| - s\|x-x_1\| \\
&= t\frac{\|x_1+(t/s)u\|-\|x_1\|}{t/s} + s\|x\| + s(\|x_1\|-\|x\|) - s\|x-x_1\| \\
&\ge td_1 + s\|x\| + s(\|x_1\|-\|x\|) - s\|x-x_1\| \\
&= td_1 + \|x\| - tm\|x\| + s(\|x_1\|-\|x\|) - s\|x-x_1\| \\
&= \|x\| + t(d_1-m) + s(\|x_1\|-\|x\|) - s\|x-x_1\| \\
&> \|x\| + t\frac{d_1-d_2}{2} - s\frac{\delta}{4} - s\frac{\delta}{4} > \|x\| + |t|\frac{\varepsilon}{2} - \delta.
\end{aligned} \tag{11.7}$$

Altogether, (11.6) and (11.7) give the conclusion for all $t \in (-1,1)$ and for the slope $\varepsilon/2$.

(iv)⇒(i): According to (iv), get $\varepsilon > 0$ such that for any $x \in S_X$ and for any $n \in \mathbb{N}$ there is $v_n \in 2B_X$ with

$$\|x+tv_n\| \ge \|x\| + \varepsilon|t| - n^{-2} \quad \text{for every} \quad |t| \le 1.$$

So, putting consecutively $t = n^{-1}$ and $t = -n^{-1}$ and adding the results, we have

$$\frac{\|x + n^{-1}v_n\| + \|x - n^{-1}v_n\| - 2}{n^{-1}} \geq \varepsilon - 2n^{-1} \quad \text{for all } n.$$

This shows that $\limsup_{h\to 0} \frac{\|x+h\|+\|x-h\|-2\|x\|}{\|h\|} \geq \varepsilon$. This happens for all $x \in S_X$, so the norm $\|\cdot\|$ is ε-rough, and (i) holds. □

Some examples of rough norms

1. First we present an instructive easy result (see, e.g., [MoZiZi15, Lemma 1044]). Note that the existence of an equivalent rough norm is a consequence of Theorem 284 below. Here we shall see that the supremum norm is such; in fact it has the worst possible roughness. Our second example —the space $\ell_1(I)$— shares the same property.

   Proposition 280. *The standard supremum norm $\|\cdot\|_\infty$ on the space $C[0,1]$ is 2-rough.*

   Proof. We provide two proofs of this result.

   The first one is purely analytical: Let f be an element of the unit sphere of $C[0,1]$. Fix $\lambda > 0$. Pick two different points t and s in $[0,1]$ such that $f(s) > 1 - \lambda^2$ and $f(t) > 1 - \lambda^2$. Pick a function g in the unit sphere of $C[0,1]$ such that $g(s) = 1$ and $g(t) = -1$. Then we have

$$\begin{aligned}\|f + \lambda g\|_\infty + \|f - \lambda g\|_\infty &\geq (f + \lambda g)(s) + (f - \lambda g)(t)\\ &= f(s) + f(t) + \lambda(g(s) - g(t))\\ &\geq 2 - 2\lambda^2 + 2\lambda.\end{aligned}$$

   This shows that

$$\frac{\|f + \lambda g\|_\infty + \|f - \lambda g\|_\infty - 2}{\lambda} \geq 2 - 2\lambda.$$

   This holds for all $\lambda > 0$ so we get, for all f in the unit sphere of $C[0,1]$,

$$\limsup_{\|h\|_\infty \to 0} \frac{\|f + h\|_\infty + \|f - h\|_\infty - 2}{\|h\|_\infty} \geq 2.$$

   Hence the norm $\|\cdot\|_\infty$ is 2-rough.

   The second one is more geometrical: Let $f \in S_{C[0,1]}$, and let $\alpha \in (0,1)$. Since $f(s) = 1$ for some $s \in [0,1]$ and f is continuous, we can find $t \in [0,1]\backslash\{s\}$ such that $f(t) > 1 - \alpha$. This shows that δ_t and δ_s are both in $S(B_{C[0,1]^*}, f, \alpha)$, where $\delta_r \in S_{C[0,1]^*}$ denotes the Dirac delta associated to $r \in [0,1]$. Since $\|\delta_t - \delta_s\|_\infty^* = 2$ for all $t \neq s$, we are done. □

2.

Proposition 281. *The norm* $\|\cdot\|_1$ *on* $\ell_1(I)$ *is 2-rough whenever* I *is an infinite set.*

Proof. Let $x \in S_X$. Find $u \in S_{\ell_\infty(I)}$ such that $\langle x, u\rangle = 1$. Given $\alpha > 0$, find an index $i_0 \in I$ such that $|x(i_0)| < \alpha$ and for $k = 1, 2$ define $u_k \in S_{\ell_\infty(I)}$ by

$$u_k(i) = \begin{cases} (-1)^k & \text{if } i = i_0, \\ u(i) & \text{otherwise.} \end{cases}$$

Then $\|u_1 - u_2\|_\infty = 2$, and $u_k \in S(B_{\ell_\infty(I)}, x, \alpha)$ for $k = 1, 2$. □

The Kurzweil, Leach, and Whitfield ingenious proof of the following Lemma is extremely important. It came before Ekeland et alter principles and will be used in this text several times in different situations:

Lemma 282 (Kurzweil, Leach, Whitfield, [Kurz54], [LeaWh72]). *Assume that a Banach space admits a rough norm* $\|\cdot\|$. *If* f *is a real-valued Fréchet differentiable function on* X *and* $f(0) = 0$, *then there exists* $x \in X$, *such that* $1 \le \|x\| < 2$ *and* $f(x) \le \|x\|$.

Proof. Assume that the norm $\|\cdot\|$ is ε-rough, for some $\varepsilon > 0$. Fix $\varepsilon' \in (0, \varepsilon)$. We choose a sequence $\{x_n\}_{n\in\mathbb{N}}$ by induction to satisfy $x_0 = 0$ and, for $n \in \mathbb{N}$,
(1) $f(x_n) < \|x_n\|$,
(2) $\|x_{n+1} - x_n\| \le 1$,
(3) $\|x_{n+1}\| \ge \|x_n\| + (\frac{\varepsilon'}{8}) \cdot \|x_{n+1} - x_n\|$,
(4) $\|x_{n+1} - x_n\| \ge \frac{1}{2} \cdot M_n$ (> 0), where

$$M_n = \sup\{\|y - x_n\| : \ y \in X \text{ and } y = x_{n+1} \text{ satisfies (1) for } n+1, \text{(2), and (3)}\}.$$

➡ We may provide some details: Assume for a moment that $f(x) \ge \|x\|$ for all $x \in B_X$. Since $\|\cdot\|$ is not Fréchet differentiable at 0, we readily get a contradiction with the Fréchet differentiability of f (see the argument in the annotation after Theorem 260). It follows that there exists $x \in B_X$ such that $f(x) < \|x\|$ (in particular, $x \ne 0$). Obviously, (3) holds for this x. Define M_0 as the supremum of $\|x\|$ for all those x's, and choose then x_1 one of them with $\|x_1\| \ge (1/2)M_0$.
Assume now that x_k has been defined for $k = 1, 2, \ldots, n$ in such a way that $f(x_k) < \|x_k\|$ for $k = 1, 2, \ldots, n$, and $\|x_{k+1} - x_k\| \le 1$, $\|x_{k+1}\| \ge \|x_k\| + (\varepsilon'/8)\|x_{k+1} - x_k\|$, and $\|x_{k+1} - x_k\| \ge (1/2)M_k$, for $k = 0, 1, 2, \ldots, n-1$, and for M_k as above.
If for some $\delta > 0$ we have

$$\frac{\|x_n + h\| - \|x_n\|}{\|h\|} < \frac{\varepsilon'}{8}, \text{ for all } h \in X,\ 0 < \|h\| < \delta,$$

we readily get a contradiction with the fact that $\|\cdot\|$ is also ε'-rough at x_n. This shows that there exists a sequence $\{y_m\}_{m\in\mathbb{N}}$ in $B(x_n,1)$ such that $y_m \to x_n$, $y_m \neq x_n$ for each $m \in \mathbb{N}$, and $\|y_m\| > \|x_n\| - (\varepsilon'/8)\|y_m - x_n\|$ for all $m \in \mathbb{N}$. Assume that $f(y_m) \geq \|y_m\|$ for all $m \in \mathbb{N}$. Since f is continuous, we get $f(x_n) \geq \|x_n\|$, a contradiction. This shows that we may find $m \in \mathbb{N}$ such that $f(y_m) < \|y_m\|$. Call x_{n+1} this element y_m. This shows that the set

$$S_n := \{y \in X :\ f(y) < \|y\|,\\ \|y - x_n\| \leq 1,\ \|y\| \geq \|x_n\| + (\varepsilon'/8)\|y - x_n\|\} \quad (11.8)$$

is non-empty. Define $M_n := \sup\{\|y - x_n\| :\ y \in S_n\}$ (> 0), and let $x_{n+1} \in S_n$, $\|x_{n+1} - x_n\| \geq (1/2)M_n$. Induction over n gives the sought sequence $\{x_n\}_{n\in\mathbb{N}}$.

In choosing $\{x_n\}_{n\in\mathbb{N}}$ one could always select $x_{n+1} = x_n$ to satisfy (1), (2), and (3). Condition (4) ensures a nontrivial choice.
We will show that, for some $n \in \mathbb{N}$, $\|x_n\| \geq 1$. If not, $\{\|x_n\|\}_{n\in\mathbb{N}}$ is a Cauchy sequence since it is (strictly) increasing. Also, by (3), if $m > n$ then

$$\begin{aligned}\|x_m - x_n\| &\leq \|x_m - x_{m-1}\| + \cdots + \|x_{n+1} - x_n\| \\ &\leq \frac{8}{\varepsilon'} \cdot \Big((\|x_m\| - \|x_{m-1}\|) + \ldots + (\|x_{n+1}\| - \|x_n\|)\Big) \\ &= \frac{8}{\varepsilon'} \cdot (\|x_m\| - \|x_n\|),\end{aligned}$$

so $\{x_n\}_{n\in\mathbb{N}}$ is a Cauchy sequence converging to $z \in X$. Observe that

$$0 < \|x_1\| < \|z\| \leq 1 \text{ and } f(z) \leq \|z\|. \quad (11.9)$$

Since f is Fréchet differentiable at z, there exists $\delta > 0$ such that

$$f(z+u) - f(z) - f'(z)u \leq \varepsilon' \cdot \frac{\|u\|}{8}, \quad (11.10)$$

provided $\|u\| < \delta$. We may also assume that $\delta < 1$ and $\delta \leq 2\|z\|$.
By Theorem 278 above applied to z, and in view of Remark 279, there exists v such that $\|v\| \leq 2$ and

$$\|z + tv\| > \|z\| + \frac{\varepsilon'}{2}|t| - \varepsilon' \cdot \frac{\delta}{8}, \quad (11.11)$$

whenever $|t| \leq 1$. Choosing $t = \pm\frac{\delta}{2}$ in (11.11) depending on the sign of $f'(z)v$, we have

$$\|z + tv\| > \|z\| + \frac{\varepsilon'}{2}\cdot\frac{\delta}{2} - \varepsilon'\frac{\delta}{8} = \|z\| + \varepsilon'\frac{\delta}{8} \geq \|z\| + \varepsilon'\frac{\|tv\|}{8}. \quad (11.12)$$

Combining (11.10), (11.9), and (11.12) in this order, and due to the fact that $f'(z)(tv) \le 0$, we get

$$\begin{aligned} f(z+tv) &\le f(z) + f'(z)(tv) + \varepsilon' \frac{\|tv\|}{8} \\ &\le f(z) + \varepsilon' \frac{\|tv\|}{8} \le \|z\| + \varepsilon \frac{\|tv\|}{8} < \|z+tv\|. \end{aligned} \tag{11.13}$$

From (11.13), the choice of δ, and (11.12), we obtain the following inequalities:
(1') $f(z+tv) < \|z+tv\|$,
(2') $\|z+tv-z\| < 1$,
(3') $\|z+tv\| - \|z\| > \varepsilon' \frac{\|tv\|}{8}$.
Because $\|\cdot\|$ and f are continuous at z, and $x_n \to z$, we see that $z+tv$ is one of the elements in the set S_n defined in (11.8) above for large n, so that $M_n \ge \|z+tv-x_n\|$. Since $\|z+tv-z\| > \varepsilon' \frac{\delta}{8}$ (see the first inequality in (11.12)), we get $M_n > \varepsilon' \cdot \frac{\delta}{8}$ for large n. By condition (4) above, we have $\|x_{n+1} - x_n\| > \varepsilon' \frac{\delta}{16}$ for large n, and this contradicts the convergence of the sequence $\{x_n\}$. This shows that $\|x_n\| \ge 1$ for some $n \in \mathbb{N}$. The smallest such n gives rise to the vector x satisfying the requirements of the lemma. □

The following result gives, in view of the implication (iv)⇒(i) in Theorem 274, a sufficient condition for Asplundness. In particular, it proves (iv)⇒(v) in Theorem 276.

Theorem 283 (Leach, Whitfield, [LeaWh72], see, e.g., [FHHMZ11, Lemma 8.9]). *If a Banach space admits a rough norm $\|\cdot\|$, then there does not exist a Fréchet differentiable bump.*

Proof. It such a function exists, there is a function f satisfying $f(0) = 2$ and $f(x) = 0$ whenever $\|x\| \ge 1$. Let

$$g(x) = 2 - f(x)$$

for $x \in X$ (see Figure 11.10). Then $g(x) - \|x\| > 0$ whenever $1 \le \|x\| < 2$ and we violate Lemma 282 above. □

The question if the reverse implication holds true in a non-separable setting, i.e., if every non-separable Banach space that does not admit a rough norm necessarily admits a Fréchet differentiable bump function, is still an open problem, one of the most important problems in this area of Banach spaces. We see in this text that there are Banach spaces that do not admit rough norms (in other words, Asplund spaces, see Theorem 274) and yet do not admit any Gâteaux differentiable norm [Hay99].

The following is an important result proved topologically. It proves (v)⇒(i) in Theorem 276.

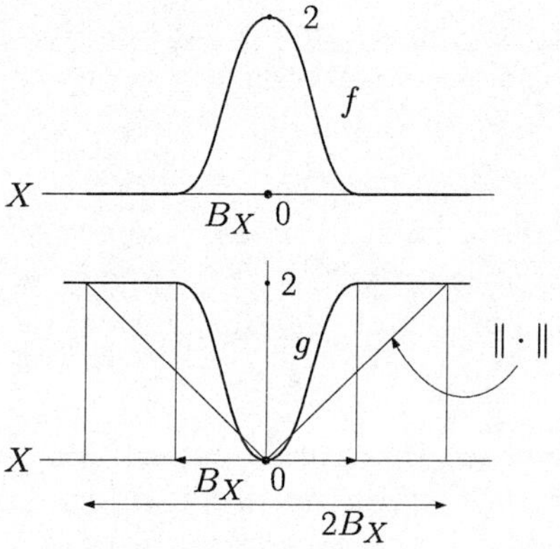

Figure 11.10: The functions f and g in the proof of Theorem 283

Theorem 284 (Leach, Whitfield, [LeaWh72], see, e.g., [FHHMZ11, Theorem 8.10]). *If a Banach space X is separable and X^* is nonseparable, then X admits a rough norm.*

Whitfield's self-contained proof of Theorem 284 is presented in the proof of [FHHMZ11, Theorem 8.10]. It uses the fact that B_{X^*} is nonseparable in norm but weak* separable to produce a lot of condensation points in the weak* topology which, by a convex argument, create the dual ball of a rough norm on X.

A separable-reduction result is the following:

Theorem 285 (Leach, Whitfield, [LeaWh72]). *If a Banach space X admits a rough norm $\|\cdot\|$, then there is a separable subspace of X where restriction of this norm $\|\cdot\|$ is also rough.*

Proof. See [DeGoZi93, Lemma I.5.8]. Compare with Lemma 277 above □

11.7 The use of variational principles

11.7.1 Ekeland's and smooth variational principles

We provide now an alternative proof of Theorem 283 based in Ekeland's Variational Principle, which is stated in the framework of Banach spaces in Theorem 286 below (see also, e.g., [FHHMZ11, Theorem 7.39]). Theorem 287 gives its abstract version in a complete metric space. We shall give two proofs. The first one of a geometric nature, the second an elegant analytic one as given in the proof of [BeLi00, Lemma 4.11]. We notice that an alternative proof of Ekeland's Variational Principle by applications of methods of smooth analysis, namely the Smooth Variational Principle (Theorem 289), is in [DeGoZi93, Chapter I, page 10]. Thus, it can be also seen as a consequence of the magic Baire category theorem.

Theorem 286 (Ekeland Variational Principle, Banach space version). [Ek74] *Let X be a Banach space and let $\varphi : X \to \mathbb{R} \cup \{+\infty\}$ be a lower semicontinuous bounded below function that is not identically equal to $+\infty$. Then given $\varepsilon > 0$ there exists $x_0 \in X$ such that $\varphi(x_0) \leq \varphi(x) + \varepsilon\|x - x_0\|$ for all $x \in X$.*

H It should be pointed out that earlier than [Ek74] was published, J. Daneš in Prague published the equivalent result known as the Drop Theorem (see Theorem 297 below). All those independently obtained results have being preceded by the proof of Bishop and Phelps of the result on the density of the set of all norm-attaining functionals (Theorem 14). See Subsection 11.7.2 below for further information.

The following is the metric-space version of Theorem 286:

Theorem 287 (Ekeland Variational Principle, metric space version). [Ek74] *Let (X, d) be a complete metric space and φ be a lower semicontinuous not always $+\infty$ function on X that is bounded below (by, say, $m_1 \in \mathbb{R}$). Then for every $\varepsilon > 0$ there is $x_0 \in X$ such that*

$$\varphi(x_0) \leq \varphi(x) + \varepsilon d(x, x_0)$$

for all $x \in X$.

In a sense, the idea for the proof of the result is based in the following trivial observation: Let $P := X \times \mathbb{R}$ be endowed with the metric $D((x,a),(y,b)) := \max\{d(x,y), |a-b|\}$ for (x,a), $(y,b) \in P$. Fix $x_0 \in X$, $K > 0$, and $\delta > 0$. Then the set $T := \{(x,t) \in P : \ \delta \geq t \geq K.d(x,x_0)\}$ (see Figure 11.11) has a D-diameter less than $2.\max\{\delta, \delta/K\}$. Indeed, $D((x_0,0),(x,t)) \leq \max\{\delta, \delta/K\}$ for all $(x,t) \in T$.

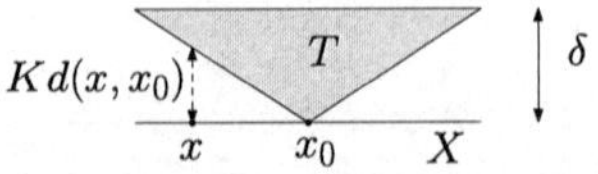

Figure 11.11: A "triangle" with small diameter

First proof of Theorem 287. The procedure is depicted in Figure 11.12. The metric space (P, D) defined above is complete. Let $\operatorname{epi}(\varphi) := \{(x,t) \in X \times \mathbb{R} : \ t \geq \varphi(x)\}$. Given $m \leq M$ in $\mathbb{R}$ and $x \in X$, define the **truncated cone** $K_\varepsilon(x, m, M) = \{(y,t) \in X \times \mathbb{R} : \ m \leq t \leq M - \varepsilon d(y,x)\}$. The previous observation gives $\operatorname{diam} K_\varepsilon(x,m,M) \leq 2\max\{M - m, (M-m)/\varepsilon\}$.

Choose $x_1 \in X$, $\varphi(x_1) < m_1 + 1/2$ and let $K_1 := K_\varepsilon(x_1, m_1, \varphi(x_1)) \cap \operatorname{epi}(\varphi)$. Put $m_2 := \inf\{\varphi(x) : (x,t) \in K_1\}$. Find $x_2 \in K_1$ such that $\varphi(x_2) < m_2 + 1/2^2$ and let $K_2 := K_\varepsilon(x_2, m_2, \varphi(x_2)) \cap \operatorname{epi}(\varphi)$. Continue this way to get a sequence $\{x_n\}$ in X. Note that $(x_n, \varphi(x_n)) \in K_n$, K_n is closed, $K_{n+1} \subset K_n$, and $\operatorname{diam} K_n \leq 2\max\{2^{-n}, 2^{-n}/\varepsilon\}$ for all $n \in \mathbb{N}$. Thus, the sequence $\{(x_n, \varphi(x_n))\}$ is Cauchy, hence convergent, say to (x_0, l_0). It is obvious that $\varphi(x_0) = l_0$ and $K(x_0, r, \varphi(x_0)) \cap \operatorname{epi}(\varphi) = \{(x_0, \varphi(x_0))\}$ for all $r \leq \varphi(x_0)$. The conclusion follows. □

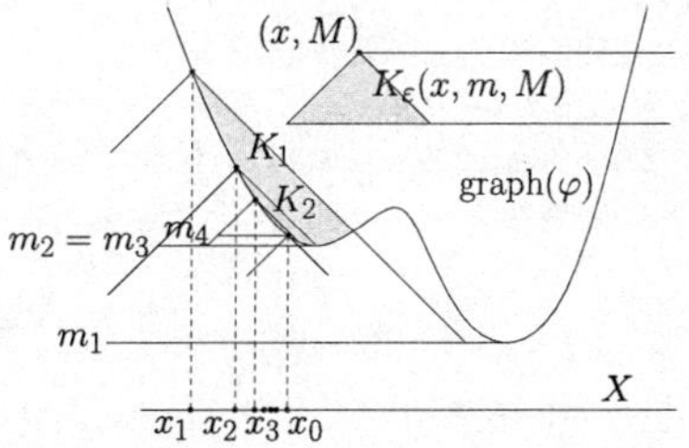

Figure 11.12: The construction in the proof of Theorem 287. The diameter of the shaded regions goes to 0.

Second proof of Theorem 287. (Benyamini, Lindenstrauss [BeLi00, proof of Lemma 4.11]; the proof there was done only for continuous functions, although it can be easily adapted to lower semicontinuous functions, as we do here). Starting with any $x_1 \in X$, define inductively a sequence $\{x_n\}$ in X. If x_n has been selected, take x_{n+1} such that

$$\varphi(x_{n+1}) + \varepsilon d(x_n, x_{n+1}) \leq \inf\{\varphi(x) + \varepsilon d(x_n, x) : \ x \in X\} + 2^{-n}. \qquad (11.14)$$

(The right-hand side of (11.14) is finite because φ is bounded below, see Figure 11.13.)
The sequence $\{\varphi(x_n)\}_{n\in\mathbb{N}}$ is **almost nonincreasing** in the sense that

$$\varphi(x_{n+1}) \leq \varphi(x_{n+1}) + \varepsilon d(x_n, x_{n+1}) \leq \varphi(x_n) + 2^{-n}.$$

Hence $\{\varphi(x_n)\}_{n\in\mathbb{N}}$ is convergent, say to L. Summing up the inequalities

$$\varepsilon d(x_n, x_{n+1}) \leq \varphi(x_n) - \varphi(x_{n+1}) + 2^{-n}$$

over n, we find from the convergence of $\{\varphi(x_n)\}_{n\in\mathbb{N}}$ that $\sum d(x_n, x_{n+1})$ converges and hence that $\{x_n\}_{n\in\mathbb{N}}$ is Cauchy. The required point x_0 is the limit of $\{x_n\}_{n\in\mathbb{N}}$. Indeed, taking $x = x_0$ and then limits over n in (11.14) we get $L \leq \varphi(x_0)$. Due to the fact that φ is lower semicontinuous, we get $L = \varphi(x_0)$. Fixing now an arbitrary x and letting $n \to \infty$ in (11.14) we get the sought inequality. □

An alternative proof of Theorem 283 *by using Theorem* 286 (see [DeGoZi93, Lemma II.5.4]). Let $\|\cdot\|$ be an ε-rough norm on X. Assume that X admits a Fréchet

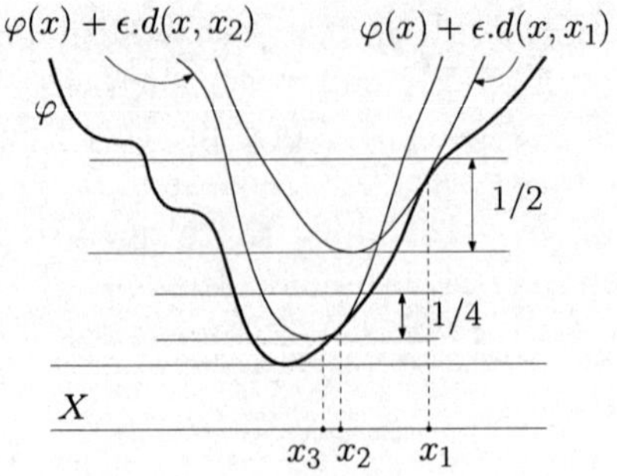

Figure 11.13: Searching for the sequence $\{x_n\}$ (Benyamini–Lindenstrauss proof of Theorem 287)

differentiable function g with bounded nonempty support. Let $f : X \to \mathbb{R} \cup \{+\infty\}$ be defined by

$$f = \begin{cases} g(x)^{-2} - \|x\| & \text{if } g(x) \neq 0; \\ +\infty & \text{otherwise.} \end{cases}$$

By Ekeland's Theorem 286, there is $x_0 \in X$ such that

$$f(x_0 + h) \geq f(x_0) - \frac{\varepsilon}{4}\|h\|, \qquad \text{for every } h \in X.$$

Write down this inequality for h and for $-h$ and add them to obtain

$$g(x_0 + h)^{-2} + g(x_0 - h)^{-2} - \|x_0 + h\| - \|x_0 - h\| \geq 2g(x_0)^{-2} - 2\|x_0\| - \frac{\varepsilon}{2}\|h\|.$$

Hence

$$\begin{aligned} &\limsup_{\|h\|\to 0} \frac{g(x_0 + h)^{-2} + g(x_0 - h)^{-2} - 2g(x_0)^{-2}}{\|h\|} \\ &\qquad \geq \limsup_{\|h\|\to 0} \frac{\|x_0 + h\| + \|x_0 - h\| - 2\|x_0\|}{\|h\|} - \frac{\varepsilon}{2} \geq \frac{\varepsilon}{2} \end{aligned}$$

(see equation (4.25)). This contradicts the Fréchet differentiability of g at x_0. □

From Proposition 280 and Theorem 284 or, alternatively, from Theorems 283 and 284, it follows:

Theorem 288 (Kurzweil, [Kurz54]). *The space $C[0, 1]$ does not admit any Fréchet smooth bump.*

H Theorem 288 was at its time a groundbreaking result of Kurzweil in [Kurz54], before any results in the vein of the Bishop–Phelps theorem. Kurzweil's proof consists of a clever comparison of the growth of a rough norm on $C[0,1]$ with the growth of a possible Fréchet differentiable bump. We recommend the reader to study Kurzweil's proof.

Here is the statement of the crucial Smooth Variational Principle:

Theorem 289 (Deville, Godefroy, Zizler). ([DeGoZi93b], Smooth Variational Principle, see also, e.g., [DeGoZi93, Theorem I.2.3] or [FHHMZ11, Theorem 7.43]) *Let X be a Banach space that admits a Lipschitzian function with bounded nonempty support that is Fréchet (Gâteaux) differentiable. Then for every lower semicontinuous and bounded below function f on X and every $\varepsilon > 0$ there exists a function g which is Lipschitzian and Fréchet (respectively, Gâteaux) differentiable on X and such that $\|g\|_\infty \le \varepsilon$, $\|g'\|_\infty \le \varepsilon$ and $f - g$ attains its minimum on X.*

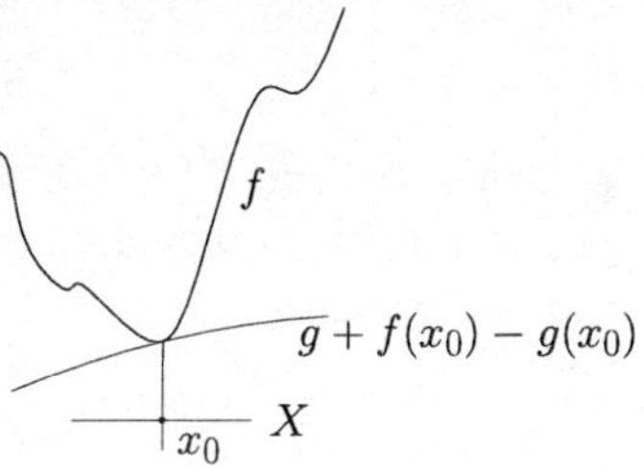

Figure 11.14: The Smooth Variational Principle (Theorem 289)

Corollary 290. *If X satisfies the assumptions of Theorem 289, then every continuous convex function on X is Fréchet (Gâteaux) differentiable at a dense set of points.*

For a proof of Corollary 290 see, e.g., [FHHMZ11, Corollary 7.44].

Remark 291. Note that Corollary 290 gives Theorem 272, as well as (iii) in Theorem 275. ®

Proof of Theorem 289 (The main idea and the Fréchet differentiability case: the Gâteaux case is similar; for details, see, e.g., [FHHMZ11, Theorem 7.43]). Let Y be the Banach space of all Lipschitz and Fréchet differentiable functions on X normed by $|||h||| = \|h\|_\infty + \|h'\|_\infty$. For a given lower semicontinuous and bounded below function f defined on X, consider the set

$$U_n := \Big\{ g \in Y : \quad \text{there is } x_0 \in X \text{ such that} \\ (f-g)(x_0) < (f-g)(x) \quad \text{for all } x \in X \setminus \operatorname{Int} B(x_0, \tfrac{1}{n}) \Big\}.$$

Then U_n is an open dense subset of Y and thus $G = \bigcap_{n=1}^{\infty} U_n$ is a dense G_δ subset of Y. If $g \in G$, then $(f - g)$ attains its minimum on X.

➪ The fact that $f - g$ attains its minimum on X at $x_0 \in X$ reads: $(f - g)(x) \geq (f - g)(x_0)$ for all $x \in X$, i.e., $f(x) \geq g(x) + f(x_0) - g(x_0)$ (see Figure 11.14).

□

H Theorem 289 had a predecessor in the ground-breaking result of Borwein and Preiss in [BoPpr87], nowadays called the **Borwein–Preiss Variational Principle**, where it was assumed that X admits a Fréchet differentiable norm. Many mathematicians tried to extend its constructive proof to the situation of bump functions, including the authors of [DeGoZi93]. The final success, thus practically custom-built for [DeGoZi93], was achieved by R. Deville. His result and its proof (a mathematical delight) appears in many references (see, e.g., those quoted in its statement). Again it was achieved by using the Baire category theorem —the "mighty good witch", as some Czech mathematicians use to say.

On the other hand, if X is separable and X^* is not separable, then any equivalent norm on X shares some portion of the rough property of Whitfield's rough norm (see Theorem 284) in the following sense.

Theorem 292 (Godefroy, Zizler, [GoZi91]). *If X is separable and X^* is not separable, then for every $\varepsilon > 0$ and for every equivalent norm $\|\cdot\|$ on X the following holds: There is a convergent sequence $\{x_n\} \subset S_X$ formed by points of Gâteaux differentiability of the norm $\|\cdot\|$ such that $\| \|x_n\|' - \|x_k\|' \| \geq 1 - \varepsilon$, where $\|x_n\|'$ denotes the Gâteaux derivative of the norm at the point x_n. In particular, the norm $\|\cdot\|$ is not Fréchet differentiable at the point* $\lim x_n$.

Theorem 293 (Fabian, Hájek, Vanderwerff, [FHV96]). *Let a Banach space X admit an equivalent Fréchet differentiable norm.*
(i) *If f is a lower semicontinuous bounded below function on X, then there is a Fréchet differentiable function φ on X such that $\varphi(x) \to -\infty$ as $\|x\| \to \infty$, and*

$$f - \varphi \text{ attains its minimum on } X.$$

(ii) *There is an equivalent Fréchet differentiable norm $|||\cdot|||$ on X such that $|||\cdot|||$ attains its minimum on the original sphere of X.*

The following is called the **Fabian–Stegall variational principle** (see [FHHMZ11, Theorem 11.6]):

Theorem 294 (Fabian [Fab83], Stegall [St78b]). *Assume that X is a reflexive Banach space and that f is a lower semicontinuous function on X such that there* *are constants $a > 0$ and $b \in \mathbb{R}$ with*

$$f(x) > a\|x\| + b \text{ for } x \in X.$$

Then for every $\varepsilon > 0$ there is $x^* \in X^*$ such that $\|x^*\| < \varepsilon$ and

$$f - x^* \text{ attains its minimum on } X.$$

For some applications see, e.g., [FHHMZ11, Exercise 7.66].

Note that the Borwein–Preiss principle [BoPpr87] creates a smooth convex variational function, while the function in Theorem 289 is not convex in general. However its use for convex problems remain powerful. This is due to the fact that if f is, say, a convex continuous function and if its graph is touched from above and a differentiable function g is such that $f \leq g$ and $f(x_0) = g(x_0)$, then f is differentiable at x_0; this is a consequence of the following elementary fact:

$$0 \leq f(x+h) + f(x-h) - 2f(x_0) \leq g(x_0+h) + g(x-h) - 2g(x_0) \to 0.$$

The following result gives that, in some cases, it is possible to get more than the mere density of the set of all Gâteaux smooth points, and it is inspired in [Klee59]:

Theorem 295 (J. Rychtář, [Rych05b]). *Let X be WCG, K be a symmetric weakly compact subset that is linearly dense in X, $x_0 \in X$ and f a continuous convex function on X. Then there is $x \in (x_0 + K)$ such that f is Gâteaux differentiable at x.*

11.7.2 Other geometric statements equivalent to the Ekeland variational principle

There is a family of results coming from different contexts (optimization, mathematical economy, fixed point theory, Banach space structure) that share a common flavour. Not by chance it happens that all of them are mutually equivalent, although they have been mostly formulated and proved independently. The root can be traced back to the method of proof in the original version of the Bishop–Phelps Theorem 14 [BiPh61]. A glance at the geometric and analytic procedure described there will illustrate the common background to all the statements. Late in the same year, Phelps isolated from it the following result, that has been known as "Phelps' Lemma":

Lemma 296 (Phelps, [Ph63], [Ph74]). (See Figure 11.16) *Let X be a Banach space. Let $A \subset X$ be a nonempty closed set and $B \subset X$ a nonempty closed convex and bounded set such that $0 \notin B$. Let K be the cone defined by B, i.e., $K := \{\alpha b : \alpha \geq 0,\ b \in B\}$. Then, given $x_0 \in A$ such that $A \cap (x_0 + K)$ is bounded, there exists $x \in A \cap (x_0 + K)$ such that $A \cap (x + K) = \{x\}$.*

The list below is a partial account of some equivalent results. Some of them have been presented earlier in this text.
Daneš result was called, by obvious reasons, the "drop theorem" (see Figure 11.15. Here, the **drop** $D(B, x)$ defined by a convex set B of X and a point $x \in X$ is the set $\operatorname{conv}(B \cup \{x\})$):

Theorem 297 (Daneš, [Dan85]). (See Figure 11.15.) *Let X be a Banach space, S a nonempty closed subset of X, x_0 a point in $X \setminus S$, $\varepsilon > 0$, and $0 < r < R :=$* $\operatorname{dist}(x_0, S)$. *Then there exists a point $s \in S$ such that*

$$\|s - x_0\| < R + \varepsilon, \text{ and } D(B(x_0, r), s) \cap S = \{s\}.$$

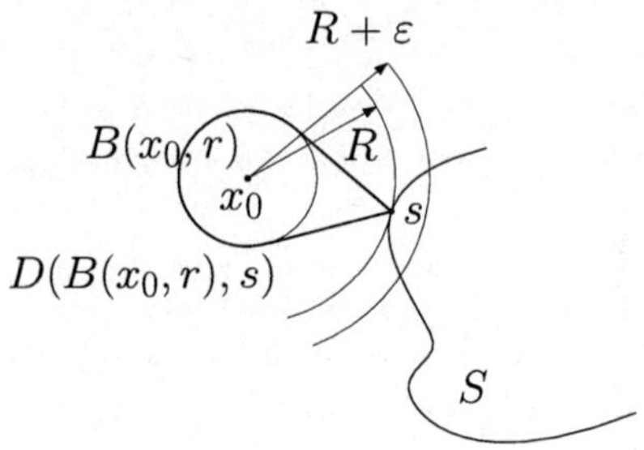

Figure 11.15: The Daneš' Drop Theorem: The drop $D(B(x_0, r), s)$ (in bold) is defined by $B(x_0, r)$ and a point $s \in S$

Observe that in case B is moreover bounded, the drop $D(B, x)$ is closed.
In [Dan85] it was proved the equivalence of several statements that have been spread in the literature. We collect those here:

(D) (The Daneš' Drop Theorem): Theorem 297 above.

(GD) (The Generalized Daneš' Drop Theorem): (See Figure 11.17.) Let X be a Banach space, S a nonempty closed subset of X, C a closed bounded convex subset of X with $\operatorname{dist}(S, C) > 0$, and s_0 a point in S. Then there exists a point $s \in S \cap D(C, s_0)$ such that

$$D(C, s) \cap S = \{s\}.$$

(KZD) (The Krasnoselskiĭ–Zabreĭko Theorem): (See Figure 11.18.) Let X be a Banach space, S a closed set in X and z_0 a point in $X \setminus S$. Let $0 < \delta < 1$. Then there exists an equivalent norm $|||\cdot|||$ on X such that the $|||\cdot|||$-distance of z_0 from S is attained at a point of S and at most at two points of S (int his case, the sum of these two points equals $2z_0$), and such that $(1 - \delta)\|\cdot\| \le |||\cdot||| \le \|\cdot\|$.

(B) (The Brézis–Browder Theorem): (See Figure 11.19.) Let S be a closed subset of a Banach space X and $z_0 \in X \setminus S$. Let $0 < r < \operatorname{dist}(z_0, S)$ and $s_0 \in S$. Then there exists a point $s \in S \cap D(B(z_0, r), s_0)$ such that

$$S \cap (s + K) \cap B(s, \delta) = \{s\}$$

for all $\delta < \operatorname{dist}(z_0, S) - r$, where $K := \bigcup_{\lambda \geq 0} \lambda D(B(z_0 - s_0), r)$.

(B') (The alternative Brézis–Browder Theorem): (See Figure 11.19.) Under the hypotheses of Theorem (B), there exists a point $s \in S \cap D(B(z_0, r), s_0)$ such that

$$S \cap (s + K) \cap B(s, \operatorname{dist}(z_0, S) - r) = \{s\}$$

(E) (The Ekeland Variational Principle): Theorem 287 above.

(C) (The Caristi–Kirk Theorem) Let (X, d) be a complete metric space, f a real-valued bounded-below lower semicontinuous function on X and $T : X \to X$ be a mapping such that $\operatorname{dist}(x, Tx) \leq f(x) - f(Tx)$ for all $x \in X$. Then T has a fixed point.

J.P. Penot [Pen86] defined, in a metric space (M, d), and for $\gamma > 0$, the **petal** associated to $\gamma > 0$ and points a and b in M as $P_\gamma(a, b) := \{x \in M : \ \gamma d(a, x) + d(x, b) \leq d(a, b)\}$ (see Figure 11.20). Analogously, for a function $f : M \to \mathbb{R}$, the set $P_\gamma(a, f) := \{x \in M : \ \gamma d(a, x) + f(x) \leq f(a)\}$ is called the **petal** associated to γ and f.

(P) (The Penot Petal Theorem) (See Figure 11.20.) Let A be a closed subset of a complete metric space (M, d). Let $x_0 \in A$ and $b \in X \setminus A$, $r \leq \operatorname{dist}(b, A)$, $s := \operatorname{dist}(b, x_0)$. Then, for every $\gamma > 0$ there exists $a \in A \cap P_\gamma(x_0, b)$ (so that, in particular, $\operatorname{dist}(a, x_0) \leq \gamma^{-1}(s - r)$) such that $P_\gamma(a, b) \cap A = \{a\}$.

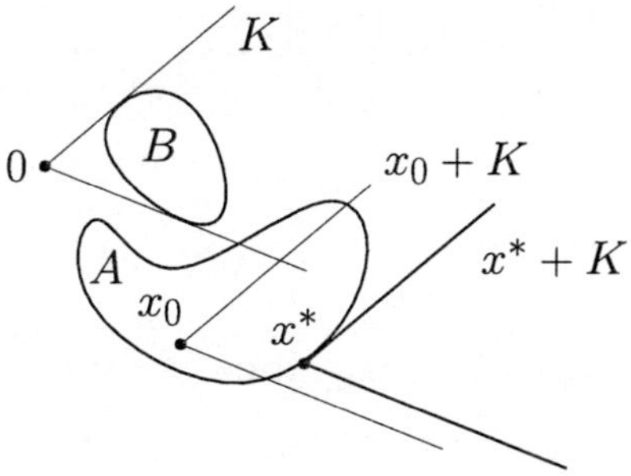

Figure 11.16: Phelps' Lemma

Based on Daneš Drop Theorem, Rolewicz introduced in [Rol87] the following concept: The norm $\|\cdot\|$ of a Banach space X has the **drop property** if given any nonempty set S such that $S \cap B_X = \emptyset$, then there exists $s \in S$ such that $D(s, B_X) \cap S = \{s\}$. He proved that *if X is a uniformly convex Banach space then its norm has the drop property*, and also that *every space whose norm has*

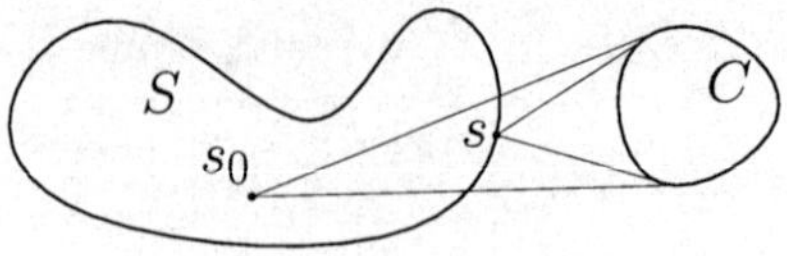

Figure 11.17: The Generalized Daneš' Drop Theorem

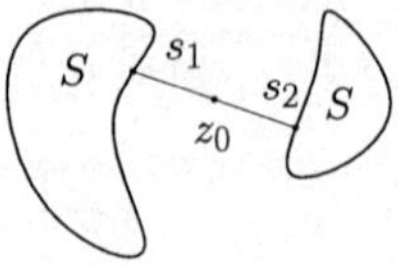

Figure 11.18: The Krasnoselskiĭ–Zabreĭko Theorem

the drop property is reflexive. The following result characterizes norms in Banach spaces with the drop property. The norm $\|\cdot\|$ has **property** (α) if given $x^* \in S_{X^*}$, the Kuratowski index of noncompactness of the slices $S(B_X, x^*, 1-\delta)$ tends to 0 whenever $\delta \to 0$. The **Kuratowski index of noncompactness** of a set A in a Banach space is the infimum of all $r > 0$ such that A can be covered by a finite number of sets with diameter less than r.

Theorem 298 (Montesinos, [Mo87]). *Let $(X, \|\cdot\|)$ be a Banach space. The following statements are equivalent:*
(i) *The norm $\|\cdot\|$ has the drop property.*
(ii) *The norm $\|\cdot\|$ has property α.*
(iii) *The norm $\|\cdot\|$ has property w-$\|\cdot\|$-SKK and the space is reflexive.*
As a consequence, X is reflexive if, and only if, there is an equivalent norm on X with the drop property.

The consequence follows from Troyanski's renorming Theorem 218.

11.7.3 The compact variational principle and related results

Extreme points of convex subsets of a Banach space were introduced in Section 2.2. The following is an extension of this concept (see, e.g., [DuSch, Definition V.8.1]):

Definition 299. *Let S be a nonempty subset of a Banach space X. A nonempty subset E of S is said to be an* **extreme subset of** S *if a proper convex combination $\lambda s_1 + (1-\lambda)s_2$ $(0 < \lambda < 1)$ of two points $s_1, s_2 \in S$ belongs to E only if both s_1 and s_2 are in E. An extreme subset of S consisting of just one point is called an* **extreme point of S**.

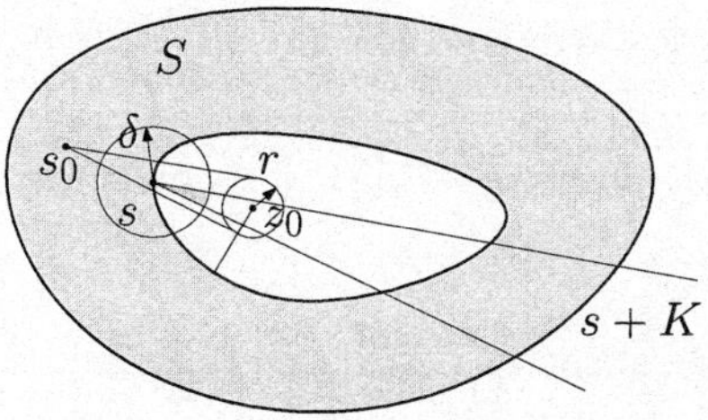

Figure 11.19: The Brézis Browder Theorem

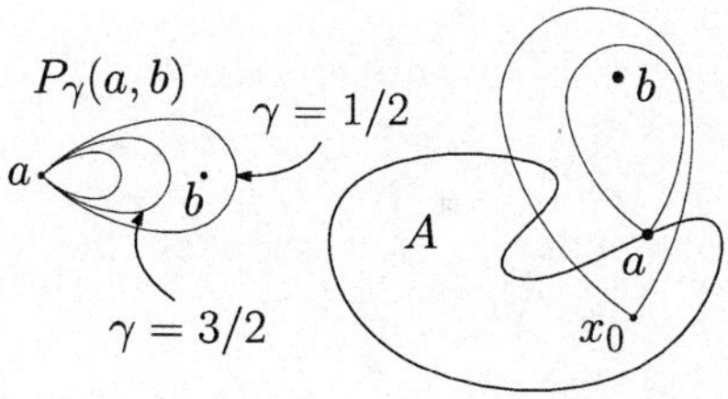

Figure 11.20: A petal and Penot's Petal Theorem

If there is a linear functional f that is bounded above on S and attains its supremum on S, the set $\{e \in S : f(e) = \sup\{f(x) : x \in S\}\}$ is clearly an example of an extreme subset of S.

➪ If C is a nonempty convex subset of X, note that the definition of extreme point given here coincides with the one provided in Section 2.2.

From now on in this subsection, the set C will be a **nonempty closed and convex subset of a Banach space**. Note that $e \in C$ is an extreme point of C if, and only if, there is no $h \in X$, $h \neq 0$, such that $e \pm h \in C$. Thus, the famous open problem already mentioned in page 215, i.e., the equivalence between the Krein–Milman (see Definition 545) and the Radon–Nikodým (see Definition 61) properties, can be formulated by asking whether a Banach space has this last property if, and only if, in any closed convex bounded $C \subset X$ there is a point $k \in C$ such that there is no $h \in X$, $h \neq 0$, with $k \pm h \subset C$ (see also the comments after Proposition 266). Whitfield and Zizler [WhZi88] considered naturally to substitute k in the previous question by a nonempty $\|\cdot\|$-compact subset of C, and so they introduced the following related concept: A nonempty subset $K \subset C$ is said to be **cylindrically extreme of** C (for short, **c-extreme**), if there is no $h \in X$, $h \neq 0$, such that $K \pm h \subset C$.

Remark 300. 1. In general, for a closed convex subset C of a Banach space X, the two concepts of being extreme and being cylindrically extreme are not related: The set K in Figure 11.21(a) (the boundary of a closed half-space C)

is extreme C but not c-extreme of C; in (b), it is c-extreme and not extreme of C.

2. For bounded sets C, however, if $K \subset C$ is extreme of C then it is c-extreme there.

   ➩ Indeed, let C be a nonempty closed convex and bounded subset of X. If a nonempty subset K of C is extreme and not c-extreme of C there exists $0 \neq h \in X$ such that $K \pm h \subset C$. Fix $k_0 \in K$. Then $k_0 \pm h \in C$, hence $k_0 + h \in K$. Thus, $(k_0 + h) \pm h$ in C, hence $k_0 + 2h \in K$. Continue in this way to get $k_0 + nh \in K$ for all $n \in \mathbb{N}$, a contradiction with the fact that C is bounded.

3. Let K be a nonempty $\|\cdot\|_\infty$-compact subset of B_{c_0}. Then there exists $n_0 \in \mathbb{N}$ such that $|a_n| < 1/2$ for every $a = (a_n) \in K$ and $n \geq n_0$ (see, e.g., [FHHMZ11, Exercise 1.50]). Thus $K \pm (1/2)e_{n_0} \subset B_{c_0}$, and so K is not c-extreme in B_{c_0}. This behaviour characterizes the existence of copies of c_0 in a Banach space, see Theorem 301 below.

4. The characteristic function of a subinterval $[a, b] \subset [0, 1]$ is denoted by $\chi_{[a,b]}$. Let $f_1 := \chi_{[0,1/2]}$ and $f_2 := \chi_{[1/2,1]}$. Then the set $\{2f_1, 2f_2\}$ is a c-extreme set in $B_{L_1[0,1]}$ that is not extreme in $B_{L_1[0,1]}$.

   ➩ To see this, for $g : [0, 1] \to \mathbb{R}$ and $[a, b] \subset [0, 1]$, put $g_{[a,b]} := g.\chi_{[a,b]}$. Assume that $h \in L_1[0, 1]$ is such that $2f_i \pm h \in B_{L_1[0,1]}$, $i = 1, 2$. Since $\|2f_1\|_1 = 1$, by convexity either $\|2f_1 + h_{[0,\frac{1}{2}]}\|_1 \geq 1$ or $\|2f_1 - h_{[0,\frac{1}{2}]}\|_1 \geq 1$. Assume the former. Then,

$$\begin{aligned} 1 \geq \|2f_1 + h\|_1 &= \|2f_1 + h_{[0,1/2]} + h_{[1/2,1]}\|_1 \\ &= \|2f_1 + h_{[0,1/2]}\|_1 + \|h_{[1/2,1]}\|_1, \end{aligned}$$

   hence $\|h_{[1/2,1]}\|_1 = 0$.

   If instead we assume the latter, then

$$\begin{aligned} 1 \geq \|2f_1 - h\|_1 &= \|2f_1 - h_{[0,1/2]} - h_{[1/2,1]}\|_1 \\ &= \|2f_1 - h_{[0,1/2]}\|_1 + \|h_{[1/2,1]}\|_1, \end{aligned}$$

   hence we get again $\|h_{[1/2,1]}\|_1 = 0$. Working similarly with f_2, we get $\|h_{[0,1/2]}\|_1 = 0$. All together, $\|h\|_1 = 0$, i.e., $h = 0$ (a.e.), and so $h = 0$ as an element in $L_1[0, 1]$.

   It is standard to see that the set $\{2f_1, 2f_2\}$ is not an extreme subset of $B_{L_1[0,1]}$.

   Indeed, you may use the functions

$$2f_1 \pm g,$$

where the function g is defined on $[0,1]$ by

$$g(x) = \begin{cases} -\frac{1}{2} & \text{for } x \in [0, \frac{1}{4}], \\ +\frac{1}{2} & \text{for } x \in (\frac{1}{4}, \frac{1}{2}], \text{ and} \\ 0 & \text{otherwise.} \end{cases}$$

For f_2 proceed similarly. ®

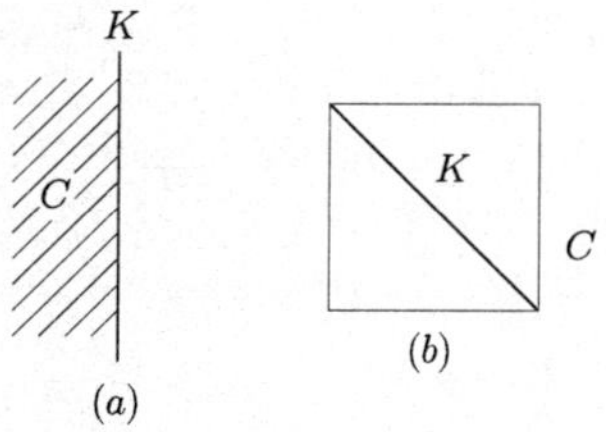

Figure 11.21: In (a), an extreme set K not c-extreme. In (b), the reverse situation

The following is the answer to the Krein–Milman versus Radon–Nikodým question for this new concept; its proof uses again Kurzweil et alt. method:

Theorem 301 (Whitfield, Zizler, [WhZi88]). *A Banach space X does not contain an isomorphic copy of c_0 if, and only if, in every nonempty closed convex bounded set $C \subset X$ there is a nonempty compact set $K \subset C$ which is c-extreme of C.*

Proof. Assume that there is a bounded closed convex set C in X such that there is no such compact set available. We use a method of Kurzweil [Kurz54] and Meshkov [Meshk78] to see that then X contains an isomorphic copy of c_0. We proceed as follows.
Pick an arbitrary element $x_1 \in C$. Since $\{x_1\}$ is a compact set, there is $x_2' \in X$ $x_2' \neq 0$ such that $x_1 \pm x_2' \in C$. Among such available x_2''s choose x_2 such that $\|x_2\| > \frac{1}{2}\sup\{\|x_2'\| : \ x_1 \pm x_2' \in C\}$.
Having chosen x_2, consider the set $\{x_1 \pm x_2\}$. Since it is compact, choose $x_3 \in X$, $x_3 \neq 0$, such that $x_1 \pm x_2 \pm x_3 \in C$ and $\|x_3\| > \frac{1}{2}\sup\{\|x_3'\| : \ x_1 \pm x_2 \pm x_3' \in C\}$. Continue this way to get a sequence $\{x_n\}_{n\in\mathbb{N}}$ in X with $x_i \neq 0$, $i \in \mathbb{N}$, such that (see Figure 11.22)

$$S := \left\{ x_1 + \sum_{i=2}^{n} \varepsilon_i x_i, \ n \in \mathbb{N}, \ \varepsilon_i = \pm 1 \right\} \text{ is bounded.} \tag{11.15}$$

Notice that the series $\sum_{i=1}^{\infty} x_i$ is weakly uniformly Cauchy.

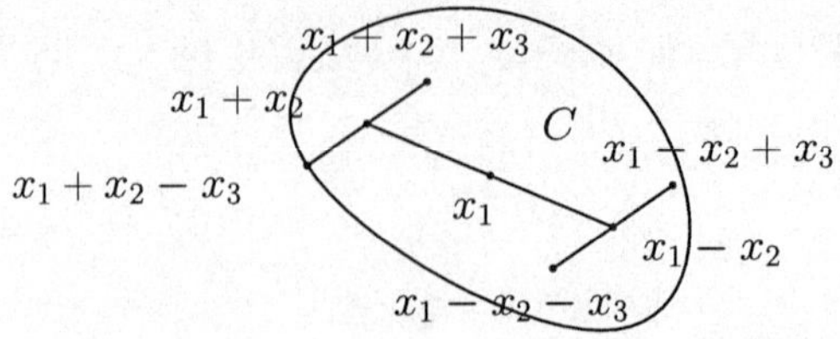

Figure 11.22: First three steps of the construction in the proof of Theorem 301

➪ If $s(F) := \sum_{i\in F} x_i$ for F a finite subset of $\mathbb{N}$, observe that $2s(F) = \sum_{i=1}^n \varepsilon_i x_i + \sum_{i=1}^n x_i$, where $\varepsilon_i \in \{-1,1\}$, $\varepsilon_i = 1$ for $i \in F$, $\varepsilon_i = -1$ otherwise. Thus, it follows from (11.15) that the set $\{s(F) : \subset \mathbb{N}, F \text{ finite}\}$ is bounded, and so the series is weakly unconditionally Cauchy (see, e.g., [Jam74, Proposition 31.9]).

Thus, X contains an isomorphic copy of c_0.
For otherwise, by a result of Bessaga and Pełczyński (see, e.g., [FHHMZ11, Corollary 4.52]), the series $\sum x_i$ would be unconditionally convergent, and therefore the set S ($\subset C$) would then be relatively compact. Thus, by the assumption, we could take $h \in X$, $h \neq 0$ such that $\overline{S} \pm h \subset C$. Since h appeared then as one of the x_n''s at each step n in the previous construction, we should have $\|x_n\| \geq \frac{1}{2}\|h\| > 0$ for each n, contradicting the convergence of $\sum x_n$. This finishes this part of the proof.
If $T : c_0 \to X$ is an isomorphism onto, then argument in Remark 300.3 shows that the set $T(B_{c_0})$ contains no nonempty $\|\cdot\|$-compact subset that is c-extreme there. □

A related result is the following (see Figure 11.18):

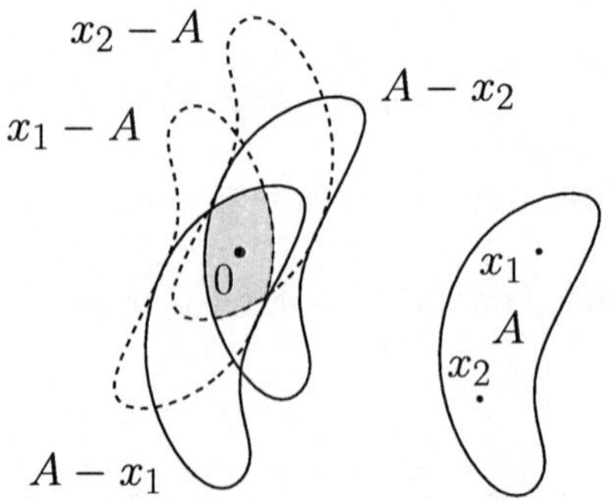

Figure 11.23: Intersecting translates of a set and their symmetrized sets (Theorem 246)

Theorem 302 (Pérez, Raja, [PerRaj17]). *A Banach space X does not contain an isomorphic copy of c_0 if, and only if, for every bounded set $A \subset X$ and each $\varepsilon > 0$, there are $\{x_1, \ldots, x_n\} \in A$ such that*

$$\operatorname{diam}\left(\bigcap_{i=1}^{n}(A - x_i) \cap (x_i - A)\right) < \varepsilon.$$

The main idea of the proof of Theorem 301 above, which is due to Kurzweil and Meshkov as mentioned, generated the following compact variational principle, which is very useful as it transfers some problems to problems on compact spaces. We will show its efficiency below, by comparing the proof of Theorem 301 with the one that employs the Compact Variational Principle.
The principle was first formulated and proved in [DeFa89]. For more on this principle see, e.g., [DeGoZi93, Theorem V.2.2], [HáJo14, Chapter 5, Lemma 47], and [Go01b].

We need first some definitions:

Definition 303. *Let U be a bounded open subset of a Banach space X, f be a real-valued continuous function on $\overline{U}$ and $K \subset U$ be a compact set. If V is a symmetric neighborhood of the origin in X such that $K + V \subset U$, then we define the function f_K on V by*

$$f_K(v) = \sup\{f(v+k) : k \in K\} \quad \text{for all } v \in V.$$

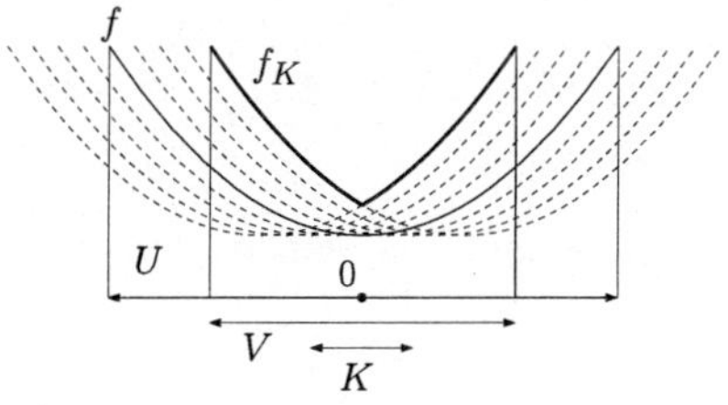

Figure 11.24: In bold, the function f_K in Definition 303

Note that the function f_K is continuous on V.

Indeed, let $\{v_n\}_{n=0}^{\infty}$ be a sequence in V such that $v_n \to v_0$. We shall prove that $f(v_n + k) \to f(v_0 + k)$ uniformly on $k \in K$. Otherwise, there exists $\varepsilon > 0$, a subsequence $\{v_{n_i}\}_{i=1}^{\infty}$ of $\{v_n\}_{n=1}^{\infty}$, and elements $k_i \in K$, $i \in \mathbb{N}$, such that

$$|f(v_{n_i} + k_i) - f(v_0 + k_i)| \geq \varepsilon \text{ for } i \in \mathbb{N}. \tag{11.16}$$

A subsequence $\{k_{i_j}\}$ of $\{k_i\}$ converges, say to k_0, so $v_{n_{i_j}} + k_{i_j} \to v_0 + k_0$.

Then, by continuity, $f(v_{n_{i_j}}+k_{i_j}) \to f(v_0+k_0)$, and $f(v_0+k_{i_j}) \to f(v_0+k_0)$, a contradiction with (11.16). The conclusion now readily follows.

As usual, ∂U will denote the boundary of the set U.

Theorem 304 (Deville, Fabian, [DeFa89]). [Compact Variational Principle] *Let X be a Banach space that does not contain a copy of c_0. Let U be a bounded symmetric open set in X containing the origin and f be a real-valued continuous function on $\overline{U}$ such that*

$$f(0) \le 0, \; f(x) = f(-x) \text{ for all } x \in \overline{U},$$

and

$$f(x) > 0 \text{ for all } x \in \partial U.$$

Then there exists a compact symmetric set $K \subset U$ and a symmetric neighborhood V of the origin in X with the following properties:
(i) $K + V \subset U$.
(ii) $f_K(0) = 0$ *and* $f_K(x) > 0$ *for* $x \in V$, $x \neq 0$.

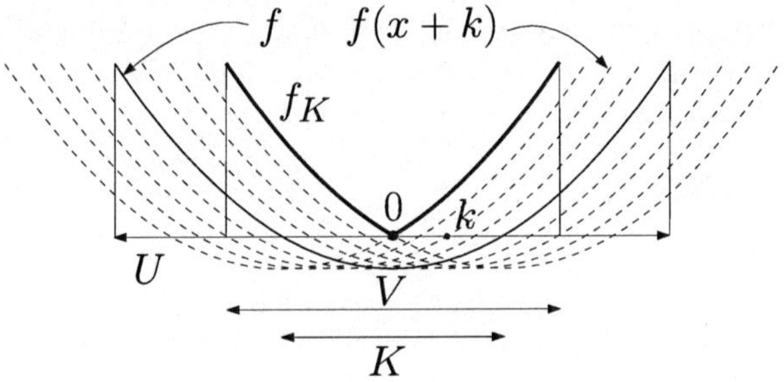

Figure 11.25: An sketch of the compact variational principle (Theorem 304). In bold, the function f_K

(See Figure 11.25.) We strongly recommend the excellent version of the proof of the result above in [Fab83, Proposition 3.5]. As an example of a simple use of this theorem, let us prove the following variant of Theorem 301 above (see, e.g., [DeGoZi93, Corollary V.2.3]).

Theorem 305. *Let X be a Banach space that does not contain a copy of c_0. Then B_X contains a compact symmetric set K such that for no $h \in X$, $h \neq 0$ we have $K + h \subset B_X$.*

Proof. Let U be a symmetric open and bounded subset of X such that $B_X \subset U$ and let f be defined for $x \in \overline{U}$ by

$$f(x) = \text{dist}\,(x, B_X).$$

By Theorem 304 above, there are a symmetric compact set K in U and a symmetric neighborhood V of the origin in X such that f_K is defined on V,

$$f_K(0) = 0, \quad \text{and } f_K(x) > 0 \ \text{ for each } x \in V, \ x \neq 0.$$

From this it follows that $K \subset B_X$ and that if $h \in V$, $h \neq 0$, then $K + h$ is not included in B_X. A simple convexity argument gives then that, for no $h \in X \setminus \{0\}$, $K + h \subset B_X$. $\square$

Further applications of the compact variational principle will be discussed, e.g., in Chapter 24.

Chapter 12

Projectional resolutions of the identity III

12.1 Introduction

Any set S of ordinal numbers can be endowed with the **order topology**, i.e., the topology on S having as a basis of open sets the family of all **open intervals** $\{s \in S : \ \alpha < s < \beta\}$, $\{s \in S : \ s < \beta\}$, $\{s \in S : \ s > \alpha\}$, where α and β are ordinal numbers. Note that every ordinal number β has an immediate successor (denoted $\beta + 1$), hence a basis of neighborhoods of β in S is given by all the sets $S \cap (\alpha, \beta]$, where α is any ordinal number such that $\alpha < \beta$. Thus, a net $\{\alpha_i : \ i \in I, \ \leq\}$ of ordinal numbers in S converges to $\beta \in S$ if (and only if) for every $\alpha < \beta$ there exists $i_0 \in I$ such that $\alpha < \alpha_i \leq \beta$ for all $i \geq i_0$. If β has an immediate predecessor (i.e., if β is not a **limit ordinal**), this can only happen if there exists $i_0 \in I$ such that $\alpha_i = \beta$ for $i \geq i_0$.

The **cofinality** of an ordinal number μ is the smallest ordinal number α such that there exists in $[0, \mu)$ a cofinal subset whose ordinal number is α.

➯ Given a partially ordered set $(S, \leq)$, a **cofinal** subset C of S is defined by the property that , for every $s \in S$, there exists $c \in C$ such that $s \leq c$.

A **regular** ordinal number is an ordinal number r such that its cofinality is precisely r. For example, ω_0 is a regular ordinal. Assuming the Axiom of Choice, we may conclude that ω_1 is regular, too. On the other side, ω_0^2 is not regular, since its cofinality is ω_0.

We shall always assume that any set of ordinal numbers is endowed with the order topology.

➯ Every strictly decreasing sequence $\{\gamma_n\}$ of ordinal numbers is finite. Indeed, the set $\{\gamma_n : \ n \in \mathbb{N}\}$ has a first element, and this gives the result. Given an ordinal number β, the set $[0, \beta]$ is compact. This follows from the previous assertion: Let $\{\Omega_i : \ i \in I\}$ be an open covering of $[0, \beta]$. We may assume, without loss of generality, that $\Omega_i = [0, \beta_i)$ for some ordinal number β_i. We have $\beta \in \Omega_{i_1}$ for some $i_1 \in I$, $\beta_{i_1} \in \Omega_{i_2}$ for some $i_2 \in I$, and so on. If the process does not stop, the sequence $\{\beta_{i_n}\}_{n=1}^{\infty}$ must have a strictly decreasing subsequence, something impossible. This shows that for some finite subset I_0 of I we have $[0, \beta] \subset \bigcup_{i \in I_0} \Omega_i$.

A. J. Guirao et al., *Renormings in Banach Spaces*, Monografie Matematyczne 75,
https://doi.org/10.1007/978-3-031-08655-7_12

12.2 Four useful properties of projectional resolutions of the identity

Let X be a Banach space. If Y is a subspace of X, then Y^* is isometrically isomorphic to $X^*/Y^\perp$ endowed with its quotient norm. In particular, if $P : X \to X$ is a projection with $\|P\| = 1$, then $(PX)^*$ is isometrically isomorphic to $X^*/(PX)^\perp$. Since $X^* = P^*X^* \oplus (PX)^\perp$, and $\|P^*\| = 1$, we get that P^*X^* is isometrically isomorphic to $(PX)^*$.

Projectional resolutions of the identity (PRI's) were introduced in Definition 130. We mention here four properties of PRI's that will be used below:

1. Let X be a Banach space and assume that $\{P_\alpha : \omega_0 \le \alpha \le \mu\}$ is a PRI for X (see Definition 130 above). Fix $x \in X$. Then the mapping $\alpha \mapsto P_\alpha x$ is continuous from $[\omega_0, \mu]$ into X.

   ➪ Indeed, the property should be checked just for any limit ordinal $\omega_0 \le l \le \mu$: Fix $x \in X$ and an arbitrary $\varepsilon > 0$. By property (iv) of a PRI, we may find $y \in X$ and $\beta \in [\omega_0, l)$ such that $\|P_l x - P_{\beta+1} y\| < \varepsilon$. Let $\beta + 1 \le \gamma < l$. Apply P_γ to obtain $\|P_\gamma P_l x - P_\gamma P_{\beta+1} y\| < \varepsilon$, i.e., $\|P_\gamma x - P_{\beta+1} y\| < \varepsilon$. Then $\|P_\gamma x - P_l x\| \le \|P_\gamma x - P_{\beta+1} y\| + \|P_{\beta+1} y - P_l x\| < 2\varepsilon$. This shows that $P_\gamma x \to P_l x$ as $\gamma \to l$.

   This continuity property may then substitute (iv) in Definition 130 of a PRI. Observe also that, fixed $x^* \in X^*$, the mapping $\alpha \mapsto P^*_\alpha x^*$ is w^*-continuous. In particular, $P^*_\alpha x^* \xrightarrow{w^*} x^*$ whenever $\alpha \to \mu$, for every $x^* \in X^*$.

2. Let X be a Banach space of density ω_1 having a PRI $\{P_\alpha : \omega_0 \le \alpha \le \omega_1\}$. Note that, in this case, $X = \bigcup_{\omega_0 \le \alpha < \omega_1} P_\alpha X$. Indeed, given $x \in X$, the function $\alpha \to (I - P_\alpha)x$ is continuous (see Item 1 above) and equal to zero at ω_1. Thus it is easy to see that there exists $\beta \in [\omega_0, \omega_1)$ (depending on x) such that this function is equal to zero for all $\alpha \in [\beta, \omega_1]$ (see, e.g., [FHHMZ11, Exercise 13.23]).

   ➪ Indeed, for every $n \in \mathbb{N}$ find, by continuity, $\alpha_n \in [\omega_0, \omega_1)$ such that $\|x - P_{\alpha_n} x\| < 1/n$. Without loss of generality we may assume that the sequence $\{\alpha_n\}$ in nondecreasing. It is enough to take $\beta := \sup\{\alpha_n : n \in \mathbb{N}\}$, and to notice that $\beta < \omega_1$, due to the fact that all α_n are countable ordinals, and so it is β.

   The same argument as above shows that $X = \bigcup_{\omega_0 \le \alpha < \mu} P_\alpha X$ for spaces X such that the first ordinal number μ associated to $\operatorname{dens}(X)$ is regular and uncountable, and having a PRI $\{P_\alpha : \omega_1 \le \alpha \le \mu\}$.

3. A related argument (compare with Item 2 above), this time by using Proposition 145 and Remark 146, shows that if $\|\cdot\|$ on X is Gâteaux differentiable,

the space X has a PRI $\{P_\alpha : \omega_0 \leq \alpha \leq \omega_1\}$, and $\operatorname{dens} X \leq \omega_1$, then $X^* = \bigcup_{\omega_0 \leq \alpha < \omega_1} P_\alpha^* X^*$. Moreover, in this case (B_{X^*}, w^*) is a Corson compact. Details are provided in the proof of the second part of Theorem 306 below.

4. The following result shows a feature of a Banach space having a PRI and whose norm has a certain differentiability property. The second part reproduces the information given in the previous item. We include this here again for a matter of contrast with the Fréchet case:

Theorem 306. (i) *Let $(X, \|\cdot\|)$ be a Banach space with a PRI $\{P_\alpha : \omega_0 \leq \alpha \leq \mu\}$. Assume that $\|\cdot\|$ is Fréchet differentiable. Then $\{P_\alpha^* : \omega_0 \leq \alpha \leq \mu\}$ is a PRI for X^*, and X is WCG.*

(ii) *Let $(X, \|\cdot\|)$ be a Banach space with density ω_1 that admits a PRI $\{P_\alpha : \omega_0 \leq \alpha \leq \omega_1\}$. Assume that $\|\cdot\|$ is Gâteaux differentiable. Then $\bigcup_{\omega_0 \leq \alpha < \omega_1} P_\alpha^*(X^*) = X^*$ and (B_{X^*}, w^*) is a Corson compact.*

Proof. (i) Obviously, properties (i) and (ii) of a PRI hold for $\{P_\alpha^* : \omega_0 \leq \alpha \leq \mu\}$. The remark at the beginning of the present Subsection and the fact that the norm is Fréchet differentiable shows that $\operatorname{dens} P_\alpha^* X^* \leq |\alpha|$ for every $\alpha \in [\omega_0, \mu]$. To prove property (iii) of a PRI we follow a variant of the argument in Remark 146 and in the previous item: Fix $\beta \in (0, \mu]$ and let us consider first the case that $P_\beta^*(x^*) \neq 0$ attains the norm. Since the net $\{P_\alpha^*(x^*) : \alpha \in [\omega_0, \beta), \leq\}$ is w^*-convergent to $P_\beta^*(x^*)$, the Šmulyan Lemma 54 shows that the net $\|\cdot\|$-converges to $P_\beta^*(x^*)$, so $P_\beta^*(x^*) \in \overline{\bigcup_{\omega_0 \leq \alpha < \beta} P_\alpha^* X^*}^{\|\cdot\|}$. The Bishop–Phelps Theorem 14, applied to the space $P_\beta X$ and its dual $P_\beta^* X^*$ proves (iii) in the general case. It is clear that (iv) also holds.

We shall provide two arguments for showing the WCG character of X:

The first one relies on Theorem 236 above. It is easy to prove that the mapping $x^* \mapsto ((P_{\alpha+1}^* - P_\alpha^*)x^*)_\alpha$ does the job.

For the second one, we shall prove first that the condition implies the existence of a shrinking M-basis $\{x_\gamma, f_\gamma\}_{\gamma \in \Gamma}$. Since the set $\{x_\gamma : \gamma \in \Gamma\}$ is always w-relatively compact (if bounded, something that may always be assumed; see the remark after Definition 128), we shall conclude that X is WCG. Let us prove the following *Claim*: If $(X, \|\cdot\|)$ is a Banach space with a PRI $\{P_\alpha : \omega_0 \leq \alpha \leq \mu\}$, and such that $\|\cdot\|$ is Fréchet differentiable, then for every $\alpha < \mu$ the space $(P_{\alpha+1} - P_\alpha)X$ has a shrinking M-basis. The proof is done by induction on the density character of X. If X is separable and the norm is Fréchet differentiable, Markushevich's theorem (see, e.g., [FHHMZ11, Theorem 4.59]) shows that, due to the fact that X^* is separable, we can find a shrinking M-basis in X. Assume now that the result has been proved for every Banach space $(X, \|\cdot\|)$ such that, for some ordinal number μ, we have $\operatorname{dens} X < \mu$, $\|\cdot\|$ is Fréchet differentiable and X has a PRI

$\{P_\alpha : \omega_0 \le \alpha \le \mu\}$. Let now $(X, \|\cdot\|)$ be a Banach space with $\operatorname{dens} X = \mu$, with a PRI $\{P_\alpha : \omega_0 \le \alpha \le \mu\}$, and such that $\|\cdot\|$ is Fréchet differentiable. For any $\alpha < \mu$, the space $P_\alpha X$ has density $\le \alpha < \mu$, and satisfies all the requirements, so $(P_{\beta+1} - P_\beta)X$ has a shrinking M-basis for each $\beta < \alpha$. Thus this happens for all $\beta < \mu$ and the Claim is proved. This applies to the Banach space $(X, \|\cdot\|)$. Putting together all those bases for $(P_{\alpha+1} - P_\alpha)X$, $\alpha < \mu$, we get a shrinking M-basis for X, and the conclusion follows.

(ii) Fix $x^* \in S_{X^*}$. Assume first that x^* attains its norm on B_X. Find then $x \in S_X$ such that $\langle x, x^*\rangle = 1$. According to Item 2 in Section 12.2, there exists $\alpha < \omega_1$ such that $x \in P_\alpha X$. Then $\langle x, P_\alpha^* x^*\rangle = \langle P_\alpha x, x^*\rangle = \langle x, x^*\rangle = 1$. Since $\|P_\alpha^* x^*\| \le 1$, we get $\|P_\alpha^* x^*\| = 1$. The Gâteaux differentiability of $\|\cdot\|$ at x shows that $x^* = P_\alpha^* x^*$, hence $x^* \in P_\alpha^* X^*$. If x^* does not attain its norm, given $n \in \mathbb{N}$ we can find, by the Bishop–Phelps Theorem 14 above, $x_n^* \in S_{X^*}$ that $\|x^* - x_n^*\| < 1/n$ and x_n^* attains its norm on B_X. By the first part of the proof, $x_n^* \in P_{\alpha_n}^* X^*$ for some $\alpha_n < \omega_1$. We may assume that the sequence $\{\alpha_n\}_{n=1}^\infty$ is increasing. Let α $(< \omega_1)$ be its supremum. Then $x^* \in P_\alpha^* X^*$.

For $\alpha < \omega_1$, fins a countable and dense subset D_α of the separable subspace $P_\alpha X$. The fact that $x^* \in B_{X^*}$ is in some $P_\alpha^* X^*$ shows that x^* is countably supported on the set $\bigcup_{\omega_0 \le \alpha < \omega_1} D_\alpha$, and this concludes the proof. □

12.3 Examples of spaces having projectional resolutions of the identity

We list now some natural examples of (nonseparable) spaces with a PRI. Let μ be the first ordinal number corresponding to a given uncountable cardinal number.

1. The spaces $c_0[0, \mu)$ and $\ell_p[0, \mu)$ for $1 \le p < \infty$, defined in Section 1.4, have a PRI. Those spaces have density $\operatorname{card}(\mu)$. For $\alpha \in [\omega_0, \mu)$ and $(a_\beta)_{\beta<\mu} \in c_0[0, \mu)$, let $P_\alpha((a_\beta)_{\beta<\mu}) = (b_\beta)_{\beta<\mu}$, where $b_\beta = a_\beta$ for $\beta < \alpha$ and $b_\beta = 0$ for $\beta \ge \alpha$. Put $P_\mu = \mathrm{Id}_X$. Then $\{P_\alpha : \omega_0 \le \alpha \le \mu\}$ is clearly a PRI for $c_0[0, \mu)$. The same applies to $\ell_p[0, \mu)$.

   Note that the support of $x \in X$ for $X := c_0[0, \mu)$ or $X := \ell_1[0, \mu)$ is countable. Hence, if μ is a regular (uncountable) ordinal number we get, without relying on the continuity of the mapping $\alpha \to P_\alpha(x)$, that $X = \bigcup_{\omega_0 \le \alpha < \mu} P_\alpha(X)$.

2. Let $C[0, \mu]$ be the Banach space of all the continuous functions on the ordinal segment $[0, \mu]$ (a compact space, see the annotation before Section 12.2), endowed with the supremum norm $\|\cdot\|_\infty$.

   ➡ Add an extra point -1 at the beginning of the transfinite sequence of all ordinal numbers $[0, \mu]$ to obtain a set $[-1, \mu]$ that, under the given order, is order-equivalent to $[0, \mu]$. The space $C[-1, \mu]$ is, under the supremum norm, isometric to $C[0,\mu]$. The hyperplane $H := \{x \in C[-1,\mu] : x(-1) = 0\}$ is isometric to $C[0,\mu]$. Since all hyperplanes of a Banach space are isomorphic, the hyperplane $C_0[0,\mu] := \{x \in C[0,\mu] : x(\mu) = 0\}$ is then isomorphic to $C[0,\mu]$.

Note that the density of $C[0,\mu]$ is card (μ).

➪ Let us consider the subset $S := \{(\kappa_\alpha :=)\ \kappa_{[0,\alpha]} :\ 0 \leq \alpha \leq \mu\}$ of $C[0,\mu]$ consisting of the characteristic functions of intervals $[0,\alpha]$. Note that $\|\kappa_\alpha - \kappa_\beta\|_\infty = 1$ if $\alpha \neq \beta$, hence the family $\{B(\kappa_\alpha, 1/3) :\ 0 \leq \alpha \leq \mu\}$ is pairwise disjoint and has cardinality card (μ). It follows that if $D \subset C[0,\mu]$ is a dense subset, then card $(D) \geq$ card (μ). On the other hand, the linear subspace $L :=$ span (S) is a sublattice, in the sense that $x \vee y$ and $x \wedge y$ belong to L whenever $x, y \in L$. Clearly, L strongly separates points of $[0,\mu]$, in the sense that given distinct points α and β of $[0,\mu]$ and real numbers a, b, there exists $x \in L$ such that $x(\alpha) = a$ and $x(\beta) = b$. Then L is dense in $C[0,\mu]$ (see, e.g., [Jam74, Proposition 26.4]). Since card $(L) =$ card (S) $(=$ card $(\mu))$ (use rational combinations and their limits), we get that dens $(C[0,\mu]) \leq$ card (μ). This shows the statement.

On $C[0,\mu]$ define the projections

$$P_\beta(x) = \begin{cases} x \text{ on } [0,\beta], \\ x(\beta) \text{ on } [\beta,\mu], \quad \text{for } x \in C[0,\mu], \end{cases}$$

where $\beta \in [\omega_0,\mu]$. Then $\{P_\beta :\ \omega_0 \leq \beta \leq \mu\}$ is a PRI on $C[0,\mu]$.

3. For K a Valdivia compact space, $C(K)$ has a PRI. More precisely (see Theorem 798), it has a PRI $\{P_\gamma :\ \omega_0 \leq \gamma \leq \mu\}$ such that each P_α is τ_p-continuous, where τ_p denotes the topology on $C(K)$ of the pointwise convergence on K ([Vald90], see also, e.g., [DeGoZi93, Theorem 7.6]). A compact space K is said to be a **Valdivia compact** if for some index set Γ there exist a homeomorphism ϕ from K into $[0,1]^\Gamma$ such that $\Sigma(\Gamma) \cap \phi(K)$ is dense in $\phi(K)$, where $\Sigma(\Gamma)$ denotes the subset of all countably supported vectors in $[0,1]^\Gamma$ (see Chapter 47 below). Of course, every Corson compact space is a Valdivia compact (see again Chapter 47). An example of a Valdivia compact space that is not Corson compact is the space $[0,\omega_1]$ (see Proposition 797). For more information on this topic see Chapter 47 and, e.g., [DeGoZi93, Section VI.7].

Proposition 307 (Fabian, Montesinos, Zizler, [FaMoZi02-b]). *There is no Gâteaux differentiable norm on $C[0,\omega_1]$ that would be $[0,\omega_1)$-lower semicontinuous.*

Proof. (Sketch) The space $Y := \overline{\text{span}}^{\|\cdot\|}\{\delta_\alpha : \alpha \in [0,\omega_1)\}$ is certainly 1-norming. Apply Remark 146 and observe that δ_{ω_1} cannot be in the w^*-closure of a countable subset of B_Y. □

This is in contrast with Proposition 232 above. Theorem 239 says that a Banach space X is a subspace of a weakly compactly generated Banach space if, and only if, (B_{X^*}, w^*) is an Eberlein compact (see also, e.g., [HMVZ07, Theorem 6.13]). Observe that, for an uncountable regular ordinal number μ, neither $X := C[0,\mu]$ nor $C_0[0,\mu]$ (they are isomorphic) are such, since $[0,\mu]$ ($\subset (B_{X^*}, w^*)$) is not an Eberlein compact (it is not angelic, see Section 6.2, due to the fact that μ is not the limit of a sequence in $[0,\mu)$).

Proposition 232 suggests to introduce the following property of a Banach space:

Definition 308. *A Banach space X is said to have* **property (P)** *if given an arbitrary norming subspace N of X^*, there exists a Gâteaux differentiable $w(X,N)$-lower semicontinuous equivalent norm $|||\cdot|||$ on X. We say that the property (P) is* **strong** *if there exists an equivalent norm $|||\cdot|||$ on X such that, for every norming subspace N of X^*, the N-lower semicontinuous envelope $|||\cdot|||_N$ of $|||\cdot|||$ is Gâteaux differentiable.*

Proposition 232 above says that *every subspace of a WCG Banach space has strong property (P).*

In [FaMoZi02-b] it was proved that *if K is Valdivia compact of density less than or equal to ω_1, and $C(K)$ has property (P), then K is a Corson compact and, more generally, $C(K)$ is WLD.* We may prove here that the result holds without any cardinality restriction. If K is a Valdivia compact, we may assume that K is a subset of a topological product $[0,1]^\Gamma$ endowed with the pointwise topology, and that the $\Sigma \cap K$ is dense in K, where $\Sigma := \{s \in [0,1]^\Gamma : \text{supp }(s) \text{ countable}\}$.

Proposition 309 (Guirao–Lissitsin–Montesinos, [GuiLiMo19]). *Let K be a Valdivia compact space. If $C(K)$ has property (P), then $C(K)$ is WLD; in particular, K is Corson compact.*

Proof. We shall prove first that K is a Corson compact. If $|\Gamma| \leq \omega_0$ there is nothing to prove. Otherwise, the subspace $N := \text{span }\{\delta(s) : s \in \Sigma\}$ of $C(K)^*$ is clearly 1-norming. By assumption, there is an equivalent Gâteaux differentiable norm $|||\cdot|||$ on C(K) such that N is still 1-norming for $|||\cdot|||$, so $N \cap B_{(C(K)^*, |||\cdot|||^*)}$ is w^*-dense in $B_{(C(K)^*, |||\cdot|||^*)}$. Fix $k \in K$ and let $u := \delta(k)/|||\delta(k)|||^*$, an element in $S_{(C(K)^*, |||\cdot|||^*)}$. By Remark 146, there exists a countable set $N_0 \subset N \cap B_{(C(K)^*, |||\cdot|||^*)}$ such that $u \in \overline{N_0}^{w^*}$. So, $\delta(k) \in \overline{|||\delta(k)|||^*.N_0}^{w^*}$. This happens for any $k \in K$.
Let us denote by p_γ the canonical projection from $[0,1]^\Gamma$ into the γ-coordinate.

Any element $\mu \in N$ is countably supported on $\{p_\gamma,\ \gamma \in \Gamma\} \subset C(K)$, i.e., $|\{\gamma \in \Gamma :\ \langle p_\gamma, \mu\rangle \neq 0\}| \leq \aleph_0$, hence the same holds for $\delta(k)$, $k \in K$. This shows that $K = \Sigma$, and so K is a Corson compact.
By Theorem 7 in [FaMoZi02-b], every element of $C(K)^*$ is a regular Borel measure on K that has a separable support. This shows that K is a Corson compact with property (M) (see [HMVZ07, Definition 5.56]). By [AMN88] (see also [HMVZ07, Theorem 5.57]) the space $C(K)$ is WLD. Another approach to obtain the same conclusion is to use [Kal00-b] (see also [HMVZ07, Theorem 5.55]): The set Σ is dense in K, and so the set $\{\mu \in C(K)^* :\ \text{supp}\ (\mu)\ \text{a separable subset of}\ \Sigma\}$ is a 1-norming Σ-subspace of $C(K)^*$. □

We may mention two open problems related to the concept of properties (P) and strong (P). First, it is apparently unknown whether every space with the strong property (P) is a subspace of a WCG Banach space. Second, assume that K is a compact topological space such that on $C(K)$ there exists an equivalent norm $|||\cdot|||$ such that, for every dense subset D of K, then the D-lower semicontinuous envelope $|||\cdot|||_D$ of $|||\cdot|||$ is Gâteaux differentiable. Is then K an Eberlein compact space?
It follows from [FaGoZi01] that every UG norm has strong property (P).

12.4 Examples of spaces without projectional resolutions of the identity

Regarding Banach spaces without PRI's, we consider a few examples below.

1. As mentioned above (see Subsection 6.1.4), the space ℓ_∞ has no PRI. Indeed, if $\{P_\alpha :\ \omega_0 \leq \alpha \leq \mu\}$ is a PRI for a space X, then for every countable ordinal number α, the subspace $P_\alpha(X)$ is separable. However, every infinite-dimensional complemented subspace of ℓ_∞ is isomorphic to ℓ_∞ (J. Lindenstrauss, [Lind67]; Lindenstrauss' result is presented in [LinTz77, Theorem 2.a.7]).
2. The example in Theorem 310 below is inspired by the fact that c_0 is not 1-complemented in c, see, e.g., [FHHMZ11, Exercise 5.28].

   ➪ The space c_0 is a hyperplane of c. In fact, it is the kernel of the continuous linear functional $f : c \to \mathbb{R}$ given by $f(x) = \lim_n x_n$, for $x = (x_n) \in c$. Let $P : c \to c_0$ be a bounded projection. Find $a \in \ker(P)$ such that $f(a) = 1$. Then obviously $P(x) = x - f(x)a$ for all $x \in c$. It is very easy to find $x \in c$ such that $\|x\|_\infty = 1$, $f(x) = 1$, and $\|x - a\|_\infty$ $(= \|P(x)\|_\infty)$ is almost 2. This shows that every bounded linear projection $P : c \to c_0$ has norm $\|P\|_\infty \geq 2$.

Theorem 310. (See also [DeGoZi93, Example VI.8.6].) *The space $C_0[0, \omega_1]$ of all real-valued continuous functions defined on $[0, \omega_1]$ and vanishing at ω_1, endowed with the supremum norm, does not admit any PRI.*

Proof. Assume that $\{P_\alpha : \omega_0 \leq \alpha \leq \omega_1\}$ is a PRI on $X := C_0[0, \omega_1]$. By the separability of P_{ω_0} and the fact that every element in $C_0[0, \omega_1]$ is zero on the interval $[\beta, \mu]$ for some $0 \leq \beta < \mu$, there is $\beta_0 < \omega_1$ such that

$$P_{\omega_0}(X) \subset X_{\beta_0} := \{f \in X : f(\gamma) = 0 \text{ if } \gamma > \beta_0\}.$$

Since X_{β_0} is separable (see Item 2 in Section 12.3), Item 2 in Section 12.2 shows that there is $\alpha_1 > \omega_0$ such that $X_{\beta_0} \subset P_{\alpha_1}(X)$. As above, there is β_1 with $\beta_0 < \beta_1 < \omega_1$ such that $P_{\alpha_1}(X) \subset X_{\beta_1}$.

Continuing in this way, we find $\alpha := \sup\{\alpha_n : n \geq 1\}$ and $\beta := \sup\{\beta_n : n \geq 1\}$ such that

$$P_\alpha(X) = \overline{\bigcup_{n \geq 1} P_{\alpha_n}(X)} = Y_\beta = \{f \in X : f(\gamma) = 0 \text{ if } \gamma \geq \beta\}.$$

Mimicking the proof of the non-1-complementability of c_0 in c given above, it is not difficult to show that Y_β is not 1-complemented in $C_0[0, \omega_1]$. This is a contradiction. □

Remark 311. (a) Theorem 310 gives an example of a Banach space with a PRI having a closed hyperplane without PRI. Since $C_0[0, \omega_1]$ and $C[0, \omega_1]$ are isomorphic (see Item 2 in Section 12.3), this shows that having a PRI is a property that is not stable under renormings.

(b) Despite the negative result in Theorem 310, note that the space $C_0[0, \omega_1]$ has the property that every separable subspace is contained in a 1-complemented separable subspace. This follows from the argument given in the proof of the aforementioned Theorem 310. We say that $C_0[0, \omega_1]$ has the **separable complementation property** (more precisely, the 1-**separable complementation property**). See, e.g., [HMVZ07, Definition 3.39]. ®

12.5 Results on projectional resolutions of the identity in the 1970s and 1980s

Let us discuss now some results that were obtained in Prague in the seventies. Even now, they may provide some inspiration for new research in this area. As we said, they were achieved by using K. John's discovery of an improvement of the Amir–Lindenstrauss construction of PRI's (see Subsections 10.2.1 and 10.2.2, and [Day73, Proof of Theorem 3.5.1]) done purely via topological methods. As we mentioned, this allowed for a construction of PRI's that would preserve a given subspace, and contained in the dual ranges of projections of prescribed differentials of the norm on X.

After Rosenthal's "negative" result for the heredity of the WCG property (see Theorem 760 below), one of the first "positive" result in this direction was the following:

Theorem 312 (John, Zizler, [JoZi74a]). *Let X be a WCG Banach space that admits a continuously Fréchet differentiable bump function. Let Y be a subspace of X. Then Y is WCG.*

Proof. Let us first consider the case that X admits a Fréchet differentiable norm $\|\cdot\|$. Then, using John's device (see the proof of Theorem 306 above), we construct on X a shrinking PRI that preserves Y, i.e., the projections P_α have the property that $P_\alpha(Y) \subset Y$. Then the restrictions of P_α to Y form a shrinking PRI. From this, it is straightforward to show that Y has a shrinking Markushevich basis (see again the proof of Theorem 306). Finally, any space with a shrinking Markushevich basis clearly is WCG (see the remark after Definition 128). □

If X admits a C^1-smooth bump function, then include "by force" its derivatives into ranges of dual projections, which is possible by John's result mentioned above, and use the fact that we get a shrinking PRI is a consequence of the following result of M. Leduc:

Theorem 313 (Leduc, [LD70]). *If X is a Banach space that admits a C^1-smooth bump function f, then $\{f'(x)/\|f'(x)\| : x \in S_X\}$ is dense in S_{X^*}.*

Next we can use the following result of Troyanski:

Theorem 314 (Troyanski, [Tr72]). *Let a Banach space X admit a shrinking Markushevich basis. Then X admits a norm whose dual norm is LUR.*

Thus we get the following result (see also Theorem 839 below):

Theorem 315 (John, Zizler, [JoZi74a]). *Let X be a Banach space. Then the following are equivalent.*
(i) *X admits a shrinking Markushevich basis.*
(ii) *X is WCG and admits a norm whose dual is LUR.*
(iii) *X is WCG and admits a Fréchet differentiable norm.*
(iv) *X is WCG and admits a C^1-smooth bump function.*

If this happens for X then any subspace of X shares these properties.

Proof. (i)⇒(ii): X is WCG, as any space with a shrinking Markushevich basis is WCG (see the comment after Definition 128), and X admit such a norm by Theorem 314 above.
(ii)⇒(iii): See Proposition 91 above.
(iii)⇒(iv): If $\|\cdot\|$ is a C^1-smooth norm, then a composition of $\|\cdot\|^2$ with a proper real-valued function creates a C^1-smooth bump function.
(iv)⇒(i): See the proof of Theorem 312. □

Corollary 316 (John, Zizler, Friedland). *For any nonempty set Γ, any subspace of $c_0(\Gamma)$ is WCG.*

Proof. The standard basis of $c_0(\Gamma)$ is a shrinking Markushevich basis. Apply now Theorem 315 above. □

Theorem 317 (John and Zizler, [JoZi74b]). *Let X admit a norming Markushevich basis. Then X admits a LUR norm.*

H Around the year of publication of [JoZi74b], John and Zizler first heard — through an anonymous referee's report— about the new notion of projectional generator. The method derived from this device was not used in the proof of [JoZi74b], where a different technique was used. Today, Theorem 317 can be obtained, too, as a consequence of a transfer technique to be presented in Section 21.2 below (see, in particular, Example 432 there).

Based on a clever idea of Tacon in [Tac70] on the uniqueness of the support functionals for differentiable norms, it was proved the following result.

Theorem 318 (John and Zizler, [JoZi75]). *If the dual norm of X^* is LUR, then X^* admits a Markushevich basis.*

Theorem 319 ([Rei76]). *Let X be a subspace of a WCG space. Then X is DENS and it admits a Markushevich basis.*

Now Theorem 318 follows from the following result.

Theorem 320 (Fabian, Godefroy [FaGo88]). *For any Asplund space X, X^* admits a PRI, a Markushevich basis, and a (non-dual, in general) LUR equivalent norm.*

The proof of Theorem 320 is much more involved than the proof of Theorem 318.

Besides Simons' Theorem 538 below, the proof of Theorem 320 involves also the following concept that is also of independent interest.

Theorem 321 (Jayne, Rogers, see, e.g. [DeGoZi93], Theorem I.4.2, [GhMauSch92]). *Let X be an Asplund space. Then there is a map $\mathcal{D} : X \to (X^*)^{\mathbb{N}}$, such that*

$$\mathcal{D}(x) = \{D_n(x) : \ n \in \mathbb{N}\} \cup \{D_\infty(x)\},$$

where

(i) $D_n : X \to X^*$ *is norm-to-norm continuous for every* $n \geq 1$.

(ii) $\lim_{n\to\infty} \|D_\infty(x) - D_n(x)\| = 0,$ *for every* $x \in X$.

(iii) $\|x\|^2 = \|D_\infty(x)\|^2 = \langle x, D_\infty(x)\rangle$ *for every* $x \in X$.

Using $\{D_n\}$ in Theorem 321 instead of derivatives of bumps in Theorem 318, Fabian obtained the following milestone result:

Theorem 322 (Fabian, [Fab87]). *If X is Vašák and Asplund then X admits a Fréchet smooth norm.*

Chapter 13

Smooth approximation of norms by norms

13.1 Smooth approximations in separable spaces

If X is a separable Asplund space, then every norm on X can be approximated uniformly on bounded sets by Fréchet differentiable norms. This is due to the fact that X^* admits a dual LUR norm $\|\cdot\|_1$, and so $(\|\cdot\|^2 + \|\cdot\|_1^2)^{1/2}$ is LUR for any dual norm $\|\cdot\|$ on X^*. This method, by taking predual norms, creates the desired approximation on X.
In general it seems to be open if such approximation can be done on X whenever we only assume that X admits a Fréchet smooth norm.

In case of higher-order differentiability, such approximation is a difficult result which uses some deep theorems of M. Zippin. It is done so far only in separable spaces. See, however, Section 13.2 below.
From the approximation of norms by higher-order smooth norms we have the following:

Theorem 323 (Deville, Fonf, Hájek, Talponen, see, e.g., [HáJo14, Chapter 7, Theorem 103]). *Let X be a separable Banach space with a C^k-smooth norm , $k \in \mathbb{N} \cup \{\infty\}$. Then every equivalent norm on X can be approximated uniformly on bounded sets by a C^k-smooth equivalent norm.*

Definition 324. *A* **real-analytic function** *is one that can locally be expressed as a Taylor series.*

The following problem is open: Does Theorem 323 hold true for nonseparable spaces as well? (Compare with Theorem 328 below.) Another related open problem is the following: Is every norm on ℓ_2 approximable uniformly on bounded sets by real-analytic norms? (See, e.g., [HáJo14, Chapter 7, Problem 112].)

The concept of polyhedral space in the statement of the next theorem has been introduced in Definition 141 above:

Theorem 325 (Deville, Fonf, Hájek, see [HáJo14, Chapter 7, Theorem 104]). *If X is a separable space isomorphic to a polyhedral space, then any norm on X can be approximated uniformly on bounded sets by real-analytic norms.*

A. J. Guirao et al., *Renormings in Banach Spaces*, Monografie Matematyczne 75,
https://doi.org/10.1007/978-3-031-08655-7_13

So we have the following open problem: Assume that a nonseparable Banach space $(X, \|\cdot\|)$ admits a C^k-smooth norm and $\varepsilon > 0$ is given. Does there exist a C^k-smooth equivalent norm $|||\cdot|||$ on X such that

$$1 \leq \frac{|||x|||}{\|x\|} < 1 + \varepsilon \text{ for all } 0 \neq x \in X?$$

The problem is solved practically only if $k = 1$ and the space belongs to some known classes, like separable spaces, WLD spaces, etc. It is open for the maximum norm on $C[0, \omega_1]$ and $k = 1$.

Quality approximations of norms by smooth norms are useful, for example, in constructing equivalent norms, as to work with smooth norms is usually much easier than with general norms. The following result goes into this direction:

Theorem 326 (Hájek and Russo, [HáRu17]). *Let $(X, \|\cdot\|)$ be a separable Banach space with Schauder basis $\{e_i\}_{i\in\mathbb{N}}$ that admits a C^k-smooth norm . Then for every sequence $\{\varepsilon_n\}_{n\geq 0}$ of positive numbers there is a C^k-smooth norm $|||\cdot|||$ on X such that for every $n \geq 0$,*

$$\big||||x||| - \|x\|\big| \leq \varepsilon\|x\| \text{ for } x \in X^{(n)},$$

where $X^{(n)} :=$ span $\{e_i\}_{i\geq n+1}$.

The following is one of the few instances in this book dealing with approximation of functions by smooth functions.

Theorem 327 (Johanis, [Joha03]). *Let X be separable, Y a Banach space, and $f : X \to Y$ be L-Lipschitz. Let $\varepsilon > 0$. Then there exist an L-Lipschitz and UG mapping $g : X \to Y$ such that $|(f - g)(x)| < \varepsilon$ for every $x \in X$.*

13.2 Smooth approximation in nonseparable spaces

In nonseparable spaces these problems are widely open and always very difficult. One of a few known results, by the way very useful, is the following result:

Theorem 328 (V. Bible, R.J. Smith, [BiSm16]). *For every Γ, every equivalent norm in $c_0(\Gamma)$ can be uniformly approximated on bounded sets by C^∞-smooth norms.*

Chapter 14

Smooth partitions of unity in nonseparable spaces

On the subject of this chapter, we recommend [DeGoZi93, Chapter VIII].

Let X be a topological space. A family $\mathcal{F}$ of subsets of X is called **locally finite** if every $x \in X$ has a neighborhood that intersects only a finite number of elements in $\mathcal{F}$. A family $\mathcal{F}$ that is a countable union of locally finite families is called **σ-locally finite**.

An **open cover** of a topological space X is a family $\mathcal{V}$ of open subsets of X such that $\bigcup_{V \in \mathcal{V}} V = X$. An open cover $\mathcal{W}$ of X is a **refinement** of $\mathcal{V}$ if every $W \in \mathcal{W}$ is contained in some $V \in \mathcal{V}$.

A **locally finite partition of unity on** X is a family $\{f_\alpha : \alpha \in A\}$ of continuous functions from X into $[0,1]$ such that
(i) the sets $\{x \in X : f_\alpha(x) > 0\}$ form a locally finite open cover of X,
(ii) $\sum_{\alpha \in A} f_\alpha(x) = 1$ for all $x \in X$.
We shall say that the partition of unity is **subordinated** to a cover $\mathcal{V}$ if the cover defined in *(i)* is a refinement of $\mathcal{V}$.

Observe that, for each point $x \in X$ there is a neighborhood $U(x)$ of x where the sum in (ii) is a finite sum.

Recall that for a real valued function f on a Banach space X, the **support of** f is denoted by supp f, i.e. supp $f = \overline{\{x \in X : f(x) \neq 0\}}$.

We say that X **admits C^k-smooth partitions of unity** if for any open covering $\mathcal{U}$ of X there is a C^k-smooth partition of unity $\{\psi_\alpha : \alpha \in \Lambda\}$ which is subordinated to $\mathcal{U}$, i.e., if $\mathcal{U} = \{U_\beta : \beta \in \mathcal{B}\}$, then the partition is of the form $\{\psi_\beta : \beta \in \mathcal{B}\}$ and supp $\psi_\beta \subset U_\beta$ for every $\beta \in \mathcal{B}$.

The following is an updated Toruńczyk result (a **cozero** set for a function $f : S \to \mathbb{R}$ is the complement of the set ker f).

Theorem 329 ([JoZi74a], [GTWZ83], see also [DeGoZi93, Theorem VIII.3.2]). *Let X be a Banach space and $k \in \mathbb{N}$ or $k = \infty$. Then the following are equivalent*
(i) *The space X admits C^k-smooth partitions of unity.*
(i′) *If p is a positive continuous function on X, Y a Banach space, and $f : X \to Y$ is a continuous mapping, then there is a C^k-smooth mapping $g : X \to Y$ such that $\|f(x) - g(x)\| < p(x)$ for every $x \in X$.*

A. J. Guirao et al., *Renormings in Banach Spaces*, Monografie Matematyczne 75,
https://doi.org/10.1007/978-3-031-08655-7_14

(ii) *There is a set* Γ *and a homeomorphic embedding* ϕ *from* X *into* $c_0(\Gamma)$ *such that the coordinate function* $\phi(\cdot)_\gamma$ *is a* C^k*-smooth function on* X*, for every* $\gamma \in \Gamma$.
(iii) *There is a* σ*-locally finite base of the topology of* X *formed by cozero sets of* C^k*-smooth functions.*

In the proof of Theorem 329 the following result is crucial (the concept of local dependence on finitely many coordinates was introduced in Definition 140):

Theorem 330 (Toruńczyk, [Tor73]). *For any set* Γ, $c_0(\Gamma)$ *has a* σ*-locally finite base of its topology formed by cozero sets of* C^k*-smooth functions that locally depend on finitely many coordinates.*

Corollary 331. *If* X *is Vašák, then* X *admits a partition of unity formed by continuous, Gâteaux differentiable functions.*

Corollary 331 was first proved for WCG spaces in [JoZi77].

Theorem 332. (i) ([DeGoZi93, Theorem VIII.3.2]) *Let* X *be WCD. Then* X *admits* C^k*-smooth partitions of unity if* X *admits a* C^k*-smooth bump function.*
(ii) (Haydon, Hájek, [HájHay07]) *Let* k *be a positive integer or* ∞. *If* $X := C(K)$ *for a compact space* K*, then* X *admits a* C^k*-smooth partition of unity if, and only if, it admits a* C^k*-smooth bump function.*

It is unknown if (i) in Theorem 332 holds true without assuming that X is WCD.

Theorem 333 (Deville, Godefroy, Zizler, [DeGoZi90]). *Assume that* Y *is a subspace of a Banach space* X *and that* $k \in \mathbb{N}$ *or* $k = \infty$. *Assume that* Y *is isomorphic to* $c_0(\Gamma)$ *and that* X/Y *admits* C^k*-smooth partitions of unity. Then* X *admits* C^k*-smooth partitions of unity.*

Therefore all spaces JL_0 , JL_2 , and Ciesielski–Pol (defined in [CiePo84], see Theorem 823, see also, e.g., [DeGoZi93, Definition VI.8.8.2] and [GPWZ88]) have C^∞-smooth partitions of unity, and thus the Ciesielski–Pol space admits a homeomorphism into $c_0(\Gamma)$ such that the coordinates are C^∞. It is shown in [DeGoZi90] that the Ciesielski–Pol space $C(K)$ is Lipschitz equivalent to $c_0(\Gamma)$.

We say that X is **Lipschitz equivalent** to Y (or **Lipschitz homeomorphic to**) if there is a one-to-one map φ from X onto Y such that both φ and φ^{-1} are Lipschitz.

Then we have

Theorem 334 (John, Toruńczyk, Zizler, [JoToZi81]). *Every superreflexive Banach space admits partitions of unity formed by functions with uniformly continuous differential.*

The proof of Theorem 335 as well as of Theorem 356 below use an idea due to V. Milman (see, e.g., [DeGoZi93, Chapter VIII. Notes, Section 3]).

Theorem 335 (Vanderwerff, [Vand92b], see, e.g., [Go01b], [DeGoZi93, Theorem VIII.3.12]). *Let* X *be a Banach space which has a LUR norm whose dual norm is LUR. Then* X *has a* C^1*-smooth partition of unity.*

Denote by $(C_0[0,\omega_1], \|\cdot\|_\infty)$ the subspace of $(C[0,\omega_1], \|\cdot\|_\infty)$ formed by all functions $f \in C[0,\omega_1]$ such that $f(\omega_1) = 0$, where $[0,\omega_1]$ is the ordinal segment for the first uncountable ordinal ω_1. Here is the idea, due to Talagrand in [Tal86], how to show that $C_0[0,\omega_1]$ admits a norm that is LFC: Let a bounded linear operator T from $C_0[0,\omega_1]$ into $c_0[0,\omega_1]$ be defined by

$$Tx(\alpha) = \begin{cases} x(\alpha+1) - x(\alpha) & \text{if } \alpha < \omega_1, \\ 0 & \text{if } \alpha = \omega_1. \end{cases}$$

➪ Let us see why $Tx \in c_0[0,\omega_1]$. Fix $\varepsilon > 0$ and assume that $I_\varepsilon := \{\alpha \in [0,\omega_1) : |Tx(\alpha)| > \varepsilon\}$ is infinite. We can find then a strictly increasing sequence $\{\alpha_n\}$ in I_ε. Let α be its limit. Then, $\alpha \in (0,\omega_1)$, and α is a limit ordinal. Obviously, $\alpha_n + 1 \to \alpha$. Then $x(\alpha_n) \to x(\alpha)$ and $x(\alpha_n+1) \to x(\alpha)$, a contradiction with the fact that $|x(\alpha_n+1) - x(\alpha_n)| > \varepsilon$ for all $n \in \mathbb{N}$.

Observe that for every $x \in C_0[0,\omega_1]$ with $\|x\|_\infty = 1$, there exists $\beta \in [0,\omega_1)$ such that $|x(\beta)| = 1$ and $Tx(\beta) \neq 0$. Indeed, it is enough to choose

$$\beta := \sup\{\alpha : \ |x(\alpha)| = 1\}. \tag{14.1}$$

So, T is an example of what is now called a **Talagrand operator** (see, e.g., [FHHMZ11, Exercises 14.54 to 14.56]).

Define an equivalent norm $\|\cdot\|_1$ on $C_0[0,\omega_1]$ by

$$\boxed{\|x\|_1 = \sup_{\alpha\in[0,\omega_1]} \{|x(\alpha)| + |Tx(\alpha)|\}}$$

Given $x \in C_0[0,\omega_1]$ with $\|x\| = 1$, let F be the finite collection of all the coordinates α such that $|Tx(\alpha)| > \frac{1}{2}|Tx(\beta)|$ (> 0). Note that $|x(\alpha)| \leq |x(\beta)| = 1$ for all $\alpha \in [0,\omega_1]$.
By the $c_0[0,\omega_1]$ behaviour and the continuity of T, it is easily seen that the norm $\|\cdot\|$ depends, on a neighborhood of x, only on coordinates from F.

➪ In order to see this, put $\varepsilon := |Tx(\beta)|/12$. Take $y \in C_0[0,\omega_1]$ such that $\|y - x\|_\infty < \varepsilon$. Then

$$|y(\beta)| + |Ty(\beta)| > |x(\beta)| - \varepsilon + |Tx(\beta)| - 2\varepsilon = |x(\beta)| + \frac{3}{4}|Tx(\beta)|.$$

Note that $\beta \in F$. If $\alpha \notin F$, we have $|Tx(\alpha)| \leq (1/2)|Tx(\beta)|$. Then, for such an α we have

$$\begin{aligned} |y(\alpha)| + |Ty(\alpha)| &< |x(\alpha)| + \varepsilon + |Tx(\alpha)| + 2\varepsilon \\ &\leq |x(\beta)| + \varepsilon + \frac{1}{2}|Tx(\beta)| + 2\varepsilon = |x(\beta)| + \frac{3}{4}|Tx(\beta)|. \end{aligned}$$

This shows that $\|y\|_1 = \sup\{|y(\alpha)| + |Ty(\alpha)| : \alpha \in F\}$.

Hence $C_0[0, \omega_1]$ admits an equivalent norm that locally depends on finitely many coordinates. As $C[0, \omega_1]$ is isomorphic to $C_0[0, \omega_1]$, we thus have that $C[0, \omega_1]$ admits such an LFC norm, too. □

By a subtle renorming one gets

Theorem 336 (Haydon, [Hay96]). *The space $C[0, \mu]$ of continuous functions on the ordinal segment $[0, \mu]$ admits a C^∞-smooth norm.*

In the end of this section let us mention the following important result:

Theorem 337 (Haydon, [Hay08]). *If X is a Banach space that admits a norm whose dual is LUR, then X admits a LUR norm. Moreover, this equivalent LUR norm on X can be chosen to have a dual LUR norm, too.*

The "moreover" statement in Theorem 337 is an immediate consequence of the Asplund averaging technique (see Item 2 in Section 8.2 above). As it is mentioned in [Hay08], it is known ([DeGoZi93, Theorem VIII.3.12]) that a Banach space with a norm which is LUR and has a LUR dual norm admits C^1-smooth partition of unity: equivalently, on such a space every continuous real-valued function may be uniformly approximated by C^1-smooth functions. We have thus the following corollary:

Corollary 338. *Let X be a Banach space whose dual norm is LUR. Then every continuous real-valued function on X may be uniformly approximated by C^1-smooth functions.*

Theorem 337 is a consequence of the following renorming result for a certain class of $C(K)$ spaces:

Theorem 339 (Haydon, [Hay08]). *Let K be a Namioka–Phelps compact space. Then there is an equivalent LUR norm on $(C(K), \|\cdot\|_\infty)$.*

If $(X, \mathcal{T})$ is a topological space and d is a metric on X inducing a topology finer than $\mathcal{T}$, we say that the **property** $P(d, \mathcal{T})$ holds if there is a sequence $\{B_n : n \in \mathbb{N}\}$ of subsets of X such that the topology generated by $\mathcal{T} \cup \{B_n : n \in \mathbb{N}\}$ is finer than the topology $\mathcal{T}_d$ induced by the metric d. If $(K, \mathcal{T})$ is compact and has $P(d, \mathcal{T})$ for some $\mathcal{T}$-lower semicontinuous metric d, then K is called a **Namioka–Phelps compact**.
Raja has shown the following result, used in the proof of Theorem 337. See also (vi) in Theorem 775, where an alternative description of the concept of Namioka–Phelps compact space appears.

Theorem 340 (Raja, [Raj02]). *If X is a Banach space, then (B_{X^*}, w^*) is a Namioka–Phelps compact if, and only if, X admits an equivalent norm with LUR dual norm.*

Chapter 15

Smooth norms in dense subspaces

Theorem 341 (Dantas, Hájek, and Russo, [DaHaRu20], Theorem A). *The following spaces have dense subspaces with an analytic renorming:*
(i) *Every separable Banach space.*
(ii) ℓ_∞.
(iii) $\ell_1(c)$.

Note that (ii) in Theorem 341 above follows by using a Preiss-like norm on the normed space of finitely valued sequences in ℓ_∞.
However,

Theorem 342 (Dantas, Hájek, and Russo, [DaHaRu20], Theorem 3.10). *No dense subspace of $c_0(\omega_1)$ admits an analytic norm.*

Theorem 343 (Dantas, Hájek, and Russo, [DaHaRu20], Theorem B). *If X is a nonseparable Banach space with uncountable Schauder basis, and Y is the linear hull of this basis of X, then Y has an equivalent C^∞-smooth norm.*

A. J. Guirao et al., *Renormings in Banach Spaces*, Monografie Matematyczne 75,
https://doi.org/10.1007/978-3-031-08655-7_15

Chapter 16

Miscellaneous applications

16.1 Difference of two convex continuous functions

A **DC function** on a Banach space X is a difference of two convex continuous functions on X.

The following results are very important for understanding the relation of linear and metric structures of Banach spaces.

Theorem 344 (Cepedello, [Cep98], see, e.g., [Go01b]). *Let X be a Banach space. The following are equivalent.*
(i) *X is superreflexive.*
(ii) *Every Lipschitz function on X is the uniform limit on X of a sequence of functions $g_n = c_n - d_n$, where c_n and d_n are convex functions which are bounded on bounded sets of X.*

Theorem 345 (Cepedello, [Cep98]). *Assume that a Banach space X admits a LUR norm. Then every Lipschitz function on X is a pointwise limit on X of DC functions.*

Theorem 346 (Cepedello, [Cep98]). *There is a Lipschitz function on ℓ_∞ that is not a pointwise limit of DC functions.*

Proof. (Sketch) The proof of Theorem 346 uses Talagrand's result in [Tal78] that there are norm-Borel sets in ℓ_∞ that are not Borel in the weak topology of ℓ_∞.□

Note that from Theorem 345 and Theorem 346 we get again that ℓ_∞ does not admit any LUR norm.

16.2 Some topological issues

The paper [Cep98], mentioned in Section 16.1, also gives a new proof to the LUR version of the first part of the following result of Edgar:

Theorem 347 (Edgar, [Edg79]). *Assume that a Banach space X admits a Kadets–Klee norm. Then the Borel fields for the norm and the weak topology coincide. Moreover, X is Borel in X^{**} in its weak star topology.*

A. J. Guirao et al., *Renormings in Banach Spaces*, Monografie Matematyczne 75,
https://doi.org/10.1007/978-3-031-08655-7_16

Proof. (Hint for the first part) If $F \subset X$ is norm-closed, then put f for the distance function to F. Apply Theorem 514 to see that f is weak Borel (continuous functions are weak-lower semicontinuous). Thus f and also its zero set are weak Borel. □

See also [FHHMZ11, Exercise 8.53].

A family of subsets of X is **discrete** if every point of X has a neighborhood that intersects at most one member of the family. A family of sets is **σ-discrete** if it is a union of a countable number of discrete families.

The following problem is open: If the Borel fields for the norm and the weak topology on a Banach space X coincide, can X be renormed by an equivalent Kadets–Klee norm?

Theorem 348 (Raja, see, e.g., [FHHMZ11, Exercise 12.42])**.** *Assume that a Banach space admits a LUR norm. Then the norm topology of X has a σ-discrete base consisting of convex sets.*

It is open whether Theorem 348 holds for all Banach spaces.

Theorem 349. *If the norm of X is LUR, then the set of norm-attaining functionals is G_δ dense in the norm topology of X^*.*

Proof. Due to the LUR property of the norm of X, it follows from Šmulyan's lemma that a nonzero element of X^* is norm-attaining if and only if it is a point of Fréchet differentiability of the dual norm. The set of Fréchet differentiability of the norm is always G_δ. □

Theorem 350 (Guirao, Montesinos, Zizler, [GuiMoZi18])**.** *Let X be a separable Banach space. Then for its norm $\|\cdot\|$ the set of all norm-attaining functionals is residual in the norm topology of X^* if and only if the dual norm $\|\cdot\|^*$ is Fréchet differentiable at a dense set of points in X^*.*

The proof of Theorem 350 is a direct consequence of the methods in [PrZa84].

In this connection, let us mention here a few results in [Kau91]:

Theorem 351 (Kaufman, [Kau91])**.** (i) *If the norm of a separable Banach space is rotund, then the set of all norm-attaining functionals is Borel even in the weak star topology of X^*.*
(ii) *Every separable nonreflexive Banach space can be so renormed that the set of all norm-attaining functionals is not Borel in the norm topology of X^*.*

Recall that a topological space is called a **Polish space** if it is homeomorphic to a complete metrizable separable space.

From the applications of LUR norms let us mention here the following.

Proposition 352 (See [FHHMZ11, Proposition 14.52]). *Let X have a w-$\mathcal{T}_{\|\cdot\|}$-KK norm. Then (B_X, w) is a Baire space.*

Proof. Assume without loss of generality that X is infinite-dimensional. (S_X, w) is a dense G_δ set in (B_X, w). Since the weak and norm topology on S_X coincide and S_X is a complete metric space, (S_X, w) is a Baire space. It is not difficult to show that thus (B_X, w) is a Baire space. □

Theorem 353 (See [FHHMZ11, Proposition 14.53]). *Let B_{c_0} be the closed unit ball of $(c_0, \|\cdot\|_\infty)$. Then (B_{c_0}, w) is not a Baire space.*

Proof. For $n \in \mathbb{N}$, let $B_n = \{x = (x_i) \in B_{c_0} : |x_i| \leq \frac{1}{2}$ for $i \geq n\}$. Then B_n is closed in (B_{c_0}, w). Moreover, B_n is nowhere dense in (B_{c_0}, w). Indeed, if $x \in B_n$ and $p \in \mathbb{N}$ and $\varepsilon > 0$ are given, there is $\tilde{x} \in B_{c_0}$ such that $|\tilde{x}_i - x_i| < \varepsilon$ for $i \leq p$ and $|\tilde{x}_{\max\{(p+1),n\}}| > \frac{1}{2}$, so $\tilde{x} \notin B_n$. Thus $B_{c_0} = \bigcup_{n=1}^\infty B_n$ is of first category in itself in its weak topology, hence not a Baire space. □

Corollary 354. *Let B be the closed unit ball of c_0 under a LUR norm. Then (B, w) is a Baire space but not a Polish space.*

Proof. Assume without loss of generality that B contains the ball B_1 under the supremum norm. If (B, w) was Polish, so it would be (B_1, w), being a closed subspace of (B, w). □

Let us mention here in passing the following result:

Theorem 355 (Godefroy, see [FHHMZ11], Theorem14.51). *If X^{**} is separable, then (B_X, w) is a Polish space.*

16.3 An application of locally uniformly convex renorming to lower semicontinuous functions

A further application of LUR renormings is the following:

Theorem 356 (Poliquin, Vanderwerff, and Zizler, [PoVaZi94]). *Let a Banach space X admit a LUR norm $\|\cdot\|$. Then every lower semicontinuous bounded below function $f : X \mapsto \mathbb{R} \cup \{\infty\}$ can be expressed in the form*

$$f(x) = \Phi(x, \|x\|^2),$$

where $\Phi : X \times \mathbb{R} \mapsto \mathbb{R} \cup \{\infty\}$ is a lower semicontinous convex function.

Chapter 17

Bumps depending locally on finitely many coordinates

The notion of dependence on a finite number of coordinates was introduced in Definition 140.
The method of proof in the second part of the following result goes back to Kurzweil [Kurz54].

Theorem 357 (Fabian, Pechanec, Whitfield, and Zizler, [PeWhZi81], [FaZi97]). *Let X be a separable Banach space. If X admits a continuous LFC bump, then X^* is separable and every subspace of X contains a subspace isomorphic to c_0.*

Proof. For the separability of X^* see, e.g., [FHHMZ11, Theorem 10.13].
For the statement about c_0, we may assume that there exists a sequence $\{y_n^*\}_{n=1}^\infty$ in B_{X^*} such that g depends locally on finitely many y_n^*'s. By composing with a convenient real-valued function on $\mathbb{R}$ we may also assume that $g(x) = 0$ for $\|x\| \leq 1/2$ and that $g(x) = 1$ for $\|x\| \geq 1$.
We will construct inductively a sequence $\{x_n\}_{n=0}^\infty \subset X$ as follows: set $x_0 = 0$ and if $x_0, x_1, \dots, x_n$ have been chosen, find x_{n+1} so that

(1) $g(\sum\limits_{i=0}^{n} \varepsilon_i x_i + \varepsilon_{n+1} x_{n+1}) = 0$ for $\varepsilon_i = \pm 1$, $i = 1, \dots, n+1$,

(2) $\|x_{n+1}\| \geq \frac{1}{2} M_n = \frac{1}{2} \sup \|y\|$,

where the supremum is taken over all $y = x_{n+1}$ satisfying (1). Since $g(\sum_{i=1}^k \varepsilon_i x_i) = 0$ for every $k \in \mathbb{N}$, we get $\|\sum_{i=1}^k \varepsilon_i x_i\| \leq 1$ for all $k \in \mathbb{N}$. If X does not contain an isomorphic copy of c_0, then we have that $\sum x_i$ is unconditionally convergent and $S = \left\{\sum_{i=1}^n \varepsilon_i x_i : \ \varepsilon_i = \pm 1, n \in \mathbb{N}\right\}$ is relatively compact in X.
From the local dependence of g on finitely many $\{y_n^*\}$ we have that for each $z \in \overline{S}$ there is $\delta_z > 0$ and a finite set K_z of natural numbers such that $g(w \pm \delta x) = g(w)$ for all $w \in B(z, \delta_z)$, $\delta \leq \delta_z$ and $x \in \bigcap\limits_{i \in K_z} (y^*)_i^{-1}(0)$. By compactness, $\overline{S} \subset \bigcup\limits_{i=1}^{k} B(z_i, \delta_{z_i})$ for some $z_i \in X$. Let $\delta = \min\{\delta_{z_i} : \ i = 1, \dots, k\}$ and $K = \bigcup\limits_{i=1}^{k} K_{z_i}$. Let $x \in \bigcap\limits_{i \in K} (y^*)_i^{-1}(0)$, $0 < \|x\| < \delta$. If $w \in S$ then $w \in B(z_i, \delta_{z_i})$ for some i and hence $g(w \pm \delta x) = g(w)$. Thus $\inf(M_n) \geq \delta > 0$ for all $n \in \mathbb{N}$, which contradicts the convergence of $\sum x_n$. □

A. J. Guirao et al., *Renormings in Banach Spaces*, Monografie Matematyczne 75,
https://doi.org/10.1007/978-3-031-08655-7_17

Proposition 358. *There exists a separable Asplund space saturated with* c_0 *and not admitting a LFC bump.*

It is an open problem whether every separable space X that admits an LFC continuous bump admits an LFC norm.
It was proved by P. Hájek [Há95] that if a separable space X admits a norm that is LFC, then X admits a norm that is C^∞-smooth (see, e.g., [FHHMZ11, Theorem 10.12]). It is an open problem whether this holds for all Banach spaces.

Chapter 18

Summary on renorming for uniformly rotund in every direction, strictly convex, and weakly uniformly rotund spaces

We mentioned in Subsection 9.3.1 that the basic Clarkson's renorming method provided R, URED, or URWC equivalent norms. We collect the aforementioned results here:

Theorem 359. (see, e.g., [DeGoZi93, Theorem II.6.8]) *Let X and Y be two Banach spaces and assume Y admits an R norm (respectively URED norm, respectively URWC norm). Assume that there is a bounded linear one-to-one operator T from X into Y. Then X admits an R norm (respectively URED norm, respectively URWC norm).*

We mentioned also in the same paragraph that, in the case that (S_{X^*}, w^*) is separable, we may use the operator T from X into ℓ_2 defined by $Tx = (f_i(x)/2^i)_{i=1}^{\infty}$, where $\{f_i\}_{i=1}^{\infty}$ is a sequence of elements of S_{X^*} that is w^*-dense in S_{X^*}. Similarly, if $\{x_i\}_{i=1}^{\infty}$ is norm dense in S_X for X separable, then the map $T : X^* \to \ell_2$ defined by $Tx^* = (\langle x_i, x^* \rangle / 2^i)_{i=1}^{\infty}$ is w^*-continuous and one-to-one. Thus, we get the following:

Theorem 360. *Let X be a separable Banach space.*
(i) *The space X admits an equivalent URWC norm (in particular, an equivalent URED norm).*
(ii) *The space X^* admits a dual W* UR norm*

Corollary 361. *Every separable Banach space admits a UG norm.*

Thus we get (i) in the following theorem.

Theorem 362. (See, e.g., [FHHMPZ01, Theorem 8.14] or [FHHMZ11, Theorem 11.21]) *Let X be a separable Banach space.*
(i) (Mazur), see, e.g., [FHHMZ11, Theorem 8.2] *For any continuous convex function on X, the set of points where f is Gâteaux differentiable is G_δ dense in X.*

A. J. Guirao et al., *Renormings in Banach Spaces*, Monografie Matematyczne 75,
https://doi.org/10.1007/978-3-031-08655-7_18

(ii) (Aronszajn, Christensen, Mankiewicz) (See, e.g., [FHHMZ11, Theorem 11.21]) *If f is a Lipschitz function on X, then the set of points where f is Gâteaux differentiable is dense in X .*
(iii) (Alexandrov, [Alex39]) *Assume that f is a convex function on a finite-dimensional Hilbert space $(H, \langle\cdot,\cdot\rangle)$. Then for almost all points $x \in H$ there is $y \in H$ and a symmetric bilinear form T_2 on H such that*

$$f(x+th) - f(x) - \langle y, th\rangle - t^2 T_2(h,h) = o(t^2), \text{ as } t \to 0.$$

Proof. (i) For an elementary proof see, e.g., [Phe93, Theorem 1.20].
(iii) See a proof in [BiCoPu96]. □

It is open if (iii) in Theorem 362 holds for separable infinite-dimensional Hilbert spaces as well.

For some more in this direction see, e.g., [BoNo94].

Theorem 363 (Kutzarova, [Kut84], see [DeGoZi93, Theorem II.7.16]). *If μ is a finite measure, then $L_1(\mu)$ admits an URED norm.*

An important information in this area was also the following result.

Theorem 364 (Kutzarova, Troyanski, [KuTr82], see, e.g., [DeGoZi93, Theorem IV.6.5], and [HMVZ07, Theorem 6.35]). *There is a reflexive Banach space with unconditional basis that cannot be renormed either by a UG norm or a URED norm.*

In the theory of injections the following result is important.

Theorem 365 (Hájek, [Há94]). *There exists a nonseparable reflexive Banach space with the property that no nonseparable subspace of which can be mapped by a one-to-one bounded linear operator into some superreflexive Banach space.*

Chapter 19

Examples on C^1-smoothness

19.1 Uniformly Gâteaux differentiable norms and related results

Before we start on examples, let us mention that we reviewed some differentiability properties of the standard norms of basic classical spaces like $C[0,1]$, c_0, ℓ_1, etc. in Subsection 4.1.4. See also [DeGoZi93, Examples I.1.6] or [FHHMZ11, Exercises 7.36–7.41]. In order to shorten some expressions, we sometimes will use G for "Gâteaux differentiable" and F for "Fréchet differentiable".

Theorem 366 (Borwein, Fabian, [BoFa93]). *Every separable infinite-dimensional Banach space can be renormed by a UG norm that fails to be Fréchet differentiable everywhere. Every WCG space can be renormed by a Gâteaux differentiable norm that is not Fréchet differentiable at some non-zero point.*

Proof. (Sketch, the detailed proof is, e.g., in [HáMoZi12]). Let $(X, \|\cdot\|)$ be a separable infinite-dimensional Banach space. Let Y be a closed hyperplane of X. We may assume that $\|\cdot\|$ on $X = Y \oplus \mathbb{R}$ satisfies $\|y + t\| = \|y\| + |t|$ for $y \in Y$ and $t \in \mathbb{R}$, so that on $(Y^* \oplus \mathbb{R})$ the dual norm is $\|(\zeta, r)\| = \max\{\|\zeta\|, |r|\}$. Let $\{y_n\}_{n=1}^{\infty}$ be a dense sequence in the unit sphere of Y, and a compact operator $T : \ell_2 \to Y$ be defined by

$$T\big((\lambda_n)\big) = \sum \frac{1}{2^n} \lambda_n y_n$$

for $(\lambda_n) \in \ell_2$.

Define a norm $|||\cdot|||$ on $Y^* \oplus \mathbb{R}$ by

$$|||(\zeta, r)|||^2 = \max\{\|\zeta\|^2, r^2\} + \|T^*(\zeta)\|_{\ell_2}^2 + r^2, \qquad (\zeta, r) \in Y^* \oplus \mathbb{R}.$$

The Josefson–Nissenzweig theorem (see, e.g., [HMVZ07, Theorem 3.27]) guarantees the existence of a sequence $\{\zeta_n\}_{n=1}^{\infty}$ in the unit sphere of $(Y^*, \|\cdot\|)$ such that $\zeta_n \to 0$ in the w^*-topology.
Put

$$\xi_n := (\zeta_n, 1)/|||(\zeta_n, 1)|||, \ \xi = \left(0, \frac{\sqrt{2}}{2}\right) \in Y^* \oplus \mathbb{R}, \ \text{and } x = (0, \sqrt{2}) \in Y \oplus \mathbb{R},$$

A. J. Guirao et al., *Renormings in Banach Spaces*, Monografie Matematyczne 75,
https://doi.org/10.1007/978-3-031-08655-7_19

for $n \in \mathbb{N}$. Then we calculate

$$|||\xi_n||| = |||\xi||| = 1 = \langle x, \xi \rangle = |||x||| \text{ for } n \in \mathbb{N},$$

$$\xi_n \to \xi \text{ weak}^*, \quad \text{and } |||\xi_n - \xi||| \to \frac{\sqrt{2}}{2}.$$

Thus, by the Šmulyan's test, the norm $|||\cdot|||$ on $Y \oplus \mathbb{R}$ is not Fréchet differentiable at x.
Using the operator T and a standard argument (see Proposition 162) we see that the norm $|||\cdot|||$ on X is UG since its dual norm is W^*UR. □

Note that it follows from the Šmulyan's test (see, e.g., [FHHMZ11, Corollary 7.22]) that *if the norm is Fréchet differentiable at a point $x \in S_X$, then x is a point of Fréchet differentiability of its second dual norm.*

The following result is also instructional.

Theorem 367 (Borwein, Fabian, [BoFa93]). *If X^* is separable, then X admits a Gâteaux differentiable norm that is Fréchet differentiable at all but one point in S_X.*

Another example of a UG norm that is not F, is as follows:

Theorem 368. *Let $\|\cdot\|$ be a WUR norm on c_0* (see Theorem 276). *Its dual norm $\|\cdot\|^*$ on ℓ_1 is UG and not Fréchet differentiable.*

Proof. If the dual norm is F, then the space is reflexive (see Theorem 105). □

If one looks just for an easy example of a point of Gâteaux and not Fréchet differentiability of a norm on ℓ_2, one can use Šmulyan's test and Lindenstrauss' example presented after Theorem 402.

R.R. Phelps provided the mathematical community with the first example of a Gâteaux differentiable but nowhere Fréchet differentiable norm in a separable Banach space (see Theorem 172 above). The two following results enlarge the view on this construction. The first one strengthens Theorem 172 (see also Theorem 371 below). The second one characterizes the Banach spaces having a Gâteaux nowhere Fréchet differentiable norm.

Theorem 369 (Wee-Kee Tang, [WKT96a], see also [HáMoZi12]). *Let X be a separable Banach space that contains an isomorphic copy of ℓ_1. Then X admits a UG norm that is nowhere very smooth .*

For a proof of the following result see, e.g., [DeGoZi93, Theorem III.1.9]. Observe that $K := B_{\ell_\infty}$ satisfies the requirements in Theorem 370, as it follows easily from the argument in the proof of Theorem 172, so we recover Theorem 172 above as a particular case.

Theorem 370 (Deville, Godefroy, Hare, Zizler, [DGHZ87]). *Let X be a separable Banach space. Then X admits a Gâteaux differentiable norm that is nowhere Fréchet differentiable if, and only if, there is a convex w^*-compact subset K of X^* and $\varepsilon > 0$ such that for every nonempty w^*-open subset O of K we have* $\operatorname{diam} O > \varepsilon$.

The following result of a former student of M.I. Kadets also belongs to this area. Compare with Theorem 369 above.

Theorem 371 (Godun, [Go85b], see also [HáMoZi12]). *Let X be a separable nonreflexive Banach space. Then X admits a UG norm that is not very smooth at some points of S_X .*

Speaking on very smooth norms, let us note in passing the following result, that was already mentioned in Section 4.2.

Theorem 372 (Singer, [Si75], see also [Di75, Theorem 2.2.2]). *Assume that the norm of a Banach space X is very smooth. Then X is Asplund.*

Proof. (Sketch). The Bishop–Phelps Theorem 14 above states that the image of the duality mapping is $\|\cdot\|$-linearly dense in X^*. □

The main idea in the proof of the following result is due to V.L. Klee ([Klee59], see, e.g., [FHHMZ11, Exercise 8.63]).

Theorem 373 (Guirao, Montesinos, Zizler, [GuiMoZi12]). *Let X be infinite-dimensional nonreflexive and separable. Then X admits a norm that is Gâteaux differentiable and LUR but its dual is not rotund. If X^* is moreover separable, then X admits a norm $\|\cdot\|$ that is F, LUR, WUR, whose dual norm has the w-$\mathcal{T}_{\|\cdot\|}$-KK property although it is not R.*

Note that the norm in the second part of the theorem above gives that the dual of this norm is Gâteaux differentiable on X^* as implied by its WUR property.

Theorem 373 is an upgraded version of a classical result of V. Klee ([Klee59]). The main idea of V. Klee was to construct a Gâteaux smooth norm on X, the quotient of which by a two-codimensional subspace is not Gâteaux differentiable. This, by duality, gives that the dual norm cannot be rotund. In the proof, the classical Šmulyan result on the existence of a nested family of closed convex bounded sets with empty intersection in nonreflexive spaces is used. Note that nowadays this Šmulyan result follows by taking a not norm-attaining functional and consider the slices generated by it (James' weak compactness Theorem 13). It is apparently open whether there is an equivalent C^2-smooth norm on c_0 whose dual is not rotund.

The first example of a separable space G the dual of which is not R was independently shown by S.L. Troyanski ([Tr70]).

Note that the dual of a G norm is R in the case of reflexive spaces (see Proposition 82 above). However,

Theorem 374 (Yost, [Yo81]). *Let X be a reflexive infinite-dimensional separable Banach space. Then X can be renormed so that its norm is Fréchet differentiable but its dual norm is not LUR.*

Thus, by the Šmulyan's lemma and duality, every separable reflexive space admits a norm that is not LUR but all points of the sphere are strongly exposed points of the ball.

Proof of Theorem 374. We will show dually that there is a norm on an infinite-dimensional reflexive separable space X that is not LUR but its dual is Fréchet smooth.
Let $\{x_n, f_n\}_{n\in\mathbb{N}}$ be a Markushevich basis for X such that $\|x_n\| = 1$ for all $n \in \mathbb{N}$ and $\{f_n\}$ is bounded (see, e.g., [HMVZ07, Theorem 1.27]).
We may assume, by Theorem 173, that the original norm $\|\cdot\|$ of X is LUR. Proposition 90 shows then that $\|\cdot\|$ is w-$\mathcal{T}_{\|\cdot\|}$-KK.
Define a norm $|||\cdot|||$ on X by

$$|||x|||^2 = (|f_1(x)| + \|x - f_1(x)x_1\|)^2 + \sum_{n=2}^{\infty} 2^{-n} f_n(x)^2. \tag{19.1}$$

Then $|||x_n||| \to |||x_1|||$ and $|||x_1 + x_n||| \to 2$, but $\{x_n\}_{n\in\mathbb{N}}$ does not converge weakly to x_1. So the norm $|||\cdot|||$ is not LUR.
We need to show that $|||\cdot|||^*$ is Fréchet differentiable. First we show that $|||\cdot|||$ is rotund. This follows from the third term of (19.1) combined with the first two terms of (19.1), using the fact that $\bigcap_{n=2}^{\infty} f_n^{-1}(0)$ is necessarily one-dimensional, in fact the linear hull of $\{x_1\}$. As X is reflexive, this means that $|||\cdot|||^*$ is Gâteaux differentiable. Next, due to the fact that $\|\cdot\|$ is w-$\mathcal{T}_{\|\cdot\|}$-KK, it follows that $|||\cdot|||$ has the w-$\mathcal{T}_{\|\cdot\|}$-KK property, too. From Corollary 147 we get that $|||\cdot|||^*$ is Fréchet differentiable. □

19.2 Uniformly Kadets–Klee smooth norms

Definition 375. *We say that the norm of a Banach space X is* **uniformly Kadets–Klee smooth (UKKS)** *if for every $\varepsilon > 0$ there exists $\delta > 0$ such that given $x^* \in X^*$ with the property that diam $(V \cap B_{X^*}) > \varepsilon$ for each w^*-neighborhood V of x^*, then $\|x^*\| < 1 - \delta$.*

The concept of separation $\operatorname{sep}\{x_n\}_{n=1}^{\infty}$ of a sequence $\{x_n\}_{n=1}^{\infty}$ was introduced in Definition 100 above.

Definition 376. *The norm is called* **uniformly Kadets–Klee (UKK)** *if for every* $\varepsilon > 0$ *there is a* $\delta < 1$ *such that if* $\{x_n\}_{n=1}^{\infty}$ *is a sequence in* B_X *such that* $\operatorname{sep}\{x_n\}_{n=1}^{\infty} \geq \varepsilon$ *and* $x_n \to x$ *weakly, then* $\|x\| \leq \delta$.

Remark 377. UKK properties originated in the paper [Hu80]. There is a more restrictive version of the concept of UKK used by Lancien [Lanc95, Definition 3.1]: The norm of a Banach space $(X, \|\cdot\|)$ is said to be **uniformly Kadets–Klee** if for every $\varepsilon > 0$ there exists $\delta \in (0,1)$ such that every $x \in B_X$ with $\|x\| > 1 - \delta$ has a w-open neighbourhood U with $\operatorname{diam}(U \cap B_X) < \varepsilon$. In the case of reflexive spaces, Lancien's UKK property coincides with Huff's UKK, and with the NUC property, too (see Theorem 421 below).
The property **uniform Kadets–Klee for the** w^***-topology** (UKK*) for the dual norm of a Banach space (Lancien, [Lanc95, Definition 3.2]) coincides with what we called "UKKS for the norm of the space" (Definition 375 above). We shall use indistinctly one or the other name along this book (the first one referred to the dual norm, the second one to the original norm). ®

Let us note that UF norms are UKKS norms (see, e.g., [FHHMZ11, Exercise 9.10]). If we replace in the definition of UKKS norms the w^*-neighborhood by w^*-slices, we arrive to the superreflexivity of the space (see [FHHMZ11, Theorem 9.22]).

UKKS norms are extensively used in questions of Lipschitz nonlinear homeomorphisms of Banach spaces (see, e.g., [FHHMZ11, Chapter 12]) and have applications in fixed point theory (see, e.g., [AyDoLo97]).

Theorem 378. (See, e.g., [FHHMZ11, Proposition 12.64]) *The canonical norm of* $c_0(\Gamma)$ *is UKKS for each* Γ.

An example of separable reflexive Banach space that cannot be renormed by UKKS norm is in [FHHMZ11, Exercise 9.9].

Finally let us mention here the following theorem.

Theorem 379. (See, e.g., [FHHMZ11, Theorem 8.24]) Let K be an infinite compact metric space. Then $C(K)$ is isomorphic to c_0 if, and only if, $C(K)$ admits an equivalent UKKS norm.

Chapter 20

Examples on Rotundity

20.1 Normal structure

Definition 380. *If C is a bounded subset of a Banach space X, a point $x \in C$ is called* **diametral** *if*

$$\sup\{\|x-y\| : y \in C\} = diam\ C \quad (=: \sup\{\|z_1 - z_2\| : \ z_1, z_2 \in C\}).$$

Definition 381. *A nonempty bounded convex subset C of a Banach space $(X, \|\cdot\|)$ is said to have* **normal structure** (a concept introduced by Brodskii and Milman [BroMi48]) *if every nonempty convex subset of C, not reduced to a point, contains a non-diametral point. A Banach space X is said to have* **normal structure** *if any nonempty convex bounded subset of X has normal structure.*

Spaces with a URED norm have normal structure (Theorem 387 below), and the notion of URED can be described by using Chebyshev centers (Theorem 383). This section explores the connection between these notions. URED norms are widely used, for instance, in Fixed Point Theory, in the study of Chebyshev (Haar) subspaces, etc. A subspace Y of a Banach space X is called a **Chebyshev subspace** if each point in X has a unique closest point in Y.

Let us mention here just a few things to start with.

1. In the space L_1, there is a nontrivial line segment on the boundary of the unit ball in each direction (see [Ph60b]). Thus, in this case, there are no one-dimensional Chebyshev subspaces.

2. In many cases, the notion of Chebyshev one-dimensional subspaces coincide with directions of URED, for example in $C[0,1]$ (see [Ziz71]).

3. There is a line segment on the unit sphere of c_0 in the direction of $z \neq 0$ if and only if $z = (z_i)$ has some coordinate 0 ([Ph60b]). However, the space c_0 is URED in no direction ([Ziz71]).

4. For any 3-dimensional ball there is a direction such that there is no nondegenerate line segment in the boundary of this direction, i.e., any 3-dimensional Banach space contains a 1-dimensional Chebyshev subspace (see references in [Ziz71], or [Klee69], or [PavZa07]). Apparently unknown is the general finite-dimensional case.

A. J. Guirao et al., *Renormings in Banach Spaces*, Monografie Matematyczne 75,
https://doi.org/10.1007/978-3-031-08655-7_20

Overall, let us note that the problems on finite-dimensional Chebyshev subspaces has been a big prosperous part of an interesting good classical Mathematics.

In the sequel, we will use the following few classical results. We need first a definition:

Definition 382. *If X is a Banach space and A is a bounded set in X, put, for $x \in X$,*

$$r_x(A) = \sup\{\|x - y\| : \ y \in A\},$$

and call it the **radius of A from x**. *Then we put*

$$\mathrm{rad}(A) = \inf\{r_x(A) : \ x \in X\}.$$

Then call $x \in X$ a **Chebyshev center of A** *if*

$$r_x(A) = \ \mathrm{rad}(A).$$

For a sketch see Figure 20.1. We have

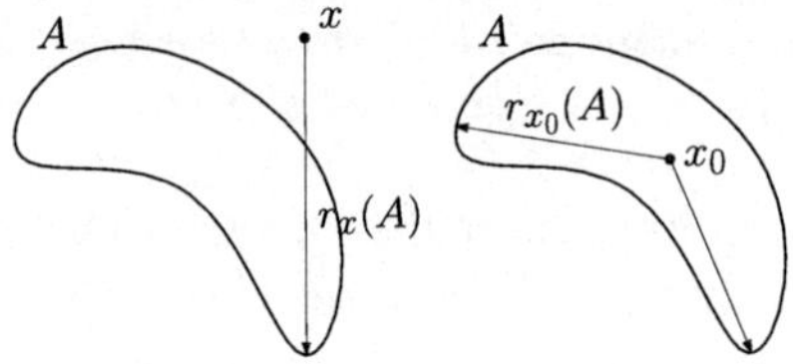

Figure 20.1: The radius form x, and a Chebyshev center x_0

Theorem 383 (Garkavi, [Gark64]). *The Banach space X is URED if, and only if, every bounded set in X has at most one Chebyshev center.*

Proof. (Partial) To see one implication, assume that for some bounded set M there are two Chebyshev centers, y_1 and y_2. Assume that $\mathrm{rad}(M) = 1$

Put

$$y_c = \frac{1}{2}(y_1 + y_2).$$

This is another Chebyshev center. Let $\{x_n : \ n \in \mathbb{N}\} \subset M$ be chosen such that

$$\lim_{n \to \infty} \|y_c - x_n\| = 1.$$

Since

$$y_c - x_n = \frac{1}{2}(y_1 - x_n) + \frac{1}{2}(y_2 - x_n),$$

$$\|y_1 - x_n\| \leq 1, \quad \text{and} \quad \|y_2 - x_n\| \leq 1,$$

we have

$$\lim_{n\to\infty} \|y_1 - x_n\| = \lim_{n\to\infty} \|y_2 - x_n\| = 1.$$

Consider the points

$$u_1^n = (y_1 - x_n), \quad u_2^n = (y_2 - x_n), \ n \in \mathbb{N}.$$

Then $\lim_n \|u_1^n\| = \lim_n \|u_2^n\| = 1$, $u_1^n - u_2^n = (y_1 - y_2)$ and $\lim_n \|u_1^n + u_2^n\| = 2\lim_n \|y_c - x_n\| = 2$. Therefore the space is not URED in the direction $y_1 - y_2$.□

As an instructive remark let us now repeat here two exercises from [FHHMZ11]: Let

$$A = \{x(t) \in C[0,1] : \ 0 = x(0) \leq x(t) \leq x(1) = 1\}.$$

Define a mapping $T : A \to A$ by

$$T(x)(t) = tx(t).$$

Then A is a bounded closed convex set in $C[0,1]$ such that every point of A is diametral [FHHMZ11, Exercise 12.11] and T is a nonexpansive mapping that has no fixed point [FHHMZ11, Exercise 12.10] (by a **nonexpansive mapping** we mean that $\|Tx - Ty\| \leq \|x - y\|$ for all x, y).

In the proof of Theorem 385 below, induction is used. Maybe the proof that every nonempty $\|\cdot\|$-compact convex subset of a Banach space has normal structure (see Definition 381) gives the right hint for understanding the Brodskii–Milman statement (Theorem 385 below):

Proposition 384. *Every nonempty $\|\cdot\|$-compact convex subset K of a Banach space $(X, \|\cdot\|)$ has normal structure.*

Proof. Arguing by contradiction, assume that K contains a nonempty convex subset C, not reduced to a point, such that every point of C is diametral in C.

➪ We may assume, without loss of generality, that C is closed. Indeed, take $\bar{c} \in \overline{C}$. Fix $\varepsilon > 0$. Find $c \in C$ such that $\|\bar{c} - c\| < \varepsilon$. Since c is diametral, we can find $c_1 \in C$ such that $\|c - c_1\| > d - \varepsilon$, where $d := \operatorname{diam} C$ $(= \operatorname{diam} \overline{C})$. Then $\|\bar{c} - c_1\| = \|c - c_1 + \bar{c} - c\| \geq \|c - c_1\| - \|\bar{c} - c\| > d - 2\varepsilon$. Since this is true for every $\varepsilon > 0$, we get that $\bar{c}$ is diametral in $\overline{C}$.

Start by taking $x_1 \in C$. Due to the fact that C is compact and that x_1 is diametral, we can find $x_2 \in C$ such that $\|x_1 - x_2\| = d$. The point $(x_1 + x_2)/2$ is also diametral, so there exists $x_3 \in C$ such that $\|\frac{x_1+x_2}{2} - x_3\| = d$. Do the same for $(x_1+x_2+x_3)/3$, finding x_4, and proceed recursively to find a sequence $\{x_n\}$ in C (see Figure 20.2) such that

$$d = \left\| \frac{x_1 + \ldots + x_n}{n} - x_{n+1} \right\| = \left\| \sum_{i=1}^{n} \frac{x_i - x_{n+1}}{n} \right\| \leq \sum_{i=1}^{n} \frac{\|x_i - x_{n+1}\|}{n} \leq d.$$

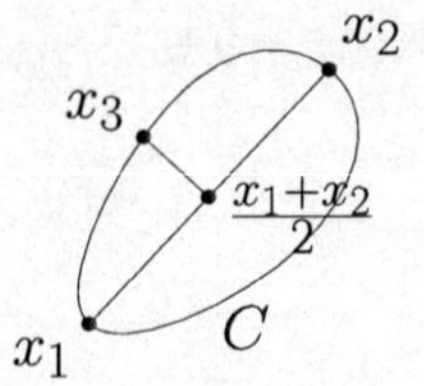

Figure 20.2: Starting the sequence $\{x_n\}_{n\in\mathbb{N}}$ in the proof of Proposition 384

Since $\|x_i - x_{n+1}\| \le d$ for all $i = 1, 2, \dots, n$, we get $\|x_i - x_{n+1}\| = d$ for all $i = 1, 2, \dots, n$. This is true for every $n \in \mathbb{N}$, and this is impossible in a compact space K. □

Theorem 385 (Brodskii–Milman, see, e.g., [Sw83]). *A Banach space X does not have normal structure if, and only if, there is a nonconstant bounded sequence $\{x_n\}$ in X such that*

$$\operatorname{dist}(x_k, \operatorname{conv}(x_1, \dots, x_{k-1})) \to \operatorname{diam}(\{x_n\}) \quad \text{when } k \to \infty.$$

Definition 386. *A sequence $\{x_n\}_{n\in\mathbb{N}}$ as in Theorem 385 is called* **diametral**.

The following theorem presents further application of the notion of URED.

Theorem 387 (Zizler [Ziz71]). (i) *Every URED norm has normal structure* (see, e.g., [FHHMZ11, Proposition 12.11]). *In particular, separable Banach spaces can be renormed to have normal structure.*
(ii) *Every separable space can be renormed to be URED* (see, e.g., [FHHMZ11, Exercise 9.20] and Proposition 167).

Proof of (i). Let C be a nonempty bounded convex subset of a URED Banach space X, not reduced to a point (see Figure 20.3). Take $x \neq y$ in C, and put $z = (1/2)(x + y)$. We claim that z is not diametral. Otherwise, there would be a sequence $\{x_n\}$ in C such that $\|z - x_n\| \to \operatorname{diam} C$.
Since $\|x_n - x\| \le \operatorname{diam} C$, $\|x_n - y\| \le \operatorname{diam} C$, and

$$\begin{aligned}\|x_n - z\| &= \left\| x_n - \frac{x+y}{2} \right\| \\ &= \frac{1}{2}\|(x_n - x) + (x_n - y)\| \le \frac{1}{2}\big(\|x_n - x\| + \|x_n - y\|\big) \le \operatorname{diam} C,\end{aligned}$$

we conclude that $\|x - x_n\| \to \operatorname{diam} C$ and $\|y - x_n\| \to \operatorname{diam} C$. Thus

$$2\|x_n - x\|^2 + 2\|x_n - y\|^2 - \|(x_n - x) + (x_n - y)\|^2 \to 0,$$

and $(x_n - x) - (x_n - y) = (x - y)$, for every $n \in \mathbb{N}$. This contradicts the definition of URED norm. □

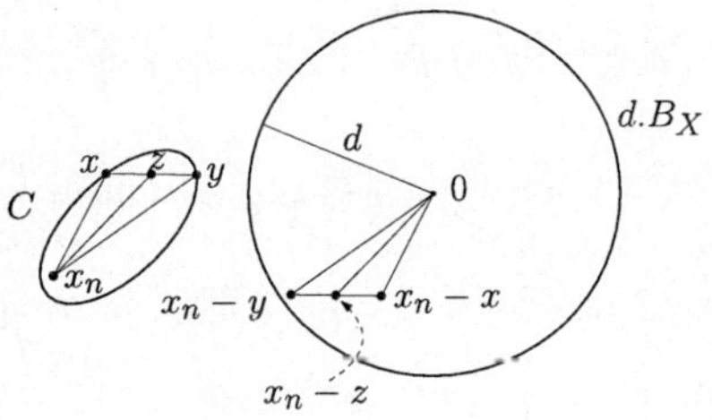

Figure 20.3: $\|x_n - z\| \to d$, $\|(x_n - x) - (x_n - y)\| \not\to 0$ (Theorem 387)

Theorem 388 (Kirk, [Ki65], see, e.g., [FHHMZ11, Theorem 12.12]). *Any non-expansive mapping from a weakly compact convex set into itself in a space with normal structure has a fixed point.*

Those are examples of how URED norms are extensively used in Fixed Point Theory. For the use of URED norms outside the area of fixed points see, e.g., [Rych05] and [Rych03].

Theorem 389 (Turett, [Tu82]). *Every UF norm has normal structure.*

However, we have

Theorem 390 (Johanis, Rychtář, [JohRy07]). *Every separable infinite-dimensional Banach space can be renormed by a UG norm that lacks normal structure.*

Proof. (Sketch) We only sketch the main idea in the definition of such norm.
Let $\{e_i, f_i\}$ be a Markushevich basis for X such that $\|e_i\| = 1$ and $\|f_i\| \leq 20$ for each i ([HMVZ07, Section 1.2]).
We put

$$C = \{x \in X : \ \|x\| \leq 2,\ 0 \leq f_n(x) \leq 1 \text{ for all } n \in \mathbb{N}\}.$$

Choose $\beta >$ diam C and define a new norm on X by

$$\|x\|_\beta = \max\left\{\|x\|, \beta \sup_{n\in\mathbb{N}} |f_n(x)|\right\}.$$

Finally define $|||\cdot|||^*_\beta$ by

$$(|||f|||^*_\beta)^2 = (\|f\|^*_\beta)^2 + \sum_{n=1}^{\infty} \frac{1}{2^n} f^2(e_n).$$

The equivalent norm $|||\cdot|||_\beta^*$ is a dual norm, and it is standard to show that it is W^*UR, which is dual notion to UG. The paper then gives a nice clear proof that the predual of the norm $|||\cdot|||_\beta^*$ on X does not have normal structure. □

The following space X_J comes from R.C. James: Let a norm on ℓ_2 be defined by

$$|||x||| = \max\left\{\|x\|_2, \frac{1}{\sqrt{2}}\|x\|_\infty\right\},$$

where $\|x\|_2$ is the standard norm on ℓ_2 and $\|x\|_\infty = \sup_{n\in\mathbb{N}} |x_n|$ for $x = \{x_n\} \in \ell_2$.

Denote ℓ_2 with this norm by X_J. Then we have

Theorem 391 (James, Karlovitz, [Kar76]). *The space X_J does not have normal structure; however every nonexpansive mapping from any closed convex bounded subset of X_J into itself has a fixed point.*

20.2 M.A. Smith's renormings of ℓ_2

We will now discuss several clever renorming constructions by M.A. Smith, a former student of M.M. Day, of norms on ℓ_2, which then became seminal in this field. Together with their dual norms they show a huge variety of equivalent norms on ℓ_2. The usual Hilbertian norm on ℓ_2 will be denoted by $\|\cdot\|_2$.
We suggest the reader try to see the geometry behind their construction.

1. For $x = (x^1, x^2, \ldots, x^n, \ldots) \in \ell_2$, an equivalent norm $\|\cdot\|_L$ on ℓ_2 is defined by

$$\|x\|_L = \left\|\left(\frac{1}{2}\|x\|_2, x^1, x^2, x^2, \ldots, x^j, x^j, \overset{(j)}{\ldots}, x^j, x^{j+1}, \ldots\right)\right\|_D, \tag{20.1}$$

where $\|\cdot\|_D$ denoted the Day norm on c_0 (see the definition in Subsection 9.4.1, and, in particular, equations (9.10), (9.11), and (9.12) there).

Theorem 392 (M.A. Smith). *The equivalent norm $\|\cdot\|_L$ on ℓ_2 is LUR and not URED* [Sm78]*; in fact, it does not have normal structure* [SmTu85].

Proof. To show that $\|\cdot\|_L$ is LUR, let $x, x_n \in \ell_2$, $\|x\|_L = 1$, $\|x_n\|_L \to 1$ and $\|x + x_n\|_L \to 2$. Since Day's norm is LUR on c_0 (Corollary 243), we get that $\|x_n\|_2 \to \|x\|_2$, and $x_n^j \to x^j$ for each j. Thus $x_n \to x$ weakly. Since the Hilbertian norm is w-$\mathcal{T}_{\|\cdot\|_2}$-KK, we have $x_n \to x$ in ℓ_2.

To see that $\|x\|_L$ is not URED, consider $x_n := e_1 + \sqrt{3}e_n$, $y_n := \sqrt{3}e_n$ for each n, where e_i are the standard unit vectors in ℓ_2. Then $\|x_n\|_L \to 1$, $\|y_n\|_L \to 1$ and $\|x_n + y_n\|_L \to 2$, while $x_n - y_n = e_1$ for all $n \in \mathbb{N}$.

To show that $\|\cdot\|_L$ does not have normal structure, it is directly calculated in [SmTu85] that the sequence $\{e_i\}$ of the standard unit vectors is a diametral sequence in the norm $\|\cdot\|_L$. □

Let us note in passing that Day's norm (see its definition in Subsection 9.4.1) is not URED on c_0: Indeed, it is enough to consider the points $(2,0,0,\dots)$ and $\sum_{j=n}^{2n} e_n$.

Also, a simple example of a R norm that is not URED is the following one on ℓ_∞ (we follow here [Gark64] or [Ziz71, Chapter 4, Example 4]):

For $x = \{x_i\} \in \ell_\infty$ consider

$$|||x||| = \|x\|_\infty + \left(\sum_{i=1}^{\infty} \frac{1}{2^i} x_i^2\right)^{\frac{1}{2}}.$$

It is simple to show that $|||\cdot|||$ on ℓ_∞ is a dual norm. By a now standard Clarkson's argument, $|||\cdot|||$ is R but it is not URED: For $n \in \mathbb{N}$, consider the elements

$$f_n = \left(\frac{1}{5}, \frac{1}{5}, \dots, \frac{1}{5}, -\frac{4}{5}, \frac{1}{5}, \dots\right)$$

where $-\frac{4}{5}$ stands on the n-position,

$$z = \left(\frac{1}{5}, \frac{1}{5}, \dots\right).$$

We have $|||z||| = 2/5$, $|||f_n||| \to 1$, $|||f_n + z||| \to 1$, $|||f_n - z||| \to 1$. For $n \in \mathbb{N}$, denote $p_n := f_n + z$, $q_n := f_n - z$. Then $|||p_n||| \to 1$, $|||q_n||| \to 1$, $|||p_n + q_n||| = 2|||f_n||| \to 2$, $p_n - q_n = 2z$ for all $n \in \mathbb{N}$, and so $|||\cdot|||$ on ℓ_∞ is not URED in the direction $z_0 := z/|||z|||$.

2. The norm $\|\cdot\|_N$ on ℓ_2 is defined as follows:

   Let $\{\alpha_n\}$ be a decreasing sequence of positive real numbers converging to 0. Let the operator T from ℓ_2 into ℓ_2 be defined by

$$T(x^1, x^2, \dots) = (\alpha_2 x^2, \alpha_3 x^3, \dots).$$

   Finally, let the norm $\|\cdot\|_N$ be defined by

$$\|x\|_N = \left(\|x\|_L^2 + \|Tx\|_2^2\right)^{\frac{1}{2}}, \tag{20.2}$$

   where $\|x\|_L$ is the norm defined in (20.1).

Theorem 393 (M.A. Smith, [Sm78]). *The equivalent norm $\|\cdot\|_N$ on ℓ_2 is LUR not URED but it has normal structure.*

Proof. The norm $\|\cdot\|_N$ is not URED: Consider $x_n = e_1 + \sqrt{3}e_n$, $y_n = \sqrt{3}e_n$. Then $\|x_n\|_N \to 1$, $\|y_n\|_N \to 1$, $\|x_n + y_n\|_N \to 2$ but $x_n - y_n = e_1$ for all $n \in \mathbb{N}$.

The fact that $\|\cdot\|_N$ is LUR follows from the fact that $\|\cdot\|_L$ is LUR.

The fact that $\|\cdot\|_N$ has normal structure follows from Smith's result in [Sm78] that the norm $\|\cdot\|$ has normal structure whenever it is uniformly rotund in all but countably many direction (the kernel of the operator T is 1-dimensional). □

3. Let us define another norm on ℓ_2.

Theorem 394 (M.A. Smith, [Sm78]). *There is an equivalent norm $\|\cdot\|_W$ on ℓ_2 that is WUR but not MLUR.*

Proof. For $x = (x^1, x^2, \ldots) \in \ell_2$ the norm $\|\cdot\|_W$ is defined as follows: Let $x' = (0, x^2, \ldots)$, and define the norm $\|\cdot\|_S$ by

$$\|x\|_S = \max\{|x^1|, \|x'\|_2\}. \tag{20.3}$$

Let $\alpha_n > 0$ be decreasing to 0 and define a linear injection T from ℓ_2 into ℓ_2 by $T(x^1, x^2, \ldots) = (x^1, \alpha_2 x^2, \alpha_3 x^3, \ldots)$.

Finally, define a norm on ℓ_2 by

$$\|x\|_W = (\|x\|_S^2 + \|Tx\|_2^2)^{\frac{1}{2}}. \tag{20.4}$$

It is standard to show that $\|\cdot\|_W$ is WUR.

The norm $\|\cdot\|_W$ is not MLUR. To see this let $\alpha = \frac{1}{\sqrt{2}}$ and let $x = \alpha e_1$, $x_n = \alpha(e_1 + e_n)$ and $y_n = \alpha(e_1 - e_n)$. Then $\|x\|_W = 1$, $\|x_n\|_W \to 1$, $\|y_n\|_W \to 1$, and $(x_n + y_n)/2 = x$ for all $n \in \mathbb{N}$, but $\|x_n - y_n\|_W \to \sqrt{2}$. □

Since Fréchet differentiability of the dual norm implies that the norm is MLUR (Theorem 105), and WUR of the norm implies that the dual norm is UG (Theorem 143), we get the following corollary of Theorem 394.

Corollary 395 (M.A. Smith, [Sm78]). *The dual norm of $\|\cdot\|_W$ on ℓ_2 is UG and not Fréchet differentiable.*

4. Next we have the following result:

Theorem 396 (Polak, Sims, Smith, [Sm78], [PolSi83]). *There is an equivalent norm $\|\cdot\|_A$ on ℓ_2 that is URED, w-$\mathcal{T}_{\|\cdot\|_A}$-SKK, MLUR, 2-R, but not WLUR.*

Proof. On ℓ_2 the norm $\|\cdot\|_A$ is defined as follows: First, if $x := (x^1, x^2, \ldots) \in \ell_2$, let $x' = (0, x^2, x^3, \ldots)$ and define the norm $\|\cdot\|_F$ on ℓ_2 by

$$\|x\|_F = |x^1| + \|x'\|_2. \tag{20.5}$$

It is easy to see that $\|\cdot\|_F$ is an equivalent norm on ℓ_2. Let $\{\alpha_n\}_{n=2}^{\infty}$ be a decreasing sequence converging to 0. Define a linear injection T from ℓ_2 into ℓ_2 by $T(x^1, x^2, \ldots) = (\alpha_2 x^2, \alpha_3 x^3, \ldots)$. Finally, define the norm $\|\cdot\|_A$ on ℓ_2 by

$$\|x\|_A = (\|x\|_F^2 + \|Tx\|_2^2)^{\frac{1}{2}}. \tag{20.6}$$

The norm $\|\cdot\|_A$ is URED (in particular, strictly convex).

➪ This is based on the following result. Here, a norm $\|\cdot\|$ in X is UR$_M$ whenever $\delta(z, \varepsilon) := \inf\{1 - \|(x_1 + x_2)/2\| : \; x_1, x_2 \in S_X, \; x_1 - x_2 \in \operatorname{span}\{z\}, \; \|x_1 - x_2\| \geq \varepsilon\} > 0$ for all $0 < \varepsilon \leq 2$ and $z \in M \setminus \{0\}$.

Theorem 397 (Zizler, [Ziz71, Proposition 1.14]). *Let $(X, \|\cdot\|_X)$ and $(Y, \|\cdot\|_Y)$ be two Banach spaces, and let $T : X \to Y$ be a continuous linear operator. Define $|||\cdot|||$ on X as $|||x||| = (\|x\|_X^2 + \|Tx\|_Y^2)^{1/2}$ for $x \in X$. Then $|||\cdot|||$ is an equivalent norm. Moreover, if $\|\cdot\|_Y$ is UR$_{T(X)}$ and $\|\cdot\|_X$ is UR$_{T^{-1}(0) \cap B_X}$, then $|||\cdot|||$ is URED.*

Thus, in view of Theorem 397, it is enough to show that $\|\cdot\|_F$ is uniformly rotund in the direction e_1. For this, let $\varepsilon > 0$ be given and let $x = \{x^j\}$ and $\{y^j\}$ be two element in ℓ_2 such that $\|x\|_F = \|y\|_F = 1$ and $x - y = \varepsilon e_1$. Then $x^j = y^j$ for all $j \geq 2$ and hence $|x^1| = |y^1|$ which, together with $x \neq y$, gives that $x^1 = -y^1$. Then

$$\left\| \frac{1}{2}(x + y) \right\|_F = \left(\sum_{j=2}^{\infty} |x^j|^2 \right)^{\frac{1}{2}} = \|x\|_F - |x_1| = 1 - \frac{\varepsilon}{2},$$

and this proves the uniform rotundity of $\|\cdot\|_F$ in the direction e_1, and so the URED property of $\|\cdot\|_A$.

We now show that $\|\cdot\|_F$ is w-$\mathcal{T}_{\|\cdot\|_F}$-SKK. For this, let $\{x_n\}_{n \in \mathbb{N}}$ be a sequence in $S_{\|\cdot\|_F}$ that w-converges to $x \in S_{\|\cdot\|_F}$. It converges coordinatewise, in particular $x_n(1) \to x(1)$ (for a moment, we write $x(1)$ for the first coordinate of x). Thus, $\|x_n'\|_2 \to \|x'\|_2$ and so $\|x_n' - x'\|_2 \to 0$. It follows that $\|x_n - x\|_F \to 0$, as we wanted to show.

From this, we get that $\|\cdot\|_A$ is w-$\mathcal{T}_{\|\cdot\|_A}$-SKK. Indeed, if $\{x_n\}_{n \in \mathbb{N}}$ is a sequence in $S_{\|\cdot\|_A}$ that w-converges to $x \in S_{\|\cdot\|_A}$, then it follows easily that $\|x_n\|_F \to$

$\|x\|_F$. From the previous paragraph we get $\|x_n - x\|_F \to 0$ and so we get $\|x_n - x\|_A \to 0$, as we wanted to show.

We can use Theorem 403 below and the previous results in this proof to see that the norm $\|\cdot\|_A$ is MLUR.

The fact that $\|\cdot\|_A$ is 2-R is proved in [PolSi83].

We finally show that $\|\cdot\|_A$ is not WLUR. Indeed, let $x := e_1$ and $x_n = e_n$ for $n = 2, 3, \ldots$ Then $\|x\|_A = 1$, $\|x_n\|_A \to 1$, and $\|x + x_n\|_A \to 2$, although $x_n \to 0$ in the weak topology. □

Note that $\|\cdot\|_A$ in Theorem 396 is strictly convex. Thus, $\|\cdot\|_A^*$ is Gâteaux differentiable (see Proposition 82). Moreover, $\|\cdot\|_A$ is w-$\mathcal{T}_{\|\cdot\|_A}$-SKK, in fact w-$\mathcal{T}_{\|\cdot\|_A}$-KK since ℓ_2 does not contain a copy of ℓ_1 (see Section 5.4). Thus, due to Corollary 147, $\|\cdot\|_A^*$ is Fréchet differentiable, and it is clear that $\|\cdot\|_A$ is not LUR. This is then a particular example of the situation in Theorem 374 above.

It is shown in [PolSi83] that the following theorem follows from Theorem 396.

Corollary 398. (See [PolSi83]) *There is a norm $\|\cdot\|$ on a reflexive space that is w-$\mathcal{T}_{\|\cdot\|}$-SKK (and then w-$\mathcal{T}_{\|\cdot\|}$-KK),* see Section 5.4*), R, not LUR, and not 2-R.*

20.3 M.A. Smith's renormings of c_0 and ℓ_1

Theorem 399 (M.A. Smith, [Sm78]). *On ℓ_1 there exists an equivalent norm $\|\cdot\|_H$ with the following properties:*
(i) *URWC.*
(ii) *R.*
(iii) *w-$\mathcal{T}_{\|\cdot\|_H}$-SKK.*
(iv) *Not MLUR.*

Proof. For $x = (x_n) \in \ell_1$ put $x' = (0, x_2, x_3, \ldots)$, and let $\|x\|_M = \max\{|x_1|, \|x'\|_1\}$. This is an equivalent norm on ℓ_1. Let $j : \ell_1 \to \ell_2$ be the inclusion mapping. Put, for $x \in \ell_1$,

$$\|x\|_H := (\|x\|_M^2 + \|jx\|_W^2)^{1/2},$$

where $\|\cdot\|_W$ is the norm on ℓ_2 given in Theorem 394. Then $\|\cdot\|_H$ is the sought equivalent norm on ℓ_1. □

Theorem 400 (M.A. Smith, [Sm78]). *On ℓ_1 there exists an equivalent norm $\|\cdot\|_E$ with the following properties:*
(i) *URWC.*
(ii) *LUR.*
(iii) *Not WUR.*

Proof. Let $j : \ell_1 \to \ell_2$ be the inclusion mapping. For $x \in \ell_1$ put

$$\|x\|_E := (\|x\|_1^2 + \|jx\|_2^2)^{1/2}. \tag{20.7}$$

This is the sought equivalent norm.
(iii) Due to the Schur (see, e.g., [FHHMZ11, Theorem 5.36]) and the non-reflexivity properties of ℓ_1, there is no equivalent WUR norm on this space. □

Theorem 401 (M.A. Smith, [Sm81]). *On c_0 there exists an equivalent norm $\|\cdot\|_G$ that is MLUR, WUR but not w-$\mathcal{T}_{\|\cdot\|_G}$-SKK.*

Proof. For each natural number k, define $V_k : c_0 \to \mathbb{R}$ by

$$V_k(x^1, x^2, \dots) = \sup_{j \geq k}\{|x^1 - x^j|\}.$$

Let $\{\alpha_n\}_{n=2}^{\infty}$ be a sequence of positive real numbers such that $\sum_{n=2}^{\infty} \alpha_n^2 = 1$. Define

$$V : c_0 \to \ell_2 \quad \text{and} \quad T : c_0 \to \ell_2$$

by

$$V(x) = (\alpha_2 V_2(x), \alpha_3 V_3(x), \dots) \text{ and } T(x) = (x^1, \alpha_2 x^2, \alpha_3 x^3, \dots)$$

for $x = (x^1, x^2, \dots)$. Finally for $x \in c_0$ define

$$\|x\|_G^2 = \|x\|_\infty^2 + \|V(x)\|_2^2 + \|T(x)\|_2^2, \tag{20.8}$$

where $\|\cdot\|_\infty$ and $\|\cdot\|_2$ denote the usual norms on c_0 and ℓ_2 respectively.

The norm $\|\cdot\|_G$ does not have property w-$\mathcal{T}_{\|\cdot\|_G}$-SKK. Indeed, let $x = e_1$ and $x_n = e_1 + e_n$ where e_i denotes the standard unit vectors in c_0. Then $\|x_n\|_G \to \|x\|_G = \sqrt{3}$ and $x_n \to x$ weakly, but x_n does not converge to x in norm.
It is standard to show that $\|\cdot\|_G$ is WUR.
The nice very instructive proof that $\|\cdot\|_G$ is MLUR is contained in [Sm81]. The proof shows implicitly why the dual version of MLUR does not imply Fréchet differentiability of the predual norm (see also [MOTV01]).
In fact, it also shows the ingenuity of the notion of LUR. We suggest that the reader try to understand the difference of the proofs for LUR and MLUR by convexity arguments. It is very well shown in [Sm81]. □

We note that Theorem 401 was upgraded in [DoHuSm95].

We continue with further examples.

Theorem 402 (Yost, [Yo81]). *Let X be a separable infinite-dimensional Banach space. Then X can be renormed by a strictly convex norm that is not weakly LUR.*

Proof. The norm is defined as follows: Let $\{x_i, f_i\}$ be a Markushevich basis for X such that $\|x_n\| = 1$ for all n and $\{f_n\}$ is bounded. Define the norm $|||\cdot|||$ as follows:

$$|||x||| := \left(\sum_{n=1}^{\infty} \frac{1}{4^n}(f_n(x))^2\right)^{\frac{1}{2}} + \max\left\{\frac{1}{2}\|x\|, \frac{1}{2}|f_1(x)| + \sup_{n=2}^{\infty}|f_n(x)|\right\}. \tag{20.9}$$

Due to the first term in (20.9), it is standard to see that $|||\cdot|||$ is strictly convex. Note that $|||x_1||| = 1$, $|||x_n||| \to 1$ and $|||x_1 + x_n||| \to 2$. However, $f_1(x_1) = 1$ and $f_1(x_n) \to 0$. Thus $\{x_n\}$ does not weakly converge to x_1. This means that $|||\cdot|||$ is not weakly LUR. □

Let us note in passing that if one needs to produce an example of an exposed point of the convex set that is not strongly exposed one can use the following argument from [Lind63]: In the space $X = \ell_2$ consider the set A to be the closed convex hull of $\bigcup_{n=2}^{\infty}\{\pm x_n\} \cup B_X$, where $x_n = (1 - \frac{1}{n}, 0, \ldots, 0, 1, 0, \ldots)$, with the number 1 in the n-th place. Then $(1, 0, 0, \ldots)$ is the only point of A which has the first coordinate 1, and hence it is an exposed point of A. It is, however not strongly exposed, since the first coordinate of x_n tends to 1.

The main idea in the following result then became seminal in the field.

Theorem 403 (M.I. Kadets, [Kad82]). *Let X be a Banach space that does not contain an isomorphic copy of ℓ_1. If X is w-$\mathcal{T}_{\|\cdot\|}$-SKK and R then X is MLUR. On the other hand, if X is separable and does contain an isomorphic copy of ℓ_1, then X can be renormed to be w-$\mathcal{T}_{\|\cdot\|}$-SKK, R but not MLUR.*

Proof. For the first part, and working by contradiction, assume that $x \in S_X$ and $\{x_n\}_{n\in\mathbb{N}}$ are such that $\|x + x_n\| \to 1$, $\|x - x_n\| \to 1$ and $\|x_n\| \geq \delta$ for all $n \in \mathbb{N}$. We need to produce y such that $\|x \pm y\| = 1$, $y \neq 0$.
As ℓ_1 is not isomorphic to a subspace of X, by Rosenthal's ℓ_1-theorem (see, e.g., [FHHMZ11, Theorem 5.37]), we may assume that $\{x_n\}_{n\in\mathbb{N}}$ is weakly Cauchy. Consider elements $x \pm \frac{1}{2}(x_n - x_m)$ for large m, n. Use the inequality

$$\|x \pm \tfrac{1}{2}(x_m - x_n)\| \leq \tfrac{1}{2}(\|x \pm x_m\| + \|x \mp x_n\|)$$

together with the lower weak semicontinuity of the norm and the fact that $\{x_n\}$ is weak Cauchy to see that $\lim \|x + \frac{1}{2}(x_m - x_n)\| = \|x\| = 1$. Since X has property w-$\mathcal{T}_{\|\cdot\|}$-SKK and $x + \frac{1}{2}(x_m - x_n) \to x$ in the weak topology, we have that $\|x_m - x_n\| \to$

0 if $m, n \to \infty$, so $\{x_m\}_{m\in\mathbb{N}}$ is norm convergent to some y. Since $\|x_n\| \geq \delta$ for all $n \in \mathbb{N}$, we get $\|y\| \geq \delta$, too. Moreover, $\|x \pm y\| = 1$ and we reach a contradiction with the property R. □

Remark 404. Observe that the second part of the statement of Theorem 403 gives an example of a w-$\mathcal{T}_{\|\cdot\|}$-SKK Banach space $(X, \|\cdot\|)$ that is not w-$\mathcal{T}_{\|\cdot\|}$-KK. Indeed, since $\|\cdot\|$ is R, the w-$\mathcal{T}_{\|\cdot\|}$-KK property would imply, by Theorem 407, that $\|\cdot\|$ is ALUR, and so MLUR, a contradiction. Ⓡ

Later on, Theorem 403 was strengthened by S. Troyanski (Theorem 407 below), identifying a geometric property of the norm called "average locally uniformly rotundness" (ALUR, for short), see Definition 406 below. Its definition needs the following concept:

Definition 405. *Let X be a Banach space. A point $x \in S_X$ is called a* **denting point** *of B_X if for every $\varepsilon > 0$ there is a slice of B_X that contains x whose diameter is less than ε. A point in S_{X^*} is said to be a* **w^*-denting point** *of B_{X^*} if it is a denting point and, for every $\varepsilon > 0$, the slice in the definition comes from an element in X.*

The notion of w^*-denting points will be used to characterize the Mazur Intersection Property (see Chapter 33 and, in particular, Theorem 567 there).

By a result of Lin–Lin–Troyanski [LLT88], the ALUR property can be equivalently described by using the notion of a denting point as follows:

Definition 406. *The norm $\|\cdot\|$ of a Banach space X is said to be* **average locally uniformly rotund** *(***ALUR***, for short) if all points in S_X are denting.*

For a chart connecting this, the LUR and MLUR notions, and some others, see Figure 20.4 below.

➪ The original definition [Tr85] of an ALUR Banach space X relies on the concept of expectation of a X-valued random variable on a probability space (see also [DeGoZi93, Section IV.2]). An alternative approach —still using integration theory— relies on the notion of a very strong extreme point: Let K be a closed convex subset of a Banach space $(X, \|\cdot\|)$. An element $x \in K$ is said to be a **very strong extreme point of** K if, given $\varepsilon > 0$, there exists $\delta > 0$ such that $\int_0^1 \|f(t) - x\| dt < \varepsilon$ whenever $f : [0, 1] \to K$ is a Bochner integrable function such that $\|\int_0^1 f(t)dt - x\| < \delta$. Troyanski [Tr85] says that $(X, \|\cdot\|)$ is ALUR whenever every point in S_X is a very strong extreme of B_X. It is proved in [LLT88] that, for K a bounded closed convex subset of a Banach space and $x \in K$, the following are equivalent: (i) x is a very strong extreme point of K, (ii) x is a denting point of K, and (iii) x is an extreme point and a

point of continuity of K (i.e., the identity mapping $I : (K, w) \to (K, \|\cdot\|)$ is continuous at x). This allows for Definition 406 and Theorem 407. For details regarding the probability approach see again [DeGoZi93, Sections IV.1 and IV.2].

The following is an alternative description of the ALUR property:

Theorem 407 (Troyanski, [Tr85]). *A space is strictly convex and has the* w-$\mathcal{T}_{\|\cdot\|}$-*KK property if, and only if, it is ALUR.*

Proof. (Rosenthal) (Sketch). It is almost obvious that ALUR implies the w-$\mathcal{T}_{\|\cdot\|}$-KK property and the strict convexity. For the reverse implication follow the hint in [FHHMZ11, Exercise 3.146]: Let $x \in S_X$. First prove that if $2x = (y^{**} + z^{**})$ for y^{**} and z^{**} in $B_{X^{**}}$, then the w^*-topology and the norm topology of $B_{X^{**}}$ coincide at y^{**} and at z^{**}. By Goldstine's theorem, we get $y^{**} \in X$ and $z^{**} \in X$ (and so $x = y^{**} = z^{**}$). This shows that x is an extreme point of $B_{X^{**}}$ (we say that x is a **preserved extreme point of** B_X). Finally, use Choquet's lemma (see, e.g., [FHHMZ11, Lemma 3.69]). □

Corollary 408. *If a space* X *is strictly convex and has the* w-$\mathcal{T}_{\|\cdot\|}$-*KK property then* X *is MLUR.*

Proof. ALUR easily implies MLUR. □

Remark 409. In this connection we showed in Section 5.4 that if X does not contain an isomorphic copy of ℓ_1, then the property w-$\mathcal{T}_{\|\cdot\|}$-SKK of X implies the w-$\mathcal{T}_{\|\cdot\|}$-KK property of X.
Note that Theorems 399 and 407 together imply that the norm from Theorem 399 does not have the w-$\mathcal{T}_{\|\cdot\|}$-KK property as otherwise would be ALUR, which implies the MLUR property, as we mentioned above. Thus we get another instance (see Remark 404) of the fact that the w-$\mathcal{T}_{\|\cdot\|}$-KK and the w-$\mathcal{T}_{\|\cdot\|}$-SKK properties are not the same. In particular it means that the sphere of the norm is not metrizable in the weak topology, which is an interesting easy reason for it. Ⓡ

Theorem 410 (Troyanski, [Tr85]). *Every ALUR space admits an equivalent LUR norm.*

An easier proof of Theorem 410 was later given by M. Raja in [Raj99], see, e.g., [FHHMZ11, Exercise 8.52]). We reproduce here this clever elegant proof:

Proof of Theorem 410 (Raja). First, let $\|\cdot\|$ be a norm on X that is both strictly convex and w-$\mathcal{T}_{\|\cdot\|}$-Kadets–Klee (Asplund averaging, see Section 10.8). Then each point of the new unit sphere is denting (see the argument in the proof of Theorem 407). For $m \in \mathbb{N}$ put

$$A_m = \left\{x \in B_X : \operatorname{diam}(B_X \cap H) > \tfrac{1}{m} \text{ for all half-spaces } H \text{ containing } x\right\}.$$

Check that A_m is closed, convex, symmetric, and that $0 \in \text{Int}(A_m)$.
Let f_m be the Minkowski functional of the set A_m, which is an equivalent norm on X. Let $a_m > 0$ be such that $a_m f_m^2(x) \leq 2^{-m}\|x\|^2$ for all $x \in X$. Define

$$\boxed{|x|^2 = \sum a_m f_m^2(x) + \|x\|^2, \quad \text{for } x \in X.} \tag{20.10}$$

Then $|\cdot|$ is an equivalent norm on X.
Let us check that $|\cdot|$ is LUR: Let $x, x_k \in X$ be such that

$$Q_{|\cdot|}(x, x_k) \to 0. \tag{20.11}$$

We have to show that $|x - x_k| \to 0$.
For $x = 0$ it is clear, so assume $x \neq 0$. Multiplying x and x_k by $\frac{1}{\|x\|}$ we assume that $\|x\| = 1$. From Lemmas 111 and 110, equation (20.11) gives
(i) $\|x_k\| \to \|x\| = 1$,
(ii) $\|x + x_k\|/2 \to 1$,
and, for $m \in \mathbb{N}$,
(iii) $f_m(x_k) \to_k f_m(x)$, and
(iv) $f_m(x + x_k) \to_k 2f_m(x)$.
Replacing x_k by $x'_k = x_k/\|x_k\|$ we get $\|x'_k\| = 1$ and $1 \geq \|x + x'_k\|/2 \to 1$. Fix any $\varepsilon > 0$. Find $m \in \mathbb{N}$ with $m > 2/\varepsilon$. Since x, x'_k are denting, $x, x'_k \notin A_m$. As A_m is closed, $f_m(x) > 1$ and $f_m(x'_k) > 1$. From (iv) above we then have $f_m\big((x+x'_k)/2\big) > 1$ for large $k \in \mathbb{N}$. Hence for these k we have $(x + x'_k)/2 \notin A_m$ and thus there is a half-space H such that $(x + x'_k)/2 \in H$ and $\text{diam}(B_X \cap H) \leq 1/m < \varepsilon/2$. As $x, x'_k \in B_X$, by a convexity argument we have $\|x - x'_k\|/2 < \varepsilon/2$, i.e. $\|x - x'_k\| < \varepsilon$. This means that $x'_k \to x$ and thus $x_k \to x$. □

It is a classical Dixmier's result that *a space is reflexive if the fourth dual norm of it is rotund.* Dixmier's result nowadays follows by duality from the following result

Theorem 411 (Giles, M.I. Kadets, Phelps). *If the dual norm of a Banach space X^* is very smooth, then the space is reflexive.*

Proof.. We may use James' Theorem 13 together with Lemma 53: Indeed, assume X is not reflexive and take $x_0^* \in S_{X^*}$ that does not attains its norm on B_X. Let $x_0^{**} \in S_{X^{**}}$ such that $\langle x_0^{**}, x_0^* \rangle = 1$. Find a w-neighbourhood W of x_0^{**} that does not cut B_X (a w-closed subset of X^{**}). Then W contains a slice S of $B_{X^{**}}$ defined by x_0^*. However, such a slice always cuts B_X, a contradiction. This is another reason for $(\ell_1, \|\cdot\|_1)$ not being very smooth. □

M.A. Smith proved that this is a sharp result in the following sense:

Theorem 412 (M.A. Smith, [Sm76]). *There is an equivalent norm on the classical James space J such that its third dual norm is rotund.*

The James' classical quasi-reflexive space J is introduced, e.g., in [FHHMZ11, Definition 4.43].

Since $\dim J^{**}/J = 1$ it is shown in [Sm76] that any norm whose dual is both W^*UR and LUR has the required property in Theorem 412. Such norm exists on J ([FHHMZ11, Chapter 8]).

Later on, P. Hájek showed the following two results:

Theorem 413 (Hájek, [Há96a]). *There is a norm on the predual of J whose third dual norm is W^*UR.*

Theorem 414 (Hájek, [Há96a]). *The predual of the James tree space JT has a norm whose third dual norm is $URED$.*

The James tree space JT is discussed e.g. in [FHHMZ11, Exercise 4.75]. It was a problem from Banach if for separable X, X is w^*-sequentially dense in X^{**} necessarily implies that X^* is separable. The space JT has this property without JT* being separable.

On the other hand we have (here, JH stands for the Hagler JH space):

Theorem 415 (Hájek, [Há96a]). *There is a separable space JH with nonseparable dual such that JH does not contain ℓ_1 and there is no norm on JH the second dual of which would be R.*

Note that the norm in Theorem 414 cannot be such that its third dual would be W^*UR, as JT is not renormable by WUR norm by a Hájek's result in [Há96a], see, e.g., [FHHMZ11, Theorem 8.12]).

Theorem 416 (Lajara, [La04]). *There is a norm on JT whose dual norm is $MLUR$.*

Note that JT does not admit any norm whose dual norm is LUR, as JT is not Asplund.

Theorem 417 (Moltó, Orihuela, Troyanski, Valdivia, [MOTV01], [LaPaTr09]). *There is a norm on the James space J whose second dual norm is $MLUR$.*

Theorem 417 shows, among others, two things:

1. The predual norm of the MLUR norm in Theorem 417 cannot be Fréchet differentiable since otherwise J would be reflexive (Theorem 105, see also [FHHMZ11, Corollary 7.26]).

2. The ball of the second dual has all points on the sphere strongly extreme (see Definition 11), and not all such points are strongly exposed (see Definition 55). Compare with the situation of exposed points.

Theorem 418 (Guirao, Montesinos, Zizler, [GuiMoZi15]). *If X is C^2-smooth and MLUR then X is superreflexive.*

Theorem 419 (Guirao, Montesinos, Zizler, [GuiMoZi15]). *There is a strictly convex C^2-smooth norm on c_0 such that no point on this sphere is an extreme point of its second dual ball.*

Theorem 419 is a C^2-smooth version of the original result of Morris ([Mo83]).

Theorem 420 (Godun, [Go85]). *A separable Banach space X is not reflexive if, and only if, it X can be renormed by such a strictly convex norm that some points of its unit sphere are not extreme in its bidual ball.*

Theorem 421 (Huff, [Hu80]). *Let X be a Banach space. Then X is NUC if, and only if, it is reflexive and UKK.*

Theorem 422 (Odell, Schlumprecht, [OdSch98]). *Let X be a separable Banach space. Then X is reflexive if and only if X admits a 2R norm.*

The proof of Theorem 422 is also in [Go01b].

It is an interesting open problem whether Theorem 422 holds for nonseparable spaces as well.

Theorem 423 (Hájek, Johanis, [HáJo04]). *Let X be a Banach space. Then X is reflexive if and only if X admits a W2R norm.*

Theorem 424 (Hájek, Johanis, [HáJo04]). *There is a norm on ℓ_2 that is W2R but it is not MLUR or 2R.*

The norm is defined by

$$\|x\|^2 = (\max\{\|x\|_2, 2|x_1|\})^2 + \sum_{i=2}^{\infty} 2^{-i} x_i^2.$$

Theorem 425 (Hájek, Johanis, [HáJo04]). *There is a norm on ℓ_2 that is LUR but not W2R.*

The norm is defined as follows:
First for $i, j \in \mathbb{N}$ put

$$|x|_{i,j}^2 = (|x_1| + |x_i| + |x_j|)^2 + \frac{1}{i+j}(x_1^2 + x_i^2 + x_j^2) + \sum_{k=2,\ k\neq i,j}^{\infty} x_k^2.$$

Then put

$$\|x\|^2 = \sup_{1<i<j} \{|x|_{i,j}^2\}.$$

20.4 A chart

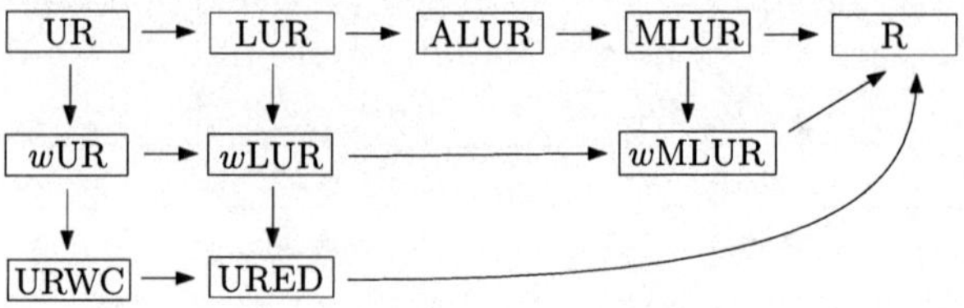

Figure 20.4: Some connections

Part III

Advances and Developments in Renorming, and Applications

Chapter 21

Nonlinear transfer techniques

21.1 Deville's master lemma and applications

Inspired by Troyanski's fundamental idea from his original [Tr71], R. Deville proved a powerful renorming theorem —reproduced below as Lemma 426— whose flexibility allows for a huge number of applications. It came as custom-built for the book [DeGoZi93]. It is used in this text in proving Lemma 427 and Theorems 220 and 428. In [DeGoZi93, Chapter VII] it is shown, among some many other results, that it gives a LUR renorming of $c_0(\Gamma)$ [DeGoZi93, Theorem VII.1.4]. See also below.

In the statement that follows, Q_p is the function of two variables defined by equation (5.2). We reproduce here this formula for the reader's convenience (p is a real-valued function defined on a set):

$$\boxed{\text{For } x, y \in X, \text{ put } Q_p(x,y) := 2p^2(x) + 2p^2(y) - p^2(x+y).} \tag{21.1}$$

Notice that the last assertion in Lemma 426 allows for dual renormings: Indeed, X ($\subset X^{**}$) is a norming subspace for X^* —in fact, 1-norming— and so the equivalent norm on X^* given by the lemma when applied to X^* is a dual norm.

Lemma 426 (Deville, see, e.g., [DeGoZi93], Lemma VII.I.1). *Let $(X, \|\cdot\|)$ be a Banach space, and $N \subset X^*$ a norming subspace. Let $\{\varphi_i : i \in I\}$ and $\{\psi_i : i \in I\}$ be two families of convex nonnegative real-valued functions defined on X and uniformly bounded on bounded sets. Then there exists an equivalent norm $|||\cdot|||$ on X such that, if $x \in X$ and $\{x_n\}$ is a sequence in X with $Q_{|||\cdot|||}(x, x_n) \to 0$, then there exists a sequence $\{i_n\}$ in I with the two following properties:*
(i) $\lim_n \varphi_{i_n}(x_n) = \lim_n \varphi_{i_n}(x) = \lim_n \varphi_{i_n}((x + x_n)/2) = \sup\{\varphi_i(x) : i \in I\}$.
(ii) $\lim_n \left[\frac{1}{2}\psi_{i_n}^2(x) + \frac{1}{2}\psi_{i_n}^2(x_n) - \psi_{i_n}^2\left(\frac{x+x_n}{2}\right)\right] = 0$.
If each φ_i and ψ_i are $w(X,N)$-lower semicontinuous, then $|||\cdot|||$ is $w(X,N)$-lower semicontinuous, too.

➡ Here is the precise definition of $|||\cdot|||$: Take $x \in X$. First, put $\varphi(x) := \sup\{\varphi_i(x) : i \in I\}$. Then, for $i \in I$ and $m \in \mathbb{N}$, let $\theta_{i,m}(x) := \varphi_i^2(x) + 2^{-m}\psi_i^2(x)$. Next, for $m \in \mathbb{N}$ put $\theta_m(x) := \sup\{\theta_{i,m}(x) : i \in I\}$. Let $\theta(x) := \|x\|^2 + \sum_{m=1}^{\infty} 2^{-m}(\theta_m(x) + \theta_m(-x))$. Finally, if $B := \{x \in X : \theta(x) \le 1\}$, $|||\cdot|||$ is defined to be the Minkowski functional of B.

A. J. Guirao et al., *Renormings in Banach Spaces*, Monografie Matematyczne 75,
https://doi.org/10.1007/978-3-031-08655-7_21

Observe that it follows from (ii) in Lemma 426 and Lemma 109 above that $|\psi_{i_n}(x_n) - \psi_{i_n}(x)| \to 0$ as $n \to \infty$.

A lattice LUR norm on $c_0(\Gamma)$

A direct application of Lemma 426 gives an equivalent lattice LUR norm on $c_0(\Gamma)$, where Γ is an arbitrary nonempty set. This is done in [DeGoZi93, Theorem VII.1.4], where all the details of the proof are given. Here we just provide a sketch:
Fix a finite nonempty set $A \subset \Gamma$. If $x \in c_0(\Gamma)$, let $x^A(\gamma) := x(\gamma)$ for $\gamma \in A$, $x^A(\gamma) = 0$ otherwise; put $\varphi_A(x) := \left(\sum_{\gamma \in A} x(\gamma)^2\right)^{1/2}$, and $\psi_A(x) := \|x - x^A\|_\infty$.
Given $p \in \mathbb{N}$, let $I_p := \{A \subset \Gamma : \ |A| = p\}$. Lemma 426, for $I := I_p$, gives an equivalent norm $\|\cdot\|_p$ on $c_0(\Gamma)$ with the properties there.
The sought lattice LUR equivalent norm $|||\cdot|||$ on $c_0(\Gamma)$ is given by

$$|||x||| := \left(\sum_{p=1}^{\infty} \beta_p \|x\|_p^2\right)^{1/2}, \ x \in c_0(\Gamma), \tag{21.2}$$

where $\{\beta_n\}_{n=1}^{\infty}$ is a sequence of positive numbers chosen to make the series in (21.2) convergent.

For a kind of survey on LUR renorming of $c_0(\Gamma)$ in the present text, see the notes after Theorem 170 above.

An application to LUR renorming

Let us provide a characterization of Banach spaces having an equivalent LUR renorming. We follow J. Orihuela and S. Troyanski lead. First we need a lemma, that the authors call, for obvious reasons, the **slice localization lemma**. Its proof is another neat example of the use of Lemma 426, so it is worth to reproduce it below.

Lemma 427 (Orihuela, Troyanski, [MOTV09]). *Let $(X, \|\cdot\|)$ be a normed space, and $N \subset X^*$ a norming subspace. Let A be a bounded subset of X and $\mathcal{H}$ a family of $w(X,N)$-open half-spaces such that $A \cap H \neq \emptyset$ for each $H \in \mathcal{H}$. Then there exists an equivalent $w(X,N)$-lower semicontinuous norm $\|\cdot\|_{\mathcal{H},A}$ on X such that, if $x_0 \in A \cap H_0$ for some $H_0 \in \mathcal{A}$ and $\{x_n\}$ is a sequence in X with $Q_{\|\cdot\|_{\mathcal{H},A}}(x_0, x_n) \to 0$, then there exists a sequence $\{H_n\}$ in $\mathcal{H}$ with the following properties:*
(i) *$x_0 \in H_n$ for all $n \in \mathbb{N}$ and, if $x_n \in A$, then $x_n \in H_n$.*
(ii) *For every $\delta > 0$ there is $n_\delta \in \mathbb{N}$ such that, for $n \geq n_\delta$,*

$$x_0, x_n \in \overline{\left(\operatorname{conv}(A \cap H_n) + \delta B_{\|\cdot\|}\right)}^{w(X,N)}.$$

Proof. If we define the function $\|\cdot\|_N : X^{**} \to \mathbb{R}$ as $\|x^{**}\|_N = \sup x^{**}(B_N)$ for $x^{**} \in X^{**}$, where $B_N := B_{X^*} \cap N$, then $\|\cdot\|_N$ is an (equivalent) norm on X (although maybe not even a norm on X^{**}).

➪ To justify that $\|\cdot\|_N$ may fail to be a norm on X^{**} select, in the non-reflexive case, $x^{**} \in X^{**} \setminus X$. Note that $N := \ker x^{**}$ is a norming subspace of X^*, as it was mentioned in Subsection 3.4.4, Item 1, although it does not separate points of X^{**}. Indeed, $\sup x^{**}(B_N) = 0$.

Now, for C a convex, w^*-closed and bounded subset of X^{**} and $x \in X$, define $d_N(x, C) := \inf\{\|x - c^{**}\|_N : \ c^{**} \in C\})$. Put

$$\varphi(x) := d_N(x, C), \text{ for } x \in X. \tag{21.3}$$

Note that φ is $w(X, N)$-lower semicontinuous, convex, and nonnegative.

➪ Maybe to check the $w(X, N)$-lower semicontinuity property needs a hint: It is necessary to prove that for any $\alpha \in \mathbb{R}$, the set $M := \{x \in X : \ \varphi(x) \le \alpha\}$ is $w(X, N)$-closed. Let $\{x_i\}$ be a net in M that $w(X, N)$-converges to x_0. Fix $\varepsilon > 0$ and i. Since $d_N(x_i, C) < \alpha + \varepsilon$, there exists $c_i^{**} \in C$ such that $\|x_i - c_i^{**}\|_N < \alpha + \varepsilon$, hence $|\langle x_i - c_i^{**}, b^*\rangle| < \alpha + \varepsilon$ for all $b^* \in B_N$. Let c_0^{**} be the w^*-limit of a convergent subnet $\{c_{i_j}^{**}\}$ of $\{c_i^{**}\}$. Then c_0^{**} belongs to C. Since $\alpha + \varepsilon > |\langle x_{i_j} - c_{i_j}^{**}, b^*\rangle| \to |\langle x_0 - c_0^{**}, b^*\rangle|$ for every $b^* \in B_N$, we get $\|x_0 - c_0^{**}\|_N \le \alpha + \varepsilon$ and so $d_N(x_0, C) \le \alpha + \varepsilon$. This holds for every $\varepsilon > 0$, hence $d_N(x_0, C) \le \alpha$, i.e., $x_0 \in M$.

For $H \in \mathcal{H}$ and $x \in X$, let $C := \overline{\operatorname{conv}(A) \cap H^c}^{w^*}$, where H^c is the complement of H in X, and define φ as in (21.3). Call the resulting function φ_H, i.e.,

$$\boxed{\varphi_H(x) := d_N(x, \overline{\operatorname{conv}(A) \cap H^c}^{w^*}), \ x \in X.} \tag{21.4}$$

Note that, if $x \in X$, then

$$\begin{cases} \text{(i) } \varphi_H(x) = 0 \Rightarrow x \in H^c. \\ \text{(ii) } \ x \in A \cap H^c \Rightarrow \varphi_H(x) = 0. \end{cases} \tag{21.5}$$

➪ (i) Assume that $\varphi_H(x) = 0$. Then we can find then a sequence $\{h_n^{**}\}_{n\in\mathbb{N}}$ in $\overline{\operatorname{conv}(A) \cap H^c}^{w^*}$ with $(\|x - h_n^{**}\|_N =) \ \sup\{\langle x - h_n^{**}, f^*\rangle : \ f^* \in B_N\} < 1/n$ for all $n \in \mathbb{N}$. Let $H = \{y \in X : \ \langle y, h^*\rangle > \alpha\}$ for some $h^* \in B_N$ and $\alpha \in \mathbb{R}$. We get, in particular, $|\langle x - h_n^{**}, h^*\rangle| < 1/n$ for all $n \in \mathbb{N}$. Since $\langle h_n^{**}, h^*\rangle \le \alpha$ for all $n \in \mathbb{N}$, we get $\langle x, h^*\rangle \le \alpha$, i.e., $x \in H^c$, as claimed.
(ii) is trivial.

We proved above that φ_H is a $w(X, N)$-lower semicontinuous function defined on X, and it is clearly convex and nonnegative.

Let $\delta > 0$. For each $H \in \mathcal{H}$, fix $a_H \in A \cap H$. Put $D_H := \operatorname{conv}(A \cap H)$, $D_H^\delta := D_H + \delta.\operatorname{Int}(B_{(X,\|\cdot\|)})$, and let p_H^δ be the Minkowski functional of $\overline{D_H^\delta}^{w(X,N)} - a_H$ (see Figure 21.1). Thus, p_H^δ is $w(X,N)$-lower semicontinuous. Finally, define the $w(X,N)$-lower semicontinuous function p_H on X by the formula

$$p_H^2(x) := \sum_{q=1}^{\infty} \frac{1}{q^2 2^q} \left(p_H^{1/q}(x)\right)^2, \ x \in X. \tag{21.6}$$

➡ Since $\delta.\operatorname{Int}B_{(X,\|\cdot\|)} \subset \overline{D_H^\delta}^{w(X,N)} - a_H$, it follows that $p_H^\delta(x) \le \|x\|/\delta'$ for all $0 < \delta' < \delta$ and $x \in X$, hence $p_H^\delta(x) \le \|x\|/\delta$ for all $x \in X$. This shows, in particular, $p_H^{1/q}(x) \le q\|x\|$ for all $x \in X$, and so the series in (21.6) converges (and gives a function $w(X,N)$-lower semicontinuous and bounded on bounded sets).

Finally, put

$$\boxed{\psi_H(x) := p_H(x - a_H), \ x \in X.} \tag{21.7}$$

Summarizing, we have two families $\{\varphi_H : \ H \in \mathcal{H}\}$ and $\{\psi_H : \ H \in \mathcal{H}\}$ (given, respectively, in (21.4) and (21.7)) of convex, nonnegative, and $w(X,N)$-lower semicontinuous functions, clearly uniformly bounded on bounded sets in X. Lemma 426 provides a $w(X,N)$-lower semicontinuous equivalent norm $\|\cdot\|_{\mathcal{H},A}$ on X with the properties listed there.

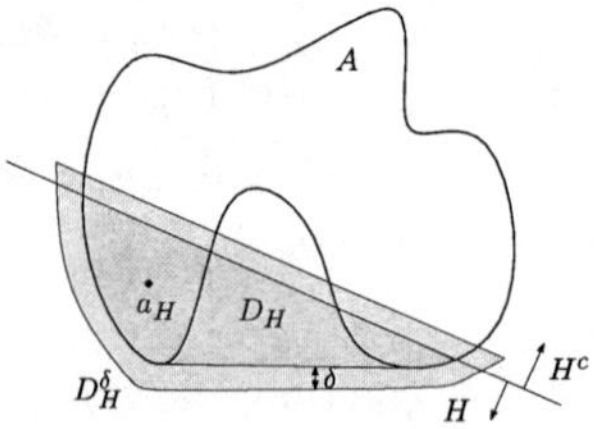

Figure 21.1: Some sets in the proof of Lemma 427

To be precise, let $x_0 \in A \cap H_0$ for some $H_0 \in \mathcal{H}$, and let $\{x_n\}_{n\in\mathbb{N}}$ be a sequence in X such that $Q_{\|\cdot\|_{\mathcal{H},A}}(x_0, x_n) \to 0$. Then there exists a sequence $\{H_n\}_{n\in\mathbb{N}}$ in $\mathcal{H}$ such that

$$\lim_n \varphi_{H_n}(x_n) = \lim_n \varphi_{H_n}(x_0) = \lim_n \varphi_{H_n}((x_0 + x_n)/2) = \sup\{\varphi_H(x_0) : \ H \in \mathcal{H}\}, \tag{21.8}$$

and

$$Q_{\psi_{H_n}}(x_0, x_n) \to 0. \tag{21.9}$$

It follows from (21.5) that $\varphi_{H_0}(x_0) > 0$, hence $\sup\{\varphi_H(x_0) : H \in \mathcal{H}\} > 0$. By (21.8), there exists $n_0 \in \mathbb{N}$ such that

$$\varphi_{H_n}(x_0) > 0 \text{ and } \varphi_{H_n}(x_n) > 0, \text{ for all } n \geq n_0. \tag{21.10}$$

Thus, from (21.10) and (21.5) we get that $x_0 \in H_n$ for all $n \geq n_0$ and, if $x_n \in A$ for some $n \geq n_0$ then $x_n \in H_n$, too. This proves (i) in Lemma 427.

In order to prove (ii), use (21.9). From Lemma 111 we get

$$Q_{p_{H_n}^{1/q}}(x_0 - a_{H_n}, x_n - a_{H_n}) \to_n 0,$$

and from Lemma 108,

$$\left| p_{H_n}^{1/q}(x_0 - a_{H_n}) - p_{H_n}^{1/q}(x_n - a_{H_n}) \right| \to_n 0, \text{ for every } q \in \mathbb{N}. \tag{21.11}$$

Fix now $0 < \delta_1 < \delta_2$, $H \in \mathcal{H}$, and $y \in A \cap H$, $y \neq a_H$. Note that

$$y - a_H + \delta_1 \frac{y - a_H}{\|y - a_H\|} \in (y - a_H) + \delta_1.B_X \subset (y - a_H) + \delta_2.\mathrm{Int}(B_X) \in D_H^{\delta_2} - a_H.$$

In other words, $(y - a_H)\left[1 + \frac{\delta_1}{\|y - a_H\|}\right] \in D_H^{\delta_2} - a_H$, and this shows that

$$p_H^{\delta_2}(y - a_H) < \left(1 + \frac{\delta_1}{\|y - a_H\|}\right)^{-1}, \tag{21.12}$$

due to the fact that $D_H^{\delta_2} - a_H$ is a $\|\cdot\|$-open set.
Fix $q \in \mathbb{N}$ such that $1/q < \delta$, and let $0 < \delta' < 1/q$. Let $n \geq n_0$. The first part of the proof showed that $x_0 \in A \cap H_n$. Assume for a moment that $x_0 \neq a_{H_n}$. We get, from (21.12), that $p_{H_n}^{1/q}(x_0 - a_{H_n}) < (1 + \frac{\delta'}{\|x_0 - a_{H_n}\|})^{-1}$. Since A is a bounded set, there exists $\xi > 0$ such that $p_{H_n}^{1/q}(x_0 - a_{H_n}) < (1 - \xi)$ for all $n \geq n_0$. Note that this inequality remains valid if $x_0 = a_{H_n}$ for some $n \geq n_0$. It follows from (21.11) that there exists $n_\delta \geq n_0$ such that, for $n \geq n_\delta$, $p_{H_n}^{1/q}(x_n - a_{H_n}) < 1 - \xi/2$. This shows that $x_n - a_{H_n} \in \overline{D_{H_n}^{1/q}}^{w(X,N)} - a_{H_n}$ $(\subset \overline{D_{H_n}^{\delta}}^{w(X,N)} - a_{H_n})$ for all $n \geq n_\delta$, and this gives the conclusion. □

Theorem 428 below gives a necessary and sufficient condition for LUR renorming. Maybe it is worth to see first that the condition there is in fact a necessary one.

➪ *Proving that a LUR and $w(X,N)$-lower semicontinuous renorming implies the condition in Theorem* 428. Let $(X, \|\cdot\|)$ be a normed space, $N \subset X^*$ a norming subspace, and assume that $\|\cdot\|$ is $w(X,N)$-lower semicontinuous (in other words, N is 1-norming) and LUR (thus, for every $\varepsilon > 0$ and $x \in S_X$ there exists $\delta := \delta(x, \varepsilon) > 0$ such that $y \in B_X$ and $\|(x + y)/2\| > 1 - \delta$ implies $\|x - y\| < \varepsilon$).

For $p, n \in \mathbb{N}$ set

$$A_{p,n} := \{x \in S_X : \ \delta(x, 1/n) > 1/p\}.$$

Note that, for $n \in \mathbb{N}$, $\bigcup_p A_{p,n} = S_X$; indeed, if $x \in S_X$, find $p \in \mathbb{N}$ such that $1/p < \delta(x, 1/n)$. Thus, $x \in A_{p,n}$.
Fix $\varepsilon > 0$ and find $n \in \mathbb{N}$ such that $1/n < \varepsilon$. Given $x \in S_X$, the argument above gives $p \in \mathbb{N}$ such that $x \in A_{p,n}$. Thus,$\delta(x, 1/n) > 1/p$. Since N is 1-norming, we can find $x^* \in S_{X^*} \cap N$ such that $\langle x, x^* \rangle > 1 - 1/p$. Let

$$H := \{y \in X : \ \langle y, x^* \rangle > 1 - 1/p\}.$$

Thus, H is a $w(X, N)$-open half-space in X. Obviously, $x \in H$. Let $y \in A_{p,n} \cap H$, so $\langle y, x^* \rangle > 1 - 1/p$, hence

$$\left(\left\|\frac{x+y}{2}\right\| \geq\right) \left\langle \frac{x+y}{2}, x^* \right\rangle > 1 - \frac{1}{p} \ (> 1 - \delta(x, 1/n)).$$

This shows that $\|y - x\| < 1/n$ $(< \varepsilon)$. Thus, $\operatorname{diam}(A_{p,n} \cap H) < 2\varepsilon$. By homogeneity, this shows that the condition in Theorem 428 below is necessary for the $w(X, N)$-lower semicontinuous LUR renorming.

Theorem 428 (Orihuela, Troyanski, [MOTV09] Theorem 1.1). *Let $(X, \|\cdot\|)$ be a normed space, and let $N \subset X^*$ be a norming subspace. The space $(X, \|\cdot\|)$ has an equivalent LUR $w(X, N)$-lower semicontinuous norm if, and only if, there exists a sequence $\{A_n\}$ of subsets of X such that, for every $\varepsilon > 0$ and $x \in X$ there exists $p \in \mathbb{N}$ and a $w(X, N)$-open half-space H with $x \in A_p \cap H$ and* $\operatorname{diam}(A_p \cap H) \leq \varepsilon$.

Proof. Without loss of generality, we may assume that $\|\cdot\| = \|\cdot\|_N$ or, in other words, that B_X is $w(X, N)$-closed (i.e., that N is 1-norming). To prove sufficiency we may assume that each A_n is bounded.

➡ Indeed, put $A_{p,k} := A_p \cap kB_X$ for all $p, k \in \mathbb{N}$. Given $\varepsilon > 0$ and $x \in X$, find $p \in \mathbb{N}$ and H as in the statement. Since $\operatorname{diam}(A_p \cap H) < \varepsilon$, there exists $k \in \mathbb{N}$ such that $A_p \cap H \subset kB_X$. Thus, $A_p \cap H = (A_p \cap kB_X) \cap H = A_{p,k} \cap H$, and the countable family $\{A_{p,k}\}_{p,k \in \mathbb{N}}$ consists of bounded sets and can be used instead of $\{A_k\}_{k \in \mathbb{N}}$.

For $m \in \mathbb{N}$, let $\mathcal{H}_{m,n}$ be the family of all $w(X, N)$-open half-spaces H such that $A_n \cap H \neq \emptyset$ and $\operatorname{diam}(A_n \cap H) \leq 1/m$ (if for some $n \in \mathbb{N}$ there is no such a slice, we do not consider it at all). By Lemma 427 applied to A_n and the family $\mathcal{H}_{m,n}$ there exists an equivalent norm $\|\cdot\|_{m,n}$ on X with the properties there. Find

$c_{m,n} > 0$ such that $\|\cdot\|_{m,n} \leq c_{m,n}\|\cdot\|$, and set

$$|||x|||_m^2 := \sum_{n=1}^{\infty} \frac{1}{c_{m,n}2^n}\|x\|_{m,n}^2, \; x \in X. \tag{21.13}$$

Find $d_m > 0$ such that $|||\cdot|||_m \leq d_m\|\cdot\|$, and set

$$|||x|||^2 := \sum_{m=1}^{\infty} \frac{1}{d_m 2^m}|||x|||_m^2, \; x \in X.$$

This norm satisfies all the requirements.

➪ Let $\{x_n\}_{n\in\mathbb{N}}$ be a sequence in X, and let $x_0 \in X$. Assume that $Q_{|||\cdot|||}(x_n, x_0) \to_n 0$. Fix $\varepsilon > 0$ and find $m \in \mathbb{N}$ such that $1/m < \varepsilon$. The assumption guarantees the existence of $p \in \mathbb{N}$ and a $w(X,N)$-open half-space H_0 such that $x_0 \in A_p \cap H_0$ and $\operatorname{diam}(A_p \cap H_0) < 1/m$. This shows that $H_0 \in \mathcal{H}_{m,p}$. We may now apply Lemma 427 to the (bounded) set A_p, the family $\mathcal{H}_{m,p}$, and the point $x_0 \in A_p \cap H_0$. Note that, thanks to Lemma 111, we have $Q_{\|\cdot\|_{m,p}}(x_n, x_0) \to 0$. Thus, we get a sequence $\{H_n\}_{n\in\mathbb{N}}$ in $\mathcal{H}_{m,p}$ with the two properties (i) and (ii) there. In particular, by (ii), there is $n_\varepsilon \in \mathbb{N}$ such that $x_0, x_n \in \overline{\operatorname{conv}(A_p \cap H_n) + \varepsilon B_{\|\cdot\|}}^{w(X,N)}$ for all $n \geq n_\varepsilon$. Since $\operatorname{diam}(A_p \cap H_n) < 1/m < \varepsilon$, we have $\operatorname{diam}(\operatorname{conv}(A_p \cap H_n)) < \varepsilon$, and so $\operatorname{diam}\big(\operatorname{conv}(A_p \cap H_n) + \varepsilon B_{\|\cdot\|}\big) < 3\varepsilon$. Due to the fact that B_X is $w(X,N)$-closed and that $x_0, x_n \in \overline{\operatorname{conv}(A_p \cap H_n) + \varepsilon B_{\|\cdot\|}}^{w(X,N)}$ for all $n \geq n_\varepsilon$, we get $\|x_0 - x_n\| < 3\varepsilon$ for $n \geq n_\varepsilon$. This shows that $x_n \to x_0$, and the conclusion follows. Note that we used (ii) in Lemma 427 since there is no guarantee that $x_n \in A_p$ for n big enough.

That the condition is also necessary was proved in the annotation before the statement of the theorem. □

There is a version of Theorem 428 that uses, instead of the diameter of the slices, their Kuratowski index of non-compactness. A required tool is the so-called "Bourgain–Namioka Superlemma" (see, e.g., [Di84, Chapter IX]). Its statement can be equivalently formulated in the following way:

Lemma 429 (Bourgain–Namioka Superlemma). *Let X be a Banach space. If $\varepsilon > 0$ and A and B are two non-empty closed convex and bounded subsets of X and* $\operatorname{diam}(B) < \varepsilon$, *then there exists a nonempty slice of* $\overline{\operatorname{conv}}(A \cup B)$ *that intersects B and has diameter less than ε.*

21.2 Nonlinear transfer

In 1979, S. Troyanski provided the first characterization of LUR renormability in linear topological terms. This result, whose origin goes back to Pisier's renorming

[Pis75] of super-reflexive Banach spaces with power type modulus of convexity, was the base for the new characterization in terms of slices of small diameter provided by A. Moltó, J. Orihuela and S. Troyanski (Theorem 428 above)

A general case providing additionally the $w(X, F)$-lower semicontinuity of the norm —for any norming subspace $F \subset X^*$— was proved by Raja in [Raj99] (for a proof using Deville's master lemma see Theorem 428), where he was able to replace Troyanski's probabilistic methods in [MOT97] in favor of geometrical ones —namely the Bourgain–Namioka Superlemma 429 above.
Theorem 428 gave rise to several developments around the classical linear transfer for rotundity, leading to a new non-linear transfer technique adapted to this characterization. The relevance of this technique follows from the fact that a big majority of the known LUR renormings can be derived from it. Sometimes the proofs could be more involved or less transparent though.
A. Moltó, J. Orihuela, S. Troyanski and M. Valdivia wrote a monograph about this so-called *nonlinear transfer technique* —see [MOTV09]— where the interested reader can access the details of the theory. We just present in this section a brief description of the main notions and results of this theory and remit the reader to [MOTV09] for further study.

In view of Theorem 428, and having as a goal transferring LUR norms in the same vein as Clarkson and Day did with rotund norms, the notion of *co-σ-continuity* is defined now.

Definition 430. *Let (X, d_X) and (Y, d_Y) be two metric spaces. A function φ from X to Y is said to be* **co-σ-continuous** *whenever for every $\varepsilon > 0$ there exists a countable cover of X, $\{X_{n,\varepsilon}\}_{n\in\mathbb{N}}$ in such a way that for every $n \in \mathbb{N}$ and for every $x \in X_{n,\varepsilon}$ there exists $\delta_n(x) > 0$ such that $d_X(x, y) < \varepsilon$ as soon as $d_Y(\varphi(x), \varphi(y)) < \delta_n(x)$ and $y \in X_{n,\varepsilon}$.*

When this definition is used in the context of normed spaces, it turns out to be equivalent to the condition that on the operator T was requested in Theorem 221 above (the proof can be found in [MOTV09, Theorem 2.32 and Proposition 2.33]). Thus, an alternative statement of Theorem 221 is the following:

Theorem 431. (See Theorem 221) *Let Y be a LUR normed space and T be a bounded linear co-σ-continuous operator from the normed space X into Y. Then X admits an equivalent LUR norm.*

Theorem 221 was already used in proving the basic Troyanski's Theorem 222: The first proof there essentially showed that, in the presence of a strong M-basis whose set of functionals is bounded, the natural evaluation mapping $T : X \to c_0(\Gamma)$ was co-σ-continuous. The LUR character of Day's norm on $c_0(\Gamma)$ plus Theorem 221 gave the LUR renormability of the space.

Let us present the following example, that provides another natural situation fitting Theorem 222. Once again, there exists a co-σ-continuous mapping into $c_0(\Gamma)$, and so it gives a second application of Theorem 221.

Example 432 ([MOTV09, Corollary 2.51]). Let X be a Banach space with an M-basis $\{e_\gamma, f_\gamma\}_{\gamma\in\Gamma}$. Assume, without loss of generality, that $\|f_\gamma\| = 1$ for all $\gamma \in \Gamma$. If

$$N := \{x^* \in X^* : \{\gamma \in \Gamma : x^*(e_\gamma) \neq 0\} \text{ is countable}\}$$

is a norming subspace of X^*, then the evaluation operator $T\colon X \to c_0(\Gamma)$, defined by $Tx = (f_\gamma(x))_{\gamma\in\Gamma}$ is co-σ-continuous. As a consequence, X has an equivalent LUR norm. Since $\overline{\text{span}}\{f_\gamma : \gamma \in \Gamma\} \subset N$, this gives a new proof of Theorem 317. Note that a Banach space with the property above can be renormed in such a way that N is 1-norming. Thus, behind this example is the fact that a Banach space with such an M-basis is a 1-Plichko space (see [FHHMZ11, Definition 5.46]), and that 1-Plichko spaces have a strong M-basis (see, e.g., [HMVZ07, Corollary 5.2]), so that Theorem 222 applies.
Every M-basis in a WLD space verifies that $\{\gamma \in \Gamma : f(e_\gamma) \neq 0\}$ is countable for every $f \in X^*$ (see Theorems 135 and 833). Thus, the evaluation operator T will be co-σ-continuous in this rather general setting. As a consequence, every WLD Banach space is LUR renormable. This is also a consequence of Theorem 222 and the fact that every WLD space has a strong M-basis (see Theorem 224). ✠

In many cases, to ask for the linearity of T may be rather restrictive. At first sight it seems that the linearity of T is necessary to transfer slices, since it is clear that if a map sends the zero vector into the zero vector and transfers slices into slices it must be linear. Nevertheless the linearity of T can be relaxed by taking advantage of the possibility to make additional countable splitting. So, and keeping in mind Theorem 428, the notion of *σ-slicely continuous maps* is introduced:

Definition 433. *Let (Y, d_Y) be a metric space, (X, τ) a linear topological space and $A \subset X$. A map $\varphi\colon A \to (Y, d_Y)$ is called* **slicely continuous at** $x \in A$ *whenever for every $\varepsilon > 0$ there exists an open half space H of X containing x with* diam$(\varphi(H \cap A)) < \varepsilon$. *Similarly, φ is called* **σ-slicely continuous** *whenever for every $\varepsilon > 0$ there exists a countable cover $\{A_{n,\varepsilon}\}_{n\in\mathbb{N}}$ of A in such a way that for every $n \in \mathbb{N}$ and for every $x \in A_{n,\varepsilon}$, there exists an open half space H of X such that* diam$(\varphi(H \cap A_{n,\varepsilon})) < \varepsilon$.

That σ-slicely continuous maps can be far from linear is evidenced, for instance, by the classical Day's map:

Example 434 ([MOTV09, Proposition 4.2]). Let $D\colon c_0(\Gamma) \to \ell^2(\Gamma)$ —called Day's map— defined by $Dx(\gamma_n) = 2^{-n}x(\gamma_n)$ if supp $x = \{\gamma_n\}_{n\geq1}$, $|x(\gamma_n)| \geq |x(\gamma_{n+1})| > 0$ for every n, and $Dx(\gamma) = 0$ if $\gamma \notin$ supp x. Then D is σ-slicely continuous and co-σ-continuous. ✠

Theorem 428 can now be rephrased in terms of σ-slicely continuous maps as follows.

Theorem 435. *A normed space X admits an equivalent LUR norm if, and only if, the identity map $I\colon X \to X$ is σ-slicely continuous.*

The following transfer theorem can be then obtained:

Theorem 436 (Nonlinear Transfer Theorem)**.** *A normed space X admits an equivalent LUR norm if, and only if, there exists a metric space (Y, d_Y) and a map $\varphi\colon X \to (Y, d_Y)$ which is σ-slicely continuous and co-σ-continuous.*

From the result that Day's map —see Example 434 above— is σ-slicely continuous and co-σ-continuous, and the use of the previous theorem, it follows at once that $c_0(\Gamma)$ is LUR renormable.

The techniques here allowed the proof of the existence of an equivalent LUR renorming of a space with a WLUR norm (Theorem 227 above).

Chapter 22

Lipschitz functions II

Recall that a function $f : M \to \mathbb{R}$, where (M, d) is a metric space, is said to be **Lipschitz** if there exists a constant $L > 0$ such that $|f(x) - f(y)| \leq L.d(x, y)$ for all $x, y \in M$. To be precise, we speak in this case of a L-Lipschitz function, and we say that L is a **Lipschitz constant for the function** f. Examples of Lipschitz functions have been provided earlier: The norm $\|\cdot\|$ on a Banach space X is a 1-Lipschitz function, a consequence of the triangle inequality. If $f : (a, b) \to \mathbb{R}$ is a differentiable function and $|f'(x)| \leq L$ for all $x \in (a, b)$, then f is an L-Lipschitz function. This is a consequence of the Mean Value Theorem 44. Obviously, every Lipschitz function is uniformly continuous. The converse is false: The function $\sqrt{x}$ on $[0, 1]$ is continuous (then uniformly continuous) and obviously not Lipschitz there.

Not every convex function on a Banach space is Lipschitz. The function $f(x) := x^2$ in $\mathbb{R}$ is a counterexample. However, every convex continuous function on a nonempty open convex subset O of a Banach space is **locally Lipschitz**, i.e., every point $x_0 \in O$ is in a neighborhood $U(x_0)$ such that $f|_{U(x_0)}$ is Lipschitz (Theorem 49). For example, the function $f(x) := x^2$ defined on $\mathbb{R}$ is Lipschitz when restricted to any nonempty bounded open subset of $\mathbb{R}$.

➪ It is useful to have in mind the following set of equivalences for a convex function f defined on a nonempty open convex subset O of a Banach space and a point $x_0 \in O$ (see Theorem 49 above and, e.g., [FHHMZ11, Lemma 7.3]): (i) f is continuous at x_0. (ii) f is locally bounded above at x_0. (iii) f is locally Lipschitz at x_0.

Another basic important result is the following

Theorem 437 (McShane, [McS34], Whitney, [Whitn34]). *Every real-valued Lipschitz function f_0 defined on a nonempty subset N of a metric space M has a Lipschitz extension to M with the same Lipschitz constant. If f_0 is bounded, the extension can be defined having also the same supremum norm.*

Proof. (Sketch, see Figure 22.1) Use the inf-convolution defined as $(E^+ f)(m) = \inf\{f(n) + \|f\|_L d(m, n) : \ n \in N\}$ for $m \in M$. Then, if $S := \sup_{n \in N} f(n)$ and $s := \inf_{n \in N} f(n)$, put $(\widehat{f})(m) := (f \wedge S) \vee s$. For details, see, e.g., [Weav18b, Theorem 1.33] (it is possible to adjust the proof to use the sup-convolution —defined symmetrically— instead). For a detailed argument see, e.g., [Weav18b, Theorem 1.33]. □

A. J. Guirao et al., *Renormings in Banach Spaces*, Monografie Matematyczne 75,
https://doi.org/10.1007/978-3-031-08655-7_22

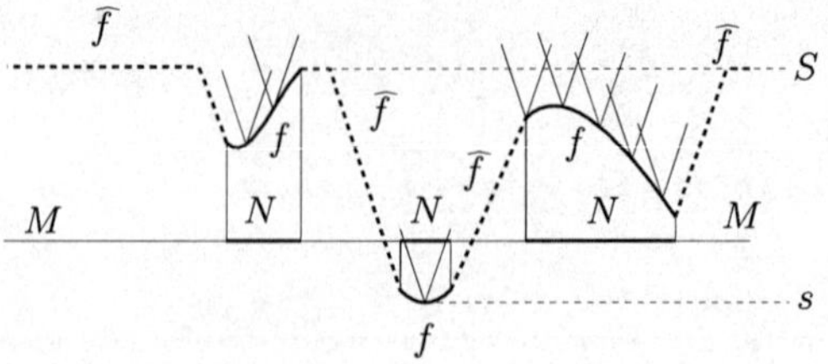

Figure 22.1: Building the McShane–Whitney extension $\widehat{f}$ (dashed in bold) of f on N (continuous, in bold) (Theorem 437)

22.1 The Rademacher theorem

One of the most important results regarding Lipschitz functions is the following one. The reader is advised to review Lemma 48 above and Theorem 443 below.

Theorem 438 (Rademacher, [Rade19]). *Every Lipschitz function from a nonempty open subset O of $\mathbb{R}^n$ into $\mathbb{R}$ is almost everywhere Fréchet differentiable. In particular, it is Fréchet differentiable at points of a dense subset of O.*

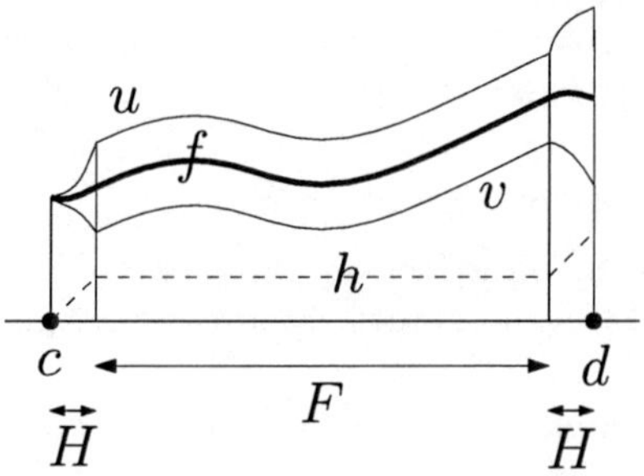

Figure 22.2: A rough idea of the construction in the second proof of Theorem 438. Note that it is impossible to have $DI_i f(x) \leq r < s \leq DI_j(x)$ (these Dini derivatives are defined below, as well as numbers r and s) in a whole interval, see Proposition 440

➪ We provide here a proof of this result in two steps. First, for the case $n = 1$. Then we shall extend the result to an arbitrary $n \in \mathbb{N}$.

1. The case $n = 1$. The basic ingredients are the so-called **Dini derivatives**: Let f be a real-valued function defined on an open interval J of $\mathbb{R}$. Let $x \in J$. Put $D^+ f(x) := \limsup_{t \to 0+} \frac{f(x+t)-f(x)}{t}$, and $D_+ f(x) := \liminf_{t \to 0+} \frac{f(x+t)-f(x)}{t}$ (eventually some of these expressions may be in-

finite). The derivatives $D^-f(x)$ and $D_-f(x)$ are defined similarly (now $t \to 0-$). For notational purposes rename those derivatives D^+f, D_+f, D^-f, and D_-f, by DI_1f, DI_2f, DI_3f, and DI_4f, respectively. Obviously, f is not differentiable at x if, for some $i, j \in \{1,2,3,4\}$, we have $DI_if(x) \neq DI_jf(x)$.

Zygmund provided a useful "one-sided" sufficient condition for monotonicity of a function. We shall prove here a simplified version, enough for our purposes:

Proposition 439 (Zygmund, see, e.g., [Saks37, Theorem VI.7.1]). *Let F be a real-valued continuous function defined on a closed interval $J \subset \mathbb{R}$. Fix $i \in \{1,2,3,4\}$. Let $E := \{x \in \operatorname{Int}(J) : DI_iF(x) \leq 0\}$. Assume that $F(E)$ contains no nondegenerate interval. Then F is nondecreasing on J.*

Proof. We present the proof for $i = 1$. Small changes give the proof of the case $i = 3$. The argument for the case $i = 2$ ($i = 4$) is included in the proof for $i = 1$ (respectively, for $i = 3$). Arguing by contradiction, assume that there exists a and b in J such that $a < b$ and $F(a) > F(b)$. By the assumption, $(F(b), F(a)) \not\subset F(E)$, hence there exists $y_0 \in (F(b), F(a)) \setminus F(E)$. Let $x_0 := \sup\{x \in [a,b] : F(x) \geq y_0\}$. Obviously, $F(x_0) = y_0$, and so $x_0 \in (a,b)$. Since $F(x) < y_0$ $(= F(x_0))$ for $x \in (x_0, b]$, we have $(F(x) - F(x_0))/(x - x_0) < 0$ for all $x \in (x_0, b]$. This shows that $DI_1F(x_0) \leq 0$, i.e., $x_0 \in E$. This contradicts $(F(x_0) =)\ y_0 \notin F(E)$. □

As a consequence, we get the following Dini's result:

Proposition 440 (Dini, see, e.g., [Saks37, page 204]). *Let f be a real-valued continuous function defined on an interval $[a,b]$. Then, for $i \in \{1,2,3,4\}$ we have*

$$\begin{aligned} m &:= \inf\{DI_if(x) : x \in [a,b]\} \\ &= \inf\left\{\frac{f(x_1) - f(x_2)}{x_1 - x_2} : x_1, x_2 \in [a,b],\ x_1 \neq x_2\right\} \end{aligned} \tag{22.1}$$

(possibly, $m = -\infty$). A similar result holds for the supremum.

Proof. Let us prove it for $i = 1$ and the infimum (proofs for the other cases are similar). Assume first that $m \neq -\infty$. Let $m' < m$; put $F(x) := f(x) - m'x$ for $x \in [a,b]$. Clearly, $DI_1F(x) > 0$ for all $x \in (a,b)$. Thus, the assumption in Proposition 439 holds trivially, hence F is non-decreasing. This shows that, for $x_1 < x_2$ in $[a,b]$, we have

$f(x_2) - f(x_1) \geq m'(x_2 - x_1)$. Thus,

$$(m \geq)\ \inf \left\{ \frac{f(x_1) - f(x_2)}{x_1 - x_2} : x_1, x_2 \in [a, b],\ x_1 \neq x_2 \right\} \geq m'.$$

Since $m' < m$ was arbitrary, we get the equality in (22.1). In case $m = -\infty$, the result holds trivially. □

Proof of the 1*-dimensional case.* We provide here two different approaches:

a) The result can be obtained from the Lebesgue differentiability theorem for monotone functions (see, e.g., [MoZiZi15, Theorem 424]). Indeed, if f if a K-Lipschitz function for some $K > 0$, then $f + Kx$ is obviously monotonically increasing.

b) We present now a very interesting elementary proof for $n = 1$, due to L. Zajíček [Zaj92], that does not use Vitali covers (an ingredient for the classical proof of the Lebesgue differentiation theorem): We may assume that f is defined on a closed and bounded non-degenerate interval $[a, b]$, that f is K-Lipschitz for some $K > 0$, and that the result fails. Thus, if λ denotes the Lebesgue measure on $\mathbb{R}$, there exists $i, j \in \{1, 2, 3, 4\}$ such that $\lambda\{x \in (a, b) : DI_i f(x) < DI_j f(x)\} > 0$ (note that the functions $x \to DI_i f(x)$, defined at the beginning of this annotation, are Lebesgue measurable). Thus, there exist rational numbers $r < s$ such that $\lambda(L) > 0$, where $L := \{x \in (a, b) : DI_i f(x) < r < s < DI_j f(x)\}$. Find $a < \alpha < \beta < b$ such that $\lambda(M) > 0$, where $M := L \cap (\alpha, \beta)$. Further, choose $\varepsilon > 0$ such that

$$s - r > 4K\varepsilon \tag{22.2}$$

and find an open set G such that $M \subset G \subset (\alpha, \beta)$ and $\lambda(M)/\lambda(G) > 1 - \varepsilon$. It is easy to see that for some component (c, d) of G we also have $\lambda(M \cap (c, d))/(d - c) > 1 - \varepsilon$. Finally, choose a closed set $F \subset M \cap (c, d)$ such that $\lambda(F)/(d - c) > 1 - \varepsilon$, and put $H := (c, d) \setminus F$. Note that $\{r, s, DI_i f(x), DI_j f(x)\} \subset [-K, K]$ for all $x \in (c, d)$. The idea is to define two functions u and v on $[c, d]$, both very similar to f there but on the small set H, that satisfy $DI_j u(x) \geq s$ and $DI_i v(x) \leq r$ for all $x \in (c, d)$. We shall apply then Proposition 440 in order to get a contradiction with equation (22.2) (see Figure 22.2). Just put $u(x) := f(x) + 2Kh(x)$ and $v(x) := f(x) - 2Kh(x)$, where $h(x) := \lambda\big((c, x) \cap H\big)$, for $x \in [c, d]$. It is simple to see that u and v satisfy the sought conditions, and that $h(d) - h(c) = \lambda(H) < \varepsilon(d - c)$. Hence, by

Proposition 440,

$$\begin{aligned} s(d-c) &\leq u(d)-u(c) \\ &= \big(f(d)-f(c)\big)+2K\big(h(d)-h(c)\big) \\ &< \big(f(d)-f(c)\big)+2K\varepsilon(d-c), \end{aligned}$$

and

$$\begin{aligned} r(d-c) &\geq v(d)-v(c) \\ &= \big(f(d)-f(c)\big)-2K\big(h(d)-h(c)\big) \\ &> \big(f(d)-f(c)\big)-2K\varepsilon(d-c), \end{aligned}$$

so $s-r<4K\varepsilon$, and this contradicts (22.2). □

2. *Proof of the n-dimensional case.* To prove the case $n \geq 2$, we shall use the one-dimensional case, Fubini's theorem, the Lebesgue Dominated Convergence Theorem, and the integration-by-parts formula. We may assume that the function f is K-Lipschitz and is defined on $\mathbb{R}^n$ (every K-Lipschitz real-valued function defined on a subset of a Banach space can be extended to a K-Lipschitz function on the whole space (see Theorem 437 above; see also, e.g., [FHHMZ11, Exercise 2.18]). The Lebesgue measure on $\mathbb{R}^n$ is denoted, again, by λ. We proceed along several steps:

   (a) First, given $0 \neq v \in \mathbb{R}^n$, we claim that $\lambda(\mathbb{R}^n \setminus D_v)=0$, being D_v the set of all $x \in \mathbb{R}^n$ where the directional derivative $D_v f(x)$ exists. Indeed, $\mathbb{R}^n \setminus D_v$ intersects any line parallel to v in a null set (thanks to the case $n=1$). This proves the claim, by Fubini's theorem.

   (b) Given again $0 \neq v=(v_1,\ldots,v_n) \in \mathbb{R}^n$, we shall prove that $\lambda(\mathbb{R}^n \setminus D_{v,1,\ldots,n})=0$, where

$$D_{v,1,\ldots,n} := \Big\{x \in \mathbb{R}^n : D_v f(x), D_1 f(x), \ldots, D_n f(x) \\ \text{exists, and } D_v f(x) = \sum_{i=1}^{n} v_i D_i f(x)\Big\}$$

   (as usual, $D_i f(x)$ denotes the i-th partial derivative of f at x; notice that if f is differentiable at x, then the formula $D_v f(x) = \sum_{i=1}^n v_i D_i f(x)$ holds). Since each $D_v f(x)$ and $D_i f(x)$, $i=1,2,\ldots$, exist a.e., it is enough to prove that $D_v f(x) = \sum_{i=1}^n v_i D_i f(x)$

holds a.e. For this, take an arbitrary C^∞-function $\phi : \mathbb{R}^n \to \mathbb{R}$ having a compact support. Then

$$\begin{aligned}
&\int_{\mathbb{R}^n} D_v f(x)\phi(x)\, dx \\
&\quad = \lim_{t\to 0} \int_{\mathbb{R}^n} \frac{f(x+tv)-f(x)}{t}\phi(x)\, dx \quad (\textit{Dom.Conv.Thm.}) \\
&\quad = \lim_{t\to 0} \int_{\mathbb{R}^n} f(x)\frac{\phi(x-tv)-\phi(x)}{t}\, dx \\
&\quad = \int_{\mathbb{R}^n} f(x)\sum_{j=1}^{n} v_j D_j\phi(x)\, dx \\
&\quad = \int_{\mathbb{R}^n} \left(\sum_{j=1}^{n} v_j D_j f(x)\right)\phi(x)\, dx \ (\textit{integration by parts}),
\end{aligned}$$

and the conclusion follows.

(c) Finally, let us show that f is differentiable a.e. Let Ω be a countable dense subset of $S_{\mathbb{R}^n}$. Using the notation of (2b) above, it is clear that $\lambda(\mathbb{R}^n \setminus \bigcap_{w\in\Omega} D_{w,1,\dots,n}) = 0$. Fix $x \in \bigcap_{w\in\Omega} D_{w,1,\dots,n}$. We shall prove that the remainder

$$r(w,t) := \frac{f(x+wt)-f(x)-t\sum_{i=1}^{n} w_i D_i f(x)}{t},\quad t\neq 0,$$

satisfies $\lim_{t\to 0} r(w,t) = 0$ uniformly on $w \in S_{\mathbb{R}^n}$. By the very definition of the differential, this will show that f is differentiable at x. Note that

$$|r(\widetilde{w},t) - r(w,t)| \le C\|\widetilde{w} - w\|,$$

where C is a constant coming from the Lipschitz behaviour of f. Fix $\varepsilon > 0$ and find a finite $\varepsilon/(2C)$-net $\Omega_0 \subset \Omega$. From (2b) above it follows that there exists $0 < t_0 = t_0(\Omega_0, \varepsilon)$ such that

$$|r(w_0,t)| < \varepsilon/2, \text{for } w_0 \in \Omega_0,\ 0 < |t| < t_0.$$

Given $w \in S_{\mathbb{R}^n}$, find $w_0 \in \Omega_0$ such that $\|w - w_0\| < \varepsilon/(2C)$. Then

$$|r(w,t)| \le |r(w_0,t)| + |r(w,t) - r(w_0,t)| < \varepsilon,\ \text{for } 0 < |t| < t_0.$$

This proves the statement. □

For another simple proof of the n-dimensional Rademacher theorem, based in the 1-dimensional case and Fubini's theorem, see [NeZa98].

Definition 441. *A function f on a Banach space X has* **finite pointwise Lipschitz constant at** $x \in X$ *if* $\limsup_{y \to x} \frac{|f(y)-f(x)|}{\|y-x\|} < \infty$.

Theorem 442 (Stepanov, see, e.g., [Mal99]). *If f on $\mathbb{R}^n$ has a.e. finite pointwise Lipschitz constant, then f is a.e. Fréchet differentiable.*

22.2 Preiss' differentiation of Lipschitz functions

We only mention the following celebrated result. It is in fact the best advertisement for renormings as it used gradual step by step renormings in the proof. Compare with Rademacher's Theorem 438 above.

Theorem 443 (Preiss, [Prei90], see also [LPT10]). *Assume that a Banach space X admits an equivalent Fréchet (resp. Gâteaux) differentiable norm. Then every Lipschitz function on X is Fréchet (resp. Gâteaux) differentiable at points of a dense subset of X.*

It should be noted that the result in Theorem 443 is new even in ℓ_2. In [LPT10], Theorem 443 is proved by a new interesting variational principle.

Comparing with Theorem 62, note that Theorem 443 is new for nonseparable spaces with Gâteaux norm.

Corollary 444. *Assume that X is an Asplund space. Then every Lipschitz function on X is Fréchet differentiable at points of a dense subset of X.*

The proof of Corollary 444 uses the separable-reduction Lemma 46 and Kadets' renorming Theorem 173 and its Corollary 178:

Proof of Corollary 444. Assume that f is a Lipschitz function on X that is nowhere Fréchet differentiable on X. Pick a separable subspace V of X. Then, by Lemma 46, there is a separable subspace $W \supset V$ of X such that f is Fréchet differentiable at each point of W where the restriction of f to W is Fréchet differentiable. Since W^* is separable (as X is Asplund), there is a Fréchet differentiable norm on W (see, e.g., [FHHMZ11, Theorem 8.6]). Thus there is a point in W where the restriction of f to W is Fréchet differentiable, and thus f is Fréchet differentiable at this point, a contradiction. □

H Theorem 443 was motivated by Phelps' suggestion to the Prague School to prove it (after he found, together with Fitzpatrick, that all available counterexamples to this statement in the literature were not correct), see the Introduction in [Prei90]. The third named author is pleased to say that he carried this message to the "differentiation unit" in Prague led by D. Preiss and L. L. Zajíček. For more information on Theorem 443 we refer to [LPT12] (please, see also Godefroy's review on this book in MathSci).

We note that measure theory in infinite dimensions can hardly be used for differentiability of even convex functions: See [MatMat99].

Finally let us point out that, unlike the set of points of Fréchet differentiability of convex functions, that is residual in a Banach space if dense, there are real-valued Lipschitz functions on the real line whose set of all points of differentiability is dense but not residual (Lebesgue differentiation theorem, Zahorski, see, e.g., [FHHMZ11, Exercise 7.34]).

This glorifies the results of Preiss, as no measure theory can be used; even Baire category cannot be used, so it must be done, as Czech mathematicians say in such situations, "by bare hands".

For functions on finite-dimensional spaces this is Rademacher Theorem 438. See also [FHHMZ11, Chapter 11] and [BoVa09].

The following problems seems to be open:

1. Assume that a separable Banach space X has the property that any Gâteaux differentiable norm on X is Fréchet differentiable somewhere. Is the same true for Lipschitz functions on X?
2. Assume that f_1, f_2, f_3 are Lipschitz functions on ℓ_2. Does there exist a point where all are Fréchet differentiable?

Remark 445. 1. Let the function g be defined by $g(t) = t$ for $t \leq 0$ and $g(t) = 2t$ for $t > 0$. Let $f : \ell_2 \to \ell_2$ be defined by $f(x_1, x_2, \dots) = (g(x_1), g(x_2), \dots)$. Then f is a Lipschitz homeomorphism of ℓ_2 onto ℓ_2 that is nowhere Fréchet differentiable (see [BeLi00, Item (i) after Problem 7.1]).

2. Note (Lindenstrauss) that the function $f : L_2[0,1] \to L_2[0,1]$ defined by $x(t) \to \sin\big(x(t)\big)$ is everywhere Gâteaux differentiable but nowhere Fréchet differentiable, see, e.g., [FHHMZ11, Exercise 11.38]. ®

Chapter 23

Spaces isomorphic to Hilbert spaces

We only list here some conditions equivalent to being isomorphic to a Hilbert space. Some other will be scattered along this text. For example, see Theorems 513, 517, 518, and 561 below. For the concept of a twice Gâteaux differentiable function, see Definition 495 below.

Theorem 446. *Let $(X, \|\cdot\|)$ be a Banach space. The following conditions are equivalent:*
(i) *X is isomorphic to a Hilbert space.*
(ii) *The square of the norm $\|\cdot\|$ is Fréchet differentiable at* 0. See, e.g., [FHHMZ11, Exercise 10.4].
(iii) (Lindenstrauss) *Every separable subspace of X is isomorphic to a Hilbert space.* See also [FaZi99b] and [FHHMZ11, Exercise 5.2].
(iv) *Both X and X^* admit bump functions with locally Lipschitzian first derivative* [FWZ83], see also [DeGoZi93, Corollary V.3.6].
(v) *Both X and X^* admit continuous twice Gâteaux differentiable bumps* [Vand93].
(vi) *Every infinite-dimensional separable subspace Y of X is isomorphic to all its infinite-dimensional subspaces.* See, e.g., [Mau03, page 1249].
(vii) (Lindenstrauss, Tzafriri), see, e.g., [FHHMZ11, Theorem 6.16] *Every subspace of X is complemented in X.*

Proof. (Hints) (ii) If $(X, \|\cdot\|)$ is a Banach space such that $\|\cdot\|^2$ is twice Fréchet differentiable at 0, then X is isomorphic to a Hilbert space. Indeed, if $\|x\|^2 = t(x) + p(x)$ is the Taylor expansion at 0, note that $t(x) = 0$ and $\|x\|^2 - p(x) = o(\|x\|^2)$. Thus there is $\varepsilon > 0$ such that $p(x) \geq (1/2)\|x\|^2$ for $\|x\| > \varepsilon$. Therefore $\sqrt{p(x)}$ defines a Hilbertian norm on X (see [FHHMZ11, Exercise 10.4]).
(v) See Theorem 498. Compare with Theorems 513, 517, and 518 below. □

A. J. Guirao et al., *Renormings in Banach Spaces*, Monografie Matematyczne 75,
https://doi.org/10.1007/978-3-031-08655-7_23

Chapter 24

Superreflexive spaces

The concept of superreflexivity (see Definition 136 above) was introduced by James in [Jame72a] and [Jame72] in terms of the concept of finite representability (see Definition 137 and Item (iv) in Theorem 447), and became one of the most important concepts in the modern Banach space theory.
Definition 136 says that superreflexivity of a Banach space is the property of having an equivalent UR norm. As this text does not specialize in uniform notions of convexity or smoothness we will mention here only few things that did not appear in [FHHMZ11] or other recent books in the area.

The following is a fundamental result:

Theorem 447 (Enflo, Fabian, James, see, e.g., [FHHMZ11], Theorem 9.14). *Let X be a Banach space. The following are equivalent:*
(i) *X is superreflexive.*
(ii) *X admits a UF norm.*
(iii) *X admits a uniformly Fréchet differentiable bump.*
(iv) *All Banach spaces that are finitely representable in X are reflexive.*

The original methods of proof of Enflo and Lancien are discussed carefully in [FHHMZ11, Chapter 9].
Note that if every separable subspace of X admits a UR equivalent norm, then X is superreflexive (see, e.g., [FHHMZ11, Exercise 9.23]). Note, too, that if Y is crudely finitely representable in a superreflexive space X, then Y is superreflexive (see, e.g., [FHHMZ11, Chapter 9]).

We mention first a couple of things on the modulus of convexity (see Definition 67 above). The moduli of convexity and smoothness of the L_p spaces are shown, e.g., in [FHHMZ11, Remark 9.4.2], where its is proved that all the L_p spaces, $1 < p < \infty$, are uniformly rotund (see Theorem 74 above).

Theorem 448 (Pisier, [Pis75], see, e.g., [DeGoZi93], Theorem IV.4.8). *Every superreflexive space can be renormed by a norm whose modulus of convexity* (see Definition 67) *is of power type, i.e., there is $q > 0$ and $K > 0$ such that*

$$\delta(\varepsilon) \geq K\varepsilon^q$$

for all $\varepsilon \in (0, 2]$.

A. J. Guirao et al., *Renormings in Banach Spaces*, Monografie Matematyczne 75,
https://doi.org/10.1007/978-3-031-08655-7_24

Theorem 449 (Liokumovich, [Lio73]). *There is a two-dimensional space with non-convex modulus of convexity.*

Theorem 450 (Figiel, [Fi76]). *If $\delta_X(\varepsilon)$ denotes the modulus of convexity of X, then*

$$\frac{\delta_X(\varepsilon)}{\varepsilon}$$

is nondecreasing on $(0,2]$.

Theorem 451 (Nordlander, [Nor60], see, e.g., [BeLi00], Proposition A.1). *For every Banach space X, $\delta_X(\varepsilon) \leq \delta_H(\varepsilon)$, where δ_H denotes the modulus of convexity of the Hilbert space* (see Proposition 75).

Note that *the norm $\|\cdot\|$ of X is LUR if and only if for every $x \in S_X$ and every $\varepsilon > 0$,*

$$0 < \delta_x(\varepsilon) := \inf\left\{1 - \left\|\frac{x+y}{2}\right\| : \ y \in S_X, \ \|x-y\| \geq \varepsilon\right\}. \tag{24.1}$$

The function δ_x, defined for $x \in S_X$ by the formula (24.1), is called the **pointwise modulus of convexity at** x of the norm $\|\cdot\|$. If there is $q > 0$ and a constant $C > 0$ such that

$$\delta_x(\varepsilon) \geq C\varepsilon^q, \quad \text{for each } \varepsilon \in (0,2],$$

we say that the norm $\|\cdot\|$ has **pointwise modulus of convexity of power type** q **at** x.

Theorem 452 (Deville, Godefroy, Zizler, [DeGoZi93a]). *Assume that for every $x \in S_X$ there is $q(x) > 0$ such that $\|\cdot\|$ has pointwise modulus of convexity of power type $q(x)$ at x. Then X is superreflexive.*

Proof. For positive integers N and q, we let

$$F_{N,q} = \{x \in S_X : \ \delta_x(\varepsilon) \geq N^{-1}\varepsilon^q, \text{ for every } \varepsilon \in (0,2]\}.$$

Then each $F_{N,q}$ is closed and

$$S_X = \bigcup_{N,q} F_{N,q}.$$

From the Baire category theorem we get that there exist positive integers N_1 and q_1 such that F_{N_1,q_1} has a nonempty interior in S_X. Hence there is an open set O in X such that $O \cap S_X \neq \emptyset$ and

$$x, y \in O \cap S_X, \ \|x-y\| \geq \varepsilon \text{ implies } \|x+y\| \leq 2 - 2N^{-1}\varepsilon^{q_1}.$$

Now a result of Day [Day43] provides a uniformly rotund norm on X of power type q_1. □

Theorem 452 shows how fine Kadets' and Troyanski's Theorems 173 and 218, respectively, are.

The following result is in [FHHMZ11], and it was achieved by using the compact variational principle (Theorem 304 above).

Theorem 453 (Fabian, Whitfield, Zizler, see, e.g., [FHHMZ11], Theorem 9.21, and [DeGoZi93], Corollary V.3.3). *Assume that a Banach space X does not contain an isomorphic copy of c_0 and admits a function with bounded nonempty support and locally uniformly continuous derivative. Then X is superreflexive.*

Theorem 454. *The Banach space $(\sum_n \ell_\infty^n)_2$ is reflexive but not superreflexive and admits no C^2-smooth norm.*

Proof. (Sketch) It is well known that the space is reflexive. It is not superreflexive, as c_0 is finitely represented in it (note to students: It is worth to show this fact precisely and meticulously, see, e.g., [FHHMZ11, Theorem 6.2]). It does not admit a C^2-smooth norm because of Theorem 453. □

Recall that a Schauder basis of the space X is **monotone** if all the basis projections have norm one.

Theorem 455 (Guirao, Hájek, [GuiHá07b]). *Assume that X^* is separable and that X has a Schauder basis. Then X is reflexive if and only if for every Schauder basis of X there is a Fréchet smooth norm on X making the basis monotone.*

Proof. (Sketch). If X is reflexive and $\{e_n, e_n^*\}_{n=1}^\infty$ is a Schauder basis of X, then $\{e_n^*, e_n\}$ is a Schauder basis of X^*. We may apply Corollary 177 to this basis to get an equivalent norm $|||\cdot|||$ on X^* that is LUR and in which $\{e_n^*, e_n\}$ is monotone. The predual norm $|||\cdot|||$ on X is then Fréchet differentiable and in this norm the basis $\{e_n, e_n^*\}$ is monotone.
On the other hand, let $\{e_n, e_n^*\}_{n=1}^\infty$ be a Schauder basis of X, and let $F := \operatorname{span}\{e_n^* : n \in \mathbb{N}\}$. Find an equivalent Fréchet differentiable norm $|||\cdot|||$ on X for which the basis is monotone. Hence, F is 1-norming for $|||\cdot|||$. Let $x^* \in S_{(X^*,|||\cdot|||)}$ a functional that attains its norm (say at $x \in S_{(X,|||\cdot|||)}$). Given $\varepsilon > 0$ find a slice $S(B_{(X^*,|||\cdot|||)}, x, 1-\delta)$ of $B_{(X^*,|||\cdot|||)}$ having a $|||\cdot|||$-diameter less than ε. The w^*-density of $B_{(X^*,|||\cdot|||)} \cap F$ in $B_{(X^*,|||\cdot|||)}$ gives a point of $B_{(X^*,|||\cdot|||)} \cap F$ in the slice, so at a $|||\cdot|||$-distance to x^* less than ε. This shows that $x^* \in \overline{B_{(X^*,|||\cdot|||)} \cap F}^{|||\cdot|||}$. By the Bishop–Phelps Theorem 14 above, we get $B_{(X^*,|||\cdot|||)} = \overline{B_{(X^*,|||\cdot|||)} \cap F}^{|||\cdot|||}$, and so the basis is shrinking. Since this happens for every Schauder basis in X, a result of Zippin [Zip68a] shows that X is reflexive. □

Theorem 456 (Guirao, Hájek, [GuiHá07b]). *Assume that X is a separable superreflexive Banach space with a Schauder basis. Then there exists an UR equivalent norm under which the basis $\{e_i\}$ is monotone.*

A result on the extension of an UR norm to the whole space from a closed subspace of a Banach space having an equivalent UR norm was presented in Theorem 122 above. It included not only uniform rotundness, but also some other notions, and a sketch of its proof for the UR case was included there.

The **girth of the unit ball** is the infimum of the lengths of centrally symmetric simple rectifiable curves on its surface.

Theorem 457 (James, Schaffer, [JaSch72]). *A Banach space X is superreflexive if, and only if, the girth of its unit ball is greater than* 4.

For more information in this direction see [Sch67] and [Sch76].

Theorem 458 (M.I. Kadets, [Kad56]). *Let $(X, \|\cdot\|)$ be a uniformly convex Banach space with modulus of convexity $\delta(\varepsilon)$. If $\sum_{j=1}^{\infty} x_j$ converges unconditionally in X, then $\sum_{j=1}^{\infty} \delta(\|x_j\|) < \infty$.*

Theorem 459 (Figiel, Pisier, [Pis75]). *If $\{x_n\}_{n=0}^{\infty}$ is a monotone basic sequence in a Banach space X with modulus of convexity δ_X such that*

$$\sup_n \left\| \sum_{i=0}^{n} x_i \right\| \le 1,$$

then

$$\|x_0\| + \sum_{i=1}^{\infty} \delta_X(\|x_i\|) \le 1.$$

Theorem 460 (Gurarii, Gurarii, James, [GuGu71], [Jame72]). *Let X be a superreflexive Banach pace with a seminormalized Schauder basis $\{x_i\}_{i=1}^{\infty}$* (see Definition 687). *Then there exist $p, q \in (1, \infty)$ and $K > 0$ such that for every $x = \sum_{i=1}^{\infty} \alpha_i x_i \in X$ we have*

$$\frac{1}{K}\Big(\sum_{i=1}^{\infty} |\alpha_i|^q\Big)^{\frac{1}{q}} \le \|x\| \le K\Big(\sum_{i=1}^{\infty} |\alpha_i|^p\Big)^{\frac{1}{p}}.$$

See also, e.g., [FHHMZ11, Theorem 9.25].

Corollary 461 (Benyamini, Starbird, [BeSta76]). *Every superreflexive Banach space injects into some $\ell_p(\Gamma)$ for some $p \in (1, \infty)$.*

On uniformization of strongly exposed points in superreflexive spaces we refer to [Fin86].

Chapter 25

The Kingdom of Tsirelson's space

This chapter presents just a short visit to the ground-breaking results that were ignited by Tsirelson's discovery of his space in 1974.
We present just a list of results together with some comments.

It has been an open problem since Banach time whether every Banach space contains an isomorphic image of an ℓ_p space for $p \in [1, \infty)$, or c_0. It was solved by B.S. Tsirelson (see Theorem 462 below). His ingenious construction is one of the gems in the modern theory of Banach spaces.

Tsirelson's idea may be formulated in the following way: Since c_0 does not admit any UR norm, it is standard to see that it cannot be crudely finitely representable (see Definition 137) in any space with a UR norm, in particular in any ℓ_p, $p > 1$. So to solve Banach problem in the negative, it is enough to find a reflexive space such that c_0 is crudely finitely representable in any infinite-dimensional subspace of it.

Theorem 462 (Tsirelson, [Tsi74], see, e.g., [FHHMZ11], Exercises 9.29–9.31). *There exists a reflexive Banach space in which no ℓ_p $(1 \le p < \infty)$ space can be isomorphically embedded.*

Definition 463. *Let $\lambda > 1$. A Banach space $(X, \|\cdot\|)$ is* **λ-distortable** *(or* **admits a λ-distortable norm***) if there exists an equivalent norm $|\cdot|$ on X so that for all infinite-dimensional subspaces Y of X,*

$$\sup\left\{\frac{|y_1|}{|y_2|} : y_1, y_2 \in S_{(Y,\|\cdot\|)}\right\} \ge \lambda.$$

X is **distortable** *if X is λ-distortable for some $\lambda > 1$. X is* **arbitrarily distortable** *if X is λ-distortable for all $\lambda > 1$.*

At the time when R.C. James showed the following result, it was an open problem about the existence of distortable spaces:

Theorem 464 (James, [Jame64c]). *Neither c_0 nor ℓ_1 is distortable.*

Milman proved the following result:

A. J. Guirao et al., *Renormings in Banach Spaces*, Monografie Matematyczne 75,
https://doi.org/10.1007/978-3-031-08655-7_25

Theorem 465 (V.D. Milman, [Milm71]). *If X does not contain an isomorphic copy of c_0 or ℓ_p for some $1 \le p < \infty$, then X contains a distortable subspace.*

The original version of Tsirelson's space is discussed, e.g., in [FHHMZ11, Exercises 9.29, 9.30, and 9.31].
Based on Tsirelson's result, soon after it, Figiel and Johnson presented an easier analytic description of the dual of Tsirelson's space in [FiJo74]. It was this crystal-clear, easy approachable explanation of Figiel and Johnson, that helped to a true revolution in Banach spaces, as we will now try to mention. Let us follow Figiel and Johnson definition of the space:

We work in c_{00}, the space of finitely supported vectors in c_0. If E and F are finite sets of integers, we write $E < F$ if $\max E < \min F$. Given finite sets of integers $E_1, E_2, \ldots, E_k$, we say that $(E_i)_{i=1}^k$ is **admissible** provided that $\{k\} \le E_1 < \ldots < E_k$.

We define a sequence $\{\|\cdot\|_n\}_{n=0}^\infty$ of norms on c_{00} recursively as follows: First, denote by $\|\cdot\|_{c_0}$ the supremum norm on c_0. Then, if $E \subset \mathbb{N}$ and $x \in c_{00}$, put $Ex(n) = x(n)$ if $n \in E$, and $Ex(n) = 0$ otherwise. For $x \in c_{00}$, let

$$\begin{cases} \|x\|_0 := \|x\|_{c_0}, \\ \|x\|_{n+1} \\ := \max\Big\{\|x\|_n, \frac{1}{2}\max\big\{\sum_{j=1}^k \|E_i x\|_n : \ k \in \mathbb{N} \text{ and } (E_i)_{i=1}^k \text{is admissible}\big\}\Big\}, \\ n = 0, 1, 2, \ldots \end{cases}$$

Then $\{\|\cdot\|_n\}_{n=0}^\infty$ is an increasing sequence of unconditional norms on c_{00}, so the norm defined by $\|x\| := \lim \|x\|_n$ is a monotonely unconditional norm on c_{00}. The completion T^* of $(c_{00}, \|\cdot\|)$ is the dual of the original Tsirelson's space T.

The norm $\|\cdot\|$ on T^* has the following property:

$$\|x\| = \max\left\{\|x\|_{c_0}, \frac{1}{2}\max\left\{\sum_{i=1}^k \|E_i x\| : \ k \in \mathbb{N} \text{ and } (E_i)_{i=1}^k \text{ is admissible}\right\}\right\}. \tag{25.1}$$

Formula (25.1) gives, in fact, $\|\cdot\|$ on c_{00} defined implicitly: $\|\cdot\|$ is the solution of (25.1) (which of course must be shown to exist).
The following result also solves in the negative Banach's problem mentioned at the beginning of this chapter:

Theorem 466 (Figiel, Johnson, [FiJo74]). *The space T^* defined in the preceding paragraphs is a reflexive separable space with unconditional basis that contains no infinite-dimensional superreflexive subspace.*

So it followed from Milman's Theorem 465 and from Tsirelson's Theorem 462 that the space T^* contained a distortable subspace. Fifteen years later, Odell proved that T^* has a $(2-\varepsilon)$-distortable norm for each $\varepsilon > 0$. The proof of Odell's result is in [OdSch03, Theorem 12]. It seems that, up to now, it is not known whether T^* is arbitrarily distortable.

Theorem 467 (Figiel, Johnson, [FiJo74]). *There is a superreflexive space with unconditional basis that isomorphically contains no ℓ_p for $p > 1$.*

Motivated by Odell's result, Thomas Schlumprecht proved the following extremely important result

Theorem 468 (Schlumprecht, [Sch91]). *There is a Banach space* (now called S) *that has the following properties: It is arbitrarily distortable, with unconditional basis, and not containing any of c_0 or ℓ_p for $1 \leq p < \infty$.*

In the notation of the Figiel–Johnson space above, and denoting the function $f(n) := \log_2(n+1)$, $n \geq 1$, the space S is defined as the completion of c_{00} under the norm defined implicitly by

$$\|x\| = \max\left\{\|x\|_{c_0}, \sup\left\{\frac{1}{f(l)}\sum_{i=1}^{l}\|E_i x\| : l \geq 2 \text{ and } E_1 < \ldots < E_l\right\}\right\}.$$

Schlumprecht's result has an enormous impact on the Banach space theory overall. Soon after Schlumprecht's talk in Jerusalem conference in 1991, Maurey, and independently Gowers, with the help of Johnson, by using the space S, showed the following ground-breaking result (a space with the properties listed there is now called **hereditarily indecomposable**):

Theorem 469 (Gowers, Johnson, Maurey, [GowMau93]). *There is a separable Banach space X such that for no subspace Y of X, there would exist a projection P from Y into Y with both PY and $(I-P)Y$ infinite-dimensional.*

Theorem 470. (a) *Every hereditarily indecomposable space is not isomorphic to any of its proper subspaces, in particular to any of its hyperplanes.*
(b) *Any hereditarily indecomposable space does not contain any infinite-dimensional subspace with unconditional basis (just consider the projection onto even coordinates).*

The result in (a) is, e.g., in Corollary 8 of the excellent paper [Mau03], that we recommend to graduate students very much.

Let us mention here just a few more results on hereditarily indecomposable spaces:

Theorem 471 (Gowers, [Gow96]). *Every infinite-dimensional Banach space contains either an infinite-dimensional subspace with unconditional basis or a hereditarily indecomposable subspace.*

Theorem 472 (Tomczak-Jaegermann, Komorowski, [KoTo95]). *If a separable Banach space X is isomorphic to all of its infinite-dimensional subspaces, then X is isomorphic to ℓ_2.*

Theorem 473 (Gowers, [Gow94]). *There is a Banach space that does not contain a reflexive subspace or a space isomorphic to* c_0 *or* ℓ_1.

The following result solved a longtime open problem which used to be one on the most important problems in Banach spaces so far.

Theorem 474 (Argyros, Haydon, [ArHay11]). *There is a hereditarily indecomposable space* X *whose dual is isomorphic to* ℓ_1 *and such that every operator from* X *into* X *is of the form a multiple of the identity plus a compact operator.*

Note that X from Theorem 474 is the first example of an infinite-dimensional space such that $\mathcal{B}(X)$ is separable, every operator has a nontrivial invariant subspace, and a hereditarily indecomposable space whose dual has an unconditional basis. We recommend to read the excellent reviews on the result in Math. Reviews MR2784662 by K. Beanland and in Zentralblatt Zbl 1223.46007 by D. Werner.
It is an open problem (G. Pisier, see, e.g., [GuiMoZi16, Problem 71]) whether in the previous result the space of compact operators can be replaced with the space of nuclear operators (for the definition see, e.g., [FHHMZ11, Definition 16.21]).

Theorem 475 (Tomczak-Jaegermann, [To96]). *Every hereditarily indecomposable space is arbitrarily distortable.*

The following result is an improvement of results of Shelah and Steprans as quoted therein:

Theorem 476 (Wark, [War01]). *There is a nonseparable reflexive Banach space on which every operator is of the form scalar multiple of the identity plus an operator with separable range.*

Theorem 477 (Koszmider, Plebánek, [Ple04]). *There is a compact space* K *such that every operator* T *from* $C(K)$ *into* $C(K)$ *is of the form* $T = gI + S$, *where* $g \in C(K)$, I *is the identity operator and* S *is a weakly compact operator.*

Theorem 478 (Argyros, Tolias, [ArTo04]). *There exists a nonseparable hereditarily indecomposable Banach space.*

Theorem 479 (Odell, Schlumprecht, [OdSch03]). *If* $1 < p < \infty$, *then* ℓ_p *is arbitrarily distortable.*

Theorem 480 (Ferenczi, [Fer97]). *There is a uniformly convex hereditarily indecomposable space.*

Remark 481. This chapter shows, in the good classical language, that the concept of Banach space is more creative than it had been thought before. In the modern business-like language, it shows that Banach spaces have more money in the bank than it had been thought before. ®

Some open problems in the area follow:

1. Is Tsirelson space arbitrarily distortable?
2. If X is of density $\geq c$, does X contain a separable subspace with unconditional basis? [Todorčević]
3. Does there exist a reflexive Banach space of density c such that c_0 is crudely finitely representable in every infinite-dimensional subspace of X?
4. Does there exist a hereditarily indecomposable space that admits a C^2-smooth norm?
5. Does there exist a hereditarily indecomposable space that is isomorphic to its dual?
6. A Banach space X is called **nonseparably hereditarily indecomposable** if whenever $Y \subset X$ is a subspace of X and $Y = Y_0 \oplus Y_1$ is a topological sum for Y_0 and Y_1 subspaces of Y, then Y_0 or Y_1 must be separable. Study this notion. In this direction, see [LopTod09].

Chapter 26

The $\mathcal{L}_\infty$ spaces

The following type of spaces were introduced by J. Lindenstrauss, A. Pełczyński, and H. P. Rosenthal in the papers [LiPe68] and [LiRo69] in attempt to get more insight into the positions of finite-dimensional subspaces of $C(K)$ spaces. We refer to [PelBe79, Chapter 5].

Definition 482. *A Banach space X is called an* **$\mathcal{L}_\infty$-space** *if there exists $\lambda > 1$ such that for every finite-dimensional subspace E of X, there exists a finite-dimensional F with $E \subset F \subset X$ so that*

$$d(F, \ell_\infty^n) \leq \lambda, \text{ where } n = \dim F,$$

and $d(.,.)$ denotes the Banach–Mazur distance (see Definition 691).
If λ works, X is called an **$\mathcal{L}_{\infty,\lambda}$ space**.

Theorem 483 (Lindenstrauss, Rosenthal, [LiRo69]). *Let K be a compact space. Then $C(K)$ is an $\mathcal{L}_{\infty,\lambda}$-space for every $\lambda > 1$.*

For a proof, see also [FHHMZ11, Proposition 5.24].

Theorem 484 (Lindenstrauss, Rosenthal, [LiRo69]). *A Banach space X is an $\mathcal{L}_{\infty,\lambda}$-space for every $\lambda > 1$ if and only if X is an $L_1(\mu)$-predual, i.e., X^* is isometric to $L_1(\mu)$ for some μ.*

Theorem 485 (Benyamini, Lindenstrauss, [BeLi72]). *There exists a Banach space X with X^* isometric to ℓ_1 such that X is not isomorphic to any $C(K)$ space.*

Theorem 486 (Zippin, [Zip69]). *Any L_1-predual contains an isometric copy of c_0.*

Theorem 487 (Bourgain, Delbaen, [BouDel81]). *There is a Banach space Y that has the following properties:*
(i) *Y is a separable $\mathcal{L}_\infty$-space.*
(ii) *Y has the RNP property, so in particular Y does not contain an isomorphic copy of c_0.*
(iii) *Y is* **somewhat reflexive**, *i.e., every infinite-dimensional subspace of Y contains an infinite-dimensional subspace that is reflexive.*
(iv) *Y^* is isomorphic to ℓ_1.*

Note that it follows from the above that the classes of $C(K)$ spaces, the preduals of L_1 and of $\mathcal{L}_\infty$ spaces are pairwise different.

A. J. Guirao et al., *Renormings in Banach Spaces*, Monografie Matematyczne 75,
https://doi.org/10.1007/978-3-031-08655-7_26

Theorem 488 (Lewis and Stegall, [St73]). *Let X be a separable $\mathcal{L}_\infty$-space. Then X^* is isomorphic either to ℓ_1 or to $C[0,1]^*$. The first case happens if, and only if, X does not contain an isomorphic copy of ℓ_1.*

Theorem 489 (Lindenstrauss and Rosenthal, [LiRo69]). *The following are equivalent.*
(i) *X is an $\mathcal{L}_\infty$-space.*
(ii) *For all Banach spaces $Z \supset Y$ and compact operators T from Y into X, T has a compact extension from Z into X.*
(iii) *For all Banach spaces Y and Z with $Z \supset X$, every compact operator T from X into Y has a compact extension from Z into Y.*

Theorem 490 (Freeman, Odell, Schlumprecht, [FOS11]). *Let X be a Banach space with separable dual. Then X embeds isomorphically into an $\mathcal{L}_\infty$-space whose dual is isomorphic to ℓ_1.*

Theorem 491 (Bourgain, Pisier, [BouPis83]). *Let $\lambda > 1$. Let X be a separable space. Then X can be isometrically embedded into a separable $\mathcal{L}_{\infty,\lambda}$-space E such that E/X has the RNP.*

Theorem 492 (Rosenthal, [Ros83]). *Let X be an $\mathcal{L}_\infty$-space which is isomorphic to a subspace of a separable Banach space with unconditional basis. Then X is isomorphic to c_0.*

Chapter 27

Higher-order smoothness

27.1 Higher-order Gâteaux smooth norms

Theorem 493. (See, e.g., [FHHMZ11, Theorem 10.8]) *Let* $p \in (1, \infty)$.
(i) *If* p *is even then the canonical norm* $\|\cdot\|_p$ *of* ℓ_p *is* C^∞*-smooth.*
(ii) *If* p *is odd then the canonical norm* $\|\cdot\|_p$ *of* ℓ_p *is* C^{p-1}*-smooth.*
(iii) *If* p *is not an integer then the canonical norm* $\|\cdot\|_p$ *of* ℓ_p *is* $C^{[p]}$*-smooth, where* $[p]$ *is the integer part of* p.

Theorem 494 (Kurzweil, [Kurz54], see also [FHHMZ11], Theorem 10.7)**.** *If* p *is odd, then* ℓ_p *does not admit a* C^p*-differentiable bump.*

Definition 495. *A function* $\varphi : X \to \mathbb{R}$ *is said to be* **twice Gâteaux differentiable at** $x \in X$ *provided that the Gâteaux derivative* $\varphi'(y)$ *exists for* y *in a neighborhood of* x*, and that the limit*

$$\varphi''(x)(h,k) = \lim_{t \downarrow 0} \frac{1}{t} \big(\varphi'(x+tk) - \varphi'(x)\big)(h)$$

exists for each $h, k \in X$*, and that* $\varphi''(x)(\cdot,\cdot)$ *is a continuous symmetric linear form.*
Higher-order Gâteaux differentiable norms are then defined inductively.

Theorem 496 (Borwein, Noll, [BoNo94])**.** *Let* f *be a continuous convex function on a Banach space* X *that is twice Gâteaux differentiable at a point* $x \in X$*. Then* f *is Fréchet differentiable at* x*.*

The smart proof depends on the Baire category theorem and the convexity.

Compare the following result with (v) in Theorem 446 above.

Theorem 497 (Vanderwerff, [Vand93])**.** *If a Banach space* X *admits a continuous bump that is twice Gâteaux differentiable, then* X *is an Asplund space.*

The following result was already recorded in Theorem 446 above. We reproduce it here because of the use of higher-order derivatives.

A. J. Guirao et al., *Renormings in Banach Spaces*, Monografie Matematyczne 75,
https://doi.org/10.1007/978-3-031-08655-7_27

Theorem 498 (Vanderwerff, [Vand93]). *If both X and X^* admit continuous twice Gâteaux differentiable bumps, then X is isomorphic to a Hilbert space.*

Theorem 499. *If $1 < p < 2$ then ℓ_p does not admit a continuous twice Gâteaux differentiable bump.*

This is because the dual of such ℓ_p admits a twice Fréchet differentiable norm. See Theorem 514 and its proof below.

In Theorem 498 it was used the following results in [MPVZ93]:

Theorem 500. *If an RNP space X admits a C^2-smooth bump, then X admits a norm with modulus of smoothness of power type 2.*

Theorem 501 (Fabian, Whitfield, and Zizler, [FWZ83]). *Let X be a separable Banach space that admits a norm with modulus of smoothness of power type 2. Then X admits a twice Gâteaux differentiable norm.*

In [DeGonJa99] it is the following result:

Theorem 502. *The space $(\sum \ell_4^n)_2$ admits a twice Gâteaux differentiable norm but admits no Fréchet twice differentiable norm.*

Theorem 503 (Troyanski, [Tr90]). *If p is an odd natural number and μ is a σ-finite measure, then $L_p(\mu)$ admits a p-times Gâteaux differentiable norm.*

Theorem 504 (Troyanski, [Tr90]). *If p is an odd natural number, then $\ell_p(\Gamma)$ admits a p-times Gâteaux differentiable norm if, and only if, Γ is countable.*

27.2 Higher-order smoothness

Theorem 505. *The following results hold.*
(i) (Preiss) [FPWZ89] *There is an equivalent real-analytic norm on c_0.*
(ii) (Pełczyński) [Pel57] *If Γ is uncountable, then $c_0(\Gamma)$ does not admit any real-analytic norm.*
(iii) (Kuiper) (see, e.g., [DeGoZi93, Theorem V.1.5]) *For any Γ, $c_0(\Gamma)$ admits a C^∞-smooth norm* (see also Theorem 506 below).

Proof. (Sketch) (i) The norm is constructed as follows: Let φ be defined on c_0 by

$$\varphi(x) = \sum_{n=1}^{\infty} (x_n)^{2n}. \tag{27.1}$$

Then φ is real-analytic on c_0 and the norm be defined as the Minkowski functional of the set $\{x : \varphi(x) \leq 1\}$ (see, e.g., [FHHMZ11, Definition 2.9 and Lemma 2.11]).

H It confirms an old good Czech mathematical saying: The utmost important is a good definition. It took D. Preiss a few minutes to show this new cornerstone result to Zizler on his request, on the corridor of CAB (Central Academic Building) at University of Alberta in 1987.

(iii) Let φ be an even C^∞-smooth real valued function on $\mathbb{R}$ such that $\varphi = 0$ on $[\frac{-1}{2}, \frac{1}{2}]$, $\varphi' > 0$ and $\varphi'' > 0$ on $(\frac{1}{2}, \infty)$, and consider, as in (i), on $c_0(\Gamma)$ the set $\{x : \sum_{\gamma \in \Gamma} \varphi(x_\gamma) \leq 1\}$. □

H [BoFra66] contains a result of N. Kuiper, where he first proved that c_0 admits a C^∞-smooth norm.
[PeWhZi81] contains a new flexible proof of (iii) in Theorem 505 above, where summations were used instead of Kuiper's products. In fact this new approach enabled to prove the following result:

Theorem 506. (See, e.g., [DeGoZi93, Theorem V.I.5]) *For any set Γ, the space $c_0(\Gamma)$ admits a norm which is Fréchet differentiable, locally uniformly rotund, and which is a limit (uniform on bounded sets) of C^∞-smooth norms. In particular, $c_0(\Gamma)$ admits a C^∞-smooth norm.*

With UG norms and the use of Minkowski functionals, one has to be careful about the use of the Implicit Function Theorem. So, some more care is needed and this was done in [WKT97], [FaZi99], and [HáJo14], and we have the following result:

Theorem 507 (Wee-Kee Tang, [WKT97]). *If a Banach space X admits a uniformly Gâteaux differentiable bump, then X admits an equivalent UG norm .*

In this connection, it should be pointed out that there are Banach spaces with C^∞-smooth functions with nonempty bounded support and yet they do not admit any G norm (see Theorem 817 below).

Theorem 508 (Fabian, Zizler, [FaZi99]). *Let X be a separable Banach space that admits a C^n-smooth norm, for $n \in \mathbb{N}$ or $n = \infty$. Then X admits a norm that is both C^n-smooth and uniformly Gâteaux differentiable.*

Theorem 508 should be compared with Theorem 810 below.

Since higher-order smoothness renormings do not allow for Asplund's averaging (see e.g. Theorem 510 below), the proof of Theorem 508 above requires a special effort. It is done by using finite-dimensional integral convolutions.

On the spaces $C[0,\alpha]$ we have the following:

Theorem 509 (Hájek, Procházka, [HáPro14]). *Let X be a WCG space that admits a C^k-smooth norm ($k \in \mathbb{N} \cup \{\infty\}$). Then X admits a C^1-smooth norm that is LUR and is a limit (uniform on bounded sets) of C^k-smooth norms. The same holds for $C[0,\mu]$ in case $k = \infty$.*

The last assertion fits into the frame because Theorem 336 above. It is unknown if every norm on $C[0,\omega_1]$ is a limit, uniform on bounded sets, of norms that are C^1-smooth.
For a separable version of Theorem 509 see [DeGoZi93, Theorem V.I.7].

Theorem 510 (Fabian, Whitfield, and Zizler, [FWZ83], see, e.g., [DeGoZi93], Proposition V.1.3). *If the norm of a space X is C^2-smooth and has the w-$\mathcal{T}_{\|\cdot\|}$-KK property, then X is superreflexive. In particular, c_0 does not admit a norm that would be LUR and C^2-smooth.*

Theorem 511 (Hájek, see, e.g., [HáJo14] Chapter 5, Theorem 132). *A $C(K)$ space admits a real-analytic norm if and only if $C(K)$ is separable.*

Proof. (partial) If $C(K)$ is separable and admits a C^∞-smooth norm, then K is scattered ([FHHMZ11, Theorem 14.25], see also Section 27.3), and since $C(K)$ is separable, K is metrizable (see Subsection 3.4.4, Item 5), so K is countable ([FHHMZ11, Lemma 14.21]). Then $C(K)$ admits a real-analytic norm (Hájek, [Há96b] or [HáJo14, Chapter 5, Theorem 132]). □

It was thus difficult to prove the following result, as the example cannot exist among $C(K)$ spaces. It was found among the so called Orlicz spaces.

Theorem 512 (Hájek, Troyanski, [HáTr01], see also [HáJo14], Chapter 5, Corollary 148). *There is a separable Banach space that admits a C^∞-smooth norm but admits no real-analytic norm.*

The following result used also methods from [Kurz54]. Compare with Theorem 498.

Theorem 513 (Meškov, see, e.g., [FHHMZ11], Theorem 11.7). *Let X be a Banach space such that both X and X^* admit C^2-smooth norms. Then X is isomorphic to a Hilbert space.*

The following statement list provides some instances of spaces that admit C^1-smooth norm but do not admit any C^2-smooth norm.

Theorem 514. (i) *If $1 < p < 2$, then ℓ_p admits no C^2-smooth norm though its standard norm is C^1-smooth.*
(ii) *Any reflexive space which is not superreflexive does not admit any C^2-smooth norm, though it admits a C^1-smooth norm.*
iii *The James quasi-reflexive space J admits a C^1-smooth norm but admits no C^2-smooth norm.*

Proof. (Sketch) To see (i), note that otherwise if q is such that $\frac{1}{p} + \frac{1}{q} = 1$, then $q > 2$ and thus both ℓ_p and $\ell_{p^*} = \ell_q$ admit C^2-smooth norms (see [FHHMZ11, Theorem 10.8]) and thus ℓ_p is isomorphic to ℓ_2, which is not the case (by Pitt's theorem, see, e.g., [FHHMZ11, Proposition 4.49]).
To see (ii), consult [DeGoZi93, Corollary V.3.3].
To see (iii), note that J does not contain an isomorphic copy of c_0, and J is not reflexive so not superreflexive; consult again [DeGoZi93, Theorems V.3.1, V.3.2, and V.3.4, and Corollary V.3.3]. □

Theorem 515 (Kurzweil, [Kurz57]). *Let f be a real-valued continuous function on ℓ_2 and $\varepsilon > 0$ be given. Then there is a real-analytic function g on ℓ_2 such that $|f(x) - g(x)| \leq \varepsilon$ for each $x \in \ell_2$.*

The result in Theorem 515 above is open for the space c_0.

The following is a milestone result of Deville. It implies, in particular, that there are no very smooth Tsirelson-like spaces:

Theorem 516 (Deville, see, e.g., [DeGoZi93], Theorem V.4.1). *Assume that an infinite-dimensional Banach space X admits a C^∞-smooth bump function and that X does not contain an isomorphic copy of c_0. Then there is an even integer $2q$ such that X contains an isomorphic copy of ℓ_{2q}.*

Theorem 517 (Makarov, [Mak79], see, e.g., [DeGoZi93], Corollary V.5.2). *Assume that a Banach space X admits a C^2-smooth bump function and that every infinite-dimensional subspace of X contains an isomorphic copy of ℓ_2. Then X is isomorphic to a Hilbert space.*

The following is an open problem: If X is saturated with ℓ_2, must the bidual norm be somewhere Fréchet differentiable? Is it true for the space JT? (See [HáMoZi12]).

A related result is the following one: We need first a definition.

A Banach space X is said to have the **Orlicz property** if $\sum_{i=1}^{\infty} \|x_i\|^2$ is convergent whenever $\sum x_i$ is unconditionally convergent.

Note that L_p for $p \leq 2$ has the Orlicz property (see e.g. [FHHMZ11, Remark 9.4.4]).

Theorem 518 (Makarov, [Mak83]). *Assume that a Banach space X has the Orlicz property and that it admits a C^2-smooth bump function. Then X is isomorphic to a Hilbert space.*

Theorem 518 above and the note preceding it show again that L_p does not admit any C^2-smooth bump function if $1 < p < 2$.

The following result allows for a C^k-smooth renorming of a Banach space under some special circumstances. Its statement will be enlarged in Theorem 657 below.

Theorem 519 ([GPWZ88], see also [DeGoZi93], Lemma V.1.9). *Let $k \in \mathbb{N}$ or $k = \infty$. Let X be a Banach space and Y be a subspace of X that is isomorphic to $c_0(\Gamma)$. Let X/Y admit C^k-smooth norm. Then the same holds for X.*

Theorem 520 (D'Alessandro, Hájek, [AlHa14]). *Let X be an infinite-dimensional Banach space with C^2-smooth norm. Then there is a separable infinite-dimensional subspace Z of X such that $X = Z \oplus Y$ for some subspace Y of X.*

In the proof the following theorem is used:

Theorem 521 (D'Alessandro, Hájek, Schlumprecht). *Let X be a Banach space with w^*-sequentially compact dual closed unit ball such that c_0 is isomorphic to a subspace Y of X. Then there is a further subspace Z of Y such that Z is isomorphic to c_0 and Z is complemented in X.*

Spaces with a w^*-sequentially compact dual closed unit ball are said to belong to the SK*-class, or to have the SK*-property. They will be the subject of Section 45.3 below.

27.3 Smooth norms in $C(K)$ spaces

Finally, we consider the problem of finding on a given $C(K)$ space an equivalent norm of the highest possible smoothness. It is a classical result (e.g., [DeGoZi93]) that the existence of an equivalent C^1-smooth norm on a Banach space implies that the space is Asplund. On the other hand, deep examples of Haydon ([Hay99], see also [DeGoZi93]) show that not every $C(K)$ Asplund space admits an equivalent Gâteaux smooth renorming (see Theorem 817 below).

So far, an equivalent C^∞-smooth renorming was constructed on $C(K)$ spaces where $K^{(\omega_0)} = \emptyset$ ([GPWZ88]), and a C^1-smooth norm is guaranteed when $K^{(\omega_1)} = \emptyset$ ([De86]). Haydon's C^∞-smooth renorming techniques work well for certain tree-like compact sets K, which may have nonempty derived sets of arbitrary large ordinal number, but their disadvantage is that they put very strong structural restrictions on K (apart from the obvious and necessary scatteredness). This is not accidental, because the above mentioned example of $C(K)$ without a Gâteaux norm (Theorem 817) has $K^{(\omega_1)}$ a singleton.

Here we show the existence of C^∞-smooth renormings whenever $K^{(\omega_1)} = \emptyset$ (Theorem 524 below). This is the best possible result without additional structural assumptions on K. However, it is really only a small step towards a desired general theorem linking the existence of C^∞-smooth renorming of $C(K)$ to some other properties of the space, such as the existence of a dual LUR renorming of $C(K)^*$ (Theorem 530 below). For background material and notation we refer to [DeGoZi93].

The concept of dependence on finitely many functionals or coordinates was introduced in Definition 140 above.

Given $1 > \delta > 0$, find $\phi_\delta : \mathbb{R} \to \mathbb{R}$ such that ϕ_δ is C^∞-smooth, even, convex, $\phi_\delta([0, 1-\delta]) = 0$, $\phi_\delta(1) = 1$.

Definition 522. *Let $f \in \ell_\infty(\Gamma)$. Put*

$$f^\infty := \inf\{t : \ \operatorname{card}\{\gamma : \ |f(\gamma)| > t\} \text{ is finite}\}.$$

Lemma 523. *Let $1 > \delta > 0$. Let $\Phi : \ell_\infty(\Gamma) \to \mathbb{R} \cup \{+\infty\}$ be a convex function defined by*

$$\Phi(f) = \sum_{\gamma\in\Gamma} \phi_\delta(f(\gamma)).$$

Then Φ restricted to $\{f \in \ell_\infty(\Gamma) : \ f^\infty < 1 - \delta\}$ is finite, LFC and C^∞-smooth.

Proof. Given f such that $f^\infty < 1-\delta$, the set $\Theta = \{\gamma \in \Gamma : \ |f(\gamma)| > f^\infty + \frac{1-\delta-f^\infty}{4}\}$ is finite. Thus for $g \in B(f, \frac{1-\delta-f^\infty}{2})$ we have $\Phi(g) = \sum_{\gamma\in\Theta} \phi_\delta(g(\gamma))$, which is a finite sum of C^∞-smooth convex functions. □

Theorem 524 (Hájek, [Há03]). *Let K be a scattered compact, $K^{(\omega_1)} = \emptyset$. Then $C(K)$ admits an equivalent LFC and C^∞-smooth norm.*

Proof. There is $\Lambda < \omega_1$ such that $K^{(\Lambda)} \neq \emptyset$ is finite and $K^{(\Lambda+1)} = \emptyset$. The space $C_0(K) = \{f \in C(K) : \ f(K^{(\Lambda)}) = 0\}$ is isomorphic to $C(K)$.
Put $L_\alpha = K^{(\alpha)} \setminus K^{(\alpha+1)}$, $\alpha \le \Lambda$, and fix $\{\delta_\alpha\}_{\alpha\le\Lambda}$, such that $\delta_\alpha > 0$ and $\prod_{\alpha=0}^{\Lambda} (1 + \delta_\alpha) < 2$. Put $\psi_\alpha = \phi_{\frac{\delta_{\alpha+1}}{8}}$, $D_\alpha = \prod_{\beta=0}^{\alpha} (1 + \delta_\beta)$. Let us define a convex function $\Psi : C_0(K) \to \mathbb{R} \cup \{+\infty\}$ by $\Psi(f) = \sum_{0\le\alpha\le\Lambda} \sum_{\gamma\in L_\alpha} \psi_\alpha(D_\alpha \cdot f(\gamma))$.
Our aim will be to show that $\Psi^{-1}([0, \frac{1}{2}])$ is the unit ball of an equivalent LFC (canonically, $C_0(K) \subset \ell_\infty(K)$) and C^∞-smooth norm on $C_0(K)$.

For every $f \in C_0(K)$, $0 \le \alpha \le \Lambda$ put

$$a^f_\alpha = (f|_{L_\alpha})^\infty, \quad b^f_\alpha = \|f|_{L_\alpha}\|_\infty.$$

Note that $\alpha \to D_\alpha$ is an increasing and thus upper semicontinuous (usc) function on $[0,\Lambda]$ (which is a compact space when considered endowed with its natural interval topology). We have

$$b^f_{\alpha+1} = a^f_\alpha, \text{ and } \alpha \to b^f_\alpha \in C[0,\Lambda], \quad b^f_\alpha = \lim_{\tau \nearrow \alpha, \tau < \alpha} a^f_\tau. \tag{27.2}$$

Therefore, for a fixed $f \neq 0$, $\alpha \to D_\alpha \cdot b^f_\alpha$ is usc on $[0,\Lambda]$, so it attains its maximum M_f at some $\beta \in [0,\Lambda)$. Clearly, $M_f \ge D_{\alpha+1} b^f_{\alpha+1} = (1+\delta_{\alpha+1}) D_\alpha a^f_\alpha$, and so $D_\alpha a^f_\alpha \le \frac{M_f}{1+\delta_{\alpha+1}} \le M_f(1-\frac{\delta_{\alpha+1}}{2})$, for all $\alpha \in [0,\Lambda]$.
First note that $D_\beta b^f_\beta > (D_\alpha a^f_\alpha)^\infty$. Indeed, otherwise there exist increasing sequences $\alpha_n \nearrow \gamma$, $\alpha_n \in [0,\Lambda)$, $D_{\alpha_n} a^f_{\alpha_n} \nearrow D_\beta b^f_\beta$. However, by (27.2), $a^f_{\alpha_n} \to b^f_\gamma$, $D_\gamma > (1+\delta_\gamma) D_{\alpha_n}$ for $n \in \mathbb{N}$, and consequently $D_\gamma b^f_\gamma \ge (1+\delta_\gamma) D_\beta b^f_\beta$, a contradiction. So there exists $\varepsilon_f > 0$ such that card $\{\alpha, D_\alpha a^f_\alpha > D_\beta b^f_\beta - 2\varepsilon_f\}$ is finite.
Next we claim that card $\{\alpha, D_\alpha b^f_\alpha > D_\beta b^f_\beta - \varepsilon_f\}$ is also finite.
Again, otherwise there exists $\alpha_n \nearrow \gamma$, $D_{\alpha_n} b^f_{\alpha_n} > D_\beta b^f_\beta - \varepsilon_f$. Using (27.2), and passing to a suitable subsequence of $\{\alpha_n\}$ we find $\beta_n \nearrow \gamma$, $\alpha_n \le \beta_n < \alpha_{n+1}$ such that

$$D_{\beta_n} a^f_{\beta_n} \ge D_{\alpha_{n+1}} b^f_{\alpha_{n+1}} - \frac{\varepsilon_f}{2}.$$

This is a contradiction with the definition of ε_f, since

$$\{\beta_n\} \subset \{\alpha, D_\alpha a^f_\alpha > D_\beta a^f_\beta - 2\varepsilon_f\}.$$

Put $O = \{f \in C_0(K), M_f < 1\} \subset 2B_{C_0(K)}$. We claim that $\Psi|_O$ is finite, C^∞-smooth and LFC. Moreover, $\Psi^{-1}([0,\frac{1}{2}]) \subset \operatorname{int} O$, which by the Implicit Function Theorem ([Dieu69], see also Theorem 77) finishes the proof.

Choose any $f \in O$. Consider the finite set $A = \{\alpha \in [0,\Lambda]$, either $D_\alpha b^f_\alpha > M_f - \varepsilon_f$ or $\delta_{\alpha+1} \ge \frac{\varepsilon_f}{4}\}$, and put $\delta_f = \min\{\frac{\delta_{\alpha+1}}{16}, \alpha \in A\}$. Let us check that $\Psi|_{B(f,\delta_f)}$ depends on finitely many coordinates (therefore it is necessarily C^∞-smooth). If $\gamma \in L_\alpha$, $\alpha \notin A$, then $4\delta_{\alpha+1} < \varepsilon_f$ and $D_\alpha b^f_\alpha \le M_f - \varepsilon_f \le 1 - \varepsilon_f \le 1 - 4\delta_{\alpha+1}$. For $g \in B(f,\delta_f)$, $|g(\gamma)| \le |f(\gamma)| + \delta_f$, so

$$|D_\alpha g(\gamma)| \le D_\alpha b^f_\alpha + 2\delta_f \le 1 - \varepsilon_f + 2\delta_f \le 1 - 2\delta_{\alpha+1}.$$

Consequently,

$$\psi_\alpha(D_\alpha g(\gamma)) = 0, \text{ and } \Psi(g) = \sum_{\alpha \in A} \sum_{\gamma \in L_\alpha} \psi_\alpha(D_\alpha g(\gamma)).$$

If $\alpha \in A$, $\gamma \in L_\alpha$, and $|f(\gamma)| < a_\alpha^f(1+\frac{\delta_{\alpha+1}}{8})$ then

$$|D_\alpha f(\gamma)| \leq D_\alpha a_\alpha^f\Big(1+\frac{\delta_{\alpha+1}}{8}\Big) \leq M_f\Big(1-\frac{\delta_{\alpha+1}}{2}\Big)\Big(1+\frac{\delta_{\alpha+1}}{8}\Big) \leq 1-\frac{\delta_{\alpha+1}}{4}.$$

If $g \in B(f, \delta_f)$, we then have

$$|D_\alpha g(\gamma)| \leq 1-\frac{\delta_{\alpha+1}}{4}+2\delta_f \leq 1-\frac{\delta_{\alpha+1}}{8}.$$

Consequently, in this case also $\psi_\alpha(D_\alpha g(\gamma)) = 0$. The remaining set $S = \{\gamma : \gamma \in L_\alpha$ for $\alpha \in A$ and $f(\gamma) \geq a_\alpha^f(1+\frac{\delta_{\alpha+1}}{8})\}$ is clearly finite, and we have

$$\Psi(g) = \sum_{\alpha\in A}\sum_{\gamma\in S\cap L_\alpha}(D_\alpha g(\gamma))$$

whenever $g \in B(f, \delta_f)$. This proves (Lemma 523) that $\Psi|_O$ is C^∞-smooth and LFC. It is obvious that $M_f \leq \frac{1}{4}$ implies $\Psi(f) = 0$ and $M_f = 1$ implies $\Psi(f) \geq 1$. Thus $B = \Psi^{-1}([0, \frac{1}{2}])$ is an equivalent unit ball of $C_0(K)$. By the Implicit Function Theorem 77, its Minkowski functional is C^∞-smooth and LFC. □

27.4 Survey on higher-order smooth norms on $C(K)$ spaces

Here we just provide an instructive brief survey of a few results discussed in this text concerning higher-order Fréchet differentiable norms on $C(K)$ spaces.

Let us start with the case the compact space K is metrizable. This is equivalent to say that $C(K)$ is separable.

Theorem 525 (Deville, Hájek, Fonf). *Let K be a metrizable compact space.*
(i) *If K is countable, then $C(K)$ admits a real-analytic norm, and thus a C^∞-smooth norm.*
(ii) *If K is uncountable, then $C(K)$ does not admit any Fréchet differentiable norm.*

Proof of (ii). Observe that, by Milyutin's Theorem (see Theorem 777), $C(K)$ is isomorphic to $C[0,1]$. This last space does not admit any Fréchet differentiable norm, as its dual is non-separable (see Remark 59 above). Another (easier) argument to show that, if K is uncountable, then $C(K)^*$ is not separable, is in Item 5 in Subsection 3.4.4 above. □

Corollary 526. *Let K be a metrizable compact space. Then, the following are equivalent:*
(i) *$C(K)$ admits a C^1-smooth norm.*
(ii) *$C(K)$ admits a C^∞-smooth norm.*
(iii) *$C(K)$ admits an analytic norm.*
(iv) *$C(K)^*$ is separable.*

From the nonseparable $C(K)$ area let us mention the following few results that we use in this text.

Theorem 527 (Deville, Pelant, Godefroy, and Zizler). *If a Banach space X contains a subspace Y isomorphic to $c_0(\Gamma)$ for some Γ, and X/Y admits a C^∞-smooth norm, then X admits a C^∞-smooth norm.*

The following result is part of Theorem 524 recorded above:

Theorem 528 (Hájek). *If $K^{(\omega_1)} = \emptyset$, then $C(K)$ admits a C^∞-smooth norm.*

Theorem 529 (Haydon). *There is a compact space K with $K^{(\omega_1+1)}$ a singleton, so that $C(K)$ does not admit any Gâteaux differentiable norm.*

Theorem 530 (Hájek, Haydon, [HájHay07]). *If $C(K)^*$ admits a dual LUR norm, then $C(K)$ admits a C^∞-smooth norm. Therefore in particular, if K is scattered and Eberlein compact, then $C(K)$ admits a C^∞-smooth norm.*

Note on the proof of the second part. The particular case in Theorem 530 above follows from the following facta: (i) If K is scattered, then $C(K)$ is Asplund (see Theorem 274). (ii) Since K is Eberlein compact, then $C(K)$ is WCG (Theorem 238). (iii) Finally, Theorem 315 concludes that $C(K)$ admits an equivalent norm whose dual norm is LUR, and we are in the situation described in the first part of the statement. □

Concerning analytic norms, we have the following result:

Theorem 531 (Hájek). *If $C(K)$ admits an analytic norm, then $C(K)$ is separable.*

We recall here Haydon's Theorem 336, where it is stated that for an ordinal μ, the space $C[0,\mu]$ admits a C^∞-smooth norm. Note that Talagrand originally constructed a Fréchet smooth norm on $C[0,\mu]$.

Theorem 532 (Haydon). *If T is a tree, then $C(T)$ has an equivalent C^∞-smooth norm if, and only if, it has an equivalent Fréchet smooth norm.*

Theorem 533 (Jiménez-Sevilla–Moreno, [JiMo97], Proposition 3.7). *If T is a tree, then $C_0(T)$ admits a norm with MIP.*

Note that a much easier proof exists showing that for any ordinal μ, $C([0,\mu])$ admits a norm with MIP (see, e.g., Theorem 572).

The following problem is open: If $C(K)$ admits a Fréchet differentiable norm, does it admit a C^∞-smooth norm?

Chapter 28

James boundaries

28.1 Introduction

The following result is in [AzDeJi03]: *If $(X, \|\cdot\|)$ is a separable Banach space, then there is a continuous Gâteaux differentiable bump f on X such that $f'(X) = X^*$.* A similar result for the norm $\|\cdot\|$ of a Banach space $(X, \|\cdot\|)$ instead is impossible except in the case that the space is reflexive. Indeed, the set $\partial\|\cdot\|(X)$ (where $\partial\|\cdot\|$ is the subdifferential of the norm, see Definition 4.1.3) consists of all the norm-attaining functionals in S_{X^*} (a James boundary, see Definition 534 below). If S_{X^*} is a James boundary then, by James' weak compactness Theorem 13 above, the space X is reflexive.
We repeat here the definition of this concept, already introduced at Item 4 in Subsection 3.4.4:

Definition 534. *Let X be a Banach space. A* **James boundary** *(for X) is a subset JB of S_{X^*} such that every $x \in X$ attains its supremum on B_{X^*} at a point in JB. More generally, if $K \subset X^*$, a subset JB of K is said to be a* **boundary of** *K if every $x \in X$ attains its supremum on K at a point of JB.*

By choosing at least an element in each set $\partial\|\cdot\|(x)$, $x \in S_X$, we get a James boundary for X. In particular, the set $\operatorname{Ext} B_{X^*}$ of all the extreme points of B_{X^*} is such (see Figure 28.1).

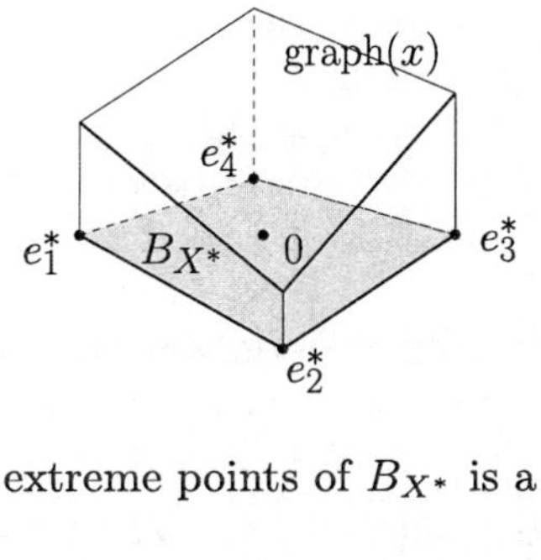

Figure 28.1: The set of extreme points of B_{X^*} is a James boundary for X

James boundaries are one of the most important and powerful concepts in Banach space theory. They lead us to many geometrical results. They are the central object for James' fundamental weak compactness Theorem 13, that gives a necessary and

A. J. Guirao et al., *Renormings in Banach Spaces*, Monografie Matematyczne 75,
https://doi.org/10.1007/978-3-031-08655-7_28

sufficient condition for reflexivity in terms of James boundaries (as was mentioned at the beginning of this chapter) and, more generally, for the w-compactness of a w-closed subset of a Banach space. Its proof will be presented below only in the separable case, as an application of Simons' Lemma 537 and Theorem 538, and Godefroy's Theorem 539. Here we just mention a few related results, and refer to [FHHMZ11, Sections 3.11.8 and 10.3], [HáJo14, Sections 5.5], and [FoLiPh01] for a more complete treatment.

28.2 The boundary problem and the strong boundary problem

If JB is a James boundary for a Banach space X, then clearly $\overline{\text{conv}}^{w^*}(JB) = B_{X^*}$ (in particular, $w(X, JB)$ is a Hausdorff locally convex topology on X). Milman's converse to the Krein–Milman theorem (Theorem 8 above) ensures that $\text{Ext}\, B_{X^*} \subset \overline{JB}^{w^*}$. For an uncountable index set Γ, the set $\{x = (x(\gamma)) \in S_{\ell_\infty(\Gamma)} :$ supp x countable$\}$ is a James boundary for $\ell_1(\Gamma)$ that is disjoint from $\text{Ext}\, B_{\ell_\infty(\Gamma)}$ (see Example 3 in Section 2.2 above).

A result of Grothendieck in 1952 said that *if K is a compact topological space, and M is a $\|\cdot\|_\infty$-bounded subset of $C(K)$ that is pointwise compact, then M is w-compact.*

➡ A streamlined argument is that if $\mathcal{T}_p$ denotes the topology of the pointwise convergence on $C(K)$, then $(C(K), \mathcal{T}_p)$ is angelic (see, e.g., [FHHMZ11, Theorem 3.58] and Definition 139 here). M is thus $\mathcal{T}_p$-sequentially compact. This subsequence is w-convergent (to the same point) by the Lebesgue Dominated Convergence Theorem. The Eberlein–Šmulyan Theorem (see, e.g., [FHHMZ11, Theorem 3.109]) finally shows that M is w-compact.

The topology $\mathcal{T}_p$ on $C(K)$ is, clearly, the topology of the pointwise convergence on the set $\text{Ext}\, B_{C(K)^*}$ (see Example 5 in Section 2.2 above), and this last set is a James boundary for $C(K)$. Thus, it is natural to ask whether the result is true for any bounded $w(X, JB)$-compact subset M of any Banach space $(X, \|\cdot\|)$ and any James boundary JB for X. This question [Godefroy [Go87]] has been known as the **boundary problem**.

Assume for a moment that a James boundary JB for $(X, \|\cdot\|)$ has the extra property that $\overline{\text{conv}}^{\|\cdot\|^*}(JB) = B_{X^*}$. Then *the boundary problem has a positive answer.*

➡ Indeed, let M be a bounded and $w(X, JB)$-compact subset of X. The set $\overline{M}^{w^*}$ is w^*-compact in X^{**}. On it, the topologies w^* and $w(X^{**}, JB)$ coincide, since the second one is coarser than the first one and Hausdorff. The result follows.

It is then highly convenient to isolate the concept of a so-called strong James boundary, already mentioned at Item 4 in Subsection 3.4.4). We repeat its definition here:

Definition 535. *A James boundary* JB *for a Banach space* X *such that* $\overline{\text{conv}}^{\|\cdot\|^*}(JB) = B_{X^*}$ *is called a* **strong James boundary**.

Not every James boundary —not even the set of extreme points of B_{X^*}— is a strong boundary. This was already mentioned at Item 4 in Subsection 3.4.4.

In view of the preceding observations, two interrelated problems arise:
(i) To know whether a given James boundary for a Banach space is strong.
(ii) The aforementioned general boundary problem.

We shown above that problem (ii) has a positive answer as soon as the James boundary is strong. Combined efforts have been devoted to provide sufficient conditions for (i) and (ii). We shall mention some of them in the History Box below. The second problem has been solved in full generality quite recently. Due to the importance of the result, let us formulate it separately. This time, the weak topologies in the statement are not supposed to be Hausdorff:

Theorem 536 (Pfitzner, [Pfit10]). *Let* X *be a Banach space. Let* $JB \subset X^*$ *be a James boundary of a bounded set* $K \subset X^*$. *Let* M *be a bounded and* $w(X, JB)$*-countably compact subset of* X. *Then* M *is* $w(X, \overline{K}^{w^*})$*-sequentially compact. In particular, if* JB *is a James boundary for* X, *then a bounded set* $M \subset X$ *is* w*-compact if, and only if, it is* $w(X, JB)$*-compact.*

Yet, it is worth to mention some of the contributions made in both related problems. After the following enumeration, we shall explicitly present some of the results listed below, and a few of the techniques used. It is very instructive to read the introduction in [Pfit10] for a discussion on these and other tools that have been used thoroughly.

H To the problem (ii):
The classical theorem of Grothendieck (1952) mentioned at the beginning of this section provides a positive answer for $X = C(K)$ and $JB = \text{Ext}\,(B_{X^*})$. The fundamental contribution of James (1957 [Jame57], 1964 [Jame64], [Jame64b]) can be formulated in terms of boundaries: If B_X is a James boundary for X^*, then B_{X^*} is w-compact (i.e., the space is reflexive). More generally, if M is a w-closed subset of X such that every $x^* \in X^*$ attains its supremum (in other words, M is a James boundary for the set $\overline{M}^{w^*} \subset X^{**}$), then M is w-compact (Theorem 13). We shall devote another History Box below to some of James' highlights in this same direction.
J. Rainwater, in 1963 [Rainw63], proved that a bounded and $w(X, JB)$-sequentially compact subset of X is w-compact if $JB = \text{Ext}\,(B_{X^*})$.

If M is bounded and $w(X, JB)$-sequentially compact, and JB is an arbitrary James boundary, then M is w-compact (Simons, 1972).
The boundary problem for convex sets was solved by M. de Wilde in 1974 [dWil74].
Bourgain and Talagrand showed in 1980 [BouTa80] that Grothendieck's result for $C(K)$ mentioned above remains valid for an arbitrary Banach space but still with $JB = \text{Ext}\,(B_{X^*})$.
Again, Grothendieck's result was generalized, still on $C(K)$ but now for an arbitrary James boundary, by B. Cascales and G. Godefroy in 1998 [CasGo98].

The presence or absence of ℓ_1 or $\ell_1(\Gamma)$ plays a relevant rôle, both in the strong and in the general boundary problems (and also in Pfitzner's proof of Theorem 536). In the direction of the boundary problem, let us mention that G. Manjabacas, G. Vera and B. Cascales showed in [CaMaVe97] that it has a positive answer if $\ell_1(\Gamma) \not\subset X$ (see also [CasShv03]).

The related strong boundary problem was treated by Haydon in 1976 [Hay76], showing tat if $\ell_1 \not\subset X$, then $\text{Ext}\, B_{X^*}$ is a strong James boundary.
The case where JB was countable was treated successfully by Rodé, in 1980 and 1981, [Rod80] and [Rod81].
I. Namioka [Nam87], in 1987, gave also a positive answer in the case that JB was a James boundary for a norm-fragmented subset K of X^*.
Rodé's result was significantly extended by Godefroy in 1987 (see Theorem 539 below), who also proved that if X is separable and $\ell_1 \not\subset X$, then every James boundary for X is strong [Go87].
A similar result, now for separable polyhedral spaces (see Definition 141), was proved by Fonf in 1999. Fonf–Lindenstrauss approach [FoLi03] to the strong boundary problem by using the so-called I-formula was seen to be equivalent to the results obtained by using Simons' inequality (Lemma 537 and Theorem 538 below).

28.3 A glimpse of some techniques for the boundary problem and James' theorem

Stegall asked whether the set of extreme points of B_{X^*}, if $\|\cdot\|^*$-separable, should be a strong James boundary. A positive answer was due independently to Fonf, Haydon, M.I. Kadets, and Rybakov. Godefroy proved the result for every separable James boundary (Theorem 539 below).

➪ **Solution of Stegall's problem.** For $\varepsilon > 0$ let $\{f_i\}$ be a norm dense subset of $\text{Ext}\, B_{X^*}$. For each i let B_i denote the intersection of the closed ball of radius $\varepsilon/3$ centered at f_i with B_{X^*}. Let $f \in B_{X^*}$ be given and let μ be a probability measure on B_{X^*} supported by $\text{Ext}\, B_{X^*}$ and having the barycenter

f, i.e. $r(\mu) = f$. Then $\mu(\bigcup B_i) = 1$. Let n be a positive integer such that if $D = \bigcup_{i=1}^n B_i$, then $\mu(D) > 1 - \frac{\varepsilon}{3}$. Then μ can be decomposed as $\mu = \lambda\mu_1 + (1-\lambda)\mu_2$, where $\lambda = \mu(D)$ and μ_1 and μ_2 are probability measures on B_{X^*} defined by

$$\lambda\mu_1 = \mu \restriction D \text{ and } (1-\lambda)\mu_2 = \mu \restriction (B_{X^*} \setminus D).$$

Then $f = r(\mu) = \lambda r(\mu_1) + (1-\lambda)r(\mu 2)$. Since $r(\mu_2) \in B_{X^*}$ we have

$$\|f - \lambda r(\mu_1)\| \leq (1-\lambda)\|r(\mu_2)\| \leq \frac{\varepsilon}{3}.$$

Due to the fact that μ_1 is a probability measure supported by $\bigcup_{i=1}^n B_i$, the point $r(\mu_1)$ lies in the convex hull of $\bigcup_{i=1}^n B_i$, which is w^*-compact. Hence $r(\mu_1) = \sum_{i=1}^n \lambda_i g_i$, where $g_i \in B_i, \lambda_i \geq 0$ and $\sum_{i=1}^n \lambda_i = 1$.

Let $h = \sum_{i=1}^n \lambda_i f_i$. This is a point of $\operatorname{conv}(\operatorname{Ext} B_{X^*})$ and $\|r(\mu_1) - h\| \leq \frac{\varepsilon}{3}$. Consequently

$$\|f - h\| \leq \|f - \lambda r(\mu_1)\| + (1-\lambda)\|r(\mu_1)\| + \|r(\mu_1) - h\| \leq \frac{\varepsilon}{3} + \frac{\varepsilon}{3} + \frac{\varepsilon}{3}.$$

Thus $\operatorname{conv}(\operatorname{Ext} B_{X^*})$ is norm dense in B_{X^*}. □

In the proof of Theorem 538 below, the following lemma, which was motivated by James' results, is used:

Lemma 537 (Simons, [Sim72], see, e.g., [FHHMZ11], Lemma 3.123). *Let B be a James boundary for a Banach space X and $\{x_n\}$ be a bounded sequence in X. Let the function u on B be defined by*

$$u(b) = \limsup_n x_n(b)$$

for $b \in B$. Then

$$\sup_B u \geq \inf\{\sup_B x : x \in \operatorname{conv}\{x_n\}\}.$$

Theorem 538 (Simons, [Sim72], see, e.g., [FHHMZ11], Theorem 3.124). *Let JB be a James boundary for a Banach space X. If $\{x_n\}$ is a bounded sequence in X, and the function u on X^* is defined by*

$$u(x^*) = \limsup_n x^*(x_n)$$

for $x^* \in X^*$, then

$$\sup_{JB} u = \sup_{B_{X^*}} u.$$

As we mentioned above, Simons' Lemma 537 and Theorem 538 originated in James' original proof of Theorem 13. Godefroy deduced from those the following result:

Theorem 539 (Godefroy, see, e.g., [FHHMZ11], Theorem 3.122). *If a Banach space X has a norm-separable James boundary JB, then JB is a strong James boundary.*

Proof. If not true, by the Hahn-Banach theorem we can find $F \in S_{X^{**}}$, $\alpha < \beta$ and $y_0^* \in B_{X^*} \setminus \overline{\text{conv}}^{\|\cdot\|}(JB)$ such that $F(f) \leq \alpha$ for all $f \in JB$ and $F(y_0^*) > \beta$. Let $S = \{x \in B_X : \ y_0^*(x) > \beta\}$. By Goldstine's theorem, we get that $F \in \overline{S}^{w^*}$. As JB is separable, on bounded sets in X^{**} the topology of pointwise convergence on JB is metrizable. Thus there is a sequence $\{x_n\}$ in S which converges to F on points of $JB \cup \{y_0^*\}$. Put $u(x^*) := \limsup x^*(x_n)$ for $x^* \in X^*$. Then, by Theorem 538, we get

$$\alpha \geq \sup_{JB} F = \sup_{JB} u = \sup_{B_{X^*}} u \geq u(y_0^*) = F(y_0^*) > \beta,$$

a contradiction. □

The separable version of James' compactness theorem follows from Theorem 539:

Proof of James' weak compactness Theorem 13 *(separable case).* If X is separable then B_X is a separable James boundary for X^* and thus by Theorem 539, $B_{X^{**}} = B_X$. □

The following is an extension of Theorem 539 above:

Theorem 540 (Contreras, Godefroy, Payá, [CoPa94], [Go00]). *If a James boundary JB of a Banach space X can be covered by a sequence of balls with a fixed radius a, $0 < a < 1$, then JB is a strong James boundary.*

The proof of Theorem 539 generated the following Godefroy's theorem, based on an Odell–Rosenthal result (see [FHHMZ11, Theorem 5.40]). It gives another instance in which a James boundary turns out to be strong.

Theorem 541 (Godefroy, [Go87]). *Let X be a separable Banach space that does not contain a copy of ℓ_1, and let JB be a James boundary of X. Then JB is a strong James boundary.*

Proof of James' compactness theorem using Theorem 536. James' reflexivity Theorem 13 follows straightforwardly from Pfitzner's Theorem 536. Indeed, assume that X is a Banach space such that every element in X^* attains its supremum on B_X. Thus, B_X is a James boundary for X^*. Since B_{X^*} is w^*-compact, it follows, by Theorem 536, that it is w^*-sequentially compact, and so, by the Eberlein–Šmulyan Theorem, w-compact. Then, the space X^* (and so X) is reflexive. The localized version of James' compactness Theorem 13 follows in a similar way by using Grothendieck's double-limit criterium for weak compactness (see, e.g., [Kothe69, §24.6.1] and [HMVZ07, Theorem 3.69]).

H It was shown in [Jame50] that a separable Banach space is reflexive if it has a Schauder basis and all continuous linear functionals attain its norm for any equivalent norm on the space.
The condition of having basis was removed in [Kl50].
It was proved in [Jame57] that a separable Banach space is reflexive if each continuous linear functional attains its norm.
It was proved in [Jame64] that a Banach space is reflexive if each continuous linear functional attains its norm.
[FoLiPh01, Theorem 5.7] shows the very interesting, unexpected, and important result that James' weak compactness Theorem 13 can be proved by using the Bishop–Phelps Theorem 14 (and the techniques of the Integral Representation Theory). This, combined with the remarks after Theorem 14, showed that James' weak compactness Theorem 13 can as well be proved by the good witch Baire category theorem (as Czech students call it).

It would be interesting if one can directly show that, if all elements of the dual of a separable space X attain their norm, then there is a LUR norm on X with the same property (without using James' weak compactness Theorem 13). This would give, using the Bishop–Phelps Theorem 14 and the Šmulyan Lemma 54, a direct proof of the aforementioned James' theorem.

Note that the completeness of the normed space in James' Theorem 13 is needed, see, e.g., [FHHMZ11, Exercise 3.160].

For some recent results in this area we refer to, e.g., [CaOrPe17].

Chapter 29

The Radon–Nikodým property

The Radon–Nikodým property was introduced in Definition 61 above by using the concept of strongly exposed point. The following is the traditional definition of this property, in terms of measure theory. Theorem 544 states the equivalence of both definitions and lists some other equivalent formulations:

Definition 542. *The space X is said to have the* **Radon-Nikodým property** *(***RNP***, for short) if for every measure space $(\Omega, \Sigma, \lambda)$ with $\lambda(\Omega) < \infty$ and every λ-continuous measure $\mu : \Sigma \to X$ of finite variation, there exists a Bochner integrable function $f : \Omega \to X$ such that $\mu(E) = \int_E f d\lambda$ for $E \in \Sigma$. The function f is then called* **the Radon-Nikodým derivative of** *μ.*

Philips proved in [Phi43] that reflexive spaces have the RNP, and in [DuPe40] it was shown that separable dual spaces have the RNP.

The following concept allows for a geometric characterization of the RNP (see the equivalence (i)$\Leftrightarrow$(iv) in Theorem 544 below):

Definition 543. *Let X be a Banach space. We say that a subset M of X is* **dentable** *if for every $\varepsilon > 0$ there are $x^* \in X^*$ and $a \in \mathbb{R}$ such that the slice $M \cap \{x^* > a\}$ is nonempty and has diameter less than ε. The Banach space X is called* **dentable** *if every nonempty bounded subset of it is dentable.*

(See, e.g., [FHHMZ11, Chapter 11 and, in particular, Definition 11.1].)

A. J. Guirao et al., *Renormings in Banach Spaces*, Monografie Matematyczne 75,
https://doi.org/10.1007/978-3-031-08655-7_29

Theorem 544. *The following are equivalent for a Banach space X.*
(i) *X is dentable.*
(ii) *The ball of every equivalent norm on X is dentable.*
(iii) *X has the RNP according to Definition 61, i.e., every bounded closed convex set in X is the $\|\cdot\|$-closed convex hull of its strongly exposed points.*
(iv) *X has the RNP according to Definition 542.*
(v) *Every Lipschitz map from $[0,1]$ into X is differentiable (Gâteaux or Fréchet, this is the same) almost everywhere on $[0,1]$.*
(vi) *The dual norm of every equivalent norm is Fréchet differentiable at points of a dense set in X^*.*
(vii) *Every weak*-lower semicontinous convex function on X^* is Fréchet differentiable at points of a dense G_δ set.*
(viii) *Each continuous convex function on X^* that is w^*-lsc is Gâteaux differentiable at least at one point with the derivative in X.*

Proof. (i) is equivalent to (iii) (see, e.g., [FHHMZ11, Theorem 11.3]), to (iv) and (v) (see, e.g., [FHHMZ11, Theorem 11.15], see also [JoLi03, p. 38]). The equivalence of (i) and (ii) is in [DaPh74]. The equivalence of (i) and (vii) is in [Coll76]. The equivalence (i)$\Leftrightarrow$(viii) is due to Bachir and Danilidis [BaDa00]. □

Definition 545. *The Banach space X is said to have the* **Krein–Milman property** *(for short,* **KMP***) if every bounded closed convex set in X is the closed convex hull of its extreme points.*

The following definition plays an important role in the theory of Asplund spaces (see the equivalence (i)$\Leftrightarrow$(viii) in Theorem 547). It will appear again in Chapter 43 related to the so-called Szlenk derivation process.

Definition 546. *Let X be a Banach space. A subset M of X^* is said to be* **w^*-fragmentable** *if for every $\varepsilon > 0$ there exists a w^*-open set $V \subset X^*$ such that $M \cap V \neq \emptyset$ and $M \cap V$ has $\|\cdot\|$-diameter less than ε. The space X is said to be* **w^*-fragmentable** *if every nonempty bounded subset of X^* is w^*-fragmentable.*

Theorem 547. *Let X be a Banach space. Then the following are equivalent.*
(i) *X is Asplund.*
(ii) *Every continuous convex function on X is Fréchet differentiable at the points of a dense set.*
(iii) *Every equivalent norm on X is Fréchet differentiable at the points of a dense set.*
(iv) *X^* has the RNP.*
(v) *X^* has the KMP.*
(vi) *X^* is dentable.*
(vii) *X^* is w^*-dentable.*
(viii) *X^* is w^*-fragmentable.*

Proof. Equivalences of (i), (ii) and (iii) and (iv) are in Theorem 283 above. Equivalence with (v) is in [HuMo75]. Equivalences of (i), (vi), (vii), and (viii) are due to Namioka and Phelps [NaPhe68] (see also, e.g., [FHHMZ11, Theorem 11.8]). □

Chapter 30

Strongly subdifferentiable norms

We will now discuss a few things on strongly subdifferentiable norms (SSD, for short). They were introduced in Definition 69. First note that due to the classical Dini theorem any norm that locally depends on finitely many coordinates is SSD [GoMoZi95], so in particular the supremum norm on c_0 is SSD. Actually, this is one of the reasons why the space c_0 is so attractive.

We have the following result of Franchetti and Payá (for the proof see, e.g., [Go00]), that is in the direction of Šmulyan's Lemma 54.

Theorem 548 (Franchetti, Payá, see, e.g., [Go00]). *The norm is SSD at $x \in S_X$ if, and only if, for every $\varepsilon > 0$ there is $\delta > 0$ such that* $\operatorname{dist}\big(x^*, \partial\|\cdot\|(x)\big) < \varepsilon$ *whenever $x^* \in B_{X^*}$ is such that $x^*(x) > 1 - \delta$.*

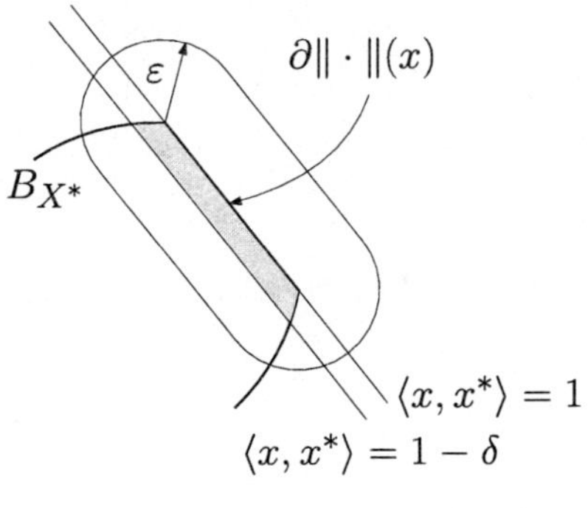

Figure 30.1: A characterization of SSD at x (Theorem 548)

Theorem 549 (Godefroy, [Go00]). *If the norm of a Banach space X is SSD norm, then X is Asplund.*

That the converse of Theorem 549 fails in a very strong sense follows from the following result (for the definition of the Kunen compact space, see Theorem 844 below):

Theorem 550 (Jiménez-Sevilla, Moreno, [JiMo97]). *If K is the Kunen compact space, then $C(K)$ is Asplund and admits no SSD norm.*

A. J. Guirao et al., *Renormings in Banach Spaces*, Monografie Matematyczne 75,
https://doi.org/10.1007/978-3-031-08655-7_30

In fact, it was proven in the same reference that the space $C(K)$ has no equivalent NS norm (see Theorem 583 below), from which the statement about the absence of an SSD renorming follows (see Corollary 584).

Theorem 551 (Godefroy, Montesinos, and Zizler, [GoMoZi95]). *If X is separable and X^* is not separable, then X admits a norm that is SSD only at the origin.*

A nonseparable version of Theorem 551 is open.

Theorem 552 (Godefroy, [Go00]). *The set of all SSD points of the sup norm on ℓ_∞ is not G_δ.*

The following problem is apparently open (see, e.g., [GuiMoZi16, Problem 162]): Assume that X admits both an SSD equivalent norm and a Gâteaux equivalent norm. Does it admit an equivalent Fréchet differentiable norm?

Chapter 31

The Banach–Saks property

Definition 553. *A Banach space X has the* **Banach–Saks property** *(***BS***) whenever every bounded sequence in X has a subsequence whose arithmetic means converge in norm, i.e. $\{\frac{1}{j}\sum_{k=1}^{j} x_{n_k}\}_{j=1}^{\infty}$ is convergent in norm. If this is required only for weakly convergent sequences, we say that X has the* **weak Banach–Saks property** *(***WBS***).*

Banach and Saks proved the following result:

Theorem 554 (Banach–Saks, [BanSa30]). *If $1 < p < \infty$, then $L_p(0,1)$ has BS.*

This was generalized by Kakutani:

Theorem 555 (Kakutani, [Kaku38]). *Every uniformly convex space has BS.*

Nishiura and Waterman proved the following result:

Theorem 556 (Nishiura, Waterman, [NiWa63]). *Every BS space is reflexive.*

Note: For an easier proof by using James' weak compactness Theorem 13 see [Di75, Chapter 3, §7, Corollary 1].

Baernstein II proved the following result:

Theorem 557 (Baernstein II, [Bae72]). *There is a reflexive space that does not have BS and there is a BS space that is not uniformly convex.*

It is not difficult to show that c_0 and ℓ_1 have WBS. See e.g. [FHHMZ11, Exercise 3.57].

Szlenk proved the following result:

Theorem 558 (Szlenk, [Sz65]). *The space $L_1[0,1]$ has WBS.*

Farnum showed the following result:

Theorem 559 (Farnum, [Far74]). *If K is a compact metric space, then $C(K)$ has WBS if, and only if, $K^{(\omega)} = \emptyset$.*

It was asked if the existence of a C^∞-smooth norm on a space can guarantee its WBS (see [GuiMoZi16, Problem 177]). From Farnum's result above and Hájek's result on C^∞-smooth norms in [Há03] it follows that the answer to this question is in the negative.

A. J. Guirao et al., *Renormings in Banach Spaces*, Monografie Matematyczne 75,
https://doi.org/10.1007/978-3-031-08655-7_31

Chapter 32

Transitive norms

Definition 560. *Let $(X, \|\cdot\|)$ be a Banach space. Its norm is called* **transitive** *if for every pair of points $x, y \in S_X$ there is a linear isometry T from X onto X such that $T(x) = y$. The norm is called* **quasitransitive** *if for every $x, y \in S_X$ and $\varepsilon > 0$ there is a linear isometry T from X onto X such that $\|T(x) - y\| < \varepsilon$.*

Theorem 561 (Auerbach, von Neumann, see, e.g., [PelBe79], Theorem 9.6). *If the norm of a finite-dimensional Banach space X is transitive, then X is linearly isometric to a Hilbert space.*

Open problem (Mazur). Does Theorem 561 hold true if finite-dimensionality is replaced by separability of X?

Theorem 562. (See, e.g., [PelBe79, Remark after Theorem 9.6]) *Let $1 \leq p < \infty$ and μ be a non-sigma-finite non-atomic measure. Then $L_p(\mu)$ admits a transitive norm.*

Open problem (Cabello). [Cab97] Does every superreflexive space admit an equivalent quasitransitive norm?

Theorem 563. (See [DeGoZi93, Corollary IV.5.7]) *Let X be a superreflexive space. Then every quasitransitive equivalent norm is uniformly convex and uniformly smooth.*

Theorem 564 (Cabello, [Cab98]). *Assume that the norm of an Asplund space X is quasitransitive. Then X is superreflexive.*

Theorem 565 (Cabello, [Cab98]). *There is a space X whose norm is quasitransitive but not transitive.*

Theorem 566 (Cabello, [Cab98]). *The space $C[0, 1]$ can be renormed by a quasitransitive norm.*

A. J. Guirao et al., *Renormings in Banach Spaces*, Monografie Matematyczne 75,
https://doi.org/10.1007/978-3-031-08655-7_32

Chapter 33

Norms with the Mazur intersection property

The Mazur intersection property (MIP, for short) was introduced in Definition 262, that we repeat here: The norm $\|\cdot\|$ of a Banach space X is said to have the **Mazur intersection property (MIP)** if each bounded closed convex set in X is the intersection of closed balls. We give below only a list of some basic results in this area. In the 40's, Mazur proved that any Fréchet differentiable norm on a reflexive space has the MIP. Since then this concept got a wide use in the Banach space theory and elsewhere (for instance in operator theory). It is favored for its elegance and usefulness.

Theorem 567 below characterizes the MIP. The concept of a w^*-denting point was introduced in Definition 405 above.

Theorem 567 (Giles, Gregory, Sims, [GiGrSi78], see, e.g., [HMVZ07], Theorem 8.32). *Let X be a Banach space. Then the norm of X has MIP if and only if the w^*-denting points of B_{X^*} form a $\|\cdot\|^*$-dense subset of S_{X^*}.*

Item (ii) in the next theorem will be strengthened by Theorem 578 (complemented with (ii) in Theorem 582):

Theorem 568 (Phelps, [Ph60a], see, e.g., [HMVZ07], Section 8.4).
(i) *If the norm on a Banach space is Fréchet differentiable, then it has MIP.*
(ii) *If the norm has MIP and the space X is separable, then X^* is separable.*

Theorem 569 (Jiménez-Sevilla, Moreno, [JiMo97], Corollary 3.4). *Every Banach space can be linearly isometrically embedded into a Banach space with the MIP.*

Theorem 570 (Jiménez-Sevilla, Moreno, [JiMo97]). *Let $(X^*, \|\cdot\|^*)$ be a dual Banach space with a biorthogonal system $\{x_i^*; x_i\}_{i\in I} \subset X^* \times X$, and let $Y_0 :=$ span $(\{x_i^* : i \in I\})$. Then X^* admits an equivalent dual norm $|\cdot|^*$ that is LUR at the points of Y_0. In particular, if Y_0 is dense in X^*, then X can be renormed to have the MIP.*

See also, e.g., [HMVZ07, Theorem 8.42].

Theorem 571 (Jiménez-Sevilla, Moreno, [JiMo97], see, e.g., [HMVZ07, Section 8.4]). *The non-Asplund space $\ell_1 \oplus \ell_2(c)$ can be renormed by a MIP norm.*

A. J. Guirao et al., *Renormings in Banach Spaces*, Monografie Matematyczne 75,
https://doi.org/10.1007/978-3-031-08655-7_33

Theorem 571 can be proved by using the following useful result:

Theorem 572 (Rychtář, [Rych04]). *Let X be a Banach space and Y be a subspace of X with a shrinking Markushevich basis $\{x_\gamma, f_\gamma\}_{\gamma\in\Gamma}$. Assume that card $\Gamma =$ dens X^*. Then X can be renormed by a MIP norm.*

Note that Theorem 572 immediately implies that $C[0,\alpha]$ admits a norm with the MIP for any α, just by considering $c_0(\Gamma) \subset C[0,\Gamma]$.

➪ Let us show, as a particular case, a linear isometry $T : c_0[0,\omega_1]$ into $C_0[0,\omega_1]$: Define

$$Tx(\alpha) := \begin{cases} x(\beta) & \text{if there is } \beta \text{ such that } \beta+1=\alpha, \\ 0 & \text{otherwise.} \end{cases}$$

The following is an interesting result:

Theorem 573 (Bačák, Hájek, Jiménez-Sevilla, Moreno, [BaHá08], [JiMo97]). *The question whether every Asplund space X of density ω_1 admits a norm with the MIP is not decidable in ZFC. If K is the Kunen compact space, then $C(K)$ admits no MIP norm .*

Theorem 574 (Jiménez-Sevilla, Moreno, [JiMo97]). *For every tree T, the space $C_0(T)$ admits a norm with MIP.*

The following result extends a theorem in [Ziz86], where it was proved for WCG spaces:

Theorem 575 (Vanderwerff, [Van98]). *Every space can be renormed so that each weakly compact convex set is in the new norm an intersection of a family of closed balls.*

Theorem 576 (Deville, Godefroy, Zizler, [DeGoZi93a]). *If X has RNP and admits a Lipschitz Fréchet differentiable bump, then X admits a norm with MIP.*

Note that it is open whether X admits a norm with MIP if it admits a Lipschitz Fréchet smooth bump (see [GuiMoZi16, Problem 187]).

Definition 577 (Whitfield, Zizler, [WhZi87]). *A Banach space $(X, \|\cdot\|)$ is said to have the* **uniform Mazur intersection property** *(***UMIP***, for short) if the following holds: For every $\varepsilon > 0$ there is $K > 0$ such that if a point $x \in X$ and a closed convex set $C \subset X$ are such that* $\operatorname{dist}(x, C) > \varepsilon$ *and* $\operatorname{diam} C < \frac{1}{\varepsilon}$, *then there is a ball B with radius $\leq K$ such that $x \notin B$ and $C \subset B$.*

Open problem. If X has the UMIP, is X necessarily superreflexive?

Chapter 34

Nicely smooth Banach spaces

Related to the existence of closed proper 1-norming subspaces of the dual X^* of a Banach space X, we introduced in Definition 33 the concept of a nicely smooth norm. Let us repeat this definition here: The norm $\|\cdot\|$ on a Banach space $(X,\|\cdot\|)$ is said to be **nicely smooth** (**NS**, for short) if there is no proper closed 1-norming subspace of X^*.

Theorem 578. *Every nicely smooth separable space is Asplund.*

Proof. Let $\{x_n\}_{n\in\mathbb{N}}$ be a sequence dense in S_X. Choose, for each $n\in\mathbb{N}$, an element $f_n\in S_{X^*}$ such that $f_n(x_n)=1$. Observe that $N:=\overline{\operatorname{span}}\{f_n:\ n\in\mathbb{N}\}$ is a closed 1-norming subspace of X^*. By the assumption, $N=X^*$, hence X^* is separable. □

Remark 579. 1. Without the separability condition, Theorem 578 may fail: Indeed, let X be a non-Asplund space. According to Theorem 569, there exists a Banach space $(Z,\|\cdot\|)$ such that $\|\cdot\|$ has MIP and X linearly isometrically embeds into Z. Theorem 582 below shows that $\|\cdot\|$ on Z is nicely smooth. However, Z is not an Asplund space.

2. The converse of Theorem 578 fails: There are separable Asplund spaces that are not nicely smooth. For example, let X be a nonreflexive Banach space whose second dual is separable (a particular instance is the space J defined by R.C. James, see, e.g., [FHHMZ11, Definition 4.43]; it is nonreflexive, separable and it has codimension 1 in its bidual space). Then X^* is a separable Asplund space and X ($\subset X^{**}$) is a proper closed 1-norming subspace of X^{**}. ®

The following result characterizes nicely smooth norms in terms of intersections of balls (see Figure 34.1).

Theorem 580 (Godefroy, Saphar, [GoSa88]). *Let $(X,\|\cdot\|)$ be a Banach space. Then, the following are equivalent:*
(i) *$\|\cdot\|$ is nicely smooth.*
(ii) *For every $x^{**}\in X^{**}$, $\bigcap_{x\in X}B(x,\|x^{**}-x\|)=\{x^{**}\}$.*
(iii) *For every $x^{**}\in X^{**}\setminus X$, $\bigcap_{x\in X}B(x,\|x^{**}-x\|)\cap X=\emptyset$.*

Proof. Obviously, (ii) implies (iii).
We shall prove now that (i) and (iii) are equivalent. Assume that N is a proper closed 1-norming subspace of X^*. By enlarging N if necessary we may assume

A. J. Guirao et al., *Renormings in Banach Spaces*, Monografie Matematyczne 75,
https://doi.org/10.1007/978-3-031-08655-7_34

that N is a proper closed 1-norming hyperplane. Then $N = \ker x^{**}$ for some $x^{**} \in X^{**} \setminus X$, and $\|x^{**} - x\| \geq \|x\|$ for all $x \in X$ (see Item 1 in Subsection 3.4.4). This shows that $0 \in \bigcap_{x \in X} B(x, \|x^{**} - x\|) \cap X$, so (iii)⇒(i).
Assume now that (iii) fails. There exists $x_0^{**} \in X^{**} \setminus X$ and a point $x_0 \in \bigcap_{x \in X} B(x, \|x_0^{**} - x\|) \cap X$. Thus, $\|x_0 - x\| \leq \|x_0^{**} - x\|$ for all $x \in X$. In other terms, if $y := x_0 - x$ and $x^{**} := x_0^{**} - x_0$, we have $\|y\| \leq \|x^{**} + y\|$, and $y \in X$ is arbitrary. This shows, again in view of Item 1 in Subsection 3.4.4, that $N := \ker x^{**}$ is a closed proper 1-norming hyperplane in X^*, so (i) fails.
For the proof that (iii)⇒(ii) we refer to [GoSa88]. □

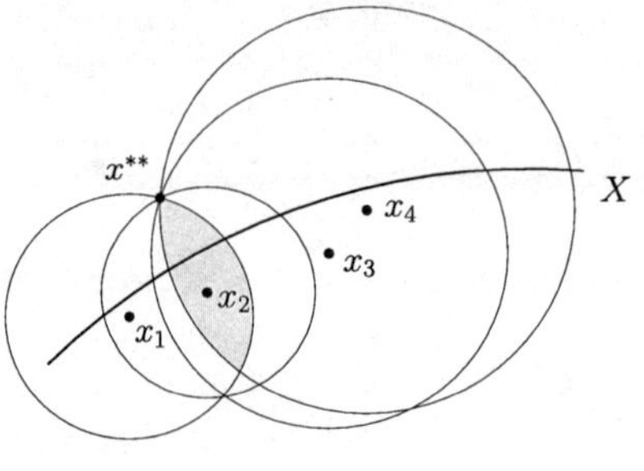

Figure 34.1: Intersection of some balls

In view of characterizations (ii) and (iii) in Theorem 580, it is natural to introduce the following notion:

Theorem 585 below is an extension of the same result for sets in reflexive spaces due to Corson and Lindenstrauss ([CorLi66]) (see, e.g., [FHHMZ11, Exercise 8.40]).

Definition 581 ([GoKa89]). *The norm $\|\cdot\|$ of a Banach space X is said to have the* **ball generated property** *(***BGP***, for short) if every bounded closed convex set in X is an intersection a finite unions of balls in $\|\cdot\|$.*

In Section 3.5, we mentioned that every Fréchet differentiable norm is nicely smooth. Item (i) in Theorem 582 below shows a stronger statement. We add some other results pertaining to nicely smooth spaces and related to MIP and BGP (for the former see Chapter 33).

Theorem 582 (Godefroy, [Go01b]).
(i) *Every norm with the SSD property is NS.*
(ii) *Every norm with MIP is NS.*
(iii) *Every norm with BGP is NS. If X is separable, then BGP is equivalent to NS.*

For the definition of the Kunen compact space, see Theorem 844 below.

Theorem 583 (Jiménez-Sevilla, Moreno, [JiMo97]). *If K is the Kunen compact space* (note that CH is needed), *then no renorming of $C(K)$ is NS.*

Corollary 584. *If K is the Kunen compact space, then no renorming of $C(K)$ is SSD , or MIP , or BGP.*

The statement about the absence of an equivalent SSD renorming of $C(K)$ was already mentioned in Theorem 550 above.

Theorem 585 (Godefroy, Kalton [GoKa89], see also [FHHMZ11], Exercise 8.40). *Every weakly compact set in a Banach space is an intersection of finite unions of balls.*

Chapter 35

Weak Hadamard differentiability

35.1 Introduction

Whether the differentiability of a convex continuous function is in the Gâteaux or the Fréchet sense depend on the limit (4.1): its existence uniformly on finite sets of directions for Gâteaux differentiability, and on bounded sets of directions for Fréchet differentiability. More generally, we may use a **bornology** on a normed space X: It is a family $\mathcal{B}$ of nonempty bounded subsets of X having the following properties: (i) Every $B \in \mathcal{B}$ is symmetric, (ii) $\mathcal{B}$ is directed upwards, and (iii) $\bigcup_{B\in\mathcal{B}} B = X$. The topology $\mathcal{T}_{\mathcal{B}}$ on X^* induced by a bornology $\mathcal{B}$ on X is the topology of the uniform convergence on all elements in $\mathcal{B}$. Typical examples are w^*, $\mu(X^*, X)$ (denoted briefly μ^*, see Section 1.3), and $\mathcal{T}_{\|\cdot\|^*}$, given by the bornologies $\mathcal{F}$, $\mathcal{K}$, and $\mathcal{B}$ of all the finite symmetric, respectively the w-compact symmetric, respectively the bounded symmetric subsets of X. We shall consider in this chapter mainly differentiability uniformly in the second sense —weak Hadamard differentiability— and its connections to the Gâteaux and Fréchet versions.

Definition 586. *Let $\mathcal{B}$ be a bornology on a normed space X. A proper extended real-valued function f on X is said to be* **$\mathcal{B}$-differentiable at a point** $x_0 \in X$ *if it is finite on a neighborhood of x_0 and there exists $f'(x_0) \in X^*$ such that the limit* (4.1) *is $f'(x_0)(h)$ uniformly on the elements of $\mathcal{B}$. It is called* **weak Hadamard differentiable at** $x_0 \in X$ *if it is $\mathcal{K}$-differentiable for the bornology $\mathcal{K}$ of all symmetric and w-compact subsets of X.*

35.2 Sequential convergence in X^*, boundedness, and differentiability

That differentiability in X is related to sequential convergence in X^* is essentially the content of the Šmulyan's Lemma 142 (see also Proposition 145 and Corollary 147). A general version for continuous convex functions and differentiability for a given bornology is presented in Theorem 587 below. It collects several statements related to the differentiability with respect to a bornology in terms of equation (4.7), continuity of the subdifferential (Subsection 4.1.3), and the Šmulyan Lemma 142, and indicates their relationship. We follow [Zal02, Theorem 3.3.2]. In the statement of Theorem 587 below, the symbol ∂_ε denotes the ε-subdifferential of a

A. J. Guirao et al., *Renormings in Banach Spaces*, Monografie Matematyczne 75,
https://doi.org/10.1007/978-3-031-08655-7_35

function f.

Theorem 587. *Let f be a proper convex function on a normed space X. Assume that f is continuous at some $x_0 \in \operatorname{dom}(f)$. Let $\mathcal{B}$ be a bornology on X. Consider the following statements:*
(i) *f is $\mathcal{B}$-differentiable at x_0.*
(ii) $\lim_{t\to 0} \frac{f(x_0+th)+f(x_0-th)-2f(x_0)}{t} = 0$ *uniformly on each element in $\mathcal{B}$.*
(iii) *Every selection of ∂f is $\mathcal{T}_{\|\cdot\|}$-$\mathcal{T}_{\mathcal{B}}$-continuous at x_0.*
(iv) *For every $x_0^* \in \partial f(x_0)$ there exists a selection γ of ∂f which is $\mathcal{T}_{\|\cdot\|}$-$\mathcal{T}_{\mathcal{B}}$-continuous at x_0 and $\gamma(x_0) = x_0^*$.*
(v) *There exists a selection of ∂f which is $\mathcal{T}_{\|\cdot\|}$-$\mathcal{T}_{\mathcal{B}}$-continuous at x_0.*
(vi) $x_n^* \overset{\mathcal{T}_{\mathcal{B}}}{\to} x_0^*$ *whenever $x_0^* \in \partial f(x_0)$, $x_n^* \in \partial f(x_n)$ for $n \in \mathbb{N}$, and $x_n \to x_0$ in X.*
(vii) $x_n^* \overset{\mathcal{T}_{\mathcal{B}}}{\to} x_0^*$ *whenever $x_0^* \in \partial f(x_0)$, $0 \le \varepsilon_n$, $x_n^* \in \partial_{\varepsilon_n} f(x_n)$ for $n \in \mathbb{N}$, $\varepsilon_n \to 0$, and $x_n \to x_0$.*
(viii) $x_n^* \overset{\mathcal{T}_{\mathcal{B}}}{\to} x_0^*$ *whenever $x_0^* \in \partial f(x_0)$, $0 \le \varepsilon_n$, $x_n^* \in \partial_{\varepsilon_n} f(x_0)$ for $n \in \mathbb{N}$, and $\varepsilon_n \to 0$.*
Then (i)$\Leftrightarrow$(ii)$\Leftrightarrow$(iii)$\Leftrightarrow$(iv)$\Leftrightarrow$(v)$\Leftrightarrow$(vi)$\Leftarrow$(vii)$\Leftrightarrow$(viii). *Moreover, when X is a Banach space and f is lower semicontinuous, then all of them are equivalent.*

The coincidence of several topologies on S_{X^*} (see Definition 89) has been considered in Section 5.4 (see, in particular Corollary 94) and Corollary 147 (see also [HMVZ07, Section 3.2]). Note that *a Banach space $(X, \|\cdot\|)$ does not contain a copy of ℓ_1 if, and only if, the topologies $\mu(X^*, X)$ and $\mathcal{T}_{\|\cdot\|^*}$ sequentially agree on* X^* ([Orn91] and [Vald93a], see, e.g., [HMVZ07, Theorem 3.17]).

In view of characterizations (vi), (vii), and (viii) of differentiability (for Banach spaces) in Theorem 587, and for the three bornologies mentioned above, we will discuss here the sequential agreement of the induced topologies on X^*. The following result gives several equivalent conditions for this. That (i)$\Rightarrow$ (iii) follows from Theorem 587 above. (iii)$\Rightarrow$(iv) is trivial. Incidentally, an equivalent condition in terms of boundedness of continuous convex functions is included.

Theorem 588. (See, e.g., [BoMoVa06, Theorem 2.3]; see also [BoFiVa94]) *Let X be a Banach space with two bornologies β_1 and β_2, $\beta_1 \subset \beta_2$. Then the following are equivalent:*
(i) *$\mathcal{T}_{\beta_1}$ and $\mathcal{T}_{\beta_2}$ agree sequentially on X^*.*
(ii) *Every continuous convex function on X that is bounded on β_1-sets is bounded on β_2-sets.*
(iii) *β_1-differentiability agrees with β_2-differentiability for continuous convex functions on X.*
(iv) *β_1-differentiability agrees with β_2-differentiability for equivalent norms on X.*

Note that in the finite-dimensional case all the three topologies w^*, μ^*, and $\mathcal{T}_{\|\cdot\|^*}$ —and so the associated differentiability notions for continuous convex functions or for equivalent norms— coincide. This is in contrast with the infinite-dimensional case:

Theorem 589. *Let X be an infinite-dimensional Banach space. Then*
(i) [Josefson–Nissenzweig, see, e.g., [HMVZ07, Theorem 3.27] or [Di84, Chapter XII]] *On X^*, the w^*-topology does not sequentially agree with the norm topology.*
(ii) (See, e.g., [FHHMZ11, Exercise 7.7]) *There is a continuous convex function f on X that is unbounded on B_X.*

Proof of (ii). (Vanderwerff) From (i), there exists a sequence $\{f_n\}_{n\in\mathbb{N}}$ in S_{X^*} that w^*-converges to 0. If τ_n is an even continuous convex function on $\mathbb{R}$ such that τ_n is non-decreasing on $[0, +\infty)$, $\tau_n = 0$ on $[0, 1/2]$, and $\tau_n(1) = n$ for all $n \in \mathbb{N}$, put then $f = \sum_n \tau_n \circ f_n$. □

Remark 590. Note that (ii) in Theorem 589 is in fact equivalent to the Josefson–Nissenzweig theorem, in view of the equivalence (i)⇔(ii) in Theorem 588. Thus, the following are equivalent (see, e.g., [BoMoVa06, Theorem 3.1]): (i) X is finite-dimensional. (ii) w^* and $\mathcal{T}_{\|\cdot\|}$-convergence coincide sequentially on X^*. (iii) Every continuous convex function on X is bounded on bounded subsets of X. (iv) Gâteaux and Fréchet differentiability coincide for continuous convex functions on X. Ⓡ

In the following chart (Figure 35.1), the name of a property (for X) located between two topologies means that they agree sequentially on X^*. Thus, for example, $\mu^* \overset{\ell_1 \not\to X}{---} \|\cdot\|^*$ means that a Banach space X does not contains a copy of ℓ_1 if, and only if, μ^* and $\mathcal{T}_{\|\cdot\|^*}$ agree sequentially on X^* (as it was mentioned above). **DP** stands for the **Dunford–Pettis property** of X, **DP*** for the so-called **Dunford–Pettis*** (of X), and **GR** for the **Grothendieck property** (of X). The sequential coincidence of the topologies w and $\mathcal{T}_{\|\cdot\|^*}$ on X^* is the **Schur property** (of X^*).

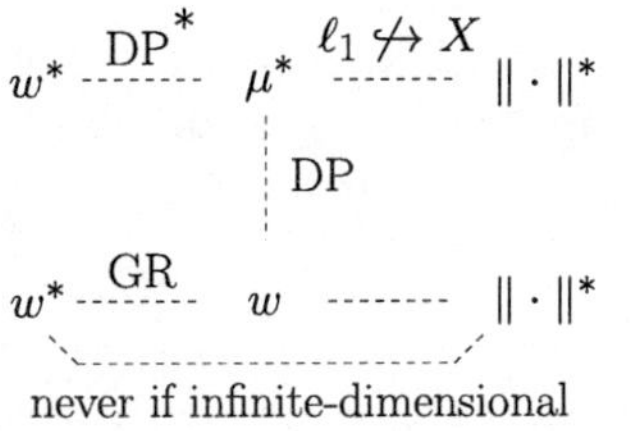

Figure 35.1: Sequential agreement of topologies in the dual

Note that $X := \ell_\infty$ has the DP* property. Indeed, ℓ_∞ is a Grothendieck space (see, e.g., [HMVZ07, Theorem 7.18]) and, at the same time, it has the Dunford–Pettis property (as it is a $C(K)$ space, see, e.g., [FHHMZ11, Theorem 13.43]).

Theorems 587 and 588 have some straightforward consequences: For example, (i) if X has the DP* property, then every continuous convex function on X that is Gâteaux differentiable is weak Hadamard differentiable.
(ii) If X has the DP* property, then every continuous convex function on X is bounded on every w-compact subset of X.

➪ This gives a proof of the celebrated Sobczyk' theorem (see, e.g., [FHHMZ11, Theorem 5.6]) on the non-complementability of c_0 in ℓ_∞. Indeed, assume that there is a continuous linear projection $P : \ell_\infty \to c_0$. Let $h(x) := \sum_{i=1}^{\infty}(x_i)^{2i}$ for $x := (x_i) \in c_0$ (this function has been used by Preiss, see equation (27.1) in Theorem 505). Then $h \circ P$ is a continuous convex function on ℓ_∞ (a space with the DP* property, see above) that is not bounded in the w-compact set $\{0, 2e_1, 2e_2, \ldots\}$, a contradiction.

(iii) Another one is (i) in the following result:

Theorem 591. *Let $(X, \|\cdot\|)$ be a Banach space.*
(i) *If X does not contain a copy of ℓ_1 and $\|\cdot\|$ is weak Hadamard differentiable at $x \in X$, then $\|\cdot\|$ is Fréchet differentiable at x.*
(ii) *If X contains a copy of ℓ_1, then there is an equivalent norm $|||\cdot|||$ on X that is weak Hadamard differentiable but not Fréchet differentiable at some point $x \in X$.*
(iii) *If $X := C(K)$ and $|||\cdot|||$ is a weak Hadamard differentiable norm on X, then $|||\cdot|||$ is Fréchet differentiable on X. In particular, $C(K)$ is then an Asplund space.*
(iv) *If $X := L_1(\mu)$ for a finite measure μ, then X admits a norm that is weak Hadamard differentiable at all nonzero points.*

Proof. (i) follows from Theorem 587 and the previously quoted characterization of spaces without copies of ℓ_1.
(ii) is proved in [BoFa93, Theorem 2].
(iii) By (i) it suffices to show that X is Asplund. If not, then X contains a copy of $C([0,1])$ (see, e.g., the argument in the proof of [FHHMZ11, Theorem 14.26 (v)]) and thus also a copy of JT [Jame74], since JT is separable. Thus, the restriction of the norm $|||\cdot|||$ to this copy of JT is weak Hadamard differentiable and thus Fréchet smooth by (i), since JT does not contain a copy of ℓ_1. This is impossible, as JT^* is nonseparable.
(iv) is proved in [BoFi93]. □

Theorem 592. *If X admits a weak Hadamard differentiable norm, then every other norm on X is weak Hadamard differentiable at a G_δ dense set of points.*

Proof. It follows from [PrPhNa90]. □

Chapter 36

Lipschitz Asplund spaces

Every Banach space with a Fréchet differentiable bump is Asplund. This is the result of Ekeland and Lebourg [EkLe76] (see (iv) in Theorem 275) that generalizes the fact that every Banach space whose norm is Fréchet differentiable is an Asplund space (see (ii) in Theorem 275). The following concept allows for a "Lipschitz" variant of the Ekeland–Lebourg result. This is Theorem 596 below.

Definition 593 (Fabian, [Fab85]). *A real-valued function f defined on an open subset U of a Banach space X is said to be* **Lipschitz smooth at a point** $x_0 \in U$ *if there exists $c > 0$, $\delta > 0$, and $x_0^* \in X^*$ such that*

$$|f(x_0 + h) - f(x_0) - \langle h, x_0^* \rangle| \leq c\|h\|^2, \text{ whenever } h \in X, \ \|h\| < \delta. \tag{36.1}$$

The function f is called **Lipschitz smooth on** U *if it is Lipschitz smooth at every point of U. A Banach space X is said to be a* **Lipschitz Asplund space** *if for every convex continuous function $f : X \to \mathbb{R}$ there exists a dense subset D_f of X such that f is Lipschitz smooth at each point of D_f.*

Remark 594. 1. Fabian called "LD-spaces" what we called here "Lipschitz Asplund spaces".

2. If $f : U \to \mathbb{R}$ is Lipschitz smooth at $x_0 \in U$, then it is obviously Fréchet smooth at x_0, and $f'(x_0) = x_0^*$ in (36.1).

3. If $f : U \to \mathbb{R}$ is Fréchet differentiable on U and $\|f'(x) - f'(x_0)\|^* \leq c\|x - x_0\|$ for some $c > 0$ and for all x in a neighborhood of x_0, then an easy application of the Mean Value Theorem 44 shows that f is Lipschitz smooth at x_0. This accounts for the name "Lipschitz smooth".

4. There is a continuous convex function on the real line such that its set of Lipschitz smooth points does not contain a G_δ-set [Fab85]. Ⓡ

For checking the Lipschitz Asplundness condition of a Banach space, it is enough to apply the following result:

Theorem 595 (Fabian, [Fab85]). *A Banach space X is Lipschitz Asplund if (and only if) every convex continuous function $f : X \to \mathbb{R}$ has at least a point of Lipschitz smoothness.*

A. J. Guirao et al., *Renormings in Banach Spaces*, Monografie Matematyczne 75,
https://doi.org/10.1007/978-3-031-08655-7_36

Theorem 596 (Fabian, [Fab85]). *Assume that X has RNP and admits a bump that is Lipschitz smooth at every point. Then X is a Lipschitz Asplund space.*

Theorem 597 (Fabian, [Fab85]). *A Banach space X is Lipschitz Asplund if every separable subspace of X is.*

We list some open problems related to this subject:

1. (M. Fabian, [Fab85]) Is c_0 a Lipschitz Asplund space? The same question is open for the James' space J.
2. Does Theorem 596 hold true if RNP is replaced by X not containing a copy of c_0?
3. Assume that X is a Lipschitz Asplund space. Is X necessarily superreflexive?
4. (M. Fabian, [Fab85]) Is every subspace of a Lipschitz Asplund space Lipschitz Asplund?

Check for updates

Chapter 37

Lipschitz-free spaces

37.1 Introduction

Lipschitz functions on a normed space and with values on a Banach space were defined in Section 4.1.2. Certainly, the definition makes sense for functions f between two metric spaces (M, d_M) and (N, d_N). The **Lipschitz constant** is

$$\|f\|_L := \sup_{x,y \in M,\ x \neq y} \frac{d_N(f(x), f(y))}{d_M(x, y)}. \tag{37.1}$$

Functions whose Lipschitz constant is $\lambda \geq 0$ are called **λ-Lipschitz**. Here we shall focus on the case where the target space N is just the real line.

This chapter presents only some basic facts about Banach spaces of Lipschitz functions and its natural preduals, the so-called **Arens–Eells** (or, under other name **Lipschitz-free** spaces), see Definition 598 below. For an authoritative account on those structures, the reader should check [Weav18b], the seminal paper [GoKa03], and the comprehensive survey [Go15]. These and some other recent sources will be quoted below.

It should be clear that $\|\cdot\|_L$ is a seminorm on the real vector space of all Lipschitz functions from M into $\mathbb{R}$ (a constant function has Lipschitz constant 0). In order to avoid this inconvenience, it is natural to fix an element in M, usually denoted by 0 (the resulting structure is called a **pointed metric space**) and to restrict ourselves to the space $\mathrm{Lip}_0(M)$ of all Lipschitz real-valued functions on M that vanish at 0. Thus, $(\mathrm{Lip}_0(M), \|\cdot\|_L)$ is a Banach space.
In what follows, (M, d) will always denote a pointed metric space.

37.2 Lipschitz-free spaces

Given a pointed metric space, the introduction of the structure $\mathcal{F}(M)$ (called the **Lipschitz-free space over** M, see Definition 598 below) accomplish the goal of "linearizing" arguments about Lipschitz functions on metric spaces. This is seen in the fundamental Theorem 600 below.
There is a natural mapping $\delta : M \to \mathrm{Lip}_0(M)^*$ that takes $x \in M$ to the element $\delta(x)$ such that $\langle \delta(x), f\rangle = f(x)$ for all $f \in \mathrm{Lip}_0(M)$. It is simple to prove that δ is a d-w^*-continuous isometric embedding that takes 0 into 0. For this and several

A. J. Guirao et al., *Renormings in Banach Spaces*, Monografie Matematyczne 75,
https://doi.org/10.1007/978-3-031-08655-7_37

other basic results it is worth to consider the 1-Lipschitz function $f_{x_0,\lambda}(x) := \max\{\lambda - d(x, x_0), 0\}$, for $x, x_0 \in M$ and $\lambda \geq 0$. Its support is $B(x_0, \lambda)$ and its graph appears in Figure 37.1. If $0 \leq \lambda \leq d(x_0, 0)$, then $f_{x_0,\lambda} \in S_{\mathrm{Lip}_0(\mathrm{M})}$. For example, it allows to show that the set $\{\delta(x) : x \in M\}$ is linearly independent.

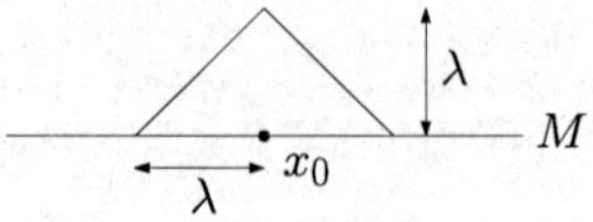

Figure 37.1: The graph of the function $f_{x_0,\lambda}$

Definition 598. *Let (M, d) be a complete pointed metric space. The* **Lipschit-free space** *$\mathcal{F}(M)$ over M is the closed linear span of $\delta(M)$ in $Lip_0(M)^*$, endowed with the restriction $\|\cdot\|$ of the dual norm of $\|\cdot\|_L$.*

Clearly, $\mathcal{F}(M)$ has the same density character as M. In particular, if M is separable so it is $\mathcal{F}(M)$. It is natural to think of $\mathcal{F}(M)$ as the completion of the set of Borel measures μ on M with finite support equipped with the norm $\|\mu\|_{\mathcal{F}} := \sup_{f \in B_{\mathrm{Lip}_0(\mathrm{M})}} \int f d\mu$. Every finite Borel measure μ supported on a compact subset K of M can be identified with an element in $\mathcal{F}(M)$: Indeed, $\mu = \int_K \delta_x d\mu(x)$ as a Bochner integral.
The basic feature about $\mathcal{F}(M)$ is that it is a (linearly isometric) predual of the space $(\mathrm{Lip}_0(M), \|\cdot\|_L)$. This is recorded in the following result:

Theorem 599. *$\mathcal{F}(M)^*$ is linearly isometric to $Lip_0(M)$ under the mapping $\Lambda \to \Lambda \circ \delta$, $\Lambda \in \mathcal{F}(M)$. Under this identification, the $w(Lip_0(M), \mathcal{F}(M))$-topology on $Lip_0(M)$ coincides on bounded sets with the topology of the pointwise convergence.*

Proof. Given $\Lambda \in \mathcal{F}(M)^*$, the mapping $\Lambda \circ \delta : M \to \mathbb{R}$ is clearly Lipschitz and $\|\Lambda \circ \delta\|_L \leq \|\Lambda\|$. Thus, the mapping $T : (\mathcal{F}(M)^*, \|\cdot\|^*) \to (\mathrm{Lip}_0(\mathrm{M}), \|\cdot\|_{\mathrm{L}})$ given by $T(\Lambda) = \Lambda \circ \delta$, for $\Lambda \in \mathcal{F}(M)^*$, is well defined and continuous. Moreover, if $\Lambda \in \mathcal{F}(M)^*$ then

$$\begin{aligned}\|\Lambda\| &= \sup_{x \in B_{\mathcal{F}(M)}} \langle x, \Lambda \rangle \\ &= \sup_{x \in B_{\mathcal{F}(M)}} \langle x, T(\Lambda) \rangle \leq \sup_{x^{**} \in B_{\mathrm{Lip}_0}(M)} \langle x^{**}, T(\Lambda) \rangle = \|T(\Lambda)\|_L,\end{aligned}$$

and so T is an isometric embedding. To see that it is onto, notice that given $f \in \mathrm{Lip}_0(\mathrm{M})$, the mapping $\lambda : \mathrm{span}\ \delta(M) \to \mathbb{R}$ given by $\lambda\left(\sum_{i=1}^n a_i \delta(x_i)\right) = \sum_{i=1}^n a_i f(x_i)$ is linear and continuous , and its (linear) continuous extension Λ to $\mathcal{F}(M)$ satisfies $T(\Lambda) = f$. □

It is an open problem whether $\mathcal{F}(M)$ is the unique (linearly isometrically) predual of $\mathrm{Lip}_0(\mathrm{M})$. The answer is positive when M is bounded [Weav18, Theorem 3.26] or "convex" (i.e., for every distinct $x, y \in M$ there exists a third distinct point z such that $d(x, y) = d(x, z) + d(z, y)$) [Weav18, Theorem 3.27]. The **metric segment** defined by two elements $x, y \in M$ is the set $[x, y] := \{z \in M : \; d(x, y) = d(x, z) + d(z, y)\}$. Thus, M is convex whenever $[x, y] \neq \{x, y\}$ for all distinct $x, y \in M$.

The following result shows the fundamental "linearization" property that justifies the introduction of the space $\mathcal{F}(M)$ (see Figure 37.2):

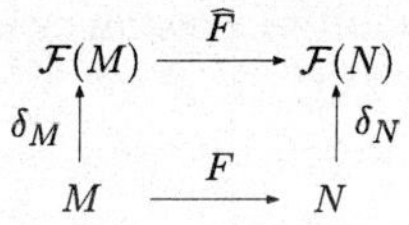

Figure 37.2: The extension Theorem 600

Theorem 600. *Let M and N be pointed metric spaces. Then, every Lipschitz function $F : M \to N$ such that $F(0) = 0$ extends to a continuous linear map $\widehat{F}$ from $\mathcal{F}(M)$ into $\mathcal{F}(N)$. Moreover, $\widehat{F}(\delta(x)) = \delta(F(x))$ for $x \in M$.*

Proof. The composition map $C_F(g) = g \circ F$ linearly and continuously maps $\mathrm{Lip}_0(\mathrm{N})$ into $\mathrm{Lip}_0(\mathrm{M})$, so its adjoint $(C_F)^*$ maps $\mathrm{Lip}_0(\mathrm{M})^*$ into $\mathrm{Lip}_0(\mathrm{N})^*$. The restriction of $(C_F)^*$ to $\mathcal{F}(M)$ gives the sought continuous linear extension. □

Notice that Theorem 600, when applied to the injection $N \to M$, for N a subspace of the metric space M, concludes that $\mathcal{F}(N)$ can be identified to a —linearly isometric— subspace of $\mathcal{F}(M)$.

The following is a list of natural examples:
(i) If $M := \{0, x_1, \ldots, x_n\}$ is a finite metric space, then $\mathrm{Lip}_0(\mathrm{M})$ is isomorphic to $\mathbb{R}^n$ via the mapping $f \in \mathrm{Lip}_0(\mathrm{M}) \to (\mathrm{f}(\mathrm{x}_1), \ldots, \mathrm{f}(\mathrm{x}_n))$ (and so $\mathcal{F}(M)$ is linearly isomorphic to $\mathbb{R}^n$, too).
(ii) If $M := \mathbb{N} \cup \{0\}$ with the usual metric, then $T : \mathrm{Lip}_0(\mathrm{M}) \to \ell_\infty$ given by $(Tf)_n := f(n) - f(n-1)$, $n \in \mathbb{N}$ is a linear isometry onto. Its preadjoint is a linear isometry from ℓ_1 onto $\mathcal{F}(M)$.
(iii) A variant of the previous example is $M = \mathbb{N} \cup \{0\}$ endowed with the following metric: $d(n, m) = 2$ for all $n \neq m \in \mathbb{N}$, and $d(n, 0) = 1$ for $n \in \mathbb{N}$. The mapping $T : \mathrm{Lip}_0(\mathrm{M}) \to \ell_\infty$ given by $(Tf)_n := f(n)$ for $n \in \mathbb{N}$ is a linear isometry onto, and its preadjoint mapping maps ℓ_1 linear isometrically onto $\mathcal{F}(M)$.
(iv) Finally, let $M := [0, 1]$ endowed with the usual metric. Then $T : L_\infty[0, 1] \to \mathrm{Lip}_0[0, 1]$ given by $(Tf)(x) := \int_0^x f(t)dt$ is a linear isometry onto. Again, its preadjoint maps linear isometrically $\mathcal{F}[0, 1]$ onto $L_1[0, 1]$.

Theorem 601 (Naor, Schechtman, [NaSch07])**.** *$\mathcal{F}(\mathbb{R}^2)$ cannot be isomorphically embedded into L_1. As a consequence, this happens to $\mathcal{F}(\mathbb{R}^n)$ for $n > 2$, and so $\mathcal{F}(\mathbb{R}^n)$ is not isomorphic to $\mathcal{F}(\mathbb{R})$ for $n > 1$.*

It is an open problem whether $\mathcal{F}(\mathbb{R}^n)$ and $\mathcal{F}(\mathbb{R}^m)$ are isomorphic for any $n > m \geq 2$.

The following result exhibits a universal property of the Lipschitz-free spaces:

Theorem 602 (Hájek, Novotný, [HáNo17]). *Any Lipschitz-free space $\mathcal{F}(M)$ contains a complemented copy of $\ell_1(\Gamma)$, where the cardinality of Γ equals the density character of M.*

For questions on embedding Lipschitz free spaces into L_1 or ℓ_1 we refer to [Goda10]. See also [AlPePr21].

37.3 The extremal structure of $B_{\mathcal{F}(M)}$

Some distinguished elements of $\mathcal{F}_M$ are the so-called **elementary molecules**, or simply **molecules**

$$m_{xy} := \frac{\delta(x) - \delta(y)}{d(x,y)}, \quad x, y \in M, \ x \neq y.$$

The set of all molecules will be denoted $\mathrm{Mol}\,(M)$. It is a subset of $S_{\mathcal{F}(M)}$. Since $\|f\|_L = \sup\{\langle m, f\rangle : m \in \mathrm{Mol}\,(M)\}$, we have $B_{\mathcal{F}(M)} = \overline{\mathrm{conv}}\,\mathrm{Mol}\,(M)$. It is natural then to wonder whether the two sets $\mathrm{Ext}\,(B_{\mathcal{F}(M)})$ and $\mathrm{Mol}\,(M)$ coincide. Below we hint at some known instances to this important question.
Not all molecules are always extreme points of $B_{\mathcal{F}(M)}$. The following result characterizes when this happens:

Theorem 603 (Aliaga and Pernecká, [AlPer20]). *Let x, y be different points of M. Then $m_{xy} \in \mathrm{Ext}\, B_{\mathcal{F}(M)}$ if, and only if, $[xy] = \{x, y\}$.*

It is an open problem whether $\mathrm{Ext}\,(B_{\mathcal{F}(M)}) \subset \mathrm{Mol}\,(M)$. This is solved in the positive in some important cases. Here we have one of them:

Theorem 604 (Aliaga, [Al21]). *If M is a compact metric space, then* $\mathrm{Ext}\,(B_{\mathcal{F}(M)}) \subset \mathrm{Mol}\,(M)$.

Figure 37.3 below summarizes some of the known results in this area (thick arrows hold in every Banach space). For the notion of exposed and strongly exposed points see Definition 55, for denting point see Definition 405, and for preserved extreme point see the proof of Theorem 407.

37.4 Lipschitz-free spaces on Banach spaces

An important aspect of the theory appears when the pointed metric space (M, d) is a Banach space $(X, \|\cdot\|)$ where the origin is the distinguished point. Even in this case, the isometric embedding $\delta : X \to \mathcal{F}(X)$ is not linear. Notice that $(X^*, \|\cdot\|^*)$ is linearly isometric to a subspace of $(\mathrm{Lip}_0(\mathrm{X}), \|\cdot\|_{\mathrm{L}})$, hence the restriction mapping $\beta : \mathcal{F}(X) \to X^{**}$ is linear and continuous. Clearly, $\beta(\sum_{i=1}^n \lambda_i \delta_{x_i}) = \sum_{i=1}^n \lambda_i x_i$

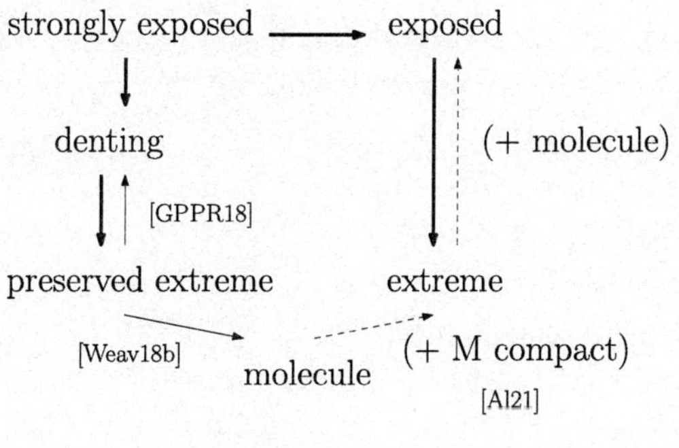

Figure 37.3: Some known connections for distinguished points in $B_{\mathcal{F}(M)}$. Thick arrows hold in any Banach space

for $\lambda_i \in \mathbb{R}$, $x_i \in X$, $i = 1, 2, \dots, n$, $n \in \mathbb{N}$, hence β maps $\mathcal{F}(X)$ into X. It is a linear left-inverse of δ; in the language of measures, as it was used above, $\beta(\mu)$ should be called the **barycenter** of $\mu \in \mathcal{F}(X)$.
If M is a metric space and X is a Banach space, then $F = \beta_X \circ \widehat{F} \circ \delta_M$ (see Figure 37.4).

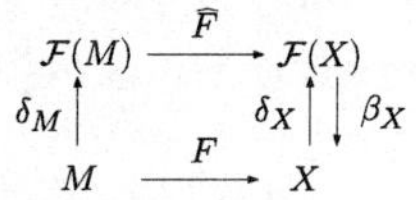

Figure 37.4: The action of the barycenter mapping

The Mazur–Ulam Theorem ensures that *every isometry from a Banach space X onto another Banach space Y that takes 0 to 0 is necessarily linear.* The surjectivity assumption cannot be omitted. Figiel proved [Fi68] the following non-surjective substitute for this result: *Let Φ be an isometry from X into Y that takes 0 to 0. Then there is a linear operator T of norm one from $\overline{\text{span}}\,\Phi(X)$ onto X such that $T \circ \Phi = Id_X$.* Godefroy and Kalton, in [GoKa03], solved a long-standing problem showing that *if X is a separable Banach space and Φ is an isometry from X into Y, then Y contains a linearly isometric copy of X.* This last result is an important application of their contributions to the theory of Lipschitz-free spaces. One technical tool is the following property: Following [GoKa03], we say that a Banach space X has the **(isometric) lifting property** if there is a (respectively, norm-1) continuous linear map $R : X \to \mathcal{F}(X)$ such that $\beta \circ R = Id_X$. This concept allows for finding examples of Banach spaces that are Lipschitz-isomorphic but not linearly isomorphic (the first example was provided by Aharoni and Lindenstrauss in [AhLi78]). Godefroy and Kalton noticed that the linear map $\widehat{\delta_X} : \mathcal{F}(X) \to \mathcal{F}(\mathcal{F}(X))$ is a linear lifting of the quotient map $\beta_{\mathcal{F}(X)}$. Thus, for any Banach space X, the free space $\mathcal{F}(X)$ has the isometric lifting property [GoKa03, Lemma

2.10], a property that passes to complemented subspaces. Then they proved the following result:

Theorem 605 (Godefroy, Kalton, [GoKa03], Theorem 2.12). *Let X be a Banach space. If $Z_X := \ker \beta_X$, then $\mathcal{G}(X) := Z_X \oplus X$ is Lipschitz isomorphic to $\mathcal{F}(X)$. Moreover, these two spaces are linearly isomorphic if, and only if, the space X has the lifting property.*

Thus, any Banach space X which fails to have the lifting property (see Theorem 607 below and notice that, besides ℓ_∞, any nonseparable reflexive Banach space or $c_0(\Gamma)$ for uncountable Γ are such) provides a couple of spaces (namely $\mathcal{F}(X)$ and $\mathcal{G}(X)$) which are Lipschitz-isomorphic but not linearly isomorphic.
The following characterization of spaces with the isometric lifting property is crucial:

Theorem 606 (Godefroy, Kalton, [GoKa03], Theorem 2.13). *Let X be a Banach space. Then X has the isometric lifting property if, and only if, whenever Y is a Banach space such that there exists a (not necessarily linear) isometry $\Phi : X \to Y$ with $\Phi(0) = 0$ and $Y : \overline{\operatorname{span}}\, \Phi(X)$, then Y contains a contractively complemented subspace linearly isometric to X.*

Theorem 607 (Godefroy, Kalton, [GoKa03]). *The following hold:*
(i) [Theorem 3.1] *Every separable Banach space has the isometric lifting property.*
(ii) [Theorem 4.3] *A WCG Banach space has the lifting property if, and only if, it is separable.*
(iii) [Proposition 4.6] *The space ℓ_∞ fails the lifting property.*

A noteworthy consequence of this result is that *in case an isometric embedding exists from a separable Banach space X into a Banach space Y, then Y contains a linear subspace which is isometric to X* [GoKa03, Corollary 3.3], solving a problem by L. Drewnowski.
Related to renorming theory, it is interesting to quote the following result [GoKa03, Proposition 4.5]: *Let X and Y be Banach spaces such that there exists a Lipschitz embedding $L : X \to Y$ such that $L(0) = 0$ and $\overline{\operatorname{span}}\, L(X) = Y$. Then L admits a continuous linear left inverse Q if, and only if, there exists an equivalent norm on Y for which L becomes an isometric embedding.*
If X is a finite-dimensional Banach space, then $\mathcal{F}(X)$ has the MAP [GoKa03, Proposition 5.1]. On the other hand, there exists a compact metric space K whose free space $\mathcal{F}(K)$ fails the approximation property [GoOz14].

Chapter 38

Polyhedral spaces

38.1 Polyhedral spaces

Let us repeat here the relevant definition. It was already introduced in Section 6.3 above:

Definition 608. *A Banach space X is called* **polyhedral** *if for every finite-dimensional subspace $F \subset X$ the ball of F is the convex hull of a finite set of points.*

We present here only a glimpse at the theory, referring to [FHHMZ11], [HáJo14], [FoLiPh01], [GuiMoZi14], and references therein. The following very useful result reflects a clever use of the compactness and the Hahn–Banach theorem. It is a core result for the understanding of polyhedrality:

Theorem 609 (Gleit, McGuigan, [GlMcG72], see also [HáJo14], Chapter 5, Lemma 101). *Let X be a Banach space. If there is a 1-norming set $B \subset B_{X^*}$ such that no w^*-accumulation point of B with norm 1 attains its norm, then B is a James boundary of X and X is polyhedral.*

Proof. To see that B is a James boundary of X, choose $x \in S_X$. There is $\{f_n\}_{n\in\mathbb{N}} \subset B$ such that $f_n(x) \to 1$. By the w^*-compactness of B_{X^*} the sequence $\{f_n\}_{n\in\mathbb{N}}$ has a w^*-cluster point $f \in B_{X^*}$. Then $f(x) = 1$ and consequently $\|f\| = 1$. By the assumptions, f is not a w^*-accumulation point of B and thus f must be one of the f_n's, so $f \in B$. This shows that B is a boundary.
Suppose further that X is not polyhedral. Then there is a finite-dimensional subspace $Y \subset X$ that is not polyhedral. Since $M = \{f \restriction_Y : f \in B\} \cap S_{Y^*}$ is a James boundary of Y, the set M is infinite as Y is not polyhedral ([HáJo14, Proposition 5.100]). By the compactness of B_{Y^*} there is a sequence $\{f_n\}_{n\in\mathbb{N}}$ of distinct elements of M such that $f_n \to g \in Y^*$. Hence $g \in S_{Y^*}$ and g attains its norm (by the compactness of B_Y). For each $n \in \mathbb{N}$ there is $h_n \in B$ such that $f_n = h_n|_Y$. By the w*-compactness of B_{X^*} the sequence $\{h_n\}_{n\in\mathbb{N}}$ has a w*-cluster point $h \in B_{X^*}$ which is also an accumulation point of B as all h_n 's are distinct. Clearly, $h|_Y = g$. Therefore $\|h\| = 1$ and h attains its norm, a contradiction with the assumptions, which proves that X is polyhedral. □

Note that the set $\{e_n : n \in \mathbb{N}\} \subset \ell_1$ can be used to show, via Theorem 609, that c_0 is polyhedral (see also Theorem 611 below).

A. J. Guirao et al., *Renormings in Banach Spaces*, Monografie Matematyczne 75,
https://doi.org/10.1007/978-3-031-08655-7_38

Theorem 610 (Fonf, see [FoLiPh01], Proposition 6.19). *If a Banach space has a countable James boundary, then for each $\varepsilon > 0$, X is $(1+\varepsilon)$-isomorphic to a polyhedral space.*

Proof. (See Figure 38.1) Let $JB = \{\pm h_i\}_{i=1}^{\infty}$ be a James boundary for X in the original norm $\|\cdot\|$. Let $\varepsilon > 0$ and let us take a sequence $\{\varepsilon_i\}_{i\in\mathbb{N}}$ of positive numbers which tends to zero and such that $\varepsilon_i < \varepsilon$ for each i. Put $B_1 := \{\pm(1+\varepsilon_i)h_i\}_{i=1}^{\infty}$ and $V^* := \overline{\operatorname{conv}}^{w^*}(B_1)$. Then $B_{X^*} \subset V^* \subset (1+\varepsilon)B_{X^*}$. Since V^* is w*-closed, it induces an equivalent norm $|||\cdot|||$ on X which is $(1+\varepsilon)$-isomorphic to the given norm. By definition, the set B_1 is 1-norming for $(X, |||\cdot|||)$.
Let $x \in X$, $x \neq 0$. Since JB is a James boundary for the original norm, there is i such that $|h_i(x)| = \|x\|$. Thus, $|||x||| \geq (1+\varepsilon_i)h_i(x) > |h_i(x)| = \|x\|$. Let f be any w*-cluster point of the set B_1. Then, since $\varepsilon_i \to 0$, we have $f \in B_{X^*}$. Suppose that for some $x \in X$ with $|||x||| = 1$, we had $f(x) = 1$. Since $f \in B_{X^*}$, we have $\|x\| \geq f(x) = 1$. Therefore $\|x\| \geq |||x|||$ which contradicts what we just proved. Thus no w*-cluster point of the set B_1 attains its nonzero norm $|||\cdot|||$, and by applying Theorem 609 above we deduce that $(X, |||\cdot|||)$ is polyhedral. $\square$

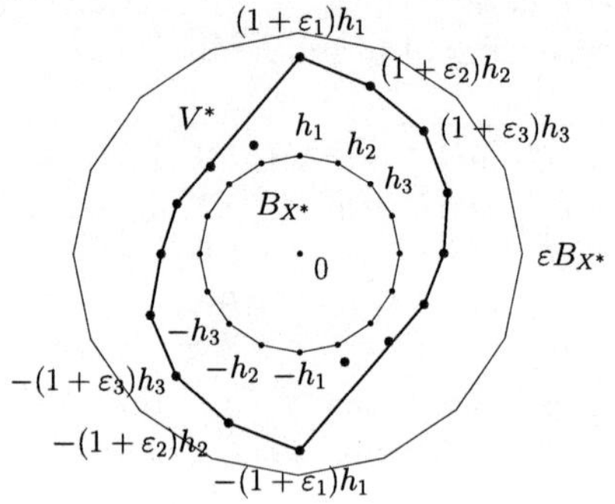

Figure 38.1: In bold, the set V^* in the proof of Theorem 610

Theorem 611. (i) *The space c_0 is polyhedral; thus, no 2-dimensional subspace of c_0 has Gâteaux smooth or strictly convex ball.*
(ii) *The space c of convergent sequences is not polyhedral.*
(iii) *If K is an infinite compact space, then $C(K)$ is not polyhedral.*

Proof. (Sketch) For (i), we mentioned an argument after the proof of Theorem 609. For a nice alternative proof see, e.g., [Go01a]; see also Theorems 609 and 610 above.
A delightful "bare hands" proof of (ii) was shown to us by Libor Veselý, see [GuiMoZi15].

(iii) Such $C(K)$ contains an isometric copy of c, so (ii) implies (iii). □

Theorem 612 (Fonf [Fon90], Hájek [Há95]). (See, e.g., [FHHMZ11, Theorem 10.12] and [HáJo14, Chapter 5, Theorem 193]) *Let X be a Banach space. Then the following are equivalent.*
(i) *X is a separable space isomorphic to a polyhedral space.*
(ii) *X has a countable James boundary.*
(iii) *X admits a norm that (outside the origin) locally depends on finitely many coordinates.*
(iv) *X admits a norm that is C^∞-smooth and locally depends on finitely many coordinates.*

Corollary 613. *If X is a separable space isomorphic to a polyhedral space, then every subspace of X contains an isomorphic copy of c_0 and X^* is separable.*

Proof. It follows from Theorem 612 and Theorem 357. □

Theorem 614 (Casazza, Kalton, Tzafriri, [CaKaTz87]). *There is a space saturated with c_0 that is not isomorphic to a polyhedral space.*

Theorem 615 (Hagler, [Hag77]). *There is a separable space with nonseparable dual that is saturated with copies of c_0.*

The following important result is a delightful gem of the theory, and we recommend the reader to enjoy it.

Theorem 616 (Deville, Fonf, Hájek, see, e.g., [HáJo14], Chapter 5, Theorem 104). *Any separable space that is isomorphic to a polyhedral space admits a real-analytic norm.*

Proof. Assume that $\{f_n : n \in \mathbb{N}\}$ is a countable James boundary of X (see Theorem 612 above). Define a function $\Phi : \mathrm{Int}B_X \to \mathbb{R}$ by

$$\Phi(x) = \sum_{n=1}^{\infty} f_n(x)^{2n}.$$

Then Φ is convex and analytic. Let $a \in (0,1)$ and set

$$A := \{x \in \mathrm{Int}B_X : \ \Phi(x) \leq a\}.$$

The set A is closed. Indeed, let $\{x_n\}_{n\in\mathbb{N}} \subset A$ be such that $x_n \to x \in X$. Then $x \in B_X$. Suppose that $x \in S_X$. Let $n \in \mathbb{N}$ be such that $f_n(x) = 1$. Then there is a neighborhood V of x such that $f_n(y) > a^{1/2n}$ for every $y \in V$.

Hence

$$\Phi(y) \geq f_n(y)^{2n} > a \text{ for } y \in V \cap \mathrm{Int}B_X,$$

which contradicts the fact that $\Phi(x_n) \leq a$ for all $n \in \mathbb{N}$. It follows that $x \in \mathrm{Int}B_X$ and so $x \in A$ by the continuity of Φ. Since $x \in \mathrm{Int}B_X$, where Φ is analytic, we can use the Implicit Function Theorem 77 again to finish the proof. □

The following problem is open: Assume X is a separable Banach space that admits a bump that locally depends of finitely many coordinates. Is X isomorphic to a polyhedral space?

38.2 Tilings

We only review several important facts in this interesting area.

Definition 617. *By a* **tiling** *of a Banach space X we will mean a covering of X by a family of closed convex bounded sets with nonempty interiors that have pairwise disjoint interiors. A tiling is* **locally finite** *if each point has a neighborhood that meets only finitely many sets from the tiling. It is called* **point-finite** *if each point lies only in a finite number of sets in the tiling.*

Theorem 618 (Corson, [Cor61b]). *Let X be an infinite-dimensional reflexive Banach space. Then any cover of X by bounded convex sets is necessarily not locally finite.*

Note that in c_0, balls with radius $1/2$ centered at points whose all coordinates are integers, form a locally finite cover of c_0.

Theorem 619 (Fonf, [Fon90]). *A separable Banach space admits a locally finite tiling if, and only if, it is isomorphic to a polyhedral space.*

The situation with point-finite tiling is as follows:

Theorem 620. (i) [Klee, [Klee86]] *For large Γ, $\ell_1(\Gamma)$ admits a tiling by pairwise disjoint shifts of the ball of $\ell_1(\Gamma)$.*
(ii) [Marchese, Zanco, [MarZa12]] *Every Banach space admits a point-finite tiling.*

Definition 621. *We say that a family $\mathcal{F}$ of sets in a Banach space X is* **star-finite** *if each member of $\mathcal{F}$ intersects only finitely many other members of $\mathcal{F}$.*

Theorem 622 (de Bernardi, Somaglia, Veselý, [dBSoVe20]). *Let $(X, \|\cdot\|)$ be LUR or F. Then X does not admit any covering by a star-finite family of closed balls. Any normed space of countable dimension admits a star-finite covering by closed balls.*

Problem. Does there exist a separable space with a star-finite covering by closed balls?

Chapter 39

Smooth functions on $c_0(\Gamma)$

Here is a list of motivations for the results in this chapter.

1. c_0 does not admit an equivalent C^2-smooth LUR norm (Theorem 510).

2. c_0 admits a strictly convex C^2-smooth norm (by using the Clarkson technique of injecting the space into ℓ_2: if T is an injection of c_0 into ℓ_2, and $\|\cdot\|$ is a C^2-smooth norm on c_0 (Theorem 418), then $|||x|||^2 := \|x\|^2 + \|Tx\|_2^2$ gives such a norm). The same is true for the nonseparable space JL_0 (see Theorem 826 below).

3. For an uncountable set Γ, $c_0(\Gamma)$ does not inject into $\ell_2(\Gamma)$. One possible argument is to use that there is no equivalent URED norm on $c_0(\Gamma)$.

Theorem 623 (Hájek, [Há98]). *Let f be a real-valued Fréchet differentiable function on B_{c_0} with uniformly continuous derivative. Then $f'(B_{c_0})$ is relatively compact in ℓ_1.*

Corollary 624 (Hájek, [Há98]). *Let f be a real-valued Fréchet differentiable function on $c_0(\Gamma)$ with locally uniformly continuous derivative. Then f depends locally on countably many coordinates. In particular, if Γ is uncountable, then there is no Fréchet differentiable function with locally uniformly continuous derivative on $c_0(\Gamma)$ which would attain its minimum at exactly one point. In particular, there is no strictly convex norm on $c_0(\Gamma)$, for Γ uncountable, that would be Fréchet smooth with locally uniformly continuous derivative (in particular, C^2-smooth).*

Theorem 625 (Hájek, [Há98]). *Let T be a (nonlinear in general) Fréchet differentiable mapping from c_0 into a superreflexive Banach space. Suppose that T' is locally uniformly continuous. Then T is locally compact.*

From the Baire category theorem, we get

Corollary 626 (Hájek, [Há98]). *There is no (nonlinear in general) C^2-smooth map from c_0 onto ℓ_p for any $p > 1$.*

The key clever idea in all of this is Hájek's smart use of the symmetrization procedure. To document its beauty, let us mention the following Hájek's result which is the core in the process of his proofs there.

Proposition 627 (Hájek, [Há98]). *Let f be a real symmetric Fréchet differentiable function on B_{c_0} with uniformly continuous derivative. Then f is constant on B_{c_0}.*

A. J. Guirao et al., *Renormings in Banach Spaces*, Monografie Matematyczne 75,
https://doi.org/10.1007/978-3-031-08655-7_39

The function f is called **symmetric** on $c_0(\Gamma)$ if $f\left(\sum a_i e_i\right) = f\left(\sum a_i e_{\pi(i)}\right)$ for any permutation of indices.
For further works in this area see [Há98b], [DeHá07], and [DeMa00].

The following problem seems to be open: If X has a C^2-smooth norm that is rotund, does X injects into $\ell_p(\Gamma)$ for some Γ? Compare with Theorem 460.

A real-valued function f on a separable infinite-dimensional Banach space X is said to be **oscillation stable** if for every infinite-dimensional subspace Z of X and every $\varepsilon > 0$ there is a further infinite-dimensional subspace Y of Z such that the oscillation of f on S_Y is at most ε, i.e., $|f(x) - f(y)| \leq \varepsilon$ for every $x, y \in S_Y$.

Theorem 628 (Gowers, Milman, Odell, Schlumprecht, see [HáNo18]). *A separable infinite-dimensional Banach space X has the property that every Lipschitz real valued function f is oscillation stable on X if, and only if, X is saturated with spaces isomorphic to c_0.*

For nonseparable spaces the situation is different:

Theorem 629 (Hájek, Novotný, [HáNo18]). *Let Γ be an uncountable cardinal. There is a Lipschitz symmetric real valued function on $S_{c_0(\Gamma)}$ such that for every nonseparable subspace Y of $c_0(\Gamma)$ there are points $x, y \in S_Y$ such that*

$$|f(x) - f(y)| > 1/4.$$

It is shown in [HáNo18] that f cannot be an equivalent norm on $c_0(\Gamma)$.

Chapter 40

Kottman-type results on separated sets

40.1 Whitley's results

The following definition was introduced by R. Whitley in [Whi68] in order to deal with Banach's problem asking whether two linearly isomorphic Banach spaces whose Banach–Mazur distance is 1 (Banach called those spaces **nearly isometric**) should be linearly isometric. By the way, the answer is negative —see, e.g., [FHHMZ11, Exercise 5.44] for a particular example—, hence the reverse.

Definition 630. *If X is a Banach space, then the* **thickness** *of X is the largest nonnegative number $T(X)$ such that any ε-net for S_X must be infinite if $0 < \varepsilon < T(X)$.*

Theorem 631 (Whitley, [Whi68]). *Let X be a Banach space.*
(i) *If X is finite-dimensional, then $T(X) = 0$.*
(ii) *If X is infinite-dimensional, then $1 \leq T(X) \leq 2$.*
(iii) *If Q is an infinite completely regular Hausdorff space, then $T(CB(Q)) = 1$ if Q contains an isolated point, and $T(CB(Q)) = 2$ otherwise. If Q is also locally compact, then $T(C_0(Q)) = T(CB(Q))$. In particular, $T(c) = T(c_0) = 1$.*
(iv) *If $1 \leq p < \infty$, then $T(\ell_p) = 2^{\frac{1}{p}}$.*
(v) *For each T in $[1, 2]$, there is a Banach space X such that $T(X) = T$.*
(vi) *Let X and Y be two isomorphic Banach spaces. If $T(X) \neq T(Y)$, then X and Y cannot be nearly isometric, i.e., the Banach–Mazur distance between X and Y is different from 1.*

Proof. (Partial.) The compactness of S_X in the case of a finite-dimensional space X (see Proposition 3) gives (i). If X is an infinite-dimensional space, $0 < \varepsilon < 1$ and $\{x_i\}_{i=1}^n$ ($\subset S_X$) is a finite ε-net for $S(X)$, then $S_X \subset \bigcup_{i=1}^n B(x_i, \varepsilon)$. Since $0 \in \overline{S_X}^w$ (see the note in page 8), a net $\{u_j : j \in J, \leq\}$ in S_X that w-converges to 0 must have a subnet in one of the $B(x_i, \varepsilon)$, and this is a contradiction since $B(x_i, \varepsilon)$ is w-closed and $x_i \in S_X$ (an alternative proof uses Riesz's Lemma 4). This shows the first inequality in (ii) in Theorem 631. The second one is obvious, since $\{x\}$ is, for every $\varepsilon > 0$, a $(2 + \varepsilon)$-net for S_X, where x is an arbitrary point in S_X. Notice that (v) follows from (iii) and (iv). □

We remark that c and c_0 are linearly isomorphic (see, e.g., [FHHMZ11, Exercise 5.16]), although the Banach–Mazur distance between c_0 and c is 3, see [PelBe79, II §2, Remark after Theorem 2.1]. However, $T(c) = T(c_0)$ according to (iii) in

A. J. Guirao et al., *Renormings in Banach Spaces*, Monografie Matematyczne 75,
https://doi.org/10.1007/978-3-031-08655-7_40

Theorem 631, showing that the converse to (vi) in the aforementioned theorem fails.

40.2 Separated and symmetrically separated sets

Let us introduce a convenient definition for another concept to be discussed in this chapter:

Definition 632. *Let $\delta \geq 0$. A subset A of a normed space $(X, \|\cdot\|)$ is said to be δ-* **separated** *(respectively,* **$(\delta+)$-separated***) if $\|x-y\| \geq \delta$ (respectively, $\|x-y\| > \delta$) for $x, y \in A$, $x \neq y$.*

The inductive use of Riesz's Lemma 4 above gives, in an infinite-dimensional normed space $(X, \|\cdot\|)$, a sequence $\{x_n\}_{n\in\mathbb{N}}$ in S_X such that $\|x_n - x_m\| \geq 1$ for all $n, m \in \mathbb{N}$, $n \neq m$. Thus, the unit sphere of every infinite-dimensional Banach space contains a 1-separated sequence. A neat improvement of this result is due to C. Kottman:

Theorem 633 (Kottman, [Kott75]). *If $(X, \|\cdot\|)$ is an infinite-dimensional normed space, then S_X contains a $(1+)$-separated sequence.*

Proof. [T. Starbird] (See, e.g., [Di84, Chapter I, Notes and Remarks]) We shall construct two sequences $\{x_n\}_{n\in\mathbb{N}}$ in S_X and $\{x_n^*\}_{n\in\mathbb{N}}$ in S_{X^*} with the following properties:
(i) $x_n^*(x_n) = 1$ for $n \in \mathbb{N}$.
(ii) $\{x_n^* : n \in \mathbb{N}\}$ is a linearly independent set.
(iii) $\|x_n - x_m\| > 1$ for $n, m \in \mathbb{N}$, $n \neq m$.
Let us start by choosing an arbitrary vector $x_1 \in S_X$. Find $x_1^* \in S_{X^*}$ such that $x_1^*(x_1) = 1$. Assume that for some $k \in \mathbb{N}$, vectors $x_1, \ldots, x_k$ and $x_1^*, \ldots, x_k^*$ that satisfy (i), (ii), and (iii) for $n = 1, 2, \ldots, k$, have been already defined. Choose $y \in X$ so that $x_1^*(y) < 0, \ldots, x_k^*(y) < 0$. Let $0 \neq x \in \bigcap_{i=1}^k \ker x_i^*$ (see Figure 40.1). Find $K > 0$ so that $\|y\| < \|y + Kx\|$. Then, for any nontrivial linear combination $\sum_{i=1}^k a_i x_i^*$ we have

$$\left|\sum_{i=1}^k a_i x_i^*(y + Kx)\right| = \left|\sum_{i=1}^k a_i x_i^*(y)\right|$$
$$\leq \left\|\sum_{i=1}^k a_i x_i^*\right\| . \|y\| < \left\|\sum_{i=1}^k a_i x_i^*\right\| . \|y + Kx\|. \tag{40.1}$$

Let $x_{k+1} := (y + Kx)\|y + Kx\|^{-1}$, and choose $x_{k+1}^* \in S_X$ so that $x_{k+1}^*(x_{k+1}) = 1$. If x_{k+1}^* is a linear combination $\sum_{i=1}^k a_i x_i^*$, equation (40.1) gives a contradiction, so $\{x_1^*, \ldots, x_{k+1}^*\}$ is a linearly independent set. Furthermore, if $1 \leq i \leq k$, then

$$\|x_i - x_{k+1}\| \geq |x_i^*(x_i) - x_i^*(x_{k+1})| = 1 - x_i^*(x_{k+1}) > 1,$$

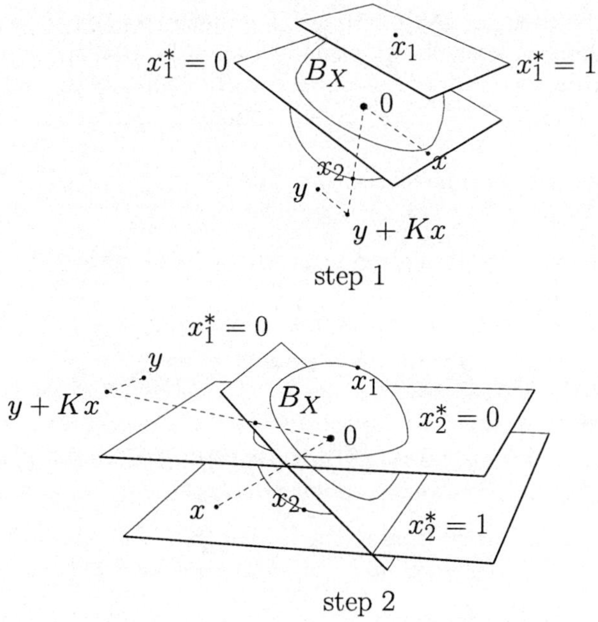

Figure 40.1: The first two steps in the proof of Kottman's theorem

since $x_i^*(x_i) = 1$ and $x_i^*(x_{k+1}) < 0$. □

J. Elton and E. Odell neatly improved Theorem 633 in the following way:

Theorem 634 (Elton, Odell, [ElOd81]). *If X is an infinite-dimensional Banach space, then there exists $\varepsilon > 0$ such that S_X contains a $(1+\varepsilon)$-separated sequence.*

For a discussion on the preceding result and for its proof see, e.g., [Di84, Chapter XIV]. A recent treatment of this topic is in [Ru18, Chapter 3]. Let us summarize its presentation: To quantify Kottman and related results, the following constant is introduced —it already appeared in [Kott75]—:

Definition 635. *The* **Kottman constant** $K(X)$ *of a normed space X is*

$$K(X) := \sup\{\sigma > 0 : \text{ there exists } \{x_n : n \in \mathbb{N}\} \subset S_X,\ \|x_n - x_k\| \geq \sigma,\ n \neq k\}.$$

Obviously, X is finite-dimensional if, and only if, $K(X) = 0$. Riesz's Lemma shows that $K(X) \geq 1$ for every infinite-dimensional normed space, and the Elton–Odell Theorem 634 can be restated as saying that $K(X) > 1$ for every infinite-dimensional normed space X. $K(\ell_p) = 2^{1/p}$ whenever $1 \leq p < \infty$, and $K(c_0) = 2$. Regarding lower estimates for $K(X)$, a result of Kryczka and Prus [KrPr01] says that for every non-reflexive Banach space X, $K(X) \geq \sqrt[5]{4}$ (~ 1.3195). Lower estimates were related to the modulus of uniform convexity δ_X by van Neerven [vNee05], Delpech [Del10], Maluta and Papini [MaPa93], giving $K(X) \geq 1 + \sqrt{2}.\delta_X(\sqrt{2})$. For upper estimates, Kottman [Kott75] already proved that $K(X) < 2$ provided X is uniformly convex or uniformly smooth. Maluta and Papini [op. cit.] gave upper estimates related to the modulus of uniform convexity and the modulus ρ_X of uniform smoothness: $K(X) \leq 2(1 - \delta_X(1))$ and $K(X) \leq 1 + 2\rho_X(1)$. See also [CGP17], [Pr10] and other references in [Ru18].

Recently, the following related concept has been investigated:

Definition 636. *Let $\delta \geq 0$. A subset A of a normed space is* **symmetrically δ-separated** *(***symmetrically $(\delta+)$-separated***) when $\|x \pm y\| \geq \delta$ (respectively, $\|x \pm y\| > \delta$) for $x, y \in A$, $x \neq y$.*

Theorem 637 (Hájek, Kania, Russo, [HáKaRu18]). *If X is an infinite-dimensional Banach space, then S_X contains a symmetrically $(1+)$-separated sequence.*

Theorem 638 (Hájek, Kania, Russo, [HáKaRu18]). *If X is an infinite-dimensional Banach space with RNP, then there exists $\varepsilon > 0$ and a $(1+\varepsilon)$-symmetrically separated sequence in S_X.*

Theorem 639 (Hájek, Kania, Russo, [HáKaRu18]). *If X is an infinite-dimensional Banach space, then there exists an equivalent norm $\|\cdot\|$ on X such that $S_{(X,\|\cdot\|)}$ contains a symmetrically 2-separated sequence.*

Theorem 640 (Kania, Kochanek, [KaKo16]). *If X is a nonseparable Banach space, and either X is reflexive or a $C(K)$-space, then S_X contains an uncountable $(1+)$-separated set.*

Theorem 641 (Hájek, Kania, Russo, [HáKaRu20]). *Let X be a Banach space. If the w^*-density of X^* is large enough, then S_X contains symmetrically $(1+)$-separated uncountable sets.*

Theorem 642 (Hájek, Kania, Russo, [HáKaRu20]). *If X is a superreflexive Banach space, then there is $\varepsilon > 0$ and a symmetrically $(1+\varepsilon)$-separated set in S_X of cardinality* $\operatorname{dens} X$.

Theorem 643 (Hájek, Kania, Russo, [HáKaRu20]). *Under CH, $c_0(\omega_1)$ can be renormed to admit no uncountable Auerbach system.*

40.3 Equilateral sets in infinite-dimensional spaces

This short section involves nice pieces of mathematics: Spreading models, distorted norms, and topology.

Definition 644. *An* **equilateral set** *S in a Banach space X is a set that has the property that there is $\lambda > 0$ such that $\|x - y\| = \lambda$ for every $x, y \in S$, $x \neq y$.*

Theorem 645 (Brass [Br99], Dekster [Dek00]). *Every infinite-dimensional Banach space contains an arbitrarily large finite equilateral set.*

Proof. This follows from Dvoretzky's theorem as shown in [Br99] and [Dek00]. □

Theorem 646 (Swanepoel, see also [MeVa15]). *Every infinite-dimensional Banach space can be renormed to contain an infinite equilateral set.*

For a proof see [MeVa15].

Theorem 647 (Swanepoel, Mercourakis and Vassiliadis, [MeVa15]). *Every renorming of c_0 contains an infinite equilateral set.*

Theorem 648 (Mercourakis and Vassiliadis, [MeVa15]). *If a Banach space has a fundamental biorthogonal system, then X can be renormed to contain a equilateral set of the size the same as the fundamental system.*

Theorem 649 (Freeman, Odell, Sari, Schlumprecht, [FOSS14]). *Let X be a Banach space with uniformly Fréchet differentiable norm. Then X contains an infinite countable equilateral set.*

Theorem 650 (Terenzi, [Ter89]). *The space ℓ_1 can be renormed so that in the new norm there is no infinite equilateral set.*

Theorem 651 (Glakousakis and Mercourakis, [GlMe16]). *Let $p > 2$ and let $Z = (\sum_{n=1}^{\infty} \oplus \ell_p^n)_1$. Then Z can be renormed not to have any infinite equilateral set.*

Theorem 652 (Koszmider, Wark, [KoWa]). *There is a Banach space of density the continuum that does not admit any infinite equilateral set.*

Theorem 653 (Terenzi, [Ter89]). *If the density of a Banach space X is larger than $2^{2^{\omega_0}}$ then X admits an uncountable equilateral set.*

Chapter 41

Three-space properties

In this chapter we refer to the book [CasGo87]. We start with the basic definition:

Definition 654. *A property P of Banach spaces is called a* **three-space property** *if whenever a Banach space X and a subspace Y of X are such that both Y and X/Y have property P, then X has property P.*

Theorem 655. (i) *Reflexivity is a three-space property.*
(ii) *Separability is a three-space property.*
(iii) *Neither WCG, nor Vašák, nor WLD, is a three-space property.*
(iv) *Renorming by a Gâteaux differentiable norm is not a three-space property. The same for a uniformly Gâteaux differentiable norm .*
(v) *Renorming by a LUR norm is a three-space property.*
(vi) *Renorming by a norm with dual LUR norm is a three-space property.*
(vii) *Being isomorphic to a Hilbert space and renorming by a C^2-smooth norm both are not a three-space property.*
(viii) *Superreflexivity is a three-space property.*
(ix) *Being a subspace of c_0 is a three-space property. For nonseparable $c_0(\Gamma)$ is not.*
(x) *Being an Asplund space is a three-space property.*
(xi) *Being a weak Asplund space is not a three-space property.*
(xii) *Renorming by R norm is not a three-space property.*
(xiii) *RNP is a three-space property.*
(xiv) *SK** (see Definition 725 below) *is not a three-space property.*

Related to (iv) and (xiv) in Theorem 655 above, we may mention that it is not known if renorming by a Fréchet smooth norm is a three-space property.

Proof of Theorem 655.
(i) See, e.g., [FHHMZ11, Exercise 3.116].
(ii) See, e.g., [FHHMZ11, Exercise 1.35].
(iii) Look at the space JL_2 (see Theorem 824 below).
(iv) For the Gâteaux version, see Theorem 841. For the UG version, Theorem 824.
(v) See Theorem 231 above.
(vi) Theorem 656 below.
(vii) See Theorems 701 and 702 below. See also [DeGoZi90].
(viii) [EnLiPi75].

A. J. Guirao et al., *Renormings in Banach Spaces*, Monografie Matematyczne 75,
https://doi.org/10.1007/978-3-031-08655-7_41

(ix) For the positive result, see [FHHMZ11, Exercise 5.58], and for the negative result, see Theorem 826.
(x) See, e.g., [CasGo87, Theorem 4.11.a] or [Fab97, Theorem 1.1.2].
(xi) Čoban, Kenderov, see, e.g., [Fab97, Theorem 2.3.1 and Subsection 2.2.3].
(xii) See Theorem 821. See also, e.g., [DeGoZi93, Theorem VII.6.10].
(xiii) See, e.g., [CasGo87, Theorem 6.5.b].
(xiv) See [Fab97, 2.4 Notes and Remarks]. □

It seems to be unknown if the renorming by a WUR norm is a three-space property.

Theorem 656 (Jiménez-Sevilla, Moreno, [JiMo97]). *If X is a Banach space and Y is a subspace of X such that both Y and X/Y have norms whose dual norms are LUR, then X has a norm whose dual is LUR.*

Theorem 657 (Deville, Godefroy, and Zizler, [DeGoZi93], Lemma V.1.9, [DeGoZi90]). *Let X be a Banach space and Y be a subspace of X such that Y is isomorphic to the space $c_0(\Gamma)$ for some Γ. Assume that X/Y admits a C^k-smooth norm (bump, partition of unity) where $k \in \mathbb{N}$ or $k = \infty$. Then X admits a C^k-smooth norm (respectively, bump, respectively, partition of unity).*

Proof. For the first part, see [DeGoZi93, Lemma V.1.9]. For the second and third parts, see [DeGoZi90]. □

Theorem 658 (Rychtář, [Rych03]). *If X contains $c_0(\Gamma)$ and $X/c_0(\Gamma)$ has a norm whose dual is URED, then X admits a norm whose dual is URED.*

Chapter 42

Polynomials on Banach spaces

The theory of polynomials on Banach spaces is an important field which until the publication of the excellent [HáJo14] lacked, in our opinion, a comprehensive modern book devoted to it. Here we touch only on a few things that we use in this text and follow the notation in the excellent [HáJo14].

We briefly recall:

Definition 659. *Let X and Y be Banach spaces. A mapping $P : X \to Y$ is called an n-***homogeneous polynomial** *if there is an n-***linear mapping** *(i.e., linear separately in each variable) M from the n-cartesian product $X \times X \times \overset{(n)}{\cdots} \times X$ into Y such that $P(x) = M(x, \ldots, x)$. A polynomial P is* **bounded** *if*

$$\|P\| := \sup\{\|Px\| : x \in B_X\} < \infty.$$

The space of all such bounded polynomials will be denoted by $\mathcal{P}(^nX, Y)$. The linear subspace of $\mathcal{P}(^nX, Y)$ consisting of all polynomials that can be written in the form

$$P(x) = \sum_{j=1}^{k} f_j(x)^n y_j, \text{ where } f_j \in X^* \text{ and } y_j \in Y,$$

will be denoted by $\mathcal{P}_f(^nX, Y)$. We let

$$\mathcal{P}_f(X, Y) = \operatorname{span} \bigcup_{n=0}^{\infty} \mathcal{P}_f(^nX, Y).$$

If Y is the scalar field, we denote $P(^nX, Y)$ by $P(^nX)$.

Let us note in passing the following two important results of independent interest:

Theorem 660 (Mazur, see, e.g., [HáJo14], Theorem 7.12). *Let X and Y be Banach spaces, $K \subset X$ be a compact set, $f \in C(K, Y)$, and $\varepsilon > 0$. Then there is a polynomial $P \in \mathcal{P}_f(X, Y)$ such that $\|f - P\|_K \leq \varepsilon$.*

Theorem 661 (Fréchet, see, e.g., [HáJo14], Theorem 7.16). *Let X and Y be Banach spaces, X separable, $\emptyset \neq \Omega \subset X$ open, and $f \in C(\Omega, Y)$. Then there is a sequence of polynomials $\{p_n\}_{n=1}^{\infty} \subset \mathcal{P}_f(X, Y)$ such that $\lim_{n\to\infty} p_n(x) = f(x)$ for every $x \in \Omega$.*

A. J. Guirao et al., *Renormings in Banach Spaces*, Monografie Matematyczne 75,
https://doi.org/10.1007/978-3-031-08655-7_42

Note that the requirement of separability in Theorem 661 is needed, see, e.g., [HáJo14, Remark after Theorem 7.16].

Theorem 662 (Semenov, [NeSe73b]). *There is a real-valued C^∞-function on ℓ_2 which is uniformly continuous and has uniformly continuous derivatives of any order which cannot be uniformly approximated on the unit ball of ℓ_2 by any polynomial.*

For a proof see, e.g., also [Ar02].

Theorem 663 (Bogdanowicz, [Bo57], Pełczyński [Pel57]). *Every bounded polynomial on c_0 is weakly sequentially continuous.*

Proof. See, e.g., [FHHMZ11, Exercise 10.5]. □

Corollary 664. *Let Γ be an uncountable set and P be a bounded polynomial on $c_0(\Gamma)$. Then P is countably supported in the sense that the set $\{\gamma \in \Gamma : P(e_\gamma) \neq 0\}$ is countable.*

Corollary 665 (Pełczyński, [Pel57]). *If Γ is an uncountable set, then $c_0(\Gamma)$ does not admit a real-analytic norm.*

Chapter 43

Szlenk derivation and applications

In this chapter we briefly discuss the concept of Szlenk index and some of its uses in problems on universal spaces connected to rotundity questions and in renorming theory. Some results are also surveyed in [HMVZ07, Chapter 2] and [Lanc06].
We shall need the following definition:

Definition 666. *A Banach space X is* **universal for a family $\mathcal{F}$ of Banach spaces** *if every member of $\mathcal{F}$ is isomorphic to a subspace of X.*

Notice the classical Mazur's theorem stating that *the space $C(\Delta)$ of all continuous functions defined on the Cantor set Δ is universal for the class of all separable Banach spaces.*

Universality in the sense of isometry is a tighter —and in this sense less flexible— requirement. For example, the classical Banach–Mazur theorem ensures that $C[0,1]$ is universal —in the sense of isometry— for all separable Banach spaces (see, e.g., [FHHMZ11, Theorem 5.8]). So it is the (nonseparable) space ℓ_∞ (see, e.g., [FHHMZ11, Proposition 5.4]). Here we shall not pursue the isometric version, but at two particular instances, Theorem 676 and the Remark after Theorem 673 below.

In the analysis of universality, in particular in the proof of (i) in Theorem 673, a powerful tool was introduced by W. Szlenk [Sz68]:

Definition 667. *Let $(X, \|\cdot\|)$ be a Banach space and let $\varepsilon > 0$. Let $\mathcal{V}$ be the family of all relatively w^*-open subsets V of B_{X^*} such that the $\|\cdot\|$-diameter of V is less than ε. Put $s_\varepsilon B_{X^*} = B_{X^*} \setminus \bigcup\{V : V \in \mathcal{V}\}$. Then we define inductively $s_\varepsilon^\alpha B_{X^*}$ for any ordinal α by $s_\varepsilon^{\alpha+1} B_{X^*} = s_\varepsilon(s_\varepsilon^\alpha B_{X^*})$ and $s_\varepsilon^\alpha B_{X^*} = \bigcap_{\beta<\alpha} s_\varepsilon^\beta B_{X^*}$ if α is a limit ordinal. We define $Sz(X,\varepsilon)$ to be the least ordinal α such that $s_\varepsilon^\alpha B_{X^*} = \emptyset$ if such an ordinal exists. Otherwise we write $Sz(X,\varepsilon) = \infty$. The* **Szlenk index of X** *is defined by $Sz(X) = \sup_{\varepsilon>0} Sz(X,\varepsilon)$.*
If we consider $\mathcal{V}$ the set of relatively w^*-open slices of B_{X^*} instead, we get the notion of **dual w^*-dentability index $\Delta(X)$ of X**.

From Theorem 547 it should be clear that a Banach space X is Asplund if, and only if, $S_z(X) < \infty$. The following result gives a more precise result in the case that the space X is separable:

Theorem 668 (Lancien, [Lanc06]). *Let X be a separable Banach space. Then the following are equivalent.*

A. J. Guirao et al., *Renormings in Banach Spaces*, Monografie Matematyczne 75,
https://doi.org/10.1007/978-3-031-08655-7_43

(i) *X^* is separable.*
(ii) $S_z(X) < \omega_1$.
(iii) *The identity map from* (B_{X^*}, w^*) *into* $(B_{X^*}, \|\cdot\|)$ *is of first Baire class.*

Lemma 669. (i) (Szlenk [Sz68]) *If Y is isomorphic to a subspace of X, then* $Sz(Y) \leq Sz(X)$.
(ii) *For every countable ordinal α there exists a separable reflexive Banach space X such that $Sz(X) \geq \alpha$.*

For a proof of (i) above see, e.g., [HMVZ07, Lemma 2.39].

The following result is the general version of a theorem proved by Knaust, Odell, and Schlumprecht [KnOdSch98] for separable spaces.

Theorem 670 (Raja, [Raj10]). *Let X be a Banach space. Then the following are equivalent.*
(i) *X admits a norm whose dual norm is w^*-UKK.*
(ii) $Sz(X) \leq \omega_0$.

Theorem 671. (See [Lanc06]) *Let K be a compact space. Then the following are equivalent.*
(i) $K^{(\omega_0)} = \emptyset$.
(ii) $Sz(C(K)) \leq \omega_0$.

Theorem 672 (Lancien, [Lanc06]). *Let X be a space such that $Sz(X) < \omega_1$. Then X can be renormed by a norm whose dual is LUR.*

Theorem 673. (i) (Szlenk, [Sz68]) *There is no separable reflexive space universal for the class of all separable reflexive Banach spaces.*
(ii) (Odell, Schlumprecht, [OdSch06]) *There is a separable reflexive space X universal for the class of all superreflexive separable spaces.*

The paper [Lind66c] gives, among other things, a beautiful short proof of the isometric version of Szlenk Theorem 673 (i) above.

Theorem 674 (Lancien, [Lanc95]). *Let X be a Banach space. Then X is superreflexive if, and only if, the dentability index $\Delta(X)$ of X is $\leq \omega_0$.*

The following result significantly improved (i) in Theorem 673.

Theorem 675 (Bourgain, [Bou80b]). *If a separable Banach space X is universal for all reflexive separable spaces, then X is universal for all separable spaces.*

For a proof see, e.g., [HMVZ07, Theorem 2.14].

Theorem 676 (Godefroy, [Go06]). *There is no separable strictly convex space that would be universal for the class of separable strictly convex space in the sense of isometry.*

Theorem 677 (Bourgain, [Bou80b]). *There is no separable superreflexive Banach space universal for all separable superreflexive Banach spaces.*

For a proof see, e.g., [HMVZ07, Proposition 2.17].

The following problem is open:

Problem. [Godefroy] Let X be a separable superreflexive Banach space. Does there exist a separable superreflexive space Y with Schauder basis such that X is isomorphic to a subspace of Y?

Notice that if the word "superreflexive" is substituted by "reflexive" in the previous question, the answer is positive by a famous Zippin's result [Zip68].

Theorem 678 (Bossard, [Bos02]). *There is a function $\psi : [1, \omega_1) \to [1, \omega_1)$ such that, if $\alpha < \omega_1$ and X is a Banach space satisfying $Sz(X) \leq \alpha$, then $\Delta(X) \leq \psi(\alpha)$.*

The next theorem requires the assumption of the Generalized Continuum Hypothesis (GCH), i.e., $\tau^+ = 2^\tau$ for all infinite cardinals τ, where τ^+ denotes the successor cardinal number to the cardinal number τ.

Theorem 679 (Yesenin-Volpin, [Yes49]). *Let τ be an uncountable cardinal. Then there is a compact space K of weight τ such that every Banach space of density τ is isometrically isomorphic to a subspace of $C(K)$.*

Proof. For a sketch of the proof see, e.g., [HMVZ07, Theorem 2.52]. □

Theorem 680 (Hájek, Lancien, Montesinos, [HáLaMo07]). *Given an infinite cardinal τ there exists no Asplund space of density τ universal for all reflexive spaces of density τ.*

Proof. For a proof see, e.g., [HMVZ07, Theorem 2.55]. □

Theorem 681. *Let Γ be an infinite set. Then*
(i) *The Szlenk index of $c_0(\Gamma)$ is ω_0.*
(ii) *The dual dentability index of $c_0(\Gamma)$ is ω_0^2.*
(iii) *The Szlenk index of $C[0, \omega_1]$ is $\omega_1 \cdot \omega_0$.*

Proof. (i) follows from Theorem 670, as $c_0(\Gamma)$ admits a w^*-UKK norm (see Theorem 378 and, e.g., [FHHMZ11, Proposition 12.64]).
For (ii) and (iii) see [HáLa07]. □

Theorem 682 (Bessaga, Pełczyński, Samuel). *The scale of spaces $C[0, \omega^{\omega^\alpha}]$, where $\alpha \in [0, \omega_1)$, contains all isomorphic types of $C(K)$ with K countable. Moreover, $S_z(C([0, \omega^{\omega^\alpha}])) = \omega^{\alpha+1}$, so the isomorphic class of $C(K)$ with K countable is determined by $S_z(C(K))$.*

This in no longer true in the nonseparable case, see [HáLa07]. In this direction see also, e.g., [GuOs75] and [Kis75].

Chapter 44

Further miscellaneous applications

Theorem 683 (Edelstein, [Ed73], John, Zizler, [Jozi78]).
(i) *There is a dentable ball in* c_0 *that has no extreme points* [Ed73].
(ii) *There is a nondentable strictly convex norm in* c_0 [Ed73].
(iii) *There is a not rough norm on* $\ell_1(\Gamma)$, *for* Γ *uncountable, that is nowhere Gâteaux differentiable* [Jozi78].
(iv) *The unit ball of* ℓ_∞ *is not dentable* [Ed73].
(v) *The unit ball of* $C[0,1]^*$ *is dentable* [Jozi78].

Theorem 684 (Lindenstrauss, see, e.g., [FHHMZ11], Exercise 7.43). *No 2-dimensional subspace of* ℓ_1 *has a smooth ball.*

J. Lindenstrauss proved in [Lind64a] that any 2-dimensional space is isometric to a subspace of L_1.
C. Bessaga showed in [Bess58] that there is no finite-dimensional space that would be isometrically universal for all 2-dimensional spaces.

The proof of the following theorem of M.I. Kadets and his former student V.P. Fonf is very instructive in this area.

Theorem 685 (Fonf, M.I. Kadets, [FonKa83]). *There is a* w^**-closed infinite-dimensional subspace of* ℓ_1 *that is MLUR. Thus there is an infinite-dimensional quotient of* c_0 *that is Gâteaux differentiable. However there is no infinite-dimensional subspace of* ℓ_1 *that is LUR.*

The origin of Theorem 685 was Lindenstrauss' idea that if (a_i) and (b_i) are in ℓ_1, and $\{a_i/b_i : i \in \mathbb{N}\}$ is dense in $\mathbb{R}$, then the 2-dimensional subspace of ℓ_1 generated by (a_i) and (b_i) is strictly convex, as the function $f(\lambda) = \sum |a_i + \lambda b_i|$ is not affine on any nondegenerate interval ([FHHMZ11, Exercise 7.44]).

Speaking on nonreflexive spaces, c_0, and biorthogonal systems, we cannot miss the following four results:

First,

Theorem 686 (Pták, [Pt59]). *Assume that* X *is a nonreflexive Banach space. Then* X *admits a biorthogonal system* $\{e_n, f_n\}_{n\in\mathbb{N}}$ *such that* $\{f_n\}$ *is bounded and the sequence* $\{(e_1 + e_2 + \ldots + e_n)\}_{n\in\mathbb{N}}$ *is bounded.*

A. J. Guirao et al., *Renormings in Banach Spaces*, Monografie Matematyczne 75,
https://doi.org/10.1007/978-3-031-08655-7_44

A new updated proof of Theorem 686 (using James' theorem) is in [FaGoMo09].

The second result to mention here is the following theorem. First we need a definition:

Definition 687. *A sequence* $\{e_n\}_{n\in\mathbb{N}}$ *in* X *is a* **seminormalized basic sequence** *if* $0 < \inf\{e_n :\ n \in \mathbb{N}\} \leq \sup\{e_n :\ n \in \mathbb{N}\} < \infty$ *and* $\{e_n\}_{n=1}^{\infty}$ *is a Schauder basis for its closed linear span.*

Theorem 688 (Milman, Milman, [MiMi64]). *Let* X *be a nonreflexive Banach space. Then there is a seminormalized basic sequence* $\{e_n\}$ *in* X *and a functional* $u^* \in X^*$ *such that* $u^*(e_n) = 1$ *for each* n.

Again, a new updated proof of Theorem 688, using (another) James' theorem, is in [AcKa11].

The third result in this group is

Theorem 689 (Pełczyński, [Pel62]). *If* X *is a nonreflexive Banach space, then* X *contains a separable nonreflexive space with Schauder basis.*

For a proof see, e.g., [Marti69, Chapter V, Theorem 10].

The fourth result in this group is:

Theorem 690 (Hagler, [Hag87]). *If* X *is a separable Banach space with nonseparable dual, then* X *contains a subspace* Y *that has a Schauder basis and has nonseparable dual.*

Another nice result (Theorem 693 below) is an application of the rotundity properties on c_0.

Recall the following definition:

Definition 691. *The* **Banach-Mazur distance** *between two isomorphic Banach spaces* X *and* Y *is*

$$\inf\{\|T\| \cdot \|T^{-1}\| :\ T \text{ isomorphism from } X \text{ onto } Y\}.$$

Notice the following result of Johnson and Odell:

Theorem 692 (Johnson and Odell, [JohOd05]). *If* X *is a separable infinite-dimensional Banach space, then* $\sup\{d(X_1, X_2) :\ X_1,\ X_2 \text{ both isomorphic to } X\} = \infty$, *where* d *is the Banach–Mazur distance.*

It is open whether Theorem 692 holds for nonseparable spaces as well. In this direction see also [Go10]

Theorem 693 (Bessaga, Pełczyński). *There are two equivalent norms on* c_0, *one is* R *and the other not* R, *that have Banach–Mazur distance* 1.

Proof. (Sketch) The first norm is

$$\|x\|_1 = (\|x\|_\infty^2 + \sum_{i=1}^{\infty} 2^{-i}|x_i|^2)^{\frac{1}{2}},$$

the second one is

$$\|x\|_2 = (\|x\|_\infty^2 + \sum_{i=2}^{\infty} 2^{-i+1}|x_i|^2)^{\frac{1}{2}}.$$

and the mappings $T_n : (X, \|\cdot\|_1) \to (X, \|\cdot\|_2)$ giving the distance 1 are

$$T_n(x) = (x_n, x_1, \dots x_{n-1}, x_{n+1}, \dots).$$

The details are in [PelBe79, Theorem 2.1] or, e.g., in [FHHMZ11, Exercise 5.44]. □

The following problem is open: [Odell] Is it true that for every infinite-dimensional separable Banach space X there exist two spaces Y and Z isomorphic to X such that the Banach–Mazur distance $d(Y, Z) = 1$ and Y is not isometric to Z? (see also, e.g., [GuiMoZi16, Problem 96]).

A further result related to this area is the following interesting result (let us mention here the classical result that every separable Banach space is isometric to a quotient of the space $(\ell_1, \|\cdot\|_1)$, see, e.g., [FHHMZ11, Theorem 5.1]):

Theorem 694 (Klee, [Klee59], see, e.g., [FHHMZ11], Exercise 5.25). *There is a strictly convex norm $\|\cdot\|$ on ℓ_1 such that each separable Banach space is isometric to a quotient of $(\ell_1, \|\cdot\|)$.*

Theorem 695 (Edelstein, Thompson, [EdTh72]). *There is an equivalent ball B of $(c_0, \|\cdot\|_\infty)$ such that no point of its complement has a closest point in B in the sense of $(c_0, \|\cdot\|_\infty)$. This means that there is no nonzero element of c_0^* that would attain its maximum on both B and on the unit ball of c_0.*

Proof. (Sketch) The main, clever idea here is to construct a convex body S in c_0 such that all continuous linear functionals that attain their maximum on S have infinite number of nonzero coordinates. Then use that all continuous linear functionals that attain their maximum on the standard ball of c_0 have only a finite number of nonzero coordinates. Finally, a standard separation argument gives the result. □

In this connection we have the following recent result:

Theorem 696 (Rmoutil, [Rm17]). *The space c_0 can be renormed so that in the new norm the set of norm-attaining functionals contains no 2-dimensional subspace.*

Theorem 696 solved a problem of G. Godefroy.

Theorem 697 (Fonf, [Fon94]). *In each incomplete separable normed space there is a closed bounded convex set V such that no continuous linear non zero functional on X attains its maximum on V.*

Related to this we would like to mention the following result:

Theorem 698 (Godun, Lin, Troyanski, [GLT92]). *Every separable Banach space can be renormed so that the unit ball has only countably many strongly extreme points.*

This should be compared with the following result:

Theorem 699 (Lindenstrauss, Phelps, [LindPh68]). *If X is infinite-dimensional and reflexive, then every bounded closed convex body in X has uncountably many extreme points.*

We note that for the case of strongly exposed points, the result in Theorem 698 was proved already in [LindPh68]. See also Theorem 259 above.

Theorem 700 (Fonf, [Fon96]). *If the space X is infinite-dimensional and reflexive, then the set of all exposed points of B_X is uncountable.*

The following is a very interesting result:

Theorem 701 (Enflo, Lindenstrauss, Pisier, [EnLiPi75]). *There is a Banach space X and a subspace Y of X such that both Y and X/Y are isometric to ℓ_2 but X is not isomorphic to a Hilbert space.*

Theorem 702. (See, e.g., [DeGoZi93, Remark V.1.10]) *There is a Banach space X and a subspace Y of X such that both Y and X/Y admit C^2-smooth norms and X does not admit any C^2-smooth norm.*

Theorem 703 (Zizler, [Ziz89], see, e.g., [DeGoZi93], Theorem II.8.3). *There is a separable Banach space X, a subspace $Y \subset X$ and a Gâteaux differentiable norm $|\cdot|$ on Y such that $|\cdot|$ cannot be extended to a Gâteaux differentiable norm on X.*

It was already mentioned after Theorem 122 that it is an open problem whether if X is a separable reflexive Banach space, Y is a subspace of X, and $\|\cdot\|$ is Fréchet differentiable on Y, this norm can be extended to a Fréchet differentiable norm on X. In this direction, see Theorems 122 and 703 above.

Theorem 704 (Jarosz, [Jar88]). *Every Banach space can be renormed to have only $\pm\mathrm{Id}_X$ as isometries.*

Theorem 705 (Godefroy, Maurey, see, e.g., [Go01b]). *Let X be a Banach space. Then the following are equivalent.*
(i) *X contains an isomorphic copy of ℓ_1.*
(ii) *There is an equivalent norm $\|\cdot\|$ on X and $t \in X^{**} \setminus \{0\}$ such that*

$$\|x + t\| = \|x\| + \|t\|$$

for all $x \in X$.
(iii) *There is an equivalent norm $\|\cdot\|$ on X such that for every finite-dimensional subspace F of X and for every $\varepsilon > 0$ there is $h \in X \setminus \{0\}$ such that for all $x \in F$ and all $\alpha \in \mathbb{R}$ one has*

$$\|x + \alpha h\| \geq (1 - \varepsilon)(\|x\| + \|\alpha h\|).$$

A norm with property (iii) in Theorem 705 is called an **octahedral norm** (see [DeGoZi93, Definition III.2.1]).

Theorem 706 (Godefroy, Kalton, Lancien, Werner, see [Go01b]). *Let X be a separable Banach space. The following are equivalent:*
(i) *X is isomorphic to a subspace of c_0.*
(ii) *There is an equivalent norm $\|\cdot\|$ on X and $k > 0$ such that if $\{f_n\}$ is a sequence in B_{X^*} which is w^*-convergent to f and $\lim \|f_n - f\| = \varepsilon > 0$, then $\|f\| \leq 1 - k\varepsilon$.*

44.1 Fabian–Preiss intermediate differentiation of Lipschitz functions

The one-sided directional derivatives were introduced by equations (4.5) and (4.6). Dini derivatives were used for dealing with continuous nondifferentiable functions (see Section 22.1). The following definition considers, as in the case of Dini derivatives, upper an lower derivatives, now in a given direction:

Definition 707. *Let f be a real valued function on a Banach space X and $x \in X$. The* **upper and lower derivatives of f at x in a direction** $h \in X$ *are defined by*

$$D^+ f(x,h) = \limsup_{t\to 0+} \frac{1}{t}(f(x+th) - f(x))$$

and

$$D_+ f(x,h) = \liminf_{t\to 0+} \frac{1}{t}(f(x+th) - f(x)),$$

respectively.
The function f is said to be **intermediately differentiable at x with intermediate derivative** $\xi \in X^*$ *if*

$$D_+ f(x,h) \le \xi(h) \le D^+ f(x,h)$$

for all $h \in X$.

Theorem 708 (Fabian, Preiss, [FaPr91])**.** *Let f be a Lipschitz real-valued function on an Asplund space X. Then f is intermediately differentiable at every point of a residual subset of X.*

There are several reasons for including this separate section in the text. First, note that the differentiability is on *residual* sets. Second, while the proof of this result is not too complicated, it can be satisfactory for many questions (see, e.g., [BeLi00, page 279]).

44.2 Bates' results, onto mappings, ranges of derivatives

Theorem 709 (Bates, [Bat97], see [FHHMZ11], Theorem 8.32)**.** *Let X and Y be two infinite-dimensional separable Banach spaces. Then there exists a C^1-Fréchet smooth and Lipschitz map from X onto Y.*

Theorem 710 (Bates, [Bat97], see [FHHMZ11], Remark 8.3)**.** *Every separable Banach space is the image, under a surjective C^∞-smooth map, of a Hilbert space.*

Concerning extensions of Bates' Theorems 709 and 710, we refer for instance to [GuiHáMo10].

Open problem. If α is a given cardinal, is it true that every two Banach spaces of density α are C^1-smooth images of each other?

It is proved among other things in [GuiHáMo10] that, under the Continuum Hypothesis, there is a relatively easy proof of this for $\alpha = c$ even for C^∞-smooth maps.

Theorem 711 (Hájek, [Há98b], see [FHHMZ11], Remark 8.2). *If X has an unconditional basis and if there is a C^2-smooth map from c_0 onto X, then X contains an isomorphic copy of c_0.*

Hájek's Theorem 711 above should be compared with Pełczyński's results [HMVZ07, Corollaries 7.7 and 7.8].
The following result should be compared with James' weak compactness theorem. See also the result mentioned after Definition 534.

Theorem 712 (Azagra, Deville, [AzDe01], see [FHHMZ11], Theorem 8.33). *Let X be an infinite-dimensional Banach space with X^* separable. Then there exists a C^1-smooth and Lipschitz bump function on X whose range of the derivative contains the whole B_{X^*}.*

In the following result, ∂U denotes the topological boundary of a given set U.

Theorem 713 ([DeGoZi93], Proposition 1.7). *Let X be a separable space with nonseparable dual and Y be a Banach space with separable dual. Let U be a bounded open subset of X and let $\varphi : \overline{U} \to Y$ be a continuous map which is Fréchet differentiable on U. Then $\varphi(\partial U)$ is norm dense in $\varphi(\overline{U})$.*

Sketch of the proof. Assume that $0 \in U$ and show that $\varphi(0) \in \overline{\varphi(\partial U)}$. To this end, proceed by contradiction: If $\varphi(0) \notin \overline{\varphi(\partial U)}$, let b be a Fréchet differentiable bump function such that $b(\varphi(0)) = 1$ and $b = 0$ on a neighborhood of $\overline{\varphi(\partial U)}$. We let

$$B(x) = \begin{cases} b(\varphi(x)) & \text{if } x \in \overline{U}, \\ 0 & \text{if } x \in X \setminus \overline{U}. \end{cases}$$

The function B is a Fréchet differentiable bump function on X, but there are not such bump functions on X, since X is not Asplund (Theorem 275). □

Theorem 714 ([BeLi00], Corollary 10.45). *Let X be a separable Banach space with nonseparable dual X^*, and let Y be a separable space with separable dual. Then there is no Fréchet differentiable uniform homeomorphism between an open set in X ad an open set in Y.*

Sketch of the proof. If such homeomorphism f between $U \subset X$ and $V \subset Y$ existed, and b is a Fréchet smooth bump supported by V, then $b(f)$ would be a Fréchet differentiable bump function supported by U. □

Chapter 45

Miscellaneous topics

45.1 Phelps' property U

The Hahn–Banach Theorem ensures that, for a subspace M of a Banach space X and for a continuous linear functional $f : M \to \mathbb{R}$, the set $H_M(f)$ (see equation (10.1) above)) of all the norm-preserving extensions of f to X is nonempty (see Subsection 10.2.3). Here we shall consider the problem of uniqueness, i.e., conditions on M (on X) to ensure that $H_M(f)$ is a singleton for every $f \in M^*$ (respectively, for all subspaces $M \subset X$ and all $f \in M^*$)

Definition 715. *Let X be a Banach space and M be a subspace of X. We say that M has* **property** *U whenever each $f \in S_{M^*}$ can be uniquely extended to $\hat{f} \in S_{X^*}$.*

➡ Let us illustrate a simple case where *all* closed subspaces M of a Banach space have property U. Let $X := H$ be a Hilbert space. Let $M \subset H$ be a closed subspace and let $q : H^* \to M^*$ be the canonical quotient mapping (the restriction-to-M mapping). Given $m^* \in S_{M^*}$, the set of *all* extensions to H is the hyperplane $x^* + M^\perp$, where x^* is a Hahn–Banach extension of m^* to X^*. Elementary computations —essentially, the Parallelogram Identity— show that $(x^* + M^\perp) \cap B_{H^*} = \{x^*\}$, what amounts to say that x^* is the (unique) closest-to-0 point in $x^* + M^\perp$ (see Figure 45.1). Despite this elementary construction, it already shows that property U is intimately related to the question of best approximation in the dual space.

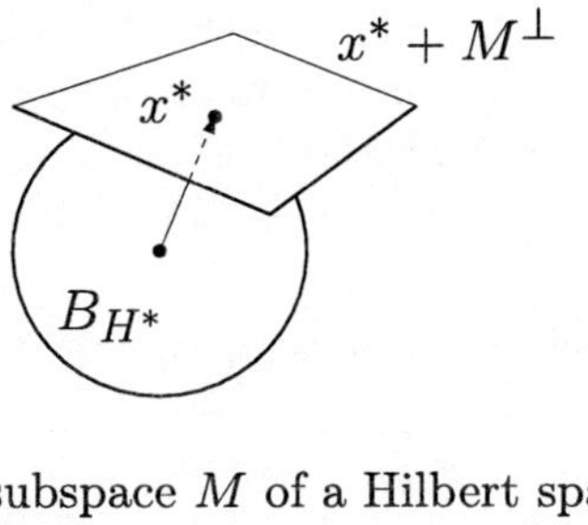

Figure 45.1: Every subspace M of a Hilbert space has property U

The notion of Chebyshev subspace (also called a **Haar subspace**) was introduced at the beginning of Chapter 20.

A. J. Guirao et al., *Renormings in Banach Spaces*, Monografie Matematyczne 75,
https://doi.org/10.1007/978-3-031-08655-7_45

Remark 716. Notice the elementary fact that if M is a closed subspace of a Banach space X and $f \in X^*$, then $\|f\|_M = \operatorname{dist}(f, M^\perp)$, where $\|f\|_M$ denotes the norm of $f|_M \in M^*$. ®

Theorem 717 (Phelps, [Ph60b]). *A subspace M of X has property U if and only if its annihilator $M^\perp$ in X^* is a Chebyshev subspace of X^*.*

Proof. First, suppose that M do not have property U. Then there exist $h \in S_{M^*}$ and $f, g \in S_{X^*}$ such that $f_{|M} = g_{|M} = h$ with $f \neq g$. Then, $0 \neq f - g \in M^\perp$ and by using Remark 716,

$$1 = \|f\| = \|f - (f - g)\| \geq \operatorname{dist}(f, M^\perp) = \|f\|_M = \|h\|_{M^*} = 1$$

So we achieve the expression $\|f\| = \|f - (f - g)\| = \operatorname{dist}(f, M^\perp)$, i.e., the distance between f and $M^\perp$ is attained at the two distinct points 0 and $f - g$, both in $M^\perp$. Thus, $M^\perp$ is not a Chebyshev subspace.
Conversely, suppose that $M^\perp$ is not a Chebyshev subspace of X^*. Then, there exist an element $f \in S_{X^*}$ and $g \in M^\perp \backslash \{0\}$ such that $\operatorname{dist}(f, M^\perp) = 1 = \|f - g\|$. By taking $h := f_{|M} = (f - g)_{|M}$ and considering that $\|h\|_M = \|f\|_M = \operatorname{dist}(f, M^\perp) = 1$ (due to Remark 716), we get that $f - g$ and f are two distinct norm-preserving extensions of h, so M does not have property U. □

We have the following consequence:

Theorem 718 (Foguel, [Fog58], Taylor, [Tay39]). *Let $(X, \|\cdot\|)$ be a Banach space. Then the following are equivalent:*
(i) *The dual norm $\|\cdot\|^*$ is strictly convex.*
(ii) *Every closed subspace of X has property U.*
(iii) *Every closed hyperplane through the origin in X has property U.*

Proof. (iii)⇒(i) follows from Theorem 717: Each line through the origin in X^* space is a Chebyshev subspace. Thus, $\|\cdot\|^*$ is clearly strictly convex.
(i)⇒(ii) follows by using the Alaoglu–Bourbaki theorem (ensuring the existence of points in $M^\perp$ at minimum distance from any other point) and the definition of strict convexity (for getting uniqueness).
(ii)⇒(iii) is trivial. □

Example 719. In [GuiMoZi12] it is proved that if X is a subspace of a weakly compactly generated Banach space, and X is moreover Asplund, then there exists an equivalent norm $|||\cdot|||$ on X that is Fréchet differentiable, LUR, and its dual norm $|||\cdot|||^*$ has property w^*-$\mathcal{T}_{\|\cdot\|^*}$-KK and is not strictly convex. This applies, in particular, to $X := c_0$. Then by Theorem 718 above there is a hyperplane in $(X, |||\cdot|||)$ that does not have property U. ✠

45.2 Pointwise uniformly rotund spaces

Definition 720. *A norm $\|\cdot\|$ on a Banach spaces X is called* **pointwise uniformly rotund (pUR)** *if there is a w^*-dense subspace $F \subset X^*$ such that $\langle x_n - y_n, f\rangle \to 0$ for all $f \in F$ whenever $\|x_n\| = \|y_n\| = 1$ in X are such $\|x_n + y_n\| \to 2$.*

This is a generalization of the notions of WUR and W*UR concepts.

Definition 721. *We say that a measure μ on a compact space K is* **strictly positive** *if $\mu(O) > 0$ for every nonempty open set O in K.*

Theorem 722 (Rosenthal, [Ros70], Rychtar, [Rych05]). *Let K be a compact space. Then the following are equivalent:*
(i) *$C(K)$ admits a* pUR *norm.*
(ii) *(Avilés, Troyanski,* [AvTr21]) *$C(K)$ admits a* URED *norm.*
(iii) *K admits a strictly positive Radon probability measure.*
(iv) *$C(K)^*$ contains a w-compact set that is linearly w^*-dense in $C(K)^*$.*

Theorem 723 (Rychtar, [Rych05]). (i) *Let X be a Banach space that contains $c_0(\Gamma)$, and such that $X/c_0(\Gamma)$ admits a norm the dual of which is pUR. Then X admits a norm whose dual is pUR.*
(ii) *Let K be a compact topological space. If $K^{(\omega_1)} = \emptyset$, then $C(K)$ admits a norm whose dual is pUR.*

Theorem 724 (Rychtar, [Rych05]). (i) *Let μ be a finite measure. Then $L_1(\mu)$ admits an equivalent pUR norm if, and only if, $L_1(\mu)$ is separable.*
(ii) *If a Banach space X admits a pUR norm, then every weakly compact set in X is a uniform Eberlein compact.*

We give now some examples:
(i) Any nonseparable $L_1(\mu)$ space admits a URED norm and does not admit any pUR norm (see Theorem 724 above and use Kutzarova's Theorem 363, see also, e.g., [DeGoZi93, Theorem II.7.16]).
(ii) JL_2 can be renormed by a norm whose dual is pUR (see Theorem 723 above and (i) in Theorem 824 below) but cannot be renormed by a norm whose dual is W*UR, as JL_2 cannot be renormed by a UG norm (indeed, every space with a UG norm is Vašák —see Lemma 752—, and JL_2 is not even WLD, see (vi) in Theorem 824 below).

45.3 Spaces with w^*-sequentially compact dual balls

Definition 725. *A Banach space X is said to have* **property** SK^* *(or that X* **belongs to the class** SK^**) if (B_{X^*}, w^*) is sequentially compact.*

For an account of the class SK^* see, e.g., [Di84, Chapter XIII]).

Remark 726. The class SK^* is quite stable: If $X \in SK^*$, Y is a Banach space, and $T : X \to Y$ is a continuous linear operator with $T(X)$ dense in Y, then $Y \in SK^*$. It is also stable by taking quotients and subspaces. ®

➡ Indeed, note that $T^* : Y^* \to X^*$ is one-to-one, so when restricted to B_{Y^*}, is a w^*-w^*-homeomorphism. Since closed subsets of sequentially compact topological spaces are again sequentially compact, this shows the first statement. The one about quotients is a straightforward consequence of this. Now, if Y is a subspace of X, and (B_{X^*}, w^*) is sequentially compact, so is its image $q(B_{X^*})$ in (Y^*, w^*) by the canonical quotient mapping q. This image is B_{Y^*} by the Hahn–Banach theorem.

Note that all separable spaces belong to SK^*.

➡ Indeed, (B_{X^*}, w^*) is metrizable and compact, hence sequentially compact.

More generally, all WCG spaces belong to SK^*.

➡ The theorem of Šmulyan–Eberlein–Grothendieck (see, e.g., [FHHMZ11, Theorem 3.58] says that if K is a nonempty topological space, then $C_p(K)$ (i.e., $C(K)$ endowed with its pointwise topology) is angelic (see Section 6.2). Now, if X is WCG, let K be a weakly compact subset of X that generates X. Then (B_{X^*}, w^*) can be considered as a topological subspace of $C_p(K)$, since the topology of the pointwise convergence on K coincides, on B_{X^*}, with the w^*-topology. It is simple to prove that on angelic spaces, the classes of compact, countably compact, and sequentially compact subsets coincide (see, e.g., [Fl80, page 31]). This shows the assertion.

By using this observation and Remark 726 above, it follows that every subspace of a WCG space belongs to SK^*. An extension of this result will be mentioned in Corollary 731 below.
On the other hand, the space ℓ_∞ does not belong to SK^*.

➡ Indeed, let a sequence $\{f_n\} \subset B_{\ell_\infty^*}$ be defined by $f_n(x) = x_n$ whenever $x = (x_i) \in \ell_\infty$. This sequence has no w^*-convergent subsequence. To see it, assume that $\{f_{n_k}\}$ is a w^*-convergent subsequence. Define a vector $x = (x_i) \in \ell_\infty$ by $x_{n_k} = (-1)^k$ for $k \in \mathbb{N}$, and $x_i = 0$ if $i \notin \{n_k : k \in \mathbb{N}\}$. Then

$\{f_{n_k}(x)\}$ is not convergent.

As a consequence of this and the previous remarks, if $X \in SK^*$, then the space X cannot contain an isomorphic copy of ℓ_∞. On the other hand, there is a Banach space X that does not contain a copy of ℓ_1 and yet, $X \notin SK^*$. For an example, see [HagOd78].

The next thing in this direction is the following lemma.

Lemma 727 (Čech, Pospíšil, [CechPo38]). *Assume the Continuum Hypothesis. If K is a compact space that is not sequentially compact, then* $\operatorname{card} K \geq 2^{\omega_1}$, *where ω_1 is the first uncountable cardinal.*

From this lemma we get the following result, that will be improved in Corollary 731 below. We still present it because of its simplicity.

Theorem 728 (Hagler and Sullivan, [HagSull80]). *Assume the Continuum Hypothesis. Let X admit a Gâteaux differentiable norm. Then $X \in SK^*$.*

Proof. Assume this is not true for X. Then there is a sequence $\{f_n\}_{n\in\mathbb{N}} \subset B_{X^*}$ that has no w^*-convergent subsequence. Thus, for each subsequence of $\{f_n\}_{n\in\mathbb{N}}$ we get a point in X at which the subsequence is not convergent. There is a number c of such points. Their closed linear span forms a subspace of X of density c whose dual unit ball is not w^*-sequentially compact. Of course, the subspace admits a Gâteaux smooth norm. Thus we can assume that our contradiction space has density c. Use the Bishop–Phelps Theorem 14 above, together with the Gâteaux smoothness to show that X^* has also $\|\cdot\|$-density c. Then we get (we use the fact that the cardinality of a Banach space equals c if the density equals c)

$$c = \operatorname{dens} X^* = \operatorname{card} X^* \geq \operatorname{card} B_{X^*} \geq 2^{\omega_1} = 2^c,$$

where the second inequality comes from Theorem 727. We get a contradiction. □

The next result, a seminal idea of Stegall, is very useful.

Lemma 729 (Stegall, [St78], see, e.g., [Fab97], Lemma 2.1.1). *Let K be a compact space such that each nonempty closed subset $Y \subset K$ has a G_δ-point y with respect to Y, that is, $\{y\} = Y \cap \bigcap_{n=1}^\infty G_n$, where each G_n is an open sets in K. Then K is sequentially compact.*

Proof. Let $\{k_n\}_{n\in\mathbb{N}}$ be a sequence in K. For $m = 1, 2, \dots$ let Y_m be the closure of $\{k_n : n \geq m\}$. Then $Y := \bigcap_{m=1}^\infty Y_m$ is precisely the set of all cluster points of the sequence $\{k_n\}_{n\in\mathbb{N}}$. By hypothesis, there are $y \in Y$ and open sets $G_n \subset K$, $n \in \mathbb{N}$, such that $\{y\} = Y \cap \bigcap_{n=1}^\infty G_n$. We may assume that $\overline{G_{n+1}} \subset G_n$ for each n. As $Y_m \cap G_n \neq \emptyset$ for all $m, n \in \mathbb{N}$, there is a sequence $n_1 < n_2 < n_3 < \dots$ such that $k_{n_i} \in G_i$ for each $i \in \mathbb{N}$. If now z is a cluster point of $\{k_{n_i}\}$, then z is a cluster point of $\{k_n\}$ and so, $z \in Y$. Clearly such a z belongs to $\bigcap_{n=1}^{\infty} \overline{G_n}$ and hence $z = y$. This proves that y is the only cluster point of the sequence $\{k_n\}$. Therefore this sequence converges to y. □

We now give two cases where Stegall's lemma can be applied.

Theorem 730 (Stegall, see, e.g., [Fab97], Theorem 2.1.2). *Let a Banach space X have the property that every continuous convex function on X is Gâteaux differentiable at some points. Then every bounded w^*-closed subset K of X^* has a w^*-G_δ-point.*

Proof. Define $\varphi : X \to \mathbb{R}$ by

$$\varphi(v) = \sup\{y(v) : y \in K\}, \text{ for } v \in X.$$

Then φ is a continuous convex function on X, so it is Gâteaux differentiable at some $v_0 \in X$. Let ξ be its Gâteaux derivative at v_0. For $n \in \mathbb{N}$ consider the sets

$$G_n = \{\eta \in K : \eta(v_0) > \varphi(v_0) - \frac{1}{n}\}$$

(see Figure 45.2). Each G_n is w^*-open in K, and it contains ξ, as $\xi(v_0) = \varphi(v_0)$. Take any $\eta \in \bigcap_{n=1}^{\infty} G_n$. Then for all $h \in X$ we have

$$\begin{aligned}(\eta - \xi)(h) &= \lim_{t \downarrow 0} \frac{1}{t}\Big(\eta(v_0 + th) - \eta(v_0)\Big) - \xi(h) \\ &\leq \lim_{t \downarrow 0} \frac{1}{t}\Big(\varphi(v_0 + th) - \varphi(v_0)\Big) - \xi(h) = 0.\end{aligned}$$

Hence $\eta = \xi$ and $\bigcap_{n=1}^{\infty} G_n = \{\xi\}$, which means that $\xi \in K$ and ξ is a G_δ-point in K. □

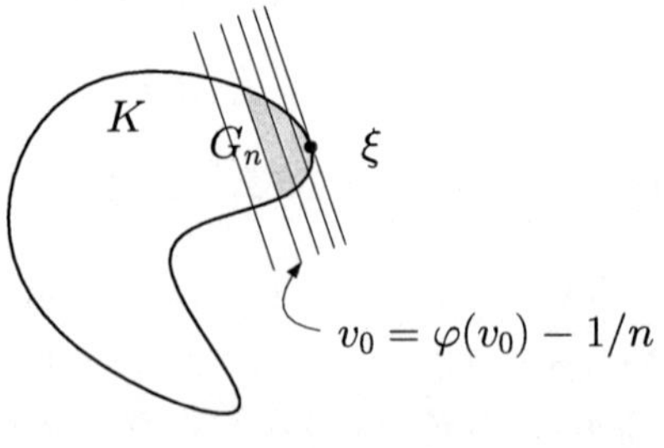

Figure 45.2: The construction in the proof of Theorem 730

Corollary 731. *Let X be a Banach space. Then $X \in SK^*$ if one of the following conditions is satisfied:*
(i) *X is WLD.*
(ii) *X admits a Gâteaux differentiable norm.*

Note that (i) in Corollary 731 improves the previous assertion on subspaces of WCG Banach spaces, and (ii) extends Theorem 728 above.

Proof of Corollary 731. (i) follows from Stegall's lemma above using the result of Shapirovskii (see [FHHMZ11, Theorem 14.41]; see also [ArMe93]). Namely, Shapirovskii proved that any Corson compact space contains a dense set formed by G_δ-points of K.
(ii) If X admits a Gâteaux differentiable norm, then every continuous convex function on X is Gâteaux differentiable at a dense set of points (use the Smooth Variational Principle). □

By using similar methods, Hagler and Sullivan proved in [HagSull80] the following result (compare with (iv) in Theorem 655):

Theorem 732 (Hagler and Sullivan, [HagSull80]). *Assume that a Banach space X has a subspace Y such that both Y and X/Y admit Gâteaux differentiable norms. Then $X \in SK^*$.*

As a consequence, we can provide some other examples of spaces in the class SK^*. For the WLD class, the result was already mentioned in Corollary 731. For the other classes see Theorems 826, 824, and 823, respectively.

Corollary 733. *The spaces JL_0, JL_2, Ciesielski–Pol , and all WLD spaces belong to SK^*.*

For a particular property of spaces in the class SK^* that contain isomorphic copies of c_0 see Theorem 521 above.

We will need the following proposition later on.

Proposition 734. *Let X be a Banach space. Then X is a subspace of ℓ_∞ if and only if for some equivalent norm on X, B_{X^*} is w^*-separable.*

Proof. Assume that X is a subspace of ℓ_∞. Since the dual ball of ℓ_∞ in the sup norm is w^*-separable by Goldstine's theorem, so is the dual ball of the restriction of this norm to X. On the other hand, if the dual ball of X is w^*-separable, and $\{f_n\}$ is a w^*-dense sequence in it, then the map $x \to \{f_i(x)\}$ gets for us the embedding into ℓ_∞. □

45.4 Injections I

Concerning injections, the following result provides the solution to the problem originally posed by S. Troyanski in one of the Czech Winter Schools.

Theorem 735 (Argyros, Mercourakis, [ArMe05]). *There is a WUR space X with no possible injection into any $c_0(\Gamma)$ space.*

Corollary 736. *There is a Vašák space X such that there is no bounded linear one-to-one operator from X^* into any $c_0(\Gamma)$.*

To prove this corollary use the fact that X from Theorem 735 has the property that its dual X^* is UG and thus Vašák (Theorem 753).

Theorem 737. *Any WUR space X admits a norm $\|\cdot\|$ such that $\|\cdot\|$ and its dual norm on X^* are both LUR. In particular, any WUR space admits a norm that is both LUR and Fréchet differentiable.*

See Theorem 746 below. Observe that, as a consequence of Theorem 737 above, Theorem 735 gives another example of space with LUR norm that does not inject into any $c_0(\Gamma)$.

Concerning injections, we would like to mention the following result of independent interest.

Theorem 738 (Argyros, Mercourakis, [ArMe93], see, e.g., [HMVZ07], Proposition 7.44). *Let Γ be uncountable. Then there is no bounded linear one-to-one operator from $\ell_\infty(\Gamma)$ into $\ell^c_\infty(\Delta)$ for any set Δ (where $\ell^c_\infty(\Delta)$ denotes the subspace of $\ell_\infty(\Delta)$ formed by countably supported vectors).*

The following is a pioneering result of Dashiell and Lindenstrauss. We need first some notation: If f is a bounded real valued function on $I := [0,1]$, and $\varepsilon > 0$, put

$$\sigma_\varepsilon(f) = \{t \in I : |f(t)| \geq \varepsilon\}.$$

Then, for a countable ordinal α, let X_α be the closed subspace of $\ell_\infty(I)$ formed by functions f such that the α-th Cantor derivative

$$\sigma_\varepsilon(f)^{(\alpha)} = \emptyset$$

for all $\varepsilon > 0$. Furthermore, put

$$Y = \bigcup_{\alpha<\omega_1} X_\alpha.$$

Then Y is a closed subspace of ℓ^c_∞ and we have

$$c_0(I) = X_1 \subset X_2 \subset \ldots \subset X_\alpha \subset X_{\alpha+1} \subset \ldots \subset Y \subset \ell^c_\infty(I).$$

In this notation we have:

Theorem 739 (Dashiell, Lindenstrauss, [DaLl73]). *Under the notation above,*
(i) *All X_α, $\alpha < \omega_1$, admit strictly convex norms.*
(ii) *Y does not admit any strictly convex norm.*
(iii) *Y is formed exactly by bounded functions with countable support that are Baire-one functions.*
(iv) *There does not exist any bounded linear one-to-one operator from X_2 into any $c_0(\Gamma)$.*
(v) *There does not exist any bounded linear one-to-one operator from ℓ_∞^c into Y.*

The space X_2 was the first example of a strictly convexifiable space with no injection into any $c_0(\Gamma)$ (see also Corollary 748). Note that it is not too difficult to note that X_2 contains an isomorphic copy of ℓ_∞ and thus cannot admit any LUR norm. For such example we go, e.g., to Theorem 823 or take Theorem 738.

In Theorem 739, besides the classical Day's technique of Day's norm used for renormings, in showing the impossibility of the injection Rosenthal's disjointization of measures (see, e.g., [HMVZ07, Theorem 7.1]) was used. In Theorem 823 below, in a similar context, the combinatorial root lemma is used instead (see, e.g., [DeGoZi93, Lemma VI.8.8.4]).

Speaking on injections let us note in passing the following result of independent interest:

Theorem 740 (Avilés, Koszmider, [AvKo13]). *There is an infinite-dimensional space X such that every bounded linear one-to-one operator from X into X is necessarily onto X.*

Another application of the LUR property is the following result. First we need a definition.

A compact space K has **the Namioka property** if for every Baire space E and every separately continuous function $f : E \times K \to \mathbb{R}$ there exists a dense G_δ subset Ω of E such that f is jointly continuous at every point of $\Omega \times K$.

Theorem 741. (See [DeGoZi93, Lemma VII.7.4]) *If K is a compact space such that $C(K)$ admits a pointwise lower semicontinuous LUR norm, then K has the Namioka property.*

The second dual ball of ℓ_1 in the weak star topology does not have the Namioka property [De89].

Proof of Theorem 741. Let f be separately continuous function on $E \times K$. Let the mapping $\varphi : E \to C(K)$ be defined by $\varphi(x)(y) = f(x, y)$ $x \in E, y \in K$. Then φ is continuous into $(C(K), \tau_p)$, where τ_p denotes the pointwise topology in $C(K)$. Let $\|\cdot\|$ be a pointwise lower semicontinuous norm on $C(K)$. Then the real-valued function ψ defined by $\psi(x) = \|\varphi(x)\|$ is lower semicontinuous on E and since E is Baire, we get that there is a dense G_δ set in E where ψ is continuous. Since on the unit sphere of $C(K)$ in the norm $\|\cdot\|$ the pointwise topology and the norm topology coincide by Proposition 88, it is not difficult to show that if x_0 is a point of continuity of ψ, then x_0 is a point of continuity for φ from E into $(C(K), \|\cdot\|)$, which gives that f is jointly continuous at (x_0, y) for every $y \in K$. $\square$

Note that by Kadets' Corollary 179, K has the Namioka property if K is a compact metric space.

Corollary 742. (See, e.g., [DeGoZi93, Theorem VII.7.5].) *If K is a Corson compact or if $K^{(\omega_1)} = \emptyset$, then K has the Namioka property.*

A related important result is the following (see, e.g., [DeGoZi93, Section VII.4, Corollary 4.5]):

Theorem 743 (Haydon and Rogers). *Let K be a compact space such that $K^{(\omega_1)} = \emptyset$. Then $C(K)$ admits a LUR equivalent norm.*

45.5 Vašák spaces II

The following result gives some equivalent formulations for the concept of Vašák (i.e., WCD) Banach space. Let $\Sigma := \mathbb{N}^{\mathbb{N}}$ be endowed with its usual product topology (i.e., the **Baire space**). Note that it is homeomorphic to the space of all irrational numbers in $[0,1]$ endowed with its standard topology.

Theorem 744. *Let X be a Banach space. The following are equivalent:*
(i) *X is Vašák.*
(ii) *There is a sequence $\{K_n\}_{n\in\mathbb{N}}$ of w^*-compact subsets of X^{**} such that for every $x \in X$ there is a subsequence $\{m_n\}_{n\in\mathbb{N}}$ of $\{1,2,\dots\}$ with $x \in \bigcap_{n=1}^\infty K_{m_n} \subset X$.*
(iii) *There is a sequence $\{K_n\}_{n\in\mathbb{N}}$ of absolutely convex w^*-compact subsets of X^{**} with nonempty norm-interior in X^{**} such that for every $x \in X$ there is a subsequence $\{m_n\}_{n\in\mathbb{N}}$ of $\{1,2,\dots\}$ with $x \in \bigcap_{n=1}^\infty K_{m_n} \subset X$.*
(iv) *There is a subset Σ' of Σ and an usco (i.e., upper semicontinuous and compact-valued) map Φ of Σ' into X such that*

$$X = \bigcup\{\Phi(\sigma) : \sigma \in \Sigma'\}.$$

Proof. (i)$\Leftrightarrow$(ii): Reindex the sequence $\{K_n\}_{n\in\mathbb{N}}$ so that every set in question repeats infinitely many times.
(ii)$\Leftrightarrow$(iii): See [FHHMZ11, Exercise 13.29].
(i)$\Leftrightarrow$(iii): See [DeGoZi93, Lemma VI.6.6]. $\square$

Let us now collect a few needed facts on the relationship of WCG and Vašák spaces.

Theorem 745. (i) *A Banach space X is Vašák if and only if there is a subset Σ' of Σ, a set Γ and a bounded linear one-to-one-operator from X^* into the Mercourakis' space $c_1(\Sigma' \times \Gamma)$ that is weak* to pointwise continuous* (Mercourakis), [Me87], see, e.g., [DeGoZi93, Theorem VI.6.7].
(ii) *Every subspace of a WCG space is Vašák* (see [FHHMZ11, Exercise 12.28]).
(iii) *Any Vašák Asplund space is WCG* ([DeGoZi93, Corollary VI.4.4], [HMVZ07, Theorem 6.3]).
(iv) *There is a Vašák space that does not contain ℓ_1 and is not a subspace of any WCG space* (this Argyros' example is presented in [Fab97, Theorem 8.4.6]).
(v) *If X is Vašák, then the weak* density of X^* equals to the norm density of X (i.e., Vašák spaces are DENS)* (see, e.g., [HMVZ07, Theorem 5.40]).
(vi) *Any closed subspace of a Vašák space is again Vašák.*

The following result lists some more facts on Vašák spaces.

Theorem 746. (i) *Any nonseparable Vašák space admits a PRI* (Vašák, [Vas81], [DeGoZi93, Theorem VI.2.5]).
(ii) *If X is Vašák, then X can be renormed by a LUR norm* (Vašák) [Vas81], see, e.g., [DeGoZi93, Theorem VII.2.3].
(iii) *If X is Vašák, then X can be renormed by a norm whose dual is rotund* (Mercourakis, [Me87], see also [DeGoZi93, Theorem VII.1.16]).
(iv) *If X is Vašák Asplund space, then X admits a norm whose dual is LUR* (Fabian, [DeGoZi93, Corollary VII.1.13]).
(v) *If X^* is Vašák, then X admits a norm whose dual is LUR* (Fabian) [DeGoZi93, Theorem VII.2.7].
(vi) *If X^* is Vašák, then X admits a LUR norm* ([DeGoZi93, Theorem VII.2.7]) or [FaTro90]).
(vii) *If X is Vašák, then there is a strong Markushevich basis for X* (see, e.g., [HMVZ07, Corollary 5.42]).
(viii) *If X is Vašák, then X is weak Asplund.*

Note that (viii) follows from (iii) via Theorem 272. Regarding (iii) it should be said that the Prague group tried to solve it quite intensively with no success. Then it was solved by S. Mercourakis in [Me87].

To prove (iii) Mercourakis used a result of his, that translated to the language used here reads:

Theorem 747 (Mercourakis). *If Z is a separable metric space and Λ is a set, then Mercourakis' space $c_1(Z \times \Lambda)$ space admits a norm that is pointwise LUR and pointwise lower semicontinuous.*

For the proof of Theorem 747 we refer to [DeGoZi93, Theorem VII.1.15] or to [MOTV09, Proposition 3.42]. □

Corollary 748. [DeGoZi93, Corollary VI.6.5] *If Δ is any set, then there is no one-to-one bounded linear operator from $c_1(\Sigma \times \Delta)$ into any $c_0(\Gamma)$.*

Note that Corollary 748 gives another example of a Banach space that can be renormed by a strictly convex norm and yet, it does not admit any bounded linear one-to-one operator into any $c_0(\Gamma)$ (see Theorem 739 and the comment after it).

45.6 Uniformly Gâteaux differentiable norms II

Theorem 749 ([FHHMZ11], Proposition 14.18). *Let M be a bounded set in a Banach space X that satisfies that*

$$\sup_{x \in M} |(f_n - g_n)(x)| \to 0$$

whenever $f_n, g_n \in S_{X^*}$ are such that $\|f_n + g_n\| \to 2$. Them M is weakly relatively compact.

Theorem 750 (Fabian, Montesinos, Zizler, [FaMoZi02]). *A bounded set in a Banach space X is weakly relatively compact if and only if for every norming subspace Y of X^* there is a Y-lower semicontinuous norm $\|\cdot\|$ on X such that*

$$\sup_{x \in M} |(f_n - f)(x)| \to 0$$

whenever $f_n, f \in S_{X^*}$ are such that $\|f_n + f\| \to 2$.

The relation of WCG property to the Gâteaux differentiability is so subtle that it was thought in the sixties that any space with Gâteaux differentiable norm must be a subspace of a WCG space. Johnson and Lindenstrauss then found a counterexample: The space JL_2 discussed below.

Actually the former conjecture was "almost right"; only one should assume uniform Gâteaux differentiability instead, see Corollary 755 below.

Theorem 751 (Benyamini, Rudin, Wage, [BRW77]). *Let K be a compact space. Then the following are equivalent.*
(i) *K is a uniform Eberlein compact.*
(ii) *$(B_{C(K)^*}, w^*)$ is uniform Eberlein compact.*
(ii) *$C(K)$ is Hilbert generated.*

For the proof, see, e.g., [FHHMZ11, Theorem 14.15].

An important step was the following Godefroy's result. It was motivated by [MMOT98] and [FaHáZi97].

Lemma 752 (Fabian, Godefroy, Zizler, [FaGoZi01], see, e.g., [FHHMZ11], Lemma 14.17). *Any space with a UG norm is a Vašák space.*

Proof (Sketch) The main thing in the proof of Lemma 752 is the definition of the sets that ensure that X is Vašák: For $\varepsilon > 0$ and $n \in \mathbb{N}$, put

$$B_n^{\varepsilon} = \left\{ x \in B_X : \ |(f-g)(x)| < \varepsilon, \text{ if } f, g \in B_{X^*} \text{ satisfy } \|f+g\| > 2 - \frac{1}{n} \right\}.$$

Then $\{\overline{B_n^{\frac{1}{m}}}^{w^*} : n, m \in \mathbb{N}\}$ ensure that X is Vašák, see [FHHMZ11, Lemma 14.17]. □

Theorem 753 (Fabian, Godefroy, Zizler, see [FHHMZ11], Theorem 14.16). (i) *A Banach space X admits a UG norm if and only if B_{X^*} in its weak* topology is a uniform Eberlein compact.*
(ii) *Let K be a compact space. Then $C(K)$ admits a UG norm if, and only if, K is a uniform Eberlein compact.*

See also [HMVZ07, Chapter 6].

Theorem 754 (Fabian, Godefroy, Zizler, see, e.g., [HMVZ07], Theorem 6.30). *For a Banach pace X the following are equivalent:*
(i) *X admits a UG norm.*
(ii) *There is a Banach space Y such that $X \subset Y$ and there is a bounded linear operator from a Hilbert space onto a dense subset of Y. In other words, if, and only if, X is a subspace of a Hilbert generated space.*

Corollary 755. *If a Banach space X admits a UG norm, then X is a subspace of a WCG space.*

Theorem 756. *There exists a reflexive Banach space X that admits a UG norm and yet X is not Hilbert generated.*

For these results see [FGHZ03].

45.7 Strongly uniformly Gâteaux differentiable norms

Here we report on the paper [FGHZ03], which is very important for getting a good insight into the relationship of weak compact generating and Gâteaux norms.

Definition 757. *Let $(X, \|\cdot\|)$ be a Banach space, and let $M \subset B_X$. We will say that the norm $\|\cdot\|$ is* M**-UG smooth** *if*

$$\sup\{\|x+th\| + \|x-th\| - 2 : \ x \in S_X, \ h \in M\} = o(t) \quad \text{when} \quad t \to 0.$$

The norm $\|\cdot\|$ is called **strongly UG-smooth** *if it is M-UG smooth for some $M \subset B_X$ such that M is linearly dense in X.*

Theorem 758 (Fabian, Godefroy, Hájek, Zizler, [FGHZ03]). (i) *Any UG norm on a separable Banach space is strongly UG.*
(ii) *Any Hilbert generated Banach space admit a strongly UG norm.*

Proof. (i) If $\|\cdot\|$ is a UG norm on a separable space X and $\{x_i : \ i \in \mathbb{N}\}$ is dense in B_X, then $M := \{\frac{1}{2^i}x_i : \ i \in \mathbb{N}\}$ is relatively norm compact. An elementary argument, that uses the Lipschitz property of the norm, gives that $\|\cdot\|$ is automatically strongly UG.
(ii) We refer to [FGHZ03]. □

Theorem 759 (Fabian, Godefroy, Hájek, Zizler, [FGHZ03]). *If a Banach space X admits a strongly UG norm, then X is WCG.*

Example. There is a Banach space R that has an unconditional basis, is not strongly UG renormable but is a subspace of a space that is strongly UG renormable.
Indeed, note that the Rosenthal example of a non-WCG space with unconditional basis is a subspace of an $L_1(\mu)$ space, that is, of course, Hilbert generated. In fact, this space has a stronger property, see Theorem 850 below.

45.8 "Pathologies" in weakly compactly generated spaces

In this section we gather a few "pathological" facts on WCG spaces.

Theorem 760. (i) *There is a nonseparable WCG space $L_1(\mu)$ for a finite measure μ having a subspace with unconditional basis that is not WCG* ([Ros74]).
(ii) *There is a nonseparable WCG space X with unconditional basis having a subspace Y with unconditional basis that is not WCG* ([ArMe04]).
(iii) *There is a nonseparable WCG space X that does not contain ℓ_1 having a subspace Y that is not WCG* ([ArMe04]).

(iv) *There is a nonseparable WCG space X of the form $C(K)$ having a subspace Y that is not WCG* (for this Argyros' example see [Fab97, Theorem 1.63]).
(v) *There is a nonseparable WCG space X having a subspace Y so that there is no bounded linear one-to-one operator from Y^* into any $c_0(\Gamma)$* (see the dual of the space in Theorem 735).

In connection with (i) in Theorem 760 we recommend to read also [MeSt06].

The following well-known problem is still open: If X is a Banach space such that X^{**} is WCG, is necessarily X WCG? See, e.g., [GuiMoZi16, Problem 114]. Observe that there are non-WCG spaces X such that X^* is WCG (see Theorem 824 below).

45.9 Unconditional bases

This section surveys some properties of nonseparable Banach spaces with unconditional bases. (Long) unconditional bases were introduced in Definition 129. We mentioned in Subsection 6.1.3 that standard examples of Banach spaces having such structure include $c_0(\Gamma)$ and $\ell_p(\Gamma)$ spaces with $p \geq 1$. If one looks for nontrivial examples we recommend, for instance, [ArMe93] or further works of the Greek school of Banach spaces. We mentioned in Subsection 6.1.3 that spaces with unconditional bases belong to the family of spaces with strong norming Markushevich bases.

Theorem 761. (See, e.g., [HMVZ07, Theorem 7.39]) *Assume that X is a nonseparable Banach space with unconditional basis $\{e_\gamma, f_\gamma\}_{\gamma\in\Gamma}$. Then the following are equivalent:*
(i) *X admits a Fréchet differentiable norm.*
(ii) *X admits a norm whose dual is LUR.*
(iii) *X is Asplund.*
(iv) *The basis $\{e_\gamma, f_\gamma\}_{\gamma\in\Gamma}$ is shrinking.*
(v) *X does not contain an isomorphic copy of ℓ_1.*

Note in particular that this implies that $C[0, \omega_1]$ does not admit an unconditional basis and that an Asplund space with unconditional basis is automatically WCG (see the observation after Definition 128).

Theorem 762. *Let a nonseparable X admit an unconditional basis. Then the following are equivalent:*
(i) *X admits a norm whose dual norm is R.*
(ii) *X admits a norm whose dual norm is W^*LUR.*

This is proved in [SmTr09].

It is an open problem whether in Theorem 762 it is equivalent also that X admits a Gâteaux differentiable norm.

Theorem 763 (Rychtář). (i) *If a Banach space X has an unconditional basis (not necessarily countable) and X^* admits a (not necessarily dual) URED norm, then X admits a UG norm* [Rych00].
(ii) *If a finite Cantor derived set of a compact K is empty, then $C(K)$ can be renormed by a norm whose dual is URED* [Rych03].

Since the dual norm to a UG norm is W*UR, and thus URED, we have the following corollary of Theorem 763:

Corollary 764. *Let a nonseparable space X admit an unconditional basis. Then X admits a UG norm if, and only if, X^* admits (a not necessarily dual) URED norm.*

For spaces with unconditional bases we have:

Theorem 765 (Johnson, see, e.g., [HMVZ07], Theorem 7.40). *Let X be a Banach space with unconditional basis such that there is a bounded linear one-to-one operator from X^* into $c_0(\Gamma)$ for some Γ. Then X is WCG.*
If this happens, then any unconditional basis in X is σ-weakly compact, i.e. its union with $\{0\}$ is the union of countably many weakly compact sets.

Thus we have

Theorem 766. *Let a nonseparable space X admit an unconditional basis. Then X is WCG if and only if there is a bounded linear injection from X^* into some $c_0(\Gamma)$. If this happens, then any unconditional basis of X is σ-weakly compact.*

An example of a Markushevich basis of a WCG space that is not σ-weakly compact is in [HMVZ07, Remark 6.8].

Theorem 767. (See, e.g., [HMVZ07, Theorem 7.43]) *Let a nonseparable X admit an unconditional basis. Then the following are equivalent:*
(i) *X is WLD.*
(ii) *X does not contain an isomorphic copy of $\ell_1(\omega_1)$.*
(iii) *There is a bounded linear injection from X^* into $\ell^c_\infty(\Delta)$ for some set Δ.*

Since $\ell_1(\omega_1)$ does not admit Gâteaux norm, we have:

Corollary 768. *Let a nonseparable space X admit an unconditional basis. Assume that X admits a Gâteaux norm. Then X is WLD.*

As it was mentioned in (ii), Theorem 760, Argyros and Mercourakis proved in [ArMe04] that there is a WCG space X with unconditional basis and a subspace $Y \subset X$ with unconditional basis such that Y is not WCG.

45.10 Fundamental biorthogonal systems

Fundamental biorthogonal systems were introduced in Subsection 6.1.1 above. They occur in nonseparable Banach spaces much more often than Markushevich bases:

Theorem 769 (Davis, Johnson, [DaJo73b], see also [HMVZ07], Theorem 4.15). *Let X be a nonseparable Banach space that has a WCG quotient of density* $\operatorname{dens} X$. *Then X has a fundamental biorthogonal system.*

From Theorem 769 and Rosenthal's Theorem 809 it follows

Theorem 770. *For any set Γ, $\ell_\infty(\Gamma)$ admits a fundamental biorthogonal system.*

Also, from Theorem 769 it follows (see Theorems 823, 824, and 826):

Theorem 771. *The spaces JL_0, JL_2, and the Ciesielski–Pol space, all have fundamental biorthogonal systems.*

Theorem 772 (Godun and M.I. Kadets, Plichko, see, e.g., [HMVZ07], Theorem 4.26). *The space $\ell_\infty^c(\Gamma)$ has a fundamental biorthogonal system if, and only if,* $\operatorname{card}\Gamma \leq c$.

Note that Theorem 844 says that under CH, there is a nonseparable space $C(K)$ with no uncountable biorthogonal system; however, under the Martin axiom MM, every nonseparable Banach space contains an uncountable biorthogonal system (Todorčević, see, e.g., [HMVZ07, Theorem 4.48]).

For applications of fundamental biorthogonal systems in the geometry of nonseparable spaces we refer, e g., to [HMVZ07, Section 8.2].

Chapter 46

Weakly compactly generated spaces and their relatives III

H Projectional resolutions of the identity were a fundamental tool for elucidating the structure of WCG spaces, and one of the fundamental contributions of Lindenstrauss first and Amir-Lindenstrauss later (see Section 10.2 above and, in particular, Lemma 196).
Then, by the result of Kamil John in Prague, PRI's were constructed in subspaces of WCG, see, e.g., [FHHMPZ01, Chapter 11] and references therein. Later on, an important goal was to get PRI's on Vašák spaces. Vašák ([Vas81]) succeeded in it by constructing PRI's on a Banach space X by working on the second dual with projections that preserve X, and using the Lindenstrauss approach, as explained in, e.g., [FHHMPZ01, Chapter 11].

L. Vašák prepared his thesis in 1975, but to have it published in a quality journal he had to wait till M. Talagrand, from Paris, found an example of a Vašák space that is not a subspace of any WCG space (see Theorem 745 above).

The class of Vašák spaces implicitly uses the topological properties of the space $\mathbb{N}^{\mathbb{N}}$ of Baire and the techniques known in the $\mathcal{K}$-analytic spaces (see, e.g., [DeGoZi93, Lemma VI.6.6]). The PRI's for Vašák spaces are explained also in [DeGoZi93, Chapter 6] and in the exercises in [FHHMZ11, Chapter 13].

The very starting point for this area in Prague was the project to show that WCG spaces are Lindelöf in the weak topology (a problem posed by Corson and Lindenstrauss). This was achieved by D. Preiss in Prague in a "field trip" with his almost dozen of students to South Bohemia and, independently, by M. Talagrand in Paris. Again, they achieved this by showing much more instead, namely, that the space is then weakly $\mathcal{K}$-analytic.

The further project was to study smooth partitions of unity in nonseparable spaces and functions that locally depend on finite number of coordinates, initiated by H. Toruńczyk from Warsaw and K. John from Prague.

The theory of the still prosperous LUR renormings of nonseparable spaces has four most efficient methods:

A. J. Guirao et al., *Renormings in Banach Spaces*, Monografie Matematyczne 75,
https://doi.org/10.1007/978-3-031-08655-7_46

1. It was an ingenious seminal idea of M.I. Kadets to use moduli of continuity and convexity to prove the existence of LUR renorming in separable spaces.
2. The method of Troyanski for spaces relative to WCG spaces that uses a quality injections of spaces into $c_0(\Gamma)$ (PRI's, Markushevich bases, etc). This gave birth to three-space renormings by LUR norms (see (v) in Theorem 655), to Deville's master lemma (Lemma 426), Moltó–Orihuela–Troyanski–Valdivia results (Subsection 10.4.2), etc.
3. The method of Godefroy, that uses instead quality injections of dual spaces into $c_0(\Gamma)$'s and the weak compactness. This gave birth to Godefroy–Fabian–Mercourakis results, etc., including a new Fabian's proof of LUR renormings of WCG spaces (see Subsection 10.4.3 above).
4. The method of projectional generators (see Definition 207 by Orihuela and Valdivia. Related to this, see Lemma 206 and Theorem 208, just to start with).
5. The new methods, that we can call "slicing topological techniques", due essentially to the Spanish, Bulgarian, and British schools.

Our discussion in the two first areas updates the book [DeGoZi93], published a quarter of century ago. Regarding the last item, its state of affairs as it was, say, 10 years ago, is described in [MOTV09]. We do not discuss it in detail in our "user's manual of basic techniques" text. We focus instead on relationship between several notions and practical applications of renormings.

The Vašák spaces turned out to be ideal in questions on smoothness in nonseparable Banach spaces (see the results of Fabian and Mercourakis in [DeGoZi93, Chapter VII]).

For example, around the year 2000 a characterization of spaces admitting UG norms in terms of uniform Eberlein compacta was found (see, e.g., [FHHMZ11, Chapter 14]).

The collaboration of these centers (Paris, Prague, Sofia, Murcia, Valencia, and Oxford) then started and keeps continuing since.

As we mentioned, the class of WLD spaces originated in Greece and Spain in the eighties. It uses properties of Corson compacts and the Mazur's exhaustion method instead of Tychonoff's compactness theorem for constructing the PRI's (see [Vald88] and [ArMe93]; we should mention also Kalenda's results in this direction (see (iv) in Theorem 834)). It is presented in [HMVZ07, Theorems 5.36 and 3.42].

46.1 Asplund spaces II

H Let us summarize here the sequence of events in the direction of the concept and results on Asplund spaces:

1. About 1963, Lindenstrauss and M.I. Kadets proved that separable reflexive spaces are strong differentiability spaces (SDS, for short) (i.e., Asplund spaces) [Lind63], [Kad58], [Kad65].
2. In 1968 Asplund [Asp68] proved that if the dual space of X has a dual LUR norm, then X is SDS.
3. In 1971 Troyanski [Tr71] proved that reflexive Banach spaces are SDS.
4. John and Zizler in 1975 [JoZi75a], proved that if X and X^* are subspaces of WCG spaces, then X is SDS.
5. In 1981, Godefroy et al. [GTWZ83] proved that if X^* is a subspace of a WCG space, then X is SDS.
6. In 1991, Fabian [Fab91] showed that if X^* is WCD, then X is SDS.

Remark 773. It is worth mentioning that, under the existence on an unconditional basis in a space X, Theorem 761 provides a set of conditions on a Banach space each of them equivalent to Asplundness. ®

Theorem 774 (Deville, Fabian, Godefroy). (i) *If X is an Asplund space, then X^* admits a LUR (not necessarily dual) norm* (Fabian, Godefroy [DeGoZi93, Corollary VII.1.12]).
(ii) *If X is an Asplund space, then there is a PRI on X^** (Fabian, Godefroy), [DeGoZi93, Theorem VI.3.4].
(iii) *If X is an Asplund space of density $\aleph_1$, then X^* admits a Markushevich basis $\{f_\alpha, g_\alpha\}$ such that for every $x \in X$, card $\{\alpha : f_\alpha(x) \neq 0\} \leq \aleph_0$* ([DeGo93]).
(iv) *If X is an Asplund space, then there is a one-to-one bounded linear operator from X^* into some $c_0(\Gamma)$* (see, e.g., [Fab97, Theorem 8.2.2]).
(v) *If X is Asplund, then X^* admits a strong M-basis* (see, e.g., [HMVZ07, Corollary 5.2]).

Compare (i) in Theorem 774 with Theorem 226. (ii) was already mentioned in Theorem 320. Notice that (i) follows from (v) and Theorem 222.
Theorem 774 was preceded by the work [JoZi75] (see Theorem 318 above) and [Tac70], see also Corollary 205.

Note that the projections in the PRI in the preceding theorem are not in general weak*-weak*-continuous, see, e.g., the space JL_2 below.

The following class of Asplund spaces —namely spaces that admit norms whose dual norms are LUR— is very wide. It contains the WCG Asplund spaces (Fabian, see Theorem 746 above), spaces whose duals are Vašák (Fabian, see again Theorem 746), spaces $C(K)$ for $K^{(\omega_1)} = \emptyset$ (Deville, Theorem 813 below), and it satisfies the three-space property (Jiménez-Sevilla, Moreno [JiMo97, Corollary 3.2], see (vi) in Theorem 655 above). Due to Talagrand's Theorem 812, the space $C([0,\omega_1])$ is not in this family.

Theorem 775 (Raja, [Raj02], [Raj03]). *Let X be an Asplund space. Then the following are equivalent.*
(i) *X^* admits an equivalent dual LUR norm.*
(ii) *X^* admits a dual norm with the w^*-$\mathcal{T}_{\|\cdot\|}$-KK property.*
(iii) *X^* admits a norm with the w^*-w-KK property.*
(iv) *X^* admits a W*LUR equivalent norm.*
In the case of $X = C(K)$, it is equivalent to say that K is σ-discrete, i.e., a union of countably many sets that are relatively discrete.

Haydon constructed a space X with dual norm rotund and such that X does not admit any Fréchet smooth norm (see, e.g., [DeGoZi93, VII Remarks, Section 6.5]).

Theorem 776 (Wee-Kee Tang, [WKT95]). *Let X be a separable Banach space and φ be a C^1-smooth function on X. Then every continuous convex function f on X such that $f \le \varphi$ is Fréchet differentiable at points of a dense set in X.*

46.2 Scattered compacta and Asplund spaces II

The space $c_0(\Gamma)$ is Asplund. Let us produce several proofs of this fact.
The first follows from Theorems 171 and 272.
Second proof: Observe that the support supp (x) of every $x \in c_0(\Gamma)$ is countable. Let S be a separable subspace of $c_0(\Gamma)$. Let D be a dense and countable subset of S. Then $\Gamma_0 := \bigcup\{\text{supp } x : x \in D\}$ is countable, and this set coincides with $\bigcup\{\text{supp } x : x \in S\}$. Thus, S is a subspace of c_0, hence S^* is separable.
The third proof seems harder, but instructive: Let Γ be an infinite set. Endow Γ with the discrete topology (it becomes a locally compact space) and let $K := \Gamma \cup \{\infty\}$ be its Alexandrov compactification. Since K is scattered, $C(K)$ is Asplund (Theorem 274).

➪ The topology on K is described in the following way: The neighborhoods of elements $\gamma \in \Gamma$ are those subsets of K containing γ, and the neighborhoods of ∞ are the complements in K of finite subsets of Γ. Obviously, we define in this way a Hausdorff topology. The resulting topological space is compact, since given an open covering $\{U_i\}$ of K, at least one of the U_i contains ∞, and leaves outside only a finite number of points in Γ, so we may extract a finite subcovering. The compact space K is scattered, since given F, a closed subset of K, if $\infty \notin F$ then F is finite, and so every point in F is isolated in F. Otherwise, F must contain elements in Γ, and each of them is isolated in

F. This shows that $C(K)$ is Asplund.

The space $C(K)$ is isomorphic to its hyperplanes determined by $\gamma \in \Gamma$.

➩ Two discrete topological spaces with the same cardinality are homeomorphic. As it is well known, if Γ is an infinite set, then Γ and the set $\Gamma \cup \mathbb{N}$ have the same cardinality. In particular, so it happens to the set Γ and the set obtained by adding an extra point to Γ. So, if $\gamma \in \Gamma$, the discrete spaces $\Gamma \setminus \{\gamma\}$ and Γ are homeomorphic, and so they are the corresponding Alexandrov compactifications. This shows that $C(K)$ is isomorphic to any of its hyperplanes $\ker \delta_\gamma$, for $\gamma \in \Gamma$.

Since any two closed hyperplanes of a Banach space are mutually isomorphic, we get that $C(K)$ is isomorphic to $H_\infty := \ker \delta_\infty$. Finally note that H_∞ is isomorphic to $c_0(\Gamma)$.

➩ Indeed, if $f \in H_\infty$ and $\varepsilon > 0$, then the set $\{\gamma \in \Gamma : \ |f(\gamma)| \geq \varepsilon\}$ must be finite. Otherwise, it will have an accumulation point, and the only candidate is ∞. This will force $|f(\infty)| \geq \varepsilon$, a contradiction.

This shows that $c_0(\Gamma)$ and $C(K)$ are mutually isomorphic, and so $c_0(\Gamma)$ is Asplund. Moreover, $c_0(\Gamma)$ is Hilbert generated, so the dual unit ball of $C(K)^*$ is, in its w^*-topology, a uniform Eberlein compact, and so it is then K (see Theorem 751).

➩ Indeed, consider the canonical injection j from $\ell_2(\Gamma)$ into $c_0(\Gamma)$. It has a $\|\cdot\|_\infty$-dense image. We saw that $c_0(\Gamma)$ and $C(K)$ are isomorphic, and so they are $\ell_1(\Gamma)$ and $C(K)^*$ (for a stronger statement, see (iii) in Theorem 778 below). Since j^* is one-to-one and w^*-w-continuous from $\ell_1(\Gamma)$ into $\ell_2(\Gamma)$, we get the conclusion.

Theorem 777. (i) [Milyutin, [Mil66]] *Let K be an uncountable compact metric space. Then $C(K)$ is isomorphic to $C[0,1]$.* (See [Ros03].)
(ii) [Mazurkiewicz, Sierpiński] , see, e.g., [HMVZ07, Theorem 2.56] *Let K be a countable compact. Then K is homeomorphic to an ordinal interval $[0,\alpha]$ for some countable ordinal α. Thus $C(K)$ is isomorphic to $C[0,\alpha]$.* (See, e.g., [HMVZ07, Theorem 2.56].)

The equivalence between (i) and (ii) in the following result was already mentioned in Theorem 274 above.

Theorem 778. (See, e.g., [HáJo14, Ch. 5, Thm. 125]) *Let K be a compact space. Then the following are equivalent.*
(i) *K is scattered.*
(ii) *$C(K)$ is an Asplund space.*
(iii) *$C(K)^*$ is isometric to $\ell_1(\Gamma)$ (in fact we can choose $\Gamma = K$).*
(iv) *$C(K)$ does not contain ℓ_1.*
(v) *$C(K)$ is c_0 saturated, i.e., each infinite-dimensional subspace contains an isomorphic copy of c_0.*

Theorem 779. *Let K be an infinite compact metric space. Then $K^{(\omega_0)} = \emptyset$ if and only if $C(K)$ is isomorphic to c_0.*

Proof. See, e.g., [FHHMZ11, Theorem 8.24]. □

This statement has the following extension.

Theorem 780 (Godefroy, Kalton, Lancien, [GoKaLa00]). *Let K be a compact space such that $C(K)$ is WLD and has density $\aleph_1$. Assume that $K^{(\omega_0)} = \emptyset$. Then $C(K)$ is isomorphic to $c_0(\Gamma)$ with* $\operatorname{card}(\Gamma) = \aleph_1$.

Proof. See [GoKaLa00]. □

We do not know if the assumption on the density in Theorem 780 is needed. On the other hand it happens that the WLD assumption cannot be omitted there. Indeed, look at the Ciesielski–Pol space (see Theorem 823) or the Aharoni–Lindenstrauss space (see Theorem 826).

The following questions are open:

1. Let $C(K)$ be Asplund. Does $C(K)$ admit a C^∞-smooth bump?
2. Let $C(K)$ be Asplund admitting a C^1-smooth norm. Does $C(K)$ admit a C^∞-smooth norm? (It is true if $C(K)$ admits a norm whose dual norm is LUR (see Theorem 530 above).
3. If K is the Kunen compact, does $C(K)$ admit a C^1-smooth bump?

Theorem 781 (Hájek, Haydon, [HájHay07]). *If K is a scattered compact and $C(K)$ admits a norm whose dual is LUR, then $C(K)$ admits a LFC C^∞-smooth norm.*

This happens for instance if $K^{(\omega_1)} = \emptyset$ by Deville's theorem (see (ii) in Theorem 813).

A related important renorming result concerning compact topological spaces K with $K^{(\omega_1)} = \emptyset$ was mentioned in Theorem 743 above.

Theorem 782 (Deville, Godefroy, Zizler, see, e.g., [Ziz03], Theorem 8.7). *For K compact, $C(K)$ is Lipschitz homeomorphic to $c_0(\Gamma)$ if and only if $C(K)$ is uniformly homeomorphic to $c_0(\Gamma)$, if and only if $K^{(\omega_0)} = \emptyset$.*

46.3 σ-Fréchet smooth norms

We discuss [FMZ07].

Definition 783. *Let $M \subset B_X$ and $\varepsilon > 0$ be given. The norm $\|\cdot\|$ of a Banach space X is said to be* **ε-M-differentiable at a point** $0 \neq x \in X$ *if*

$$\lim_{t\downarrow 0} \frac{1}{t} \sup\{\|x+th\| + \|x-th\| - 2\|x\| : \ h \in M\} < \varepsilon.$$

The norm $\|\cdot\|$ is called **ε-M-smooth** *if it is ε-M-differentiable at each $0 \neq x \in X$.*
The norm $\|\cdot\|$ is called **σ-Fréchet differentiable** *if for every $\varepsilon > 0$ there is a decomposition $B_X = \bigcup_{n=1}^{\infty} A_n^{\varepsilon}$ such that the norm $\|\cdot\|$ is ε-A_n^{ε}-differentiable at every $0 \neq x \in X$ for every $n \in \mathbb{N}$.*

Theorem 784 (Fabian, Montesinos, Zizler, [FMZ07]). (i) *If $\|\cdot\|$ is a Gâteaux differentiable norm on a separable Banach space, then $\|\cdot\|$ is σ-Fréchet differentiable.*
(ii) *If $\|\cdot\|$ is a norm on any Banach space that is σ-Fréchet differentiable, then $\|\cdot\|$ is Gâteaux differentiable.*
Thus a norm $\|\cdot\|$ on a separable Banach space is σ-Fréchet differentiable if, and only if, it is Gâteaux differentiable.

Proof. (sketch) (i) Let $\{x_n : \ n \in \mathbb{N}\}$ be a dense subset of B_X. For $n \in \mathbb{N}$ and $\varepsilon > 0$ put

$$A_n^{\varepsilon} = \big(x_n + (\varepsilon/3)B_X\big) \cap B_X$$

Then $\bigcup_{n=1}^{\infty} A_n^{\varepsilon} = B_X$. Moreover, for every $0 \neq x \in X$, $n \in \mathbb{N}$, and $\varepsilon > 0$,

$$\begin{aligned}&\lim_{t\downarrow 0} \frac{1}{t} \sup\{\|x+th\| + \|x-th\| - 2\|x\| : h \in A_n^{\varepsilon}\}\\ &\quad\leq \lim_{t\downarrow 0} \frac{1}{t}\{\|x+tx_n\| + \|x-tx_n\| - 2\|x\|\} + \frac{2\varepsilon}{3} = \frac{2\varepsilon}{3} < \varepsilon.\end{aligned}$$

(ii) Let the norm on X be σ-Fréchet differentiable. Given x and h in S_X, and $m \in \mathbb{N}$, get $A_n^{\frac{1}{m}}$ with $h \in A_n^{\frac{1}{m}}$. Then we have $\lim_{t\downarrow 0}\{\|x+th\| + \|x-th\| - 2\} \leq \frac{1}{m}$. This gives the Gâteaux differentiability. □

Theorem 785 (Fabian, Godefroy, Montesinos, Zizler, [FGMZ04]). *If $\|\cdot\|$ is a uniformly Gâteaux differentiable norm on a Banach space, then $\|\cdot\|$ is σ-Fréchet differentiable.*

Thus the class of σ-Fréchet differentiable norms represent a common roof over both uniformly Gâteaux differentiable norms and Fréchet differentiable norms.

Proof of Theorem 785. (See [FGMZ04, (ii)⇔(iv) in Theorem 6].) *Hint*: For $\varepsilon > 0$ and $n \in \mathbb{N}$ put

$$A_n^{\varepsilon} := \left\{h \in B_X : \ \left\|x + \frac{1}{n}h\right\| + \left\|x - \frac{1}{n}h\right\| - 2 < \frac{\varepsilon}{n} \ \text{ whenever } \ x \in S_X\right\}.$$

The uniform Gâteaux smoothness guarantees that $B_X = \bigcup_{n=1}^{\infty} A_n^{\varepsilon}$.
Also it is clear that the norm is ε-A_n^{ε}-Fréchet smooth for all $x \in S_X$, and thus for all $0 \neq x \in X$ by a homogeneity argument. □

Theorem 786 (Fabian, Montesinos, Zizler, [FMZ07]). (i) *Let X be a subspace of a WCG Banach space. Then X admits a σ-Fréchet differentiable norm.*
(ii) *If a WLD space admits a σ-Fréchet differentiable norm, then X is a subspace of a WCG Banach space.*

Examples. (i) Let $\|\cdot\|$ be a Gâteaux differentiable norm on ℓ_1 whose dual is not strictly convex (see, e.g., Theorem 373 above). Then $\|\cdot\|$ is not UG, and cannot be Fréchet differentiable as ℓ_1 is not Asplund. However, $\|\cdot\|$ is σ-Fréchet differentiable by Theorem 784 above.
(ii) Let X be a Vašák Banach space that is not a subspace of a WCG space (see Theorem 745 above), and let $\|\cdot\|$ be a Gâteaux differentiable norm on X (see (iii) in Theorem 746). Then $\|\cdot\|$ is not σ-Fréchet differentiable by Theorem 786 above, as X is WLD but not a subspace of a WCG space.

46.4 The Daugavet property

Definition 787. *A Banach space X has the* **Daugavet property** *if for every rank-one operator T on X the following equality holds true*

$$\|I + T\| = 1 + \|T\|,$$

where I denotes the identity operator on X. We say that T **satisfies the Daugavet equation** *above.*

The notion of the Daugavet property has provided a new insight into many questions of representation of convex sets, approximation properties and many others.

In this section we just provide a brief information on some results in this area.

In 1963, Daugavet proved the following result:

Theorem 788 (Daugavet, [Dau63]). *If $X = C[0,1]$ and T is a compact operator on X, then $\|I + T\| = 1 + \|T\|$.*

Note that if a compact space K contains an isolated point, then Theorem 788 does not hold (see, e.g., [Wer01]). The reader is advised to check the proof of [FHHMZ11, Exercise 9.27] for the main idea in the proof of Daugavet's result.

In 1966, Lozanovskii proved

Theorem 789 (Lozanovskii, [Loz66]). *If $X = L_1[0,1]$ and T is a compact operator on X, then $\|I + T\| = 1 + \|T\|$.*

In 1986 Holub proved the following result:

Theorem 790 (Holub, [Ho86]). *If $X = C[0,1]$ and T is a weakly compact operator on X, then $\|I + T\| = 1 + \|T\|$.*

In 1996, V.M. Kadets [Kad96] used the Daugavet property to show that neither $C[0,1]$ nor $L_1[0,1]$ have an unconditional Schauder basis. Kadets' proof is a real little gem in Banach space theory.

An excellent overview of the state of affairs in this direction to the year 2001 is Werner's paper [Wer01], including the references there, where the following theorems can be found.

Theorem 791. *The space X has the Daugavet property if, and only if, for each $x \in S_X$ and each $\varepsilon > 0$ $B_X = \overline{\operatorname{conv}}\, \Delta_\varepsilon(x)$, where $\Delta_\varepsilon(x) = \{y \in B_X : \|y - x\| \geq 2 - \varepsilon\}$.*

Theorem 792. *Let X have the Daugavet property. Then*
(i) *Every slice of B_X has diameter 2, in particular, X does not have the RNP property.*
(ii) *X contains an isomorphic copy of ℓ_1.*
(iii) *Every weakly compact operator from X into X satisfies the Daugavet equation.*

Theorem 793 (V.M. Kadets, [Kad21]). *X has the Daugavet property if, and only if, for every convex combination C of slices of B_X, every $x \in C$ and every $\varepsilon > 0$, there is $y \in C$ such that $\|y - x\| > 1 + \|x\| - \varepsilon$.*

Theorem 794 (Abramovich, Aliprantis, and Burkinshaw, [AbAlBu91]). *Let the norm of a space be LUR. Then a compact operator T from X into X satisfies the Daugavet equation if, and only if, $\|T\|$ is an eigenvalue of T.*

Chapter 47

Valdivia compacta

Valdivia compacta were introduced in Section 12.3. We repeat here their definition.

Definition 795. *A compact space K is called a* **Valdivia compact** *if K is homeomorphic to a compact set K_0 in some $[0,1]^\Gamma$ in its pointwise topology such that the set of points in K_0 that have countable support is dense in K_0.*

Here we shall mention only a few facta on the class of Valdivia compact spaces. For more information, the reader may consult, e.g., [Kal00-b].

As we already mentioned, all Corson compacta as well as $[0,1]^\Gamma$ itself (for any index set Γ) are Valdivia compacta, the latter one since finitely supported vectors in $[0,1]^\Gamma$ are dense in $[0,1]^\Gamma$.

Angelic space were introduced in Section 6.2 above. For a compact space K to be angelic it is enough to show that, for every nonempty subset A of K and for every $a \in \overline{A}$, there is sequence $\{a_n\} \subset A$ such that $a_n \to a$.

Proposition 796. (i) *Every Corson compact is angelic.*
(ii) *Every separable Corson compact is metrizable.*

Proof. (i) Let $K \subset [0,1]^\Gamma$ be a compact space such that, for every $f \in K$, its support $\operatorname{supp} f := \{\gamma \in \Gamma : \ f(\gamma) \neq 0\}$ is countable. Let $H \subset K$ and $h \in \overline{H}$. Let $\operatorname{supp} h := \{\gamma_i^0 : \ i \in \mathbb{N}\}$. Find $h_1 \in H$ such that $|(h-h_1)(\gamma_1^0)| < 1$. Let $\operatorname{supp} h_1 = \{\gamma_i^1 : \ i \in \mathbb{N}\}$. Find $h_2 \in H$ such that $|(h-h_2)(\gamma_j^i)| < \frac{1}{2}$, $i = 0,1$, $j = 1,2$, etc. The sequence $\{h_i\}_{i=1}^\infty$ converges to h at every point of $\operatorname{supp} h \cup \bigcup_{i=1}^\infty \operatorname{supp} h_i$. Outside this set, h and all h_j are zero.
(ii) If $K \subset [0,1]^\Gamma$ is a Corson compact space, then every element $k \in K$ has a countable support. Let $\{k_n : \ n \in \mathbb{N}\}$ be dense in K. If Γ_1 denotes the union of the supports of all k_n's, use the fact that $[0,1]^{\Gamma_1}$ is metrizable. □

Proposition 797. (i) *The segment $[0,\omega_1]$ in its interval topology is a Valdivia compact but not angelic, hence not Corson.*
(ii) *The segment $[0,\omega_2]$, where ω_2 is the ordinal of the least cardinality strictly greater than $\aleph_1$, is not Valdivia compact.*
(iii) (Deville, Godefroy [DeGo93]) *A compact space K is Valdivia but not Corson if, and only if, K contains a set homeomorphic to $[0,\omega_1]$.*

Proof. (i) Define the mapping $\varphi : [0,\omega_1] \to [0,1]^{[0,\omega_1]}$ by

$$\varphi(\eta) = \mathbf{1}_{[0,\eta)}, \ \eta \in [0,\omega_1].$$

A. J. Guirao et al., *Renormings in Banach Spaces*, Monografie Matematyczne 75,
https://doi.org/10.1007/978-3-031-08655-7_47

Then φ is continuous and injective, and hence it is a homeomorphism of $[0,\omega_1]$ onto a subset F in $[0,1]^{[0,\omega_1]}$. All but $\varphi(\omega_1)$ have countable support, and they form a dense subset of F. That $[0,\omega_1]$ is not angelic follows from the fact that ω_1 is not the limit of a sequence in $[0,\omega_1)$.
(ii) The proof is presented in [DeGo93].
(iii) The proof is presented in [DeGo93], see also [HMVZ07, Theorem 5.31]. □

Thus $B_{C[0,\omega_1]^*}$ is Valdivia compact in its w^*-topology (indeed, J. Orihuela noted in [Ori92] that $B_{C(K)^*}$ is Valdivia compact if K is). The space $C_0[0,\omega_1]$ is isomorphic to $C[0,\omega_1]$ (look at the mapping T from $C[0,\omega_1]$ into $C_0[0,\omega_1]$ defined by $Tx = (x(\omega_1), x(0) - x(\omega_1), ..., x(t) - x(\omega_1), ..., x(\omega_1) - x(\omega_1)))$. However the dual ball in $C_0[0,\omega_1]^*$ is not Valdivia compact [Kal99]. This shows several things. First, the dual ball of $C[0,\omega_1]^*$ is not Corson in its w*-topology since $[0,\omega_1]$ (a subspace of it) is not angelic. Then, it shows that you can have two equivalent norms on the same space $C_0[0,\omega_1]$ the dual ball of one being Valdivia and the second one not. Next the dual ball of $C_0[0,\omega_1]^*$ is a restriction of the dual ball of $C[0,\omega_1]^*$. This shows that the continuous image of Valdivia may not be Valdivia (the first such example is due to Valdivia, see e.g. [HMVZ07, Proposition 5.30]). From these all it follows also that a closed subspace of a Valdivia compact may not be Valdivia.

We include only a glimpse on Valdivia compacts.

Theorem 798. (i) [Valdivia] *If K is Valdivia compact space, then $C(K)$ admits a PRI and a pointwise lower semicontinuous LUR norm* (see e.g. [DeGoZi93, Theorem VI.7.6]).
(ii) [Kalenda] *A Banach space X is WLD if, and only if, for every equivalent norm $|||\cdot|||$ on X the dual ball of $|||\cdot|||$ is Valdivia in its w^*-topology* (see , e.g., [HMVZ07, Theorem 5.51]).
(iii) *Let the norm of a Banach space be Gâteaux differentiable and assume that its dual unit ball in its w^*-topology is Valdivia compact. Then X is WLD* (see, e.g., [HMVZ07, Exercise 5.4]).

Chapter 48

Checking renormability in some classical spaces

This chapter provides a list of classical spaces with some possible and impossible renormings. Risking some redundancies, it is maybe better to present in many cases different approaches to the same result; they will provide different techniques useful in various circumstances.

48.1 Spaces of bounded functions

For reference, let us start by listing some properties of the space ℓ_∞.

Theorem 799. (i) *ℓ_∞ is not a subspace of any space that admits a LUR norm. In particular, of a WLD space or $C^*[0,1]$.*
(ii) *ℓ_∞ is not a subspace of any space that admits a Gâteaux differentiable norm, in particular of JL_0 or JL_2 or the Ciesielski–Pol space.*
(iii) *ℓ_∞ is a subspace of any nontrivial Mercourakis space or Dashiell space.*
(iv) *ℓ_∞ contains JL_0 but not JL_2.*
(v) *Any space with an unconditional basis of density $\leq c$ is Lipschitz homeomorphic to a subset of ℓ_∞.*
(vi) *Nonseparable WLD spaces do not even inject into ℓ_∞.*
(vii) *$\ell_1(c)$ is a subspace of ℓ_∞.*
(viii) *ℓ_∞ does not have Markushevich basis, but is a complemented subspace of a space with Markushevich basis.*
(ix) *ℓ_∞ has a fundamental biorthogonal system.*

Proof. (i) See Theorem 802 below.
(ii) See Corollary 807 below.
For (v), see [Ka11].
(vi) See Theorem 834 below.
(vii) $\ell_1(c)$ is a subspace of $(C[0,1])^*$.
(viii) See Theorem 827 below.
(ix) See Theorem 770 above. □

Note that it follows from (v) in Theorem 799 that if Γ is uncountable then there is a countable family of Lipschitz functions on $c_0(\Gamma)$ that separates points of $c_0(\Gamma)$,

A. J. Guirao et al., *Renormings in Banach Spaces*, Monografie Matematyczne 75,
https://doi.org/10.1007/978-3-031-08655-7_48

though there does not exist a countable family of linear continuous functionals that separates the points of $c_0(\Gamma)$, as this space is WCG.

Let us recall that, for some index set Γ, $\ell^c_\infty(\Gamma)$ is the subspace of $\ell_\infty(\Gamma)$ consisting of all countable supported vectors. The **support** $\sigma(x)$ of an element $x \in \ell_\infty(\Gamma)$ is the set $\{\gamma \in \Gamma : x(\gamma) \neq 0\}$.

48.1.1 Strict convexity

The reader is advised to complement the information in this Subsection with the one in Chapter 50 below.

H In 1955, M.M. Day published a seminal paper [Day55], where, among many other results, he proved that for an uncountable index set Γ, the space $\ell_\infty(\Gamma)$ cannot be renormed to become strictly convex . The technique of proof was used by Lindenstrauss [Lind72]. Troyanski [Tr72] showed that there is no Kadets–Klee renorming of the space ℓ_∞ (see Corollary 804 below, see also, e.g., [DeGoZi93, Theorem II.7.10]). Later, Hu, Moors and Smith pushed further Day's and Lindenstrauss' construction to show in [HMS97] that ℓ_∞ has no equivalent weakly midpoint locally uniformly rotund norm, a result that was independently proved by G.A. Aleksandrov and I.P. Dimitrov [AlDi85]. As a byproduct, it is proven in [HMS97] Bourgain's result [Bou80] that ℓ_∞/c_0 has no equivalent strictly convex norm.

Theorem 800 (Day, Partington, [Par80], [Day55], see, e.g., [DeGoZi93], Theorem II.7.12). *Let Γ be uncountable. Then $\ell^c_\infty(\Gamma)$ does not admit an equivalent strictly convex norm. As a consequence, $\ell_\infty(\Gamma)$ shares the same property.*

➡ On the other hand,

Theorem 801. *For every Γ, the space $\ell_\infty(\Gamma)$ contains a dense linear subspace that admits a LUR norm.*

Proof.: By Theorem 770, $\ell_\infty(\Gamma)$ admits a fundamental biorthogonal system. Then use Theorem 223. □

Day's method of proof of Theorem 800 is similar to the theory of the so called "distorted norms" (Definition 463), which became a powerful industry in Banach spaces. The question here is how much an equivalent norm must share properties of the initial norm. This area includes also questions on the smooth variational principles, etc. (see, e.g., [FHHMZ11, Theorem 7.43]).
It is worth to reproduce here Day's proof, as it was the seed of several other related constructions. Later on, we shall also present Hu, Moors, and Smith modification.

Proof of Theorem 800 *(Day).*
Let Γ be an uncountable set. Put $X := \ell^c_\infty(\Gamma)$. For any equivalent norm $|||\cdot|||$ on X, we shall search for a "flat" set in some positive multiple of $(S_{|||\cdot|||})$. This flat set will be of the form

$$F_x := \{y \in S_{(X,\|\cdot\|_\infty)} : \ y(\gamma) = x(\gamma) \text{ for } x(\gamma) \neq 0\}, \tag{48.1}$$

for some $x \in S_{(X,\|\cdot\|_\infty)}$. Figure 48.1 below tries to convey the aspect of F_x for several points $x \in S_{(X,\|\cdot\|_\infty)}$.

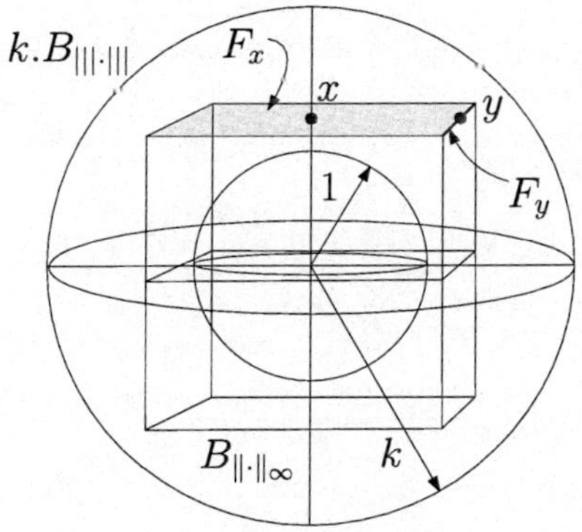

Figure 48.1: Some of the sets F_x. The picture is misleading: If $x(\gamma) \neq 0$ for all $\gamma \in \Gamma$, then $F_x = \{x\}$. However, $\Gamma \setminus \sigma(x)$ is uncountable for every x, so F_x is in fact always huge (it contains an uncountable number of segments of length 2)

Observe that if $y \in F_x$, then $\sigma(x) \subset \sigma(y)$ and $F_y \subset F_x$. Note, too, that $2x - F_x = F_x$.
Let $|||\cdot|||$ be an equivalent norm on X. We may assume, without loss of generality, that $\|x\|_\infty \leq |||x|||$ for all $x \in X$. Put $k := \sup\{|||x||| : \ x \in S_{(X,\|\cdot\|_\infty)}\}$, so $\|x\|_\infty \leq |||x||| \leq k\|x\|_\infty$ for all $x \in X$. In case $k = 1$, we have $|||\cdot||| = \|\cdot\|_\infty$, and certainly $\|\cdot\|_\infty$ is not strictly convex. So we may assume $k > 1$.

➪ The rough idea of the proof is in Figure 48.2 below: By taking $x_1 \in S_{(X,\|\cdot\|_\infty)}$ with $|||x_1|||$ almost the supremum k, the oscillation of $|||\cdot|||$ on the set F_{x_1} (a flat set) is small, so F_{x_1} is "almost" in $kS_{(X,|||\cdot|||)}$. In the next step we choose $x_2 \in F_{x_1}$ with $|||x_2|||$ almost the supremum there, getting that the oscillation of $|||\cdot|||$ on F_{x_2} (again a flat set) is smaller than before, so F_{x_2} is even closer to be in some $|||\cdot|||$-sphere. After a countable number of steps, still we have a substantial (it contains many segments of $\|\cdot\|_\infty$-length 2, since I is uncountable) flat set that lies on some $|||\cdot|||$-sphere, and so $|||\cdot|||$ cannot be strictly convex.

Let us continue with the proof. For $x \in S_{(X,\|\cdot\|_\infty)}$ put (see Figure 48.3)

$$m_x := \inf\{|||y||| : \ y \in F_x\}, \ M_x := \sup\{|||y||| : \ y \in F_x\}. \tag{48.2}$$

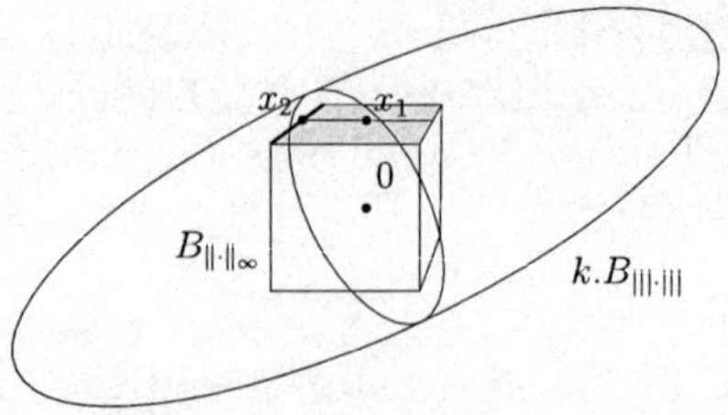

Figure 48.2: Searching for a flat set on some sphere $r.S_{|||\cdot|||}$

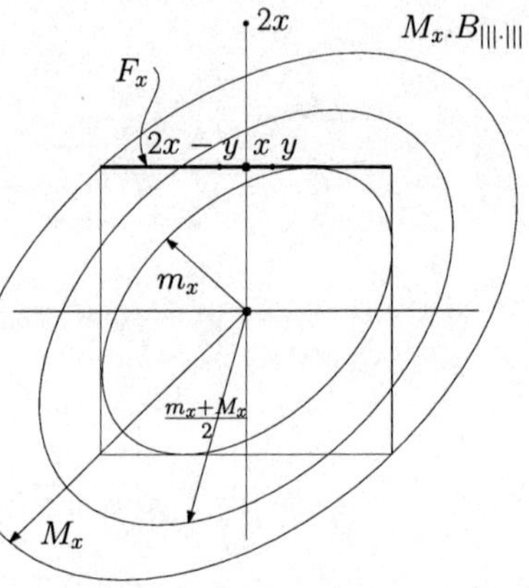

Figure 48.3: Inequality (48.3)

Claim: For every $x \in S_{(X,\|\cdot\|_\infty)}$ (see again Figure 48.3),

$$|||x||| \leq \frac{m_x + M_x}{2}, \text{ i.e., } m_x \geq 2|||x||| - M_x. \tag{48.3}$$

➪ *Proof of the Claim*: For $\varepsilon > 0$ find $y \in F_x$ such that $|||y||| < m_x + \varepsilon$. Since $2x - y \in F_x$, we have $|||2x - y||| \leq M_x$. Then

$$2|||x||| \leq |||y||| + |||2x - y||| \leq m_x + \varepsilon + M_x.$$

Letting $\varepsilon \to 0$ we get (48.3).

We shall select by induction a sequence $\{x_n\}$ in $S_{(X,\|\cdot\|_\infty)}$ with the property that $\lim M_{x_n} = \lim m_{x_n}$.
To choose x_1, note that $(3k+1)/4 < k$. Find then $x_1 \in S_{(X,\|\cdot\|_\infty)}$ so that $|||x_1||| \geq (3k+1)/4$. Thus, by (48.3) above and the fact that $M_{x_1} \leq k$, we get

$$m_{x_1} \geq 2|||x_1||| - M_{x_1} \geq \frac{3k+1}{2} - k = \frac{k+1}{2}. \tag{48.4}$$

We may now estimate the oscillation of $|||\cdot|||$ on F_{x_1}:

$$M_{x_1} - m_{x_1} \leq M_{x_1} - \frac{k+1}{2} \leq k - \frac{k+1}{2} = \frac{k-1}{2}. \tag{48.5}$$

To choose x_2, we repeat the former construction with M_{x_1} and F_{x_1} instead of k and $S_{(X,\|\cdot\|_\infty)}$, respectively: Note that $(3M_{x_1} + |||x_1|||)/4 < M_{x_1}$. Find then $x_2 \in F_{x_1}$ such that $|||x_2||| \geq (3M_{x_1} + |||x_1|||)/4$. Thus, by (48.3) above and the fact that $M_{x_2} \leq M_{x_1}$, we get

$$m_{x_2} \geq 2|||x_2||| - M_{x_2} \geq \frac{3M_{x_1} + |||x_1|||}{2} - M_{x_1} = \frac{M_{x_1} + |||x_1|||}{2}. \tag{48.6}$$

We may now estimate the oscillation of $|||\cdot|||$ on F_{x_2}:

$$\begin{aligned} M_{x_2} - m_{x_2} &\leq M_{x_2} - \frac{M_{x_1} + |||x_1|||}{2} \\ &\leq M_{x_1} - \frac{M_{x_1} + |||x_1|||}{2} = \frac{M_{x_1} - |||x_1|||}{2} \leq \frac{M_{x_1} - m_{x_1}}{2} \leq \frac{k-1}{2^2}. \end{aligned}$$

Continue in this way to get a sequence $\{x_n\}$. The following properties should now be clear: For all $n \in \mathbb{N}$,
(i) $F_{x_{n+1}} \subset F_{x_n}$,
(ii) $x_{n+1} \in F_{x_n}$,
(iii) $\sigma(x_n) \subset \sigma(x_{n+1})$,
(iv) $1 \leq m_{x_n} \leq m_{x_{n+1}} \leq M_{x_{n+1}} \leq M_{x_n} \leq k$, and
(v) $M_{x_n} - m_{x_n} \leq (k-1)/2^n$.
By (v), $\lim_n m_{x_n} = \lim_n M_{x_n}$ $(= r)$. The set $\bigcup_n \sigma(x_n)$ is countable. Put $x(\gamma) := \lim_n x_n(\gamma)$ for $\gamma \in \Gamma$ (this limit is well defined). Then obviously $F_x = \bigcap_n F_{x_n}$. The set F_x contains an uncountable number of segments of length 2. Note that if $y \in F_x$, then $m_{x_n} \leq |||y||| \leq M_{x_n}$ for all $n \in \mathbb{N}$, so $|||y||| = r$. This shows that $F_x \subset rS_{(X,|||\cdot|||)}$, so $|||\cdot|||$ is not strictly convex. □

Based on Day's argument in Theorem 800, Lindenstrauss proved in [Lind72] the following result:

Theorem 802 (Lindenstrauss, [Lind72]). *The space ℓ_∞ does not admit any WLUR equivalent norm.*

Nowadays, by using similar ideas to those in the proof of Theorem 800, it is possible to prove something more, namely that the space ℓ_∞ does not admit any equivalent WMLUR norm. As a byproduct of the construction, it is seen that the space ℓ_∞/c_0 does not admit any equivalent strictly convex norm, a result due to Bourgain [Bou80] (see (ii) in Theorem 803 below).

Our proof of Theorem 803 follows the original one and introduces some little simplifications in the argument. The reader will notice the similarities with the proof of Theorem 800 above (to stress this, we even kept the name of the symbols used when possible).

➪ The proof of Theorem 800 works because the support of any vector in $\ell_\infty^c(\Gamma)$ leaves out a huge number of coordinates (for the final step, in fact one dimension left is enough). In [HMS97], Hu, Moors, and Smith adapted the construction to ℓ_∞ (proving even that there is no WMLUR equivalent norm on it) by building a sequence of vectors in ℓ_∞ whose supports again leave out an infinite number of coordinates, and being sure that in the final step still this happens. This is obtained by keeping out of the supports two sequence of vectors.

Theorem 803. (i) [Aleksandrov, Dimitrov, [AlDi85], [Hu, Moors, Smith, [HMS97]] *The space ℓ_∞ does not admit any WMLUR equivalent norm.*
(ii) [Bourgain, [Bou80]] *The space ℓ_∞/c_0 does not admit any strictly convex norm.*

Proof. (i) Let $|||\cdot|||$ be an equivalent norm on ℓ_∞.
We shall select inductively a sequence $\{F_n\}_{n=0}^\infty$ of subsets of ℓ_∞, points $x_{n+1}\in F_n$ for all $n=0,1,2,\ldots$, and distinct natural numbers i_n, j_n, $n\in\mathbb{N}$. To simplify the notation, put $\sigma_n:=\sigma(x_n)$ for all $n\in\mathbb{N}$. For $A\subset\mathbb{N}$ and $x,y\in\ell_\infty$, we shall say that $x=y$ **on** A whenever $x(n)=y(n)$ for all $n\in A$.
Once F_n would be defined, put

$$m_n:=\inf\{|||y|||:\ y\in F_n\},\ M_n:=\sup\{|||y|||:\ y\in F_n\},\ n=0,1,2,\ldots$$

Note that $m_0\le m_1\le m_2\le\ldots\le M_2\le M_1\le M_0$.
Let $F_0:=\{x\in S_{(\ell_\infty,\|\cdot\|_\infty)}:\ \mathbb{N}\setminus\sigma(x)\text{ is infinite}\}$. Choose $x_1\in F_0$ such that $|||x_1|||\ge(3M_0+m_0)/4$ (note that in case $m_0=M_0$, then any point in F_0 would do the job). Choose $i_1\ne j_1$ in $\mathbb{N}\setminus\sigma_1$. Let

$$F_1:=\{y\in F_0:\ y=x_1\text{ on }\sigma_1,\ y(i_1)=y(j_1)=0\}.$$

Choose $x_2\in F_1$ such that $|||x_2|||\ge(3M_1+m_1)/4$ (the same remark as above applies). Note that $\sigma_1\subset\sigma_2$ and that $i_1,j_1\notin\sigma_2$. Find $i_2\ne j_2\in\mathbb{N}\setminus(\sigma_1\cup\{i_1,j_1\})$. Put

$$F_2:=\{y\in F_0:\ y=x_2\text{ on }\sigma_2,\ y(i_k)=y(j_k)=0,\ k=1,2\}.$$

Continue this way.
The following scheme (Figure 48.4) may help to understand the position of several of the elements to be built in the process.
Some remarks, some of them already mentioned, are in order:
(i) $0<m_0\le m_1\le\ldots\le M_1\le M_0$.
(ii) $F_0\supset F_1\supset F_2\supset\ldots$, and $x_n\in F_{n-1}$ for all $n\in\mathbb{N}$.

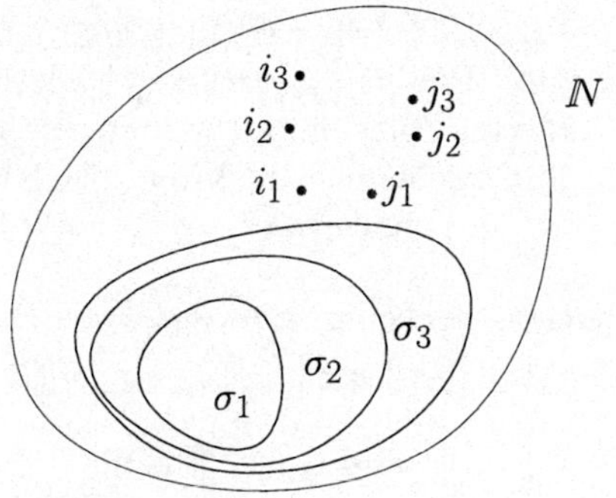

Figure 48.4: Relative positions of some elements in $\mathbb{N}$

(iii) $\sigma_1 \subset \sigma_2 \subset \ldots$.
(iv) For every $n \in \mathbb{N}$, $2x_n - F_n \subset F_n$.
(v) For every $n \in \mathbb{N}$, $(M_n - m_n) \leq (M_{n-1} - m_{n-1})/2$.

➪ From (iv) above it follows that if $y \in F_n$, then $|||2x_n - y||| \leq M_n$. Thus, $|||2x_n||| \leq |||2x_n - y||| + |||y||| \leq M_n + |||y|||$. By letting $|||y|||$ approach m_n we get $|||2x_n||| \leq M_n + m_n$. Hence

$$m_n \geq 2|||x_n||| - M_n \geq \frac{3M_{n-1} + m_{n-1}}{2} - M_n = \frac{M_{n-1} + m_{n-1}}{2}.$$

Thus,

$$\begin{aligned} M_n - m_n &\leq M_n - \frac{M_{n-1} + m_{n-1}}{2} \\ &\leq M_{n-1} - \frac{M_{n-1} + m_{n-1}}{2} = \frac{M_{n-1} - m_{n-1}}{2}. \end{aligned}$$

It follows from (v) above that $\lim_n m_n = \lim_n M_n \; (= r)$.
Put x_∞ for the pointwise limit of the sequence $\{x_n\}$. This is well defined in view of (iii) above. Note that $x_\infty \in \bigcap_{n=0}^{\infty} F_n$, hence $|||x_\infty||| = r$. For $n \in \mathbb{N} \cup \{0\}$, let $h_n = \chi(\{i_k : \; k \geq n\})$, where $\chi(A)$ denotes the characteristic function of a set $A \subset \mathbb{N}$. Obviously, $x_\infty \pm h_{n+1} \in F_n$ for all $n \in \mathbb{N} \cup \{0\}$, hence $|||x_\infty \pm h_{n+1}||| \to r \; (= |||x_\infty|||)$.
However, $\{h_n\}$ does not w-converge to 0.

➪ Indeed, the dual space of ℓ_∞/c_0 is $c_0^\perp$ ($\subset \ell_\infty^*$. Let $q : \ell_\infty \to \ell_\infty/c_0$ be the canonical quotient mapping. Note that in the quotient norm of $\|\cdot\|_\infty$, we have $\|q(h_1)\| = 1$. Thus there exists $f \in S_{c_0^\perp}$ such that $\langle h_1, f\rangle = 1$, and we have obviously $\langle h_n, f\rangle = 1$ for all $n \in \mathbb{N}$.

This shows that $|||\cdot|||$ is not WMLUR. □

(ii) As above, let $q : \ell_\infty \to \ell_\infty/c_0$ be the canonical quotient mapping. Let $|||\cdot|||$ be an equivalent norm on ℓ_∞/c_0. Put $\|x\| := \|x\|_\infty + |||q(x)|||$ for $x \in \ell_\infty$. This is an equivalent norm on ℓ_∞. The construction in the first part of the proof, now for the norm $\|\cdot\|$, gives $x_\infty \in \ell_\infty$ and a sequence $\{h_n\}$ in ℓ_∞ such that $\|x_\infty \pm h_n\| \to \|x_\infty\|$. Note that $\|x_\infty \pm h_n\|_\infty = 1$, and that $q(h_n) = q(h_1) \neq 0$ for all $n \in \mathbb{N}$. Then

$$\begin{aligned}\|x_\infty \pm h_n\| &= \|x_\infty \pm h_n\|_\infty + |||q(x_\infty \pm h_n)||| \\ &= 1 + |||q(x_\infty) \pm q(h_1)||| \to \|x_\infty\| = \|x_\infty\|_\infty + |||q(x_\infty)||| = 1 + |||x_\infty|||.\end{aligned}$$

Thus, $|||q(x_\infty) \pm q(h_1)||| = |||x_\infty|||$, and so $|||\cdot|||$ is not strictly convex. □

Corollary 804 (Lindenstrauss, [Lind72], Troyanski, [Tr72]). *The space ℓ_∞ does not admit any equivalent LUR norm. In fact, it does not admit any equivalent norm having the Kadets–Klee property.*

Proof. The first assertion follows from Theorem 803 and from the fact that every LUR norm is, obviously, WMLUR. For the second, more general statement (every LUR norm has the Kadets–Klee property, see Proposition 88), observe first that ℓ_∞ has an equivalent strictly convex norm. This was mentioned in Paragraph 9.3.1. In case ℓ_∞ admits an equivalent norm with the Kadets–Klee property, use Theorems 407 and 410 to get a contradiction. □

For a different approach to Corollary 804 see, e.g., [DeGoZi93, Theorem II.7.10].

48.1.2 Smoothness

Mazur's Theorem 362(i) says that every separable Banach space is a weak Asplund space. In particular, this holds for the space ℓ_1. However, this is no longer true in a dramatic form —in the sense that there is no equivalent Gâteaux equivalent norm on it— for the space $\ell_1(\Gamma)$, where Γ is an uncountable set (see Theorem 805 below). This property is shared by some other unconditional spaces. This is recorded in the following theorem, a relative of Theorem 800. Let us mention here, too, that the space ℓ_∞ does not admit any Gâteaux differentiable bump (see Theorem 806 and Corollary 807 below).

Theorem 805 (Day, [Day55]). *If Γ is an uncountable set, then the space $\ell_1(\Gamma)$ has no equivalent Gâteaux smooth norm. If Γ is infinite, then neither the space $\ell_\infty^c(\Gamma)$ nor $\ell_\infty^c(\Gamma)/c_0(\Gamma)$ have any equivalent Gâteaux smooth norm (so the same is true for $\ell_\infty(\Gamma)$ and $\ell_\infty(\Gamma)/c_0(\Gamma)$).*

In the contemporary arguments, this can be proved in the following way: In case $\ell_1(\Gamma)$ for an uncountable set Γ should admit an equivalent Gâteaux differentiable norm, Preiss, Phelps, and Namioka Theorem 273 will conclude that it is a weak

Asplund space, and this is not the case since $\|\cdot\|_1$ on $\ell_1(\Gamma)$ is nowhere Gâteaux differentiable (see Examples, Item 6, in Subsection 4.1.4). By the way, it is not necessary to use the full strength of Preiss, Phelps, and Namioka result: The Smooth Variational Principle (Theorem 289) gives, as a consequence, that the presence of an equivalent Gâteaux differentiable norm implies that the space is a Gâteaux differentiability space.
For the space ℓ_∞, use the previous argument together with the fact that the continuous seminorm $x \mapsto \limsup |x_n|$ on ℓ_∞ is nowhere Gâteaux differentiable (see, e.g., [FHHMZ11, Exercise 7.42]).

We shall provide several other approaches to the proof of Theorem 805.

Day's original proof of Theorem 805. The arguments for ℓ_1 and ℓ_∞ share some common ground. To avoid duplicities, we denote by $\|\cdot\|$ the canonical norm ($\|\cdot\|_1$, $\|\cdot\|_\infty$, $\|\cdot\|_\infty$) of the space X (respectively, $\ell_1(\Gamma)$, $\ell_\infty^c(\Gamma)$, $\ell_\infty(\Gamma)$).
Let $|\cdot|$ be a Gâteaux smooth equivalent norm on X. We may always assume that $m\|\cdot\| \le |\cdot| \le \|\cdot\|$ for some $0 < m$ (≤ 1). For $x \in S_{(X,\|\cdot\|)}$, define

$$\boxed{E_x := \{y \in S_{(X,\|\cdot\|)} : \ |y| \le |x| - m\|x-y\|/4\}.} \tag{48.7}$$

Figure 48.5 pretends to illustrate how the set E_x looks like for a particular selection of $\|\cdot\|$ and $|\cdot|$ on $\mathbb{R}^2$.

➪ Figure 48.5 deserves an explanation: We consider $X := \ell_1(\{1,2\})$, $\|\cdot\| := \|\cdot\|_1$, and $|\cdot| := \|\cdot\|_2$. All lengths in the picture are measured in the norm $|\cdot|$. Clearly, $m = |(x+z)/2|$. Thales' theorem was used to compute $m/4$ and $(m/4)\|x-y\|$ for y with $\|y\| = 1$ in the picture on top. The small circle around y in this picture has radius $(m/4)\|x-y\|$. So $y \in E_x$ as long as this circle remains in $B_{|\cdot|}$. This can be seen in the picture at the bottom, where different y's and the corresponding circles are drawn.

Let us list some of the properties of sets E_x for $x \in S_{(X,\|\cdot\|)}$.
(i) $\emptyset \ne E_x \subset S_{(X,\|\cdot\|)}$. Indeed, $x \in S_x$.
(ii) Each E_x is clearly closed.
(iii) If $y \in E_x$, then $E_y \subset E_x$, and $|y| \le |x|$.
(iv) $m \le |y|$ for all $y \in E_x$.
(v) If we select a sequence $\{z_n\}_{n\in\mathbb{N}}$ in X such that $z_1 \in S_{(X,\|\cdot\|)}$ is arbitrary and $z_{n+1} \in E_{z_n}$ for all $n \in \mathbb{N}$, the sequence $\{z_n\}_{n\in\mathbb{N}}$ converges, say to z.

➪ Indeed, by (iii) and (iv) above, the sequence $\{|z_n|\}$ converges. Let $\mu := \lim_n |z_n|$. Fix $\varepsilon > 0$. Find $n \in \mathbb{N}$ such that $|z_n| < \mu + \varepsilon$, and let $k > n$. Then $\mu \le |z_k| \le |z_n| < \mu + \varepsilon$, so $\|z_k - z_n\| < 4\varepsilon/m$.

It follows from (ii) above and the completeness of $(X, \|\cdot\|)$ that z defined in (v) belongs to E_{z_n} for each $n \in \mathbb{N}$.

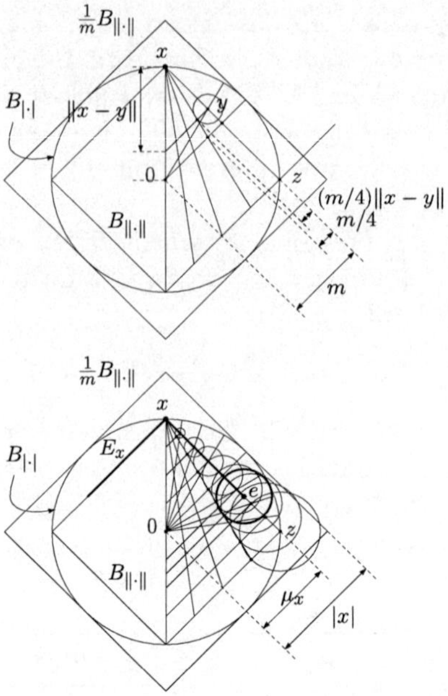

Figure 48.5: The set E_x (thick straight lines at the bottom picture) in the proof of Theorem 805 for $(X, \|\cdot\|) := (\ell_1(\Gamma), \|\cdot\|_1)$ and $|\cdot| := \|\cdot\|_2$. The value μ_x is also represented there.

1. **The case $(\ell_1(\Gamma), \|\cdot\|_1)$, for Γ uncountable**. For $x \in S_{(X,\|\cdot\|)}$, put (see Figure 48.5)

$$\boxed{\mu_x := \inf\{|y| : y \in E_x\}.} \tag{48.8}$$

Claim (see again Figure 48.5):

$$\boxed{\mu_x < |x| \text{ for all } x \in S_{(X,\|\cdot\|)}.} \tag{48.9}$$

To prove the Claim, let $z \in S_{\|\cdot\|}$ be such that supp $z \cap$ supp $x = \emptyset$ (this point depends, obviously, on x). Then

$$\|tx + t'z\|_1 = |t| + |t'| \text{ for } t, t' \in \mathbb{R}. \tag{48.10}$$

Let us consider the 2-dimensional plane Π generated by x and z and the closed smooth curve $S := |x|.S_{(X,|\cdot|)} \cap \Pi$.

*Subclaim: There is an open arc C on the curve S with x as one of its end-points, and entirely lying outside the parallelogram P:=*conv $\{x, -x, 2z, -2z\}$ (see Figure 48.6, where now we consider $\|\cdot\|$ the ℓ_1 norm, as above, and $|\cdot|$ an "arbitrary" smooth norm).

➡ If not, the smooth ball $B_{(X,|\cdot|)} \cap \Pi$ is entirely inside P. This forces the restriction of $|\cdot|$ to Π not to be differentiable at x. Indeed, we can define two continuous linear functionals u_1^* and u_2^* on Π satisfying the following two properties: $\langle x, u_i^* \rangle = |x|$ and $\langle (-1)^i 2z, u_i^* \rangle = |x|$, $i = 1, 2$. This is easy, just by recalling that x and $2z$ have disjoint supports. It is obvious then that $|\langle p, u_i^* \rangle| \leq |x|$ for $p \in P$, in particular for $y \in B_{(X,|\cdot|)}$ (so $|u_i^*| \leq |x|$), $i = 1, 2$. Since $\langle x, u_i^* \rangle = |x|$ we get $|u_i^*| = |x|$, $i = 1, 2$, and we conclude that $u_1^*/|x|$ and $u_2^*/|x|$ are distinct elements in the subdifferential at x of the restriction of $|\cdot|$ to Π, a contradiction to the differentiability of $|\cdot|$. This proves the Subclaim.

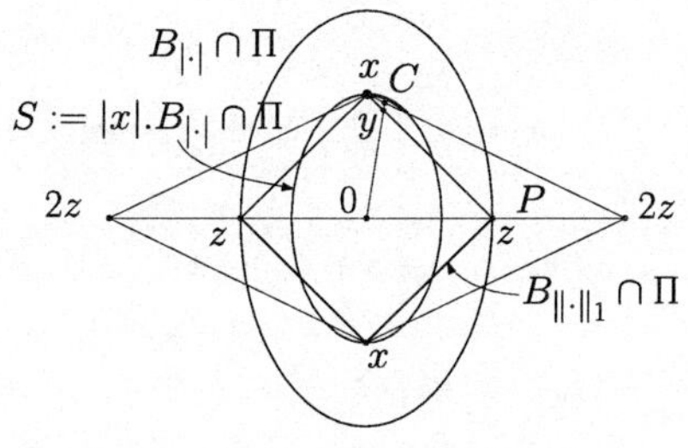

Figure 48.6: The arc C (in bold) on the curve S is out of the parallelogram P, (the Subclaim in the proof of Theorem 805)

Assume that C is "to the right of x" (see Figure 48.6) (otherwise, change z to $-z$ in the following argument). Put $y := tz + (1-t)x$ for some $t \in (0, 1)$. By (48.10), $\|y\| = 1$. Find $\lambda > 1$ such that $\lambda y = s(2z) + (1-s)x$. It is easy to prove that $\lambda = (1 - t/2)^{-1}$ and $s = t(2-t)^{-1}$. According to the Subclaim, we may find $t \in (0, 1)$ small enough such that $|\lambda y| < |x|$ (see the position of y and its multiple λy in Figure 48.6). This shows that $|y| < (1 - t/2)|x|$. So we have

$$|y| < |x|(1 - t/2) \leq |x| - mt/2 = |x| - m\|x - y\|/4 < |x|.$$

Thus, $y \in E_x$, and so $\mu_x < |x|$, as claimed.

➡ Note that if $x_0 := (x + z)/2$ in Figure 48.5 then obviously $E_{x_0} = \{x_0\}$ (since $|y| > |x|$ for all $y \in S_{(X,\|\cdot\|)}$, $y \neq x$), so (48.9) fails. This is not a contradiction, but an illustration of why the set I should be uncountable. Since for $x \in S_{(X,\|\cdot\|)}$ we may always find $z \in S_{(X,\|\cdot\|)}$ with supp $(z) \cap$ supp $(x) = \emptyset$, x appears in the plane $\Pi :=$ span $\{x, z\}$ *as a vertex* of the ball $B_{(X,\|\cdot\|)} \cap \Pi$, and Figure 48.5 is, in the case $\|\cdot\| := \|\cdot\|_1$ and $|\cdot| := \|\cdot\|_2$, quite accurate.

We shall define now a particular sequence $\{x_n\}$ by induction (we are searching for a point —its limit— were $\mu_x = |x|$, placing $|x|B_{|\cdot|}$ "below"

$S_{\|\cdot\|}$ locally at x, something impossible due to the lack of differentiability of $\|\cdot\|$ at x). Choose an arbitrary $x_1 \in S_{(X,\|\cdot\|)}$. By (48.9) we know that $\varepsilon := |x_1| - \mu_{x_1} > 0$. Find $x_2 \in E_{x_1}$ such that $|x_2| < \mu_{x_1} + \varepsilon/4$. Assume that x_n has been already defined. Then choose $x_{n+1} \in E_{x_n}$ such that $|x_{n+1}| < \mu_{x_n} + \varepsilon/4^n$. Note that $\{\mu_{x_n}\}$ is non-decreasing and bounded above, hence it converges. By (v) above, the sequence $\{x_n\}$ converges, too. Let x be its limit. Since $x \in E_{x_n}$ for all $n \in \mathbb{N}$, we have, by (iii) above,

$$\mu_x \geq \lim_n \mu_{x_n} = \lim_n |x_n| = |x|.$$

This inequality contradicts (48.9), and proves the result in case $(X, \|\cdot\|) := (\ell_1(\Gamma), \|\cdot\|_1)$. $\square$

2. **The case $(\ell^c_\infty(\Gamma), \|\cdot\|_\infty)$ and $(\ell_\infty(\Gamma), \|\cdot\|_\infty)$ (and their quotients by $c_0(\Gamma)$) for Γ infinite.**

The proof for $\ell^c_\infty(\Gamma)$ works also for $\ell^c_\infty(\Gamma)/c_0(\Gamma)$, and it follows the previous one with some small changes. The result for the overspaces $\ell_\infty(\Gamma)$ and $\ell_\infty(\Gamma)/c_0(\Gamma)$, respectively, is then a consequence.

Put $X := \ell^c_\infty(\Gamma)$ and $\widehat{X} := X/c_0(\Gamma)$. Let $q : X \to \widehat{X}$ be the canonical quotient mapping. Put $F := q^{-1}(S_{(\widehat{X}\|\cdot\|)}) \cap S_{(X,\|\cdot\|)}$ (we denote by $\|\cdot\|$ also the quotient norm of $\|\cdot\|$). Note that $x \in F$ if, and only if, there exists a sequence $\{\gamma_n\}$ of distinct points in Γ such that $|x(\gamma_n)| \to \|x\| = 1$. For $x \in S_{(X,\|\cdot\|)}$ let $F_x := F \cap E_x$, where E_x was defined in (48.7). Obviously, properties (i) to (v) of E_x are also shared by F_x. For $x \in S_{(X,\|\cdot\|)}$, define μ'_x instead of μ_x (see (48.8)) by

$$\boxed{\mu'_x = \inf\{|y| : y \in F_x\}}. \tag{48.11}$$

The corresponding Claim to be proved is now

$$\boxed{\mu'_x < |x| \text{ for all } x \in F.} \tag{48.12}$$

The proof would be the same as soon as we may reproduce the two-dimensional construction there without leaving F. Precisely, for $x \in F$ we should find $z \in F$ such that $\Pi := \text{span}\,\{x, z\}$ cuts out the sphere $S := \{y : |y| = |x|\}$ a curve in F. For this it is enough to define z as follows: If $|x(\gamma_n)| \to \|x\| = 1$ for an infinite sequence $\{\gamma_n\}$ in Γ, put

$$z(\gamma) := \begin{cases} -x(\gamma) & \text{if } \gamma = \gamma_{2n-1},\ n \in \mathbb{N}, \\ x(\gamma) & \text{otherwise.} \end{cases}$$

Note that for $y := tz + t'x$, $t, t' \in \mathbb{R}$, we have $y(\gamma_{2n-1}) = (-t + t')x(\gamma_{2n-1})$ for $n \in \mathbb{N}$, and $y(\gamma) = (t + t')x(\gamma)$ for other γ's in Γ. This shows that

$\|y\| = |t| + |t'|$, in particular $y \in F$ whenever $|t| + |t'| = 1$. This is what was needed for proving the Subclaim and the Claim above. For building $\{x_n\}$ as before, start with an arbitrary $x_1 \in F$ and proceed in a similar way as we did for $\ell_1(\Gamma)$ to get a contradiction.

Note that the proof is also showing the statement not only for X but also for $\widehat{X}$, since elements in F are mapped by q to elements in $\widehat{X}$ with the same norm. □

An alternative way for proving Theorem 805 for ℓ_∞ is the following result.

Theorem 806 (Fabian, [Fab87-2], see also [DeGoZi93], Proposition II.5.2). *Let φ be a continuous and Gâteaux differentiable bump function on a Banach space X. Then $\text{dens}\, X^* \leq \text{card}\, X$ and if φ is C^1-smooth, then $\text{dens}\, X^* \leq \text{dens}\, X$.*

Corollary 807 (John, Zizler, Fabian, [Jozi78]). *The space ℓ_∞ does not admit any continuous Gâteaux smooth bump function.*

Proof. Use that $\text{card}\,(\ell_\infty) = c < 2^c = \text{card}\,(\beta\mathbb{N}) = \text{dens}\,(\ell_\infty^*)$ (Pospíšil, see [Du66, XI.8 Example 1 (a)]). □

Theorem 808. (i) [Leonard, Whitfield] *The standard norm of ℓ_∞/c_0 is nowhere Gâteaux differentiable. Thus this space admits no Gâteaux differentiable norm* ([LeWhi83].
(ii) [Bourgain] *The space ℓ_∞/c_0 admits no rotund norm* ([Bou80]).
(iii) *The unit ball of ℓ_∞/c_0 is the norm-closed convex hull of its extreme points* ([LeWhi83]).
(iv) [Parovičenko] *The space ℓ_∞/c_0 linearly isometrically contains any Banach space of density $\aleph_1$, the least uncountable cardinal* ([Pa63]).
(v) *Thus, from* (iii), *the space ℓ_∞/c_0 does not have property C.*

Proof. (Sketch) (i) Note that $\ell_\infty/c_0 = C(\beta\mathbb{N} \setminus \mathbb{N})$, where $\beta\mathbb{N}$ stands for the Čech–Stone compactification of $\mathbb{N}$.
(ii) A proof, different from the original one and based on the concept of WMLUR, is in Theorem 803.
Note that this is an example of the sup norm on the classical space $C(\beta\mathbb{N} \setminus \mathbb{N})$ that is nowhere Gâteaux differentiable ([LeWhi83]).
(iii) Parovičenko proved that any compact space of topological weight $\aleph_1$ is a continuous image of the space $\beta\mathbb{N}\setminus\mathbb{N}$ and thus any Banach space of the density $\aleph_1$ is linearly isometric to a subspace of ℓ_∞/c_0 ([Pa63]). The fact that ℓ_∞/c_0 contains an isometric image of any separable Banach space is in [FHHMZ11, Proposition 5.5] and the fact that this space contains an isometric image of $c_0(c)$ is in [FHHMZ11, Exercise 5.33]. □

Let us note in passing the following important result:

Theorem 809 (Rosenthal, [Ros68], see [HMVZ07], Theorem 4.22). *For any set* Γ, *the space* $\ell_2(2^\Gamma)$ *is a quotient of* $\ell_\infty(\Gamma)$.

We will also need to review results on the spaces $c_0(\Gamma)$, $C[0,\mu]$ of continuous functions on the ordinal segment $[0,\mu]$, $\ell_\infty(\Gamma)$, spaces of continuous functions on tree spaces, and the Ciesielski–Pol spaces of continuous functions on ladder system compacts (see Theorem 823).

Theorem 810 (Hájek, Johanis, [HáJo09]). *For any set* Γ, $c_0(\Gamma)$ *admits an equivalent norm that is at the same time* C^∞*-smooth and UG.*

Theorem 811 (Day, James, Swaminathan, (see, e.g., [FHHMZ11], Exercise 9.18)). *If* Γ *is uncountable, then* $c_0(\Gamma)$ *does not admit any URED norm.*

The following proof is very instructive for "negative results" for rotund renormings:

Proof of Theorem 811. Let $\|\cdot\|_\infty$ be the supremum norm of $c_0(\Gamma)$ and assume that $\|\cdot\|$ is an equivalent norm on $c_0(\Gamma)$. Set $M = \sup_{\|x\|_\infty \le 1} \|x\|$. Let $x_n \in c_0(\Gamma)$ satisfy $\|x_n\|_\infty = 1$ and $\|x_n\| \to M$. Let z be such that $\|z\|_\infty = 1$ and its support is disjoint from all the supports of x_n. Then $\|x_n \pm z\|_\infty = 1$ for every n.
Since $\|(x_n + z + x_n - z)/2\| \to M$, we have $\|x_n \pm z\| \to M$ and also $2\|x_n + z\|^2 + 2\|x_n - z\|^2 - \|x_n + z + x_n - z\|^2 \to 0$. Thus $\|\cdot\|$ is not URED. □

Let Γ be an uncountable set. We recall here that $\ell_1(\Gamma)$ admits a dual LUR norm (Theorem 171), and that $\ell_1(\Gamma)$ does not admit any G norm (Theorem 805) .

The following is now a classical important result:

Theorem 812 (Talagrand). *If* μ *is an uncountable ordinal, then* $C[0,\mu]$ *does not admit any norm the dual of which is R* ([Tal86], see, e.g., [DeGoZi93, Theorem VII.5.2]).

Theorem 812 should be compared with Theorems 336 and 509 above.

Theorem 813. (i) [Hájek, [Há03], see, e.g., [HáJo14], Chapter 5, Corollary 130] *If* K *is a compact space such that* $K^{(\omega_1)} = \emptyset$, *then* $C(K)$ *admits a* C^∞*-smooth norm that is LFC.*
(ii) [Deville, [De86]] *Under the same assumptions,* $C(K)$ *has a norm whose dual is LUR.*

48.1.3 Spaces of bounded functions with countable support

We will denote the space of bounded functions with countable support on the ordinal interval $[0,\omega_1]$, endowed with the supremum norm, by $\ell_\infty^c(\omega_1)$. Similarly we denote by $\ell_\infty(\omega_1)$ the space of bounded functions on $[0,\omega_1]$ and $c_0(\omega_1)$ the space $c_0([0,\omega_1])$.

Theorem 814 (Pełczyński and Sudakov, [PelSu62]). *The space $\ell_\infty^c(\omega_1)$ is not complemented in $\ell_\infty(\omega_1)$.*

Theorem 815 (Johnson and Kania, [JoKa19]). *Let X be a WLD subspace of $\ell_\infty^c(\omega_1)$. Then the following are equivalent:*
(i) *X is separable.*
(ii) *X embeds in ℓ_∞.*
(iii) *There is a bounded linear injection from X into ℓ_∞.*
(iv) *X does not contain a subspace isometrically isomorphic to $c_0(\omega_1)$.*
In particular, every reflexive subspace of $\ell_\infty^c(\omega_1)$ is separable.

Theorem 816 (Johnson and Kania, [JoKa19]). *There is a subspace Z of $\ell_\infty^c(\omega_1)$ with unconditional basis such that Z does not embed into ℓ_∞, $c_0(\omega_1)$ does not embed into Z and Z admits a bounded linear injection into ℓ_∞.*

The first example of a space not embeddable but injectable into ℓ_∞ was JL_2. However, this space has no Markushevich basis.

48.2 "Haydon's forest"

Topological spaces called "tree spaces" are defined, for example, in [DeGoZi93, VI.9]. They are locally compact spaces T, and the Alexandroff one-point compactifications of T, $\hat{T} = T \cup \{\infty\}$ are scattered compacts. We denote by $C_0(T)$ the space of continuous functions f on $\hat{T}$ such that $f(\infty) = 0$.

Note that all such $C_0(T)$ spaces are Asplund spaces. Thus the following is a striking result. The ingenious Haydon's key argument in this result, based on the idea that there does not exist "too flat" Gâteaux differentiable norm on $C(0, \omega_1)$, is described in [DeGoZi93, Chapter 7].

Theorem 817 (Haydon, [Hay90], see also [DeGoZi93], Theorem VII.6.4). *There is a tree T such that $C_0(T)$ admits no Gâteaux differentiable norm.*

Here are some other results from the invaluable paper of R. Haydon [Hay99] on "Haydon's forest" (see [DeGoZi93, Chapter 7, Section 6]).

Theorem 818 (Haydon, [Hay99]). *For every tree T, $C_0(T)$ admits a C^∞-smooth function with bounded support.*

Theorem 819 (Haydon, [Hay99]). *The following properties for a tree T are equivalent:*
(i) *$C_0(T)$ admits a C^∞-smooth norm.*
(ii) *$C_0(T)$ admits an F norm.*
(iii) *$C_0(T)$ admits a LUR norm.*
(iv) *$C_0(T)$ admits a $WLUR$ norm.*

It is an open problem if (ii) and (iii) in Theorem 819 are equivalent for every Asplund space.

Theorem 820 (Haydon, [Hay99]). *For a tree T, the following properties are equivalent:*
(i) *$C_0(T)$ admits an R norm.*
(ii) *$C_0(T)$ admits an $MLUR$ norm.*
(iii) *There is a bounded linear one-to-one operator from $C_0(T)$ into some $c_0(\Gamma)$.*

Theorem 821 (Haydon, [Hay99]). (i) *There is a tree T such that $C_0(T)$ admits a w-$\mathcal{T}_{\|\cdot\|}$-KK renorming but no R renorming.*
(ii) *There is a tree T such that $C_0(T)$ admits an equivalent R norm and admits no w-$\mathcal{T}_{\|\cdot\|}$-KK renorming.*
(iii) *There is a tree T such that $C_0(T)$ admits an equivalent G norm and admits no R renorming.*
(iv) *There is a tree T such that $C_0(T)$ has no w-$\mathcal{T}_{\|\cdot\|}$-KK equivalent norm, although it has an equivalent $MLUR$ norm.*
(v) *There is a Banach space X with a subspace Y such that Y admits a LUR norm, X/Y admits an R norm and X does not admit any R norm* (see, e.g., [DeGoZi93, Theorem VII.6.10]).
(vi) *There is a tree T such that $C_0(T)$ has no Fréchet differentiable equivalent norm although it is a quotient of a space with a C^∞-norm.*

Corollary 822 (Haydon, [Hay99]). *There is a tree T such that $C_0(T)$ admits a Gâteaux differentiable norm but does not admit any Fréchet differentiable equivalent norm.*

The research on trees discussed here had been partly motivated by the seminal question of M.I. Kadets: Assume that a Banach space X does not contain an isomorphic copy of ℓ_∞. Must X necessarily admit a LUR norm?, and by a suggestion of J. Lindenstrauss to Zizler to contact Haydon in this direction. The first (negative) answer to Kadets' question was published in [HayZiz89]. In this direction note that the following is open: Assume that for separable space X, the space X^* cannot be renormed by a LUR norm. Does X necessarily contain a copy of ℓ_1? (See, e.g., [GuiMoZi16, Problem 152].)

Since for T being a tree, $C_0(T)$ is Asplund (see, e.g., [FHHMZ11, Theorem 14.25]) and thus cannot contain a copy of ℓ_∞ (since ℓ_∞ contains, e.g., ℓ_1), the results above also provide a negative answer to Kadets' question.

48.3 Injections II

The Ciesielski–Pol ladder system compact space is defined in [CiePo84], see Theorem 823 and, e.g., [DeGoZi93, Definition VI.8.8.2] and [GPWZ88]).

Theorem 823 (Ciesielski, Pol). *Let K be the ladder system compact defined in* [CiePo84], [DeGoZi93, Definition VI.8.8.2], *and* [Ziz03]. *Then*
(i) *The third Cantor derived set of K is empty.*
(ii) *There is a subspace Y of $C(K)$ that is isomorphic to $c_0(\Gamma_1)$ for some uncountable Γ_1 such that the quotient space $C(K)/Y$ is isomorphic to $c_0(\Gamma_2)$ for some uncountable Γ_2.*
(iii) *There is no bounded linear one-to-one map from $C(K)$ into any $c_0(\Gamma)$.*
(iv) ((Hájek), Theorem 813) *$C(K)$ admits a C^∞-smooth norm.*
(v) ((Deville), [DeGoZi93, Theorem VII.4.7]) *$C(K)^*$ admits a dual LUR norm*
(vi) *$C(K)$ can be renormed by a norm whose dual is URED* (Theorem 763).
(vii) *The dual ball of $C(K)$ is w^*-sequentially compact.*
(viii) *$C(K)$ admits a LUR norm (although $C(K)$ does not admit a WUR norm,* see (ii) above and Theorem 231).
(ix) *$C(K)$ is Lipschitz homeomorphic to $c_0(\Gamma)$* ([DeGoZi90]).

The first example of a space with LUR norm and no injection into $c_0(\Gamma)$ was in [De86].

48.4 Johnson–Lindenstrauss spaces

The space JL_2 below was the first example which admits a Gâteaux differentiable norm and is not a subspace of a WCG space.

Theorem 824 (Johnson, Lindenstrauss, [JoLi74], see, e.g., [Ziz03]). *There is a nonseparable Banach space JL_2 with the following properties:*
(i) c_0 is linearly isometric to a subspace of JL_2 and JL_2/c_0 is linearly isometric to $\ell_2(\Gamma)$, where Γ has the cardinal of the continuum. In particular, JL_2 is not separable.
(ii) JL_2^* is weak-separable. JL_2 is not WCG (it is not even WLD, see* (vi) *below).*
(iii) JL_2^* is WCG and admits a norming Markushevich basis, but JL_2^* has an equivalent norm in which it has no 1-norming Markushevich basis.
(iv) JL_2 admits a norm that is both LUR and whose dual norm is LUR.
(v) JL_2 admits a C^∞-smooth norm.
(vi) JL_2 does not admit any Markushevich basis, and it is not WLD.
(vii) JL_2 is not weakly Lindelöf.
(viii) JL_2 has C^∞-smooth partitions of unity ([GPWZ88], see [DeGoZi90]).
(ix) The dual ball is w^*-sequentially compact.
(x) For no equivalent norm on JL_2, its dual ball is w^*-separable. Thus JL_2 is not a subspace of ℓ_∞.
(xi) [Ka11] *JL_2 is Lipschitz homeomorphic to $c_0 \oplus \ell_2(\Gamma)$.*

Proof. (Sketch)
We start by a description of JL_2: It is well known that there exists a collection $\{N_\gamma\}_{\gamma\in\Gamma}$ of infinite subsets of the integers so that $N_{\gamma_1}\cap N_{\gamma_2}$ is finite for $\gamma_1 \neq \gamma_2$, and Γ has the cardinality of the continuum. For each $\gamma \in \Gamma$ let ϕ_γ be the characteristic function of N_γ, an element in ℓ_∞. Let

$$U_0 := \operatorname{span}\{c_0 \cup \{\phi_\gamma :\ \gamma \in \Gamma\}\}\ \ (\subset \ell_\infty). \tag{48.13}$$

Observe that for $x \in U_0$, say $x = y + \sum_{i=1}^k a_{\gamma_i}\phi_{\gamma_i}$, the finite sequence $\{a_{\gamma_i}\}_{i=1}^k$ is uniquely determined, since $a_{\gamma_i} = \lim_{n\to\infty,\ n\in N_{\gamma_i}} x(n)$ (so $y \in c_0$ is also uniquely defined). We define a norm on U_0 by

$$|||x||| = \max\{\|x\|_\infty, \|(a_{\gamma_i})\|_2\},\quad \text{where } x = y + \sum_{i=1}^k a_{\gamma_i}\phi_{\gamma_i} \in U_0, \tag{48.14}$$

$\|\cdot\|_\infty$ is the usual sup norm in ℓ_∞, $\|\cdot\|_2$ is the Euclidean norm in ℓ_2^k, $y \in c_0$, and $\gamma_i \neq \gamma_j$ if $i \neq j$. It is easily seen that $|||\cdot|||$ is indeed a norm on U_0. On c_0, $|||\cdot|||$ obviously coincides with $\|\cdot\|_\infty$.
Let

$$U := \{x \in \ell_\infty :\ \forall\gamma \in \Gamma,\ \exists \lim_{n\to\infty,\ n\in N_\gamma} x(n)\ (=: a_\gamma);\ (a_\gamma)_{\gamma\in\Gamma} \in \ell_2(\Gamma)\}. \tag{48.15}$$

For $x \in U$ put $T(x) := (a_\gamma)_{\gamma\in\Gamma}$, where a_γ is defined as in (48.15) for $\gamma \in \Gamma$, and define

$$|||x||| := \max\{\|x\|_\infty, \|T(x)\|_2\}, \tag{48.16}$$

where $\|\cdot\|_2$ is the canonical norm of $\ell_2(\Gamma)$.
We claim that $(U, |||\cdot|||)$ is a Banach space.

➪ In order to show this, first note that $\|\cdot\|_\infty \le |||\cdot|||$ on U, that U is a linear subspace of ℓ_∞, that U_0 is a subspace of U, and that obviously $|||\cdot|||$ defined by (48.16) on U induces $|||\cdot|||$ defined by (48.14) on U_0. Let $\{x_m\}_{m=1}^\infty$ be a Cauchy sequence in $(U, |||\cdot|||)$. Then it is $\|\cdot\|_\infty$-Cauchy, hence it $\|\cdot\|_\infty$-converges to some $x \in \ell_\infty$ due to the fact that $(\ell_\infty, \|\cdot\|_\infty)$ is complete. Fix $\gamma \in \Gamma$. Note that, for all $m \in \mathbb{N}$, $x_m|_{N_\gamma} \in c$, where $(c, \|\cdot\|_\infty)$ is the Banach space of all convergent sequences equipped with the supremum norm. The sequence $\{x_m|_{N_\gamma}\}_{m=1}^\infty$ is Cauchy in $(c, \|\cdot\|_\infty)$, so it has a $\|\cdot\|_\infty$-limit in c, clearly $x|_{N_\gamma}$, so $x|_{N_\gamma} \in c$. The mapping $L : c \to \mathbb{R}$ that to $y \in c$ associates $\lim_{n\to\infty} y(n)$ is linear and $\|\cdot\|_\infty$-continuous. This implies that $L(x_m|_{N_\gamma}) \to L(x|_{N_\gamma})$. This happens for every $\gamma \in \Gamma$, hence $T(x_m) \to T(x)$ coordinatewise. Note that $T(x_m) \in \ell_2(\Gamma)$ and that $\|T(x_m)\|_2 \le |||x_m|||$ for all m. The sequence $\{|||x_m|||\}_{m=1}^\infty$ is bounded, due to the fact that $\{x_m\}_{m=1}^\infty$ is $|||\cdot|||$-Cauchy. This shows that the sequence $\{T(x_m)\}$ is $\|\cdot\|_2$-bounded in $\ell_2(\Gamma)$, hence it has a w-cluster point in $\ell_2(\Gamma)$. This point must be then $T(x)$. This shows that $T(x) \in \ell_2(\Gamma)$ (in particular, that $x \in U$) and that the sequence $\{T(x_m)\}_{m=1}^\infty$ is w-convergent to Tx in $\ell_2(\Gamma)$. Note that $\|T(u)\|_2 \le |||u|||$ for all $u \in U$. Thus $T : U \to \ell_2(\Gamma)$ is $|||\cdot|||$-$\|\cdot\|_2$-continuous, so $\{T(x_m)\}_{m=1}^\infty$ is $\|\cdot\|_2$-Cauchy in $\ell_2(\Gamma)$. Since it is w-convergent to $T(x)$, we get that $T(x_m) \to T(x)$ in $\|\cdot\|_2$. All together, $x_m \to x$ in $(U, |||\cdot|||)$, and so this last space is complete.

The $|||\cdot|||$-closure of U_0 in U, endowed with the norm induced by $|||\cdot|||$, is the completion of $(U_0, |||\cdot|||)$, and is denoted by JL_2. Note that JL_2, as a vector space, is a linear nonclosed subspace of ℓ_∞. We shall prove below the stated properties of JL_2:

(i) That c_0 is linearly isometric to a subspace of JL_2 follows from the construction above. In order to prove that JL_2/c_0 is linearly isometric to $\ell_2(\Gamma)$ we prove first that $|||q(u_0)||| = \|T(u_0)\|_2$ for all $u_0 \in U_0$, where q is the canonical quotient mapping from JL_2 onto JL_2/c_0.

➪ Given $u_0 := y_0 + \sum_{i=1}^k a_{\gamma_i}\phi_{\gamma_i} \in U_0$, where $y_0 \in c_0$, we have $|||q(u_0)||| = \inf_{y\in c_0} |||u_0+y|||$, where $|||\cdot|||$ denotes, too, the quotient norm in $(JL_2, |||\cdot|||)/c_0$. By definition,

$$\left|\left|\left|\sum_{i=1}^k a_{\gamma_i}\phi_{\gamma_i} + y\right|\right|\right| \ge \left(\sum_{i=1}^k |a_{\gamma_i}|^2\right)^{\frac{1}{2}}$$

for every $y \in c_0$, so we get $|||q(u_0)||| \ge \left(\sum_{i=1}^k |a_{\gamma_i}|^2\right)^{1/2}$.
On the other hand, given $\{a_{\gamma_i}\}_{i=1}^k$ it is possible to find a finitely supported

vector s so that

$$\left\|\sum_{i=1}^{k} a_{\gamma_i}\phi_{\gamma_i} + s\right\|_{\infty} = \max_{1\le i\le k} |a_{\gamma_i}| \quad \left(\le \left(\sum_{i=1}^{k} |a_{\gamma_i}|^2\right)^{1/2}\right),$$

and thus

$$|||u_0 + (s - y_0)||| = \left|\left|\left|\sum_{i=1}^{k} a_{\gamma_i}\phi_{\gamma_i} + s\right|\right|\right| \le \left(\sum_{i=1}^{k} |a_{\gamma_i}|^2\right)^{\frac{1}{2}},$$

so finally $|||q(u_0)||| = (\sum_{i=1}^{k} |a_{\gamma_i}|^2)^{1/2}$ $(= \|T(u_0)\|_2)$.

We can conclude now that the same is true for every $x \in JL_2$.

➪ Indeed, if $x \in JL_2$ we can get a sequence $\{u_n\}_{n=1}^{\infty}$ in U_0 that $|||\cdot|||$-converges to x. In particular, $\{q(u_n)\}_{n=1}^{\infty}$ converges to $q(x)$ in $|||\cdot|||$, so $|||q(u_n)||| \to |||q(x)|||$. We mentioned above that $T : (U, |||\cdot|||) \to (\ell_2(\Gamma), \|\cdot\|_2)$ is continuous. Thus, $T(u_n) \to T(x)$ in $\|\cdot\|_2$, so $\|T(u_n)\|_2 \to \|Tx\|_2$. Since $|||q(u_n)||| = \|T(u_n)\|_2$ for all $n \in \mathbb{N}$, we finally get $|||q(x)||| = \|Tx\|_2$.

This shows that JL_2/c_0 is linearly isometric to $\ell_2(\Gamma)$. It follows that JL_2 is not separable (in (ii) we shall see more, namely that it is not WCG, and in (vi) that, it is not even WLD).

(ii) JL_2 is a nonclosed linear subspace of ℓ_∞. The n-th coordinate functional δ_n is linear and $|||\cdot|||$-continuous on JL_2, since $|\langle x, \delta_n\rangle| = |x_n| \le |||x|||$ for all $x = (x_n) \in JL_2$. Then, if $x_n = 0$ for all $n \in \mathbb{N}$, we get $x = 0$. In other words, the subset $\{\delta_n : n \in \mathbb{N}\}$ of JL_2^* is w^*-linearly dense, and so JL_2^* is w^*-separable. Assume that JL_2 is WCG, say generated by the weakly compact subset K. Then (K, w) is metrizable, so separable. It will follow then that JL_2 is w-separable (i.e., $|||\cdot|||$-separable), and this was shown to be false. By the way, this shows that being WCG is not a three-space property: c_0 is a separable subspace of JL_2 and JL_2/c_0 is reflexive, although JL_2 itself is not WCG.

For (iii) use the fact that JL_2^* is isomorphic to $\ell_1 \oplus \ell_2(\Gamma)$ by the lifting property of ℓ_1.

➪ A Banach space Z has the **lifting property** if, for every pair of spaces X and Y and every operator S from X onto Y, the following condition holds: Whenever T is an operator form Z into X, there is an operator $\widetilde{T} : Z \to Y$ such that $T = S \circ \widetilde{T}$ (see Figure 48.7). It is easy to prove that *the space ℓ_1 has the lifting property* (see Exercise 825). Figure 48.8 below is a diagram for the relevant mappings and spaces in the proof. Since c_0 is a subspace of JL_2 (let j be the canonical injection), we get that ℓ_1 is isometric to a quotient of JL_2^* via the quotient mapping $j^* : JL_2^* \to \ell_1$. The identity mapping id from ℓ_1 into itself can be "lifted" to the quotient mapping j^* by a mapping $\widetilde{id}: \ell_1 \to JL_2^*$, clearly an isomorphism into a subspace $L1$ of JL_2^*, and so $L1$ is closed in JL_2. On the other hand, $\ell_2(\Gamma)$ is isometric to a quotient $q(JL_2) = JL_2/c_0$ of JL_2, so it follows that $\ell_2(\Gamma)^*$ $(= \ell_2(\Gamma))$ is isometric to the subspace $q^*(c_0^\perp)$ of JL_2^*, that we denote by $L2$. Assume that $x^* \in L1 \cap L2$ $(= q^*(c_0^\perp) \cap \widetilde{id}(\ell_1) \subset JL_2^*)$. Let $x^* = q^*(z^*) = \widetilde{id}(y^*)$ for some $z^* \in c_0^\perp$ and $y^* \in \ell_1$. Then we have, for $y \in c_0$,

$$\langle y, j^*(x^*)\rangle = \langle j(y), x^*\rangle = \langle j(y), q^*(z^*)\rangle = \langle q \circ j(y), z^*\rangle = 0,$$

since $q \circ j(y) = 0$. On the other hand,

$$(0 =)\ \langle y, j^*(x^*)\rangle = \langle y, j^* \circ \widetilde{id}(y^*)\rangle = \langle y, y^*\rangle.$$

Since this holds for all $y \in c_0$, we get $y^* = 0$, so $x^* = 0$. Moreover, $L1 + L2 = JL_2^*$. Indeed, given $x^* \in JL2$, put $x_1^* := \widetilde{id}(id^{-1}(j^*(x^*)))$. Then, if $x_2^* := x^* - x_1^*$ we have, for $y \in c_0$,

$$\langle j(y), x_2^*\rangle = \langle j(y), x^* - x_1^*\rangle = \langle y, j^*(x^*) - j^*(x_1^*)\rangle = 0,$$

since clearly $j^*(x_1^*) = j^*(x^*)$. This proves that $x_2^* \in L_2$. The conclusion follows.

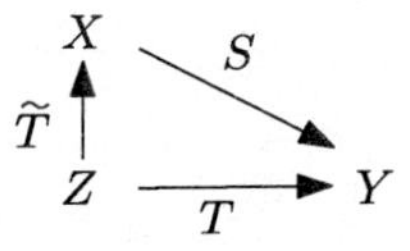

Figure 48.7: The lifting property of Z

Exercise 825. *Prove that for every nonempty set Γ, the space $\ell_1(\Gamma)$ has the lifting property.*

Hint: Let X and Y be Banach spaces, and $S : X \to Y$ be a linear and continuous mapping from X onto Y. This mapping is open, so there exists $C > 0$ such that given $y \in Y$ we can find $x \in X$ with $Sx = y$ and $\|x\| \le C\|y\|$. If $\{e_\gamma\}$ is the canonical basis of $\ell_1(\Gamma)$, find for $\gamma \in \Gamma$ a vector $x_\gamma \in X$ accordingly. Define $\widetilde{T} : \ell_1(\Gamma) \to X$ by $\widetilde{T}(\sum_\gamma a_\gamma e_\gamma) := \sum_\gamma a_\gamma x_\gamma$. It is easily proven that $\widetilde{T}$ is well defined and satisfies all the requirements. ◇

Since JL_2^* is isomorphic to $\ell_1 \oplus \ell_2(\Gamma)$, JL_2^* has a norming M-basis by [Go95] (see, e.g., [HMVZ07, Proposition 5.20]). The last statement in (iii) is the result of Troyanski (see, e.g., [HMVZ07, Theorem 5.22].
Observe then that a Banach space may fail to be WCG even if its dual space is WCG.

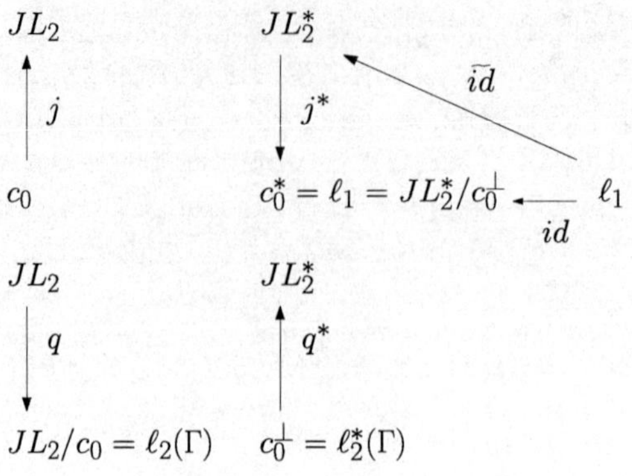

Figure 48.8: Subspaces and quotients of JL_2 and its dual JL_2^*

For (iv) we use [DeGoZi93, Theorem VII.2.7].
For (v) we use [DeGoZi93, Lemma V.1.9].
For (vi) use the following arguments: JL_2 is not WLD as it is nonseparable and its dual is weak* separable (use Theorem 834). As JL_2 has property C, it cannot admit a Markushevich basis, as we proved above that it is not WLD (use Theorem 833).
(vii) is in [Edg79].
(viii) See [DeGoZi90].
(ix) See Corollary 733.
(x) This is in [JoLi74], see Proposition 734 above. □

Theorem 826 (Aharoni, Johnson, Lindenstrauss). *There is a nonseparable space JL_0 with the following properties* ([JoLi74], [AhLi78], see, e.g., [FHHMZ11, Theorem 14.54]).
(i) *JL_0 is isomorphic to a subspace of ℓ_∞.*
(ii) *$c_0 \subset JL_0$ and JL_0/c_0 is isomorphic to $c_0(\Gamma)$ for some uncountable set Γ.*
(iii) *JL_0 is isomorphic to $C(K)$ for a separable scattered uncountable compact K of the third Cantor derived set empty.*
(iv) *JL_0 admits a norm that is LUR and its dual norm is LUR. Thus JL_0 is an Asplund space.*
(v) *JL_0 admits C^∞-smooth norm that is LFC.*
(vi) *JL_0 has property C and is not WLD. Thus it admits no Markushevich basis and JL_0 admits no UG norm. In fact JL_0 does not contain any nonseparable subspace with Markushevich basis* [FHHMZ11, Theorem 14.54].
(vii) *JL_0 does not admit any WUR norm.*
(viii) *JL_0 can be renormed by a norm whose dual is URED.*
(ix) *JL_0 is not weakly Lindelöf.*
(x) *JL_0 is Lipschitz homeomorphic to $c_0(\Gamma)$* (Aharoni, Lindenstrauss).
(xi) *The space $c_0(\Gamma)$ does not inject into JL_0 if Γ is uncountable.*
(xii) *The space JL_0 admits C^∞-smooth partitions of unity* ([GPWZ88], see [DeGoZi90]).
(xiii) *JL_0^* is not WLD.*
(xiv) *$B_{JL_0^*}$ is w^*-separable. However, there is an equivalent norm on JL_0 whose dual ball is not w^*-separable.*
(xv) *$B_{JL_0^*}$ is w^*-sequentially compact.*
(xvi) *JL_0 is saturated with c_0 and yet contains no nonseparable $c_0(\Gamma)$. JL_0 contains a complemented subspace isomorphic to c_0.*

Proof. (Notes) For (i) and (iii) use, e.g., [Yo97].
For (ii) use [JoLi74].
For (iv) use, e.g., [DeGoZi93, Theorem VII.4.8].
For (v) see Theorem 813.
For (vi) use the fact that JL_0^* is w^*-separable and JL_0 is nonseparable and use Theorem 834 to see that JL_0 is not WLD. JL_0 has property C since C has the three-space property. Thus JL_0 cannot admit any Markushevich basis by Theorem 834.
For (vii) use that otherwise JL_0^* would have an equivalent UG norm, and this is impossible as JL_0^* is isomorphic to $\ell_1(\Gamma)$ due to the lifting property of ℓ_1 (see [JoLi74]).
For (viii) use Theorem 763.
For (ix) use Pol's result that if the ω_1-derived set of K is empty, K is separable and $C(K)$ is weakly Lindelöf, then K is countable ([Po77]).
(x) is proved in [FHHMZ11, Theorem 14.54].
(xi) follows from the fact that JL_0^* is weak star separable but $c_0(\Gamma)^*$ is not.

(xii) See [DeGoZi90].
Note that, due to (vi) and (viii), JL_0^* admits a dual norm that is URED but not W*UR.
(xiii) Every subspace of a WLD space is itself WLD. This is a consequence of the deep result of Gul'ko stating that the continuous image of a Corson compact space is itself Corson compact. By (ii), JL_0^* has a subspace isomorphic to $\ell_1(\Gamma)$ for uncountable Γ. If $\ell_1(\Gamma)$ was WLD, then $B_{\ell_\infty(\Gamma)}$ would be Corson compact in the w^*-topology, hence angelic, and this is not the case, as $B_{\ell_\infty(\Gamma)}$ is the w^*-closure of the subset of all countably supported vectors in $B_{\ell_\infty(\Gamma)}$, and the vector with all coordinates equal to 1 is not the w^*-limit of a sequence of countably supported vectors.
(xiv) Put $X := JL_0$. Since K is separable, B_{X^*} is w^*-separable by the Krein–Milman Theorem 7. Let $\|\cdot\|$ be an equivalent norm on X such that its dual norm $\|\cdot\|^*$ is LUR. Let B be the unit ball and S the unit sphere for $\|\cdot\|^*$. Assume that there is a countable set $C \subset B$ such that $\overline{C}^{w^*} = B$. By the LUR property it follows that $S \subset \overline{C}$ and thus X^* would be separable, a contradiction.
(xv) See Corollary 733.
(xvi) See (vi) above and [FHHMZ11, Theorem 14.26]. □

Let us note that Talagrand proved [Tal80] that under the CH there is a nonseparable compact K such that $(B_{C(K)^*}, w^*)$ is separable.

Note that, due to (x) and (xi), JL_0 and $c_0(\Gamma)$ are Lipschitz homeomorphic but not linearly isomorphic. JL_0 and $c_0(\Gamma)$ were the first found example of such couple of spaces in [AhLi78]. It is interesting that the existence of a WCG space X that would be Lipschitz homeomorphic to $c_0(\Gamma)$ but not linearly isomorphic to is undecidable in ZFC (see [GoLaZi14]). The existence of such a separable couple of spaces is still a well-known important open problem.
Note also the following theorem [Mazur, Ulam, see, e.g., [FHHMZ11, Theorem 12.54]]: *Any isometry from a Banach space X onto a Banach space Y that takes 0 to 0 is necessarily linear.*

48.5 Markushevich bases II

From the area of Markushevich bases we additionally mention the following things.

Theorem 827.
(i) *Any space X with an M-basis admits a bounded linear one-to-one operator from X into $c_0(\Gamma)$ for some Γ* (see, e.g., [HMVZ07, Theorem 5.3]).
(ii) *The space ℓ_∞ does not admit any M-basis* (Johnson, see, e.g., [FHHMZ11, Theorem 4.61]).
(iii) *If Γ is uncountable, then $\ell_\infty(\Gamma)$ is not a subspace of a space with an M-basis* (Theorem 800).
(iv) *The space ℓ_∞ is a complemented subspace of a space with an M-basis* (Plichko, see, e.g., [FHHMZ11, Theorem 4.63]).
(v) *Any space with an M-basis admits a rotund norm.*
(vi) *There is a space with an M-basis that does not admit any LUR norm.*
(vii) *If X is separable, then X has a strong norming M-basis* [HMVZ07, Theorem 1.42].
(viii) *If X^* is separable, then X has an M-basis that is strong, and its dual coefficients form a strong M-basis for the dual.*
(ix) *Any separable infinite-dimensional space has an M-basis that is not strong* (Johnson, see, e.g., [HMVZ07, Proposition 1.34]).

Proof. For (v), use (i) in the Theorem. For (vi), use (iv). □

The following is a deep difficult result.

Theorem 828 (Aleksandrov, Plichko, see, e.g., [HMVZ07], Theorem 5.25). *The Banach space $C[0, \omega_1]$ does not admit any norming Markushevich basis.*

Let us recall here the recent result on the non-existence of norming M-bases on some Hilbert generated spaces formulated in Theorem 241 above: *There exists a Hilbert generated Banach space without norming M-basis.*

Theorem 829 (Edgar, [Edg77]). *Let X^* be a weakly Lindelöf space. Then X is an Asplund space.*

Sketch of the proof. Stegall [St75] showed that if X is not Asplund then there is an uncountable set in X^* that is closed and discrete in the weak topology. □

We sum up now some properties of $c_0(\Gamma)$ and $\ell_1(\Gamma)$ for uncountable Γ that we discussed above:

Theorem 830. *If Γ is uncountable, then $c_0(\Gamma)$ has the following properties:*
(i) *$c_0(\Gamma)$ admits a LUR norm that is the limit, uniform on bounded sets, of C^∞-smooth norms.*
(ii) *$c_0(\Gamma)$ admits a UG norm that is C^∞-smooth and LFC.*
(iii) *$c_0(\Gamma)$ does not admit any C^2-smooth strictly convex norm.*
(iv) *$c_0(\Gamma)$ does not admit any real-analytic norm.*
(v) *$c_0(\Gamma)$ does not admit any URED norm.*
(vi) *$c_0(\Gamma)$ and its all subspaces are WCG.*
(vii) *$c_0(\Gamma)$ is weakly Lindelöf.*
(viii) *$c_0(\Gamma)$ cannot be injected into a reflexive space.*
(ix) *Every space with a Markushevich basis can be injected into some $c_0(\Gamma)$. In particular, every WLD space can be injected into some $c_0(\Gamma)$.*

Theorem 831. *If the cardinality of Γ is the continuum, then $\ell_1(\Gamma)$ has the following properties.*
(i) *$\ell_1(\Gamma)$ is a subspace of $C[0,1]^*$, so it is a subspace of ℓ_∞.*
(ii) *$\ell_1(\Gamma)$ admits a LUR norm which is URED. It admits a dual LUR norm.*
(iii) *$\ell_1(\Gamma)$ does not admit any G norm.*
(iv) *$\ell_1(\Gamma)$ injects into a Hilbert space.*
(v) *No nonseparable reflexive space or $c_0(\Gamma)$ for Γ uncountable injects into $\ell_1(\Gamma)$* (the Grothendieck property).
(vi) *$\ell_1(\Gamma)$ is not weakly Lindelöf, neither it has the property C of Corson. So the dual ball of $\ell_\infty(\Gamma)$ is not weak*-Corson.*

48.6 Weakly Lindelöf determined spaces I

We mention here only a few things needed for presenting a list of results and examples related to the class of WLD spaces. First, we reproduce here a theorem of Pol that will also be needed:

Theorem 832 (Pol, see, e.g., [FHHMZ11], Theorem 14.33)**.** *If X is a Banach space and Y is a subspace of X such that both Y and X/Y have property C, then X has property C.*

This compares with the fact that the weak Lindelöf property is not a three-space property (see Theorem 826 above).

There is an example of a weakly Lindelöf space that is not WLD [Po79].

Theorem 833. *Let X be a Banach space. Then the following are equivalent.*
(i) *X is WLD.*
(ii) *The dual unit ball of X^* in its weak*-topology is a Corson compact* (see, e.g., [HMVZ07, Theorem 5.37 (iii)]).
(iii) *X has a Markushevich basis and X is weakly Lindelöf* ([HMVZ07, Theorem 5.37 (iv)]).

(iv) *X has a Markushevich basis and X has the property C* ([VWZ94]).
(v) *X has a Markushevich basis* $\{x_\alpha, f_\alpha\}$ *such that for every* $f \in X^*$, card $\{\alpha : f(x_\alpha) \neq 0\} \leq \aleph_0$ (see Theorem 135; see also, e.g., [HMVZ07, Theorem 5.37 (ii)]).
(vi) *X has a full projectional generator* (see Item (iii) after Definition 207).

We also will need the following result:

Theorem 834. (See, e.g., [HMVZ07], Section 5.4) *Let X be a WLD space. Then*
(i) *X has a PRI.*
(ii) *X is a DENS space.*
(iii) *X admits a LUR norm.*
(iv) *If* dens $X = \omega_1$, *then a Banach space X is WLD if, and only if, X admits a PRI in every equivalent norm* ([Kal00], see, e.g., [HMVZ07, Theorem 5.51]).

The following is an important result in understanding the relationship of Vašák and WLD properties.

Theorem 835 (Argyros, Mercourakis, [ArMe93]). (i) *There is a WLD space X with unconditional basis such that its dual admits a nondual rotund norm, and that X is not weak Asplund. In particular this space X does not admit any Lipschitz Gâteaux differentiable bump, in particular X admits no Gâteaux differentiable norm.*
(ii) *There is a WLD space X with* X^* *admitting a dual rotund norm, (so X admits a Gâteaux smooth norm) such that X is not Vašák.*
(iii) *There is a WLD space X with unconditional basis such that* X^* *does not admit any rotund norm.*

The following problem is open: If X is WLD, is every continuous convex real-valued function defined on it differentiable at the points of a dense set? (see [Ziz03, page 1800]).

We need the following couple of things on Mercourakis spaces. See (i) in Theorem 745.

Lemma 836 (Mercourakis, see [DeGoZi93], Lemma VI.6.2). *Let* Γ *be a set. Let* $h \in c_1(\Sigma' \times \Gamma)$ *and put* $a_0 = (p_0, q_0) \in \Sigma' \times \Gamma$. *Then, for every* $\varepsilon > 0$ *there is an open neighborhood V in* Σ' *such that* $h((V \setminus \{p_0\}) \times \Gamma) \subset (-\varepsilon, \varepsilon)$, *and thus* $\{a \in V \times \Gamma : \ |h(a)| \geq \varepsilon\}$ *is finite.*

Proof. If not, there exist$\varepsilon > 0$ and a sequence $a_n = (p_n, q_n)$ in $\Sigma' \times \Gamma$ such that $p_n \neq p_0$ for every n and $\lim p_n = p_0$ $p_n \neq p_m$ if $m \neq n$ and $|h(a_n)| \geq \varepsilon$. But $K = \{p_n : \ n \in \mathbb{N}\}$ is a compact subset of Σ' and $\{a_n : \ n \in \mathbb{N}\}$ is an infinite subset $K \times \Gamma$ such that $|h(a_n)| \geq \varepsilon$. This contradicts $h \in c_0(K \times \Gamma)$. The last statement in Lemma 836 follows from the fact that the set

$$\{a \in V \times \Gamma : \ |h(a)| \geq \varepsilon\} = \{a \in \{p_0\} \times \Gamma : \ |h(a)| \geq \varepsilon\}$$

is finite. □

From the Lindelöf property of Σ' it thus follows:

Theorem 837 (Mercourakis). *If $h \in c_1(\Sigma' \times \Gamma)$, then the support of h is countable.*

Theorem 838. (i) *Every Vašák space is WLD.*
(ii) *There is a space that is WLD but not Vašák.*

Proof. (i) If X is Vašák, then X in the weak topology is Lindelöf ([Vas81], see, e.g., [FHHMZ11, Exercise 13.30]. Thus, a fortiori, X has property C. If X is Vašák, then X has Markushevich basis by Theorem 746. Therefore (i) follows from (iv) in Theorem 833.
(ii) Theorem 835, Theorem 746. □

The following result appears in [HMVZ07, Theorem 6.3]. See also Theorem 315 above.

Theorem 839. *Let X be a Banach space. Then the following are equivalent:*
(i) *X admits a shrinking Markushevich basis.*
(ii) *X is WCG and Asplund.*
(iii) *X is WLD and Asplund.*
(iv) *X is WLD and has an equivalent norm whose dual norm is LUR.*
(v) *X is WLD and admits an equivalent Fréchet differentiable norm.*

Proof. (i)⇒(ii) follows from the fact that, for a shrinking Markushevich basis $\{x_\gamma, f_\gamma\}$, the set $\{x_\gamma\} \cup \{0\}$ is weakly compact if bounded and from the remark after [Fab97, Definition 6.2.3].
(ii)⇒(iii) is trivial.
(iii)⇒(i) is in [Fab97, Theorem 8.3.3].
(ii) ⇒(iv) is in [DeGoZi93, Corollary VII.1.13].
(iv) ⇒(v) is trivial.
(v) ⇒(iii) follows from the fact that every Banach space with a Fréchet smooth norm is Asplund. □

Theorem 840 (Argyros and Mercourakis, [ArMe93], see also [HMVZ07], Theorem 7.43). *Let X be a Banach space with an unconditional basis $\{e_\gamma; f_\gamma\}_{\gamma\in\Gamma}$. Then the following are equivalent:*
(i) *X is WLD.*
(ii) *X does not contain an isomorphic copy of $\ell_1(\omega_1)$.*
(iii) *For every uncountable subset $\Delta \subset \Gamma$, $\{e_\gamma : \gamma \in \Delta\}$ is not equivalent to the unit vector basis of $\ell_1(\Delta)$.*
(iv) *There is a bounded linear one-to-one operator T from X^* into $\ell_\infty^c(\Delta)$ for some set Δ.*

48.7 Double arrow space

Let D be the space of bounded functions on $[0,1]$ that are left continuous and have right limits at each point, endowed with the supremum norm. This space can be represented as a space of continuous functions on the so called "two arrow compact space". Then the quotient space $D/C[0,1]$ is isomorphic to $c_0([0,1))$ (see, e.g., [DeGoZi93, Section VII.3], [Fab97, Theorem 2.3.1], or [FHHMZ11, Theorem 14.39]).

Then we have

Theorem 841.
(i) (Talagrand)([Tal86], see, e.g., [DeGoZi93, Theorem VII.3.5]) *For every equivalent norm $|\cdot|$ on D there is a continuous function x on $[0,1]$ such that $|\cdot|$, as a norm on D, is not Gâteaux differentiable at x. Thus D admits no Gâteaux differentiable norm.*
(ii) *D admits a LUR norm* (Theorem 231).
(iii) *D does not admit a Markushevich basis* (it is not DENS, see (iv)) *but it has property C* (see (vi)).
(iv) *D^* is w^*-separable* ([FHHMZ11, Theorem 14.39]).
(v) *D admits a URED norm* (see (iv) here and Corollary 169 above).
(vi) *D is not weakly Lindelöf but it has property C* (see [FHHMZ11, Theorem 14.39]).

48.8 Space of continuous functions on ordinals

Theorem 842. *Let μ be an uncountable ordinal. Then the space $C[0,\mu]$ has the following properties:*
(i) *The space $C[0,\omega_1]$ does not have property C of Corson* (see [FHHMZ11, Theorem 14.36]).
(ii) (Haydon)([Hay92]) *$C[0,\mu]$ admits an C^∞-smooth norm that is LFC* (see also Theorem 336 above).
(iii) (Troyanski) (see, e.g., [DeGoZi93, Theorem VII.5.1]) *$C[0,\mu]$ admits a LUR norm.*
(iv) *$C[0,\mu]$ admits a Markushevich basis* (see, e.g., [HMVZ07, Proposition 4.8]).
(v) *$C[0,\mu]$ does not admit a URED norm* (it contains $c_0(\Gamma)$ for uncountable Γ, see Theorem 811).
(vi) (Semadeni), [Sem60] *The space $C[0,\omega_1]$ is not isomorphic to its Cartesian square.*

In this context let us note in passing that the James' space J is clearly not isomorphic to its Cartesian square, and Figiel showed [Fi72] that there exists an infinite-dimensional separable reflexive space that is not isomorphic to its square.

48.9 Weakly Lindelöf determined spaces II

We now can sum up some remaining examples concerning WLD spaces that we discussed above. For simplicity, let us collect them in a theorem.

Theorem 843.
(i) *The Ciesielski–Pol space is not WLD.*
(ii) *The space D on the two arrow compact space is not WLD.*
(iii) *The space ℓ_∞/c_0 is not WLD.*
(iv) *ℓ_∞ is not weakly Lindelöf.*
(v) *The Johnson–Lindenstrauss space JL_0 is not WLD.*
(vi) *The Johnson–Lindenstrauss space JL_2 is not WLD.*

Proof.
(i) Theorem 833, Theorem 823.
(ii) The space D is not a DENS space (Theorem 841, Theorem 834).
(iii) ℓ_∞/c_0: Theorem 834, Theorem 808.
(iv) Otherwise, $\ell_1(c) \subset \ell_\infty$ would be weakly Lindelöf, impossible as it has a Markushevich basis and is not a DENS space (the w^*-density of $\ell_1(c)^*$ is $\aleph_0$).
(v) and (vi): See Theorem 834 above. □

48.10 Kunen compact space

A topological space is said to be **hereditarily separable** if all its subspaces are separable.

Theorem 844 (Kunen). *Assuming the Continuum Hypothesis, there is an uncountable scattered compact space K such that $C(K)$ is hereditarily weakly Lindelöf and $C(K)^*$ is hereditarily w^*-separable. Moreover, K is not metrizable and thus $C(K)$ is not separable and $C(K)$ admits no uncountable biorthogonal system* (see, e.g., [HMVZ07, Theorem 4.41]).

The compact space K in Theorem 844 is nowadays called the **Kunen compact space**. Note that we have

Corollary 845. *If K is the Kunen compact space in Theorem 844, then $C(K)$ admits no LUR norm and admits no Fréchet smooth norm* (see, e.g., [Ziz03]).

Proof. Otherwise, in case of the existence of an equivalent LUR norm, the w-$\mathcal{T}_{\|\cdot\|}$-KK property together with the hereditarily Lindelöf property would imply the norm separability of the sphere. Similarly for the dual sphere in the case of existence of an equivalent Fréchet norm, due to the Šmulyan and Bishop–Phelps theorems. □

Some other results in the direction of impossible renormings of the space $C(K)$ were given in Theorems 550, 583, and Corollary 584 above.

Let us proceed with coordinates of a few examples from the weak topology of Banach spaces (for definitions, see the Introduction).

First, an example of an Eberlein compact space that is not uniform Eberlein is, e.g., in [FHHMZ11, Exercise 14.13].

An example of a Corson compact that is not Eberlein can be derived from Theorems 835 and 746.

For the results that the classes of Eberlein compacts, uniform Eberlein compacts and Corson compacts are, respectively, stable for continuous images we refer to [HMVZ07, Chapter 6] and [Vald91], where the quotations to the original papers are mentioned.

48.11 Strongly weakly compactly generated spaces

We mentioned in Section 10.1 that a Banach space is WCG if, and only if, there exists a w-compact subset K (called a "generator") of X that "almost absorbs" (see the definition of this term there) every $\|\cdot\|$-compact subset of X or, alternatively, that almost absorbs every null sequence in X. The space $(\ell_1, \|\cdot\|_1)$, being a separable Banach space, is WCG. Since w-compact subsets of ℓ_1 are $\|\cdot\|_1$-compact (the Schur property of ℓ_1), a generator of ℓ_1 almost absorbs every w-compact subset. This property is shared by the Banach space $L_1(\mu)$, where μ is a σ-finite measure (see, e.g., [HMVZ07, Proposition 6.41]). This, and the technique of proof of Theorem 191, motivated G. Schlüchtermann and R.F. Wheeler in [SchW88] to introduce the following subclass of the class of all WCG spaces:

Definition 846. *A Banach space* $(X, \|\cdot\|)$ *is called* **strongly weakly compactly generated** *(*β**WCG**, *for short) if there exists a weakly compact subset* K *of* X *that* **strongly generates** X *, i.e., it almost absorbs every* w*-compact subset* W *of* X.

We may easily translate to this case what was said for the class of WCG spaces in Section 10.1 above. Note that, thanks to Krein's theorem (see, e.g., [FHHMZ11, Theorem 3.133]), a generator for a space can always be taken absolutely convex:

Proposition 847. *Let* $(X, \|\cdot\|)$ *be a Banach space, and let* K *be a generator of* X. *Under the assumption that* K *is closed and absolutely convex, the following are equivalent:*
(i) K *is a strong generator of* X.
(ii) K *almost absorbs any* w*-null sequence in* X.

➪ *Proof.* (Sketch) Obviously (i)⇒(ii). Assume now that K absorbs every w-null sequence in X, and that there exists a w-compact subset A of X and $\varepsilon > 0$ such that, for every $n \in \mathbb{N}$, $A \not\subset nK + \varepsilon B_X$. There exists then $a_n \in A \setminus (nK + \varepsilon B_X)$. The Eberlein–Šmulyan theorem allows for a w-convergent subsequence $\{a_{n_j}\}_{j\in\mathbb{N}}$ of $\{a_n\}_{n\in\mathbb{N}}$. Let $a \in A$ be its w-limit. The fact that $\{a_{n_j} - a\}_{j\in\mathbb{N}}$ is w-null, and so that K almost absorbs $\{a_{n_j} - a\}_{j\in\mathbb{N}}$ and a, readily gives a contradiction. □

Note that Theorem 191 shows that X is βWCG if, and only if, there exists a reflexive Banach space R and a bounded linear mapping $T : R \to X$ such that TB_R strongly generates X (we say then that X is **strongly reflexive generated**). Another description of this class is given by the following statement: A Banach space X is βWCG if, and only if, the topology $\mu(X^*, X)$ on X^* of the uniform convergence on the family of all absolutely convex and w-compact subsets of X^* (see Section 2.1) is, when restricted to B_{X^*}, metrizable. Observe that a separable Schur space is βWCG (the apparently more general statement that every WCG Schur space is βWCG reduces to the former one, since such space is already separable). For a proof of the former assertions and a comprehensive study of this class see [SchW88] and, e.g., [HMVZ07, Section 6.4] and [FaMoZi09b].

We summarize below some other results on the class of βWCG spaces:

Theorem 848 (Schlüchtermann and Wheeler, [SchW88]). *Let $(X, \|\cdot\|)$ be a βWCG space. Then*
(i) *X is WCG.*
(ii) *X is w-sequentially complete.*
(iii) *If X does not contain a copy of ℓ_1, then X is reflexive.*

Notice that (iii) follows from (ii) and Rosenthal's dichotomy theorem (see, e.g., [FHHMZ11, Theorem 5.37]). Indeed, if $\{x_n\}_{n=1}^{\infty}$ is a sequence in B_X, then it must have a w-Cauchy subsequence, that converges because of (ii). The Eberlein–Šmulyan theorem concludes that B_X is w-compact (see, e.g., [HMVZ07, Corollary 6.39]).
Notice that from (ii) in Theorem 848, it follows that c_0 is a WCG space that is not βWCG.

An even more restrictive class consists of all the **strongly superreflexive generated Banach spaces**, i.e., those Banach spaces X for which there exists a superreflexive Banach space SR and a bounded linear mapping $T : SR \to X$ such that TB_{SR} strongly generates X. Similarly we define the smaller class of the **strongly Hilbert generated**, as we did in Subsection 6.1.8 above. Theorem 850 shows a classical space that is strongly Hilbert generated.
For the class of the strongly superreflexive generated spaces we mention two theorems:

Theorem 849. *Let X be strongly superreflexive generated Banach space. Then*
(i) [See, e.g., [FaMoZi09b, Proposition 6]] *If X does not contain a copy of ℓ_1, then X is superreflexive.*
(ii) [[FaMoZi09b]], see also [HMVZ07, Proposition 6.41] *There is an equivalent norm $\|\cdot\|$ on X such that for its dual norm on X^* the following holds: $f_n - g_n \to 0$ uniformly on each weakly compact set in X, whenever $f_n, g_n \in S_{X^*}$ are so that $\|f_n + g_n\| \to 2$.*

Theorem 850. (See, e.g., [HMVZ07, Proposition 6.41]) *If μ is a finite measure, then $L_1(\mu)$ is strongly Hilbert generated.*

This gives a new proof to the following classical result of Rosenthal (see, e.g., [HMVZ07, Corollary 6.46]):

Theorem 851 (Rosenthal, [Ros73]). *If μ is a probability measure and X is a reflexive subspace of $L_1(\mu)$, then X is superreflexive.*

Let us note that Mercourakis and Stamati proved in [MeSt06] that *there is a subspace of $L_1[0,1]$ that is not strongly superreflexive*, in fact, *not strongly reflexive generated.*

The following is an extension of the concept of strong weakly compact generation:

Definition 852. *A Banach space X is called* **strongly WCD** *(β***WCD***, for short) if the following holds: There is a sequence $\{K_n\}_{n\in\mathbb{N}}$ of w^*-compact subsets of X^{**} such that for each w-compact subset K of X, there is a subsequence $\{m_n\}_{n\in\mathbb{N}}$ of $\{1,2,\ldots\}$ with $K \subset \bigcap_{n=1}^{\infty} K_{m_n} \subset X$.*

Theorem 853 (Wilkins, [Wil95]). *Let X be a Banach space.*
(i) *If X is βWCG, then X is βWCD.*
(ii) *c_0 is βWCD and yet it is not βWCG.*

48.12 Effros–Borel structure

This short section provides a glimpse to the huge, interesting part of mathematics, the so-called "Descriptive Set Theory", discussed, e.g., in [Kec95] and [ArGoRo03].

Let $\mathcal{F}(C[0,1])$ be the set of all closed subsets of $C[0,1]$. The **Effros–Borel structure in** $\mathcal{F}(C[0,1])$ is the σ-algebra generated by all elements $F \in \mathcal{F}(C[0,1])$ such that $F \cap U \neq \emptyset$ for any nonempty open U in $C[0,1]$.

It is very interesting nice mathematics to study which classes of sets in $\mathcal{F}(C[0,1])$ are Borel in the Effros–Borel sense. We start with the following result:

Theorem 854 (Bossard, [Bos02]). *The set of all Banach spaces isomorphic to ℓ_2 is Borel in the Effros–Borel sense.*

The proof uses Kwapien's characterization of such spaces in terms of the notion of type and cotype. By using deep results in geometry of Banach spaces and descriptive set theory, this was extended to the following result:

Theorem 855 (Godefroy, [Go17]). *If $1 < p < \infty$, then the class of spaces isomorphic to ℓ_p is Borel.*

Theorem 855 is not true for spaces L_p. This is in [BouRoSch81]. The following result is worth to be quoted here:

Theorem 856 (Kurka, [Kur19]). *The class of spaces isomorphic to c_0 is not Borel.*

It is proved in [Go17b] that $(\sum \ell_{2+(\ln n)^{-1}}^{2^n})_2$, which is not isomorphic to ℓ_2, has a Borel isomorphic class; this provides the first example of this kind.
The paper [CDDK] goes in this direction.

Chapter 49

Symmetric norms

Recall that two Schauder bases $\{e_n\}_{n\in\mathbb{N}}$ and $\{f_n\}_{n\in\mathbb{N}}$ of a Banach space X are **equivalent** if for every sequence $\{a_n\}_{n\in\mathbb{N}}$ of scalars, $\sum a_n e_n$ is convergent if, and only if, $\sum a_n f_n$ is convergent. A basis $\{e_n\}_{n\in\mathbb{N}}$ of a Banach space X is called **symmetric** if for any permutation π of $\mathbb{N}$, $\{e_{\pi(n)}\}_{n\in\mathbb{N}}$ is equivalent to $\{e_n\}_{n\in\mathbb{N}}$. Every symmetric basis is certainly unconditional. The previous definition applies to an unconditional long Schauder basis $\{e_\gamma\}_{\gamma\in\Gamma}$: For any permutation π of Γ, $\sum_\gamma a_\gamma e_\gamma$ converges if, and only if, $\sum_\gamma a_\gamma e_{\pi(\gamma)}$ converges. In such a case, and if $x = \sum_\gamma x(\gamma)e_\gamma$, put $x_\pi := \sum_\gamma x(\gamma)e_{\pi(\gamma)}$. A norm $|||\cdot|||$ on a Banach space X with a symmetric basis $\{e_\gamma\}$ is called **symmetric** if for every $x \in X$ and every permutation π of Γ, $|||x||| = |||x_\pi|||$.
Not too much is known on symmetric bases in nonseparable spaces. We have the following.

Theorem 857 (Troyanski, [Tr75]). *Let X be a nonseparable Banach space with a symmetric basis $\{e_\gamma\}_{\gamma\in\Gamma}$. Then*
(i) *X admits an equivalent UG norm if and only the basis is not equivalent to the canonical basis of $\ell_1(\Gamma)$.*
(ii) *X admits an equivalent URED norm if and only if the basis is not equivalent to the canonical basis of $c_0(\Gamma)$.*

Theorem 858 (Troyanski, [Tr75]). *There is a nonseparable nonreflexive space X with symmetric basis such that X does not contain any subspace isomorphic to $c_0(\Gamma)$ for Γ uncountable while every infinite-dimensional subspace of X contains a subspace isomorphic to c_0.*

This should be compared with the famous James' theorem that *any separable nonreflexive space with unconditional basis must contain either c_0 or ℓ_1* (see, e.g., [FHHMZ11, Corollary 4.42].

We also have the following result.

Theorem 859 (Troyanski, [Tr75]). *The space $c_0(\omega_1) \times \ell_1(\omega_1)$ is not isomorphic to a subspace of a Banach space with symmetric basis.*

Theorem 859 is in contrast with Lindenstrauss' result that any separable Banach space with unconditional basis is isomorphic to a complemented subspace of a separable Banach space with symmetric basis ([LinTz77, Thm 3.b.1]).

A. J. Guirao et al., *Renormings in Banach Spaces*, Monografie Matematyczne 75,
https://doi.org/10.1007/978-3-031-08655-7_49

It is interesting that the following separable results follow from Theorem 857.

Theorem 860 (Hájek and Zizler, [HáZi95]). *Let X be a separable Banach space with symmetric basis. Then X admits an equivalent symmetric Gâteaux differentiable norm if, and only if, X admits an equivalent symmetric UG norm, if, and only if, X is not isomorphic to ℓ_1.*

Proof. (Sketch) Let $(X, \|\cdot\|)$ be a separable Banach space with a symmetric basis $\{e_i\}_{i\in\mathbb{N}}$. If θ is an injection of $\mathbb{N}$ into $\mathbb{N}$, we denote

$$x_\theta = \sum_{i=1}^{\infty} x_i e_{\theta(i)}.$$

Clearly, $\sum x_i e_{\theta(i)}$ is convergent.
We note that if π is a permutation of $\mathbb{N}$, θ is an injection of $\mathbb{N}$ into $\mathbb{N}$, and $x \in X$, then

$$\|x_\pi\| = \|x_\theta\| = \|x\|.$$

Let Γ be an infinite set. We introduce the space $X(\Gamma)$ with symmetric basis $\{e_\gamma, \gamma \in \Gamma\}$ as the completion of $c_{00(\Gamma)}$ under the norm

$$\Big\|\sum a_{\gamma_i} e_{\gamma_i}\Big\|_{X(\Gamma)} = \Big\|\sum a_{\gamma_i} e_i\Big\|.$$

We check that $X(\Gamma)$ in the norm $\|\cdot\|_{X(\Gamma)}$ is a well defined "nonseparable version" of X. □

Theorem 861 (Hájek and Zizler, [HáZi95]). *Let X be a separable Banach space with symmetric basis. Then X admits an equivalent symmetric F norm if and only if X^* is separable.*

Theorem 862 (Hájek and Zizler, [HáZi95]). *Let X be a separable Banach space with symmetric norm. Then X admits an equivalent symmetric URED norm if and only if X is not isomorphic to c_0.*

Theorem 863 (Hájek and Zizler, [HáZi95]). *The space ℓ_∞ admits no equivalent symmetric rotund norm.*

Theorem 864 (Hájek and Zizler, [HáZi95]). *If X is a separable Banach space with a symmetric Schauder basis and X^* is nonseparable, then X admits a symmetric rough norm.*

Figiel and Johnson proved in [FiJo74] that there exists a reflexive space with symmetric basis that does not contain any isomorphic copy of ℓ_p, $p > 1$.

Chapter 50

Strictly convex renorming

This chapter follows in part the excellent introduction and reproduces some results in [OriSmTr12], where a more complete information is provided. It complements previous information given in Sections 5.1, 9.2, and Subsection 48.1.1 above. It mentions also results in [GuiLaTr20].

50.1 Introduction

Sufficient conditions for the existence of a strictly convex renorming of a Banach space X were given at several stages of this text. An earlier result is Clarkson's contribution —Propositions 152 and 161, two simple yet effective transfer methods that allow, for example, to renorm in a strictly convex way a space with a w^*-separable dual (Theorem 154), see also Theorem 359—. Day, by following Clarkson's proof of the aforementioned result, showed that the dual space of a separable Banach space can be renormed by a *dual* strictly convex norm (Theorem 157), and that the space $c_0(\Gamma)$ admits a strictly convex norm (Theorem 170).
WCG spaces can be renormed to be strictly convex (in fact, LUR, see Theorem 218). Amir and Lindenstrauss showed that they admit a norm whose dual norm is strictly convex. Both statements about strict convexity (not the one about a LUR equivalent norm) are the consequence of Proposition 152, the existence of a bounded linear one-to-one mapping from the space (and from the dual space) into some $c_0(\Gamma)$ (Theorems 201 and 236, respectively), and the fact, already mentioned, that $c_0(\Gamma)$ can be renormed to be strictly convex (Proposition 152). Thus, as a byproduct of Theorem 148 (mentioned in Subsection 9.2.3, Item 1), we get that a WCG space has an equivalent strictly convex norm whose dual norm is also strictly convex.
The existence of a dual strictly convex norm on the dual space of any WCG Banach space prompted Lindenstrauss to conjecture that if X admits an equivalent Gâteaux differentiable norm then it must embed as a subspace of a WCG space; this conjecture was answered in the negative by Johnson and Lindenstrauss (see Theorem 824 above).
Godefroy, Troyanski, Whitfield, and Zizler [GTWZ] provided another sufficient condition for strictly convex renorming (Theorem 231 above). Its formulation is close to a three-space property for strictly convex renorming (a literal three-space behaviour for this property is impossible —see (xii) in Theorem 655).
On the other hand, some spaces *do not* have strictly convex renormings. In this text we mentioned $\ell_\infty^c(\Gamma)$ for Γ uncountable (Theorem 800) (and so $\ell_\infty(\Gamma)$ shares

A. J. Guirao et al., *Renormings in Banach Spaces*, Monografie Matematyczne 75,
https://doi.org/10.1007/978-3-031-08655-7_50

the same property), and the space ℓ_∞/c_0 (Item (ii) in Theorem 803). See also Subsection 48.1.1 above.

50.2 Topological sufficient conditions for strictly convex renorming and a characterization

The results mentioned in Section 50.1 showed to some extent the limits of linear-transfer methods in the problem of characterization of strictly convex renorming. They are far to provide a characterization of the existence of a strictly convex renorming.

Topological properties of the closed unit ball (in its w^*-topology) entered the scene. For example, it was shown by Ribarska that if X^* admits a strictly convex dual norm then (B_{X^*}, w^*) is fragmentable (see Definition 546). This necessity condition is far from be sufficient. For example, Talagrand [Tal86] (see Theorem 812) showed that $C(\omega_1+1)^*$ does not admit a strictly convex dual norm; however, $(B_{C(\omega_1+1)^*}, w^*)$ is fragmentable. The renorming theory for $C(K)$ spaces allows for results on general Banach spaces (after all, any Banach space X is isometrically isomorphic to a closed subspace of $C((B_{X^*}, w^*))$ endowed with the supremum norm. A subclass of the class of fragmentable compact spaces consists of the descriptive spaces, a concept introduced below:

Definition 865. *A compact space K is* **descriptive** *if it admits a σ-isolated network, i.e., a family $\mathcal{N} = \bigcup_{n=1}^{\infty}$ of subsets of K with the following properties:*
(i) $N \cap \overline{\bigcup \mathcal{N}_n \setminus \{N\}} = \emptyset$ *whenever $N \in \mathcal{N}_n$ and $n \in \mathbb{N}$, and*
(ii) *if $x \in U \subset K$, where U is open, then there exists $n \in \mathbb{N}$ and $N \in \mathcal{N}_n$ such that $x \in N \subset U$.*

Every descriptive compact space is fragmentable [Raj03]. A compact scattered space is descriptive if, and only if, it is σ-discrete (see Theorem 775). Another Raja's result, proved in [Raj03], is that *if K is a descriptive compact space, then $C(K)^*$ admits a strictly convex dual norm.* Prior to these results, Haydon built strictly convex dual norms on spaces of the form $C(K)^*$, where K are 1-point compactifications of certain trees in their interval topology [Hay99]. Some of those spaces are not descriptive, so Haydon's results were not covered by Raja's ones. R. Smith characterized in [Smi06] those trees for which the associated spaces $C(K)^*$ have an equivalent strictly convex dual norm, and in [Smi09] (Theorem 867 below) a sufficient condition for a dual strictly convex renorming of $C(K)^*$-spaces was provided, where K was a particular kind of compacta, that we define now:

Definition 866 (Gruenhage, [Gru87]). *A compact space K is called* **Gruenhage** *if there exists a sequence $\{\mathcal{U}_n\}_{n=1}^{\infty}$ of families of open subsets of K such that if $x, y \in K$ are distinct, then there exists $n \in \mathbb{N}$ and $U \in \mathcal{U}_n$ with the following two properties:*
(i) *$U \cap \{x, y\}$ is a singleton, and*
(ii) *either x lies in finitely many $U' \in \mathcal{U}_n$ or y lies in finitely many $U' \in \mathcal{U}_n$.*

Then, the following result holds:

Theorem 867 (R.J. Smith, [Smi09]). *Let K be a compact space. Then the following statement hold:*
(i) *if K is Gruenhage then $C(K)^*$ admits a dual strictly convex lattice norm.*
(ii) *If K is the 1-point compactification of a tree and $C(K)^*$ admits a dual strictly convex norm, then K is Gruenhage.*

It turned out that a convenient generalization of the concept of Gruenhage compactum (the so-called Property (*), see Definition 868 below) provided the sought strictly convex renorming characterization (Theorem 869). This definition, and all the other "topological" ones mentioned above, display a common "separation" flavour, natural in the context of the strict convexity property treated here.

Definition 868. *A topological space X has* **property (*)** *if there exists a sequence $\{\mathcal{U}_n\}_{n=1}^{\infty}$ of families of open subsets of X with the following property:*
Given $x, y \in X$, there exists $n \in \mathbb{N}$ such that
(i) *$\{x, y\} \cap \bigcup \mathcal{U}_n \neq \emptyset$, and*
(ii) *$\{x, y\} \cap U$ is at most a singleton for all $U \in \mathcal{U}_n$.*
Any sequence $\{\mathcal{U}_n\}_{n=1}^{\infty}$ satisfying (i) and (ii) will be called a **(*)-sequence for** *X.*
If, in addition, X is a locally convex space and $A \subset X$, then we say that A has **(*) with slices** *if A admits a (*)-sequence $\{\mathcal{U}_n\}_{n=1}^{\infty}$ with the property that every element of $\bigcup_{n=1}^{\infty} \mathcal{U}_n$ is an open slice of A.*

The main result in [OriSmTr12], a characterization of Banach spaces having a strictly convex renorming, follows:

Theorem 869 (Orihuela, Smith, Troyanski, [OriSmTr12]). *Let X be a Banach space. Let $N \subset X^*$ be a 1-norming subspace. Then the following are equivalent:*
(i) *X admits an equivalent $w(X,N)$-lower semicontinuous strictly convex norm.*
(ii) *$(X, w(X,N))$ has property (*) with slices.*
(iii) *$(S_X, w(X,N))$ has property (*) with slices.*
(iv) *There is a sequence $\{X_n\}_{n=1}^{\infty}$ of subsets of X such that*

$$\{(x,y) \in X^2 : \; x \neq y\} \subset \bigcup_{n=1}^{\infty} X_n^2,$$

and where each $(X_n, w(X,N))$ has property (*) with slices.

An application of the previous result is the following

Theorem 870 (Guirao-Lajara-Troyanski, [GuiLaTr20]). *Let X be a Banach space, let $M \subset X^*$ be a 1-norming subspace, let Y be a $w(X,M)$-closed subspace of X and let $Q : X \to X/Y$ denote the canonical quotient mapping. Assume that Y has an equivalent $w(X,M)$-lower semicontinuous strictly convex norm and there exists a 1-norming subspace $N \subset (X/Y)^*$ such that*
(i) *$Q^*(N) \subset M$, and*
(ii) *X/Y has an equivalent $w(X/Y,N)$-lower semicontinuous norm.*
Then X admits an equivalent $w(X,M)$-lower semicontinuous strictly convex norm.

Theorem 231 above follows from Theorem 870.

The three-space problem for strictly convex dual renormings has a negative answer. In fact, as we mentioned, Talagrand [Tal86] gave an example of a Banach space X that contains $C[0,1]$ as a closed subspace, and such that X admits no equivalent Gâteaux smooth norm and $X/C[0,1]$ is isomorphic to $c_0([0,1])$ (Section 48.7 and Theorem 841 above). The spaces $(C[0,1])^*$ and $(X/C[0,1])^*$ have, respectively, an equivalent dual strictly convex norm and an equivalent dual LUR norm, while X admits no equivalent Gâteaux smooth renorming (hence X^* admits no dual strictly convex equivalent norm, see Proposition 82). So, the following consequence of Theorem 870 is essentially optimal:

Corollary 871 (Guirao–Lajara–Troyanski, [GuiLaTr20]). *Let Y be a closed subspace of a Banach space X such that Y admits an equivalent norm whose dual norm is LUR. Then X admits an equivalent norm with dual strictly convex norm if, and only if, so does X/Y.*

Chapter 51

A concise list of coordinates for some relationships

We collect here a number of examples, counterexamples, and results, classified according to some basic features, and formulated in a telegraphic way. Most entries, if not all, are again in the indices —the General, the Renormings and the Impossible Renormings ones— in an even more condensed appearance.
For example, Item 51.1.1.5, meaning that there is no equivalent Gâteaux differentiable equivalent norm on the space ℓ_∞, refers to Theorem 805, where this result appears. In the Impossible Renormings Index, and under the entry ℓ_∞, the reader will find the subentry "no ⟨G⟩", together with the page where the result is located. In this chapter, some extra information is added to some or many of the items.

Following M.M. Day in [Day73], if P is one of the properties discussed here, we will say that a space has property ⟨P⟩ if it can be equivalently renormed by a norm with property P. Some other symbols used here are described in the Symbol Index and at the beginning of the part devoted to indices (page 553).

51.1 Differentiability

51.1.1 Gâteaux differentiability (G)

1. G not SG: Theorem 372.

2. There is a separable Banach space X, and a subspace Y of X which admits a Gâteaux differentiable norm that cannot be extended to a Gâteaux differentiable norm on X: Theorem 703.

3. G $\not\Rightarrow$ ⟨R⟩: Theorem 821.

4. ⟨G⟩ plus Asplund does not imply ⟨F⟩: Corollary 822. Note that ⟨UG⟩ plus Asplund implies ⟨F⟩ : Theorem 746.

5. No ⟨G⟩ on ℓ_∞: Theorem 805.

6. No ⟨G⟩ on ℓ_∞/c_0: Theorem 808.

7. No ⟨G⟩ on D (the double arrow space): Theorem 841.

A. J. Guirao et al., *Renormings in Banach Spaces*, Monografie Matematyczne 75,
https://doi.org/10.1007/978-3-031-08655-7_51

51.1.2 Uniform Gâteaux differentiability (UG)

1. UG not F: Theorems 366 and 368.
2. Separable reflexive UG not F: Theorems 172 and 366.
3. UG no ⟨F⟩: Any UG norm on ℓ_1.
4. UG no ⟨SG⟩: Theorem 372.
5. UG does not imply ⟨URED⟩: $c_0(\Gamma)$ for uncountable Γ: Theorem 811, Theorem 810.
6. UG does not imply ⟨NS⟩: Renorm ℓ_1 to be UG. Note that ℓ_1 cannot be ⟨NS⟩ since it is not Asplund (Theorem 578).
7. UG does not imply normal structure: Theorem 390.
8. Example of second dual UG and nonreflexive: Theorem 413.
9. There is a Banach space X that admits no G norm but there is a subspace Y of X such that both Y and X/Y admit UG norms: Theorem 841.
10. UG implies subspace of WCG: Corollary 755. UG does not imply WCG (Rosenthal's example of a SWCG space that is not WCG, see Theorem 760).
11. Example of reflexive space not ⟨UG⟩: Theorem 364.
12. No ⟨UG⟩ on JL_2: By Lemma 752, the existence of an equivalent UG norm implies that the space is Vašák, and the space JL_2 is not such (see Theorem 824).
13. UG and reflexive not Hilbert generated: Theorem 756.

51.1.3 Very smoothness (VS)

1. VS not F: Theorem 366 (take X reflexive).
2. VS implies Asplund: Theorem 372.
3. If the dual of a Banach space is VS, the space is reflexive: Theorem 411.

51.1.4 Strong uniform Gâteaux differentiability (SUG)

1. SUG implies WCG: Theorem 759.

51.1.5 Nice smoothness (NS)

1. NS does not imply Asplund: Otherwise MIP would imply Asplund (Theorem 582). In particular, NS does not imply ⟨F⟩.
2. NS does not imply ⟨SSD⟩, otherwise NS would imply Asplund (Theorem 549), and this is not true.

3. NS does not imply ⟨G⟩: Any $C_0(T)$, where T is a tree, implies ⟨MIP⟩, which implies ⟨NS⟩. There is tree T such that $C_0(T)$ does not imply ⟨G⟩.

51.1.6 Fréchet differentiability (F)

1. F not UG: Theorem 373.
2. F plus LUR and WUR whose dual norm is not R: Theorem 373. Note: Cannot be reflexive (duality).
3. Reflexive, F and dual not LUR: Theorem 374.
4. ⟨F⟩ does not imply ⟨UG⟩: Theorem 824, Theorem 823, Theorem 826.

51.1.7 C^2-smoothness

1. The existence of a C^2-smooth norm on X and the existence of R norm on X does not imply the existence of a norm on X with both of these properties simultaneously. See Corollary 624.
2. There is a Banach space X and its subspace Y such that both Y and X/Y admit C^2-smooth norms and X does not admit a C^2-smooth norm: Theorem 702.

51.1.8 C^∞-smoothness

1. C^∞-smooth $\not\Rightarrow$ ⟨URED⟩: Theorem 811.
2. There is a space X that admits C^∞-smooth norm but there is no norm on X the dual of which on X^* is R: Theorems 812 and 842.
3. ⟨C^∞-smooth⟩ $\not\Rightarrow$ ⟨UG⟩: Theorem 823.
4. A separable X admitting C^∞-smooth norm but no real-analytic norm: Theorem 512.

51.2 Rotundity

51.2.1 Strict convexity (R)

1. R not MLUR: Theorem 403.
2. R not w-$\mathcal{T}_{\|\cdot\|}$-SKK, in particular not w-$\mathcal{T}_{\|\cdot\|}$-KK: Theorem 401.
3. R not WLUR: Theorem 402.
4. R $\not\Rightarrow$ ⟨WMLUR⟩: Theorem 803.
5. ⟨R⟩ does not imply ⟨w-$\mathcal{T}_{\|\cdot\|}$-KK⟩: Theorem 821.

6. Talagrand theorem on $C[0,\alpha]$ for uncountable α: There is no equivalent norm with R dual norm: Theorem 812.
7. There is a Banach space X that is not $\langle R\rangle$ but has a subspace Y such that both Y and X/Y are $\langle R\rangle$: Theorem 821.
8. No $\langle R\rangle$ on ℓ_∞/c_0: Theorem 808.
9. No $\langle R\rangle$ on $\ell_\infty^c(\Gamma)$: Day–Partington Theorem 800.

51.2.2 Local uniform convexity (LUR)

1. LUR, reflexive no normal structure: Theorem 392.
2. LUR norm with dual LUR does not imply $\langle$WUR$\rangle$: Theorem 826.
3. LUR not URED: Theorem 392.
4. Three-space problem for LUR: Theorem 231.
5. LUR on dense subspace only: Theorem 801.
6. LUR norm the dual of which is not G: Theorem 373. Note that it cannot be reflexive (use duality). Such is, e.g., any LUR norm on ℓ_1 as ℓ_∞ does not have any G norm.
7. Reflexive LUR not 2WR: Theorem 425.
8. The space $C[0,1]^*$ admits a LUR norm (Troyanski), so it does not contain ℓ_∞. This implies that ℓ_1 is not a quotient of $C[0,1]$ (see, e.g., [HMVZ07, Exercise 3.18]).
9. LUR $\not\Rightarrow$ $\langle$G$\rangle$: Theorem 805.

51.2.3 2-rotundness (2R)

1. 2R not WLUR, see Theorem 396.
2. 2R $\not\Rightarrow$ $\langle$UR$\rangle$: $(\sum \ell_\infty^n)_2 \not\Rightarrow \langle$UR$\rangle$ and Theorem 422, see [FHHMZ11, Section 9.2].

51.2.4 Weak 2-rotundness (W2R)

1. W2R not 2R: Theorem 424.

51.2.5 Average local uniform convexity (ALUR)

1. ALUR not LUR, see [Tr85].

The example there is the following norm on ℓ_1:

$$|||x||| = \sum_{i=1}^{\infty} |x_i| + \left(\sum_{i=1}^{\infty} 2^{-i+1} x_i^2\right)^{\frac{1}{2}}.$$

Another norm $\|\cdot\|$ on ℓ_1 with the same property is given by $\|\cdot\| := \|\cdot\|_1 + \|\cdot\|_2$ (see [DeGoZi93, Remark II.7.6]). The norm $\|\cdot\|$ is a dual norm on ℓ_1, which is R. It is easy to show that it has property w^*-$\mathcal{T}_{\|\cdot\|}$-KK. Thus, it is ALUR (see Theorem 407). On the other hand, it is not LUR. Indeed, if $x : (1/2, 0, \ldots)$ and $x_n = (0, 1/n, \ldots, 1/n, 0, \ldots)$, where $1/n$ is repeated n times, then $\|x\| = 1$, $\|x_n\| \to 1$, and $\|x + x_n\| \to 2$, while $\|x - x_n\| \geq 1$ for each $n \in \mathbb{N}$.

51.2.6 Midpoint local uniform convexity (MLUR)

1. MLUR not w-$\mathcal{T}_{\|\cdot\|}$-SKK, in particular not w-$\mathcal{T}_{\|\cdot\|}$-KK: Theorem 401.
2. MLUR not ALUR: see [Tr85].

   The example there is the following norm on the space c of convergent sequences:

   $$\|x\| = \left(\sum_{1}^{\infty} 2^{-i-1}(x_i^2 + |x|_i^2)\right)^{\frac{1}{2}},$$

   where $|x|_i = \sup_{k \geq i} |x_k|$. Note that this is an example of a ball with all points on S_X strongly extreme and some points not being strongly exposed. This should be compared with the easily seen fact that each point on S_X of a rotund norm is exposed.
3. MLUR not LUR: Theorem 396.
4. MLUR $\not\Rightarrow$ $\langle w$-$\mathcal{T}_{\|\cdot\|}$-KK$\rangle$: Theorem 821.
5. $\langle$MLUR$\rangle$ does not imply $\langle$LUR$\rangle$: Theorems 820 and 821.
6. Not F but dual MLUR: Theorem 417.

51.2.7 Near uniform convexity (NUC)

1. NUC not superreflexive: $(\sum \ell_n^n)_2$ (see [Hu80]).

51.2.8 Uniformly rotund in every direction (URED)

1. URED not WLUR: Theorem 396.
2. Dual URED does not imply X $\langle$UG$\rangle$: Theorem 826.
3. Example of reflexive space not $\langle$URED$\rangle$: Theorem 364.

51.2.9 Weak uniform convexity (WUR)

1. WUR not LUR: Theorem 394.
2. WUR does not imply w-$\mathcal{T}_{\|\cdot\|}$-SKK: Theorems 394 and 403.
3. WUR does not imply $\langle$UG$\rangle$: Theorem 824 or Theorem 735.

51.3 Some extra properties

51.3.1 Kadets–Klee and sequential Kadets–Klee (KK) (SKK)

1. No w-$\mathcal{T}_{\|\cdot\|}$-KK norm on ℓ_∞: Corollary 804.
2. Example of an w-$\mathcal{T}_{\|\cdot\|}$-SKK norm on ℓ_1 that is not w-$\mathcal{T}_{\|\cdot\|}$-KK in [Sm78]: Remarks 404 and 409.
3. w-$\mathcal{T}_{\|\cdot\|}$-SKK, R not MLUR: Theorem 403.
4. w-$\mathcal{T}_{\|\cdot\|}$-KK not R: On ℓ_2, take the norm defined for $x = (x^1, x^2, \ldots)$ by $\|x\| = |x^1| + \|(x^2, x^3, \ldots)\|_2$ where $\|\cdot\|_2$ denotes the Hilbertian norm. Then it is standard to see that $\|\cdot\|$ is w-$\mathcal{T}_{\|\cdot\|}$-KK as the Hilbertian norm is, and not R ($\|e_1 + e_2\| = \|e_1\| + \|e_2\|$).
5. $\langle w$-$\mathcal{T}_{\|\cdot\|}$-KK$\rangle$ does not imply $\langle R\rangle$: Theorem 821.

51.3.2 Uniform Kadets–Klee (UKK)

1. UKK not isomorphic to NUC: ℓ_1, see [Hu80].
2. Reflexive not isomorphic to UKK: $(\sum \ell_n)_2$ ([Hu80]). So it gives by renormings, w-$\mathcal{T}_{\|\cdot\|}$-KK not isomorphic to UKK.

51.3.3 Normal structure

1. Reflexive, not URED but normal structure: Theorem 393.
2. Reflexive no normal structure but fixed points for nonexpansive: Theorem 391.

51.3.4 Asplundness

1. Asplund implies LUR on dual: Theorem 774.
2. Asplund does not imply either NS, either SSD, either MIP (Theorems 582 to 584).
3. Asplund space and no injection into $c_0(\Gamma)$, Theorem 821.

4. Asplund space with no ⟨F⟩: Corollary 845 or Corollary 822, even with Gâteaux norm.
5. Example of Asplund, not ⟨G⟩: Theorem 817.
6. Example of Asplund, not ⟨WUR⟩: Theorem 826.

51.3.5 Mazur intersection property (MIP)

1. MIP does not imply ⟨SSD⟩: otherwise it would imply Asplund (Theorem 582).

51.4 Weak compact generation (WCG) and relatives

1. WCG implies Vašák: Theorem 745.
2. Subspace of WCG implies Vašák: Theorem 745.
3. Not hereditary WCG. Rosenthal counterexample: Theorem 760.
4. Example of a WCG space (in fact reflexive space) that does not admit UG norm: Theorem 364.
5. No operator from $L_1(\mu)$ for finite μ onto a dense set in ℓ_∞ (the latter is not WCG).
6. There is a Vašák space which is not even a subspace of WCG: Theorem 745.
7. Vašák Asplund implies WCG: Theorems 812, 745, 839, and 843.
8. Vašák implies DENS: Theorem 745.
9. Vašák implies PRI: Theorem 746.
10. Vašák implies an equivalent LUR norm: Theorem 746.
11. Vašák Asplund implies an equivalent norm whose dual norm is LUR: Theorem 746.
12. X^* Vašák implies an equivalent norm on X whose dual norm is LUR: Theorem 746.
13. If X^* is Vašák then X has an equivalent LUR norm: Theorem 746.
14. Vašák implies strong Markushevich basis: Theorem 746.

51.4.1 Weak Lindelöf determinacy (WLD)

1. WLD does not imply Vašák (Theorem 835).
2. WLD with no G norm: Theorem 835.

3. $\ell_1(\Gamma)$ not weakly Lindelöf [Edg79], showing that it does not have property C.

4. None of the spaces, D, $C[0,\omega_1]$, JL_0, JL_2, Ciesielski–Pol space, ℓ_∞/c_0 is WLD: Theorem 839.

51.5 Injections

1. C^∞-smooth $\not\Rightarrow$ injection into $c_0(\Gamma)$: Theorem 823.
2. LUR$\not\Rightarrow$ injection into $c_0(\Gamma)$: Theorem 823.
3. WUR $\not\Rightarrow$ injection into $c_0(\Gamma)$: Theorem 735. The first example that R $\not\Rightarrow$ injection into $c_0(\Gamma)$ was in Theorem 739.
4. X Vašák does not imply injection of X^* into $c_0(\Gamma)$ even not weak*-weak continuous: Theorem 735.
5. $c_0(\Gamma)$ for uncountable Γ cannot be injected to either $\ell_p(\Gamma)$, $p \geq 1$ or to ℓ_∞: These spaces admits URED norms and $c_0(\Gamma)$ not.
6. If $p < q$ and Γ is uncountable then $\ell_q(\Gamma)$ does not inject into $\ell_p(\Gamma)$ (Pitt's theorem).
7. Only separable WLD spaces inject into ℓ_∞: Theorem 745.
8. The space ℓ_∞/c_0 does not inject into $c_0(\Gamma)$: Theorem 808.
9. The space D injects into ℓ_∞ : Theorem 841.
10. The space $C[0,\omega_1]$ does not inject into ℓ_∞: It does not admit URED norm as it contains a copy of $c_0(\omega_1)$.
11. $c_0(\Gamma)$ for uncountable Γ does not inject into ℓ_∞: The dual of $c_0(\Gamma)$ is not weak* separable, or $c_0(\Gamma)$ does not admit a URED norm.
12. $\ell_1(\omega_1)$ injects into ℓ_∞: $\ell_1(\omega_1)$ is a subspace of $C[0,1]^*$.
13. Nonseparable $L_1(\mu)$, μ finite, does not inject into $\ell_1(\omega_1)$. Otherwise it would inject into ℓ_∞. Use that $L_1(\mu)^*$ is not w^*-separable, as $L_1(\mu)$ is WCG.
14. To be a Vašák space does not imply injection of the dual into $c_0(\Gamma)$: Corollary 736.
15. To be an Asplund space implies injection of the dual into $c_0(\Gamma)$: Theorem 774.
16. Nonseparable $L_1(\mu)$ for finite measure μ does not inject into any reflexive space, see [Ros70].
17. For nonseparable Γ, $c_0(\Gamma)$ does not inject into $L_1(\mu)$ since the latter admits an URED norm (Kutzarova, Troyanski, see [DeGoZi93, Theorem II.7.16]) and $c_0(\Gamma)$ does not have an URED renorming (Theorem 811).

18. No injection from $c_0(c)$ into ℓ_∞: No w^*-separability of $c_0(c)^*$.

19. There is no injection from $c_0(\omega_1)$ into a reflexive pace. Indeed, Otherwise the dual operator to such injection would be w-compact so $\|\cdot\|$-compact (Schur). Thus, $c_0(\omega_1)^*$ would be w^*-separable, and this is not the case.

20. The reflexivity of the space does not imply the possibility of injection into a superreflexive space: Theorems 363 and 365.

51.6 Markushevich bases

1. Every separable space admits M-basis that is not Schauder, see Theorem 827.

2. Injection into $c_0(\Gamma)$ and no M-basis: ℓ_∞ (Theorem 827).

3. M-basis no $\langle$LUR$\rangle$: Theorem 827.

4. Johnson's example of a space with M-basis and no strong M basis: Theorem 827.

5. Example of WCG and a non weakly compact M-basis: [HMVZ07, Remark 6.8].

6. Strong M-basis does not imply PRI. See Theorem 225. Compare with Theorem 222, so we have LUR without working with a PRI.

51.7 Miscellanea

1. Bishop–Phelps not simultaneously for two sets: Theorem 695.

2. There is a Banach space X and a subspace Y of X such that both Y and X/Y are isometric to Hilbert spaces and X is not even isomorphic to a Hilbert space: Theorem 701.

3. $(\ell_\infty/c_0)^*$ is not weak* separable: Theorem 808, or, $c_0(c)$ is in ℓ_∞/c_0, and $c_0(c)^*$ is not w^*-separable.

4. $C[0,\alpha]$ does not admit any unconditional basis if α uncountable: Theorems 746, 765, and 812.

5. If Γ is uncountable, then $\ell_1(\Gamma)$ is not isomorphic to a subspace of any $L_1(\mu)$ for finite measure μ.

   To prove it, note that $L_1(\mu)$ admits then a Gâteaux differentiable norm (see [FHHMZ11, 13.1, Example (v)], but $\ell_1(\Gamma)$ does not admit Gâteaux differentiable norm (see, e.g., [DeGoZi93, 13.1, Example (iv)]).

6. $\ell_1(\Gamma)$ does not contain ℓ_∞ (renormings by LUR, see Theorem 171). Alternatively, $\ell_1(\Gamma)$ is weakly sequentially complete (in fact, it has the Schur

property, see, e.g., [FHHMZ11, Subsection 5.2.1]), but ℓ_∞ is not weakly sequentially complete (indeed, it contains c_0, which is not weakly sequentially complete —and this property is hereditary—: consider the sequence $\{e_1 + \ldots + e_n\}_{n=1}^\infty$, where e_n is the n-th vector of the canonical basis of c_0. Another, more sophisticated argument, is that *every weakly sequentially complete Banach space either is reflexive or it contains an isomorphic copy of* ℓ_1 —see. e.g., [FHHMZ11, Exercise 5.85]. The space c_0 fails this).

7. Operator from ℓ_∞ onto a dense set of $L_1(\mu)$ for finite μ: Composition of the Rosenthal quotient Theorem 809 with embedding from L_2 into L_1.
8. Operator from ℓ_∞ onto dense subspace of $c_0(c)$: Rosenthal quotient Theorem 809.
9. No operator from $c_0(c)$ onto a dense subspace of ℓ_∞ (the latter not WCG).
10. There is no operator from JL_0 onto a dense set in ℓ_∞ (the latter does not admit a Gâteaux norm, see [FHHMZ11, Exercise 7.64]).
11. If μ is finite then there is no operator from $L_1(\mu)$ onto a dense set in $\ell_1(\omega_1)$ (as the latter is not WCG).

Chapter 52

Open questions spread along the book and some additional ones

We gather in this chapter a list of open questions. The ones already mentioned along the text start with the reference to its location. Those explicitly related to renorming appear in gray boxes. Some other problems, not in the text above, are also included. For a more systematic collection we refer to [GuiMoZi16].

52.1 Differentiability

52.1.1 Gâteaux differentiability

1. Assume that X admits a Gâteaux differentiable norm and has an unconditional basis. Does X^* admit a dual rotund norm? Compare with Theorems 761 and 762.

2. Assume that X is a Banach space an Z is a subspace of X such that X/Z is separable. Assume that Z admits a Gâteaux differentiable norm. Does X also admit a Gâteaux differentiable norm?

3. Assume that X is a separable Banach space such that X does not contain a copy of ℓ_1.

   (i) Does X^* admit a not necessarily dual Gâteaux differentiable norm?

   (ii) Is the second dual norm on X^{**} Fréchet differentiable somewhere?

4. Assume that X is WLD and consider
   (i) X admits a Gâteaux differentiable bump.
   (ii) X admits a Gâteaux differentiable norm.
   (iii) X admits a norm whose dual is rotund. Does (i)⇒(ii)? Does (ii)⇒(iii)?

5. Does every infinite-dimensional separable space admit a norm that is R and G but not LUR?

6. Which separable spaces can be renormed by a G norm that is nowhere SG?

   For the Fréchet case (instead of SG) it is solved in e.g. [DeGoZi93, Theorem III.1.9].

A. J. Guirao et al., *Renormings in Banach Spaces*, Monografie Matematyczne 75,
https://doi.org/10.1007/978-3-031-08655-7_52

7. Which separable Banach spaces can be renormed by a norm that is nowhere SG? See Theorems 369 and 371.

8. (Page 430) It is an open problem whether in Theorem 762 it is equivalent also that X admits a Gâteaux differentiable norm.

52.1.2 Fréchet differentiability

1. Can every infinite-dimensional separable reflexive be renormed by F that is not UG?

2. Can every space with a Fréchet smooth norm be renormed by a LUR norm? (see [MOTV09, Question 6.14]).

3. (Pages 140 and 204) Assume X admits a LUR norm and also a Fréchet smooth norm. Does X admit a norm with both of these properties?

4. Can every infinite-dimensional superreflexive separable space be renormed by a norm that is Fréchet smooth and not UF?

5. (Page 399) It is not known if renorming by a Fréchet smooth norm is a three-space property. This is related to (iv) and (xiv) in Theorem 655.

6. (Page 410) Assume that Y is a subspace of a reflexive space X and $\|\cdot\|$ is an equivalent Fréchet smooth norm on Y. Can $\|\cdot\|$ be extended to an equivalent Fréchet smooth norm on X? The problem is open even in the separable setting. Related to this, see Theorems 122 and 703.

7. (Page 230) The question if every non-separable Banach space that does not admit a rough norm necessarily admits a Fréchet differentiable bump function, is still an open problem, one of the most important problems in this area of Banach spaces.

8. (Page 261) It seems to be open if every norm on a space X that admits a Fréchet smooth norm can be approximated uniformly on bounded sets by Fréchet differentiable norms. This happens if X is a separable Asplund space (see the proof at the mentioned location).

9. (Page 362) A nonseparable version of Theorem 551 is open.

10. (Page 362) The following problem is apparently open (see, e.g., [GuiMoZi16, Problem 162]): Assume that X admits both an SSD equivalent norm and a Gâteaux equivalent norm. Does it admit an equivalent Fréchet differentiable norm?

52.1.3 Higher-order smoothness

1. (Page xxii) Assume that a separable space X admits a C^2-smooth bump. Does it admit a C^2-smooth norm? Please check Theorems 500, 517, and 518.

2. Assume a separable X admits a C^k-smooth norm for each k. Does X admit a C^∞-smooth norm? [Pełczyński]

3. Is there a norm on c_0 that is C^∞-smooth, dentable, and has no extreme points on the sphere?

4. (Page xxii) Assume that X admits a C^k-smooth bump. Does every continuous function on X approximates by C^k-smooth functions?

5. Assume X admits a C^k-smooth norm. Do all equivalent norms approximate by C^k-smooth norms?

6. Assume that X admits a C^2-Fréchet smooth norm. Does X admit a LUR norm?

7. (Page 261) Does Theorem 323 hold true for nonseparable spaces as well? (Compare with Theorem 328.)

8. (Page 262) Assume that a nonseparable Banach space $(X, \|\cdot\|)$ admits a C^k-smooth norm and $\varepsilon > 0$ is given. Does there exist a C^k-smooth equivalent norm $|||\cdot|||$ on X such that

$$1 \leq \frac{|||x|||}{\|x\|} < 1 + \varepsilon \text{ for all } 0 \neq x \in X?$$

   The problem is solved practically only if $k = 1$ and the space belongs to some known classes, like separable spaces, WLD spaces, etc.

9. (Page xxii) If X admits a C^1-smooth bump, does it allow for smooth approximations of continuous functions?

10. (Page 392) The following problem seems to be open: If X has a C^2-smooth norm that is rotund, does X injects into $\ell_p(\Gamma)$ for some Γ? Compare with Theorem 460.

11. (Page 276) It is unknown whether a Banach space that admits a norm that is LFC admits a norm that is C^∞-smooth.

    The separable case has a positive answer [Há95], see also, e.g., [FHHMZ11, Theorem 10.12]).

12. (Page 264) It is unknown if (i) in Theorem 332 holds true without assuming that X is WCD.

52.1.4 Space c_0

1. Does c_0 admit a C^∞-smooth norm whose dual is LUR? Note that the answer is yes if LUR replaced by R (see Theorem 508).
2. Does there exist a C^2-smooth norm on c_0 whose ball is dentable?
3. (Page 341) Does every continuous function on c_0 approximates by real-analytic functions? Related to this, see Theorem 515.
4. Does there exist a C^∞-smooth norm on c_0 that is not dentable but strictly convex?
5. Is every continuous real-valued function on c_0 uniformly approximable by real-analytic functions?
6. (Page 282) It is apparently open whether there is an equivalent C^2-smooth norm on c_0 whose dual is not rotund.
7. (Page xxii) Assume that a separable space X admits a C^2-smooth bump. Does it admit a C^2-smooth norm?

52.1.5 Space ℓ_1

1. Can Theorem 694 be proven for a LUR norm on ℓ_1?

52.1.6 Space ℓ_2

1. (Page xxii) Does every equivalent norm on ℓ_2 can be approximated by real-analytic norms?
2. (Page 261) Does every continuous function on ℓ_2 approximates by real-analytic functions even if Γ is uncountable? See Theorem 515.
3. Does the space of compact operators on ℓ_2 admit a real-analytic norm?
4. (Page 261) Is every norm on ℓ_2 approximable uniformly on bounded sets by real-analytic norms? (See, e.g., [HáJo14, Chapter 7, Problem 112].)
5. (Page 341) If X is saturated with ℓ_2, must the bidual norm be somewhere Fréchet differentiable? Is it true for the space JT? (See [HáMoZi12]).

52.1.7 $C(K)$ and related spaces

1. (Page 438) If K is the Kunen compact, does $C(K)$ admit a C^1-smooth bump?
2. Is D a Gâteaux differentiability space? (see Theorem 841).

3. (Page 471) If X is WLD, is every continuous convex real function defined on it differentiable at the points of a dense set? (see [Ziz03, page 1800].

4. (Page 204) It is open if in $C[0, \omega_1]$, the set of all Fréchet differentiable norms is dense in the set of all equivalent norms. However, there is in $C[0, \omega_1]$ a Fréchet differentiable LUR norm [HáPro14].

5. (Page 340) It is unknown if every norm on $C[0, \omega_1]$ is a limit, uniform on bounded sets, of norms that are C^1-smooth.

6. (Page 347) If $C(K)$ admits a Fréchet differentiable norm, does it admit a C^∞-smooth norm?

7. (Page 262) Let $\varepsilon > 0$. Does there exist a C^1-smooth equivalent norm $|||\cdot|||$ on $C[0, \omega_1]$ such that

$$1 \leq \frac{|||x|||}{\|x\|_\infty} < 1 + \varepsilon \text{ for all } 0 \neq x \in C[0, \omega_1]?$$

8. Find a straightforward short proof that $C[0, \omega_1]$ cannot be renormed to have Phelps' property U without using Talagrand's Theorem 812 and Theorem 718.

9. (Page 255) Assume that K is a compact topological space such that on $C(K)$ there exists an equivalent norm $|||\cdot|||$ such that, for every dense subset D of K, then the D-lower semicontinuous envelope $|||\cdot|||_D$ of $|||\cdot|||$ is Gâteaux differentiable. Is then K an Eberlein compact space?

52.2 Asplund and weak Asplund spaces

1. (Page 460) It is an open problem if (ii) and (iii) in Theorem 819 are equivalent for every Asplund space.

2. (Page 221) It is open whether (iv) implies (iii) in Theorem 275.

3. (Page 219) It is not known whether every Asplund space admits a C^1-differentiable bump.

4. (Page 219) It is unknown whether every weak Asplund space X^* can be renormed by a rotund not necessarily dual norm, see, e.g., [Fab97, 5.4. Notes and Remarks].

5. (Page 221) It is open whether every Asplund space must admit a Fréchet differentiable bump.

6. (Page 219) It is an open question whether every subspace of a weak Asplund space is again weak Asplund.

7. (Page 438) Let $C(K)$ be Asplund. Does $C(K)$ admit a C^∞-smooth bump?

8. (Page 438) Let $C(K)$ be Asplund admitting a C^1-smooth norm. Does $C(K)$ admit a C^∞-smooth norm? (It is true if $C(K)$ admits a norm whose dual norm is LUR [HájHay07]. See Theorem 530.)

9. (Page 438) If K is a scattered compact space, does $C(K)$ admit a C^∞-smooth bump?

10. (Page 221) Does every Asplund space admit a Fréchet smooth bump? See also Theorem 272 and pages 219 and 221.

11. (Page 219) If X is an Asplund space, does X admit a C^1-smooth bump?

12. Assume that X is a separable space such that in its norm $\|\cdot\|$, every subspace contains a point of Fréchet differentiability of the restricted norm $\|\cdot\|$. Is X necessarily Asplund?

52.3 Rotundity

1. Does every infinite-dimensional separable superreflexive space admit a norm that is LUR and WUR and not UR?

2. Can the norm in Theorem 402 be URED?

3. Can every infinite-dimensional separable space be renormed to be R but not MLUR? See Theorems 394, 399, 403, and 424.

4. Can every infinite-dimensional separable space be renormed to be LUR and not URED? See Theorems 392 and 393.

5. Can every infinite-dimensional separable space be renormed by URED that is not LUR? See Theorems 392, 393, 396, 841, and 842.

6. Can every infinite-dimensional space with separable dual be renormed by WUR and not MLUR?

7. Can every infinite-dimensional separable Asplund space be renormed to be URED and not WUR?

8. Does every reflexive URED space admit a WUR norm ([Sm77])?

9. Find a straightforward proof of Theorem 221 for the case of $Y = c_0(\Gamma)$ by using only ideas from the original Troyanski's paper [Tr71].

10. (Page 460) Assume that for separable space X, the space X^* cannot be renormed by a LUR norm. Does X necessarily contain a copy of ℓ_1? (See, e.g., [GuiMoZi16, Problem 152].)

11. (Page 400) It seems to be unknown if the renorming by a WUR norm is a three-space property.

12. (Page 215) It is unknown whether the RNP property implies the existence of an equivalent LUR norm.

13. (Page 460) Assume that X is a separable Banach space such that X does not contain a copy of ℓ_1. Does X^* admit a LUR norm (not necessarily dual)?

14. (Pagc 301) It is an interesting open problem whether Theorem 422 holds for nonseparable spaces as well.

52.4 Weakly compactly generated spaces and relatives

1. Assume that X is a subspace of a WCG (or even Vašák) space. Can we renorm X by a Gâteaux norm so that we do not change the set of norm-attaining functionals? See Proposition 232.

2. Let X be a WCG space. Can we give a "Banach space proof" that it is homeomorphic to a Hilbert space? It was done for reflexive spaces (Bessaga) and $c_0(\Gamma)$ (Troyanski).

3. (Page 429) Is X WCG if X^{**} is?

4. (Page 255) It is apparently unknown whether every space with the strong property (P) is a subspace of a WCG Banach space.

52.5 Auerbach and Markushevich bases

1. Let X admit a LUR norm and a Markushevich basis. Does X admit a strong Markushevich basis? See Theorem 222.

2. (Page 120) Does every separable space have an Auerbach basis? What about $C[0,1]$?

52.6 Miscellanea

1. Let X be separable and S denote the set of norm-attaining functionals. For what kind of S can we produce a LUR norm which preserves (is contained, or contains) S ? If $S = X^*$, can we do it without James' weak compactness theorem?

2. (Page xxii) It is unknown whether every nonseparable Banach space has a infinite-dimensional separable quotient, see, e.g., [GuiMoZi16, Problem 32].

3. (Page 388) Assume X is a separable Banach space that admits a bump that locally depends of finitely many coordinates. Is X isomorphic to a polyhedral space? See Theorem 612.
4. (Page 408) It is open whether Theorem 692 holds for nonseparable spaces as well. In this direction see also [Go10].
5. (Page 409) The following problem is open: Is it true that for every infinite-dimensional separable Banach space X there exist two spaces Y and Z isomorphic to X such that the Banach–Mazur distance $d(Y, Z) = 1$ and Y is not isometric to Z? (see also, e.g., [GuiMoZi16, Problem 96]). [Odell]
6. (Page 412) If α is a given cardinal, is it true that every two Banach spaces of density α are C^1-smooth images of each other?
7. (Page 215) Is it true that in every nonreflexive Banach space there is a norm in which some closed convex bounded set has no farthest point? (see, e.g., [DeGoZi93, Open problems. Problem III.4]).
8. (Pages 215 and 241) It is not known whether the KMP property implies the RNP property.
9. (Page 272) If the Borel fields for the norm and the weak topology on a Banach space X coincide, can X be renormed by an equivalent Kadets–Klee norm?
10. (Page 285) It is apparently unknown whether for any n-dimensional ball there is a direction such that there is no nondegenerate line segment in the boundary of this direction. In other words, does any n-dimensional Banach space contains a 1-dimensional Chebyshev subspace? (see references in [Ziz71], or [Klee69], or [PavZa07]). For $n = 3$ it is solved.
11. (Page 170) It is an open problem whether every normed space is homeomorphic to an inner product (non complete in general) space [PelBe79, page 278, Notes after Theorem 13.9].
12. (Page 278) It is open if (iii) in Theorem 362 holds for separable infinite-dimensional Hilbert spaces as well.
13. (Page 276) It is an open problem whether every separable space X that admits an LFC continuous bump admits an LFC norm.
14. (Page 365) Does Theorem 561 hold true if finite-dimensionality is replaced by separability of X? [Mazur].
15. (Page 405) Let X be a separable superreflexive Banach space. Does there exist a separable superreflexive space Y with Schauder basis such that X is isomorphic to a subspace of Y? [Godefroy].
16. (Page 272) It is open whether Theorem 348 holds for all Banach spaces.

17. (Page 365) Does every superreflexive space admit an equivalent quasitransitive norm? [Cabello] [Cab97].

18. (Page 389 Does there exist a separable space with a star-finite covering by closed balls?

52.7 Distortion, hereditarily indecomposable spaces

1. (Page 331) It is not known whether T^* is arbitrarily distortable.

2. (Page 332) It is an open problem whether in Theorem 474 the space of compact operators can be replaced with the space of nuclear operators (for the definition see, e.g., [FHHMZ11, Definition 16.21]) [Pisier], see, e.g., [GuiMoZi16, Problem 71].

3. (Page 333) Is Tsirelson space arbitrarily distortable?

4. (Page 333) If X is of density $\geq c$, does X contain a separable subspace with unconditional basis? [Todorčević]

5. (Page 333) Does there exist a reflexive Banach space of density c such that c_0 is crudely finitely representable in every infinite-dimensional subspace of X?

6. (Page 333) Does there exist a hereditarily indecomposable space that admits a C^2 smooth norm?

7. (Page 333) Does there exist a hereditarily indecomposable space that is isomorphic to its dual?

8. (Page 333) A Banach space X is called nonseparably hereditarily indecomposable if whenever $Y \subset X$ is a subspace of X and $Y = Y_0 \oplus Y_1$ is a topological sum for Y_0 and Y_1 subspaces of Y, then Y_0 or Y_1 must be separable. Study this notion. In this direction, see [LopTod09].

52.8 Mazur intersection property

1. (Page 368) It is open whether X admits a norm with MIP if it admits a Lipschitz Fréchet smooth bump (see [GuiMoZi16, Problem 187]).

2. (Page 368) If X has the UMIP, is X necessarily superreflexive?

3. Assume that X admits a Fréchet differentiable bump function. Does X have the MIP? See Theorem 568. It is even open in the case that the bump is Lipschitz and Fréchet differentiable (see page 368).

52.9 Lipschitz functions and Lipschitz Asplund spaces

52.9.1 Lipschitz functions

1. (Page 322) Assume that a separable Banach space X has the property that any Gâteaux differentiable norm on X is Fréchet differentiable somewhere. Is the same true for Lipschitz functions on X?

2. (Page 322) Assume that f_1, f_2, f_3 are Lipschitz functions on ℓ_2. Does there exist a point where all are Fréchet differentiable?

3. (Page 468) Is there a couple of separable spaces that are Lipschitz homeomorphic but no linearly isomorphic?

52.9.2 Lipschitz-free spaces

1. (Page 382) It is an open problem whether $\text{Ext}\,(B_{\mathcal{F}(M)}) \subset \text{Mol}\,(M)$. Related to this, see Theorem 604.

2. (Page 380) It is an open problem whether $\mathcal{F}(M)$ is the unique (linearly isometrically) predual of $\text{Lip}_0(\text{M})$. The answer is positive when M is bounded [Weav18, Theorem 3.26] or "convex".

3. (Page 382) It is an open problem whether $\mathcal{F}(\mathbb{R}^n)$ and $\mathcal{F}(\mathbb{R}^m)$ are isomorphic for any $n > m \geq 2$.

52.9.3 Lipschitz Asplund spaces

1. (Page 378) Is c_0 a Lipschitz Asplund space? The same question is open for the James' space J.

2. (Chapter 36) If H is a separable infinite-dimensional Hilbert space, does (iii) in Theorem 362 hold?

3. (Page 378) Does Theorem 596 hold true if RNP is replaced by X not containing a copy of c_0?

4. (Page 378) Assume that X is a Lipschitz Asplund space. Is X necessarily superreflexive?

5. (Page 378) Is every subspace of a Lipschitz Asplund space Lipschitz Asplund?

6. (Page 378) Is c_0 a Lipschitz Asplund space? [Fabian]

7. (Page 378) Is the James space J a Lipschitz Asplund space? [Fabian]

Bibliography

[AaLu74] J. Aarts and D. Lutzer, *Completeness properties designed for recognizing Baire space*, Dissert. Math. **116** (1974), 1–43.

[AHNT18] T. Abrahamsen, P. Hájek, O. Nygaard, and S.L. Troyanski, *Strongly extreme points and approximation properties*, Canad. Math. Bull. **61** , 3 (2018), 449–459.

[AHT20] T. Abrahamsen, P. Hájek, S.L. Troyanski, *Almost square dual Banach spaces*, J. Math. Anal. Appl. **487**, 2 (2020).

[AbAlBu91] Y.A. Abramovich, C.D. Aliprantis, and O. Burkinshaw, *The Daugavet equation in uniformly convex Banach spaces*, J. Functional Anal. **97** (1991), 215–230.

[AcKa11] M.D. Acosta and V.M. Kadets, *A characterization of reflexive spaces*, Math. Ann. **349** (2011), 577-588.

[AhLi78] I. Aharoni and J. Lindenstrauss, *Uniform equivalence between Banach spaces*, Bull. Amer. Math. Soc. **84** (1978), 281–283.

[AlKa06] F. Albiac and N.J. Kalton, *Topics in Banach Space Theory.* Graduate texts in Mathematics, vol. 233, Springer 2006.

[Alex39] A.D. Alexandrov, *Almost everywhere existence of the second differential of a convex function and some properties of convex surfaces connected with it*, Uchenye Zapiski Leningrad. Univ. **6** (1939), 3–35.

[AlDi85] G.A. Aleksandrov and I.P. Dimitrov, *On the equivalent weakly midpoint locally uniformly rotund renorming of the space* ℓ^∞, (in Russian), in Mathematics and Mathematical Education (Sunny Beach, 1985), pp. 189–191.

[AlePli06] G.A. Aleksandrov and A. Plichko, *Connection between strong and norming Markushevich bases in nonseparable Banach spaces*, Mathematika **53** (2), 2006, 2007, 321–328.

[AlHa14] S. D'Alessandro and P. Hájek, *Polynomial algebras and smooth functions in Banach spaces*, J. Functional Analysis, **266** (2014), 1627–1646.

[AlPer20] R.J. Aliaga and E. Pernecka, *Supports and extreme points in Lipschitz-free spaces*, Rev. Mat. Iberoam. **6**, 7 (2020), 2073–2089.

[Al21] R.J. Aliaga, *Extreme points in Lipschitz-free spaces over compact metric spaces*, arXiv:2102.01219 [math.FA].

A. J. Guirao et al., *Renormings in Banach Spaces*, Monografie Matematyczne 75,
https://doi.org/10.1007/978-3-031-08655-7

[AlPePr21] R.J. Aliaga, C. Petitjean, and A. Procházka, *Embedding Lipschitz-free spaces into ℓ_1*, J. Functional Anal. **280**, 6 (2021), Article 108916.

[Am86] D. Amir, *Characterizations of Inner Product spaces, Operator theory: Advances and Applications*, **20**, Birkhäuser, (1986).

[AmLi68] D. Amir and J. Lindenstrauss, *The structure of weakly compact sets in Banach spaces*, Ann. Math. **88** (1968), 35–44.

[And60] K.W. Anderson, *Midpoint local uniform convexity, and other geometric properties of Banach spaces.* Dissertation, University of Illinois 1960.

[ArGoRo03] S.A. Argyros, G. Godefroy, and H.P. Rosenthal, *Descriptive set theory and Banach spaces*, Handbook of Banach spaces II, Editors W.B. Johnson and J. Lindenstrauss, Elsevier Science B.V., 2003, 1007–1071.

[ArHay11] S.A. Argyros, R.G. Haydon, *A hereditarily indecomposable $\mathcal{L}_\infty$-space that solves the scalar-plus-compact problem*, Acta Math. **206** (2011), 1–54.

[ArMe93] S.A. Argyros and S. Mercourakis, *On weakly Lindelöf Banach spaces*, Rocky Mountain J. Math. **23** (1993), 395–446.

[ArMe04] S.A. Argyros and S. Mercourakis, *Examples concerning heredity problems of WCG Banach spaces*, Proc. Amer. Math. Soc. **133** (3), (2004), 773–785.

[ArMe05] S.A. Argyros and S. Mercourakis, *A note on the structure of WUR spaces*, Comment. Math. Univ. Carolin. **46** (3) (2005), 399–408.

[AMN88] S.A. Argyros, S. Mercourakis, and S. Negrepontis, *Functional-analytic properties of Corson-compact spaces*, Studia Math. **89** (1988), 197–229.

[ArTo04] S.A. Argyros and A. Tolias, *Methods in the theory of hereditarily indecomposable Banch spaces*, Mem. Amer. Math. Soc. **170** (2004).

[Ar02] R. Aron, *An introduction to polynomials on Banach spaces*, Extracta Mathematicae **17** , no. 3 (2002), 303–329.

[Asp66] E. Asplund, *Farthest points in reflexive locally uniformly rotund Banach spaces*, Israel J. Math **4** (1966), 213–216.

[Asp67] E. Asplund, *Averaged norms*, Israel J. Math. **5** (1967), 227–233.

[Asp68] E. Asplund, *Fréchet differentiability of convex functions*, Acta Math. **121** (1968), 31–47.

[Av09] A. Avilés, *Compact space that does not map onto finite products*, Fund. Math. **202** (2009), 81–96.

[AvKo13] A. Avilés and P. Koszmider, *A Banach space in which every injective operator is surjective*, Bull. London Math. Soc. **45** (2013),1065–1074.

[AvTr21] A. Avilés and S.L. Troyanski, *Spaces $C(K)$ with an equivalent URED norm*, Proc. Am. Math Soc, **149**, 4 (2021), 1569–1579.

[AyDoLo97] J.M. Ayerbe Toledano, T. Domínguez Benavides, and G. López Acedo, Measures of noncompactness in metric fixed point theory, Operator Theory: Advances and Applications, **99**, Birkhuser Verlag, Basel, 1997.

[AzDe01] D. Azagra and R. Deville, *James' theorem fails for starlike bodies*, J. Funct. Anal. **180** (2) (2001), 328–346.

[AzDeJi03] D. Azagra, R. Deville, and M. Jiménez-Sevilla, *On the range of the derivatives of smooth functions between Banach spaces*, Proc. Cambridge Math. Soc. **134** (2003), 163–185.

[BaDa00] M. Bachir and A. Danilidis, *A dual characterization of the Radon–Nikodým property*, Bull. Austral. Math. Soc. **62** (2000), 379–387.

[Ban32] S. Banach, *Théorie des opérations linéaires*, Monografie Matematyczne, Warszawa 1932.

[BanSa30] S. Banach and S. Saks, *Sur la convergence forte dans les champs L^p*, Studia Math. **2** (1930), 51–57.

[BaHá08] M. Bačák and P. Hájek, *Mazur intersection property for Asplund spaces*, J. Functional Anal. **255** (2008), 2090–2094.

[Bae72] A. Baernstein II, *On reflexivity and summability*, Studia Math. **42** (1972), 91–94.

[Bat97] S.M. Bates, *On smooth nonlinear surjections of Banach spaces*, Israel J. Math. **100** (1997), 209–220.

[Batu87] D.P. Baturov, *On subspaces of function spaces*, Vestnik Moskov. Univ. Ser. Mat. 4 (1987), 66-69. English transl. in Moscow Univ. Math. Bull. **42** (1987).

[BeLi72] Y. Benyamini and J. Lindenstrauss, *A predual of ℓ_1 which is not isomorphic to a $C(K)$ space*, Israel J. Math. **13** (1972), 246–259.

[BeLi00] Y. Benyamini and J. Lindenstrauss, *Geometric nonlinear functional analysis*, **Vol. 1**, Colloquium Publications **48**, American Mathematical Society, 2000.

[BRW77] Y. Benyamini, M.E. Rudin, and M. Wage, *Continuous images of weakly compact subsets of Banach spaces*, Pacific J. Math. **70**, 2 (1977), 309–324.

[BeSta76] Y. Benyamini and T. Starbird, *Embedding weakly compact sets into Hilbert space*, Israel J. Math. **23** (1976), 137–141.

[Bess58] C. Bessaga, *A note on universal Banach spaces of finite dimension*, Bull. Pol. Acad. **6** (1958), 97–101.

[BessPe60] C. Bessaga and A. Pełczyński, *Some remarks on homeomorphisms of B- spaces*, Bull. Acad. Polon. Sci. **8** (1960), 757–760.

[BessPe66] C. Bessaga and A. Pełczyński, *On extreme points in separable conjugate spaces*, Israel J. Math. **4** (1966), 262–264.

[BessPe75] C. Bessaga and A. Pełczyński, *Selected topics in infinite dimensional topology*, Polish Scientific Publishers, Warszawa, 1975.

[BiCoPu96] G. Bianchi, A. Colesanti, and C. Pucci, *On the second differentiability of convex surfaces*, Geometricae Dedicata **60** (1996), 39–48.

[BiSm16] V. Bible and R.J. Smith, *Smooth and polyhedral approximation in Banach spaces*, J. Math. Anal. Appl. **435** (2) (2016), 1262–1272.

[BiPh61] E. Bishop and R.R. Phelps, *A proof that every Banach space is subreflexive*, Bull. Amer. Math. Soc. **67** (1961), 97–98.

[BogShk88] V.I. Bogachevand and S.A. Shkarin, *Differentiable and Lipschitzian mappings of Banach spaces*, Math. Notes **44** (1988), 790–798.

[Bo57] W.M. Bogdanowicz, *Weak continuity of polynomial functionals on the space* c_0 (Russian), Bull. Acad. Polon. Sci **5** (1957), no. 3, 243–246.

[Bol70] B. Bollobás, *An extension to the theorem of Bishop and Phelps*, Bull. London Math. Soc. **2** (1970), 181–182.

[Bol99] B. Bollobás, *Linear Analysis* (second edition), Cambridge University Press, 1999.

[BoFra66] R. Bonic and J. Frampton, *Smooth functions on Banach manifolds*, Journal of Mathematics and Mechanics, **15** (5), (1966), 877–898.

[BoFa93] J.M. Borwein and M. Fabian, *On convex function s having points of Gateaux differentiability which are not points of Fréchet differentiability*, Can. J. Math. **45** (6) (1993), 1121–1134.

[BoFi93] J.M. Borwein and S. Fitzpatrick, *A weak Hadamard smooth renorming of* $L_1(\Omega, \mu)$, Canad. Math. Bull. **36**, 4 (1993), 407–413.

[BoFiVa94] J.M. Borwein, S. Fitzpatrick, and J. Vanderwerff, *Examples of convex functions and classifications of normed spaces*, J. Convex Anal. **1** (1) (1994), 61–73.

[BoMoVa06] J.M. Borwein, V. Montesinos, and J. Vanderwerff, *Boundedness, differentiability and extensions of convex functions*, J. Convex Anal. **13** (2006), 3-4, 587–602.

[BoNo94] J.M. Borwein and D. Noll, *Second order differentiability of convex functions in Banach spaces*, Trans. Amer. Math. Soc. **342** (1), (1994), 43–81.

[BoPpr87] J.M. Borwein and D. Preiss, *A smooth variational principle with applications to subdifferentiability and differentiability of convex functions*, Trans. Amer. Math. Soc. **303** (1987), 517–527.

[BoVa09] J.M. Borwein and J. Vanderwerff, *Convex functions: Constructions, characterizations and counterexamples*, Encyclopedia of Mathematics and Its Applications **109**, Cambridge University Press, 2009.

[Bos02] B. Bossard, *A coding of separable Banach spaces. Analytic and coanalytic families of Banach spaces*, Fund. Math. **172**, 117–152.

[BoGoKa96] B. Bossard, G. Godefroy, and R. Kaufman, *Hurewicz's Theorems and Renormings of Banach spaces*, J. Funct. Anal. **140** (1996), 142–150.

[Bou80] J. Bourgain, *ℓ_∞/c_0 has no equivalent strictly convex norm*, Proc. Amer. Math. Soc. **78** (1980), 225–226.

[Bou80b] J. Bourgain, *On separable Banach spaces universal for all separable reflexive spaces*, Proc. Amer. Math. Soc. **79** (1980), 241–246.

[BouDel81] J. Bourgain and F. Delbaen, *A class of special $\mathcal{L}_\infty$-spaces*, Acta Math. **145** (1981), 155–176.

[BouPis83] J. Bourgain and G. Pisier, *A construction of of $\mathcal{L}_\infty$-spaces and related Banach spaces*, Bol. Soc. Bras. Mat. **14** (2), (1983), 109–123.

[BouRoSch81] J. Bourgain, H.P. Rosenthal, and G. Schechtman, *An ordinal L^p-index for Banach spaces, with applications to complemented subspaces of L^p*, Ann. Math. **114**, 2 (1981), 193–228.

[BouTa80] J. Bourgain and M. Talagrand, *Compacité extremal*, Proc. Amer. Math. Soc. **80** (1980), 68–70.

[Bour83] J. Bourgin, *Geometric Aspects of Convex Sets with the Radon–Nikodým Property*, Lecture Notes in Mathematics **993**, Springer Verlag 1983.

[Br99] P. Brass, *On equilateral simplices in normed spaces*, Beitrage Algebra Geom. **40** (1999), 303–307.

[BroMi48] M.S. Brodskii and D.P. Mil'man, *On the center of a convex set.* Dokl. Akad. Nauk SSSR (N. S.) **59**, 837–840 (1948).

[BroIt80] L. Brown and T. Ito, *Classes of Banach spaces with unique isometric preduals*, Pacific J. Math., **90**, 2 (1980), 261–283.

[Cab97] F. Cabello Sánchez, *Regards sur le problème des rotations de Mazur*, Extracta Mathematicae **12** (1997), 97–116.

[Cab98] F. Cabello Sánchez, *Transitivity of M-spaces and Wood's conjecture*, Math. Proc. Camb. Phil. Soc. **214**(1998), 513–520.

[Car67] H. Cartan, *Calcul Différentiel. I Calcul différentiel dans les espaces de Banach. II Équations différentielles*, Collection Méthodes, Hermann, Paris, 1967.

[CasGo98] B. Cascales and G. Godefroy, *Angelicity and the boundary problem.* Mathematika **45** (1998), 105–112.

[CaMaVe97] B. Cascales, G. Manjabacas, and G. Vera, *A Krein–Šmulian type result in Banach spaces*, Quart. J. Math. Oxford **48**, 2 (1997), 161–167.

[CaOrPe17] B. Cascales, J. Orihuela, and A. Pérez, *One-sided James' compactness theorem*, J. Math. Anal. and Appl. **445**, 2 (2017), 1267–1283.

[CasShv03] B. Cascales and R. Shvydkoy, *On the Krein-Šmulian theorem for weaker topologies*, Ill. J. Math. **47** (2003), 957–976.

[Casz77] P.G. Casazza, *James' quasi-reflexive space is primary*, Israel J. Math. **26** (3-4) (1977), 294–305.

[CaJohLin84] P.G. Casazza, W.B. Johnson and L. Tzafriri, *On Tsirelson's space*, Israel J. Math. **47** (2-3) (1984), 81–98.

[CaKaTz87] P.G. Casazza, N.J. Kalton, and L. Tzafriri, *Decompositions of Banach lattices into direct sums*, Trans. Amer. Math. Soc. **304** (1987), 2, 771–800.

[CaSh89] P.G. Casazza and T. Shura, *Tsirelson's space*, Springer Lecture Notes, **1363** (1989).

[CasGo87] J.M.F. Castillo and M. González, *Three-Space Problems in Banach Space Theory*, Lecture Notes in Mathematics, vol. 1667 (Springer, Berlin, 1997).

[CGKP20] J.M.F. Castillo, M. González, T. Kania, and P.L. Papini, *The isomorphic Kottman constant of a Banach space*, Proc. Amer. Math. Soc. **148**, 10 (2020), 4361–4375.

[CGP17] J.M.F. Castillo, M. González, and P.L. Papini, *New results on Kottman constant*, Banach J. Math. Anal. **11** (2), (2017), 348–352.

[CechPo38] E. Čech and B. Pospíšil, *Sur les espaces compacts*, Publ. Fac. Sci. Univ. Masaryk **258** (1938), 1–14.

[Cep98] M. Cepedello, *Approximation of Lipschitz functions by* Δ*- convex functions in Banach spaces*, Israel J. Math. **106** (1998), 269-284.

[Cho69] G. Choquet, *Lectures on Analysis I, II, III*, W. A. Benjamin, New York, 1969.

[CiePo84] K. Ciesielski and R. Pol, *A weakly Lindelöf function space* $C(K)$ *without any continuous injection into* $c_0(\Gamma)$, Bull. Pol. Acad. Sci. Math. **32** (1984), 681–688.

[Clar36] J.A. Clarkson, *Uniformly convex spaces.* Trans. Amer. Math. Soc. **40**, 396–414 (1936).

[CobKen85] M. Čoban and P.S. Kenderov, *Dense Gâteaux differentiability of the sup-norm in* $C(T)$ *and the topological properties of* T, C. R. Acad. Bulgare Sci. **38** (1985), 1603–1604.

[Coll76] J.B. Collier, *The dual of a space with the Radon-Nikodým property*, Pacific J. Math. **64** (1), (1976), 103–106.

[CoEd74] J.B. Collier and M. Edelstein, *On strongly exposed points and Fréchet differentiability*, Israel J. Math. **17** (1974), 66–68.

[CoPa94] M.D. Contreras and R. Payá, *On uppersemicontinuity of duality mappings*, Proc. Amer. Math. Soc. **121** (1994), 451–459.

[Cor61] H.H. Corson, *The weak topology of a Banach space*, Trans. Amer. Math. Soc. **101** (1961), 1–15.

[Cor61b] H.H. Corson, *Collections of convex sets which cover a Banach space*, Fundamenta Math. **49** (1961) 143–145.

[Cor65] H.H. Corson, *A compact convex set in* E^3 *whose exposed points are of the first category*, Proc. Amer. Math. Soc. **16** (5) (1965), 1015–1021.

[CorLi66] H.H. Corson and J. Lindenstrauss, *On weakly compact subsets of Banach spaces*, Proc. Amer. Math. Soc. **17** (1966), 407-412.

[Cud64] D.F. Cudia, *The geometry of Banach spaces. Smoothness*, Trans. Amer. Math. Soc. **110** (1964), 284–314.

[CuFa17] M. Cúth and M. Fabian, *Rich families and projectional skeletons in Asplund WCG spaces*, J. Math. Anal. Appl. **448** (2017) 1618–1632.

[CDDK] M. Cúth, M. Dolezal, M. Doucha, and O. Kurka, *Polish spaces of Banach spaces. Complexity of isometry classes and generic properties*, arXiv: 1912.03994 [math.FA], 2019.

[Dan72] J. Daneš, *A geometric theorem useful in nonlinear functional analysis*, Boll. Un. Mat. Ital. **6** (1972), 369–375.

[Dan85] J. Daneš, *Equivalence of some geometric and related results of nonlinear functional analysis*, Commentationes Mathematicae Universitatis Carolinae, **26**, 3 (1985), 443–454.

[DaHaRu20] S. Dantas, P. Hájek and T. Russo, *Smooth norms in dense subspaces of Banach spaces*, J. Math. Anal. Appl. **487** (2020).

[DaLl73] F.K. Dashiell and J. Lindenstrauss, *Some examples concerning strictly convex norms on $C(K)$ spaces*, Israel J. Math. **16** (1973), 329-342.

[Dau63] I.K. Daugavet, *A property of completely continuous operators in the space C*, Uspekhi Mat. Nauk **18** (5) (1963) 157–158.

[DFJP74] W.J. Davis, T. Figiel, W.B. Johnson, and A. Pełczyński, *Factoring weakly compact operators*, J. Funct. Anal. **17** (1974), 311–327.

[DaJo73] W.J. Davis and W.B. Johnson, *A renorming of nonreflexive Banach spaces*, Proc. Amer. Math. Soc. **37** (1973), 486-488.

[DaJo73b] W.J. Davis and W.B. Johnson, *On the existene of fundamental and total bounded biorthogonal systems in Banach spaces*, Studia Math **45** (1973), 173–179.

[DaLi72] W.J. Davis and J. Lindenstrauss, *On total nonnorming subspaces*, Proc. Amer. Math. Soc. **31** (1) (1972), 109–111.

[DaPh74] W.J. Davis ad R.R. Phelps, *The Radon-Nikodým property and dentable sets in Banach spaces*, Proc. Amer. Math. Soc. **45** (1) (1974), 119–122.

[Day55] M.M. Day, *Strict convexity ans smoothness of normed spaces*, Trans. Amer. Math. Soc. **78** (1955), 516-528.

[Day43] M.M. Day, *Uniform convexity III*, Bull. Amer. Math. Soc. **49** (1943), 745–750.

[Day73] M.M. Day, *Normed linear spaces*, Third Edition, Springer-Verlag (1973).

[DaJaSw72] M.M. Day, R.C. James, and S. Swaminathan, *Normed linear spaces that are uniformly convex in every direction*, Canadian J. Math. **83** (1972), 1051–1059.

[dBSoVe20] C.A. De Bernardi, J. Somaglia, and L. Veselý, *Star-finite coverings of Banach spaces*, J. Math. Anal. Appl. **491** (2), (2020), 124384.

[dWil74] M. de Wilde, *Pointwise compactness in spaces of functions and R.C. James theorem*, Math. Ann. **208** (1974), 33–47.

[Del10] S. Delpech, *Separated sequences in asymptotically uniformly convex Banach spaces*, Colloq. Math. **119** (2010), 123–125.

[Dek00] B.V. Dekster, *Simplexes with prescribed edge lengths in Minkowski and Banach spaces*, Acta Math. Hungar. **86**, 4 (2000), 343–358-.

[De86] R. Deville, *Problemes de renormages*, J. Functional Anal. **68** (1986), 117-129.

[De89] R. Deville, *Convergence puntuelle et uniforme sur un espace compact*, Bull. Acad. Polon. **37** (1989), 507–515.

[DeFa89] R. Deville and M. Fabian, *Principes variationnels et différentiabilité d'applications définies sur un espace de Banach*, Publ. Math. Besançon, Analyse non Linéaire, No. **10** (1989), 79–102.

[DeFoHá98] R. Deville, V.P. Fonf, and P. Hájek, *Analytic and polyhedral approximation of convex bodies in separable polyhedral Banach spaces*, Israel J. Math. **105** (1) (1998), 139–154.

[DeGhou01] R. Deville and N. Ghoussoub, *Perturbed minimization problems*, Handbook of Banach spaces I, Editors W.B. Johnson and J. Lindenstrauss, Elsevier Science B.V., 2001, 393–435.

[DeGo93] R. Deville and G. Godefroy, *Some applications of projectional resolutions of identity*, **67** (1993), 183–199.

[DGHZ87] R. Deville, G. Godefroy, D.E.G. Hare, and V. Zizler, *Differentiability of convex functions and the convex point of continuity property in Banach spaces*, Israel J. Math. **59** (2) (1987), 245–255.

[DeGoZi90] R. Deville, G. Godefroy and V. Zizler, *The three space problem for smooth partitions of unity and $C(K)$ spaces*, Math. Ann. **288** (1990), 613–625.

[DeGoZi91] R. Deville, G. Godefroy and V. Zizler, *Un principle variationnel utilisant des functions bosses*, C.R. Acad. Sci. Paris, Serie I **312** (1991), 281–286.

[DeGoZi93] R. Deville, G. Godefroy, and V. Zizler, *Smoothness and renormings in Banach spaces*, Pitman Monographs **64**, London, Longman, 1993.

[DeGoZi93a] R. Deville, G. Godefroy, and V. Zizler, *Smooth bump functions and geometry of Banach spaces*, Mathematika **40** (1993), 305–321.

[DeGoZi93b] R. Deville, G. Godefroy, and V. Zizler, *A smooth variational principle with applications to Hamilton–Jacobi equations in infinite-dimensions*, Journal of Functional Analysis **111**, 1 (1993), 197–212.

[DeGonJa99] R. Deville, R. Gonzalo, and J.A. Jaramillo, *Renormings of $L_p(L_q)$*, Mat. Proceedings of the Cambridge Philosophical Society **126** (1) (1999), 155–169.

[DeHá07] R. Deville and P. Hájek, *Smooth noncompact operators from $C(K)$, K scattered*, Israel J. Math. **162** (2007), 29–56.

[DeMa00] R. Deville and E. Matheron, *Pyramidal vectors and smooth functions on Banach spaces*, Proc. Amer. Math. Soc. **128**, 12 (2000), 3601–3608.

[DeZiz88] R. Deville and V. Zizler, *Farthest points in weak* compact sets*, Bull. Australian Math. Soc. **38** (1988), 433–439.

[Di75] J. Diestel, *Geomety of Banach spaces: Selected topics*, Lecture Notes in Mathematics **485**, Springer, 1975.

[Di84] J. Diestel, *Sequences and Series in Banach Spaces*, GTM **92**, Springer-Verlag, 1984.

[DiFa74] J. Diestel and B. Faires, *On vector measures*, Trans. Amer. Math. Soc. **198** (1974), 253–271.

[DiUh77] J. Diestel and J.J. Uhl, *Vector Measures*, Math. Surveys **15**, Amer. Math. Soc., 1977.

[Dieu69] J. Dieudonné, *Foundations of Modern Analysis*, Enlarged and corrected Printing, Academic Press, 1969.

[DiGiKu95] S.J. Dilworth, M. Girardi, and D. Kutzarova, *Banach spaces which admit a norm with the uniform Kadets–Klee property*, Studia Math. **112** (1995), 267–277.

[Dix48] J. Dixmier, *Sur un theoreme de Banach*, Duke Math. J. **15** (1948), 1057–1071.

[DoHuSm95] P.N. Dowling, Z. Hu, and M.A. Smith, *MLUR renormings of Banach spaces*, Pacific J. Math. **170** (2), (1995), 473-482.

[Du66] J. Dugundji, *Topology*, Allyn and Bacon, Inc., Boston 1966.

[DuGr03] J. Dugundji and A. Granas, *Fixed Point Theory*, Springer-Verlag 2003.

[Du82] D. van Dulst, *Equivalent norms and fixed point property for nonexpansive mappings*, J. London Math. Soc. **25** (1982), 139–144.

[DuSi75] D. van Dulst and I. Singer, *On Kadets–Klee norms on Banach spaces*, Studia Math. **54** (1975), 215–211.

[DuPe40] N. Dunford and B.J. Pettis, *Linear operations on summable functions*, Trans. Amer. Math. Soc. **47** (1940), 323–392.

[DuSch] N. Dunford and J.T. Schwartz, *Linear Operators; Part I: General Theory*, Interscience Pub., Inc., New York, John Wiley and Sons, Inc., New York, London, Sidney, 1957.

[Ed66] M. Edelstein, *Farthest points in uniformly convex spaces*, Israel J. Math. **4** (1966), 171–176.

[Ed73] M. Edelstein, *Concerning dentability*, Pacific J. Math. **46** (1), (1973), 111–114.

[EdTh72] M. Edelstein and A.C. Thompson, *Some results on nearest points and support properties of convex sets in* c_0, Pacific . J. Math **40** (3), (1972), 553-560.

[Edg77] G.A. Edgar, *Measurability in a Banach space*, Indiana Univ. Math. J.**26** (1977), 663-677.

[Edg79] G.A. Edgar, *Measurability in Banach spaces II*, Indiana Univ. Math. J. **28** (1979), 559-579.

[Edw81] H.M. Edwards, *Read the Masters!*, Steen L.A. (eds.), Mathematics Tomorrow, Springer, New York, NY., 1981.

[Ek74] I. Ekeland, *On the variational principle*, J. Math. Anal. Appl. **47** (1974), 324–353.

[EkLe76] I. Ekeland and G. Lebourg, *Generic Fréchet differentiability and perturbed optimization problems in Banach spaces*, Trans. Amer. Math. Soc. **224** (1976), 193–216.

[Enf72] P. Enflo, *Banach spaces which can be given an equivalent uniformly convex norm*, Israel J. Math. **13** (1972), 281–288.

[EnLiPi75] P. Enflo, J. Lindenstrauss, and G. Pisier, *On the "Three space problem"*, Math. Scand. **36** (1975), 189-210.

[Eng85] R. Engelking, *General Topology*, PWN, Warszawa, 1985.

[ElOd81] J. Elton and E. Odell, *The unit ball of every infinite dimensional normed space contains a* $(1+\varepsilon)$ *sparated sequence*, Colloq. Math. **44** (1) (1981), 105–109.

[Fab83] M. Fabian, *On minimum principles*, Acta Polytechnica **20** (1983), 109–118.

[Fab85] M. Fabian, *Lipschitz smooth points of convex functions and isomorphic characterizations of Hilbert spaces*, Proc. London Math. Soc. **51** (1985), 113–126.

[Fab87] M. Fabian, *Each weakly countably determined Asplund space admits a Fréchet differentiable norm*, Bull. Austr. Math. Soc **36** (1987), 367–374.

[Fab87-2] M. Fabian, *On projectional resolution of identity on the duals of certain Banach spaces*, Bull. Austral. Math. Soc. **35** (1987), 363–371.

[Fab97] M. Fabian, *Gâteaux differentiability of convex functions and topology —Weak Asplund spaces*, Wiley, New York, 1997.

[Fab91] M. Fabian, *On a dual locally uniformly rotund norm on a dual of Vašák space*, Studia Math. **101** (1991), 69-81.

[Fab95] M. Fabian, *On an extension of norms from a subspace to the whole Banach space keeping their rotundity*, Studia Math. **112** (1995), no 3, 203-211.

[FaGo88] M. Fabian and G. Godefroy, *The dual of every Asplund space admits a projectional resolution of the identity*, Studia Math. **91** (1988), 141–151.

[FGHZ03] M. Fabian, G. Godefroy, P. Hájek, and V. Zizler, *Hilbert-generated spaces*, J. Funct. Anal. **200** (2003), 301–323.

[FGMZ04] M. Fabian, G. Godefroy, V. Montesinos, and V. Zizler, *Inner characterizations of weakly compactly generated Banach spaces and their relatives,* J. Math. Anal. Appl. **297** (2004), 419–455.

[FaGoZi01] M. Fabian, G. Godefroy, and V. Zizler, *The structure of uniformly Gâteaux smooth Banach spaces*, Israel J. Math. **124** (2001), 243-252.

[FaGoMo09] M. Fabian, A. González, and V. Montesinos, *A note on biorthogonal systems in weakly compacly generated Banach spaces*, Annales Acad. Sci. Fennicae Mathematica **34** (2009), 555-564.

[FHHMPZ01] M. Fabian, P. Habala, P. Hájek, V. Montesinos, J. Pelant, and V. Zizler, *Functional Analysis and Infinite-Dimensional Geometry,* CMS, Springer-Verlag, 2001.

[FHHMZ11] M. Fabian, P. Habala, P. Hájek, V. Montesinos, and V. Zizler, *Banach Space Theory: The Basis for Linear and Nonlinear Analysis,* CMS, Springer-Verlag, 2011.

[FHV96] M. Fabian, P. Hájek, and J. Vanderwerff, *On Smooth Variational Principles in Banach Spaces*, Journal of Math. Analysis and Applications, **197** (1996), 153–172.

[FaHáZi97] M. Fabian, P. Hájek and V. Zizler, *Uniform Eberlein compacta and uniformly Gâteaux smooth norms*, Serdica Math. J. **23** (3-4), (1997), 351–362.

[FaMo18] M. Fabian and V. Montesinos, *WCG spaces and their subspaces grasped by projectional skeletons.* Funct. Approx. Comment. Math. **59**, 2 (2018), 231–250.

[FaMoZi02] M. Fabian, V. Montesinos, and V. Zizler, *Weakly compact sets and smooth norms in Banach spaces*, Bull. Australl. Math. Soc. **62** (2002), 223–230.

[FaMoZi02-b] M. Fabian, V. Montesinos, and V. Zizler, *Pointwise semicontinuous smooth norms*, Arch. Math. **78** (2002), 459–464.

[FMZ07] M. Fabian, V. Montesinos, and V. Zizler, *Weak compactness and $\sigma-$ Asplund generated Banach spaces*, Studia Math. **181** (2) (2007), 125–152.

[FaMoZi09] M. Fabian, V. Montesinos, and V. Zizler, *Sigma-finite dual dentability indices*, J. Math. Anal. Appl. **350** (2009), 498-507.

[FaMoZi09b] M. Fabian, V. Montesinos, and V. Zizler, *On weak compactness in L_1 spaces*, Rocky Mountain J. of Math. **39**, 6 (2009), 1885–1893.

[FaPr91] M. Fabian and D. Preiss, *On intermediate differentiability of Lipschitz functions on certain Banach spaces*, Proc. Amer. Math. Soc. **113** (3) (1991), 733–740.

[FPWZ89] M. Fabian, D. Preiss, J.H.M. Whitfield, and V. Zizler, *Separating polynomials on Banach spaces*, Quart. J. Math. Oxford (2), **40**, (1989), 409-422.

[FaTro90] M. Fabian and S.L. Troyanski, *A Banach space admits a locally uniformly rotund norm if its dual is a Vašák space*, Israel J. Math. **69** (1990), 214-224.

[FaZaZi81] M. Fabian, L. Zajíček, and V. Zizler, *On residuality of the set of rotund norms on a Banach space*, Math. Ann. **258** (1981/82), 349-351.

[FaZi97] M. Fabian and V. Zizler, *A note on bump functions that locally depend on finitely many coordinates*, Bull. Austral. Math. Soc. **56** (1997), 447–451.

[FaZi99] M. Fabian and V. Zizler, *On uniformly Gâteaux smooth C^n-smooth norms on separable Banach spaces*, Czech. J. Math. **49** (124) (1999), 657-672.

[FaZi99b] M. Fabian and V. Zizler, *An elementary approach to some questions in higher order smoothness in Banach spaces*, Extracta Math. **14** (3), (1999), 275–317.

[FWZ83] M. Fabian, J.H.M. Whitfield, and V. Zizler, *Norms with locally Lipschitzian derivatives*, Israel J. Math. **44** (1983), 262–276.

[Far74] N.R. Farnum, *The Banach–Saks theorem in $C(S)$*, Canad. J. Math. **26** (1974), 91–97.

[Fer97] V. Ferenczi, *A uniformly convex hereditarily indecomposable Banach space*, Israel J. Math. **102** (1997), 199–225.

[FeOrRa19] S. Ferrari, J. Orihuela, and M. Raja, *Generalized metric properties of spheres and renorming of Banach spaces*, RACSAM, **113**, 3 (2019), 2655–2663.

[Fi68] T. Figiel, *On non linear isometric embeddings of normed linear spaces*, Bull. Acad. Polon. Sci. Math. Astro. Phys. **16** (1968), 185–188.

[Fi72] T. Figiel, *An example of an infinite dimensional Banach space not isomorphic to its Cartesian square*, Studia Math. **42** (1972), 295–306.

[Fi76] T. Figiel, *On the moduli of convexity and smoothness*, Studia Math. **66** (1976), 121–155.

[FiJo74] T. Figiel and W.B. Johnson, *A uniformly convex Banach space which contains no ℓ_p*, Compositio Math. **29** (1974), 179–190.

[Fin86] C. Finet, *Uniform convexity properties of norms on superreflexive Banach spaces*, Israel J. Math **53**, 1 (1986), 81–92.

[FiSch89] C. Finet and W. Schachermayer, *Equivalent norms on separable Asplund spaces*, Studia Math. **89** (1989), 275–283.

[Fl80] K. Floret, *Weakly compact sets*, Lecture Notes in Mathematics **801**, Springer, 1980.

[Fog58] S.R. Foguel, *On a theorem by A.E. Taylor*, Proc. Amer. Math. Soc. **9** (1958), 325.

[Fon79] V.P. Fonf, *One property of Lindenstrauss–Phelps spaces*, Funktsional Anal. i Prilozh. **13**, 1 (1979), 79–80. (Functional Anal. Appl. **13**, 1 (1979), 66–67).

[Fon90] V.P. Fonf, *Three characterizations of polyhedral spaces*, Ukrainskii Math. Zhurnal **42** (1990), 1286–1290.

[Fon94] V.P. Fonf, *On supportless convex sets in incomplete normed spaces*, Proc. Amer. Math. Soc. **120** (1994), 1173–1176.

[Fon96] V.P. Fonf, *On exposed and smooth points of of convex bodies in Banach spaces*, Bull. London Math. Soc. **28** (1996), 51–58.

[FoKa76] V.P. Fonf and M.I. Kadets, *Some properties of the set of extreme points of the unit ball of a Banach space*, Mat. Zam. **20** (1976), 315-320.

[FonKa83] V.P. Fonf and M.I. Kadets, *Subspaces of ℓ_1 with strictly convex norm*, Mat. Zametki **33** (2), (1983), 41–422 Math. Notes of the Academy of Sc. USSR **33** (1983), 213-215.

[FLTZ18] V.P. Fonf, S. Lajara, S.L. Troyanski, and C. Zanco, *Norming subspaces of Banach spaces*, Proc. Amer. Math. Soc. **147** (2019), 3039–3045.

[FoLi03] V.P. Fonf and J. Lindenstrauss, *Boundaries and generation of convex sets.* Israel J. Math. **136** (2003), 157–172.

[FoLiPh01] V.P. Fonf, J. Lindenstrauss, and R.R. Phelps, *Infinite-dimensional convexity*, Handbook of Banach spaces I, Editors W.B. Johnson and J. Lindenstrauss, Elsevier Science B.V., 2001, 599–670.

[Fos92] M. Fosgerau, *A Banach space with Lipschitz Gâteaux smooth bump has weak* fragmentable dual*, Part of the author's Thesis, University College, London.

[Fre28] M. Fréchet, *Les espaces abstraits*, Gauthier–Villars, Paris, 1928.

[FOS11] D. Freeman, E. Odell and Th. Schlumprecht, *The universality of ℓ_1 as a dual space*, Math. Ann. **351** (2011), 149–186.

[FOSS14] D. Freeman, E. Odell, B. Sari, and Th. Schlumprecht, *Equilateral sets in uniformly smooth Banach spaces*, Mathematika **60**, 1, (2014), 219–231.

[Fro87] Z. Frolik, *A note on countably determined and distinguishable sets.* Comment. Math. Univ. Carolinae **28**, 2 (1987), 301–310.

[GPPR18] L. García-Lirola, C. Petitjean, A. Procházka, and A. Rueda-Zoca, *Extremal structure and duality of Lipschitz-free spaces*, Mediterr. J. Math., **15**, 69 (2018).

[Gark64] A.L. Garkavi, *On the best net and the best cross section of a set in a normed space*, Izv. Akad. Nauk USSR **26** (1962), 86-106, Eng. translation Amer. Math. Soc. Transl. **2** (1964), 111–137.

[Geo91] P.G. Georgiev, *On the residuality of the set of norms having Mazur's intersection property*, Math. Balkanica **5** (1991), 20-26.

[Geo05] P.G. Georgiev, *Parametric Borwein–Preiss variational principle and applications*, Proc. Amer. Math. Soc. **133** (2005), 3211–3225.

[GhMauSch92] N. Ghoussoub, B. Maurey, and W. Schachermayer, *Slicing, selections and their applications*, Canadian J. Math. **44** (1992), 483-504.

[Gil82] J.R. Giles, *Convex Analysis with Applications in Differentiation of Convex Functions*, Research Notes in Mathematics **58**, Pitman, 1982.

[GiGrSi78] J.R. Giles, D.A. Gregory, and B. Sims, *Characterization of normed linear spaces with Mazur's intersection property*, Bull. Austral. Math. Soc. **18** (1978), 471–476.

[GiGrSi78b] J.R. Giles, D.A. Gregory, and B. Sims, *Geometrical implications of upper semicontinuity of the duality mapping on a Banach space*, Pacific Journal of Mathematics, **79**, 1 (1978), 99–109.

[GlMe16] E. Glakousakis and S. Mercourakis, *Examples of infinite dimensional Banach spaces without infinite equilateral sets*, Serdica Math. **42**, no 1 (2016), 65–88.

[GlMcG72] A. Gleit and R. McGuigan, *A note on polyhedral Banach spaces*, Proc. Amer. Math. Soc. **33** (1972), (2), 398–404.

[Goda10] A. Godard, *Tree metrics and their Lipschitz-free spaces*, Proc. Amer. Math. Soc. **138**, 12 (2010), 4311–4320.

[Go78] G. Godefroy, *Espaces de Banach: existence et unicité de certains préduaux*, Ann. Inst. Fourier **28**, 3 (1978), 87–105.

[Go82] G. Godefroy, *Existence de normes tres lisses sur certains espaces de Banach*, Bull. Sci. Math. 2 **106** (1982), 63-68.

[Go87] G. Godefroy, *Boundaries of convex sets and interpolation sets*, Math. Ann. **277** (1987), 173-184.

[Go95] G. Godefroy, *Decomposable Banach spaces*, Rocky Mountain J. Math. **25** (1995), 1013-1024.

[Go00] G. Godefroy, *Some applications of Simons inequality*, Serdica Math. J **26** (2000), 59-78.

[Go01a] G. Godefroy, *The Banach space* c_0, Extracta Math. **16** (1) (2001), 1–25.

[Go01b] G. Godefroy, *Renormings of Banach spaces*, Handbook of Banach Spaces I, Editors W.B. Johnson and J. Lindenstrauss, Elsevier Science B.V., 2001, 781–835.

[Go06] G. Godefroy, *Universal spaces for strictly convex Banach spaces*, RACSAM **100** (2006), 136–146.

[Go10] G. Godefroy, *On the diameter of the Banach-Mazur set*, Czechoslovak Mathematical Journal **60** (2010), 95–100.

[Go15] G. Godefroy, *A survey on Lipschitz-free Banach spaces*, Commentationes Mathematicae **55**, 2 (2015), 89–118.

[Go17] G. Godefroy, *The isomorphism classes of* ℓ_p *are Borel*, Houston J. Math. **43**, 3 (2017), 947–951.

[Go17b] G. Godefroy, *The complexity of the isomorphism class of some Banach spaces*, J. of Nonlinear and Convex Anal. **18**, 1 (2017), 231–240.

[GoKa89] G. Godefroy and N.J. Kalton, *The ball topology and its applications*, Contemporary Mathematics **85** (1989), 195-237.

[GoKa03] G. Godefroy and N.J. Kalton, *Lipschitz-free Banach spaces*, Studia Math. **159**, I (2003), 121–141.

[GoKa07] G. Godefroy and N.J. Kalton, *Isometric embeddings and universal spaces*, Extracta Mathematicae **22**, 2 (2007), 179–189.

[GoKaLa00] G. Godefroy, N.J. Kalton, and G. Lancien, *Subspaces of* $c_0(\mathbb{N})$ *and Lipschitz isomorphisms*, GAFA (Geometric Functional Analysis) **10** (2000), 798–820.

[GoLaZi14] G. Godefroy, G. Lancien, and V. Zizler, *The non-linear geometry of Banach spaces after Nigel Kalton*, Rocky Mt. J. Math. **44** (5) (2014), 1529–1583.

[GoMoZi95] G. Godefroy, V. Montesinos, and V. Zizler, *Strong subdifferentiability of norms and geometry of Banach spaces*, Comment. Math. Univ. Carol. **36** (3), (1995), 493–502.

[GoOz14] G. Godefroy and N. Ozawa, *Free Banach spaces and the approximations properties*, Proc. Amer. Math. Soc. **142**, 5 (2014), 1681–1687.

[GPWZ88] G. Godefroy, J. Pelant, J.H.M. Whitfield, and V. Zizler, *Banach space properties of Ciesielski–Pol* $C(K)$ *space*, Proc. Amer. Math. Soc. **103** (1988), 1087–1094.

[GoSa88] G. Godefroy and P.D. Saphar, *Duality in spaces of operators and smooth norms on Banach spaces*, Illinois J. Math. **32** (4) (1988), 672–695.

[GTWZ83] G. Godefroy, S.L. Troyanski, J.H.M. Whitfield, and V. Zizler, *Smoothness in weakly compactly generated Banach spaces*, J. Funct. Anal. **52** (1983), 344–352.

[GTWZ84] G. Godefroy, S.L. Troyanski, J.H.M. Whitfield, and V. Zizler, *Locally uniformly rotund renormings and injections into* $c_0(\Gamma)$. Canad. Math. Bull. **27** (1984), 494–500.

[GTWZ] G. Godefroy, S.L. Troyanski, J.H.M. Whitfield, and V. Zizler, *Three space problem for locally uniformly rotund renormings of Banach spaces*, Proc. Amer. Math. Soc. **94** (1985), 647–652.

[GoZi91] G. Godefroy and V. Zizler, *Roughness Properties of norms on Non-Asplund spaces*, Michigan Math. J. **38** (1991), 461–466.

[Go80] B.V. Godun, *Renormings of quasi-reflexive Banach spaces*, Rev. Roumaine Math. Pures Appl. **25** (7) (1980), 1033–1038.

[Go85] B.V. Godun, *Preserved extreme points*, Funct. Anal. i Pril. **19** (1985), 75–76.

[Go85b] B.V. Godun, *Points of smoothness of convex bodies in separable Banach spaces*, Matem.Zametki **38** (1985), 713–716.

[GLT92] B.V. Godun, Bor-Luh Lin, and S.L. Troyanski, *On the strongly extreme points of convex bodies in separable Banach spaces*, Proc. Amer. Math. Soc. **114** (1992), 673–675.

[Gow94] W.T. Gowers, *A Banach space not containing c_0, ℓ_1 or a reflexive subspace*, Trans. Amer. Math. Soc. **344** (1994), 407–420.

[Gow96] W.T. Gowers, *A new dichotomy for Banach spaces*, Geom. Funct. Anal. **6** (1996), 1083–1093.

[GowMau93] W.T. Gowers and B. Maurey, *The unconditional basic sequence problem*, J. Amer. Math. Soc. **6** (1993), 851–874.

[GJM04] A.S. Granero, M. Jiménez-Sevilla, and J.P. Moreno, *Intersections of closed balls and geometry of Banach spaces*, Extracta Mathematicae **19** (1) (2004), 55–92.

[Gru87] G. Gruenhage, *A note on Gul'ko compact spaces*, Proc. Amer. Math. Soc. **100** (1987), 371–376.

[GuiHá07a] A.J. Guirao and P. Hájek, *On the moduli of convexity*, Proc. Amer. Math. Soc. **10** (2007), 3233–3240.

[GuiHá07b] A.J. Guirao and P. Hájek, *Schauder bases under uniform renormings*, Positivity **11** (2007), 627–638.

[GuiHáMo10] A.J. Guirao, P. Hájek, and V. Montesinos, *Ranges of operators and derivatives*, J. Math. Anal and Appl. **367** (2010), 29–33

[GuiLaTr20] A.J. Guirao, S. Lajara, and S.L. Troyanski, *Three-space problem for strictly convex renormings.* The Mathematical Legacy of Victor Lomonosov: Operator Theory, edited by Richard M. Aron, Eva A. Gallardo Gutiérrez, Miguel Martín, Dmitry Ryabogin, Ilya M. Spitkovsky and Artem Zvavitch, Berlin, Boston, De Gruyter, 2020, pp. 149–156.

[GuiLiMo19] A.J. Guirao, A. Lissitsin, and V. Montesinos, *A note on lower semicontinuous smooth norms*, Revista de la Real Academia de Ciencias Exactas, Físicas y Naturales, Serie A. Matemáticas (RACSAM), **113** (2019), 2081–2085.

[GuiMoZi12] A.J. Guirao, V. Montesinos, and V. Zizler, *On a classical renorming construction of V. Klee*, J. Math. Anal. Appl. **385** (2012), 458–465.

[GuiMoZi14] A.J. Guirao, V. Montesinos, and V. Zizler, *On preserved and unpreserved extreme points*, in Ferrando, J., López-Pellicer M, (eds) Descriptive Topology and Functional Analysis, Springer Proceedings in Mathematics and Statistics, **80**, 163–194, Springer, Cham, 2014.

[GuiMoZi15] A.J. Guirao, V. Montesinos, and V. Zizler, *A note on extreme points of C^∞-smooth balls in polyhedral spaces*, Proc. Amer. Math. Soc. **143** (8), (2015), 3413–3420.

[GuiMoZi16] A.J. Guirao, V. Montesinos, and V. Zizler, *Open Problems in the Geometry and Analysis on Banach Spaces*, Springer, 2016.

[GuiMoZi18] A.J. Guirao, V. Montesinos, and V. Zizler, *Remarks on the set of norm-attaining functionals and differentiablity*, Studia Mathematica, **241**, 1 (2018), 71–86.

[GuOs75] S.P. Gulko and A.V. Oskin, *Isomorphic classification of spaces of continuous functions on totally ordered compact sets*, Funct. Anal. Appl., **9**, 1 (1975), 56–57.

[GuGu71] V.I. Gurarii and N.I. Gurarii, *On bases in uniformly convex and uniformly smooth spaces*, Iz. Akad. Nauk SSSR Ser. Mat. **35** (1971), 210–215.

[HHZ96] P. Habala, P. Hájek, and V. Zizler, *Introduction to Banach Spaces I, II*, Matfyzpress Faculty of Mathematics, Charles University, Prague, 1996.

[Hag77] J. Hagler, *A counterexample to several questions about Banach spaces*, Studia Math. **60** (1977) 289–308.

[Hag87] J. Hagler, *A note on separable Banach spaces with nonseparable dual*, Proc. Amer. Math. Soc. **99** (3), (1987), 452–454.

[HagOd78] J. Hagler and E. Odell, *A Banach space not containing ℓ_1 whose dual ball is not weak* sequentially compact.* Illinois J. of Math. **22** (2) (1978), 290–294.

[HagSull80] J. Hagler and F.E. Sullivan, *Smoothness and weak star sequential compactness*, Proc. Amer. Math. Soc. **78** (1980), 497–503.

[Há94] P. Hájek, *Polynomials and injections of Banach spaces into superreflexive spaces*, Arch. Math. **63** (1994), 39–44.

[Há95] P. Hájek, *Smooth norms that depend locally on a finite number of coordinates*, Proc. Amer. Math. Soc. **123** (1995), 3817–3821.

[Há96a] P. Hájek, *Dual renormings of Banach spaces*, Comment. Math. Univ. Carol. **37** (1996), 241–253.

[Há96b] P. Hájek, *Analytic renormings of $C(K)$ spaces*, Serdica Math. J. **22** (1) (1996), 25–28.

[Há98] P. Hájek, *Smooth functions on c_0*, Israel J. Math. **104** (1998), 17–28.

[Há98b] P. Hájek, *Smooth functions on $C(K)$* , Israel J. Math. **107** (1998), 237–252.

[Há03] P. Hájek, *Smooth norms on certain $C(K)$ spaces*, Proc. Amer. Math. Soc. **131** (7) (2003), 2049–2051.

[Há20] P. Hájek, *Hilbert generated Banach spaces need not have a norming Markushevich basis*, Adv. Math. **351** (2019), 702–717.

[HájHay07] P. Hájek and R. Haydon, *Smooth norms and approximation in Banach spaces of the type $C(K)$*, Quartely J. Math. **58** No. 2, (2007), 221–228.

[HáJo04] P. Hájek and M. Johanis, *Characterization of reflexivity by equivalent renorming*, J. Funct. Anal. **211** (2004), 163–172.

[HáJo09] P. Hájek and M. Johanis, *Uniformly Gâteaux smooth approximations in $c_0(\Gamma)$*, J. Math. Anal. Appl. **350** (2009), 623–629.

[HáJo14] P. Hájek and M. Johanis, *Smooth Analysis in Banach spaces*, De Gruyter 2014.

[HáKaRu18] P. Hájek, T. Kania, and T. Russo, *Symmetrically separated sequences in the unit sphere of a Banach space*, J. Funct. Anal. **275**, 11 (2018), 3148–3168.

[HáKaRu20] P. Hájek, T. Kania, and T. Russo, *Separated sets and Auerbach systems in Banach spaces*, Trans. Amer. Math. Soc. **373** (2020), 6961–6998.

[HáLa07] P. Hájek and G. Lancien, *Various slicing indices on Banach spaces*, Mediterr. J. Math. **4** (2007), 179–190.

[HáLaMo07] P. Hájek, G. Lancien, and V. Montesinos, *Universality of Asplund spaces*, **135** (7), (2007), 2031–2035.

[HMVZ07] P. Hájek, V. Montesinos, J. Vanderwerff, and V. Zizler, *Biorthogonal systems in Banach spaces*, CMS Books in Mathematics, Canadian Mathematical Society, Springer, 2007.

[HáMoZi12] P. Hájek, V. Montesinos, and V. Zizler, *Geometry and Gâteaux smoothness in separable Banach spaces*, Oper. Matrices **6** (2012), (2), 201–232.

[HáNo17] P. Hájek and M. Novotný, *Some remarks on the structure of Lipschitz-free spaces*, Bull. Belg. Math. Soc. Simon Stevin **24**, (2017), 283–304.

[HáNo18] P. Hájek and M. Novotný, *Distortion of Lipschitz functions on* $c_0(\ggg)$, Proc. Amer. Math. Soc. **146**, 5 (2018), 2173–2180.

[HáPro14] P. Hájek and A. Procházka, C^k*-smooth approximations of LUR norms*, Trans. Amer. Math. Soc. **366** (4) (2014), 1973–1992.

[HáRu17] P. Hájek and T. Russo, *Some remarks on smooth renormings of Banach spaces*, J. Math. Anal. Appl. **455** (2017), 1272–1284.

[HáRu2019] P. Hájek and T. Russo, *An uncountable version of Pták combinatorial lemma*, J. Math. Anal. Appl. **470**, 2 (2019), 1070–1080.

[HRST20] P. Hájek, T. Russo, J. Somaglia, and S. Todorčević, *An Asplund space with norming Markushevich basis that is not weakly compactly generated*, arXiv: 2007.14201v2 [math.FA]

[HáSchSz19] P. Hájek, Th. Schlumprecht, and A. Szak, *Generalization of Zippin's theorem on perturbing Banach spaces with separable dual*, Studia Math. **245**, 2 (2019), 169–183.

[HáTr01] P. Hájek and S.L. Troyanski, *Analytic norms in Orlicz spaces*, Proc. Amer. Math. Soc.**129** (3), (2001), 713–717.

[HáZi95] P. Hájek and V. Zizler, *Remarks on symmetric smooth norms*, Bull. Austral. Math. Soc. **52** (1995), 225–229.

[HáZi06] P. Hájek and V. Zizler, *Functions locally dependent on finitely many coordinates*, RACSAM **100** (1-2) (2006), 147–154.

[Hal50] P.R. Halmos, *Measure Theory*, Springer-Verlag New York, Heidelberg, Berlin, GTM **18**, 1950.

[Hay76] R. Haydon, *Some more characterizations of Banach spaces containing* ℓ_1, Math. Proc. Cambridge Philos. Soc. **80**, 2 (1976), 269–276.

[Hay90] R. Haydon, *A counterexample to several questions about scattered compact spaces*, Bull. London Math. Soc. 22 (1990), 261–268.

[Hay92] R. Haydon, *Normes infiniment differentiables sur certain Banach spaces*, C. R. Acad. Sci. Paris Ser. 1 **315** (1992), 1175–1178.

[Hay96] R. Haydon, *Smooth functions and partitions of unity on certain Banach spaces*, Quart. J. Math. **47** (1996), 455–468.

[Hay99] R. Haydon, *Trees in renorming theory*, Proc. London Math. Soc. **78** (1999), 541–584.

[Hay08] R. Haydon, *Locally uniformly rotund norms in Banach spaces and their duals*, J. Funct. Anal. **254** (2008), 2023–2039.

[HayMoOr07] R. Haydon, A. Moltó, and J. Orihuela, *Spaces of functions with countably many discontinuities*, Israel J. Math. **158** (2007), 19–39.

[HayZiz89] R. Haydon and V. Zizler, *A new space with no locally uniformly rotund renorming*, Canad. Math. Bull. **32** (1) (1989), 122–128.

[HeiMa82] S. Heinrich and P. Mankiewicz, *Applications of ultrapowers to the uniform and Lipschitz classification of Banach spaces*, Studia Math. **73** (1982), 225–251.

[HSZ98] P. Holický, M. Šmídek, and L. Zajíček, *Convex functions with non Borel set of Gâteaux differentiability points*, Comment. Math. Univ. Carolinae **39** (1998), 469–482.

[Ho86] J.A. Holub, *A property of weakly compact operators on* $C[0,1]$, Proc. Amer. Math. Soc. **97** (3), (1986), 396–398.

[HMS97] Z. Hu, W.B. Moors, and M.A. Smith, *On a Banach space without a weak mid-point locally uniformly rotund norm*, Bull. Austral. Math. Soc. **56** (1997), 193–196.

[Hu80] R.E. Huff, *Banach spaces which are nearly uniformly convex*, Rocky Mountain J. Math. **10** (4) (1980), 743–750.

[HuMo75] R.E. Huff and P.B. Morris, *Dual spaces with the Krein-Milman property have the Radon-Nikodým property*, Proc. Amer. Math. Soc. **49** (1) (1975), 104–108.

[Jame50] R.C. James, *Bases and reflexivity of Banah spaces*, Bull. Amer. Math. Soc. **56** (1950), 518–527.

[Jame57] R.C. James, *Reflexivity and the supremum of linear functionals*, Ann. of Math. **66** (1), (1957), 159–169.

[Jame64] R.C. James, *Characterizations of reflexivity*, Studia Math. **23** (1964), 205–216.

[Jame64b] R.C. James, *Weak compactness and reflexivity*, Israel J. Math. **2** (1964), 101–119.

[Jame64c] R.C. James, *Uniformly nonsquare Banach spaces*, Ann. of Math. (2) **80** (1964), 542–550.

[Jame71] R.C. James, *A counterexample for the sup theorem in normed spaces*, Israel J. Math. **9** (1971), 511–512.

[Jame72a] R.C. James, *Superreflexive Banach spaces*, Canadian J. Math. **24** (1972), 896–904.

[Jame72] R.C. James, *Superreflexive spaces with bases*, Pacific J. Math. **41–42** (1972), 409–419.

[Jame72b] R.C. James, *Reflexivity and the supremum of linear functionals*, Israel J. Math. **13** (1972), 289–300.

[Jame74] R.C. James, *A separable somewhat reflexive Banach space with nonseparable dual*, Bull. Amer. Math.Soc. **80** (1974), 738–743.

[JaSch72] R.C. James and J.J. Schäffer, *Superreflexivity and the girth of spheres*, Israel J. Math. **11** (1972), 398–404.

[Jam74] G.J.O. Jameson, *Topology and Normed Spaces*, Chapman and Hall, 1974.

[Jarc81] H. Jarchow, *Locally Convex Spaces*, B. G. Teubner, Stuttgart 1981.

[Jar88] K. Jarosz, *Any Banach space has an equivalent norm with trivial isometries*, Israel J. Math. **64** (1988), 49–56.

[JaRo85] J.E. Jayne and C.A. Rogers, *Borel selectors for upper semicontinuous maps*, Acta Math. **155** (1985), 41–79.

[JiMo97] M. Jiménez-Sevilla and J.P. Moreno, *Renorming Banach spaces with the Mazur intersection property*, J. Funct. Anal. **144** (1997), 486–504.

[Joha03] M. Johanis, *Approximation of Lipschitz mappings*, Serdica Math. J. **29** (2003), 141–148.

[Joha17] M. Johanis, *Smooth partitions of unity on Banach spaces*, J. Funct. Anal. **273**, 2 (2017), 3226–3260.

[JohRy07] M. Johanis and J. Rychtář, *On uniformly Gâteaux smooth norms and normal structure*, Proc. Amer. Math. Soc. **135** (5), (2007), 1511–1514.

[JoToZi81] K. John, H. Toruńczyk, and V. Zizler, *Uniformly smooth partitions of unity on superreflexive Banach spaces*, Studia Math. **70** (1981), 129–137.

[JoZi72] K. John and V. Zizler, *A renorming of dual spaces*, Israel J. Math. **12** (1972), 331–336.

[JoZi74a] K. John and V. Zizler, *Smoothness and its equivalents in weakly compactly generated Banach spaces*, J. Funct. Anal. **15** (1974), 1–11.

[JoZi74b] K. John and V. Zizler, *Some remarks on nonseparable Banach spaces with Markushevich basis*, Comment. Math. Univ. Carolinae **15** (4) (1974), 679–691.

[JoZi75] K. John and V. Zizler, *Markushevich bases in some dual spaces*, Proc. Amer. Math. Soc. **50** (1), (1975), 293–296.

[JoZi75a] K. John and V. Zizler, *On the heredity of weak compact generating*, Israel J. Math. **20**, 3-4 (1975), 228–236.

[JoZi75/76] K. John and V. Zizler, *Weak compact generating in duality*, Studia Math. **55** (1975/6), 1–20.

[JoZi76] K. John and V. Zizler, *On extension of rotund norms*, Bull. Acad. Polon. Sci. Ser. Sc. Math. Astron. Phys. **24** (1976), 705–707.

[JoZi77] K. John and V. Zizler, *Gâteaux smooth partitions of unity on weakly compactly generated Banach spaces*, Studia Math. **60** (1977), 131–135.

[Jozi78] K. John and V. Zizler, *On rough norms on Banach spaces*, Comment. Math. Univ. Carolinae, **19** (1978) (2), 335–347.

[JoZi78b] K. John and V. Zizler, *Topological linear spaces*, Lecture Notes, Charles University, Prague, 1978 (in Czech).

[JoZi79] K. John and V. Zizler, *A short proof of a version of Asplund averaging theorem*, Proc. Amer. Math. Soc. **73** (1979), 277–278.

[JoKa19] W.B. Johnson and T. Kania, *Embedding Banach spaces into the space of bounded functions with countable support*, Mathematische Nachrichten **292** (2019), 2028–2031.

[JoLi74] W.B. Johnson and J. Lindenstrauss, *Some remarks on on weakly compactly generated Banach spaces*, Israel J. Math. **17** (1974), 219–230.

[JoLi03] W.B. Johnson and J. Lindenstrauss, *Basic concepts in the geometry of Banach spaces*,Handbook of the Geometry of Banach spaces II, Editors W.B. Johnson and J. Lindenstrauss, Elsevier Science B.V., 2003, 1–85.

[JohOd05] W.B. Johnson and E. Odell, *The diameter of the isomorphism class of a Banach space*, Annals of Mathematics **162** (2005), 423–437.

[Kad53] M.I. Kadets, *On homeomorphism of certain Banach spaces*, (Russian) Dokl. Akad. Nauk SSSR (N. S.) **92** (1953), 465–468.

[Kad55] M.I. Kadets, *On topological equivalence of uniformly convex spaces* (Russian), Usp. Mat. Nauk **10** (1955), 137–141.

[Kad56] M.I. Kadets, *Unconditional convergence of series in uniformly convex spaces*, Uspechi Mat. Nauk SSSR **11** (1956), 185–190.

[Kad58] M.I. Kadets, *On strong and weak convergence*, Dokl. Akad. Nauk SSSR (N. S.) **122** (1958), 13–16.

[Kad65] M.I. Kadets, *Conditions on the differentiability of the norm of a Banach space*, Uspechi Mat. Nauk SSSR **20** (1965), 183–187.

[Kad66] M.I. Kadets, *On topological equivalence of separable Banach spaces*, Soviet Math. Dokl. **7** (1966), 319–322.

[Kad67] M.I. Kadets, *A proof of topological equivalence of all separable infinite dimensional Banach spaces*, Funktsional Anal. i Prilozen **1** (1) (1967), 53–62.

[Kad82] M.I. Kadets, *Relation between certain convexity properties of the unit ball of a Banach space*, Funktsional. Anal. i. Prilozen **16** (1982), 58–60.

[KadTo75] M.I. Kadets and E.V. Tokarev, *Conjugate Banach spaces with a countable set of extreme points on the unit sphere*, Izv. Vyssh. Ucebn. Zaved. Mat. **4** (1975), 98–99. (Soviet Math. (Iz VUZ) **19**, 4 (1975), 77–78).

[Kad96] V.M. Kadets, *Some remarks concerning the Daugavet equation.* Questions Math. **19** (1996), 225–235.

[Kad21] V.M. Kadets, *The diametral strong diameter 2 property of Banach spaces is the same as the Daugavet property*, Proc Amer. Math. Soc. **149** (6), (2021), 2579–2582.

[KKLP13] J. Kąkol, W. Kubiś, and M. López-Pellicer, *Descriptive Topology in Selected Topics of Functional Analysis*, Developments in Mathematics, Springer, 2013.

[Kaku38] S. Kakutani, *Weak convergence in uniformly convex spaces*, Tohoku Math. J. **45** (1938), 188–193.

[Kaku39] S. Kakutani, *Some characterizations of Euclidean space*, Japan J. Math. **16** (1939), 93–97.

[Kal99] O.F.K. Kalenda, *An example concerning Valdivia compact spaces*, Serdica Math. J. **25** (1999), 131–140.

[Kal00] O.F.K. Kalenda, *Valdivia compacta and equivalent norms*, Studia Math. **138** (2) (2000), 179–191.

[Kal00-b] O.F.K. Kalenda, *Valdivia compact spaces in topology and Banach space theory*, Extracta Mathematicae **15** (2000), 1–85.

[Kal] O.F.K. Kalenda, *Projectional skeletons and Markushevich bases*, Proc. London Math. Soc. (3) **120** (2020), no. 4, 514–586.

[Kal03] N.J. Kalton, *Quasi-Banach Spaces*, Handbook of Banach Spaces II, Editors W.B. Johnson and J. Lindenstrauss, Elsevier Science B.V., 2003, 1101–1130.

[Ka11] N.J. Kalton, *Lipschitz and uniform embeddings into* ℓ_∞, Fund. Math. **212** (2011), 53–69.

[Kam50] E. Kamke, *Theory of Sets*, Dover, 1950.

[KaKo16] T. Kania and T. Kochanek, *Uncountable sets of unit vectors that are separated by more than 1*, Studia Math. **232** (1), (2016), 19–44.

[Kar76] L.A. Karlovitz, *Existence of fixed points of nonexpansive mappings in a space without normal structure*, Pacific J. Math. **66** (1) (1976), 153–159.

[Kau91] R. Kaufman, *Topics on analytic sets*, Fundmenta Math. **139** (1991), 215–229.

[Kec95] A.S. Kechris, *Classical Descriptive Set Theory*, Graduate texts in Mathematics **156**, Springer-Verlag, 1995.

[Kel55] J.L. Kelley, *General Topology*, Van Nostrand, Newyork, 1955.

[Ki65] W.A. Kirk, *A fixed point theorem for mappings which do not increase distances*, Amer. Math. Monthly **72** (1965), 1000–1006.

[Kis75] V.E. Kislyakov, *Classification of spaces of continuous functions of ordinals*, Sib. Math. J., **16**, 2 (1975), 226—231.

[Kl50] V.L. Klee, *Some characterizations of reflexivity*, Revista de Ciencias, **52**, (1950), 15–23.

[Klee59] V.L. Klee, *Some new results on smoothness and rotundity in normed linear spaces*, Math. Ann. **139** (1959), 51–63.

[Klee69] V.L. Klee, *Can the boundary of a d-dimensional convex body containsegments in all directions?*, Amer. Math. Monthly **76** (1969), 408–410.

[Klee86] V.L. Klee, *Do infinite dimensional Banach spaces admit nice tilings*, Studia Sci. Math. Hungarica **21** (1986), 415–427.

[KnOdSch98] H. Knaust, E. Odell, and Th. Schlumprecht, *On asymptotic structure, the Szlenk index and UKK properties in Banach spaces*, Positivity **3**, (1998), 173–199.

[KoTo95] R. Komorowski and N. Tomczak-Jaegermann, *Banach spaces without local unconditional structure*, Israel J. Math. **89** (1995), 205–226.

[KoWa] P. Koszmider and H. Wark, *Large Banach spaces with no infinite equilateral sets*, arXiv: 2104.05330v4 [math.FA]

[Kothe69] G. Köthe, *Topological Vector Spaces I*, Springer-Verlag 1969.

[Kott75] C. Kottman, *Subsets of the unit ball that are separated by more than one*, Studia Math. **53** (1975), 15–27.

[KrPr01] A. Kryczka and S. Prus, *Separated sequences in nonreflexive Banach spaces*, Proc. Amer. Math. Soc. **129** (1) (2001), 155–163.

[Kub09] W. Kubiś, *Banach spaces with projectional skeletons*, J. Math. Anal. Appl. **350** (2009), 758–776.

[Kuo75] T.H. Kuo, *On conjugate Banach spaces with the Radon-Nikodým property*, Pacific J. Math. **59** (1975) (2), 497–503.

[Kur19] O. Kurka, *The isomorphic class of c_0 is not Borel*, Israel J. Math. **231** (2019), 243–268.

[Kurz54] J. Kurzweil, *On approximation in real Banach spaces*, Studia Math. **14** (1954) 213–231.

[Kurz57] J. Kurzweil, *On approximation in real Banach spaces by analytic operations*, Studia Math. **16** (1957), 124–129.

[Kut84] D.N. Kutzarova, *On an equivalent norm in L_1 which is uniformly convex in every direction*, Constructive Theory of Functions, Sofia **84** (1984), 507–512.

[KuTr82] D.N. Kutzarova and S.L. Troyanski, *Reflexive Banach spaces without equivalent norms which are uniformly convex or uniformly differentiable in every direction*, Studia Math. **72** (1982), 91–95.

[KutTro83] D.N. Kutzarova and S.L. Troyanski, *On equivalent lattice norms which are uniformly convex or uniformly differentiable in every direction in Banach lattices with a weak unit*, Serdica Bulgaricae Math. Publ. **9** (1983), 249–262.

[La04] S. Lajara, *ALUR dual renorming of Banach spaces*, J. Math. Anal. Appl. **289** (2004), 221–226.

[LaPaTr09] S. Lajara, A.J. Pallarés, and S.L. Troyanski, *Bidual renormings of Banach spaces*, J. Math. Anal. Appl. **350** (2009), 630–639.

[Lanc93] G. Lancien, *Dentability indexes and locally uniformly convex renorming*, Rocky Mount. J. Math. **23** (1993), 633–647.

[Lanc95] G. Lancien, *On uniformly convex and uniformly Kadets–Klee renormings*, Serdica Math. J. **21** (1995), 1–18.

[Lanc96] G. Lancien, *On the Szlenk index and the weak* dentability index*, Quart. j. Math. Oxford **47** (1996), 59–71.

[Lanc06] G. Lancien, *A survey on the Szlenk index and some of its applications*, RACSAM **100** (1-2), 2006, 209–235.

[Lau75] K.S. Lau, *Farthest points in weakly compact sets*, Israel J. Math. **22** (1975), 168–174.

[LeaWh72] E.B. Leach and J.H.M. Whitfield, *Differentiable functions and rough norms on Banach spaces*, Proc. Amer. Math. Soc. **33** (1972), 120–126.

[LD70] M. Leduc, *Densité de certaines familles d'hyperplanes tangents*, C. R. Acad. Sci. Paris **270**, A (1970), 326–328.

[LeWhi83] I.E. Leonard and J.H.M. Whitfield, *A classical Banach space: ℓ_∞/c_0*, Rocky Mount. J. Math. **13** (3) (1983), 531–539.

[LLT88] Bor-Luh Lin, Pei-Kee Lin, and S.L. Troyanski, *Characterisations of denting points*, Proc. Amer. Math. Soc. **102** (1988), 526–528.

[Lind63] J. Lindenstrauss, *On operators which attain their norms*, Israel J. Math. (1963), 139–148.

[Lind64a] J. Lindenstrauss, *On the extension of operators with a finite dimensional range*, Illinois J. Math. **8** (1964), 488–499.

[Lind64b] J. Lindenstrauss, *Extension of compact operators*, Mem. Amer. Math. Soc. **48** (1964).

[Lind66a] J. Lindenstrauss, *On extreme points in ℓ_1*, Israel J. Math. **4** (1966), 59–61.

[Lind66b] J. Lindenstrauss, *On nonseparable reflexive Banach spaces*, Bull. Amer. Math. Soc. **72** (1966), 967–970.

[Lind66c] J. Lindenstrauss, *Notes on Klee's paper "Polyhedral sections of convex bodies"*, Israel J. Math. **4** (1966), 235–242.

[Lind67] J. Lindenstrauss, *On complemented subspaces of m*, Israel J. Math. **5**, (1967), 153–156.

[Lind72] J. Lindenstrauss, *Weakly compact sets —their topological properties and the Banach spaces they generate*, Ann. of Math. Studies **69** (1972), 235–273.

[LiPe68] J. Lindenstrauss and A. Pełczyński, *Absolutely summing operators in $\mathcal{L}_p$ spaces and their applications*, Studia Math. **29** (1968), 275–326.

[LindPh68] J. Lindenstrauss and R.R. Phelps, *Extreme points properties ofconvex bodies in reflexive Banach space*, Israel J. Math. **6** (1968), 39–48.

[LPT10] J. Lindenstrauss, D. Preiss, and S. Tišer, *Fréchet differentiability of Lipschitz functions via a variational principle*, J. Eur. Math. Soc. **12** (2010), 395–412.

[LPT12] J. Lindenstrauss, D. Preiss, and J. Tišer, *Fréchet differentiability of Lipschitz functions and porous sets in Banach spaces*, Annales of Mathematics Studies, **vol 179**, Princeton Unversity Press, Princeton 2012.

[LiRo69] J. Lindenstrauss and H.P. Rosenthal, *The $\mathcal{L}_p$ spaces*, Israel J. Math. **7** (1969), 325–349.

[LinSt75] J. Lidenstrauss and Ch. Stegall, *Examples of separable spaces which do not contain ℓ_1 and whose duals are noseparable*, Studia Math. **54** (1975), 81–105.

[LinTz77] J. Lindenstrauss and L. Tzafriri, *Classical Banach spaces I. Sequence spaces*, Springer, Berlin 1977.

[LinTz79] J. Lindenstrauss and L. Tzafriri, *Classical Banach spaces vol. II; Function Spaces*, Springer, **97**, 1979.

[Lio73] V.I. Liokumovich, *Existence of B- spaces with a nonconvex modulus of convexity*, Izv. Vyssh. Uchebn. Zav. Mat. **12** (1973), 43–49.

[LopTod09] J. López-Abad and S. Todorčević, *A c_0-saturated Banach space with no long unconditional basic sequence*, Trans. Amer. Math. Soc. **361**, 4541–4560 (2009).

[Lov55] A.R. Lovaglia, *Locally uniformly convex Banach spaces.* Trans. Amer. Math. Soc. **78**, 225–238 (1955).

[Loz66] G.Ya. Lozanovskii, *On almost integral operators on KB-spaces*, Vestnik Leningr. Univ. **21.7** (1966), 35–55 (in Russian).

[LjuSo61] L.A. Ljusternik and B.I. Sobolev, *Elements of Functional Analysis*, New Delhi, 1961.

[Mak79] B.M. Makarov, *A characterization of Hilbert space*, Mat. Zametki **26** (1979), 739–746.

[Mak83] B.M. Makarov, *A condition for the isomorphism of a Banach space having the Orlicz property to a Hilbert space*, Zap. Nauč. Sem. Lenigrad **126** (1983), 138–142.

[MaPa93] E. Maluta and P.L. Papini, *Relative centers and finite nets for the unit ball and its finite subsets.* Boll. Un. Mat. Ital. B (7) **7** (1993), 451–472.

[Mal99] J. Malý, *A simple proof of the Stepanov theorem on differentiability almost everywhere on $\mathbb{R}^n$*, Expositio Math. **17** (1999), 59–61.

[MarZa12] A. Marchese and C. Zanco, *On a question by Corson about pointfinite coverings*, Israel J. Math. **189** (2012), 55–63.

[Marti69] J.T. Marti, *Introduction to the Theory of Bases*, Springer Tracts in Natural Philosophy **18**, Springer-Verlag 1969.

[MatMat99] J. Matoušek and E. Matoušková, *A highly nonsmooth norm on Hilbert space*, Israel J. Math. **112** (1999), 1–27.

[Mau03] B. Maurey, *Banach spaces with few operators*, Handbook of the Geometry of Banach spaces II, Editors W.B. Johnson and J. Lindenstrauss, Elsevier Science B.V., 2003, 1247–1297.

[Maz29] S. Mazur, *Une remarque sur l'homéomorphie des champs fonctionnels*, Studia Math. **1** (1929), 83–85.

[Maz33] S. Mazur, *Über schwache Konvergentz in den Raumen ℓ^p*, Studia Math. **4** (1933), 128–133.

[MazSt33] S. Mazur and L. Sternbach, *Über die Borelschen Typen von linearenMengen*, Studia Math. **4** (1933), 48–53.

[McL92] D. McLaughlin, *Smooth paritions of unity in preduals of WCG spaces*, Math. Z. **211** (1992), 189–194.

[MPVZ93] D. McLaughlin, R. Poliquin, J. Vanderwerff, and V. Zizler, *Second order Gâteaux differentiable bump funcions and approximations in Banach spaces*, Canad. J. Math. **45** (3) (1993), 612–625.

[McS34] E.J. McShane, *Extension of range of functions*, Bull. Amer. Math. Soc. **40** (1934), 837–842.

[Meg98] R.E. Megginson, *An introduction to Banach space theory*, GTM **183**, Springer, 1998.

[Me87] S. Mercourakis, *On weakly countably determined Banach spaces*, Trans. Amer. Math. Soc. **300** (1987), 307–327.

[MeSt06] S. Mercourakis and E. Stamati, *A new class of weakly $\mathcal{K}$-analytic Banach spaces*, Comment. Math. Univ.Carolin. **42** (2) (2006), 291–312.

[MeVa15] S.K. Mercourakis and G. Vassiliadis, *Equilateral sets in infinite dimensional Banach spaces*, Studia Math. **231**, 3 (2015), 241–255.

[Meshk78] V.Z. Meshkov, *Smoothness properties in Banach spaces*, Studia Math. **63** (1978), 111–123.

[Milm71] V.D. Milman, *Geometric theory of Banach spaces II: Geometry of the unit sphere*, Russian Math. Surveys **26** (1971), 79–163.

[MiMi64] D.P. Milman and V.D. Milman, *Some properties of nonreflexive Banach spaces*, Mat. Sb. **65** (1964), 486–497.

[Mil66] A.A. Milyutin, *Isomorphisms of spaces of continuous functions on compacts of power continuum*, Teorija Func. (Kharkov) **2** (1966), 150–156 (in Russian).

[MMOT98] A. Moltó, V. Montesinos, J. Orihuela, and S.L. Troyanski, *Weakly uniformly rotund Banach spaces*, Comment. Math. Univ. Carolinae **39** (1998), 749–753.

[MOT97] A. Moltó, J. Orihuela, and S.L. Troyanski, *Locally uniformly rotund renorming and fragmentability*, Proc. London Math. Soc. **75** (1997), 619–640.

[MOTV99] A. Moltó, J. Orihuela, S.L. Troyanski and M. Valdivia, *On weakly locally uniformly rotund Banach spaces*, J. Functional Analysis **163** (1999), 252–271.

[MOTV01] A. Moltó, J. Orihuela, S.L. Troyanski, and M. Valdivia, *Midpoint locally uniform rotundity and a decomposition method for renorming,* Quart. J. Math. **52** (2001), 181–193.

[MOTV09] A. Moltó, J. Orihuela, S.L. Troyanski, and M. Valdivia, *A Nonlinear Transfer Technique for Renorming*, Lecture Notes in Mathematics **1951**, Springer, 2009.

[Mo87] V. Montesinos, *Drop property equals reflexivity*, Studia Math. **87** (1987), 93–100.

[MoZiZi15] V. Montesinos, P. Zizler, and V. Zizler, *An introduction to Modern Analysis*, Springer, 2015.

[MoSo06] W.B. Moors and S. Somasundaram, *A Gâteaux differentiability space that is not weak Asplund*, Proc. Amer. Math. Soc. **134** (2006), 2745–2754.

[Mo83] P. Morris, *Dissapearance of extreme points*, Proc. Amer. Math. Soc. **88**, 2 (1983), 244–246.

[Nam87] I. Namioka, *Radon–Nikodým compact spaces and fragmentability*, Mathematika **34**, 2 (1987), 258–281.

[NaPhe68] I. Namioka and R.R. Phelps, *Banach spaces which are Asplund spaces*, Duke Math. J. **42** (1968), 735–750.

[NaSch07] A. Naor and G. Schechtman, *Planar earthmover is not in L_1*, SIAM J. Comput. **37** (2007), 804–826.

[Nat64] I.P. Natanson, *Theory of Functions of a Real Variable I, II*, Frederick Ungar Publishing Co. , New York, 1964.

[NeZa98] A. Nekvinda and L. Zajíček, *A simple proof of the Rademacher theorem*, Časopis pro pěstování matematiky, **113** (1998), 337–341.

[NeSe73] A.S. Nemirovskii and S.M. Semenov, *On polynomial approximation in function spaces*, Mat. Sbornik **21** (1973), 255–277.

[NeSe73b] A.S. Nemirovskii and S.M. Semenov, *The polynomial approximation of functions on Hilbert space*, Mat. Sb. (NS) **92** (134) (1973), 257–281.

[NiWa63] T. Nishiura and D. Waterman, *Reflexivity and summability*, Studia Mat. **23** (1963), 53–57.

[Nor60] G. Nordlander, *The modulus of convexity in normed linear spaces*, Ark. Math. **4** (1960), 15–17.

[OdSch98] E.W. Odell and Th. Schlumprecht, *On asymptotic properties of Banach spaces under renormings*, J. Amer. Math. Soc. **11** (1998), 175–188.

[OdSch03] E.W. Odell and Th. Schlumprecht, *Distortion and Asymptotic Structure*, Handbook of the Geometry of Banach spaces II, Editors W.B. Johnson and J. Lindenstrauss, Elsevier Science B.V., 2003, 1333–1360.

[OdSch06] E.W. Odell and Th. Schlumprecht, *A universal reflexive space for the class of uniformly convex Banach spaces*, Math. Ann. **335** (2006), 901–916.

[Ore74] O. Ore, *Niels Henrik Abel, Mathematician Extraordinary*, Chelsea Pub. Co. 1974.

[Ori92] J. Orihuela, *On weakly Lindelöf Banach space*, Progress in Funct. Anal., K. D. Bierstedt, J. Bonet, J. Horvath, and M. Maestre, ed., Elsevier (1992).

[OriSmTr12] J. Orihuela, R. Smith, and S.L. Troyanski, *Strictly convex norms and topology*, Proc. London Math. Soc. **104**, 1 (2012), 197–222.

[OrVa90] J. Orihuela and M. Valdivia, *Projective generators and resolutions of identity in Banach spaces*, Rev. Mat. Univ. Complutense, Madrid, **2** Suppl. Issue (1990), 179–199.

[Orn91] P. Ørno, *On J. Borwein's concept of sequentially reflexive Banach spaces*, Banach Space Bulletin Board, 1991.

[Pa63] I.I. Parovičenko, *A universal bicompactum of weight* $\aleph$, Dokl. Akad. Nauk SSSR **150** (1963), 36–39.

[Par80] J.R. Partington, *Equivalent norms on on spaces of bounded functions*, Israel J. Math. **35** (1980), 205–209.

[PavZa07] D. Pavlica and L. Zajíček, *On the directions of segments and τ-dimensional balls on a convex surface,* J.Conv. Anal. **14** (1) (2007), 149–167.

[PeWhZi81] J. Pechanec, J.H.M. Whitfield, and V. Zizler, *Norms locally dependent on finitely many coordinates*, An. Acad. Brasil Ci. **53** (1981), 415–417.

[Pel57] A. Pełczyński, *A property of multilnear transformations*, Studia Math. **16** (2) (1957), 173–182.

[Pel62] A. Pełczyński, *A note on the paper of I. Singer "Basic sequences and reflexivity of Banach spaces"* Studia Math. **21** (1962), 371–374.

[Pel65] A. Pełczyński, *On strictly singular and strictly cosingular operators I*, Bull. Polish Acad. Sci. Math. **13** (1965), 31–36.

[PelBe79] A. Pełczyński and C. Bessaga, *Some aspects of the present theory of Banach spaces*, in 1979 Edition of S. Banach: Travaux sur L'Analyse Fonctionnelle, Warszaw, 1979.

[PelSu62] A. Pełczyński and V.N. Sudakov, *Remarks on noncomplemented subspaces of the space* $m(S)$, Collog. Math. **9** (1962), 85–88.

[Pen86] J.P. Penot, *The drop theorem, the petal theorem and Ekeland's variational principle*, Nonlinear Analysis: Theory, Methods and Applications, **10**, 9 (1986), 813–822.

[PerRaj17] A. Pérez and M. Raja, *A Bourgain-like property of Banach spaces with no copies of c0*, Revista de la Real Academia de Ciencias (RACSAM) **111** (2017), 205–211.

[Pet64] Ju.I. Petunin, *Conjugate Banach spaces containing subspaces of zero characteristic*, Soviet Math. Dokl. **5** (1964), 131–134.

[Pfit10] H. Pfitzner, *Boundaries for Banach spaces determine weak compactness*, Invent. Math. **182** (2010), 585–604.

[Ph60a] R.R. Phelps, *A representation theorem for bounded convex sets*, Proc. Amer. Math. Soc. **11** (1960), 976–983.

[Ph60b] R.R. Phelps, *Uniqueness of Hahn-Banach extensions and unique best approximation* , Trans. Amer. Math. Soc. **95** (1960), 23–255.

[Ph63] R.R. Phelps, *Support cones and their generalization*, in V. Klee, editor, *Convexity*, Vol. VII of Proceedings of Symposia in Pure Mathematics, pp. 393–401. American Mathematical Society, 1963.

[Ph74] R.R. Phelps, *Support cones in Banach spaces and their applications*, Advances in Mathematics **13** (1974), 1–19.

[Phe74b] R.R. Phelps, *Dentability and extreme points in Banach spaces*, J. of Functional Analysis, **6** (1974), 78–90.

[Phe93] R.R. Phelps, *Convex Functions, Monotone Operators and Differentiability* (2nd Ed.), LNM **1364**, Springer, 1993.

[Phe01] R.R. Phelps, *Lectures on Choquet's theorem*, 2nd Edition, Springer 2001.

[Phe02] R.R. Phelps, "Biography of John Rainwater". Melvin Henriksen, ed. Topological Commentary. Topology Atlas (York University, Canada) **7**, 2 (2002).

[Phi43] R.S. Philips, *On weakly compact subsets of a Banach space*, Amer. J. Math. **65** (1943), 108–136.

[Pie07] A. Pietsch, *History of Banach spaces and linear operators*, Birkhäuser, Boston, Inc. Boston MA, 2007.

[Pis75] G. Pisier, *Martingales with values in uniformy convex spaces*, Israel J. Math. **20** (1975), 326–350.

[Ple04] G. Plebánek, *A construction of a Banach space $C(K)$ with few operators*, Topology and Applications **143** (2004), 127–129.

[PlYo01] A.N. Plichko and D. Yost, *The Radon–Nikodym property does not imply the separable complementation property*, J. Functional Analysis, **180** (2001), 481–487.

[Po77] R. Pol, *Concerning function spaces on separable compact spaces*, Bull. Acad. Polon. Ser. Sci. Math. Astronom. Phys. **25** (1977), 993–997.

[Po79] R. Pol, *A function space $C(X)$ which is weakly Lindelöf but not weakly compactly generated*, Studia Math. **64** (1979), 279–285.

[PolSi83] T. Polak and B. Sims, *A Banach space which is fully 2–rotund but not locally uniformly rotund*, Canad. Math. Bull. **26** (1), (1983), 118–120.

[PoVaZi94] R. Poliquin, J. Vanderwerff, and V. Zizler, *Renormings and convex composite representations of functions*, Bulletin of the Polish Academy of Science Mathematics, **4** (1), (1994), 9–19.

[Prei84] D. Preiss, *Gâteaux differentiable functions are somewhere Fréchet differentiable*, Rendiconti del Circolo Matematico di Palermo, **33** (2) (1984), 122–133.

[Prei90] D. Preiss, *Differentiability of Lipschitz functions on Banach spaces*, J. Funct. Anal. **91** (1990), 312–345.

[Prei10] D. Preiss, *Differentiability problems in Banach spaces*, https://www.mat.univie.ac.at/ anstap10/slides/Preiss.pdf, 2010.

[PrPhNa90] D. Preiss, R.R. Phelps, and I. Namioka, *Smooth Banach spaces, weak Asplund spaces, and monotone or USCO mappings*, Israel. J. Math. **72** (1990), 257–279.

[PrZa84] D. Preiss and L. Zajíček, *Fréchet differentiation of convex functions in a Banach space with a separable dual*, Proc. Amer. Math. Soc. **91** (1984), 202–204.

[Pru83] S. Prus, *Finite dimensional decompositions with p-estimates*, Bull. Ac. Pol. Sci. **31** (5-8), (1983), 281–288.

[Pru87] S. Prus, *Finite-dimensional decompositions of Banach spaces with (p, q)-estimates*, Dissertationes Math. (Rozprawy Mat.) **263** (1987), 1–41.

[Pr10] S. Prus, *Constructing separated sequences in Banach space*, Proc. Amer. Math. Soc. **138** (1) (2010), 225–234.

[Pt59] V. Pták, *Biorthogonal systems and reflexivity of Banach spaces*, Czechoslovak Math. J. **9** (84), (1959), 319–326.

[Rade19] H. Rademacher, *Über partielle und totale Differenzierbarkeit I*, Math. Ann. **89** (1919), 340–359.

[Rainw63] J. Rainwater, *Weak convergence of bounded sequences*, Proc. Amer. Math. Soc. **14** (1963), 999.

[Rainw69] J. Rainwater, *Local uniform convexity of Day's norm on* $c_0(\Gamma)$, Proc. Amer. Math. Soc. **22** (1969), 335–339.

[Raj99] M. Raja, *On locally uniformly rotund norms*, Mathematika **46** (1999), 343–358.

[Raj02] M. Raja, *On dual locally uniformly rotund norms*, Israel J. Math. **129**, 1, (2002), 77–91.

[Raj03] M. Raja, *Weak* locally uniformly rotund norms and descriptive compact spaces*, J. of Functional Analysis **197** (2003), 1–13.

[Raj10] M. Raja, *On weak* uniformly Kadets-Klee renormings*, Bull. London Math. Soc. **42** (2010), 221–228.

[Rei74] J. Reif, *A note on Markushevich bases in weakly compactly generated Banach spaces*, Comment. Math. Univ. Carolinae **15** (1974), 83–111.

[Rei76] J. Reif, *Some remarks on subspaces of weakly compactly generated Banach spaces*, Comment. Math. Univ. Carolinae **16** (1976), 787–793.

[Res64] G. Restrepo, *Differentiable norms in Banach spaces*, Bull. Amer. Math. Soc. **70** (1964), 413–414.

[Reth68] J.R. Retherford, *A semishrinking basis which is not shrinking*, Proc. Amer. Math. Soc. **19** (1968), 766.

[Rez90] E.A. Rezniczenko, *Normality and collectionwise normality of function spaces*, Vestnik Moskov. Univ. Ser. Mat. (1990), 56–58. English transl. Moscow Univ. Math. Bull. **45**, no 6 (1990), 25–26.

[Rib92] N.K. Ribarska, *The dual of a Gâteaux smooth Banach space is* w^**-fragmentable*, Proc. Amer. Math. Soc. **114** (1992), 1003-1008.

[Rie67] M.A. Rieffel, *Dentable subsets of Banach spaces, with applications to Radon–Nikodým theorem*, Functional Analysis, Proc. Conference, Thompson Books C. Washington, D. C. 1967, 71–77.

[Rm17] M. Rmoutil, *Norm attaining functionals need not contain* 2*-dimensional subspaces*, J. Funct. Anal. **272** (2017), 918–928.

[Rod80] G. Rodé, *Superkonvexe Analysis.* Arch. Math. **34** (1980), 452–462.

[Rod81] G. Rodé, *Superkonvexität und schwache Kompaktheit.* Arch. Math. **36** (1981), 62–72.

[Rol87] S. Rolewicz, *On drop property*, Studia Math. **85** (1987), 27–35.

[RoRo64] A.P. Robertson and W. Robertson, *Topological Vector Spaces*, Cambridge Tracts in Mathematics and Mathematical Physics **53**, Cambridge at the University Press, 1964.

[Rock96] R.T. Rockafellar, *Convex Analysis*, Princeton University Press, 1996.

[Ros68] H.P. Rosenthal, *On quasicomplemented subspaces of Banach spaces*, Proc. Nat. Acad. Sci. USA **59** (1968), 361–364.

[Ros70] H.P. Rosenthal, *On injective Banach spaces and the spaces* $L^\infty(\mu)$ *for finite measure* μ, Acta Math. **124** (1970), 205–248.

[Ros74] H.P. Rosenthal, *The heredity problem for weakly compactly generated Banach spaces*, Compositio Math. **28** (1974), 83–111.

[Ros73] H.P. Rosenthal, *On subspaces of* L^p, Ann. Math. **97** (1973), 344–373.

[Ros70] H.P. Rosenthal, *On injective Banach spaces and the spaces* $L^\infty(\mu)$ *for finite measure* μ, Acta Math. **124** (1970), 205–248.

[Ros83] H.P. Rosenthal, *A characterization of* c_0 *and some remarks concerning the Grothendieck property* C, Longhorn Notes, The University of Texas Functional Anlysis Seminar (1982-83), 95–108.

[Ros03] H.P. Rosenthal, *The Banach spaces* $C(K)$, in Handbook of the Geometry of Banach spaces II, Editors W.B. Johnson and J. Lindenstrauss, Elsevier Science B.V. (2003), 1547–1602.

[Ru18] T. Russo, *On some open problems in Banach space theory*, PHD dissertation, Università degli Studi di Milano, Facoltà di Scienze e Tecnologie, Milano, 2017–2018.

[Rych00] J. Rychtář, *Uniformly Gâteaux differentiable norms in spaces with unconditional basis*, Serdica Math. J. **28** (2000), 353–358.

[Rych03] J. Rychtář, *Renorming of* $C(K)$ *spaces*, Proc. Amer. Math. Soc. **131** (7), (2003), 2063–2070.

[Rych04] J. Rychtář, *On biorthogonal systems and Mazur's intersection property*, Bull. Austral. Math. Soc. **69** (2004), 107–111.

[Rych05] J. Rychtář, *Pointwise uniformly rotund norms*, Proc. Amer. Math. Soc. **133** (8), (2005), 2259–2266.

[Rych05b] J. Rychtář, *On Gâteaux differeniability of convex functions in WCG spaces*, Canad. Math. Bull. **48** (3) (2005), 455–459.

[Saks37] S. Saks, *Theory of the Integral*, Second Revised Edition, Monografie Matematyczne, Hafner Publishing Company, New York, 1937.

[Sch83] W. Schachermayer, *Norm attaining operators and Renormings of Banach spaces*, Israel J. Math. **44** (3), (1983), 201–212.

[Sch89] W. Schachermayer, *Some more remarkable properties of the James Tree space*, Contemp. Math. **89** (1989), 465-496.

[Schae71] H.H. Schaefer, *Topological Vector Spaces*, Third Printing Corrected, GTM **3**, Springer-Verlag, 1971

[Sch67] J.J. Schäffer, *Inner diameter, perimeter and girth of spheres*, Math. Annalen **173** (1967), 59–82.

[Sch76] J.J. Schäffer, *Geometry of spheres in Normed Spaces*, Marcel Dekker, 1976.

[SchW88] G. Schlüchtermann and R.F. Wheeler, *On strongly WCG Banach spaces*, Math. Z. **199** (1988), 273–289.

[SchW91] G. Schlüchtermann and R.F. Wheeler, *The Mackey dual of a Banach space*, Note Math. **11** (1991), 273–287.

[Sch91] Th. Schlumprecht, *An arbitrarily distortable Banach space*, Israel J. Math. **76** (1991), 81–95.

[Sem60] Z. Semadeni, *Banach spaces not isomorphic to their Cartesian squares, II*, Bull. Polon. Sci. Ser. Sc. Math. Astronom. Physics, **8** (1960), 81–84.

[Sin63] I. Singer, *On bases in quasi-reflexive Banach spaces*, Rev. Math. Pures Appl. (Bucarest) **8** (1963), 309–311.

[Sim72] S. Simons, *A convergence theorem with boundary* , Pacific J. Math. **40** (1972), 703–708.

[SiYo89] B. Sims and D. Yost, *Linear Hahn-Banach extension operators*, Proc. Edingburgh Math. Soc. **3** (1989), 53–57.

[Si75] I. Singer, *On the problem of non-smoothness of non-reflexive second conjugate space*, Bull. Austral. Math. Soc.**12** (1975), 404–416.

[Sio58] M. Sion, *On general minimax theorems*, Pacific J. Math. **8** (1958), 171–176.

[Sm76] M.A. Smith, *Rotundity and smoothness in conjugate spaces*, Proc. Amer. Math.Soc. **61** (2), 1976), 232–234.

[Sm77] M.A. Smith, *Banach spaces that are uniformly rotund in weakly compact sets of directions*, Canadian J. Math. **29** (5) (1977), 963–970.

[Sm78] M.A. Smith, *Some examples concerning rotundity in Banach spaces*, Math. Ann. **233** (1978) 155–161.

[Sm81] M.A. Smith, *A Banach space that is MLUR but not HR*, Math. Ann. **256** (1981), 277–279.

[SmTu85] M.A. Smith and B. Turett, *A reflexive LUR Banach space that lacks normal structure*, Canad. Math. Bull. **28** (4) (1985), 492–494.

[Smi06] R.J. Smith, *On trees and dual rotund norms.* J. Funct. Anal. **231** (2006), 177–194.

[Smi09] R.J. Smith, *Gruenhage compacta and strictly convex dual norms*, J. Math. Anal. Appl. **350** (2009), 745–757.

[SmTr09] R.J. Smith and S.L. Troyanski, *Unconditional bases and strictly convex dual renormings*, Bull. London Math. Soc. **41** (2009), 831–840.

[SmTr19] R.J. Smith and S.L. Troyanski, *Approximation of norms on Banach spaces*, J. Funct. Anal. **276**, 10 (2019), 3226–3260.

[Smu40] V.L. Šmul'yan, *Sur la derivabilité de la norme dans l'espace de Banach.* C. R. (Doklady) Akad. Sci. URSS **27**, 643–648 (1940).

[St73] Ch. Stegall, *Banach spaces whose duals contain $\ell_1(\Gamma)$, with applications to the study of dual $L_1(M)$-spaces*, Trans. Amer. Math. Soc.**176** (1973), 463–477.

[St75] Ch. Stegall, *The Radon–Nikodým property property in conjugate Banach spaces*, Trans. Amer. Math. Soc.**206** (1975), 213–223.

[St78] Ch. Stegall, *The Radon–Nikodým property in conjugate spaces II*, Trans. Amer. Math. Soc. **264** (2), (1978), 290–294.

[St78b] Ch. Stegall, *Optimization of functions on certain subsets of Banach spaces*, Math. Ann. **236** (1978), 171–176.

[St78c] Ch. Stegall, *The duality between Asplund spaces and spaces with the Radon-Nikodým property*, Israel J. Math. **29** (4) (1978), 408–412.

[Str35] S. Straszewicz, *Uber exponierte Punkte abgeschlossener Punktmengen*, Fund. Math. **24** (1935), 139–143.

[Sw83] S. Swaminathan, *Normal structure in Banach spaces and its generalisations*, Contemporary Math. **18** (1983), 201–215.

[Sz65] W. Szlenk, *Sur les suites faiblemente convergents dans l'espace L*, Studia Math. **25** (1965), 337–341.

[Sz68] W. Szlenk, *The nonexistence of a separable reflexive universal for all separable reflexive Banach spaces*, Studia Math. **30** (1968), 53–61.

[Tac70] D.G. Tacon, *The conjugate of a smooth Banach space*, Bull. Austral. Math. Soc. **2** (1970), 415–425.

[Tal78] M. Talagrand, *Comparison des boréliens dún espace de Banach pour les topologies fortes et faibles*, Indiana University Math. J. **27**, 6 (1978), 1001–1004.

[Tal79] M. Talagrand, *Deux exemples de fonctions convexes*, C. R. Acad. Sc. Paris **288** (1979), 461–464.

[Tal80] M. Talagrand, *Separabilité vague dans l'espace des measures sur un compact*, Israel J. Math. **37** (1980), 171–180.

[Tal86] M. Talagrand, *Renormages de quelques $C(K)$*, Israel J. Math. **54** (1986), 327–334.

[WKT95] Wee-Kee Tang, *On Fréchet differentiability of convex functions on Banach spaces*, Comment. Math. Univ. Carolinae **36** (1995), 249–253.

[WKT96a] Wee-Kee Tang, *A note on preserved smoothness*, Serdica Math. J. **22** (1996), 29–32.

[WKT97] Wee-Kee Tang, *Uniformly differentiable bump functions*, Arch. Math. **68** (1997), 55–59.

[WKT96b] Wee-Kee Tang, *On the extension of rotund norms*, Manuscripta Math. **91** (1996), 73–82.

[Tay39] A.E. Taylor, *The extension of linear functionals*, Duke Math. J. **5** (1939), 538–547.

[Ter89] P. Terenzi, *Equilateral sets in Banach spaces*, Boll. Un. Math. Ital. A (7), **13** , (1) (1989), 119–124.

[Ter94] P. Terenzi, *Every separable Banach space has a bounded biorthogonal sequence which is also a Steinitz basis*, Studia Math. **111** (1994), 207–222.

[To96] N. Tomczak-Jaegermann, *Banach spaces of type p have arbitrarily distortable subspaces*, GaFa **6** (1996), 1074–1082.

[ToKo98] N. Tomczak-Jaegermann, R. Komorowski, *Erratum to: Banach spaces without local unconditional structure*, Israel J. Math. **105** (1998), 85–92.

[Tor73] H. Toruńczyk, *Smooth partitions of unity on some nonseparable Banach spaces*, Studia Math. **46** (1973), 43–51.

[Tor81] H. Toruńczyk, *Characterizing Hilbert Space topology*, Fund. Math **111** (1981), 247–262.

[Tr67] S.L. Troyanski, *On topological equivalence of spaces* $c_0(\aleph)$ *and* $\ell_1(\aleph)$. Bull. Acad. Polon. Sci. Ser. Sci. Math. Astr. et Phys. **15** (1967), 389–396.

[Tr68] S.L. Troyanski, *Equivalent norms in nonseparable B-spaces with an unconditional basis*, Teor. Funkcii Funkcional Anal. i Prilozen. **6** (1968), 59–65 (in Russian).

[Tr70] S.L. Troyanski, *An example of a smooth space, the dual of which is not strictly convex*, (Russian), Studia Math. **35** (1970), 305–309.

[Tr71] S.L. Troyanski, *On locally uniformly convex and differentiable norms in certain nonseparable Banach spaces*, Studia Math. **37** (1971), 173–180.

[Tr72] S.L. Troyanski, *On equivalent norms and minimal systems in nonseparable Banach space*, Studia Math. **43** (1972), 125–138.

[Tr75] S.L. Troyanski, *On nonseparable Banach spaces with a symmetric basis*, Studia Math. **53**, 3 (1975), 253–263.

[Tr85] S.L. Troyanski, *On a property of the norm which is close to local uniform convexity*, Math. Ann. **271** (1985), 305–313.

[Tr90] S.L. Troyanski, *Gâteaux differentiable norms in* L_p, Math. Ann. **287** (1990), 221–227.

[Tsi74] B.S. Tsirelson, *Not every Banach space contains* ℓ_p *or* c_0, Functional Anal. and Its Appl. **8** (1974), 138–141.

[Tu82] B. Turett, *A dual view of a theorem of Baillon*, Proc. Conf. Nonlinear Analysis and Applications, Lecture Notes in Math. **80** (1982), 279–286.

[Vald77] M. Valdivia, *On a class of Banach spaces*, Studia Math. **60** (1977), 11–13.

[Vald88] M. Valdivia, *Resolutions of the identity in certain Banach spaces*, Collect. Math. **39** (1988), 127–140.

[Vald90] M. Valdivia, *Projective resolutions of the identity in* $C(K)$ *spaces*, Arch. Math., **54** (1990), 493–498.

[Vald91] M. Valdivia, *Simultaneous resolutions of the identity operator in normed spaces* , Collect. Math. **42** (1991), 265–284.

[Vald93a] M. Valdivia, *Fréchet spaces with no subspaces isomorphic to* ℓ_1. Math. Japonica **38** (1993), 397–411.

[vNee05] J.M.A.M. van Neerven, *Separated sequences in uniformly convex Banach spaces*, Colloq. Math. **102** (2005), 147–153.

[Vand92a] J. Vanderwerff, *Fréchet differentiable norms on spaces of countable dimension*, Arch. Math. **58** (1992), 471–476.

[Vand92b] J. Vanderwerff, *Smooth approximations in Banach spaces*, Proc. Amer. Math. Soc. **115** (1992), 113–120.

[Vand93] J. Vanderwerff, *Second order Gâteaux differentiability and an isomorphic characterization of Hilbert spaces*, Quart. J. Math. Oxford (2), **44** (1993), 249–255.

[Van98] J. Vanderwerff, *Mazur intersecion properties for compact and weakly compact sets*, Canad. Math. Bull. **41** (2) (1998), 225–230.

[VWZ94] J. Vanderwerff, J.H.M. Whitfield, and V. Zizler, *Markushevich bases and Corson compacta in duality*, Can. J. Math. **46** (1), (1994), 200–211.

[Vas81] L. Vašák, *On one generalization of weakly compactly generated Banach spaces*, Studia Math. **70** (1981), 11–19.

[Ver00] R. Vershynin, *On constructions of strong and uniformly minimal M-bases in Banach spaces*, Arch. Math. **74** (2000), 50–60.

[Ves99] L. Veselý, *A geometric proof of a theorem about non-dual renormings*, Proc. Amer. Math. Soc. **127** (1999), 2807–2809.

[Ves09] L. Veselý, *A parametric smooth variational principle and support properties of convex sets and functions*, J. Math. Anal. and Appl. **350** (2009), 550–561.

[VesZa89] L. Veselý and L. Zajíček, *Delta-convex mappings between Banach spaces and applications*, Dissertationes Math. **289** (1989), 1–48.

[War01] H.M. Wark, *A nonseparable reflexive Banach space on which there are few operators*, J. London Math. Soc. (2) **64** (2001), 675–698.

[Weav18] N. Weaver, *On the unique predual problem for Lipschitz spaces*, Math. Proc. Cambridge Phil. Soc. **165** (2018), 467–473.

[Weav18b] N. Weaver, *Lipschitz Algebras*, Second Ed., World Scientific, 2018.

[Wer01] D. Werner, *Recent progress on the Daugavet property*, Irish Math. Soc. Bulletin **46** (2001), 77–97.

[WhZi87] J.H.M. Whitfield and V. Zizler, *Uniform Mazur intersection property for balls*, Canad. Mat. Bull. **30** (4) (1987), 455–460.

[WhZi88] J.H.M. Whitfield and V. Zizler, *Extremal structure of convex sets in spaces not containing* c_0, Math. Z. **197** (1988), 219–221.

[Whi68] R. Whitley, *The size of the unit sphere*, Canad. J. Math. **20** (1968), 450–455.

[Whitn34] H. Whitney, *Analytic extensions of differentiable functions defined in closed sets*, Trans. Amer. Math. Soc. **36**, 1 (1934), 63–89.

[Wil95] D. Wilkins, *The strong WCD property for Banach spaces*, Internat. J. Math & Math. Sci. **18** (1) (1995), 67–70.

[Woj91] P. Wojtaszczyk, *Banach Spaces for Analysts*, Cambridge Studies in Advanced Mathematics **25** (1991).

[Yes49] A.C. Yesenin-Volpin, *On the existence of a universal bicompact of arbitrary weight*, Dokl. Akad. Nauk USSR **68** (1949), 649–652.

[Yo81] D. Yost, *M-ideals, their strong 2-property and some renorming theorems*, Proc. Amer. Math. Soc. **81** (1981), 289–303.

[Yo93] D. Yost, *Asplund spaces for beginners*, Acta Univ. Carolinae **34** (1993), 159–177.

[Yo97] D. Yost, *The Johnson–Lindenstrauss space*, Extracta Math. **12** (1997), 185–192.

[Yu12] Yu, Y.-L., *Arithmetic dualitiy for norm*, Department of Computing Science, University of Alberta, Canada, 2012.

[Zah46] Z. Zahorski, *Sur l'ensemble des points de non-derivabilité d'une fonction continue*, Bull. Soc. Math. France **74** (1946), 147–178.

[Zaj88] L. Zajíček, *Porosity and σ-porosity*, Real Anal. Exchange **13** , 2 (1987/88) 314–350.

[Zaj92] L. Zajíček, *An elementary proof of the one-dimensional Rademacher theorem*, Mathematica Bohemica **117**, 2 (1992), 133–136.

[Zal02] C. Zălinescu, *Convex Analysis in General Vector Spaces*, World Scientific, 2002.

[Zip68] M. Zippin, *Banach spaces with separable duals*, Trans. Amer. Math. Soc. **310** (1) (1968), 371–379.

[Zip68a] M. Zippin, *A remark on bases and reflexivity in Banach spaces*, Israel J. Math **6** (1968), 74–79.

[Zip69] M. Zippin, *On some subspaces of Banach spaces whose duals are* L_1 *spaces*, Proc. Amer. Math. Soc.**23** (1969), 378–385.

[Ziz71] V. Zizler, *On some rotundity and smoothness properties of Banach spaces*, Dissert. Math **87** (1971), 1–33.

[Ziz73] V. Zizler, *On some extremal problems in Banach spaces*, Math. Scand. **32** (1973), 214–224.

[Ziz75] V. Zizler, *A note on duality mapping in Banach spaces*, Arch. Math. (Basel) **26** (1975), 94–97.

[Ziz84] V. Zizler, *Locally uniformly rotund renorming and decomposition of Banach spaces*, Bull. Austr. Math. Soc. **29** (1984), 259–265.

[Ziz86] V. Zizler, *Renorming concerning Mazur's intersection property for weakly compact convex sets*, Math. Ann. **276** (1986), 61–66.

[Ziz89] V. Zizler, *Smooth extension of norms and complementability of subspaces*, Arch. Math. **53** (1989), 585–589.

[Ziz03] V. Zizler, *Nonseparable Banach spaces*, Handbook of Banach spaces II, Editors W.B. Johnson and J. Lindenstrauss, Elsevier Science B.V., 2003.

List of Figures

A. J. Guirao et al., *Renormings in Banach Spaces*, Monografie Matematyczne 75,
https://doi.org/10.1007/978-3-031-08655-7

How to use the indices (and some notations)

In this text there are four indices. Some instructions for their use follow:

General Index

It contains a list of concepts. We tried to include not only their location in the text but also a short information regarding their function at a particular spot. Uses in and results on renormings are in general not included —those will be postponed to the "Renormings" and "Impossible Renormings" indices.

Index of Symbols

It lists symbols appearing in the text. Each one comes with a little description of the object. Pages point to the first moment the symbol pops out.

Index of Authors

We list authors together with pages where we mention them. Of course, S. Banach does not appear any time the word "Banach" pops up, and this applies to some other proper names, too. A reference to a paper does not provoke invoking the authors' name in this list. The Bibliography is to some extent independent of this list.

Renormings

Here the code turns to be more complicated, and deserves a longer explanation, better understood —we hope— looking at some examples. P and Q are norm properties, X is a Banach space, and X^* its dual space. Some of the next examples are redundant; yet, we prefer redundancy that lack of understanding.

We shall omit the reference X at an entry as $X\langle \mathrm{P} \rangle$ if the space X is understood beyond any doubt.

A. J. Guirao et al., *Renormings in Banach Spaces*, Monografie Matematyczne 75,
https://doi.org/10.1007/978-3-031-08655-7

1	$X\langle \mathrm{P}\rangle$	X has an equivalent norm with property P.
2	$X\langle \mathrm{P+Q}\rangle$	X has an equivalent norm with properties P and Q simultaneously.
3	$X\langle \mathrm{P}\rangle + \langle \mathrm{Q}\rangle$	X has an equivalent norm with property P and another one with property Q.
4	$X\langle \mathrm{P}^*\rangle$	X has an equivalent norm whose dual norm has property P.
5	$X\langle \mathrm{P}^{**}\rangle$	X has an equivalent norm whose second dual norm has property P (etc.).
6	$X^*\langle \mathrm{P}\rangle^*$	X^* has an equivalent *dual* norm with property P.
7	$X^*\langle \mathrm{P}\rangle$	X^* has an equivalent norm —not necessarily dual— with property P.
8	$X\langle \mathrm{P+Q}^*\rangle$	X has an equivalent norm with property P whose dual norm has property Q.
9	$X\langle \mathrm{P}\rangle + \langle \mathrm{Q}^*\rangle$	X has an equivalent norm with property P and another equivalent norm whose dual norm has property Q.
10	$X\langle \mathrm{P}^* + \mathrm{Q}^*\rangle$	X has an equivalent norm whose dual norm has properties P and Q simultaneously.
11	$X\langle \mathrm{P}^*\rangle + \langle \mathrm{Q}^*\rangle$	X has an equivalent norm whose dual norm has property P and another equivalent norm whose dual norm has property Q.

Observe that 4 and 6 say exactly the same thing. Yet, it is convenient to have both versions at hand.

There are several ways to look at this list:

(i) Searching for renormings of a space having a particular property —for example, being Asplund separable. Then, under the entry “Asplund, separable”, we shall collect possible renormings for this class of spaces.

(ii) Searching for renormings of a particular space —say $C[0,1]$—. Even in the case that it can be renormed by a norm having properties, say, P, Q, and R simultaneously —and so an entry will appear as $C[0,1]\langle \mathrm{P+Q+R}\rangle$—, we may refer to a particular equivalent norm having just, say, property P —and then the reader will find also $C[0,1]\langle \mathrm{P}\rangle$. This will be done for several reasons. (i) Maybe the less ambitious renorming is easier to get or depends on different techniques. (ii) Maybe its relevance is due to historical reasons and it will be unfair to omit it. (iii) Maybe the “better” renorming is just a consequence of the first, seminal one. This applies to the entire list.

(iii) If we look for a particular kind of norm, say Fréchet differentiable (F, by using our acronym), then we shall collect under the entry $\langle \mathrm{F}\rangle$ a list of spaces or properties of spaces admitting such a norm.

Impossible Renormings

Here we include a list of spaces and classes of spaces mentioned in the text together with properties of norms that those spaces *do not accept.*

General Index

A. J. Guirao et al., *Renormings in Banach Spaces*, Monografie Matematyczne 75,
https://doi.org/10.1007/978-3-031-08655-7

Index of Symbols

A. J. Guirao et al., *Renormings in Banach Spaces*, Monografie Matematyczne 75,
https://doi.org/10.1007/978-3-031-08655-7

Index of Authors

A. J. Guirao et al., *Renormings in Banach Spaces*, Monografie Matematyczne 75,
https://doi.org/10.1007/978-3-031-08655-7

Renormings

A. J. Guirao et al., *Renormings in Banach Spaces*, Monografie Matematyczne 75,
https://doi.org/10.1007/978-3-031-08655-7

Impossible Renormings

A. J. Guirao et al., *Renormings in Banach Spaces*, Monografie Matematyczne 75,
https://doi.org/10.1007/978-3-031-08655-7